T0181814

Lecture Notes in Computer Science   12681

More information about this subseries at http://www.springer.com/series/7409

Christian S. Jensen · Ee-Peng Lim ·
De-Nian Yang · Wang-Chien Lee ·
Vincent S. Tseng · Vana Kalogeraki ·
Jen-Wei Huang · Chih-Ya Shen (Eds.)

Database Systems for Advanced Applications

26th International Conference, DASFAA 2021
Taipei, Taiwan, April 11–14, 2021
Proceedings, Part I

 Springer

Editors
Christian S. Jensen ⓘ
Aalborg University
Aalborg, Denmark

De-Nian Yang
Academia Sinica
Taipei, Taiwan

Vincent S. Tseng
National Chiao Tung University
Hsinchu, Taiwan

Jen-Wei Huang ⓘ
National Cheng Kung University
Tainan City, Taiwan

Ee-Peng Lim ⓘ
Singapore Management University
Singapore, Singapore

Wang-Chien Lee
The Pennsylvania State University
University Park, PA, USA

Vana Kalogeraki
Athens University of Economics
and Business
Athens, Greece

Chih-Ya Shen
National Tsing Hua University
Hsinchu, Taiwan

ISSN 0302-9743   ISSN 1611-3349   (electronic)
Lecture Notes in Computer Science
ISBN 978-3-030-73193-9   ISBN 978-3-030-73194-6   (eBook)
https://doi.org/10.1007/978-3-030-73194-6

LNCS Sublibrary: SL3 – Information Systems and Applications, incl. Internet/Web, and HCI

This Springer imprint is published by the registered company Springer Nature Switzerland AG
The registered company address is: Gewerbestrasse 11, 6330 Cham, Switzerland

Preface

Welcome to DASFAA 2021, the 26th International Conference on Database Systems for Advanced Applications, held from April 11 to April 14, 2021! The conference was originally planned to be held in Taipei, Taiwan. Due to the outbreak of the COVID-19 pandemic and the consequent health concerns and restrictions on international travel all over the world, this prestigious event eventually happens on-line as a virtual conference, thanks to the tremendous effort made by the authors, participants, technical program committee, organization committee, and steering committee. While the traditional face-to-face research exchanges and social interactions in the DASFAA community are temporarily paused this year, the long and successful history of the events, which established DASFAA as a premier research conference in the database area, continues!

On behalf of the program committee, it is our great pleasure to present the proceedings of DASFAA 2021, which includes 131 papers in the research track, 8 papers in the industrial track, 8 demo papers, and 4 tutorials. In addition, the conference program included three keynote presentations by Prof. Beng Chin Ooi from National University of Singapore, Singapore, Prof. Jiawei Han from the University of Illinois at Urbana-Champaign, USA, and Dr. Eunice Chiu, Vice President of NVIDIA, Taiwan.

The highly selective papers in the DASFAA 2021 proceedings report the latest and most exciting research results from academia and industry in the general area of database systems for advanced applications. The quality of the accepted research papers at DASFAA 2021 is extremely high, owing to a robust and rigorous double-blind review process (supported by the Microsoft CMT system). This year, we received 490 excellent submissions, of which 98 full papers (acceptance ratio of 20%) and 33 short papers (acceptance ratio of 26.7%) were accepted. The selection process was competitive and thorough. Each paper received at least three reviews, with some papers receiving as many as four to five reviews, followed by a discussion, and then further evaluated by a senior program committee (SPC) member. We, the technical program committee (TPC) co-chairs, considered the recommendations from the SPC members and looked into each submission as well as the reviews and discussions to make the final decisions, which took into account multiple factors such as depth and novelty of technical content and relevance to the conference. The most popular topic areas for the selected papers include information retrieval and search, search and recommendation techniques; RDF, knowledge graphs, semantic web, and knowledge management; and spatial, temporal, sequence, and streaming data management, while the dominant keywords are network, recommendation, graph, learning, and model. These topic areas and keywords shed light on the direction in which the research in DASFAA is moving.

Five workshops are held in conjunction with DASFAA 2021: the 1st International Workshop on Machine Learning and Deep Learning for Data Security Applications (MLDLDSA 2021), the 6th International Workshop on Mobile Data Management,

Mining, and Computing on Social Networks (Mobisocial 2021), the 6th International Workshop on Big Data Quality Management (BDQM 2021), the 3rd International Workshop on Mobile Ubiquitous Systems and Technologies (MUST 2021), and the 5th International Workshop on Graph Data Management and Analysis (GDMA 2021). The workshop papers are included in a separate volume of the proceedings, also published by Springer in its Lecture Notes in Computer Science series.

We would like to express our sincere gratitude to all of the 43 senior program committee (SPC) members, the 278 program committee (PC) members, and the numerous external reviewers for their hard work in providing us with comprehensive and insightful reviews and recommendations. Many thanks to all the authors for submitting their papers, which contributed significantly to the technical program and the success of the conference. We are grateful to the general chairs, Christian S. Jensen, Ee-Peng Lim, and De-Nian Yang for their help. We wish to thank everyone who contributed to the proceedings, including Jianliang Xu, Chia-Hui Chang and Wen-Chih Peng (workshop chairs), Xing Xie and Shou-De Lin (industrial program chairs), Wenjie Zhang, Wook-Shin Han and Hung-Yu Kao (demonstration chairs), and Ying Zhang and Mi-Yen Yeh (tutorial chairs), as well as the organizers of the workshops, their respective PC members and reviewers.

We are also grateful to all the members of the Organizing Committee and the numerous volunteers for their tireless work before and during the conference. Also, we would like to express our sincere thanks to Chih-Ya Shen and Jen-Wei Huang (proceedings chairs) for working with the Springer team to produce the proceedings. Special thanks go to Xiaofang Zhou (DASFAA steering committee liaison) for his guidance. Lastly, we acknowledge the generous financial support from various industrial companies and academic institutes.

We hope that you will enjoy the DASFAA 2021 conference, its technical program and the proceedings!

February 2021

Wang-Chien Lee
Vincent S. Tseng
Vana Kalogeraki

Organization

Organizing Committee

Honorary Chairs

Philip S. Yu	University of Illinois at Chicago, USA
Ming-Syan Chen	National Taiwan University, Taiwan
Masaru Kitsuregawa	University of Tokyo, Japan

General Chairs

Christian S. Jensen	Aalborg University, Denmark
Ee-Peng Lim	Singapore Management University, Singapore
De-Nian Yang	Academia Sinica, Taiwan

Program Committee Chairs

Wang-Chien Lee	Pennsylvania State University, USA
Vincent S. Tseng	National Chiao Tung University, Taiwan
Vana Kalogeraki	Athens University of Economics and Business, Greece

Steering Committee

BongHee Hong	Pusan National University, Korea
Xiaofang Zhou	University of Queensland, Australia
Yasushi Sakurai	Osaka University, Japan
Lei Chen	Hong Kong University of Science and Technology, Hong Kong
Xiaoyong Du	Renmin University of China, China
Hong Gao	Harbin Institute of Technology, China
Kyuseok Shim	Seoul National University, Korea
Krishna Reddy	IIIT, India
Yunmook Nah	DKU, Korea
Wenjia Zhang	University of New South Wales, Australia
Guoliang Li	Tsinghua University, China
Sourav S. Bhowmick	Nanyang Technological University, Singapore
Atsuyuki Morishima	University of Tsukaba, Japan
Sang-Won Lee	SKKU, Korea

Industrial Program Chairs

Xing Xie   Microsoft Research Asia, China
Shou-De Lin   Appier, Taiwan

Demo Chairs

Wenjie Zhang   University of New South Wales, Australia
Wook-Shin Han   Pohang University of Science and Technology, Korea
Hung-Yu Kao   National Cheng Kung University, Taiwan

Tutorial Chairs

Ying Zhang   University of Technology Sydney, Australia
Mi-Yen Yeh   Academia Sinica, Taiwan

Workshop Chairs

Chia-Hui Chang   National Central University, Taiwan
Jianliang Xu   Hong Kong Baptist University, Hong Kong
Wen-Chih Peng   National Chiao Tung University, Taiwan

Panel Chairs

Zi Huang   The University of Queensland, Australia
Takahiro Hara   Osaka University, Japan
Shan-Hung Wu   National Tsing Hua University, Taiwan

Ph.D Consortium

Lydia Chen   Delft University of Technology, Netherlands
Kun-Ta Chuang   National Cheng Kung University, Taiwan

Publicity Chairs

Wen Hua   The University of Queensland, Australia
Yongxin Tong   Beihang University, China
Jiun-Long Huang   National Chiao Tung University, Taiwan

Proceedings Chairs

Jen-Wei Huang   National Cheng Kung University, Taiwan
Chih-Ya Shen   National Tsing Hua University, Taiwan

Registration Chairs

Chuan-Ju Wang	Academia Sinica, Taiwan
Hong-Han Shuai	National Chiao Tung University, Taiwan

Sponsor Chair

Chih-Hua Tai	National Taipei University, Taiwan

Web Chairs

Ya-Wen Teng	Academia Sinica, Taiwan
Yi-Cheng Chen	National Central University, Taiwan

Finance Chair

Yi-Ling Chen	National Taiwan University of Science and Technology, Taiwan

Local Arrangement Chairs

Chien-Chin Chen	National Taiwan University, Taiwan
Chih-Chieh Hung	National Chung Hsing University, Taiwan

DASFAA Steering Committee Liaison

Xiaofang Zhou	The Hong Kong University of Science and Technology, Hong Kong

Program Committee

Senior Program Committee Members

Zhifeng Bao	RMIT University, Vietnam
Sourav S. Bhowmick	Nanyang Technological University, Singapore
Nikos Bikakis	ATHENA Research Center, Greece
Kevin Chang	University of Illinois at Urbana-Champaign, USA
Lei Chen	Hong Kong University of Science and Technology, China
Bin Cui	Peking University, China
Xiaoyong Du	Renmin University of China, China
Hakan Ferhatosmanoglu	University of Warwick, UK
Avigdor Gal	Israel Institute of Technology, Israel
Hong Gao	Harbin Institute of Technology, China
Dimitrios Gunopulos	University of Athens, Greece
Bingsheng He	National University of Singapore, Singapore
Yoshiharu Ishikawa	Nagoya University, Japan

Nick Koudas	University of Toronto, Canada
Wei-Shinn Ku	Auburn University, USA
Dik-Lun Lee	Hong Kong University of Science and Technology, China
Dongwon Lee	Pennsylvania State University, USA
Guoliang Li	Tsinghua University, China
Ling Liu	Georgia Institute of Technology, USA
Chang-Tien Lu	Virginia Polytechnic Institute and State University, USA
Mohamed Mokbel	University of Minnesota Twin Cities, USA
Mario Nascimento	University of Alberta, Canada
Krishna Reddy P.	International Institute of Information Technology, India
Dimitris Papadias	The Hong Kong University of Science and Technology, China
Wen-Chih Peng	National Chiao Tung University, Taiwan
Evaggelia Pitoura	University of Ioannina, Greece
Cyrus Shahabi	University of Southern California, USA
Kyuseok Shim	Seoul National University, Korea
Kian-Lee Tan	National University of Singapore, Singapore
Yufei Tao	The Chinese University of Hong Kong, China
Vassilis Tsotras	University of California, Riverside, USA
Jianyong Wang	Tsinghua University, China
Matthias Weidlich	Humboldt-Universität zu Berlin, Germany
Xiaokui Xiao	National University of Singapore, Singapore
Jianliang Xu	Hong Kong Baptist University, China
Bin Yang	Aalborg University, Denmark
Jeffrey Xu Yu	The Chinese University of Hong Kong, China
Wenjie Zhang	University of New South Wales, Australia
Baihua Zheng	Singapore Management University, Singapore
Aoying Zhou	East China Normal University, China
Xiaofang Zhou	The University of Queensland, Australia
Roger Zimmermann	National University of Singapore, Singapore

Program Committee Members

Alberto Abelló	Universitat Politècnica de Catalunya, Spain
Marco Aldinucci	University of Torino, Italy
Toshiyuki Amagasa	University of Tsukuba, Japan
Ting Bai	Beijing University of Posts and Telecommunications, China
Spiridon Bakiras	Hamad Bin Khalifa University, Qatar
Wolf-Tilo Balke	Technische Universität Braunschweig, Germany
Ladjel Bellatreche	ISAE-ENSMA, France
Boualem Benatallah	University of New South Wales, Australia
Athman Bouguettaya	University of Sydney, Australia
Panagiotis Bouros	Johannes Gutenberg University Mainz, Germany

Stéphane Bressan	National University of Singapore, Singapore
Andrea Cali	Birkbeck University of London, UK
K. Selçuk Candan	Arizona State University, USA
Lei Cao	Massachusetts Institute of Technology, USA
Xin Cao	University of New South Wales, Australia
Yang Cao	Kyoto University, Japan
Sharma Chakravarthy	University of Texas at Arlington, USA
Tsz Nam Chan	Hong Kong Baptist University, China
Varun Chandola	University at Buffalo, USA
Lijun Chang	University of Sydney, Australia
Cindy Chen	University of Massachusetts Lowell, USA
Feng Chen	University of Texas at Dallas, USA
Huiyuan Chen	Case Western Reserve University, USA
Qun Chen	Northwestern Polytechnical University, China
Rui Chen	Samsung Research America, USA
Shimin Chen	Chinese Academy of Sciences, China
Yang Chen	Fudan University, China
Brian Chen	Columbia University, USA
Tzu-Ling Cheng	National Taiwan University, Taiwan
Meng-Fen Chiang	Auckland University, New Zealand
Theodoros Chondrogiannis	University of Konstanz, Germany
Chi-Yin Chow	City University of Hong Kong, China
Panos Chrysanthis	University of Pittsburgh, USA
Lingyang Chu	Huawei Technologies Canada, Canada
Kun-Ta Chuang	National Cheng Kung University, Taiwan
Jonghoon Chun	Myongji University, Korea
Antonio Corral	University of Almeria, Spain
Alfredo Cuzzocrea	Universitá della Calabria, Italy
Jian Dai	Alibaba Group, China
Maria Luisa Damiani	University of Milan, Italy
Lars Dannecker	SAP SE, Germany
Alex Delis	National and Kapodistrian University of Athens, Greece
Ting Deng	Beihang University, China
Bolin Ding	Alibaba Group, China
Carlotta Domeniconi	George Mason University, USA
Christos Doulkeridis	University of Piraeus, Greece
Eduard Dragut	Temple University, USA
Amr Ebaid	Purdue University, USA
Ahmed Eldawy	University of California, Riverside, USA
Sameh Elnikety	Microsoft Research, USA
Damiani Ernesto	University of Milan, Italy
Ju Fan	Renmin University of China, China
Yixiang Fang	University of New South Wales, Australia
Yuan Fang	Singapore Management University, Singapore
Tao-yang Fu	Penn State University, USA

Yi-Fu Fu	National Taiwan University, Taiwan
Jinyang Gao	Alibaba Group, China
Shi Gao	Google, USA
Wei Gao	Singapore Management University, Singapore
Xiaofeng Gao	Shanghai Jiaotong University, China
Xin Gao	King Abdullah University of Science and Technology, Saudi Arabia
Yunjun Gao	Zhejiang University, China
Jingyue Gao	Peking University, China
Neil Zhenqiang Gong	Iowa State University, USA
Vikram Goyal	Indraprastha Institute of Information Technology, Delhi, India
Chenjuan Guo	Aalborg University, Denmark
Rajeev Gupta	Microsoft India, India
Ralf Hartmut Güting	Fernuniversität in Hagen, Germany
Maria Halkidi	University of Pireaus, Greece
Takahiro Hara	Osaka University, Japan
Zhenying He	Fudan University, China
Yuan Hong	Illinois Institute of Technology, USA
Hsun-Ping Hsieh	National Cheng Kung University, Taiwan
Bay-Yuan Hsu	National Taipei University, Taiwan
Haibo Hu	Hong Kong Polytechnic University, China
Juhua Hu	University of Washington, USA
Wen Hua	The University of Queensland, Australia
Jiun-Long Huang	National Chiao Tung University, Taiwan
Xin Huang	Hong Kong Baptist University, China
Eenjun Hwang	Korea University, Korea
San-Yih Hwang	National Sun Yat-sen University, Taiwan
Saiful Islam	Griffith University, Australia
Mizuho Iwaihara	Waseda University, Japan
Jiawei Jiang	ETH Zurich, Switzerland
Bo Jin	Dalian University of Technology, China
Cheqing Jin	East China Normal University, China
Sungwon Jung	Sogang University, Korea
Panos Kalnis	King Abdullah University of Science and Technology, Saudi Arabia
Verena Kantere	National Technical University of Athens, Greece
Hung-Yu Kao	National Cheng Kung University, Taiwan
Katayama Kaoru	Tokyo Metropolitan University, Japan
Bojan Karlas	ETH Zurich, Switzerland
Ioannis Katakis	University of Nicosia, Cyprus
Norio Katayama	National Institute of Informatics, Japan
Chulyun Kim	Sookmyung Women's University, Korea
Donghyun Kim	Georgia State University, USA
Jinho Kim	Kangwon National University, Korea
Kyoung-Sook Kim	Artificial Intelligence Research Center, Japan

Seon Ho Kim	University of Southern California, USA
Younghoon Kim	HanYang University, Korea
Jia-Ling Koh	National Taiwan Normal University, Taiwan
Ioannis Konstantinou	National Technical University of Athens, Greece
Dimitrios Kotzinos	University of Cergy-Pontoise, France
Manolis Koubarakis	University of Athens, Greece
Peer Kröger	Ludwig-Maximilians-Universität München, Germany
Jae-Gil Lee	Korea Advanced Institute of Science and Technology, Korea
Mong Li Lee	National University of Singapore, Singapore
Wookey Lee	Inha University, Korea
Wang-Chien Lee	Pennsylvania State University, USA
Young-Koo Lee	Kyung Hee University, Korea
Cheng-Te Li	National Cheng Kung University, Taiwan
Cuiping Li	Renmin University of China, China
Hui Li	Xidian University, China
Jianxin Li	Deakin University, Australia
Ruiyuan Li	Xidian University, China
Xue Li	The University of Queensland, Australia
Yingshu Li	Georgia State University, USA
Zhixu Li	Soochow University, Taiwan
Xiang Lian	Kent State University, USA
Keng-Te Liao	National Taiwan University, Taiwan
Yusan Lin	Visa Research, USA
Sebastian Link	University of Auckland, New Zealand
Iouliana Litou	Athens University of Economics and Business, Greece
An Liu	Soochow University, Taiwan
Jinfei Liu	Emory University, USA
Qi Liu	University of Science and Technology of China, China
Danyang Liu	University of Science and Technology of China, China
Rafael Berlanga Llavori	Universitat Jaume I, Spain
Hung-Yi Lo	National Taiwan University, Taiwan
Woong-Kee Loh	Gachon University, Korea
Cheng Long	Nanyang Technological University, Singapore
Hsueh-Chan Lu	National Cheng Kung University, Taiwan
Hua Lu	Roskilde University, Denmark
Jiaheng Lu	University of Helsinki, Finland
Ping Lu	Beihang University, China
Qiong Luo	Hong Kong University of Science and Technology, China
Zhaojing Luo	National University of Singapore, Singapore
Sanjay Madria	Missouri University of Science & Technology, USA
Silviu Maniu	Universite Paris-Sud, France
Yannis Manolopoulos	Open University of Cyprus, Cyprus
Marco Mesiti	University of Milan, Italy
Jun-Ki Min	Korea University of Technology and Education, Korea

Jun Miyazaki	Tokyo Institute of Technology, Japan
Yang-Sae Moon	Kangwon National University, Korea
Yasuhiko Morimoto	Hiroshima University, Japan
Mirella Moro	Universidade Federal de Minas Gerais, Brazil
Parth Nagarkar	New Mexico State University, USA
Miyuki Nakano	Tsuda University, Japan
Raymond Ng	The University of British Columbia, Canada
Wilfred Ng	The Hong Kong University of Science and Technology, China
Quoc Viet Hung Nguyen	Griffith University, Australia
Kjetil Nørvåg	Norwegian University of Science and Technology, Norway
Nikos Ntarmos	University of Glasgow, UK
Werner Nutt	Free University of Bozen-Bolzano, Italy
Makoto Onizuka	Osaka University, Japan
Xiao Pan	Shijiazhuang Tiedao University, China
Panagiotis Papapetrou	Stockholm University, Sweden
Noseong Park	George Mason University, USA
Sanghyun Park	Yonsei University, Korea
Chanyoung Park	University of Illinois at Urbana-Champaign, USA
Dhaval Patel	IBM TJ Watson Research Center, USA
Yun Peng	Hong Kong Baptist University, China
Zhiyong Peng	Wuhan University, China
Ruggero Pensa	University of Torino, Italy
Dieter Pfoser	George Mason University, USA
Jianzhong Qi	The University of Melbourne, Australia
Zhengping Qian	Alibaba Group, China
Xiao Qin	IBM Research, USA
Karthik Ramachandra	Microsoft Research India, India
Weixiong Rao	Tongji University, China
Kui Ren	Zhejiang University, China
Chiara Renso	Institute of Information Science and Technologies, Italy
Oscar Romero	Universitat Politècnica de Catalunya, Spain
Olivier Ruas	Inria, France
Babak Salimi	University of California, Riverside, USA
Maria Luisa Sapino	University of Torino, Italy
Claudio Schifanella	University of Turin, Italy
Markus Schneider	University of Florida, USA
Xuequn Shang	Northwestern Polytechnical University, China
Zechao Shang	Univesity of Chicago, USA
Yingxia Shao	Beijing University of Posts and Telecommunications, China
Chih-Ya Shen	National Tsing Hua University, Taiwan
Yanyan Shen	Shanghai Jiao Tong University, China
Yan Shi	Shanghai Jiao Tong University, China
Junho Shim	Sookmyung Women's University, Korea

Hiroaki Shiokawa	University of Tsukuba, Japan
Hong-Han Shuai	National Chiao Tung University, Taiwan
Shaoxu Song	Tsinghua University, China
Anna Squicciarini	Pennsylvania State University, USA
Kostas Stefanidis	Tampere University, Finland
Kento Sugiura	Nagoya University, Japan
Aixin Sun	Nanyang Technological University, Singapore
Weiwei Sun	Fudan University, China
Nobutaka Suzuki	University of Tsukuba, Japan
Yu Suzuki	Nara Institute of Science and Technology, Japan
Atsuhiro Takasu	National Institute of Informatics, Japan
Jing Tang	National University of Singapore, Singapore
Lv-An Tang	NEC Labs America, USA
Tony Tang	National Taiwan University, Taiwan
Yong Tang	South China Normal University, China
Chao Tian	Alibaba Group, China
Yongxin Tong	Beihang University, China
Kristian Torp	Aalborg University, Denmark
Yun-Da Tsai	National Taiwan University, Taiwan
Goce Trajcevski	Iowa State University, USA
Efthymia Tsamoura	Samsung AI Research, Korea
Leong Hou U.	University of Macau, China
Athena Vakal	Aristotle University, Greece
Michalis Vazirgiannis	École Polytechnique, France
Sabrina De Capitani di Vimercati	Università degli Studi di Milano, Italy
Akrivi Vlachou	University of the Aegean, Greece
Bin Wang	Northeastern University, China
Changdong Wang	Sun Yat-sen University, China
Chaokun Wang	Tsinghua University, China
Chaoyue Wang	University of Sydney, Australia
Guoren Wang	Beijing Institute of Technology, China
Hongzhi Wang	Harbin Institute of Technology, China
Jie Wang	Indiana University, USA
Jin Wang	Megagon Labs, Japan
Li Wang	Taiyuan University of Technology, China
Peng Wang	Fudan University, China
Pinghui Wang	Xi'an Jiaotong University, China
Sen Wang	The University of Queensland, Australia
Sibo Wang	The Chinese University of Hong Kong, China
Wei Wang	University of New South Wales, Australia
Wei Wang	National University of Singapore, Singapore
Xiaoyang Wang	Zhejiang Gongshang University, China
Xin Wang	Tianjin University, China
Zeke Wang	Zhejiang University, China
Yiqi Wang	Michigan State University, USA

Raymond Chi-Wing Wong	Hong Kong University of Science and Technology, China
Kesheng Wu	Lawrence Berkeley National Laboratory, USA
Weili Wu	University of Texas at Dallas, USA
Chuhan Wu	Tsinghua University, China
Wush Wu	National Taiwan University, Taiwan
Chuan Xiao	Osaka University, Japan
Keli Xiao	Stony Brook University, USA
Yanghua Xiao	Fudan University, China
Dong Xie	Pennsylvania State University, USA
Xike Xie	University of Science and Technology of China, China
Jianqiu Xu	Nanjing University of Aeronautics and Astronautics, China
Fengli Xu	Tsinghua University, China
Tong Xu	University of Science and Technology of China, China
De-Nian Yang	Academia Sinica, Taiwan
Shiyu Yang	East China Normal University, China
Xiaochun Yang	Northeastern University, China
Yu Yang	City University of Hong Kong, China
Zhi Yang	Peking University, China
Chun-Pai Yang	National Taiwan University, Taiwan
Junhan Yang	University of Science and Technology of China, China
Bin Yao	Shanghai Jiaotong University, China
Junjie Yao	East China Normal University, China
Demetrios Zeinalipour Yazti	University of Cyprus, Turkey
Qingqing Ye	The Hong Kong Polytechnic University, China
Mi-Yen Yeh	Academia Sinica, Taiwan
Hongzhi Yin	The University of Queensland, Australia
Peifeng Yin	Pinterest, USA
Qiang Yin	Alibaba Group, China
Man Lung Yiu	Hong Kong Polytechnic University, China
Haruo Yokota	Tokyo Institute of Technology, Japan
Masatoshi Yoshikawa	Kyoto University, Japan
Baosheng Yu	University of Sydney, Australia
Ge Yu	Northeast University, China
Yi Yu	National Information Infrastructure Enterprise Promotion Association, Taiwan
Long Yuan	Nanjing University of Science and Technology, China
Kai Zeng	Alibaba Group, China
Fan Zhang	Guangzhou University, China
Jilian Zhang	Jinan University, China
Meihui Zhang	Beijing Institute of Technology, China
Xiaofei Zhang	University of Memphis, USA
Xiaowang Zhang	Tianjin University, China
Yan Zhang	Peking University, China
Zhongnan Zhang	Software School of Xiamen University, China

Pengpeng Zhao	Soochow University, Taiwan
Xiang Zhao	National University of Defence Technology, China
Bolong Zheng	Huazhong University of Science and Technology, China
Yudian Zheng	Twitter, USA
Jiaofei Zhong	California State University, East, USA
Rui Zhou	Swinburne University of Technology, Australia
Wenchao Zhou	Georgetown University, USA
Xiangmin Zhou	RMIT University, Vietnam
Yuanchun Zhou	Computer Network Information Center, Chinese Academy of Sciences, China
Lei Zhu	Shandong Normal Unversity, China
Qiang Zhu	University of Michigan-Dearborn, USA
Yuanyuan Zhu	Wuhan University, China
Yuqing Zhu	California State University, Los Angeles, USA
Andreas Züfle	George Mason University, USA

External Reviewers

Amani Abusafia	Sujatha Das Gollapalli
Ahmed Al-Baghdadi	Panos Drakatos
Balsam Alkouz	Venkatesh Emani
Haris B. C.	Abir Farouzi
Mohammed Bahutair	Chuanwen Feng
Elena Battaglia	Jorge Galicia Auyon
Kovan Bavi	Qiao Gao
Aparna Bhat	Francisco Garcia-Garcia
Umme Billah	Tingjian Ge
Livio Bioglio	Harris Georgiou
Panagiotis Bozanis	Jinhua Guo
Hangjia Ceng	Surabhi Gupta
Dipankar Chaki	Yaowei Han
Harry Kai-Ho Chan	Yongjing Hao
Yanchuan Chang	Xiaotian Hao
Xiaocong Chen	Huajun He
Tianwen Chen	Hanbin Hong
Zhi Chen	Xinting Huang
Lu Chen	Maximilian Hünemörder
Yuxing Chen	Omid Jafari
Xi Chen	Zijing Ji
Chen Chen	Yuli Jiang
Guo Chen	Sunhwa Jo
Meng-Fen Chiang	Seungwon Jung
Soteris Constantinou	Seungmin Jung
Jian Dai	Evangelos Karatzas

Enamul Karim
Humayun Kayesh
Jaeboum Kim
Min-Kyu Kim
Ranganath Kondapally
Deyu Kong
Andreas Konstantinidis
Gourav Kumar
Abdallah Lakhdari
Dihia Lanasri
Hieu Hanh Le
Suan Lee
Xiaofan Li
Xiao Li
Huan Li
Pengfei Li
Yan Li
Sizhuo Li
Yin-Hsiang Liao
Dandan Lin
Guanli Liu
Ruixuan Liu
Tiantian Liu
Kaijun Liu
Baozhu Liu
Xin Liu
Bingyu Liu
Andreas Lohrer
Yunkai Lou
Jin Lu
Rosni Lumbantoruan
Priya Mani
Shohei Matsugu
Yukai Miao
Paschalis Mpeis
Kiran Mukunda
Siwan No
Alex Ntoulas
Sungwoo Park
Daraksha Parveen
Raj Patel
Gang Qian
Jiangbo Qian
Gyeongjin Ra

Niranjan Rai
Weilong Ren
Matt Revelle
Qianxiong Ruan
Georgios Santipantakis
Abhishek Santra
Nadine Schüler
Bipasha Sen
Babar Shahzaad
Yuxin Shen
Gengyuan Shi
Toshiyuki Shimizu
Lorina Sinanaj
Longxu Sun
Panagiotis Tampakis
Eleftherios Tiakas
Valter Uotila
Michael Vassilakopoulos
Yaoshu Wang
Pei Wang
Kaixin Wang
Han Wang
Lan Wang
Lei Wang
Han Wang
Yuting Xie
Shangyu Xie
Zhewei Xu
Richeng Xuan
Kailun Yan
Shuyi Yang
Kai Yao
Fuqiang Yu
Feng (George) Yu
Changlong Yu
Zhuoxu Zhang
Liang Zhang
Shuxun Zhang
Liming Zhang
Jie Zhang
Shuyuan Zheng
Fan Zhou
Shaowen Zhou
Kai Zou

Contents – Part I

Spatial and Temporal Data

Contents – Part II

Data Mining

Machine Learning

Information Retrieval and Search

Social Network

Contents – Part III

Recommendation

Industrial Papers

Demo Papers

Ph.D Consortium

Tutorials

Big Data

Learning the Implicit Semantic Representation on Graph-Structured Data

Likang Wu[1], Zhi Li[1], Hongke Zhao[2], Qi Liu[1], Jun Wang[1], Mengdi Zhang[3], and Enhong Chen[1(✉)]

[1] Anhui Province Key Laboratory of Big Data Analysis and Application, University of Science and Technology of China, Hefei, China
{wulk,zhili03}@mail.ustc.edu.cn, {qiliuql,cheneh}@ustc.edu.cn
[2] Tianjin University, Tianjin, China
hongke@tju.edu.cn
[3] Meituan-Dianping Group, Beijing, China
zhangmengdi02@meituan.com

Abstract. Existing representation learning methods in graph convolutional networks are mainly designed by describing the neighborhood of each node as a perceptual whole, while the implicit semantic associations behind highly complex interactions of graphs are largely unexploited. In this paper, we propose a Semantic Graph Convolutional Networks (SGCN) that explores the implicit semantics by learning latent semantic-paths in graphs. In previous work, there are explorations of graph semantics via meta-paths. However, these methods mainly rely on explicit heterogeneous information that is hard to be obtained in a large amount of graph-structured data. SGCN first breaks through this restriction via leveraging the semantic-paths dynamically and automatically during the node aggregating process. To evaluate our idea, we conduct sufficient experiments on several standard datasets, and the empirical results show the superior performance of our model (Our code is available online at https://github.com/WLiK/SGCN_SemanticGCN).

Keywords: Graph neural networks · Semantic representation · Network analysis

1 Introduction

The representations of objects (nodes) in large graph-structured data, such as social or biological networks, have been proved extremely effective as feature inputs for graph analysis tasks. Recently, there have been many attempts in the literature to extend neural networks to deal with representation learning of graphs, such as Graph Convolutional Networks (GCN) [15], GraphSAGE [12] and Graph Attention Networks (GAT) [34].

In spite of enormous success, previous graph neural networks mainly proposed representation learning methods by describing the neighborhoods as a perceptual

C. S. Jensen et al. (Eds.): DASFAA 2021, LNCS 12681, pp. 3–19, 2021.
https://doi.org/10.1007/978-3-030-73194-6_1

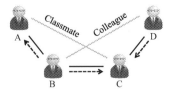

Fig. 1. Example of implicit semantic-paths in a scholar cooperation network. There are not explicit node (relation) types. Behind the same kind of relation (black solid edge), there are implicit factors (dotted line, A is the student of B, B is the advisor of C). So, the path A-B-C expresses "Student-Advisor-Student", A and C are "classmates". B-C-D expresses "Advisor-Student-Advisor", B and D are "colleagues".

whole, and they have not gone deep into the exploration of semantic information in graphs. Taking the movie network as an example, the paths based on composite relations of "Movie-Actor-Movie" and "Movie-Director-Movie" may reveal two different semantic patterns, i.e., the two movies have the same actor (director). Here the semantic pattern is defined as a specific knowledge expressed by the corresponding path. Although several researchers [30,35] attempt to capture these graph semantics of composite relations between two objects by meta-paths, existing work relies on the given heterogeneous information such as different types of objects and distinct object connections. However, in the real world, quite a lot of graph-structured data do not have the explicit characteristics. As shown in Fig. 1, in a scholar cooperation network, there are usually no explicit node (relation) types and all nodes are connected through the same relation, i.e., "Co-author". Fortunately, behind the same relation, there are various implicit factors which may express different connecting reasons, such as "Classmate" and "Colleague" for the same relation "Co-author". These factors can further compose diverse semantic-paths (e.g. "Student-Advisor-Student" and "Advisor-Student-Advisor"), which reveal sophisticated semantic associations and help to generate more informative representations. Then, how to automatically exploit comprehensive semantic patterns based on the implicit factors behind a general graph is a non-trivial problem.

In general, there are several challenges to solve this problem. Firstly, it is an essential part to adaptively infer latent factors behind graphs. We notice that several researches begin to explore desired latent factors behind a graph by disentangled representations [18,20]. However, they mainly focus on inferring the latent factors by the disentangled representation learning while failing to discriminatively model the independent implicit factors behind the same connections. Secondly, after discovering the latent factors, how to select the most meaningful semantics and aggregate the diverse semantic information remain largely unexplored. Last but not the least, to further exploit the implicit semantic patterns and to be capable of conducting inductive learning are quite difficult.

To address above challenges, in this paper, we propose a novel Semantic Graph Convolutional Networks (SGCN), which sheds light on the exploration of implicit semantics in the node aggregating process. Specifically, we first propose a latent factor routing method with the DisenConv layer [20] to adaptively infer the probability of each latent factor that may have caused the link from a given node to one of its neighborings. Then, for further exploring the diverse semantic information, we transfer the probability between every two connected nodes to the corresponding semantic adjacent matrix, which can present the semantic-paths in a graph. Afterwards, most semantic strengthen methods like the semantic level attention module can be easily integrated into our model and aggregate the diverse semantic information from these semantic-paths. Finally, to encourage the independence of the implicit semantic factors and conduct the inductive learning, we design an effective joint loss function to maintain the independent mapping channels of different factors. This loss function is able to focus on different semantic characteristics during the training process.

Specifically, the contributions of this paper can be summarized as follows:

- We first break the heterogeneous restriction of semantic representations with an end-to-end framework. It automatically infers the independent factor behind the formation of each edge and explores the semantic associations of latent factors behind a graph.
- We propose a novel Semantic Graph Convolutional Networks (SGCN), to learn node representations by aggregating the implicit semantics from the graph-structured data.
- We conduct extensive experiments on various real-world graphs datasets to evaluate the performance of the proposed model. The results show the superiority of our proposed model by comparing it with many powerful models.

2   Related Works

Graph neural networks (GNNs) [10,26], especially graph convolutional networks [13], have been proven successful in modeling the structured graph data due to its theoretical elegance [5]. They have made new breakthroughs in various tasks, such as node classification [15] and graph classification [6]. In the early days, the graph spectral theory [13] was used to derive a graph convolutional layer. Then, the polynomial spectral filters [6] greatly reduced the computational cost than before. And, Kipf and Welling [15] proposed the usage of a linear filter to get further simplification. Along with spectral graph convolution, directly performing graph convolution in the spatial domain was also investigated by many researchers [8,12]. Among them, graph attention networks [34] has aroused considerable research interest, since it adaptively specify weights to the neighbors of a node by attention mechanism [1,37].

For semantic learning research, there have been studies explored a kind of semantic-path called meta-path in heterogeneous graph embedding to preserve structural information. ESim [28] learned node representations by searching the user-defined embedding space. Based on random walk, meta-path2vec [7] utilized skip-gram to perform a semantic-path. HERec [29] proposed a type constraint strategy to filter the node sequence and captured the complex semantics reflected in heterogeneous graph. Then, Fan et al. [9] suggested a meta-graph2vec model for malware detection, where both the structures and semantics are preserved. Sun et al. [30] proposed meta-graph-based network embedding models, which simultaneously considers the hidden relations of all meta information of a meta-graph. Meanwhile, there were other influential semantic learning approaches in some studies. For instance, many models [4,17,25] were utilized to various fields because of their latent semantic analysis ability.

In heterogeneous graphs, two objects can be connected via different semantic-paths, which are called meta-paths. It depends on the characteristic that this graph structure has different types of nodes and relations. One meta-path Φ is defined as a path in the form of $A_1 \xrightarrow{R_1} A_2 \xrightarrow{R_2} \cdots \xrightarrow{R_l} A_{l+1}$ (abbreviated as $A_1 A_2 \cdots A_{l+1}$), it describes a composite relation $R = R_1 \circ R_2 \circ \cdots \circ R_l$, where $\circ$ denotes the composition operator on relations. Actually, in homogeneous graph, the relationships between nodes are also generated for different reasons (latent factors), so we can implicitly construct various types of relationships to extract various semantic-paths correspond to different semantic patterns, so as to improve the performance of GCN model from the perspective of semantic discovery.

3   Semantic Graph Convolutional Networks

In this section, we introduce the Semantic Graph Convolutional Networks (SGCN). We first present the notations, then describe the overall network progressively.

3.1   Preliminary

We focus primarily on undirected graphs, and it is straightforward to extend our approach to directed graphs. We define $G = (V, E)$ as a graph, comprised of the nodes set V and edges set E, and $|V| = N$ denotes the number of nodes. Each node $u \in V$ has a feature vector $\mathbf{x}_u \in \mathbb{R}^{d_{in}}$. We use $(u, v) \in E$ to indicate that there is an edge between node u and node v. Most graph convolutional networks can be regarded as an aggregation function $f(\cdot)$ that outputs the representations of nodes when given features of each node and its neighbors:

$$\mathbf{y} = f(\mathbf{x}_u, \mathbf{x}_v : (u, v) \in E \mid u \in V),$$

where the output $\mathbf{y} \in \mathbb{R}^{N \times d_{out}}$ denotes the representations of nodes. It means that neighborhoods of a node contains rich information, which can be aggregated to describe the node more comprehensively. Different from previous studies [12,15,34], in our work, proposed $f(\cdot)$ would automatically learn the semantic-path from graph data to explore corresponding semantic pattern.

3.2   Latent Factor Routing

Here we aim to introduce the disentangled algorithm that calculates the latent factors between every two objects. We assume that each node is composed of K independent components, hence there are K latent factors to be disentangled. For the node $u \in V$, the hidden representation of u is $\mathbf{h_u} = [\mathbf{e_{u,1}}, \mathbf{e_{u,2}}, ..., \mathbf{e_{u,K}}] \in \mathbb{R}^{K \times \frac{d_{out}}{K}}$, where $\mathbf{e}_{u,k} \in \mathbb{R}^{\frac{d_{out}}{K}} (k = 1, 2, ..., K)$ denotes corresponding aspect of node u that is pertinent to the k-th disentangled factor.

In the initial stage, we project its feature vector $\mathbf{x}_u$ into K different subspaces:

$$\mathbf{z}_{u,k} = \frac{\sigma(\mathbf{W_k x}_u + \mathbf{b}_k)}{\|\sigma(\mathbf{W_k x}_u + \mathbf{b}_k)\|_2}, \tag{1}$$

where $\mathbf{W}_k \in \mathbb{R}^{d_{in} \times \frac{d_{out}}{K}}$ and $\mathbf{b}_k \in \mathbb{R}^{\frac{d_{out}}{K}}$ are the mapping parameters and bias of k-th subspace, the nonlinear activation function σ is ReLU [23]. To capture aspect k of node u comprehensively, we construct $\mathbf{e}_{u,k}$ from both $\mathbf{z}_{u,k}$ and $\{\mathbf{z}_{v,k} : (u, v) \in E\}$, which can be utilized to identify the latent factors. Here we learn the probability of each factor by leveraging neighborhood routing mechanism [18,20], it is a DisenConv layer:

$$\mathbf{e}_{u,k}^t = \frac{\mathbf{z}_{u,k} + \sum_{v:(u,v)\in E} \mathbf{p}_{u,v}^{k,t-1} \mathbf{z}_{v,k}}{\left\|\mathbf{z}_{u,k} + \sum_{v:(u,v)\in E} \mathbf{p}_{u,v}^{k,t-1} \mathbf{z}_{v,k}\right\|_2}, \tag{2}$$

$$\mathbf{p}_{u,v}^{k,t} = \frac{\exp(\mathbf{z}_{v,k}^\top \mathbf{e}_{u,k}^t)}{\sum_{k=1}^K \exp(\mathbf{z}_{v,k}^\top \mathbf{e}_{u,k}^t)}, \tag{3}$$

where iteration $t = 1, 2, ..., T$, $\mathbf{p}_{u,v}^k$ indicates the probability that factor k indicates the reason why node u reaches neighbor v, and satisfies $\mathbf{p}_{u,v}^k \geq 0, \sum_{k=1}^K \mathbf{p}_{u,v}^k = 1$. The neighborhood routing mechanism will iteratively infer $\mathbf{p}_{u,v}^k$ and construct $\mathbf{e}_k$. Note that, there are total L DisenConv layers, $\mathbf{z}_{u,k}$ is assigned the value of $\mathbf{e}_{u,k}^T$ finally in each layer $l \leq L - 1$, more detail can refer to Algorithm 1.

3.3   Discriminative Semantic Aggregation

For the data that various relation types between nodes and their corresponding neighbors are explicit and fixed, it is easily to construct multiple sub-semantic graphs as the input data for multiple GCN model. As shown in Fig. 2(a), a heterogeneous graph G contains two different types of meta-paths (meta-path 1, meta-path 2). Then G can be decomposed to multiple graphs $\tilde{G}$ consisting of single semantic graph G_1 and G_2, where u and its neighbors are connected by path-relation 1(2) for each node u in $G_1(G_2)$.

However, we cannot simply transfer the pre-construct multiple graph method to all network architectures. In detail, for a graph with no different types of edges,

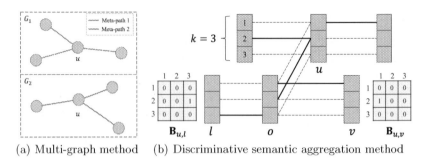

(a) Multi-graph method   (b) Discriminative semantic aggregation method

Fig. 2. A previous meta-paths representation on heterogeneous graph and our discriminative semantic aggregation method.

we have to judge implicit connecting factors of these edges to find semantic-paths. And the probability of each latent factor is calculated in the iteratively running process as mentioned in last section. To solve this dilemma, we propose a novel algorithm to automatically represent semantic-paths during the model running.

After the latent factor routing process, we get the soft probability matrix of node latents $\mathbf{p} \in \mathbb{R}^{N \times N \times K}$, where $0 \leq \mathbf{p}_{i,j}^k \leq 1$ means the possibility that node i connects to j because of the factor k. In our model, the latent factor should identify the certain connecting cause of each connected node pair. Here we transfer the probability matrix $\mathbf{p}$ to an semantic adjacent matrix $\mathbf{A}$, so the element in $\mathbf{A}$ only has binary value (0 or 1). In detail, for every node pair i and j, $\mathbf{A}_{i,j}^k = 1$ if $\mathbf{p}_{i,j}^k$ denotes the biggest value in $\mathbf{p}_{i,j}$. As shown in Fig. 2(b), each node is represented by K components. In this graph, every node may connect with others by one relationship from K types, e.g., the relationship between node u and o is R_2 (denotes $\mathbf{A}_{u,o}^2 = 1$). For node u, we can find that it has two semantic-path-based neighbors l and v. And, the semantic-paths of (u, l) and (u, v) are two different types which composed by $\Phi_{u,o,l} = (\mathbf{A}_{u,o}^2, \mathbf{A}_{o,l}^3) = R_2 \circ R_3$ and $\Phi_{u,o,v} = (\mathbf{A}_{u,o}^2, \mathbf{A}_{o,v}^1) = R_2 \circ R_1$ respectively. We define the adjacent matrix $\mathbf{B}$ for virtual semantic-path-based edges,

$$\mathbf{B}_{u,v} = \sum_{[(u,o),(o,v)] \in E} \mathbf{A}_{u,o}^\top \mathbf{A}_{o,v}, \quad \{u, v\} \subset V, \tag{4}$$

where $\mathbf{A}_{u,o} \in \mathbb{R}^K$, $\mathbf{A}_{o,v} \in \mathbb{R}^K$, and $\mathbf{B}_{u,v} \in \mathbb{R}^{K \times K}$. For instance, in Fig. 2(b), $\mathbf{A}_{u,o} = [0, 1, 0]$, $\mathbf{A}_{o,v} = [1, 0, 0]$, and $\mathbf{A}_{o,l} = [0, 0, 1]$, in this way two semantic-paths start from node u can be expressed as $\mathbf{B}_{u,l}^{2,3} = 1$ and $\mathbf{B}_{u,v}^{2,1} = 1$.

In the semantic information aggregation process, we aggregate the latent vectors connected by corresponding semantic-path as:

$$\begin{aligned}
\mathbf{h}_u &= [\mathbf{e}_{u,1}, \mathbf{e}_{u,2}, ..., \mathbf{e}_{u,K}] \in \mathbb{R}^{K \times \frac{d_{out}}{K}}, \\
\tilde{\mathbf{h}}_v &= [\mathbf{z}_{v,1}, \mathbf{z}_{v,2}, ..., \mathbf{z}_{v,K}] \in \mathbb{R}^{K \times \frac{d_{out}}{K}}, \\
\mathbf{y}_u &= \mathbf{h}_u + \underset{v \in V, v \neq u}{\text{MeanPooling}}(\mathbf{B}_{u,v} \tilde{\mathbf{h}}_v), \quad u \in V,
\end{aligned} \tag{5}$$

where we just use MeanPooling to avoid large values instead of $\sum_{v \in \mathbf{V}}$ operator, and $\mathbf{h}_u, \tilde{\mathbf{h}}_v \in \mathbb{R}^{K \times \frac{d_{out}}{K}}$ are both returned from the last layer of Disen-Conv operation, in this time that factor probabilities would be stable since the representation of each node considers the influence from neighbors. According to Eq. (5), the aggregation of two latent representations (end points) of one certain semantic-path denotes the mining result of this semantic relation, e.g., Pooling($\mathbf{e}_{u,2}, \mathbf{z}_{v,1}$) and Pooling($\mathbf{e}_{u,2}, \mathbf{z}_{l,3}$) express two different kinds of semantic pattern representations in Fig. 2(b), $R_2 \circ R_1$ and $R_2 \circ R_3$ respectively. And, for all types of semantic-paths start from node u, the weight of each type depends on its frequency. Note that, although the semantic adjacent matrix $\mathbf{A}$ neglects some low probability factors, our semantic paths are integrated with the node states of DisenGCN, which would not lose the crucial information captured by basic GCN model. The advantage of this aggregation method is that our model can distinguish different semantic relations without adding extra parameters, instead of designing various graph convolution networks for different semantic-paths. That is to say, the model does not increase the risk of over fitting after the graph semantic-paths learning. Here we only consider 2-order-paths in our model, however, it can be straightly extended to longer path mining.

3.4   Independence Learning for Mapping Subspaces

In fact, one type of edge in a meta-path tries to denote one unique meaning, so the K latent factors in our work should not overlap. So, the assumption of using latent factors to construct semantic-paths is that these different factors extracted by latent factor routing module can focus on different connecting causes. In other words, we should encourage the representations of different factors to be of sufficient independence. Before the probability calculating, on our features, the focused point views of K subspaces in Eq. (1) should keep different. Our solution considers that the distance between independence factor representations $\mathbf{z}_{i,k}, k \leq K$ should be sufficient long if they were projected to one subspace.

First, we project the input values $\mathbf{z}$ in Eq. (1) into an unified space to get vectors $\mathbf{Q}$ and $\mathbf{K}$ as follow:

$$\mathbf{Q} = \mathbf{zw}, \mathbf{K} = \mathbf{zw}, \qquad (6)$$

where $\mathbf{w} \in \mathbb{R}^{\frac{d_{out}}{K} \times \frac{d_{out}}{K}}$ is the projection parameter matrix. Then, the independence loss based on distances between unequal factor representations could be calculated as follow:

$$\mathcal{L}_i = \frac{1}{M} \sum \text{softmax}\left(\frac{\mathbf{Q}\mathbf{K}^{\top}}{\sqrt{\frac{d_{out}}{K}}}\right) \odot (1 - \mathbf{I}), \qquad (7)$$

where $\mathbf{I} \in \mathbb{R}^{K \times K}$ denotes an identity matrix, $\odot$ is element-wise product, $M = K^2 - K$. Specifically, we learn a lesson from [33] that scaling the dot products by $1/\sqrt{d_{out}/K}$, to counteract the gradients disappear effect for large values. As long as $\mathcal{L}_i$ is minimized in the training process, the distances between different factors tend to be larger, that is, the K subspaces would capture sufficient different information to encourage independence among learned latent factors.

Next, we would analyze the validity of this optimization. Latent Factor Routing aims to utilize the disentangled algorithm to calculate the latent factors between every two objects. However, this approach is a variant of von Mises-Fisher (vMF) [2] mixture model, such an EM algorithm cannot optimize the independences of latent factors within the iterative process. And random initialization of the mapping parameters is also not able to promise that subspaces obtain different concerns. For this shortcoming, we give an assumption:

Assumption 31. *The features in different subspaces keep sufficient independent when the margins of their projections in the unified space are sufficiently distinct.*

This assumption is inspired by the Latent Semantic Analysis algorithm (LSA) [16] that projects multi-dimensional features of a vector space model into a semantic space with less dimensions, which keeps the semantic features of the original space in a statistical sense. So, our optimization approach is listed below:

$$
\begin{aligned}
\mathbf{w} &= \arg\min \sum \mathrm{softmax}(\mathbf{Q}\mathbf{K}^{\mathrm{T}}) \odot (1 - \mathbf{I}), \\
&= \arg\min \sum_{u}^{V} \mathrm{softmax}((\mathbf{z_u}\mathbf{w})(\mathbf{z_u}\mathbf{w})^{\mathrm{T}}) \odot (1 - \mathbf{I}), \\
&= \arg\min \sum_{u}^{V} \frac{\sum_{k_1 \neq k_2} \exp(\mathbf{z}_{u,k_1}\mathbf{w} \cdot \mathbf{z}_{u,k_2}\mathbf{w})}{\sum_{k_1,k_2} \exp(\mathbf{z}_{u,k_1}\mathbf{w} \cdot \mathbf{z}_{u,k_2}\mathbf{w})}, \\
&= \arg\max \sum_{u}^{V} \sum_{k_1 \neq k_2} \mathrm{distance}(\mathbf{z}_{u,k_1}\mathbf{w}, \mathbf{z}_{u,k_2}\mathbf{w}).
\end{aligned}
\tag{8}
$$

$$S.t. : 1 \leq k_1 \leq K, \ 1 \leq k_2 \leq K.$$

In the above equation, $\mathbf{w}$ denotes the training parameter to be optimized. We ignore the $1/M$ and $1/\sqrt{d_{out}/K}$ in Eq. (7), because they do not affect the optimization procedure. With the increase of Inter-distances of K subspaces, the IntraVar of factors in each subspace would not larger than the original level (as the random initialization). The InterVar/IntraVar ratio becomes larger, in other word, we get more sufficient independence of mapping subspaces.

3.5   Algorithm Framework

In this section, we describe the overall algorithm of SGCN for performing node-related tasks. For graph G, the ground-truth label of node u is $\dagger_u \in \{0,1\}^{\mathcal{C}}$, where $\mathcal{C}$ is the number of classes. The details of our algorithm are shown in Algorithm 1. First, we calculate the independence loss $\mathcal{L}_i$ after factor channels capture features. Then, L layers of DisenConv operations would return the stable

---

Algorithm 1. Semantic Graph Convolutional Networks

Input: the feature vector matrix $\mathbf{x} \in \mathbb{R}^{N \times d_{in}}$, the graph $G = (V, E)$, the number of iterations T, and the number of disentangle layers L.
Output: the representation of node u by $\mathbf{y}_u \in \mathbb{R}^{d_{out}}, \forall u \in V$

1: **for** $i \in V$ **do**
2:      **for** $k = 1, 2, ..., K$ **do**
3:           $\mathbf{z}_{i,k} \leftarrow \sigma(\mathbf{W_k x}_i + \mathbf{b}_k)/\| \sigma(\mathbf{W_k x}_i + \mathbf{b}_k) \|_2$
4: $\mathbf{Q} \leftarrow \mathbf{z w}_q, \mathbf{K} \leftarrow \mathbf{z w}_k$
5: $\mathcal{L}_i = \frac{1}{M} \sum \mathrm{softmax}(\mathbf{Q K}^\top / \sqrt{\frac{d_{out}}{K}}) \odot (1 - \mathbf{I})$
6: **for** disentangle layer $l = 1, 2, ..., L$ **do**
7:      $\mathbf{e}_{u,k}^{t=1} \leftarrow \mathbf{z}_{u,k}, \forall k = 1, 2, ..., K, \forall u \in V$
8:      **for** routing iteration $t = 1, 2, ..., T$ **do**
9:           Get the soft probability matrix $\mathbf{p}$, where calculating $p_{u,v}^{k,t}$ by Eq. (3)
10:          Update the latent representation $e_{u,k}^t, \forall u \in V$ by Eq. (2)
11:     $\mathbf{e}_u \leftarrow \mathrm{dropout}(\mathrm{ReLU}(\mathbf{e}_u)), \mathbf{z}_{u,k} \leftarrow \mathbf{e}_{u,k}^{t=T}, \forall k = 1, 2, ..., K, \forall u \in V$ ◁ when $l \leq L - 1$
12: Transfer $\mathbf{p}$ to hard probability matrix $\mathbf{A}$
13: $\mathbf{B}_{u,v} \leftarrow \sum_{[(u,o),(o,v)] \in E} \mathbf{A}_{u,o}^\top \mathbf{A}_{o,v}, \{u, v\} \subset V$
14: Get each aggregation $\mathbf{y}_u^k$ of the latent vectors on semantic-paths by Eq. (5)
15: **return** $\{\mathbf{y}_u, \forall u \in V\}, \mathcal{L}_i$

---

probability matrix $\mathbf{p}$. After that, the automatic graph semantic-path representation $\mathbf{y}$ is learned based on $\mathbf{p}$. To apply $\mathbf{y}$ to different tasks, we design the final layer by a fully-connected layer $\mathbf{y}' = \mathbf{W}_y \mathbf{y} + \mathbf{b}_y$, where $\mathbf{W}_y \in \mathbb{R}^{d_{out} \times \mathcal{C}}$, $\mathbf{b}_y \in \mathbb{R}^{\mathcal{C}}$. For instance, for the semi-supervised node classification task, we implement

$$\mathcal{L}_s = - \sum_{u \in V^L} \frac{1}{\mathcal{C}} \sum_{c=1}^{\mathcal{C}} \dagger_u(c) \ln(\hat{\mathbf{y}}_u(c)) + \lambda \mathcal{L}_i \qquad (9)$$

as the loss function, where $\hat{\mathbf{y}}_u = \mathrm{softmax}(\mathbf{y}'_u)$, V^L is the set of labeled nodes, and $\mathcal{L}_i$ would be joint training by sum up with the task loss function. For the multi-label classification task, since the label $\dagger_u$ consists of more than one positive bits, we define the multi-label loss function for node u as:

$$\mathcal{L}_m = -\frac{1}{\mathcal{C}} \sum_{c=1}^{\mathcal{C}} [\dagger_u(c) \cdot \mathrm{sigmoid}(\mathbf{y}'_u(c)) + (1 - \dagger_u(c)) \cdot \mathrm{sigmoid}(-\mathbf{y}'_u(c))] + \lambda \mathcal{L}_i. \qquad (10)$$

Moreover, for the node clustering task, $\mathbf{y}'$ denotes the input feature of K-Means.

3.6 Time Complexity Analysis and Optimization

We should notice a problem in Sect. 3.3 that the time complexity of Eq. (4–5) by matrix calculation is $O(N(N-1)(N-2)K^2 + N((N-1)K^2 \times \frac{d_{out}}{K} + 2K\frac{d_{out}}{K})) \approx O(N^3 K^2 + N^2 K^2)$. Such a complex time complexity will bring a lot of computing load, so we optimize this algorithm in the actual implementation. For real-world datasets, one node connects to neighbors that are far less than the total number of nodes in the graph. Therefore, when we create the semantic-paths based adjacent matrix, the matrix $\tilde{\mathbf{A}} \in \mathbb{R}^{N \times C \times K}$ is defined to denote 1-order neighbor relationships, C is the maximum number of neighbors that we define, and $\tilde{\mathbf{A}}_u^k$ is the id of a neighbor if they are connected by R_k, else $\tilde{\mathbf{A}}_u^k = 0$. Then the semantic-path relations of type (R_{k_1}, R_{k_2}) of $u \in V$ are denoted by $\tilde{\mathbf{B}}_u^{k_1,k_2} = \tilde{\mathbf{A}}[\tilde{\mathbf{A}}[u,:,k_1],:,k_2] \in R^{C \times C}$, and the pooling of this semantic pattern is the mean pooling of $\mathbf{z}[\tilde{\mathbf{B}}_u^{k_1,k_2}, k_2, :]$. According to the analysis above, the time complexity can be reduced to $O(K^2(NC^2 + NC^2\frac{d_{out}}{K})) \approx O(2NK^2C^2)$.

Table 1. The statistics of datasets.

Dataset	Type	Nodes	Edges	Classes	Features	Multi-label
Pubmed	Citation network	19,717	44,338	3	500	False
Citeseer	Citation network	3,327	4,732	6	3,703	False
Cora	Citation network	2,708	5,429	7	1,433	False
Blogcatalog	Social network	10,312	333,983	39	–	True
POS	Word co-occurrence	4,777	184,812	40	–	True

4   Experiments

In this section, we empirically assess the efficacy of SGCN on several node-related tasks, includes semi-supervised node classification, node clustering and multi-label node classification. We then provide node visualization analysis and semantic-paths sampling experiments to verify the validity of our idea.

4.1   Experimental Setup

Datasets. We conduct our experiments on 5 real-world datasets, Citeseer, Cora, Pubmed, POS and BlogCatalog [11,27,32], whose statistics are listed in Table 1. The first three citation networks are benchmark datasets for semi-supervised node classification and node clustering. For graph content, the nodes, edges, and labels in these three represent articles, citations, and research areas, respectively. Their node features correspond a bag-of-words representation of a document.

POS and BlogCatalog are suitable for multi-label node classification task. Their labels are part-of-speech tags and user interests, respectively. In detail, BlogCatalog is a social relationships network of bloggers who post blogs in the BlogCatalog website. These labels represent the blogger's interests inferred from the text information provided by the blogger. POS (Part-of-Speech) is a co-occurrence network of words appearing in the first million bytes of the Wikipedia dump. The labels in POS denote the Part-of-Speech tags inferred via the Stanford POS-Tagger. Due to the two graphs do not provide node features, we use the rows of their adjacency matrices in place of node features for them.

Baselines. To demonstrate the advantages of our model, we compare SGCN with some representative graph neural networks, including the graph convolution network (GCN) [15] and the graph attention network (GAT) [34]. In detail, GCN [15] is a simplified spectral method of node aggregating, while GAT weights a node's neighbors by the attention mechanism. GAT achieves state of the art in many tasks, but it contains far more parameters than GCN and our model. Besides, ChebNet [6] is a spectral graph convolutional network by means of a Chebyshev expansion of the graph Laplacian, MoNet [22] extends CNN architectures by learning local, stationary, and compositional task-specific features. And IPGDN [18] is the advanced version of DisenGCN. We also implement other non-graph convolution network method, including random walk based network embedding DeepWalk [24], link-based classification method ICA [19], inductive embedding based approach Planetoid [38], label propagation approach LP [39], semi-supervised embedding learning model SemiEmb [36] and so on.

In addition, we conduct the ablation experiments into nodes classification and clustering to verify the effectiveness of the main components of SGCN: SGCN-path is our complete model without independence loss, and SGCN-indep denotes SGCN without the semantic-path representations.

In the multi-label classification experiment, the original implementations of GCN and GAT do not support multi-label tasks. We therefore modify them to use the same multi-label loss function as ours for fair comparison in multi-label tasks. We additionally include three node embedding algorithms, including DeepWalk [24], LINE [31], and node2vec [11], because they are demonstrated to perform strongly on the multi-label classification. Besides, we remove IPGDN since it is not designed for multi-label task.

Implementation Details. We train our models on one machine with 8 NVIDIA Tesla V100 GPUs. Some experimental results and the settings of common baselines that we follow [18,20], and we optimize the parameters of models with Adam [14]. Besides, we tune the hyper-parameters of both our model and baselines using hyperopt [3]. In detail, for semi-supervised classification and node clustering, we set the number of iterations $T = 6$, the layers $L \in \{1, 2, ..., 8\}$, the number of components $K \in \{1, 2, .., 7\}$ (denotes the number of mapping channels. Therefore, for our model, the dimension of a component in the SGCN model is $[d_{out}/K] \in \{10, 12, ..., 8\}$), dropout rate $\in \{0.05, 0.10, ..., 0.95\}$, trade-off $\lambda \in \{0.0, 0.5, ..., 10.0\}$, the learning rate $\sim$ loguniform $[e-8, 1]$, the l_2 regularization term $\sim$ loguniform $[e-10, 1]$. Besides, it should be noted that, in the multi-label node classification, the output dimension d_{out} is set to 128 to achieve better performance, while setting the dimension of the node embeddings to be 128 as well for other node embedding algorithms. And, when tuning the hyper-parameters, we set the number of components $K \in \{4, 8, ...28\}$ in the latent factor routing process. Here $K = 8$ makes the best result in our experiments.

Table 2. Semi-supervised classification.

Models	Cora	Citeseer	Pubmed
MLP	55.1	46.5	71.4
SemiEmb	59.0	59.6	71.1
LP	68.0	45.3	63.0
DeepWalk	67.2	43.2	65.3
ICA	75.1	69.1	73.9
Planetoid	75.7	64.7	77.2
ChebNet	81.2	69.8	74.4
GCN	81.5	70.3	79.0
MoNet	81.7	–	78.8
GAT	83.0	72.5	79.0
DisenGCN	83.7	73.4	80.5
IPGDN	84.1	74.0	81.2
SGCN-indep	84.2	73.7	82.0
SGCN-path	84.6	**74.4**	81.6
SGCN	**85.4**	74.2	**82.1**

Table 3. Node clustering with double metrics.

Models	Cora		Citeseer		Pubmed	
	NMI	ARI	NMI	ARI	NMI	ARI
SemiEmb	48.7	41.5	31.2	21.5	27.8	35.2
DeepWalk	50.3	40.8	30.5	20.6	29.6	36.6
Planetoid	52.0	40.5	41.2	22.1	32.5	33.9
ChebNet	49.8	42.4	42.6	41.5	35.6	38.6
GCN	51.7	48.9	42.8	42.8	35.0	40.9
GAT	57.0	54.1	43.1	43.6	35.0	41.4
DIsenGCN	58.4	60.4	43.7	42.5	36.1	41.6
IPGDN	59.2	61.0	44.3	43.0	37.0	42.0
SGCN-indep	60.2	59.2	44.7	42.8	37.2	42.3
SGCN-path	60.5	60.7	**45.1**	44.0	37.3	**42.8**
SGCN	**60.7**	**61.6**	44.9	**44.2**	**37.9**	42.5

4.2 Semi-Supervised Node Classification

For semi-supervised node classification, there are only 20 labeled instances for each class. It means that the information of neighbors should be leveraged when predicting the labels of target nodes. Here we follow the experimental settings of previous works [15,34,38].

We report the classification accuracy (ACC) results in Table 2. The majority of nodes only connect with those neighbors of the same class. According to Table 2, it is obvious that SGCN achieves the best performance amongst all baselines. Here SGCN outperforms the most powerful baseline IPGDN with 1.55%, 0.47% and 1.1% relative accuracy improvements on three datasets, compared with the increasing degrees of previous models, our model express obvious improvements in the node classification task. And our proposed model achieves the best ACC of 85.4% on Cora dataset, it is a great improvement on this dataset. On the other hand, in the ablation experiment (the last three rows of Table 2), the complete SGCN model is superior to either algorithm in at least two datasets. Moreover, we can find that SGCN-indep and SGCN-path are both perform better than previous algorithms to some degree. It reveals the effectiveness of our semantic-paths mining module and the independence learning for subspaces.

4.3 Multi-label Node Classification

In the multi-label classification experiment, every node is assigned one or more labels from a finite set $\mathcal{L}$. We follow node2vec [11] and report the performance of each method while varying the number of nodes labeled for training from $10\%\ |V|$ to $90\%\ |V|$, where $|V|$ is the total number of nodes. The rest of nodes are split equally to form a validation set and a test set. Then with the best hyper-parameters on the validation sets, we report the averaged performance of 30 runs on each multi-label test set. Here we summarize the results of multi-label node classification by Macro-F1 and Micro-F1 scores in Fig. 3. Firstly, there is an obvious point that proposed SGCN model achieves the best performances in both two datasets. Compared with DisenGCN model, SGCN combines with semantic-paths can achieve the biggest improvement of 20.0% when we set 10% of labeled nodes in POS dataset. The reason may be that the relation type of POS dataset is Word Co-occurrence, there are lots of regular explicit or implicit semantics amongst these relationships between different words. In the other dataset, although SGCN does not show a full lead but achieves the highest accuracy on both indicators. We find that the GCN-based algorithms are usually superior to the traditional node embedding algorithms in overall effect. Although for the Micro-F1 score on Blogcatalog, GCN produces the poor results. In addition, the SGCN algorithm can make both Macro-F1 and Micro-F2 achieve good results at the same time, and there will be no bad phenomenon in one of them. Because this approach would not ignore the information provided by the classes with few samples but important semantic relationships.

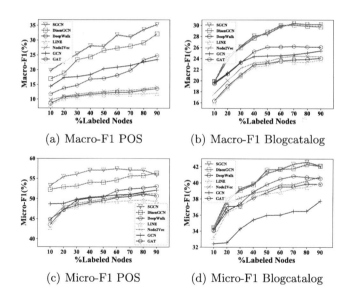

(a) Macro-F1 POS   (b) Macro-F1 Blogcatalog

(c) Micro-F1 POS   (d) Micro-F1 Blogcatalog

Fig. 3. Results of multi-label node classification.

4.4   Node Clustering

To further evaluate the embeddings learned from the above algorithms, we also conduct the clustering task. Following [18], for our model and each baseline, we obtain its node embedding via feed forward when the model is trained. Then we input the node embedding to the K-Means algorithm to cluster nodes. The ground-truth is the same as that of node classification task, and the number of clusters K is set to the number of classes. In detail, we employ two metrics of Normalized Mutual Information (NMI) and Average Rand Index (ARI) to validate the clustering results. Since the performance of K-Means is affected by initial centroids, we repeat the process for 20 times and report the average results in Table 3. As can be seen in Table 3, SGCN consistently outperforms all baselines, and GNN-based algorithms usually achieve better performance. Besides, with the semantic-path representation, SGCN and SGCN-path performs significantly better than DisenGCN and IPGDN, our proposed algorithm gets the best results on both NMI and ARI. It shows that SGCN captures a more meaningful node embedding via learning semantic patterns from graph.

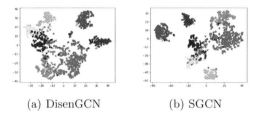

(a) DisenGCN	(b) SGCN

Fig. 4. Node representation visualization of Cora.

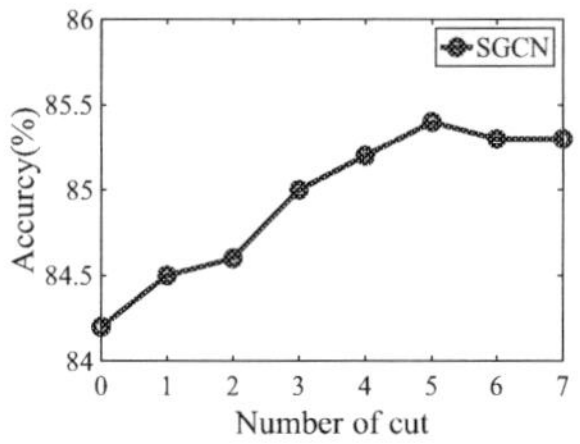

Fig. 5. Semantic-paths sampling.

4.5 Visualization Analysis and Semantic-Paths Sampling

We try to demonstrate the intuitive changes of node representations after incorporating semantic patterns. Therefore, we utilize t-SNE [21] to transform feature representations (node embedding) of SGCN and DisenGCN into a 2-dimensional space to make a more intuitive visualization. Here we visualize the node embedding of Cora (actually, the change of representation visualization is similar in other datasets), where different colors denote different research areas. According to Fig. 4, there is a phenomenon that the visualization of SGCN is more distinguishable than DisenGCN. It demonstrates that the embedding learned by SGCN presents a high intra-class similarity and separates papers into different research areas with distinct boundaries. On the contrary, DisenGCN dose not perform well since the inter-margin of clusters are not distinguishable enough. In several clusters, many nodes belong to different areas are mixed with others.

Then, to explore the influence of different scales of semantic-paths on our model performance, we implement a semantic-paths sampling experiment on Cora. As mentioned in the Sect. 3.6, for capturing different numbers of semantic paths, we change the hyper-parameter of cut size C to restrict the sampling size on each node's neighbors. As shown in Fig. 5, the SGCN model with the path representation achieves higher performances than the first point $(C = 0)$. From the perspective of global trend, with the increase of C, the classification accuracy of SGCN model is also improved steady, although it get the highest score when $C = 5$. It means that GCN model combines with more sufficient scale semantic-paths can really learn better node representations.

5 Conclusion

In this paper, we proposed a novel framework named Semantic Graph Convolutional Networks which incorporates the semantic-paths automatically during the node aggregating process. Therefore, SGCN provided the semantic learning ability to general graph algorithms. We conducted extensive experiments on various real-world datasets to evaluate the superior performance of our proposed model. Moreover, our method has good expansibility, all kinds of path-based algorithms in the graph embedding field can be directly applied in SGCN to adapt to different tasks, we will take more explorations in future work.

Acknowledgements. This research was partially supported by grants from the National Key Research and Development Program of China (No. 2018YFC0832101), and the National Natural Science Foundation of China (Nos. U20A20229 and 61922073). This research was also supported by Meituan-Dianping Group.

References

1. Bahdanau, D., Cho, K., Bengio, Y.: Neural machine translation by jointly learning to align and translate. arXiv preprint arXiv:1409.0473 (2014)
2. Banerjee, A., Dhillon, I.S., Ghosh, J., Sra, S.: Clustering on the unit hypersphere using von Mises-Fisher distributions. J. Mach. Learn. Res. **6**, 1345–1382 (2005)
3. Bergstra, J., Yamins, D., Cox, D.D.: Hyperopt: a python library for optimizing the hyperparameters of machine learning algorithms. In: Proceedings of the 12th Python in Science Conference, pp. 13–20. Citeseer (2013)
4. Blei, D.M., Ng, A.Y., Jordan, M.I.: Latent dirichlet allocation. J. Mach. Learn. Res. **3**, 993–1022 (2003). http://jmlr.org/papers/v3/blei03a.html
5. Bronstein, M.M., Bruna, J., LeCun, Y., Szlam, A., Vandergheynst, P.: Geometric deep learning: going beyond Euclidean data. IEEE Sig. Process. Mag. **34**(4), 18–42 (2017)
6. Defferrard, M., Bresson, X., Vandergheynst, P.: Convolutional neural networks on graphs with fast localized spectral filtering. In: Advances in Neural Information Processing Systems, pp. 3844–3852 (2016)
7. Dong, Y., Chawla, N.V., Swami, A.: metapath2vec: scalable representation learning for heterogeneous networks. In: Proceedings of the 23rd ACM SIGKDD International Conference on Knowledge Discovery and Data Mining, pp. 135–144 (2017)
8. Duvenaud, D.K., Maclaurin, D., Iparraguirre, J., Bombarell, R., Hirzel, T., Aspuru-Guzik, A.: Convolutional networks on graphs for learning molecular fingerprints. In: Advances in Neural Information Processing Systems, pp. 2224–2232 (2015)
9. Fan, Y., Hou, S., Zhang, Y., Ye, Y., Abdulhayoglu, M.: Gotcha-Sly Malware! Scorpion a metagraph2vec based malware detection system. In: Proceedings of the 24th ACM SIGKDD, pp. 253–262 (2018)

10. Gori, M., Monfardini, G., Scarselli, F.: A new model for learning in graph domains. In: Proceedings of the 2005 IEEE International Joint Conference on Neural Networks, vol. 2, pp. 729–734. IEEE (2005)
11. Grover, A., Leskovec, J.: node2vec: scalable feature learning for networks. In: Proceedings of the 22nd ACM SIGKDD, pp. 855–864 (2016)
12. Hamilton, W., Ying, Z., Leskovec, J.: Inductive representation learning on large graphs. In: NIPS, pp. 1024–1034 (2017)
13. Henaff, M., Bruna, J., LeCun, Y.: Deep convolutional networks on graph-structured data. arXiv preprint arXiv:1506.05163 (2015)
14. Kingma, D.P., Ba, J.: Adam: a method for stochastic optimization. In: 3rd International Conference on Learning Representations, ICLR 2015, San Diego, CA, USA, 7–9 May 2015, Conference Track Proceedings (2015)
15. Kipf, T.N., Welling, M.: Semi-supervised classification with graph convolutional networks. arXiv preprint arXiv:1609.02907 (2016)
16. Landauer, T.K., Foltz, P.W., Laham, D.: An introduction to latent semantic analysis. Discourse Process. **25**(2–3), 259–284 (1998)
17. Li, Z., Wu, B., Liu, Q., Wu, L., Zhao, H., Mei, T.: Learning the compositional visual coherence for complementary recommendations. In: IJCAI-2020, pp. 3536–3543 (2020)
18. Liu, Y., Wang, X., Wu, S., Xiao, Z.: Independence promoted graph disentangled networks. In: Proceedings of the AAAI Conference on Artificial Intelligence (2020)
19. Lu, Q., Getoor, L.: Link-based classification. In: Proceedings of the 20th International Conference on Machine Learning (ICML-2003), pp. 496–503 (2003)
20. Ma, J., Cui, P., Kuang, K., Wang, X., Zhu, W.: Disentangled graph convolutional networks. In: International Conference on Machine Learning, pp. 4212–4221 (2019)
21. van der Maaten, L., Hinton, G.: Visualizing data using t-SNE. J. Mach. Learn. Res. **9**, 2579–2605 (2008)
22. Monti, F., Boscaini, D., Masci, J., Rodola, E., Svoboda, J., Bronstein, M.M.: Geometric deep learning on graphs and manifolds using mixture model CNNs. In: IEEE Conference on Computer Vision and Pattern Recognition, pp. 5115–5124 (2017)
23. Nair, V., Hinton, G.E.: Rectified linear units improve restricted Boltzmann machines. In: Proceedings of the 27th International Conference on Machine Learning (ICML-2010), pp. 807–814 (2010)
24. Perozzi, B., Al-Rfou, R., Skiena, S.: DeepWalk: online learning of social representations. In: Proceedings of the 20th ACM SIGKDD, pp. 701–710 (2014)
25. Qiao, L., Zhao, H., Huang, X., Li, K., Chen, E.: A structure-enriched neural network for network embedding. Expert Syst. Appl. **117**, 300–311 (2019)
26. Scarselli, F., Gori, M., Tsoi, A.C., Hagenbuchner, M., Monfardini, G.: The graph neural network model. IEEE Trans. Neural Netw. **20**(1), 61–80 (2008)
27. Sen, P., Namata, G., Bilgic, M., Getoor, L., Galligher, B., Eliassi-Rad, T.: Collective classification in network data. AI Mag. **29**(3), 93–93 (2008)
28. Shang, J., Qu, M., Liu, J., Kaplan, L.M., Han, J., Peng, J.: Meta-path guided embedding for similarity search in large-scale heterogeneous information networks. arXiv preprint arXiv:1610.09769 (2016)
29. Shi, C., Hu, B., Zhao, W.X., Philip, S.Y.: Heterogeneous information network embedding for recommendation. IEEE Trans. Knowl. Data Eng. **31**(2), 357–370 (2018)
30. Sun, L., et al.: Joint embedding of meta-path and meta-graph for heterogeneous information networks. In: 2018 IEEE International Conference on Big Knowledge, pp. 131–138. IEEE (2018)

31. Tang, J., Qu, M., Wang, M., Zhang, M., Yan, J., Mei, Q.: LINE: large-scale information network embedding. In: Proceedings of the 24th International Conference on World Wide Web, pp. 1067–1077 (2015)
32. Tang, L., Liu, H.: Leveraging social media networks for classification. Data Min. Knowl. Disc. **23**(3), 447–478 (2011). https://doi.org/10.1007/s10618-010-0210-x
33. Vaswani, A., et al.: Attention is all you need. In: Advances in Neural Information Processing Systems, pp. 5998–6008 (2017)
34. Veličković, P., Cucurull, G., Casanova, A., Romero, A., Lio, P., Bengio, Y.: Graph attention networks. arXiv preprint arXiv:1710.10903 (2017)
35. Wang, X., et al.: Heterogeneous graph attention network. In: The World Wide Web Conference, pp. 2022–2032 (2019)
36. Weston, J., Ratle, F., Mobahi, H., Collobert, R.: Deep learning via semi-supervised embedding. In: Montavon, G., Orr, G.B., Müller, K.-R. (eds.) Neural Networks: Tricks of the Trade. LNCS, vol. 7700, pp. 639–655. Springer, Heidelberg (2012). https://doi.org/10.1007/978-3-642-35289-8_34
37. Wu, L., Li, Z., Zhao, H., Pan, Z., Liu, Q., Chen, E.: Estimating early fundraising performance of innovations via graph-based market environment model. In: AAAI, pp. 6396–6403 (2020)
38. Yang, Z., Cohen, W.W., Salakhutdinov, R.: Revisiting semi-supervised learning with graph embeddings. arXiv preprint arXiv:1603.08861 (2016)
39. Zhu, X., Ghahramani, Z., Lafferty, J.D.: Semi-supervised learning using gaussian fields and harmonic functions. In: Proceedings of the 20th International Conference on Machine Learning (ICML-2003), pp. 912–919 (2003)

Multi-job Merging Framework and Scheduling Optimization for Apache Flink

Hangxu Ji[1], Gang Wu[1(✉)], Yuhai Zhao[1], Ye Yuan[2], and Guoren Wang[2]

[1] School of Computer Science and Engineering, Northeastern University,
Shenyang, China
wugang@mail.neu.edu.cn
[2] School of Computer Science and Technology, Beijing Institute of Technology,
Beijing, China

Abstract. With the popularization of big data technology, distributed computing systems are constantly evolving and maturing, making substantial contributions to the query and analysis of massive data. However, the insufficient utilization of system resources is an inherent problem of distributed computing engines. Particularly, when more jobs lead to execution blocking, the system schedules multiple jobs on a first-come-first-executed (FCFE) basis, even if there are still many remaining resources in the cluster. Therefore, the optimization of resource utilization is key to improving the efficiency of multi-job execution. We investigated the field of multi-job execution optimization, designed a multi-job merging framework and scheduling optimization algorithm, and implemented them in the latest generation of a distributed computing system, Apache Flink. In summary, the advantages of our work are highlighted as follows: (1) the framework enables Flink to support multi-job collection, merging and dynamic tuning of the execution sequence, and the selection of these functions are customizable. (2) with the multi-job merging and optimization, the total running time can be reduced by 31% compared with traditional sequential execution. (3) the multi-job scheduling optimization algorithm can bring 28% performance improvement, and in the average case can reduce the cluster idle resources by 61%.

Keywords: Multi-job merging · Scheduling optimization · Distributed computing · Flink

1 Introduction

The IT industry term "Big Data" has existed for more than a decade and is a household term. To provide improved support for massive data computing, researchers have developed various distributed computing systems and are constantly releasing new versions of them to improve the system performance and enrich system functions.

Apache Flink [2] is the latest generation of distributed computing systems and exhibits high throughput and low latency when processing massive data.

© Springer Nature Switzerland AG 2021
C. S. Jensen et al. (Eds.): DASFAA 2021, LNCS 12681, pp. 20–36, 2021.
https://doi.org/10.1007/978-3-030-73194-6_2

It can cache intermediate data and support incremental iteration using its own optimizer. Many experimental studies, optimization technologies, and application platforms based on Flink are emerging because of its numerous advantages. For example, in the early days of Flink's birth, most of the research focused on the comparison between Flink and Spark [10,11,15], and pointed out that Flink is more suitable for future data computing. With the popularity of Flink, recent researches include testing tools based on Flink [9], multi-query optimization technology [16], and recommender systems [5], etc.

However, almost the distributed computing systems exhibit insufficient utilization of the hardware resources. Although Flink maximizes resource utilization by introducing TaskSlot to isolate memory, idle resources also exist because of the low parallelism of some Operators during traditional sequential execution. Moreover, when a user submits multiple jobs, Flink only run them on a first-come-first-executed (FCFE) basis, which cannot make jobs share the Slots. In a worse-case scenario, if job A is executing and job B after it does not meet the execution conditions because of insufficient remaining resources, job C cannot be executed in advance even though job C after job B meets the execution conditions, causing severe wastage of resources. These FCFE strategies of running multiple jobs only ensure fairness at the job level, but are not desired by users. In most cases, users only desire the minimum total execution time for all jobs. The above problems can be solved by simultaneously executing multiple jobs and dynamically adjusting the job execution sequence so that jobs that meet the execution conditions can be executed in advance.

In this study, we review the problem of insufficient utilization of system resources due to the fact that Flink does not support simultaneous execution of multiple jobs and optimization of execution sequence, and then focus on the multi-job efficiency improvement brought about by increasing Slot occupancy rate. The basic idea is to make simultaneously executing through multi-job merging and dynamically adjusting the execution sequence through multi-job scheduling, and the contributions of this paper are summarized below.

(1) We propose a groundbreaking framework that can support multi-job merging and scheduling in Flink. It can collect and parse multiple jobs to be executed, and generate new job execution plans through two optimization methods of multi-job merging and scheduling, and submit them to Flink for execution.

(2) To simultaneously execute multiple jobs, we propose multi-job merging algorithms based on subgraph isomorphism and heuristic strategies to enable multiple jobs to share the Slots. Both two algorithms can improve the efficiency and adapt to different job scenarios during the experiment.

(3) To dynamically adjusting the job execution sequence, we propose multi-job scheduling algorithm based on maximum parallelism to make jobs that satisfy the remaining resources execute in advance. Experimental results demonstrate that the algorithm can enhance the efficiency and reduce free resources.

The remainder of this paper is organized into 5 sections. Section 2 introduces the Flink DAGs, Flink Slots, and summarizes the current contributions toward improving resource utilization in distributed computing. Section 3 introduces the multi-job collection and execution agent, including the components, implementation method, and function. Section 4 describes multi-job execution optimization algorithms, including merging optimization and scheduling optimization. Section 5 presents the performance evaluation with respect to the running time and the number of Slots idle. Section 6 presents a brief conclusion.

2 Background and Related Work

In this section, we first summarized some of the implementation principles in Flink, including the composition and generation process of Flink DAGs, the functions and advantages of Flink Slot, to verify the feasibility of our work. Then, the related work of distributed job generation optimization and scheduling optimization is explained, and the advantages and deficiencies of existing work are pointed out.

2.1 Flink DAGs

Flink uses DAGs (Directed Acyclic Graphs) to abstract operations, which are more able to express the data processing flow than the traditional MapReduce [7]. According to the job submission and deployment process, Flink DAGs mainly includes JobGraph and ExecutionGraph. Figure 1 depicts the process of generating Flink DAGs. First, the system create JobVertexIDs based on the Operators in the job, chains the Operators through the Optimizer, and then generates a JobGraph by adding JobEdges and setting attributes. JobGraph is composed of three basic elements: JobVertex, JobEdge and IntermediateDataSet,

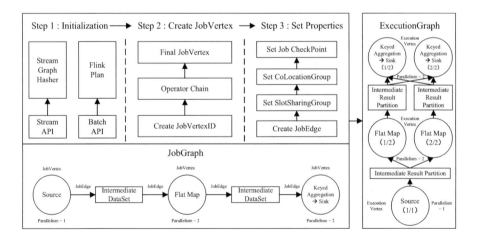

Fig. 1. The process of generating Flink DAGs (JobGraph and ExecutionGraph)

and contains all the contents of a job by assigning various attributes to these three elements. Finally, JobManager divides JobGraph into ExecutionVertex, IntermediateResultPartition and ExecutionEdge equal to its number according to the degree of parallelism to generate the final ExecutionGraph. Therefore, the research on JobGraph generation process is the core of Flink optimization, and it is also the focus of this work.

2.2 Flink Slot

In order to control the number of subtasks run by internal threads, Flink introduced TaskSlot as the minimum resource unit. The advantage of Slot is that it isolates memory resources, so jobs transmitted from JobMaster can be independently executed in different Slot, which can improve the utilization of cluster resources. As shown in Fig. 2, TaskManagers receive the task to be deployed from JobManager. If a TaskManager has four Slots, it will allocate 25% of memory for each Slot. One or more threads can be in each Slot, and threads in the same Slot share the same JVM. When subtasks belong to the same job, Flink also allows sharing Slot, which can not only quickly execute some tasks that consume less resources, but also logically remove redundant calculations that consume resources. It is precisely because of the existence of shared Slot that the multi-job merging and optimization techniques we will introduce in Sect. 4 are possible.

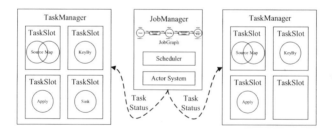

Fig. 2. Flink task deployment

2.3 Related Work

Current distributed computing systems, such as Spark [23] and Flink, are executed by converting complex programming logic into simple DAGs. The complex programming logic is mainly reflected in the user-defined function (UDF) in the Operators, so most of the research is to analyze UDF and construct optimization technology. Mainstream DAGs generation and optimization strategies include nested query decomposition technologies involving UDF [13], Operator reuse method [17,19], and Operator rearrangement algorithms [18], etc. In addition, part of the research is based on UDF code analysis to seek optimization opportunities [1,8,12,17]. In terms of distributed job scheduling optimization and load balancing, each distributed computing system has its own scheduler as its core component [3,21]. At the same time, due to the increasing complexity of distributed operations and the continuous expansion of node scale, a large number of optimization technologies have been born. For example, in the research on Hadoop, researchers have proposed scheduling strategies based on job size [22], resource quantity [14], and deadline awareness [4]. In Spark based on memory computing, current research includes interference-aware job scheduling algorithm [21], job scheduling algorithm based on I/O efficiency [20], etc. Although the above works have improved the job efficiency, they are all oriented to a single job, without considering the mutual influence between multiple jobs.

3 Framework Structure

3.1 Model

We propose a framework, which is an Agent implemented between the Flink Client and Flink JobManager, and capable of supporting multi-job merging and scheduling optimization. The ultimate goal of the framework is to generate optimized Flink DAGs. The composition of this framework is illustrated in Fig. 3. In the following, we will introduce Collector, Parser and Generator respectively. The Optimizer will be described in detail in Sect. 4.

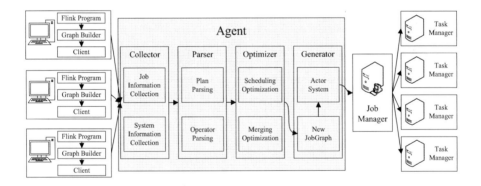

Fig. 3. The structure of the framework

Collector. In order to improve the multi-job efficiency by improving the utilization of system resources, the Optimizer must receive job information and system information as the data to be analyzed. Therefore, the Agent first provides a Collector, which collects the following information:

- **Jar Files:** As mentioned above, Flink abstracts computational logic in the form of DAGs. DAGs contain important information such as operators and UDFs, which are encapsulated in the Jar files.

- **Flink Plan:** The Flink Plan contains the deployment and strategy for job execution, mainly including the following information: GenricDataSinkBase in a collective form, which contains data terminal attributes; Cache files in the form of a HashMap, where the key is the name of the file and the value is the path to the file; ExecutionConfig is the configuration information for executing a job, which includes the adopted execution policy and the recovery method for the failed restart.
- **System Resources:** These mainly include the number of CPU cores and the memory size of each node in the cluster, as well as specific information in the Flink configuration file, including parallelism, Slot quantity, etc.

Parser. Because there are nested composite attributes inside Flink Plan and Operator, they cannot be directly converted to transferable byte streams, so a Parser is built in the Agent to serialize them. Algorithm 1 shows the process of serialization. First, gets the Sinks collection in Flink Plan (lines 1–3), and then assign the Operator and the user code encapsulation properties in it to the new object (lines 4–5). Finally, CatchFile, ExecutionConfig, JobID and the user code encapsulation properties are serialized respectively (lines 6–9). The final result is written on the Agent (line 10). At this point, the Agent has obtained all the information the Optimizer needs.

Algorithm 1: Serialization

Input: Flink Plan

Output: Job information of byte stream type

1 get DataSinkBase and GenericDataSinkBase from Plan;
2 get Operator op from GenericDataSinkBase;
3 read input Operator from DataSinkBase;
4 **if** *UserCodeWraper is not null* **then**
5 | Set UserCodeWraper to OperatorEx;

6 Serialize(CatchFile); Serialize(ExecutionConfig); Serialize(OperatorEx);
7 **if** *length > Buffer_Size* **then**
8 | add Buffer_Size;

9 Serialize(JobID);
10 writeToAgent;

Generator. The Generator first receives the optimization strategy sent from the Optimizer, including the jobs that can be merged and the optimal execution sequence of the jobs. The multi-job JobGraph is then generated by calling the implemented multiJobGraphGenerator. Since the multi-job scheduling optimization is an optimization in job execution order, when the jobs do not need to be merged, JobGraphGenerator in Flink is directly called by the Generator to generate JobGraph. Finally, similar to the functionality of the Flink Client, the Actor System is responsible for submitting jobs to the cluster for execution.

3.2 Advantages

In the traditional Flink, the Client submits a job to the JobManager, and the JobManager schedules tasks to each TaskManager for execution. Then, the TaskManager reports heartbeat and statistics to the JobManager. When a user submits multiple jobs, Flink schedules each job on a FCFE basis. The proposed framework establishes multi-job collection, merging and scheduling optimization functions between Client and JobManager, without modifying and deleting Flink's own source code, which has two advantages. Firstly, since the framework and Flink are completely independent of each other, no matter which version is updated, they will not be affected by each other. In addition, the framework adds a switch to Flink, which allows users to choose whether to turn on the multi-job merging and scheduling optimization functions, because the traditional FCFE process ensures fairness at job level, which is also what some users require.

4 Multi-job Merging and Scheduling

This section introduces the two modules in Optimizer in detail, including two multi-job merging and optimization algorithms, and a multi-job scheduling optimization algorithm.

4.1 Multi-job Merging

Multi-job merging is suitable for situations where the cluster nodes are not large and the maximum parallelism of a single job reaches or approaches the maximum parallelism of the cluster. When the user submits multiple jobs and expects the shortest total execution time of the them, the multi-job merging and optimization module will be turned on. For jobs that can be merged, Optimizer merges the execution plans of these jobs into one execution plan, and the internal

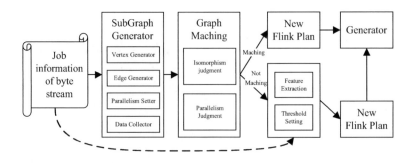

Fig. 4. Multi-job merging module

structure of these execution plans still guarantees their connection sequence. In addition, the user can choose whether to enable the function and the maximum number of jobs to be merged. As shown in Fig. 4, the module first obtains the job information, and uses the subgraph isomorphism algorithm to select jobs with higher similarity to merge. For dissimilar jobs, a heuristic method is used to set thresholds to determine the merged jobs.

Problem Description. Flink Slot uses a memory isolation method to allocate and reclaim resources, which greatly facilitates the management of cluster memory resources, but when users submit multiple jobs, memory resources cannot be shared between jobs. A job in the execution process has all its allocated resources, and the resources will not be recycled until the end of the job. Especially when the parallelism of some Operators is too low, the waste of resources will be more obvious. Therefore, the proposed multi-job merging algorithm aims at merging multiple jobs so that they can use Slot resources together, thereby improving the resource utilization.

Subgraph-Based Merging. In order to find out which jobs are merged first, we give a definition as follow:

- **Definition 1: Job Isomorphism.** A job can be represented by triples: $job = (G, P, D)$, where G is the JobGraph, P represents the maximum degree of parallelism, and D is the amount of data input. G_{sub} is a subgraph generated based on the key information in JobGraph. If $job1.G_{sub}$ and $job2.G_{sub}$ are isomorphic, $job1.P = job2.P$, and $job1.D \approx job2.D$, we determine that the two jobs are isomorphic and can be merged first.

The algorithm first choose the isomorphic jobs with similar data size for merging as much as possible, because they have similar task deployment and data deployment in TaskManager. It will make the merged jobs have fewer thread switching during execution, with better utilization of system resources. Since Flink uses JobGraphs to abstract jobs, the subgraph isomorphism algorithm can be used to judge the similarity of jobs. We only need to find a classic exact subgraph isomorphism algorithm to solve it because the vertex scale of JobGraphs is very small. We choose VF2 [6] among a large number of classic exact subgraph isomorphism algorithms, because it is more suitable for solving subgraph isomorphism of directed graphs and has a smaller space consumption.

Algorithm 2 shows the process of subgraph-based merging. First, select a job from the jobs to be executed (line 1), and use the reverse traversal method to generate a subgraph by reading its job information (line 2). Then use VF2 to perform subgraph isomorphism calculation, find all mergeable jobs, and add them to a collection (lines 3–7). If there are jobs that can be merged, the job merge algorithm will be executed and the merged job information will be deleted (lines 8–10). Finally, generate the new Flink Plan (line 11).

Algorithm 2: Subgraph-based Merging

Input: Job information of byte stream type from Parser
Output: The merged Flink Plan

1 **for** *job in jobs* **do**
2 | generate subgraph for *job*;
3 | *collection*.add(*job*);
4 | *rjobs = jobs*.remove(*job*);
5 | **for** *j in rjobs* **do**
6 | | **if** *isomorphism(j, job)* **then**
7 | | | *collection*.add(*j*);

8 | **if** *collection.size ≥ 2* **then**
9 | | *mergeJob(j, job)*;
10 | | *jobs = jobs*.remove(*job and collection*);

11 write(new Plan);

Heuristic-Based Merging. For jobs with different structures and input data sizes, we propose a heuristic-based merging strategy to find the jobs with the highest "similarity" to merge. Through a large number of experimental results, we selected the jobs with better results after merging, and performed feature extraction on them. The following are some definitions:

- **Definition 2: Job Similarity.** The value obtained by weighting the ratio of the feature parameters between the two jobs extracted, the specific parameters will be introduced later.
- **Definition 3: GlobalOperator.** GlobalOperator refers to operators that need to obtain data from other nodes for processing in a multi-node cluster, such as Join and Reduce.

Table 1. Parameters of job feature

Definition	Formula	Description	Threshold
Task size ratio	$F = \frac{size(m)}{size(n)}$ (1)	The ratio of the total data size processed by two different jobs (m is larger)	[1, 1.8]
DAG depth ratio	$D = \frac{dept(m)}{dept(n)}$ (2)	The ratio of the length of the longest Operator chain in the JobGraph of the two jobs (m is larger)	[1, 2]
GlobalOperator ratio	$G = \frac{gol(m)}{gol(n)}$ (3)	The ratio of the number of GlobalOperator in the two jobs (m is larger)	[1, 1.5]
Parallelism ratio	$P = \frac{parallelism(m)}{parallelism(n)}$ (4)	The ratio of the parallelism of the two jobs (m is larger)	[1, 2]
DAG similarity	$S = \sqrt{\sum_{i=1, j\leq i}^{n}(M_{ij} - N_{ij})^2}$ (5)	Euclidean distance of JobGraph of the two jobs	/

- **Definition 4: LocalOperator.** LocalOperator refers to operators that do not need to obtain data from other nodes for processing, but only process local node data, such as Map and Filter.

Table 1 shows the job feature parameters that need to be extracted and have a greater impact on the efficiency of the merged job execution, Through a large number of experiments, the threshold range of the parameters that meet the combined conditions is estimated. In the case of a small cluster node, the parallelism setting of most jobs adopts the default value, which is equal to the maximum parallelism of Flink. At the same time, we also found that if the parallelism is the same and the thresholds of **F**, **D**, and **G** are all within the threshold range, the merged jobs can bring satisfactory performance improvement. Since subgraph-based merging has filtered out most similar jobs, the number of jobs to be merged is not large and most of them have larger differences. Therefore, in heuristic-based merging, two jobs are selected for merging.

Algorithm 3 shows the process of heuristic-based merging. First, select one of the submitted jobs to compare with other jobs (lines 1–3). If the jobs' parallelism is the same, merge the two jobs by calculating the threshold (lines 4–5) and selecting the job with the highest score (lines 6–9). Then, perform the same threshold calculation and scoring operations as above in the remaining jobs with different parallelism (lines 10–16). Finally merge the jobs that meet the conditions (line 17) and generate the new Flink Plan (line 18).

4.2   Multi-job Scheduling

Multi-job scheduling is suitable for the large scale of cluster nodes so that the parallelism setting of jobs is less than the maximum parallelism of Flink. Through the scheduling optimization strategy, the system can not only improve operating efficiency by making full use of resources, but also maintain a balanced state of cluster load.

Problem Description. The upper limit of Flink's parallelism is the total number of Slots, which means that the total parallelism of running jobs must be less than or equal to the total number of Slots. When the parallelism of a job to be submitted is greater than the number of remaining Slots, the job will be returned. If the total number of Slots in Flink is n, the parallelism of a job being executed is $0.5n$, and the parallelism of the next job to be executed is $0.6n$, then half of the cluster resources will be idle because the job cannot be submitted. If the parallelism of the job being executed and the job to be submitted are both small, the resource usage of the cluster will be higher. The above situation will make the resource usage of the cluster unstable, which neither guarantees the cluster to run Flink jobs efficiently for long time, nor can it maintain a stable load.

Algorithm 3: Heuristic-based Merging

Input: Job information of byte stream type that does not satisfy the subgraph isomorphism condition

Output: The merged Flink Plan

1 **for** *job in jobs* **do**

2      $rjobs = jobs$.remove(job);

3   {**for** *j in rjobs* **do**

4   **if** *job.parallelism == j.parallelism* **then**

5   calculate(F); calculate(D); calculate(G);

6   **if** *meet the threshold* **then**

7   score = $F \times 0.8 + D \times 0.5 + G \times 0.3$;

8   Find *j* with the largest score;

9   *mergeJob*(j, job); $jobs = jobs$.remove(job);

10 **for** *job in jobs* **do**

11      $rjobs = jobs$.remove(j and job);

12   **for** *j in rjobs* **do**

13   calculate(F); calculate(D); calculate(G); calculate(P);

14   **if** *meet the threshold* **then**

15   calculate(S);

16   Find *j* with the smallest S;

17   *mergeJob*(j, job); $jobs = jobs$.remove(j and job);

18 write(new Plan);

Scheduling Based on Maximum Parallelism. Our solution is to give priority to the execution of the job with the highest degree of parallelism that meets the remaining resources of the system, so as to avoid system resource idleness to the greatest extent. In addition, try to make long-running and short-running jobs run at the same time to avoid excessive thread switching caused by intensive short jobs at a certain time. Therefore, the algorithm first extracts the characteristics of each job, and divides it into three groups of long, medium, and short jobs through KMeans clustering algorithm, and finally uses a round-robin scheduling method to submit to the cluster the job with the highest degree of parallelism that meets the execution conditions in each group.

Algorithm 4 describes the process of scheduling optimization. First, extract the features of each job (line 1), including the amount of data, the number of GlobalOperators, the degree of parallelism of each Operator, and the DAG depth. A monitor is placed to monitor whether the job to be executed is empty (line 2). If there are jobs to be submitted, the number of Slot remaining in

---

Algorithm 4: Multi-job Scheduling Based on Maximum Parallelism

---

Input: Jobs information of byte stream type from Parser
Output: Flink Plan for the next job to be executed

1 Extract the features of each job;
2 **while** *jobs.size ≠ null* **do**
3   slot = the number of remaining Slots;
4   **for** *i = 1 to 3* **do**
5        $\lfloor$ job[i] = k_means(features);

6     $i = (i{+}{+})$ % 3;
7   **for** *job in job[i]* **do**
8   **if** *job.parallelism > slot* **then**
9           $\lfloor$ continue;

10   Find the *job* with maximum parallelism;

11   **if** *job.exsist* **then**
12   *job*.execute();
13   *jobs = jobs*.remove(*job*);

14   continue;

---

the system will be obtained (line 3). According to the extracted job features, the KMeans clustering algorithm is used to divide the job into three groups: long-time running, medium-time running, and short-time running (lines 4–5). Then find a job by group, the job that satisfy the remaining Slot number of the system and have the greatest degree of parallelism is selected (lines 6–10). If such a job exists, it is submitted to the cluster for execution and removed from the queue of jobs to be executed (lines 11–13). Finally, regardless of whether a job is submitted for execution, the search for jobs that can be submitted for execution will continue according to the above criteria (line 14).

5 Evaluation Results

In this section, we describe the performance evaluation of the proposed multi-job merging algorithms and the scheduling optimization algorithm in our framework. The data sets is used to test the running time and the number of Slots occupied.

5.1 Experimental Setup

We run experiments on a 7-nodes OMNISKY cluster (1 JobManager & 6 TaskManagers), and all nodes are connected with 10-Gigabit Ethernet. Each node has two Intel Xeon Silver 4210 CPUs @ 2.20 GHz (10 cores × 2 threads, 20

TaskNumberSlots), 128 GB memory, and 1 TB SSD. Hadoop version 2.7.0 (for storing data on HDFS) and Flink version 1.8.0 are chosen as the experimental environment, and their configuration files are configured according to the hardware environment as mentioned above.

We select three distributed jobs to run the experiment, namely WordCount, Table Join and KMeans, from the perspectives of the type of Operators included in the job and whether the job includes iterative tasks. All running time measurements include the generation time of new JobGraphs. We use a large number of real-world data sets and generated data sets to evaluate the experimental results, and all the experiments are tested more than 10 times. The specific information are as follows:

- **WordCount:** The WordCount job contains almost only LocalOperators, and there is not too much data exchange between TaskManagers. We select the text from the literary work Hamlet as the data set and manually expand it to 500 MB–5 GB.
- **Table Join:** As Join is a GlobalOperator in Flink, it needs to obtain intermediate data from each TaskManager, so a large number of data exchanges between nodes will occur in Table Join job, especially in multi-Table Join. We choose multiple relational tables generated by the big data test benchmark TPC-H, with the size range 500 MB–10 GB.
- **KMeans:** It is a clustering job with iterative tasks in which both LocalOperators and GlobalOperators are iterated. We use the UCI standard data set Wine, and manually expand the number of data samples to reach 500 MB–1 GB.

Table 2. Effect of merging two identical jobs on data sets of different sizes

WordCount			Table join			KMeans		
Size	Running time		Size	Running time		Size	Running time	
	FCFE	Merge		FCFE	Merge		FCFE	Merge
500 MB & 500 MB	12.8 s	**10.2 s**	1 GB & 1 GB	56.7 s	**46.3 s**	500 MB & 500 MB	76.5 s	**65.0 s**
1.5 GB & 1.5 GB	33.5 s	**26.6 s**	1.2 GB & 1.2 GB	79.2 s	**65.0 s**	700 MB & 700 MB	121.6 s	**105.5 s**
3 GB & 3 GB	62.6 s	**48.2 s**	1.5 GB & 1.5 GB	122.0 s	**99.8 s**	850 MB & 850 MB	155.7 s	**131.1 s**
5 GB & 5 GB	110.7 s	**84.5 s**	2 GB & 2 GB	213.5 s	**170.8 s**	1 GB & 1 GB	212.4 s	**177.7 s**
500 MB & 5 GB	68.6 s	**61.2 s**	1 GB & 2 GB	132.1 s	**121.8 s**	500 MB & 1 GB	142.7 s	**128.1 s**
3 GB & 5 GB	85.5 s	**72.0 s**	1.5 GB & 2 GB	160.9 s	**136.4 s**	700 MB & 1 GB	165.0 s	**144.7 s**

5.2 Testing of Multi-job Merging

We first test the merging effect of two identical jobs under different scale data sets, and the specific results are shown in Table 2. The first 3 rows are the effects of merging data sets of the same size (sorted by data set size), and the rest are the effects of merging data sets of different sizes. It can be found that the three types of jobs can bring about efficiency improvement after merging and executing. Among them, WordCount job has better improvement effect than the other two, with the improvement efficiency reaching 31% in the best case, because both Table Join and KMeans have a large number of data exchanges between nodes during execution. In addition, the efficiency of merged jobs with the same size data set is better, because when the data set sizes are different, the smaller job will complete the data processing in advance, and the efficiency improvement brought by sharing resources between the two jobs cannot be maintained for a long time.

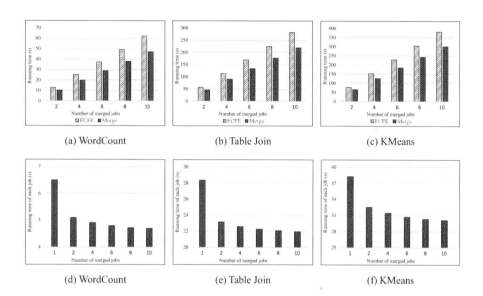

(a) WordCount   (b) Table Join   (c) KMeans

(d) WordCount   (e) Table Join   (f) KMeans

Fig. 5. Effect of the number of merged jobs on efficiency

Next we show the effect of the number of merged jobs on efficiency, which is shown in Fig. 5. We use the same job to measure the experimental results, and we can see that the more jobs are merged, the more obvious the efficiency improvement, because when the number of jobs is large, the extraction of CPU and memory resources will be more sufficient.

For merging different types of jobs, we use data sets of the same size to evaluate efficiency. As shown in Fig. 6, merging different types of jobs can still bring good performance improvement, ranging from 15% to 21%.

5.3   Testing of Scheduling Optimization

Finally, we show the effect of the multi-job scheduling optimization algorithm. According to the hardware environment described in Sect. 5.1, the maximum parallelism of Flink is set to 240, which is the same as the total number of CPU

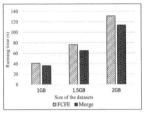 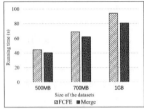 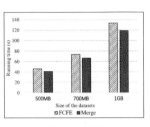

(a) WordCount & Table Join   (b) WordCount & KMeans   (c) Table Join & KMeans

Fig. 6. Effect of merging different jobs

cores in 6 TaskManagers. We generated 10 jobs in WordCount, Table Join and KMeans respectively, and randomly set the degree of parallelism for these 30 jobs, ranging from 50 to 180. Then 30 submissions are randomly generated for these 30 jobs, and the total execution time of these 30 submissions is measured and compared with the scheduling optimization method proposed by us.

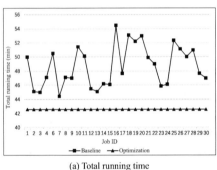 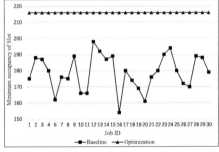

(a) Total running time   (b) Minimum occupancy of Slot

Fig. 7. Effect of different Job execution sequences

Figure 7(a) shows the running time of each set of jobs by random submission. It can be seen that the total time of executing jobs in the sequence after scheduling optimization is shorter than the total time of executing jobs in the 30 randomly generated sequences, and the performance improvement of 28% can be achieved in the best case. Slot occupancy is shown in Fig. 7(b). Since the Slot usage of the last executed job may be too low due to the job parallelism being set too small, we only measure the Slot usage when the first 25 jobs are executed. It can be found that when jobs are executed in an unoptimized order, Slot usage will be too low for a certain period of time during each execution of the jobs. On average, executing jobs in the optimized order will reduce cluster idle resources by 61%.

6   Conclusion and Discussion

In this paper, we propose the groundbreaking framework that support multi-job merging and scheduling. Based on these two functions, optimization strategies are proposed to improve the efficiency of multi-job by making full use of the cluster resources. In order to verify the effectiveness of the proposed algorithms, we conduct many experiments to prove the superiority of our work. Since Flink is a "unify batch & streaming" system, a particularly interesting direction for future work is to extend our proposed framework and optimization algorithms to the streaming jobs, which can improve the function of the framework.

Acknowledgments. This research was supported by the National Key R&D Program of China under Grant No. 2018YFB1004402; and the NSFC under Grant No. 61872072, 61772124, 61932004, 61732003, and 61729201; and the Fundamental Research Funds for the Central Universities under Grant No. N2016009 and N181605012.

References

1. Borkar, V., Carey, M., Grover, R., Onose, N., Vernica, R.: Hyracks: a flexible and extensible foundation for data-intensive computing. In: Proceedings of the International Conference on Data Engineering, pp. 1151–1162 (2011)
2. Carbone, P., et al.: Apache flink: stream and batch processing in a single engine. IEEE Data Eng. Bull. **38**, 28–38 (2015)
3. Chakraborty, R., Majumdar, S.: A priority based resource scheduling technique for multitenant storm clusters. In: International Symposium on Performance Evaluation of Computer and Telecommunication Systems, pp. 1–6 (2016)
4. Cheng, D., Rao, J., Jiang, C., Zhou, X.: Resource and deadline-aware job scheduling in dynamic Hadoop clusters. In: IEEE International Parallel and Distributed Processing Symposium, pp. 956–965 (2015)
5. Ciobanu, A., Lommatzsch, A.: Development of a news recommender system based on apache flink, vol. 1609, pp. 606–617 (2016)
6. Cordella, L.P., Foggia, P., Sansone, C., Vento, M.: A (sub)graph isomorphism algorithm for matching large graphs. IEEE Trans. Pattern Anal. Mach. Intell. **26**, 1367–1372 (2004)
7. Dean, J., Ghemawat, S.: MapReduce. Commun. ACM **51**(1), 107–113 (2008)
8. Eaman, J., Cafarella, M.J., Christopher, R.: Automatic optimization for MapReduce programs. Proc. VLDB Endow. (2011)
9. Espinosa, C.V., Martin-Martin, E., Riesco, A., Rodriguez-Hortala, J.: FlinkCheck: property-based testing for apache flink. IEEE Access **99**, 1–1 (2019)
10. Falkenthal, M., et al.: OpenTOSCA for the 4th industrial revolution: automating the provisioning of analytics tools based on apache flink, pp. 179–180 (2016)
11. Garca-Gil, D., Ramrez-Gallego, S., Garca, S., Herrera, F.: A comparison on scalability for batch big data processing on apache spark and apache flink. Big Data Anal. **2** (2017)
12. Hueske, F., Krettek, A., Tzoumas, K.: Enabling operator reordering in data flow programs through static code analysis. In: XLDI (2013)
13. Kougka, G., Gounaris, A.: Declarative expression and optimization of data-intensive flows. In: Bellatreche, L., Mohania, M.K. (eds.) DaWaK 2013. LNCS, vol. 8057, pp. 13–25. Springer, Heidelberg (2013). https://doi.org/10.1007/978-3-642-40131-2_2

14. Pandey, V., Saini, P.: An energy-efficient greedy MapReduce scheduler for heterogeneous Hadoop YARN cluster. In: Mondal, A., Gupta, H., Srivastava, J., Reddy, P.K., Somayajulu, D.V.L.N. (eds.) BDA 2018. LNCS, vol. 11297, pp. 282–291. Springer, Cham (2018). https://doi.org/10.1007/978-3-030-04780-1_19
15. Perera, S., Perera, A., Hakimzadeh, K.: Reproducible experiments for comparing apache flink and apache spark on public clouds. arXiv:1610.04493 (2016)
16. Radhya, S., Khafagy, M.H., Omara, F.A.: Big data multi-query optimisation with apache flink. Int. J. Web Eng. Technol. **13**(1), 78 (2018)
17. Rumi, G., Colella, C., Ardagna, D.: Optimization techniques within the Hadoop eco-system: a survey. In: International Symposium on Symbolic and Numeric Algorithms for Scientific Computing, pp. 437–444 (2015)
18. Simitsis, A., Wilkinson, K., Castellanos, M., Dayal, U.: Optimizing analytic data flows for multiple execution engines. In: Proceedings of the ACM SIGMOD International Conference on Management of Data, pp. 829–840 (2012)
19. Tian, H., Zhu, Y., Wu, Y., Bressan, S., Dobbie, G.: Anomaly detection and identification scheme for VM live migration in cloud infrastructure. Future Gener. Comput. Syst. **56**, 736–745 (2016)
20. Tinghui, H., Yuliang, W., Zhen, W., Gengshen, C.: Spark I/O performance optimization based on memory and file sharing mechanism. Comput. Eng. (2017)
21. Wang, K., Khan, M.M.H., Nguyen, N., Gokhale, S.: Design and implementation of an analytical framework for interference aware job scheduling on apache spark platform. Cluster Comput. **22**, 2223–2237 (2019). https://doi.org/10.1007/s10586-017-1466-3
22. Yao, Y., Tai, J., Sheng, B., Mi, N.: LsPS: a job size-based scheduler for efficient task assignments in Hadoop. IEEE Trans. Cloud Comput. **3**, 411–424 (2015)
23. Zaharia, M., et al.: Apache spark: a unified engine for big data processing. Commun. ACM **59**, 56–65 (2016)

CIC-FL: Enabling Class Imbalance-Aware Clustered Federated Learning over Shifted Distributions

Yanan Fu[1], Xuefeng Liu[1(✉)], Shaojie Tang[2], Jianwei Niu[1,3],
and Zhangmin Huang[3]

[1] State Key Laboratory of Virtual Reality Technology and Systems, School of
Computer Science and Engineering, Beihang University, Beijing 100191, China
{fuyanan,liu_xuefeng,niujianwei}@buaa.edu.cn
[2] Jindal School of Management, The University of Texas at Dallas,
Dallas, TX 75080-3021, USA
shaojie.tang@utdallas.edu
[3] Hangzhou Innovation Institute of Beihang University, Hangzhou 310051, China
zmhuang15@fudan.edu.cn

Abstract. Federated learning (FL) is a distributed training framework where decentralized clients collaboratively train a model. One challenge in FL is concept shift, i.e. that the conditional distributions of data in different clients are disagreeing. A natural solution is to group clients with similar conditional distributions into the same cluster. However, methods following this approach leverage features extracted in federated settings (e.g., model weights or gradients) which intrinsically reflect the joint distributions of clients. Considering the difference between conditional and joint distributions, they would fail in the presence of class imbalance (i.e. that the marginal distributions of different classes vary in a client's data). Although adopting sampling techniques or cost-sensitive algorithms can alleviate class imbalance, they either skew the original conditional distributions or lead to privacy leakage. To address this challenge, we propose CIC-FL, a class imbalance-aware clustered federated learning method. CIC-FL iteratively bipartitions clients by leveraging a particular feature sensitive to concept shift but robust to class imbalance. In addition, CIC-FL is privacy-preserving and communication efficient. We test CIC-FL on benchmark datasets including Fashion-MNIST, CIFAR-10 and IMDB. The results show that CIC-FL outperforms state-of-the-art clustering methods in FL in the presence of class imbalance.

Keywords: Federated learning · Clustering · Concept shift · Class imbalance

1 Introduction

Federated learning (FL) [10,13,19] is a distributed training framework in which multiple clients (e.g., organizations, data centers or mobile devices) collaboratively

C. S. Jensen et al. (Eds.): DASFAA 2021, LNCS 12681, pp. 37–52, 2021.
https://doi.org/10.1007/978-3-030-73194-6_3

train a global model under the orchestration of a central server. Not surprisingly, to obtain a high-quality global model that can fit data of all participating clients, FL requires data of all clients to be independent and identically distributed (IID).

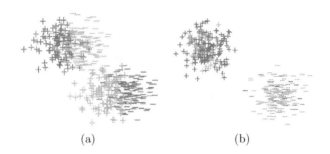

(a)   (b)

Fig. 1. (a) An example of concept shift. FL would fail in the presence of concept shift. (b) In the presence of class imbalance, two clients with the same conditional distribution would be classified into two different clusters.

However in real conditions, it is very common that data of different clients are heterogeneous. In this situation, many works have shown that conventional FL would fail to produce a good global model [8,18,28].

One common type of statistical heterogeneity is called concept shift [13]. For a supervised classification task with feature x and labels y, concept shift means that conditional distributions $\varphi(x|y)$ vary among different clients. Concept shift is quite common in many applications such as recommender systems[23], image recognition [27], and smart keyboards [10]. For example, when FL is utilized to train the model of a recommender system involving data from different groups of people, different groups generally have different preferences, generating concept shift. A toy example of concept shift is shown in Fig. 1(a), in which two clients (whose samples are respectively colored as blue and orange) collaboratively train a model that categorizes a sample as either positive (denoted as '+') or negative (denoted as '−'). The conditional distributions of two clients vary significantly. In this situation, FL generally would fail to give a single global model that can fit data distributions of all clients [7,8,13,18,20,22].

To address this problem, an intuitive approach is to group clients into clusters and train one personalized model in each cluster. Clients in the same cluster should have similar *conditional distributions* $\varphi(x|y)$ to generate an appropriate model in FL. There are recently many methods that follow this approach like [2,8,9,18,20,27]. In these clustered federated learning methods, the clustering is generally based on some features of clients. Note that as these features need to be collected in federated settings without sharing local data, they are restricted to model weights [9,27], gradients [2,20], local optima [8,18], etc., all of them computed based on *joint distributions* $\varphi(x, y)$. As we know, joint distribution $\varphi(x, y)$ is determined by both conditional distribution $\varphi(x|y)$ and marginal distribution $\varphi(y)$ (also called *label distributions*), i.e., $\varphi(x, y) = \varphi(x|y)\varphi(y)$.

We can see that these clustering methods only work well when $\varphi(y)$ for all clients are uniform among different classes.

However, in real applications, the label distributions $\varphi(y)$ can vary significantly in a client's data, which is called *class imbalance*. Moreover, concept shift and class imbalance often appear simultaneously, posing challenges to clustering-based methods in FL. An example is illustrated in Fig. 1(b), where two clients have similar conditional distributions but they both have class imbalance. In this example, class imbalance leads to different joint distributions and the existing clustering methods would erroneously divide them into different clusters.

There are two straightforward approaches that can naturally handle this problem. One approach is to first adopt some undersampling [15,17] or data augmentation [3,11] techniques to alleviate class imbalance of each client, and then apply clustering techniques to deal with concept shift. However, unless all clients have significant amount of data samples, these methods would generally skew the original conditional distributions, especially in the extreme class imbalance. Another approach is to apply new loss functions [16,26] or cost-sensitive learning techniques [14,24] to mitigate the effect of class imbalance during FL. However, these methods require clients to upload their local label distributions to the server, which can lead to privacy leakage [25].

To address the challenge, we propose CIC-FL, a Class Imbalance-aware Clustered Federated Learning method to deal with concept shift in FL in the presence of class imbalance. CIC-FL estimates a particular feature $LEGLD$ that is sensitive to concept shift but robust to class imbalance and iteratively utilizes the feature to bipartition clients into clusters. In addition, CIC-FL is privacy-preserving as the process of acquiring $LEGLD$ does not need clients to share additional information about their raw data and data distributions. Moreover, CIC-FL has a low communication cost as the feature for clustering can be obtained after a few rounds of FL. This is in contrast with some existing approaches like [2,20] in which features for clustering can only be obtained after the global model has converged. We test CIC-FL on different benchmark datasets including Fashion-MNIST, CIFAR-10, and IMDB. The results show that CIC-FL outperforms the state-of-the-art clustered federated learning methods like *ClusteredFL* [20] and *IFCA* [8] in the presence of class imbalance.

The contributions of this paper are as follows:

- We observe that the existing clustering methods in FL generally fail to deal with concept shift in the presence of class imbalance.
- We propose a clustering approach for FL called CIC-FL. CIC-FL iteratively conducts bi-partitioning using a particular feature sensitive to concept shift but robust to class imbalance. CIC-FL is privacy-preserving and communication efficient.
- We test CIC-FL on benchmark datasets created based on Fashion-MNIST, CIFAR-10, and IMDB. The results show that CIC-FL outperforms state-of-the-art clustered federated learning methods in the presence of class imbalance.

2   Related Work

In FL, one important research area is to address the heterogeneity of data in participating clients. One common type of statistical heterogeneity is concept shift, and existing solutions share the same idea: instead of learning a single global model for all clients, FL trains personalized models for different clients [13].

To address concept shift, one line of research [6,22] adopts a multi-task learning (MTL) framework: where the goal is to consider fitting separate but related models simultaneously. Another approach is to adopt a meta-learning approach [4,7]. In this setup, the objective is to first obtain a single global model, and then each client fine-tunes the model using its local data. The third approach is clustering: clients whose data have similar conditional distributions are grouped, and one personalized model is trained for clients in each cluster [2,8,9,18,20,27].

The research work of clustering approach can be further classified into two categories. The first category is based on the partition, in which clustering algorithms iteratively update centers of clusters. In [9], model weights of all clients are utilized to cluster different clients. In addition, Ghosh et al. [8], Xie et al. [27] and Mansour et al. [18], utilize some proposed loss functions related to the joint distribution to guide the clustering process. Another category of clustering is based on hierarchy. The basic idea of this kind of clustering algorithms is to construct the hierarchical relationship among clients in FL. In the clustering process, either the most neighboring clients are merged into a new cluster, or a reverse process is implemented in which distant clients are iteratively split into two clusters. Typical algorithms of this kind of clustering include [2,20,21].

However, for the clustering methods above, as clustering also needs to be implemented in federated settings without sharing local data, features extracted for clustering are based on the joint distributions of clients. Considering joint distribution is determined by both conditional distribution and label distribution, they would fail in the presence of class imbalance.

3   The CIC-FL

In this section, we first introduce an overview of CIC-FL, followed by an introduction about the feature we designed for clustering. At last, we introduce how to use the feature to cluster.

3.1   System Overview

CIC-FL is a top-down hierarchical clustering process in FL where initially all the clients belong to a single cluster. Then CIC-FL keeps partitioning clusters by leveraging a particular feature sensitive to concept shift but robust to class imbalance. Figure 2 shows an iteration at the beginning of CIC-FL. In particular, CIC-FL implements the following steps for clients in a cluster:

1. **Federated Learning.** The standard FL algorithm *FedAvg* [19] is implemented in the cluster for t rounds, and each round generates a global model $W_{Avg}^{(t)}$.

2. **Calculation of label-wise gradients.** For client i with data Z_i, samples of each label q are sorted (denoted as $Z_{i,q}$) and respectively fed into the global model $W_{Avg}^{(t-1)}$ obtained at $(t-1)^{th}$ round. For each label q, we can obtain a label-wise gradient vector denoted as $G_{i,q}$. Label-wise gradients for all Q labels $q = 1, \cdots, Q$ form a set denoted as $G_i = \{G_{i,1}, \cdots, G_{i,Q}\}$.

3. **Calculation of the feature for clustering.** Based on the weight updates of global model $\Delta W_{Avg}^{(t)}$ (i.e. that $W_{Avg}^{(t)} - W_{Avg}^{(t-1)}$) obtained in step 1 and the label-wise gradients G_i from step 2, every client calculates a feature, called the locally estimated global label distribution (*LEGLD*) and sends it to the server.

4. **Bipartition.** For the cluster, the server utilizes *LEGLDs* of the clients and partitions them into two clusters.

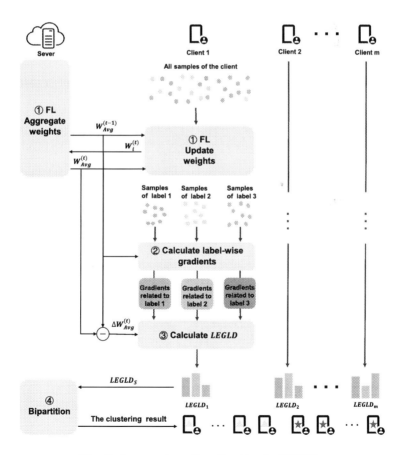

Fig. 2. An overview of an iteration in CIC-FL.

The process is iterated until clients in all clusters have similar *LEGLD*s. The overall procedures of CIC-FL are shown in Algorithm 1. In the following sections, we will introduce *LEGLD*, the feature we proposed for clustering, in detail.

Algorithm 1: CICFL

Input: a set of clients $C = \{1, \cdots, m\}$, the initial global model $W_{Avg}^{(0)}$, the round t.

Output: the cluster structure $\{C_1, \cdots, C_K\}$.

1   $W_{Avg}^{(t-1)}, W_{Avg}^{(t)} \leftarrow \text{FedAvg}(C, W_{Avg}^{(0)}, t)$

2   $\Delta W_{Avg}^{(t)} \leftarrow W_{Avg}^{(t)} - W_{Avg}^{(t-1)}$

3   **for** i *in* C **do**

4   **for** q *in* Q **do**

5         $G_{i,q} \leftarrow \text{CalculateLabelwiseGradients}(W_{Avg}^{(t-1)}, Z_{i,1}, \cdots, Z_{i,Q})$

6   **end**

7      $LEGLD_i \leftarrow \text{CalculateLEGLD}(\Delta W_{Avg}^{(t)}, G_i)$

8   **end**

9   $C_1, C_2 \leftarrow \text{Bipartition}(C, LEGLD_1, \cdots, LEGLD_m)$

10 **if** $LEGLD_1, \cdots, LEGLD_m$ *are similar* **then**

11   **return** C

12 **end**

13 **else**

14     $\text{CICFL}(C_1, W_{Avg}^{(0)}, t)$

15     $\text{CICFL}(C_2, W_{Avg}^{(0)}, t)$

16 **end**

3.2   Requirements of the Feature for Clustering

We first introduce the requirements for the feature selected for clustering in the presence of class imbalance. The feature should satisfy the following requirements.

1. The feature should be sensitive to concept shift but robust to class imbalance.
2. The feature should be able to be estimated in the framework of FL. This means that communication between clients and the server is restricted to model weights and gradients without sharing local data or other additional information of clients.
3. It is preferable that the estimation of the feature has a low communication cost.

3.3   Locally Estimated Global Label Distribution (*LEGLD*)

Before we introduce the feature we proposed, we first define a term called the global label distribution (*GLD*). We consider a supervised Q-classification task in FL involving m clients with features $\boldsymbol{x}$ and labels y, where $\boldsymbol{x} \in X$, X is the features space, and $y \in Y$, $Y = \{1, \cdots, Q\}$.

The local data of client i, denoted as Z_i, can be sorted according to the labels and therefore can be represented as $Z_i = \{Z_{i,1}, \cdots, Z_{i,Q}\}$, where $Z_{i,q}$ ($q = 1, \cdots, Q$) is the local data of label q. Then the *GLD* is a vector defined as

$$GLD = [\frac{\mathcal{N}_1}{\sum_{q=1}^{Q} \mathcal{N}_q}, \cdots, \frac{\mathcal{N}_Q}{\sum_{q=1}^{Q} \mathcal{N}_q}] \tag{1}$$

where $\mathcal{N}_q = \sum_{i=1}^{m} N_{i,q}$, $N_{i,q}$ is the number of samples in $Z_{i,q}$. We can see that for each element in *GLD*, its numerator $\mathcal{N}_q$ is the total sample number in class q across all clients, and the denominator is the number of overall training data. The *GLD* thus represents label distribution of data across all clients.

In FL settings, the calculation of *GLD* using Eq. 1 needs clients to transmit its label information and therefore is not privacy-preserving [25]. In the following part of this section, we will show that *GLD* can be estimated purely based on the weight updates in FL.

Let $f(W; z) : \Theta \to \mathbb{R}$ be the loss function associated with a sample $z = (\boldsymbol{x}, y)$, where W is the model weights, $W \in \Theta$ and Θ is the model weights space. For client i, the gradient of its loss function $L_i(W)$, is defined as:

$$\nabla L_i(W) = \nabla \mathbb{E}_{z \sim \varphi_i(\boldsymbol{x}, y)}[f(W; z)]. \tag{2}$$

$\nabla L_i(W)$ is calculated using its local data Z_i. Without loss of generality, we divide Z_i into class q and $\bar{q}$, with $\bar{q}$ including all the classes except q. Then based on Eq. 2, $\nabla L_i(W)$ can be written as:

$$\nabla L_i(W) = \varphi_i(q) \nabla L_{i,q}(W) + \varphi_i(\bar{q}) \nabla L_{i,\bar{q}}(W), \tag{3}$$

where $\varphi_i(q)$ and $\varphi_i(\bar{q})$ are the label distributions of label q and $\bar{q}$. $\nabla L_{i,q}(W)$ and $\nabla L_{i,\bar{q}}(W)$ are the gradients updated using data of label q and $\bar{q}$, respectively.

In FL, the gradients from all m clients are aggregated in the server by FedAvg [19]:

$$\nabla L_{Avg}(W) = \frac{\sum_{i=1}^{m} N_i \nabla L_i(W)}{\sum_{i=1}^{m} N_i} \tag{4}$$

where N_i is the number of samples in client i.

Substituting Eq. 3 into Eq. 4, we have

$$\nabla L_{Avg}(W) = \frac{\sum_{i=1}^{m} N_i(\varphi_i(q) \nabla L_{i,q}(W) + \varphi_i(\bar{q}) \nabla L_{i,\bar{q}}(W))}{\sum_{i=1}^{m} N_i}. \tag{5}$$

Assuming that $\frac{N_{i,q}}{N_i}$ is the unbiased estimate of $\varphi_i(q)$, then we have

$$\nabla L_{Avg}(W) = \frac{\sum_{i=1}^{m} N_{i,q} \nabla L_{i,q}(W) + \sum_{i=1}^{m} N_{i,\bar{q}} \nabla L_{i,\bar{q}}(W)}{\sum_{i=1}^{m} N_i}. \tag{6}$$

For simplicity, we define

$$
\begin{cases}
\nabla \widetilde{L}_q(W) &= \dfrac{\sum_{i=1}^{m} N_{i,q} \nabla L_{i,q}(W)}{\mathcal{N}_q}, \\
\nabla \widetilde{L}_{\bar{q}}(W) &= \dfrac{\sum_{i=1}^{m} N_{i,\bar{q}} \nabla L_{i,\bar{q}}(W)}{\mathcal{N}_{\bar{q}}}.
\end{cases}
\tag{7}
$$

Then Eq. 6 can be presented as:

$$
\nabla L_{Avg}(W) = \frac{\nabla \widetilde{L}_q(W) \mathcal{N}_q + \nabla \widetilde{L}_{\bar{q}}(W) \mathcal{N}_{\bar{q}}}{\sum_{i=1}^{m} N_i}.
\tag{8}
$$

Based on Eq. 8, we can calculate $\mathcal{N}_q$ as:

$$
\mathcal{N}_q = \frac{(\nabla L_{Avg}(W) - \nabla \widetilde{L}_{\bar{q}}(W)) \sum_{i=1}^{m} N_i}{\nabla \widetilde{L}_q(W) - \nabla \widetilde{L}_{\bar{q}}(W)}
\tag{9}
$$

Using Eq. 9, GLD can be calculated according to Eq. 1.

However, according to Eq. 7, we can see that the calculation of both $\nabla \widetilde{L}_q(W)$ and $\nabla \widetilde{L}_{\bar{q}}(W)$ require each client to upload its label distribution to the server, which violates the second requirement of the feature for clustering described in Sect. 3.2.

To address the problem, we directly replace $\nabla \widetilde{L}_q(W)$ and $\nabla \widetilde{L}_{\bar{q}}(W)$ with $\nabla L_{i,q}(W)$ and $\nabla L_{i,\bar{q}}(W)$, respectively. The latter two terms can be calculated by client i in FL settings. Then a client i can estimate a $\mathcal{N}_q$, which is denoted $\hat{\mathcal{N}}_q^{(i)}$ as:

$$
\hat{\mathcal{N}}_q^{(i)} = \frac{(\nabla L_{Avg}(W) - \nabla L_{i,\bar{q}}(W)) \sum_{i=1}^{m} N_i}{\nabla L_{i,q}(W) - \nabla L_{i,\bar{q}}(W)}.
\tag{10}
$$

Each client can estimate a Locally Estimated GLD ($LEGLD$), which is defined as

$$
LEGLD_i = [\frac{\hat{\mathcal{N}}_1^{(i)}}{\sum_{q=1}^{Q} \hat{\mathcal{N}}_q^{(i)}}, \cdots, \frac{\hat{\mathcal{N}}_Q^{(i)}}{\sum_{q=1}^{Q} \hat{\mathcal{N}}_q^{(i)}}]
\tag{11}
$$

The $LEGLD$ will be utilized as the feature for clustering.

In essence, $LEGLD$ calculated by a client is the estimation of GLD based on the gradients with respect to each label in the client. We will show that $LEGLD$ meets the requirements described in Sect. 3.2. First, $LEGLD$ can be estimated by each client in FL settings without sharing any information about its raw data or data distribution, which is privacy-preserving. We now analyze why $LEGLD$ is sensitive to concept shift but robust to class imbalance.

From Eq. 10, we can see that $\hat{\mathcal{N}}_q^{(i)}$ is determined by $\nabla L_{Avg}(W)$, $\sum_{i=1}^{m} N_i$, $\nabla L_{i,q}(W)$, and $\nabla L_{i,\bar{q}}(W)$. $\nabla L_{Avg}(W)$ reflects the overall gradient of clients participating in FL and $\sum_{i=1}^{m} N_i$ is the number of samples in all clients. Therefore, $\nabla L_{Avg}(W)$ and $\sum_{i=1}^{m} N_i$ are neither affected by concept shift nor by class imbalance.

While $\nabla L_{i,q}(W)$ and $\nabla L_{i,\bar{q}}(W)$ are related to concept shift. To explain this relatedness, we express $\nabla L_{i,q}(W)$ as

$$\nabla L_{i,q}(W) = \nabla \mathbb{E}_{z \sim \varphi_i(\boldsymbol{x}|q)}[f(W; z)], \tag{12}$$

which implies that the gradient for a particular label can reflect the information of its conditional distribution $\varphi(\boldsymbol{x}|y)$.

If there is no concept shift between clients, the data of same class possessed by different clients will result in same local gradients (i.e. that $\nabla L_{i,q}(W) = \nabla L_{j,q}(W)$, for $i,j \in \{1,\cdots,m\}$). In this situation, $\hat{\mathcal{N}}_q^{(i)}$ of any client i is equal to $\mathcal{N}_q$. Therefore, all clients have the same $LEGLD$. On the contrary, considering the condition when concept shift exists between client i and client j (i.e., $\nabla L_{i,q}(W) \neq \nabla L_{j,q}(W)$), they will obtain different $LEGLDs$.

Based on the above analysis, $LEGLD$ is sensitive to concept shift but robust to class imbalance.

To calculate $\hat{\mathcal{N}}_q^{(i)}$ in Eq. 10, we utilize the empirical loss associated with Z_i as an unbiased estimate of $L_i(W)$ (i.e., $L_i(W) = \frac{1}{|Z_i|}\sum_{z \in Z_i} f(W; z)$). In addition, we utilize weight updates to replace gradients when computation and communication budgets are limited.

Furthermore, although for any weight $w \in W$, we can obtain the same $\hat{\mathcal{N}}_q^{(i)}$ in Eq. 10, the results obtained by some weights are not reliable. Directly averaging all $\hat{\mathcal{N}}_q^{(i)}$s can also suffer from outliers. Many researches [1,25] have discovered that the weights between the hidden layer and the output layer are more sensitive to the conditional distribution of the training data. Therefore, the average of the updates of these weights will be utilized to calculate $\hat{\mathcal{N}}_q^{(i)}$s.

Another advantage of $LEGLD$ is that the calculation of this feature is communication-efficient. The feature can be obtained after a few rounds in FL [25]. This is in contrast with some existing approaches [2,20] in which features for clustering can only be obtained after the global model has converged.

3.4   Bipartition

After acquiring $LEGLDs$ of clients in the current cluster, the server will partition clients into two clusters.

The server first computes the cosine similarity matrix S based on $LEGLDs$ of all clients in cluster C, and its element $S_{i,j}$ is defined as:

$$S_{i,j} = \frac{<LEGLD_i, LEGLD_j>}{\|LEGLD_i\|\|LEGLD_j\|}, \quad i,j \in C. \tag{13}$$

Based on S, two candidate clusters are generated according to the following objective function:

$$C_1, C_2 = \arg \min_{C_1 \cup C_2 = C} (\max_{i \in C_1, j \in C_2} S_{i,j}). \tag{14}$$

The similarity between the two candidate clusters is defined as

$$SIM(C_1, C_2) = \max_{i \in C_1, j \in C_2} S_{i,j}. \tag{15}$$

If $SIM(C_1, C_2)$ is lower than a threshold ϵ, then the cluster C will be partitioned into two cluster C_1, C_2.

4   Experiments

4.1   Experimental Settings and Evaluation Metrics

We evaluate the proposed CIC-FL on different benchmark datasets (Fashion-MNIST (FMNIST), CIFAR-10 and IMDB) and on three common types of deep network architectures. Specifically, we test CIC-FL using a single-layer feedforward neural network for FMNIST, a convolutional neural network (consisting 4 convolutional layers followed by a fully connected layer) for CIFAR-10, and a recursive neural network (including two Long Short Term Memory layers followed by a fully connected layer) for IMDB.

In all experiments, we set the number of clients as m. Then we generate K clusters, and each cluster has the same number of clients. The training and testing data of each client are generated based on the following procedures. For a given dataset, each client randomly samples about 10,000 samples according to a pre-defined class imbalance. Then based on the sampled samples, concept shift among multiple clients is generated by re-assigning labels according to different permutations of a label sequence. For example, for a 5-classification task with 2 clusters, two permutations of the label sequence, namely, $[2, 1, 4, 5, 3]$ and $[3, 1, 5, 4, 2]$, are randomly generated. Then they are utilized respectively as new labels for all the clients in each of the two clusters. The generated data in each client is then evenly divided for training and testing.

We define a metric for class imbalance. First, the level of class imbalance for a single clients i [12], denoted as α_i, is defined as:

$$\alpha_i := \frac{max_{q \in Q} N_{i,q}}{min_{q \in Q} N_{i,q}} \tag{16}$$

Then αs are averaged to represent the level of class imbalance across all clients.

Then, we define a metric $\beta_{k,k'}$ to evaluate the level of concept shift between cluster C_k and $C_{k'}$. Let the label sequences generated by random permutation be P_k and $P_{k'}$ respectively, then $\beta_{k,k'}$ is defined as:

$$\beta_{k,k'} := (1 - \frac{\sum_{q=1}^{Q} I(P_k(q), P_{k'}(q))}{Q}) \times 100\% \tag{17}$$

where $I(P_k(q), P_{k'}(q)) = 1$ if $P_k(q) = P_{k'}(q)$, and $I(P_k(q), P_{k'}(q)) = 0$ if otherwise. Then βs of any two cluster pairs are averaged to represent the level of class imbalance across all clients.

We use two metrics to evaluate CIC-FL's performance. One is the RandIndex [5], denoted as RI, which is defined as follows:

$$RI := (1 - \frac{\sum_{i<j} |\mathbf{H}_{i,j} - \mathbf{H}'_{i,j}|}{\binom{m}{2}}) \times 100\% \tag{18}$$

where $\mathbf{H}$ and $\mathbf{H}'$ are matrix representation of clustering results of the ground truth and of the estimated results. They are defined to be an $m * m$ matrix where the entires for $i, j \in \{1, 2, \cdots, m\}$ are given by

$$\mathbf{H}_{i,j}, \mathbf{H}'_{i,j} = \begin{cases} 1 & \text{if client } i \text{ and } j \text{ are in the same cluster,} \\ 0 & \text{if otherwise.} \end{cases}$$

The other one is the classification accuracy (denoted as Acc) defined as

$$Acc = \frac{\sum_{i=1}^{m} Acc_i}{m} \tag{19}$$

where Acc_i is the classification accuracy of client i 's model.

We compare CIC-FL with the following four methods:

– **Local model scheme (Local).** The model in each client performs gradient descent only on local data available.
– **Global model scheme (Global).** The algorithm learns a single global model to fit data of all clients [19].
– **ClusteredFL.** One of the most state-of-the-art clustered federated learning algorithms that hierarchically separates clients into different groups based on the cosine similarities between their gradients [20].
– **IFCA.** An efficient framework for clustered federated learning that estimates the cluster identities of the clients and optimizes model parameters for the clusters via gradient descent [8].

In all the training processes of FL, we set the size of the local training batch is 32, and the learning rate is 0.001. The standard SGD optimizer is utilized to optimize the loss function. For the bipartition process, we set the threshold $\epsilon = 0.95$.

4.2 Experimental Results of CIC-FL

In the experiments of this section we consider the following FL setup: All experiments are performed on all datasets using $m = 20$ clients. Particularly, for the FMNIST or CIFAR-10 dataset, we generate $K = 4$ clusters. For IMDB, we generate $K = 2$ clusters. The experimental results on different levels of class imbalance α and concept shift β for various datasets are shown in Table 1.

We can see neither using global model based on all clients nor using local models for each client is a good choice. Correctly grouping clients into clusters and train one personalized model in each cluster is better option. In addition, CIC-FL outperforms the ClusteredFL and IFCA in our experimental scenarios in the presence of both concept shift and class imbalance.

4.3 CIC-FL at Different Levels of Class Imbalance

In this section, under different levels of class imbalance, we compare RI of CIC-FL with ClusteredFL and the following two-staged methods:

Table 1. The comparison with four baseline algorithms.

Datasets		FMNIST	CIFAR10	IMDB
(α,β)		(11,50%)	(8.5,50%)	(4,100%)
Acc (%)	Local	62.6	56.8	65.0
	Global	68.6	59.7	53.7
	ClusteredFL	75.9	62.3	75.7
	IFCA	69.5	64.9	79.6
	CIC-FL (ours)	**78.6**	**67.4**	**83.2**
RI (%)	ClusteredFL	89.4	87.3	75.0
	IFCA	82.3	91.1	89.5
	CIC-FL (ours)	**96.2**	**93.6**	**95.6**

- **Undersampling + ClusteredFL (USClusteredFL).** Each client randomly discards samples from the classes whose size is larger than the smallest one until class imbalance is eliminated. Then ClusteredFL is applied to group clients into clusters.
- **Oversampling + ClusteredFL (OSClusteredFL).** Each client randomly generates samples (via data augmentation techniques) for the classes whose size is smaller than the largest one until class imbalance is eliminated. Then ClusteredFL is applied to group clients into clusters.

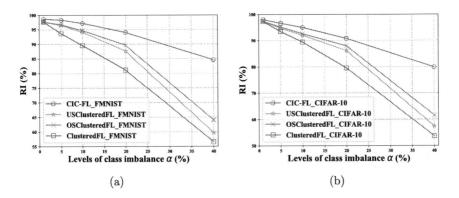

(a)   (b)

Fig. 3. The comparison of *RI* among CIC-FL, USClusteredFL, OSClusteredFL and ClusteredFL on FMNIST (as shown in (a)) and CIFAR-10 (as shown in (b)) at different levels of the class imbalance, with a fixed level of concept shift.

Specially, the experiments in this section are based on FMNIST and CIAFR-10, where we set $m = 40$ and $K = 5$. The level of class imbalance on each client is set to be changed from 1, 5, 10, 20 to 40. The level of concept shift is fixed as 60%, and other settings are described in Sect. 4.1.

RI of these clustering methods on two different datasets are shown in Fig. 3(a) (for FMINST) and Fig. 3(b) (for CIFAR-10) respectively. In Fig. 3, we can see that the performance of all 4 methods decreases with the increase in level of class imbalance, which indicates that the presence of class imbalance generally has a negative effect on clustering methods for concept shift. In addition, compared to ClusteredFL, the two-staged methods (namely, the USClusteredFL and OSClusteredFL) have better performance, which shows that mitigating the effect of class imbalance does help clustered methods generate better results. The proposed CIC-FL has consistently higher RI than other methods, especially with high level of class imbalance. Therefore, CIC-FL is more robust to class imbalance than others.

4.4   CIC-FL at Different Levels of Concept Shift

We conduct experiments to evaluate the effectiveness of CIC-FL at different levels of concept shift. We choose FMNIST and CIFAR-10 datasets and use similar experiment settings as Sect. 4.1 except that two clusters are simulated ($K = 2$), each of which has 10 clients.

Then we test the RI of CIC-FL and OSClusteredFL at different levels of concept shift, namely 20%, 40%, 60%, 80% and 100%, respectively. At the same time, the level of class imbalance is fixed to be 15. The results are shown Fig. 4. We can observe that on both datasets, the performance of CIC-FL and OSClusteredFL increases with the increase in concept shift. Furthermore, with the decrease of the level of concept shift, the gap between the performance of CIC-FL and of OSClusteredFL increases.

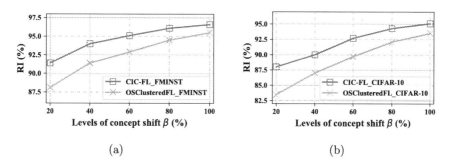

(a)   (b)

Fig. 4. The comparison of RI between CIC-FL and OSClusteredFL on FMNIST (as shown in (a)) and CIFAR-10 (as shown in (b)) at different levels of the concept shift, with a fixed level of class imbalance.

To further illustrate that *LEGLD* is sensitive to concept shift, we display the heatmaps of cosine similarity matrix S for the 20 clients on FMNIST and CIFAR-10, respectively shown in Fig. 5. The 20 clients are denoted by $0 \sim 19$ and every ten clients have a same permutation of label sequences, generating a cluster structure $\{\{0, \cdots, 9\}, \{10, \cdots, 19\}\}$.

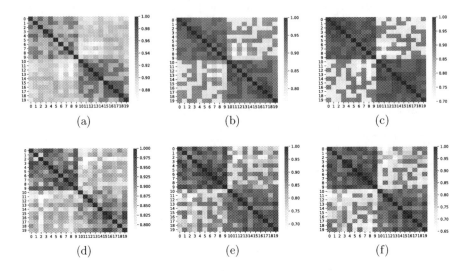

(a)   (b)   (c)

(d)   (e)   (f)

Fig. 5. The cosine similarity matrices of *LEGLD* for 20 clients. Specially, figures in the first row are on FMNIST at three different levels of concept shift (corresponding to 20%, 60%, and 100% from left to right). Similar to those the first row, figures in the second row are on CIFAR-10.

Figures in the first row in Fig. 5 show the heatmaps of cosine similarity matrix on FMNIST at three different levels of concept shift (corresponding to 20%, 60%, and 100% from left to right). We can see that initially when the concept shift is light, the heatmap does not show the cluster structure obviously, and when the concept shift rises up to 60%, and 100%, the cluster structure is shown more obviously. The above phenomenon can also be discovered in figures of the second row, which show the heatmaps of cosine similarity matrix on CIFAR-10 at three different levels of concept shift (corresponding to 20%, 60%, and 100% from left to right). The above results show that *LEGLD* is a good feature for clustering in FL for concept shift.

5   Conclusion

In this paper, we address the problem of concept shift among clients in the presence of class imbalance in FL. We propose a Class Imbalance-aware Clustered Federated Learning method (CIC-FL). CIC-FL iteratively group clients into two clusters by leveraging *LEGLD*, a feature sensitive to concept shift but robust to the class imbalance. In addition, CIC-FL is privacy-preserving and communication efficient. We test CIC-FL on benchmark datasets including Fashion-MNIST, CIFAR-10, and IMDB. The results show that CIC-FL outperforms the state-of-the-art clustered federated learning methods in the presence of the class imbalance. Then, an immediate future work is to research how to choose a better threshold ϵ for clustering.

Acknowledgement. This work is supported by the National Natural Science Foundation of China under Grant Nos. 61976012 and 61772060, and China Education and Research Network Innovation Project under Grant No. NGII20170315.

References

1. Anand, R., Mehrotra, K.G., Mohan, C.K., Ranka, S.: An improved algorithm for neural network classification of imbalanced training sets. IEEE Trans. Neural Networks **4**(6), 962–969 (1993)
2. Briggs, C., Fan, Z., Andras, P.: Federated learning with hierarchical clustering of local updates to improve training on non-IID data. arXiv preprint arXiv:2004.11791 (2020)
3. Chawla, N.V., Bowyer, K.W., Hall, L.O., Kegelmeyer, W.P.: SMOTE: synthetic minority over-sampling technique. J. Artif. Intell. Res. **16**, 321–357 (2002)
4. Chen, F., Luo, M., Dong, Z., Li, Z., He, X.: Federated meta-learning with fast convergence and efficient communication. arXiv preprint arXiv:1802.07876 (2018)
5. Collins, L.M., Dent, C.W.: Omega: a general formulation of the rand index of cluster recovery suitable for non-disjoint solutions. Multivar. Behav. Res. **23**(2), 231–242 (1988)
6. Corinzia, L., Buhmann, J.M.: Variational federated multi-task learning. arXiv preprint arXiv:1906.06268 (2019)
7. Fallah, A., Mokhtari, A., Ozdaglar, A.: Personalized federated learning: a meta-learning approach. arXiv preprint arXiv:2002.07948 (2020)
8. Ghosh, A., Chung, J., Yin, D., Ramchandran, K.: An efficient framework for clustered federated learning. arXiv preprint arXiv:2006.04088 (2020)
9. Ghosh, A., Hong, J., Yin, D., Ramchandran, K.: Robust federated learning in a heterogeneous environment. arXiv preprint arXiv:1906.06629 (2019)
10. Hard, A., et al.: Federated learning for mobile keyboard prediction. arXiv preprint arXiv:1811.03604 (2018)
11. Hensman, P., Masko, D.: The impact of imbalanced training data for convolutional neural networks. Degree Project in Computer Science, KTH Royal Institute of Technology (2015)
12. Johnson, J.M., Khoshgoftaar, T.M.: Survey on deep learning with class imbalance. J. Big Data **6**(1), 1–54 (2019). https://doi.org/10.1186/s40537-019-0192-5
13. Kairouz, P., et al.: Advances and open problems in federated learning. arXiv preprint arXiv:1912.04977 (2019)
14. Khan, S.H., Hayat, M., Bennamoun, M., Sohel, F.A., Togneri, R.: Cost-sensitive learning of deep feature representations from imbalanced data. IEEE Trans. Neural Netw. Learn. Syst. **29**(8), 3573–3587 (2017)
15. Lee, H., Park, M., Kim, J.: Plankton classification on imbalanced large scale database via convolutional neural networks with transfer learning. In: 2016 IEEE International Conference on Image Processing (ICIP), pp. 3713–3717. IEEE (2016)

16. Lin, T.Y., Goyal, P., Girshick, R., He, K., Dollár, P.: Focal loss for dense object detection. In: Proceedings of the IEEE International Conference on Computer Vision, pp. 2980–2988 (2017)
17. Mani, I., Zhang, I.: KNN approach to unbalanced data distributions: a case study involving information extraction. In: Proceedings of Workshop on Learning from Imbalanced Datasets, vol. 126 (2003)
18. Mansour, Y., Mohri, M., Ro, J., Suresh, A.T.: Three approaches for personalization with applications to federated learning. arXiv preprint arXiv:2002.10619 (2020)
19. McMahan, B., Moore, E., Ramage, D., Hampson, S., y Arcas, B.A.: Communication-efficient learning of deep networks from decentralized data. In: Artificial Intelligence and Statistics, pp. 1273–1282. PMLR (2017)
20. Sattler, F., Müller, K.R., Samek, W.: Clustered federated learning: Model-agnostic distributed multitask optimization under privacy constraints. IEEE Trans. Neural Netw. Learn. Syst. (2020)
21. Sattler, F., Müller, K.R., Wiegand, T., Samek, W.: On the byzantine robustness of clustered federated learning. In: ICASSP 2020–2020 IEEE International Conference on Acoustics, Speech and Signal Processing (ICASSP), pp. 8861–8865. IEEE (2020)
22. Smith, V., Chiang, C.K., Sanjabi, M., Talwalkar, A.S.: Federated multi-task learning. In: Advances in Neural Information Processing Systems, pp. 4424–4434 (2017)
23. Tan, B., Liu, B., Zheng, V., Yang, Q.: A federated recommender system for online services. In: Fourteenth ACM Conference on Recommender Systems, pp. 579–581 (2020)
24. Wang, H., Cui, Z., Chen, Y., Avidan, M., Abdallah, A.B., Kronzer, A.: Predicting hospital readmission via cost-sensitive deep learning. IEEE/ACM Trans. Comput. Biol. Bioinform. 15(6), 1968–1978 (2018)
25. Wang, L., Xu, S., Wang, X., Zhu, Q.: Eavesdrop the composition proportion of training labels in federated learning. arXiv preprint arXiv:1910.06044 (2019)
26. Wang, S., Liu, W., Wu, J., Cao, L., Meng, Q., Kennedy, P.J.: Training deep neural networks on imbalanced data sets. In: 2016 International Joint Conference on Neural Networks (IJCNN), pp. 4368–4374. IEEE (2016)
27. Xie, M., Long, G., Shen, T., Zhou, T., Wang, X., Jiang, J.: Multi-center federated learning. arXiv preprint arXiv:2005.01026 (2020)
28. Zhao, Y., Li, M., Lai, L., Suda, N., Civin, D., Chandra, V.: Federated learning with non-IID data. arXiv preprint arXiv:1806.00582 (2018)

vRaft: Accelerating the Distributed Consensus Under Virtualized Environments

Yangyang Wang[1,2] and Yunpeng Chai[1,2(✉)]

[1] Key Laboratory of Data Engineering and Knowledge Engineering,
MOE, Beijing, China
ypchai@ruc.edu.cn
[2] School of Information, Renmin University of China, Beijing, China

Abstract. In recent years, Raft has been gradually widely used in many distributed systems (e.g., Etcd, TiKV, PolarFS, etc.) to ensure the distributed consensus because it is effective and easy to implement. However, because the performance of the virtual nodes in cloud environments is usually heterogeneous and fluctuant due to the "noisy neighbor" problem and the cost efficiency, the strong leader mechanism makes the Raft protocol encounter a serious performance challenge. Specifically, when the performance of the leader node is low, the whole system performance will descend accordingly since both the write and the read requests serving will be blocked by the slow leader processing. Aiming to solve this problem, we proposed a modified version of Raft specially optimized for virtualized environments, i.e., vRaft. It breaks Raft's strong leader restriction and can fully utilize the temporarily fast followers to accelerate both the write and the read requests processing in a virtualized cloud environment, without affecting the linearizability guarantee of Raft. The experiments based on the virtual nodes in Tencent Cloud indicate that vRaft improves the throughput by up to 64.2%, reduces average latency by 38.1%, and shortens the tail latency by 88.5% in a typical read/write-balanced workload compared with Raft.

1 Introduction

For distributed systems, the consensus algorithm is a key component to guarantee data consistency and system reliability, especially in the presence of system faulty processes. Traditionally, the Paxos [1] protocol is employed by many distributed systems to achieve the distributed consensus. However, Paxos is particularly difficult to understand and implement in practical distributed systems. In this case, the Raft protocol [2,3], which was proposed in 2014, is easy to be comprehended and realized, and thus soon has been widely adopted by many practical distributed systems like Etcd [4], TiKV [5], and PolarFS [6]. Although the sequential execution limitation weakens the performance of Raft compared

© Springer Nature Switzerland AG 2021
C. S. Jensen et al. (Eds.): DASFAA 2021, LNCS 12681, pp. 53–70, 2021.
https://doi.org/10.1007/978-3-030-73194-6_4

with Paxos, the multiple Raft groups (see Sect. 2.1) or the ParallelRaft [6] mechanism can improve the parallelism of operation processing and promote the performance. According to statistics from Raft's official website, as of November 2020, Raft has been used in 117 projects [7].

Motivation. In recent years, more and more distributed systems are deployed in cloud environments, i.e., in virtual machines (e.g., KVM [8], Xen [9], etc.) or containers (e.g., Docker [10]). And the CPU, memory, I/O, and network resources are isolated by tools like *cgroup* [11]. However, the nowadays technique cannot guarantee accurate performance isolation, the performance of a virtual node is highly affected by the other virtual nodes located on the same physical machine; this is called the noisy neighbor problem [12]. In addition, the emerging storage devices (e.g., SSDs or non-volatile memory (NVM)) have obvious performance advantages over the traditional ones. However, these new devices are usually much more expensive, so we may deploy them in only a subset of the clusters for cost efficiency. Therefore, the virtual nodes, even with the same configurations, often have different performance, and the performance of any node may fluctuate frequently. For example, when the same program runs 300 times in a virtualized environment, the performance difference is up to 60× or more [13]. Moreover, we rent two virtual nodes with exactly the same configuration from Tencent Cloud [14], but the I/O performance of these two nodes has 3 to 10 times difference, as Fig. 1 shows.

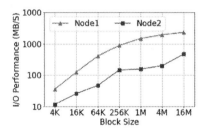

Fig. 1. The I/O performance gap between two virtual nodes with completely the same configuration. The tests were performed by using *fio* [15] with the block size setting ranging from 4 KB to 16 MB.

Considering the heterogeneous and unstable performance of virtual nodes, the distributed systems based on Raft have an important performance challenge: Raft adopts a strong leader mechanism to ensure the data consistency, i.e., the leader undertakes much more jobs compared with the followers and is the most critical part for performance. Once the leader locates on a slow node in a time period, the performance of the whole system will be slowed down (see Sect. 2.1 for more about Raft). For example, we have made some comparative experiments by forcing the leader to locate on the fastest node or the slowest node. The system throughput gap between the two configurations reaches 62.8% (see Sect. 2.2 for details).

Basic Idea. For the above problem appeared in virtualized environments, there are no existing solutions; for example, a common-sense method of migrating the leader to fast nodes introduces other additional problems (see Sect. 2.2 for more). Therefore, in this paper, we propose an improved version of the Raft protocol under the virtualized environment, i.e., vRaft, to solve the above problem and improve the performance. vRaft breaks the strong leader limitation on the basis of maintaining the same level data consistency, allowing fast followers to boost both the write and the read request processing. Two new mechanisms called Fastest Return (Sect. 3.1) and Optimal Read (Sect. 3.1) are proposed in vRaft, in order to fully take the advantage that some followers located on temporarily fast virtual nodes has fast progress and strong processing ability.

The comparison experiments between vRaft and Raft in the Tencent Cloud environment indicate that vRaft improves the throughput by 64.2% in a read/write-balanced workload, reduces the average latency by 38.1%, and shortens the tail latency by 88.5% at the same time. Furthermore, more experiments under different configurations (e.g., the numbers of replicas, system loads, and system scales) exhibit that vRaft is effective in various environments.

Our contributions in this paper are summarized as follows:

(1) *We identify the important new performance problem of Raft introduced by the virtualized environment.* Due to the heterogeneous and unstable performance of virtual nodes, if the performance of the node where the Raft leader locates is temporarily poor, the system performance will be deteriorated.
(2) *We solve the above problem of Raft by proposing a modified version of Raft, called vRaft.* vRaft breaks Raft's strong leader mechanism and thus can fully utilize the fast follower(s) to accelerate the request processing in a virtualized environment. And we prove that vRaft does not break the linear consistency guaranteed by Raft.
(3) *We improve the performance of an industrial-grade distributed key-value storage systems (i.e., TiKV) by incorporating vRaft to demonstrate its effectiveness.* Compared with Raft, vRaft promotes the throughput by 64.2%, shrinks the average latency by 38.1%, and reduces 88.5% tail latency for typical workloads.

The rest of this paper is organized as follows. Section 2 introduces the background of our research and the motivation of this paper. In Sect. 3, we present the design of our proposed vRaft. The implementation details and the evaluations of vRaft are described in Sect. 4, followed by the related work presented in Sect. 5. Finally, we conclude this paper in Sect. 6.

2   Background and Motivation

2.1   The Raft Protocol for Distributed Consensus

Traditionally, the Paxos [1] protocol is classical to ensure data consistency in distributed systems. However, Paxos is particularly difficult to understand and implement. In this case, the Raft protocol [2], which is easy to be comprehended and realized, has been widely adopted by many practical distributed systems like Etcd [4], TiKV [5], and PolarFS [6] since it was proposed in 2014.

Write Process of Raft. Among the N copies of any data segment, one of them is elected as the leader replica according to Raft, while the other $N-1$ ones become followers. Raft's processing of write requests includes three key operations, i.e., *append, commit,* and *apply,* as shown in Fig. 2.

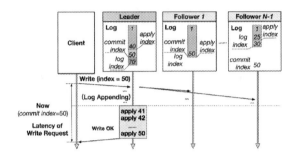

Fig. 2. Raft's procedure of processing write requests.

First, when a client sends a write request to the leader, the leader *appends* the content of the write request into the persistent log, and meanwhile the leader sends the content of the write request to all the other followers in parallel. When a follower receives the write request from the leader, it will also *append* the content into the log and then notify the leader after its appending procedure succeeds.

Subsequently, when the leader finds out that more than half of the replicas (maybe not including the leader itself) have successfully finished the append operations, the log is in the *committed* status and the *commit index* (i.e., the version id) of the replica is increased (e.g., from C to $C+1$). Note that the version of the latest log (a.k.a., *log index*) on a node is often newer than the *commit index*; this means that the latest content is not safe enough because only minority nodes have this content. For instance, as the example in Fig. 2 illustrates, the *log index* of the leader has already reached 70, while the *commit index* is only 50.

Then, after a written data is committed, the leader and all the followers start *applying* the log into the state machine. And after the leader's *apply* operation succeeds, the leader can return success to the client, and the *apply index* is increased. According to Raft, the log appending should be performed sequentially according to requests' arrival order. So do the log committing and the data applying. Therefore, the leader has to apply the previous updates (e.g., index 41–49 in Fig. 2) first before processing the target data (e.g., index 50).

Read Process of Raft.[1] In Raft, all the read requests are processed by the leader to ensure that the client would not get the out-of-date data. When the leader receives a read request from a client, it records the current *commit index* as the *read index* of the read request [3].

When the leader's *apply index* is no less than the *read index* of the read request, the leader can execute the read request immediately and return the result to the client. However, when its *apply index* is lower than the *read index*, some additional time-consuming operations should be performed first before processing the read request. As Fig. 3 plots, the *read index* is 50 while the *apply index* of the leader is only 40. So the leader has to apply the contents of 41–50 first and then process the read request, in order to ensure linear consistency.

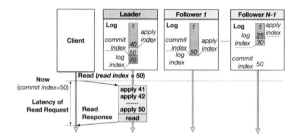

Fig. 3. Raft's procedure of processing read requests.

The four kinds of indexes used in Raft are summarized in Table 1.

Table 1. The four kinds of indexes used in Raft.

Index name	Descriptions
log index	Index of log that has been appended on a node
commit index	Index of log that has been appended by majority nodes
apply index	Index of log that has been written into status machine
read index	The commit index at the time a read request arrives

Multiple Raft Groups. If all the data are put into one Raft group, the system scalability is poor, because only N nodes can be used for N copies. In addition, all the Raft's operations will be executed sequentially, without parallelism. Therefore, practical systems usually adopt the solution of *multiple Raft groups*. i.e., the data are divided into many segments and the replicas of each data segment compose one independent Raft group. Figure 4 is an example of

---

[1] Raft's read process is not detailed described in the original paper [2], but in the doctoral thesis [3] of the author.

multiple Raft groups with the 3-copies setting, 6 nodes, and 4 Raft groups, where the operations of different Raft groups can be processed in parallel for higher performance.

Coupled with the multiple Raft groups, we can solve the performance problem of Raft caused by sequential processing of requests, making Raft comparable with Paxos in performance. Thus most Raft-based practical distributed systems adopt the multiple Raft groups such as TiKV and PolarFS. Note that all experiments in this paper are based on multiple Raft groups, and the size of each Raft group is usually about 100 MB.

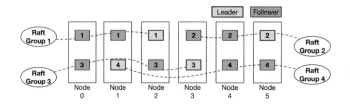

Fig. 4. An example of multiple Raft group.

2.2   Motivation

Raft Is Not Performing Well in a Virtualized Environment. The reason lies in that the leader processing affects the performance the most in Raft and the node performance is often heterogeneous and unstable in a virtualized environment. When a leader is locating on a temporarily weak node, it will slow down the whole Raft group significantly.

Specifically, for write operations, when the progress of the leader is slow, even if most nodes have already finished writing, we have to wait for the leader's accomplishment before replying to the client. For read requests, if the leader's *apply index* falls behind the *commit index* due to the poor performance of the leader node, the read request will not be executed until the *apply index* of the leader reaches the *read index* of the read request, even though other follower nodes can serve the read request already.

To illustrate the impacts of the leader node's performance. We wrote 10GB of data into a three-node cluster, forcing the leader to locate on the fastest node or the slowest node, respectively. The results exhibit the former leads to a 62.8% higher throughput and a 42.9% lower latency compared with the latter, indicating that the slow progress of the leader can result in significant performance declining (Fig. 5).

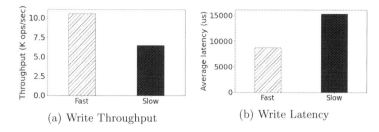

(a) Write Throughput

(b) Write Latency

Fig. 5. The performance when the leader locates on the fastest or the slowest node.

The Leader Migration Solution Does Not Work. A common-sense idea of solving the slow leader problem is to migrate the leader replica to a fast node. However, there are some problems for this solution: First, it is different to measure the node performance accurately in real-time, because the software (e.g., a key-value engine) performance on a node is usually affected by multiple factors.

Second, the leader migration causes significant additional overhead, such as the latency brought by the leader election or the new leader fetching its missing logs from others. Especially in virtualized environments, the virtual node performance often fluctuates, leading to frequent leader switches and much overhead.

Finally, another problem of leader migration lies in the possible excessive leader concentration, which will weaken the parallelism of the read request processing and make the fast node become overloaded, resulting in performance degradation. For example, as Fig. 6 shows, we read 10GB of data in a real physical cluster, comparing the performance of the evenly distributed leaders and the concentrated leaders on the fastest node. The results illustrate that the evenly distributed leader solution has a 101% higher throughput and a 50.4% lower latency than the other one.

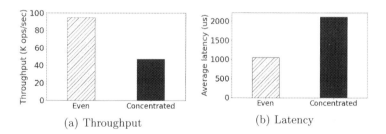

(a) Throughput

(b) Latency

Fig. 6. The performance gap between the even and concentrated leader distributions.

Summary. According to the above discussions, we cannot solve the slow leader problem through the leader migration. Therefore, we should change the direction, i.e., fully utilizing the fast follower to accelerate the request processing in a virtualized environment. We should make the follower replace part of the leader's work, breaking Raft's strong leader mechanism, thereby improving performance, but at the same time not destroying the linear consistency of Raft.

3   Design of vRaft

In this section, we will present the design of our improved Raft protocol, i.e., vRaft, which aims to boost both the write and the read request processing under the virtualized environments. We first present the basic idea of vRaft in Sect. 3.1. Then the algorithm design of vRaft will be given in Sect. 3.2 and the linearizability of vRaft will be discussed in Sect. 3.3.

3.1   Overview

In order to solve the slow leader problem under virtualized environments, vRaft boosts both the write and read processing by breaking the roadblock of the leader and creating new paths for request processing, without influence the linear consistency. Specifically, vRaft introduces two key components, i.e., Fastest Return and Optimal Read, to boost writing and reading, respectively.

Fastest Return. For the write request processing of the original Raft, if the progress of the leader is slow, even if most other nodes have finished writing, the client has to wait until the slow leader completes writing. Recall the example shown in Fig. 2, i.e., for a write operation with the index 50, since the *apply index* of the leader is only 40, older than the *write index*, we have to apply logs 41–50 to the state machine in order to finish this write operation. The massive applying operations slow down the write operation processing significantly.

However, a Raft group contains multiple nodes. Some of the follower nodes may be faster than the leader at the current time period in virtualized cloud environments. In this case, the follower should send its accomplished apply index to the leader when its apply index changes. Therefore, when the leader knows one of the followers has finished the applying phase, it can return success to the client ahead of time compared with the original Raft, even if the leader itself has not finished applying.

Example 3.1. *An write operation processing example of vRaft.* As Fig. 7 illustrates, the follower 1 notifies the leader that its apply index is 50, so the leader can return the response of the write request with index 50 to the client, even if the leader's own apply index is less than 50.

Optimal Read. Because we need to read the target data from the state machine, only the nodes with newer apply indexes compared with the read index can serve the read request. For reading, the performance problem of the original Raft lies in that when the leader node is temporarily slow in the virtualized environment and has a low apply index, we have to increase the leader's apply

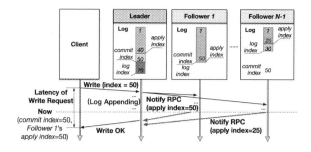

Fig. 7. vRaft's procedure of processing write requests.

index first before the request serving. Recall the example shown in Fig. 3, the read index is 50, and the temporarily slow leader which has an apply index of 40. So the read request will not be executed until the leader has applied the logs 41–50 to the state machine.

However, if there is a follower whose apply index is greater than or equal to the read index, it may be faster to redirect the read request to the follower for processing. Therefore, when the leader cannot serve the request immediately due to the low apply index, the leader checks whether there is a follower with a high enough apply index who can immediately process the read request. And the leader also needs to judge whether the time it waits for the apply index increment is greater than the time to redirect the read request to such a follower. If there are multiple followers that meet the condition, the leader will redirect the read request to the follower with the lowest pressure for processing.

Furthermore, the redirected read request must include the read index, and the follower who receives the read request must make sure the apply index not lower than the read index before executing the read request. In this way, even if an error occurs in the redirection, linear consistency can be guaranteed (See Sect. 3.3 for details).

Example 3.2 *An read operation processing example of vRaft.* As Fig. 8 plots, the leader finds that the time waiting for the apply index increment of itself is larger than the time to redirect the read request to the follower 1, so the follower 1 will process this request.

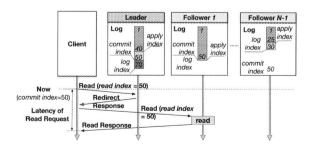

Fig. 8. vRaft's procedure of processing read requests.

3.2   Algorithm Design

Fastest Return. Algorithm 1 exhibits the key functions of FR, i.e., the newly added *sendNotify* for the followers and *handleNotify* for the leader. When the apply index of any follower has changed, this follower will send a *Notify RPC* (including its apply index) to the leader, shown as Lines 1–3 in Algorithm 1.

When the leader receives a *Notify RPC*, it updates the record of the corresponding follower' apply index, and calculates the maximum apply index of all the nodes in the Raft group, as shown in Lines 4–6. Through the variable *status* in Line 5, the leader records the status of all the nodes including their apply indexes, log indexes, etc.

Finally, the leader will check all the write requests that have not finished (i.e., the list *unresponsiveWrites* in Line 7). If the updated *maxApplyIndex* is newer than the index of a waiting write request, the write request is considered as an accomplished one and the leader can directly return success to the client, as shown in Lines 7–12. Note that the list *unresponsiveWrites* is a FIFO queue, so when the first request's write index is newer than the current max apply index, the other request will also have to wait.

Algorithm 1. Fastest Return

1: **function** *Follower* :: *sendNotify()* :
2:     *message.set(leaderID, this.applyIndex)*;
3:     *send(message)*;
4: **function** *Leader* :: *handleNotify(message)* :
5:     *status.update(message)*;
6:     $maxApplyIndex \leftarrow status.getMaxApplyIndex()$;
7:     **while** *unresponsiveWrites.len()* > 0 **do**
8:         $writeIndex \leftarrow unresonsiveWrites[0].index$;
9:         **if** $maxApplyIndex >= writeIndex$ **then**
10:             $write \leftarrow unresonsiveWrites.remove()$;
11:             *respond(write)*;
12:         **else** *break*;

Optimal Read. The key functions of Optimal Read are described in Algorithm 2. All the incoming read requests are put into a list of the leader called *pendingReads*. The leader checks all the pending read requests when a new request enters the pending request queue, or when the leader's apply index changes, or when it receives the follower's *Notify RPC*. If the apply index of the leader is no less than the read index of the request, the leader can immediately execute the read operation, as shown in Lines 1–5.

Otherwise, if the apply indexes of some followers are newer or equal to the read index, we need to compare the cost of redirecting the request to one follower and the overhead of waiting for the leader finishing the apply operations. Assume that the average time for a leader to apply a log is a, and the additional time of network transmission caused by request redirecting is b. If $(read\ index - leader's\ apply\ index) * a > b$, the redirecting plan is faster. Let $c = b/a$; when $the\ leader's\ apply\ index + c < read\ index$ and $all\ followers'\ maxApplyIndex >= read\ index$ are satisfied, we should redirect the read request to a follower to process, as shown in Lines 6–7.

Then, if there are multiple followers that satisfy the apply index condition, the leader will choose the follower with the minimal read load, as shown in Lines 8–9, where F is the set of the followers that satisfy the apply index condition. In this case, the leader will send a redirect response to the client (including the read index and the follower id) and update the follower's read load record, as shown in Lines 10–11. Note that each node records its own read load and the followers report their records to the leader periodically.

After that, the follower receives the redirected read request. If its apply index is greater than or equal to the read index of the request, the follower can directly execute the read request, as shown in Lines 12–15.

For the client, it first sends the read request to the leader and gets a response. If the response is a redirect message, the client sends the read request to the corresponding follower to get the target data, as shown in Lines 16–22.

Algorithm 2. Optimal Read

1: **function** $Leader :: checkReads()$:
2:     **for** each $readReq \in pengdingReads$ **do**
3:         $readIndex \leftarrow readReq.readIndex$;
4:         **if** $applyIndex >= readIndex$ **then**
5:             $execute(readReq)$;
6:         **else if** $applyIndex + c < readIndex$ **then**
7:             **if** $maxApplyIndex >= readIndex$ **then**
8:                 $F \leftarrow getFollowers(readIndex)$;
9:                 $followerId \leftarrow F.minReadLoadNode()$;
10:                 $redirect(readReq, readIndex, followerId)$;
11:                 $status.update(followerId)$;
12: **function** $Follower :: handleRedirectRead(readReq)$:
13:     $readIndex \leftarrow readReq.readIndex$;
14:     **if** $applyIndex >= readIndex$ **then**
15:         $execute(readReq)$;
16: **function** $Client :: Read(readReq)$:
17:     $response \leftarrow sendNrecv(readReq, leaderID)$;
18:     **if** $response.type = Redirect$ **then**
19:         $readIndex, followerId \leftarrow reponse.get()$;
20:         $readReq.readIndex \leftarrow readIndex$;
21:         $readReq.type \leftarrow Redirect$;
22:         $response \leftarrow sendNrecv(readReq, followerId)$;

3.3   Linearizability of vRaft

Although vRaft changes both the write and the read procedures compared with Raft, it does not break the linearizability [16] guaranteed by Raft.

Theorem 1. *vRaft does not break the linearizability, which means once a new value has been written or read, all the subsequent reads see the new value, until it is overwritten again.*

Proof (sketch): Once writing a new value is finished, it triggers that the apply index is updated to a_1 and we assume the current commit index is c_1. Thus the commit index must be larger than or equal to the apply index (i.e., $c_1 \geq a_1$). The read index r of any subsequent read request will be equal to or larger than the current commit index c_1 (i.e., $r \geq c_1 \geq a_1$). Because vRaft also guarantees that only when the apply index is greater than or equal to the read index, the read request can be executed, the apply index (i.e., a_2) when serving a subsequent read request, which has the same or larger read index than r, needs to be equal to or greater than r, i.e., $a_2 \geq r \geq c_1 \geq a_1$. Therefore, the state machine of the version a_2 definitely contains the written value in the version a_1, which will make sure the newly written value can be read by all the subsequent read requests.

When a new value has been read, assuming the current *apply index* is a_1. A subsequent read request's *read index* r_2 is equal to the current *commit index* c_2, which is larger than a_1, i.e., $r_2 = c_2 \geq a1$. And the new read request with the read index r_2 will also be served by a node with the *apply index* a_2 which is larger or equal to r_2. So we can get $a_2 \geq r_2 = c_2 \geq a_1$. Similar to the above case, because of $a_2 \geq a_1$, we can make sure the subsequent read requests can get the new value.

4   Implementation and Evaluation

Raft has been widely implemented in the industrial community, such as famous open-source systems like Etcd [4] and TiKV [5]. Etcd is based on a memory-based state machine, adopting a single Raft group; it is designed to store a small amount of data such as metadata. Different from Etcd, TiKV adopts multiple Raft groups and the disk-based state machine for massive data. Therefore, we implemented our proposed vRaft and integrated it into TiKV for evaluations.

As a distributed key-value storage system based on Raft, TiKV utilizes RocksDB [17] as the key-value engine for the state machine on each node, and employs another system called Placement Driver (PD) [18] to manage the data distribution of TiKV. In fact, TiKV contains more than 100K LOC of *Rust*, which is already one of the largest open-source projects in the *Rust* community.

4.1   Experimental Setup

The experiments were performed in a cluster consisted of eight virtual nodes in Tencent Cloud; each virtual node is coupled with Linux Centos 7.4.1708, 8

GB DRAM, and a 200 GB Solid State Drive (SSD). Six of the virtual nodes serve as TiKV nodes, one as PD, and the last one runs the benchmark tool, i.e., go-YCSB [19].

Go-YCSB is a *Go* language version of the widely used YCSB benchmark [20] for evaluating key-value systems. In the experiments, the used workloads are listed in Table 2, including *Load* (insert-only), *Workload A* (50:50 read/update), *Workload B* (95:5 read/update), and *Workload C* (read-only). Each key-value pair contains a 16-B key and a 1-KB value, and each data block has 3 replicas in TiKV. The default thread number of the go-YCSB client is 200. Other configurations all adopt the default ones in the go-YCSB specification [21].

Table 2. The YCSB workloads used in evaluations.

Name	Description
Load	Insert only
Workload A	50:50 Read/Update
Workload B	95:5 Read/Update
Workload C	Read only

In the following experiments, we adopt the system throughput (i.e., operations per second or ops/sec), the average latency, and the 99th percentile latency as the performance metrics.

4.2   Overall Results

In the overall experiments, we first load 100-GB data into the TiKV cluster, and then perform the workloads *A*, *B*, and *C* of YCSB, respectively, accessing 100 GB of data respectively.

As Fig. 9 plots, vRaft achieves higher throughput than Raft in most cases, i.e., 80% higher for *Load*, 64.2% higher for *Workload A*, and 2.7% higher for *Workload B*. This is because more writes make the differences in the apply indexes of different nodes be greater, thereby vRaft gains more performance improvement compared with Raft. For the read-only workload (i.e., *C*), there is no difference for the apply indexes of the nodes, so vRaft achieves almost the same throughput as Raft in *Workload C*.

Figure 10 exhibits the results of the average latency and the 99th percentile latency. vRaft achieves lower average latency than Raft in most cases, i.e., 44.2% lower for *Load*, 38.1% lower for *Workload A*, and 1.7% lower for *Workload B*. And the reduction on the 99th percentile latency of vRaft is more significant compared with Raft, e.g., 86.3% lower for *Load*, 88.5% lower for *Workload A*, and 13.9% lower for *Workload B*. Because when the leader writes slowly, Raft has to wait for the requests to be written one by one on the leader node, while vRaft can return the read and write results to clients through the faster progress of

followers. In *Workload C*, vRaft's average latency and the 99th percentile latency
are close to Raft's, since the apply indexes of all the nodes do not change at all
due to no new writes coming.

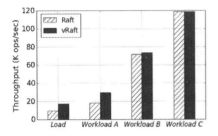 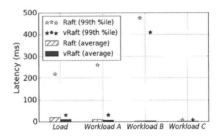

Fig. 9. Overall throughput results.   **Fig. 10.** Overall latency results.

4.3   Impacts of the Number of Replica

In this part, we measure the performance of vRaft and Raft under different
numbers of replica configurations, including 3, 5, and 7 replicas. We adopt the
read/write-balanced workload, i.e., *Workload A*, to perform the experiments.
Specifically, we first load 10 GB of data into the cluster, and then perform
Workload A of 10-GB data. The throughput and the latencies of performing
Workload A are exhibited in Fig. 11 and Fig. 12.

As Fig. 11 plots, vRaft can achieve 46.2%–63.5% higher throughput compared
with Raft under all different numbers of replica. In addition, Fig. 12 exhibits the
results of the average latency and the 99th percentile latency. vRaft shortens
the average latency by 30.4% to 38.9% and reduces the tail latency by 5.6% on
average compared to Raft under all these cases. Because the loaded data is 10
GB, less than the amount of 100 GB in the overall results, the tail latency is not
reduced as significantly as the above experiments.

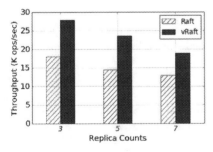 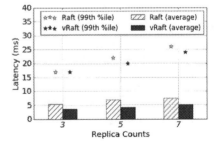

Fig. 11. Throughput for different replica
counts.

Fig. 12. Average and tail latencies for
different number of replica.

4.4   Impacts of System Load

In this part, we measure vRaft and Raft under different system loads. The low, medium, and high system loads are configured by setting different numbers of client threads of *go-YCSB*. The thread number of low system load is only 10, the thread number of medium load is 50, and the number of high load is up to 200. In the experiments, we perform 10-GB *Workload A* based on an existing data set of 10 GB.

As Fig. 13 exhibits, no matter how much the system load is, vRaft can increase the system throughput significantly (i.e., 41.7% to 67.8% higher). Figure 14 indicates that vRaft can reduce the average latency by 30.4% to 46.1% and at the same lower the 99th percentile latency by 3.6% on average under all kinds of system loads. Of course, under the medium or the high system load, the advantage of vRaft can be fully exploited.

4.5   Scalability Evaluation

In order to evaluate the scalability of vRaft, we performed experiments on clusters with different counts of TiKV nodes (i.e., 3, 6, 12, 18, 24, or 30 TiKV nodes). All nodes here indicate virtual nodes. In the evaluation, we perform 10-GB *Workload A* based on an existing data set of 10 GB.

Figure 15 exhibits the results of the relative throughput, the relative average latency and the relative 99th percentile latency between vRaft and Raft under different system scales. As Fig. 15 shows, vRaft increases the throughput by 37.4% to 54.4% compared with Raft. vRaft reduces the average latency by 25.5% to 35.5% and reduces the tail latency by 9.6% on average compared with Raft. The results indicate that vRaft has good scalability and can improve the performance compared with Raft stably under different system scales.

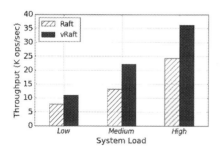
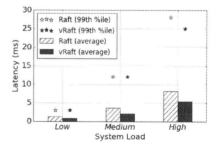

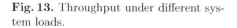

Fig. 13. Throughput under different system loads.

Fig. 14. Average and tail latencies under different system loads.

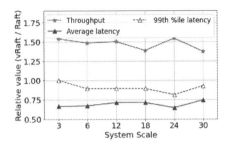

Fig. 15. Performance comparison under different system scales.

5   Related Work

Raft Optimization.   Due to the importance of the Raft protocol for distributed systems, there are many existing works for optimizing Raft from different aspects. Some work optimizes the leader election mechanism of Raft, tuning the parameters about the election to make the election procedure faster [22,23]; some other work speeds up the leader election when some failures happen [24]. In addition, some researchers combine Raft with Software Defined Networking (SDN) [25–27].

As the number of nodes in the cluster increases, the throughput may decline because the only leader becomes the bottleneck of communication. In consequence, Sorensen et al. [28] proposed Obiden, a variation of the Raft protocol for distributed consensus with a president and a vice president to provide higher performance in large clusters. Besides, PolarFS [6] implements a parallel Raft to allow Raft's logs to be processed out of order, breaking Raft's strict limitation that logs have to be submitted in sequence, with the benefit of increasing concurrency.

Hanmer et al. [29] found that Raft may not work properly under overloaded network conditions. A request such as a heartbeat cannot be returned within a specified time, thereby being considered as a failure. The heartbeat failure may cause the leader to be converted to a follower, restarting the slow leader election procedure. Furthermore, the leader election may be repeated again and again under the poor network condition, delaying the request processing seriously. Therefore, they proposed DynRaft [29], a dynamic extension of Raft to optimize the performance when the network is overloaded.

Copeland et al. [30] proposed BFTRaft, a Byzantine fault tolerant variant of the Raft consensus protocol. BFTRaft maintains the safety, the fault tolerance, and the liveness properties of Raft in the presence of Byzantine faults, and keeps the modified protocol simple and readily comprehensible, just as Raft does.

Paxos Optimization. In order to reduce the high latency of the Paxos protocol, Wang et al. proposed APUS [31], the first RDMA-based Paxos protocol that aims to be fast and scalable to client connections and hosts. Ho et al. [32] proposed a Fast Paxos-based Consensus (FPC) algorithm which provides strong consistency.

FPC adopts a controller priority mechanism to guarantee that a proposal must be elected in each round and no additional round is needed, even more than two proposers get the same votes.

In summary, the existing optimization work about the Raft-based distributed consensus does not consider the performance heterogeneity and fluctuation problem of virtual nodes in the cloud environment. Our proposed vRaft solution is the first method to solve this new problem under virtualized environments.

6   Conclusion

In a virtualized cloud environment, the performance of each virtual nodes may be heterogeneous, and they often affected seriously by the behavior of other virtual nodes located on the same physical node, thus keeps fluctuating. Therefore, the Raft protocol, which has been widely used in many distributed systems to achieve consensus, will encounter new performance problems when the leader node is temporarily slow, because both the read and the write requests have to wait for the leader's processing to be finished in Raft, even if some follower nodes are obviously faster.

In order to break the too strict leader limitation in Raft and to fully utilize the fast follower to accelerating the request processing, we propose a new version of Raft for performance optimization in virtualized environments, called vRaft. vRaft contains two new mechanisms, i.e., Fastest Return and Optimal Read, to accomplish the processing of both the write and the read requests ahead of time compared with Raft, respectively, through involving fast followers in the processing. Besides, we have implemented our proposed vRaft in an industrial level distributed key-value systems (i.e., TiKV). And the experiments based on the Tencent Cloud platform indicate that vRaft can effectively and stably improve all the key performance metrics at the same time, including the throughput, the average latency, and the tail latency.

Acknowledgement. This work is supported by the National Key Research and Development Program of China (No. 2019YFE0198600), National Natural Science Foundation of China (No. 61972402, 61972275, and 61732014).

References

1. Lamport, L.: Paxos made simple. ACM SIGACT News **32**(4), 18–25 (2001)
2. Ongaro, D., Ousterhout, J.: In search of an understandable consensus algorithm. In: 2014 USENIX Annual Technical Conference (USENIXATC 2014), pp. 305–319 (2014)
3. Ongaro, D.: Consensus: bridging theory and practice. Stanford University (2014)
4. Etcd. https://github.com/etcd-io/etcd
5. TiKV. https://github.com/pingcap/tikv
6. Cao, W., Liu, Z., Wang, P., et al.: PolarFS: an ultra-low latency and failure resilient distributed file system for shared storage cloud database. Proc. VLDB Endow. **11**(12), 1849–1862 (2018)

7. Where can I get Raft? https://raft.github.io/#implementations
8. Kernel-based Virtual Machine. https://en.wikipedia.org/wiki/Kernel-based_Virtual_Machine
9. Xen. https://en.wikipedia.org/wiki/Xen
10. docker. https://www.docker.com/
11. cgroups. https://en.wikipedia.org/wiki/Cgroups
12. Performance interference and noisy neighbors. https://en.wikipedia.org/wiki/Cloud_computing_issues#Performance_interference_and_noisy_neighbors
13. Misra, P.A., Borge, M.F., Goiri, Í., et al.: Managing tail latency in datacenter-scale file systems under production constraints. In: Proceedings of the Fourteenth EuroSys Conference, p. 17. ACM (2019)
14. Tencent Cloud. https://intl.cloud.tencent.com/
15. Flexible I/O Tester. https://github.com/axboe/fio
16. Kleppmann, M.: Designing Data-intensive Applications: The Big Ideas Behind Reliable, Scalable, and Maintainable Systems. O'Reilly Media Inc., Sebastopol (2017)
17. RocksDB. http://rocksdb.org/
18. PD. https://github.com/pingcap/pd
19. go-ycsb. https://github.com/pingcap/go-ycsb
20. Cooper, B.F., Silberstein, A., Tam, E., et al.: Benchmarking cloud serving systems with YCSB. In: Proceedings of the 1st ACM Symposium on Cloud Computing, pp. 143–154. ACM (2010)
21. go-ycsb workloads. https://github.com/pingcap/go-ycsb/tree/master/workloads
22. Howard, H., Schwarzkopf, M., Madhavapeddy, A., et al.: Raft refloated: do we have consensus? ACM SIGOPS Oper. Syst. Rev. **49**, 12–21 (2015)
23. Howard, H.: ARC: analysis of Raft consensus. Computer Laboratory, University of Cambridge (2014)
24. Fluri, C., Melnyk, D., Wattenhofer, R.: Improving raft when there are failures. In: 2018 Eighth Latin-American Symposium on Dependable Computing (LADC), pp. 167–170. IEEE (2018)
25. Sakic, E., Kellerer, W.: Response time and availability study of RAFT consensus in distributed SDN control plane. IEEE Trans. Netw. Serv. Manage. **15**(1), 304–318 (2017)
26. Zhang, Y., Ramadan, E., Mekky, H., et al.: When raft meets SDN: how to elect a leader and reach consensus in an unruly network. In: Proceedings of the First Asia-Pacific Workshop on Networking, pp. 1–7. ACM (2017)
27. Kim, T., Choi, S.G., Myung, J., et al.: Load balancing on distributed datastore in opendaylight SDN controller cluster. In: 2017 IEEE Conference on Network Softwarization (NetSoft), pp. 1–3. IEEE (2017)
28. Sorensen, J., Xiao, A., Allender, D.: Dual-leader master election for distributed systems (Obiden) (2018)
29. Hanmer, R., Jagadeesan, L., Mendiratta, V., et al.: Friend or foe: strong consistency vs. overload in high-availability distributed systems and SDN. In: 2018 IEEE International Symposium on Software Reliability Engineering Workshops (ISSREW), pp. 59–64. IEEE (2018)
30. Copeland, C., Zhong, H.: Tangaroa: a byzantine fault tolerant raft (2016)
31. Wang, C., Jiang, J., Chen, X., et al.: APUS: fast and scalable paxos on RDMA. In: Proceedings of the 2017 Symposium on Cloud Computing, pp. 94–107. ACM (2017)
32. Ho, C.C., Wang, K., Hsu, Y.H.: A fast consensus algorithm for multiple controllers in software-defined networks. In: 2016 18th International Conference on Advanced Communication Technology (ICACT), pp. 112–116. IEEE (2016)

Secure and Efficient Certificateless Provable Data Possession for Cloud-Based Data Management Systems

Jing Zhang[1], Jie Cui[1(✉)], Hong Zhong[1], Chengjie Gu[2], and Lu Liu[3]

[1] School of Computer Science and Technology, Anhui University, Hefei, China
cuijie@mail.ustc.edu.cn, zhongh@ahu.edu.cn
[2] Security Research Institute, New H3C Group, Hefei, China
gu.chengjie@h3c.com
[3] School of Informatics, University of Leicester, Leicester, UK
l.liu@leicester.ac.uk

Abstract. Cloud computing provides important data storage, processing and management functions for data owners who share their data with data users through cloud servers. Although cloud computing brings significant advantages to data owners, the data stored in the cloud also faces many internal/external security attacks. Existing certificateless data provider schemes have the following two common shortcomings, i.e., most of which use plaintext to store data and use the complex bilinear pairing operation. To address such shortcomings, this scheme proposes secure and efficient certificateless provable data possession for cloud-based data management systems. In our solution, the data owners and cloud servers need to register with the key generation center only once. To ensure the integrity of encrypted data, we use the public key of the cloud server to participate in signature calculation. Moreover, the third-party verifier can audit the integrity of ciphertext without downloading the whole encrypted data. Security analysis shows that our proposed scheme is provably secure under the random oracle model. An evaluation of performance shows that our proposed scheme is efficient in terms of computation and communication overheads.

Keywords: Cloud data management · Provable data possession (PDP) · Certificateless cryptography · Security · Efficient

1 Introduction

With the rapid development of cloud computing, more and more people outsource their data to cloud servers [1,13], which brings three main advantages. Firstly, resource-constrained users no longer need to process and store a large amount of data, so that a lot of computing and storage costs can be saved.

C. S. Jensen et al. (Eds.): DASFAA 2021, LNCS 12681, pp. 71–87, 2021.
https://doi.org/10.1007/978-3-030-73194-6_5

Secondly, users can access data anytime and anywhere without requiring high-performance hardware. Thirdly, users can share data conveniently.

Although cloud services bring many benefits to people's lives, many challenges [3,5] need to be solved properly. Firstly, user loses the direct control of their outsourced data, i.e., whether the data has been modified or deleted is unknown. Secondly, the leakage of data may damage the privacies of users, such as the time when users are not at home and the routes that users frequently travel. In the worst cases, the property safety of users may be threaten. Therefore, how to ensure the confidentiality and integrity of outsourced data has become a great concern to users.

At present, some researchers have proposed provable data possession (PDP) schemes for the integrity of outsourced data [2,4–6,9,11,12,14,15,17–22]. Although the existing schemes ensure the integrity of cloud storage data, they do not consider the confidentiality of data. Moreover, due to the usage of complex bilinear pairing operation, these schemes also bring heavy computation and communication costs to the third-party verifier (TPV). Therefore, it is urgent to design a secure and efficient provable data possession scheme for cloud data management systems.

1.1 Related Work

To ensure the integrity of outsourced data, Ateniese et al. [2] first introduced the concept of PDP in 2007, and further considered public validation. Many PDP schemes [8,9,12,17,22] that follow Ateniese et al.'s work have been introduced to protect the integrity of outsourced data. Unfortunately, these schemes have a common drawback, i.e., most of which rely on trusted third parties to generate certificates for users, so that users have serious certificate management problems and heavy computing costs.

To solve the certificate management issues, Wang et al. [15] proposed an identity-based PDP scheme and provided a corresponding security model. To improve performance and security, some identity-based PDP schemes have also been proposed [14,18,20]. However, these schemes have a common disadvantage, that is, they need the secret key generation center to generate a series of private keys for users, which brings the key escrow problem.

To overcome the key escrow problem, a series of certificateless provable data possession (CL-PDP) schemes have been proposed [4–6,11,19,21]. Unfortunately, there are still many security issues in these schemes. Zhang et al. [19] pointed out that schemes [6,11] cannot guarantee the privacy of data. He et al. [5] discovered that scheme [19] had a malicious server attack and proposed an improved scheme. Recently, Zhou et al. [21] discovered that scheme [5] is vulnerable to tag forging and data loss hiding attacks. In addition, these schemes use complex bilinear pairing operations, which bring deficiencies in terms of computation and communication.

1.2   Contribution

To achieve the security of outsourced data and further reduce the waste of resources, this paper proposes a secure and efficient certificateless provable data possession scheme for cloud-based data management systems. There are three main contributions of the proposed scheme.

1. We propose to use a symmetric and asymmetric encryption algorithms simultaneously, which cannot only realize the security of data sharing, but also further ensure the confidentiality of outsourced data.
2. The proposed scheme can resist the attack of Type I and Type II adversaries, and can resist the tag forgery attack. The security analysis reveals that our scheme is provably secure under the random oracle model.
3. The detailed comparisons with the existing related schemes in terms of computational and communication overhead on the Tag Generation, Generate-Proof and Verify-Proof Algorithms, demonstrates that our scheme provides better performance.

The outline of the rest study is as follows: In Sect. 2, we introduce the background of this study. In Sect. 3, we put forward the proposed scheme. The security analysis is proved in Sect. 4. The performance evaluation is outlined in Sect. 5. Lastly, we present the conclusion of this study in Sect. 6.

2   Background

In this section, we introduce the preliminary knowledge and network model.

2.1   Elliptic Curve Cryptosystem (ECC)

Let E_p: $y^2 = x^3 + ax + b(\bmod p)$ be a non-singular elliptic curve over the finite field F_p, where $p > 3$ is a large prime, $a, b \in F_p$, and $4a^3 + 27b^2(\bmod p) \neq 0$. Let G be a cyclic group on E_p of prime order q.

Discrete Logarithm (DL) Problem: Given two random points $P, Q \in G$, where $Q = xP$, $x \in Z_q^*$, and $Z_q^* = \{1, 2, ..., q-1\}$, it is difficult to calculate x from Q in a probabilistic polynomial time (PPT).

2.2   Network Model

The system architecture comprises a key generation center KGC, a cloud server CS, a data owner DO and a third-party verifier TPV. As shown in Fig. 1, the details of each component are described as follows:

- **KGC:** It is a trusted third party, which is in charge of generating and publishing system parameters. It also generates a partial key for each DO and delivers these sensitive information to them via secure channels.

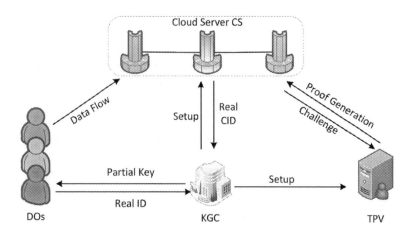

Fig. 1. A network model of the CL-PDP protocol

- **CS:** It is an honest but curious entity that is assumed to have sufficient computing and storage capabilities.
- **DO:** It is a resource constrained data owner, who outsources their data to CS and entrusts TPV to verify the integrity of cloud storage data.
- **TPV:** It verifies the integrity of cloud storage data when users need it, and is responsible for the verification results.

3   Proposed Scheme

In this section, we describe the proposed CL-PDP scheme based on ECC to solve the security problem and reduce the time cost.

3.1   Setup Algorithm

Given a security parameter λ, KGC generates a cyclic group G with prime order q and generator P. Then, KGC randomly chooses $s \in Z_q^*$ and computes the system public key $P_{pub} = s \cdot P$. KGC selects six one-way hash function $H_0 : \{0,1\}^* \to \{0,1\}^q$, $H_k : \{0,1\}^* \to Z_q^*, k = 1, 2, 3, 4, 5$. Finally, KGC publishes system parameters $params = \{q, Z_q^*, P_{pub}, H\}$ and saves the master key s secretly.

3.2   Key Generation Algorithm

Given the real identities $ID_i, CID_k \in Z_q^*$ of DO_i and CS_k, KGC performs as follows:

- KGC randomly selects $\alpha_k \in Z_q^*$ as CS_k's secret key $sk_k = \alpha_k$, and computes $PK_k = \alpha_k P$ as CS_k's public key. Then, KGC sends the key $\{sk_k, PK_k\}$ to CS_k via a secure channel.

- KGC randomly picks $\alpha_i \in Z_q^*$ and computes $A_i = \alpha_i P$, $h_{i,1} = H_1(ID_i \| A_i)$ and $sk_1 = \alpha_i + h_{i,1}s(\mathrm{mod}q)$. Then, KGC sends the partial key $\{sk_1, A_i\}$ to DO_i via a secure channel.
- DO randomly chooses $\beta \in Z_q^*$ as their secret value $sk_2 = \beta$ and computes $PK_i = \beta P$ as their public key.

3.3 Store Algorithm

Encrypt Data. DO first divides their data M into n blocks: $M = \{M_l\}_{l=1}^n$. DO then generates a corresponding signature for each block of data.

- DO randomly picks $x_M \in Z_q^*$, $\delta \in \{0,1\}^q$, computes $X_M = x_M P$, and saves $\{x_M, X_M\}$ as a one-time-use signing key and verification key, respectively.
- DO computes $h_{i,2} = H_2(E_K(M_1)\|...\|E_K(M_n)\|S_1\|...\|S_n\|\delta\|X_M)$, $Z = h_{i,2}PK_k$, $Y = \delta + H_0(h_{i,2}P)$, $h_{i,3} = H_3(ID_i\|A_i\|PK_i)$, and $S_M = h_{i,2}x_M + h_{i,3}sk_2 + sk_1$. Note that authorized users can utilize the secret key K to decrypt data $M_l = D_K(E_K(M_l))$.

Tag Generation. Through the execution of this algorithm, DO produces a Tag for each block of data and stores the encrypted data into the cloud.

- DO randomly picks $x_l \in Z_q^*$, computes $X_l = x_l P$, $h_{i,4}^l = H_4(ID_i \|name_l\|X_l\|PK_i)$, $h_{i,5}^l = H_5(name_l\|X_l\|A_i)$, and $S_l = E_K(M_l)x_l + h_{i,4}^l sk_2 + h_{i,5}^l sk_1$. Note that $name_l$ denotes the unique name of data M_l.
- DO outputs $T_l = \{X_l, S_l, E_K(M_l)\}$ as M_l's tag.
- Finally, DO sends $\{X_M, S_M, Z, Y, \{T_l\}_{l=1}^n\}$ to CS.

Store. After receiving the request from the DO, CS computes $\delta = Y + H_0(Z')$ by decrypting $Z' = h_{i,2}P = Z sk_k^{-1}$.

- CS computes $h_{i,1} = H_1(ID_i\|A_i)$, $h_{i,3} = H_3(ID_i\|A_i\|PK_i)$ and $h_{i,2} = H_2(E_K(M_1)\|...\|E_K(M_n)\|S_1\|...\|S_n\|\delta\|X_M)$. CS then checks whether the following condition is true.

$$S_M P = h_{i,2}X_M + h_{i,3}PK_i + A_i + h_{i,1}P_{pub} \tag{1}$$

- If it is not true, CS immediately stops the session. Otherwise, CS computes $h_{i,4}^l = H_4(ID_i\|name_l\|X_l\|PK_i)$, $h_{i,5}^l = H_5(name_l\|X_l\|A_i)$ and verifies the condition.

$$\sum_{l=1}^n S_l P = \sum_{l=1}^n [E_K(M_l)X_l] + \sum_{l=1}^n h_{i,4}^l PK_i + \sum_{l=1}^n h_{i,5}^l(A_i + h_{i,1}P_{pub}) \tag{2}$$

If the verification holds, CS stores the encrypted data; otherwise, CS rejects the request.

3.4   Challenge Algorithm

Through the execution of this algorithm, a TPV produces a challenging message to verify the data integrity of data.

1. TPV chooses a random subset $I \in \{1, 2, ..., n\}$ and a small number v_j for each $j \in I$.
2. TPV outputs $\{j, v_j\}_{j \in I}$ as a challenging message and returns it to CS.

3.5   Generate-Proof Algorithm

When CS receives the TPV's auditing challenge $\{j, v_j\}_{j \in I}$, CS produces the following steps to complete the proof.

1. CS calculates $S_{cs} = \sum_{j \in I} v_j S_j P$ and $C_{cs} = \sum_{j \in I} [v_j E_K(M_j) X_j]$.
2. CS outputs the proof $\{S_{cs}, C_{cs}\}$ and returns it to TPV.

3.6   Verify-Proof Algorithm

Upon receiving the proof $\{S_{cs}, C_{cs}\}$, TPV executes the following steps to check the correctness.

1. TPV calculates $h_{i,1} = H_1(ID_i \| A_i)$, $h_{i,4}^j = H_4(ID_i \| name_j \| X_l \| PK_i)$ and $h_{i,5}^j = H_5(name_j \| X_l \| A_i)$.
2. TPV checks whether the following equation holds.

$$S_{cs} = C_{cs} + \sum_{j \in I} (v_j h_{i,4}^j) PK_i + \sum_{j \in I} (v_j h_{i,5}^j)(A_i + h_{i,1} P_{pub}) \qquad (3)$$

If the equation holds, the TPV outputs "Accept"; otherwise, TPV outputs "Reject".

4   Security Analysis

In this section, we firstly present a security model for the proposed scheme. Ane then, we analyze and prove the security of the proposed CL-PDP scheme.

4.1   Security Model

There are two types of unbounded adversaries namely $\mathcal{A}_1$ and $\mathcal{A}_2$. Type I adversary $\mathcal{A}_1$ can replace the public key of the user but doesn't access the master key. Type II adversary $\mathcal{A}_2$ cannot access replace the public key of the user but has ability to access the master key. The adversary $\mathcal{A}_1$ and $\mathcal{A}_2$ and the challenger $\mathcal{C}$ could make the following queries in the game.

Setup. In this query, $\mathcal{C}$ inputs the master key and public parameters. $\mathcal{C}$ keeps the master key secretly and sends the public parameters $\mathcal{A}$. $\mathcal{C}$ also sends the master key to $\mathcal{A}$ if $\mathcal{A}$ is a Type II adversary.

Query. In this query, $\mathcal{A}$ can make some queries and $\mathcal{C}$ answers back:

1. *Create Data Owner*: $\mathcal{C}$ executes the key generation algorithm to generate the DO's partial private key and secret value, and returns the DO's public key to $\mathcal{A}$.
2. *Extract Partial Private Key*: $\mathcal{C}$ returns a partial private key of DO to $\mathcal{A}$ as an answer.
3. *Public Key Replacement*: $\mathcal{A}$ can replace the public key of DO with a new value chosen by $\mathcal{A}$.
4. *Extract Secret Value*: $\mathcal{C}$ returns a secret value of ID to $\mathcal{A}$ as an answer.
5. *Generate Tag*: $\mathcal{C}$ generates a Tag of a block and returns it to $\mathcal{A}$.

Forge. $\mathcal{A}$ outputs a one-time-use verification key S^* and a Tag X^* corresponding the challenging identity ID^*.

$\mathcal{A}$ wins the game if the following requirements are satisfied:

1. T^* is the corresponding valid tag of the challenging identity ID^*.
2. T^* is not generated by querying *Generate Tag*.
3. ID^* is independent of algorithm of *Extract Partial Private Key/Extract Secret Value* if $\mathcal{A}$ is Type I/Type II adversary.

Definition 1. The proposed certificateless provable data possession (CL-PDP) scheme is secure against forging Tag attack, if there is no any adversary $\mathcal{A} \in \{\mathcal{A}_1, \mathcal{A}_2\}$ which wins the above-mentioned game with a non-negligible probability.

4.2 Security Theorem

Theorem 1. *According to the assumption of the difficulty of the DL problem, the proposed CL-PDP scheme is secure against Type I adversary.*

Proof. Assuming given $P, Q = aP$, where P, Q are two points on elliptic curve E_q, $\mathcal{A}_1$ can forge a one-time-use verification key S^* and a Tag X^* corresponding the challenging identity ID^*. We have built a game between $\mathcal{A}_1$ and a challenger $\mathcal{C}_1$, and $\mathcal{C}_1$ has the ability to run $\mathcal{A}_1$ with a non-negligible probability as a subroutine to solve DL problem.

Setup: The master key s is randomly selected by challenger $\mathcal{C}_1$. And $\mathcal{C}_1$ then calculates the corresponding public key $P_{pub} = sP$. Next, $\mathcal{C}_1$ sends the system parameters $params = \{q, Z_q^*, P_{pub}, H\}$ to $\mathcal{A}_1$. $\mathcal{C}_1$ chooses a challenging identity ID^* and answers the following queries from $\mathcal{A}_1$.

H_i Queries: When $\mathcal{A}_1$ uses the elements m_i for H_i query, $\mathcal{C}_1$ checks whether the elements (m_i, τ_{h_i}) already exists in the hash list $L_{h_i} (i = 0, 1, ..., 5)$. If it is, $\mathcal{C}_1$ sends $\tau_{h_i} = H_1(m_i)$ to $\mathcal{A}_1$. Otherwise, $\mathcal{C}_1$ picks $\tau_{h_i} \in Z_q^*$ randomly and adds the elements (m_i, τ_{h_i}) to the hash list L_{h_i}, then $\mathcal{C}_1$ sends $\tau_{h_i} = H_1(m_i)$ to $\mathcal{A}_1$.

Create Data Owner Query: When $\mathcal{A}_1$ performs a create data owner query on the challenging identity ID^*, $\mathcal{C}_1$ checks whether the form $(ID_i, sk_2, sk_1, \alpha_i, PK_i, A_i)$ exists in L_6. If exists, $\mathcal{C}_1$ replies (PK_i, A_i) to $\mathcal{A}_1$. Otherwise, $\mathcal{C}_1$ works as following:

- If $ID_i = ID^*$, $\mathcal{C}_1$ picks three elements $sk_1, sk_2, \tau_{h_1} \in Z_q^*$ randomly and computes $PK_i = sk_2P$ and $A_i = sk_1P - \tau_{h_1}P_{pub}$. $\mathcal{C}_1$ inserts the tuple (ID_i, A_i, τ_{h_1}) and $(ID_i, sk_2, sk_1, \perp, PK_i, A_i)$ into L_{h_1} and L_6, respectively. Note that $\perp$ denotes null.
- Otherwise, $ID_i \neq ID^*$, $\mathcal{C}_1$ picks three elements $\alpha_i, sk_2, \tau_{h_1} \in Z_q^*$ randomly and computes $PK_i = sk_2P$ and $A_i = \alpha_iP$. $\mathcal{C}_1$ inserts the tuple (ID_i, A_i, τ_{h_1}) and $(ID_i, sk_2, \perp, \alpha_i, PK_i, A_i)$ into L_{h_1} and L_6, respectively.

Extract Partial Private Key Query: Upon receiving $\mathcal{A}_1$'s query, $\mathcal{C}_1$ checks whether ID_i already exists in hash list L_{h_2}. If $\mathcal{C}_1$ cannot find the corresponding tuple, $\mathcal{C}_1$ makes H_1 query on ID_i itself to produce τ_{h_1}. Then, $\mathcal{C}_1$ works as following:

- If $ID_i \neq ID^*$, $\mathcal{C}_1$ first checks whether ID_i exists in L_6. If exists, $\mathcal{C}_1$ searches the tuple $(ID_i, sk_2, sk_1, \alpha_i, PK_i, A_i)$ and returns (A_i, sk_1) to $\mathcal{A}_1$. Otherwise, $\mathcal{C}_1$ picks two element $sk_1, \tau_{h_1} \in Z_q^*$ and computes $A_i = sk_1P - \tau_{h_1}P_{pub}$. Then, $\mathcal{C}_1$ returns (A_i, sk_1) to $\mathcal{A}_1$ and stores $(ID_i, sk_2, sk_1, \alpha_i, PK_i, A_i)$ to L_6.
- Otherwise, $ID_i = ID^*$, $\mathcal{C}_1$ stops the game.

Public Key Replacement Query: When $\mathcal{A}_1$ performs a public key replacement query on (ID_i, A_i^*, PK_i^*), $\mathcal{C}_1$ first checks whether ID_i exists in L_6. $\mathcal{C}_1$ answers as following:

- If list L_6 contains ID_i, $\mathcal{C}_1$ replaces the tuple $(ID_i, sk_2, sk_1, \alpha_i, PK_i, A_i)$ with $(ID_i, sk_2, sk_1, \alpha_i, PK_i^*, A_i^*)$.
- Otherwise, $\mathcal{C}_1$ inserts the tuple $(ID_i, \perp, \perp, \perp, PK_i^*, A_i^*)$ to L_6.

Extract Secret Value Query: Upon receiving $\mathcal{A}_1$'s extract secret value query on ID_i, $\mathcal{C}_1$ answers as following:

- If list L_6 involves $(ID_i, sk_2, sk_1, \alpha_i, PK_i, A_i)$, $\mathcal{C}_1$ checks whether $sk_2 = \perp$ is true. If $sk_2 = \perp$, $\mathcal{C}_1$ sends sk_2 to $\mathcal{A}_1$. Otherwise, $\mathcal{C}_1$ performs a create data owner query to generate $PK_i = sk_2P$. After that, $\mathcal{C}_1$ sends sk_2 to $\mathcal{A}_1$ and updates (sk_i, PK_i) to list L_6.
- If list L_6 does not involve $(ID_i, sk_2, sk_1, \alpha_i, PK_i, A_i)$, $\mathcal{C}_1$ performs a create data owner query and sends sk_2 to $\mathcal{A}_1$. After that, $\mathcal{C}_1$ sends sk_2 to $\mathcal{A}_1$ and updates (ID_i, sk_i, PK_i) to list L_6.

Generate Tag Query: $\mathcal{A}_1$ performs a generate tag query on $(name_l, M_l)$ under (ID_i, PK_i, A_i). $\mathcal{C}_1$ first checks whether ID_i exists in L_6, L_{h_1}, L_{h_4} and L_{h_5}. $\mathcal{C}_1$ answers as following:

- If $ID_i = ID^*$, $\mathcal{C}_1$ stops the game.
- Otherwise, $\mathcal{C}_1$ picks three elements $S_l, \tau_{h_1}, \tau_{h_4}, \tau_{h_5} \in Z_q^*$ randomly and computes $X_l = E_K(M_l)^{-1}(S_lP - \tau_{h_4}PK_i - \tau_{h_5}(A_i + \tau_{h_1}P_{pub}))$. Then, $\mathcal{C}_1$ returns (S_l, X_l) to $\mathcal{A}_1$. Note that if τ_{h_4} or τ_{h_5} already exists in hash list L_{h_4} or L_{h_5}, $\mathcal{C}_1$ picks an element S_l and works again.

Forgery: At last, $\mathcal{A}_1$ outputs a M_l's Tag $\{X_M, S_M^*, Z, Y, X_l, S_l^*, E_K(M_l)\}$ under (ID_i, PK_i, A_i). If $ID_i \neq ID^*$, $\mathcal{C}_1$ aborts the game. Otherwise, on the basis of the forking lemma [10], $\mathcal{C}_1$ has the ability to get two different valid Tags $T_l = (X_l, S_l)$ and $T_l^* = (X_l, S_l^*)$ in polynomial time through $\mathcal{A}_1$, if $\mathcal{C}_1$ repeat the process with a different choice of H_1. We have the following equation:

$$S_l P = E_K(M_l)X_l + h_{i,4}^l PK_i + h_{i,5}^l(A_i + h_{i,1}P_{pub}) \tag{4}$$

$$S_l^* P = E_K(M_l)X_l + h_{i,4}^l PK_i + h_{i,5}^l(A_i + h_{i,1}^* P_{pub}) \tag{5}$$

Hence, we can get that

$$
\begin{aligned}
(S_l - S_l^*)P &= S_l P - S_l^* P \\
&= E_K(M_l)X_l + h_{i,4}^l PK_i + h_{i,5}^l(A_i + h_{i,1}P_{pub}) \\
&\quad - E_K(M_l)X_l + h_{i,4}^l PK_i + h_{i,5}^l(A_i + h_{i,1}^* P_{pub}) \\
&= (h_{i,1} - h_{i,1}^*)h_{i,5}^l P_{pub} \\
&= a(h_{i,1} - h_{i,1}^*)h_{i,5}^l P
\end{aligned}
\tag{6}
$$

and

$$a = \frac{S_l - S_l^*}{(h_{i,1} - h_{i,1}^*)h_{i,5}^l} \tag{7}$$

Thus, $\mathcal{C}_1$ could solve the DL problem. However, this is in contradiction with the difficulty of DL problem.

Similarly, if $\mathcal{A}_1$ could correctly guess the output of H_2, $\mathcal{C}_1$ also has the ability to get two different valid signatures $\{X_M, S_M, Z, Y\}$ and $\{X_M, S_M^*, Z, Y\}$ based on the forking lemma [10]. $\mathcal{C}_1$ also repeat the process with a different choice of H_1 and we have the following equation:

$$S_M P = h_{i,2}X_M + h_{i,3}PK_i + A_i + h_{i,1}P_{pub} \tag{8}$$

$$S_M^* P = h_{i,2}^* X_M + h_{i,3}PK_i + A_i + h_{i,1}^* P_{pub} \tag{9}$$

In the same way, if $h_{i,2} = h_{i,2}^*$, we can get $a = \frac{S_M - S_M^*}{h_{i,1} - h_{i,1}^*}$.

Unfortunately, the premise of this equation is not only that $\mathcal{A}_1$ can correctly guess the output of H_2, but also that $\mathcal{C}_1$ can solve the DL problem.

Analysis: The probability that $\mathcal{A}_1$ can correctly guess the output of H_2 is $\frac{1}{2^q}$. Assume $\mathcal{C}_1$ can solve the DL problem with negligible advantage ε. The following three events are used to analyze the probability that $\mathcal{C}_1$ can solve the DL problem.

- Event E_1: $\mathcal{A}_1$ can forge a valid Tag $\{X_M^*, S_M^*, Z^*, Y^*, X_l^*, S_l^*, E_K(M_l)^*\}$ under (ID_i, PK_i, A_i).
- Event E_2: $\mathcal{C}_1$ does not abort when $\mathcal{A}_1$ performs extract partial private key query and generate tag query.
- Event E_3: $ID_i = ID^*$.

Under the random oracle model, a probabilistic polynomial-time adversary $\mathcal{A}_1$ forges a Tag in an attack modeled by the forking lemma after making q_{H_i} ($i = 1, 2, 3, 4, 5$) times queries, q_{ppk} times extract partial private key queries, and q_{tag} times generate tag queries. We can achieve that $Pr(E_1) = \eta$, $Pr(E_2|E_1) = (1 - \frac{1}{q_{H_1}})^{q_{ppk}+q_{Tag}}$ and $Pr(E_3|E_1 \wedge E_2) = \frac{1}{q_{H_1}}$. The probability that $\mathcal{C}_1$ can solve the DL problem is

$$
\begin{aligned}
\varepsilon &= Pr(E_1 \wedge E_2 \wedge E_3) \\
&= Pr(E_3|E_1 \wedge E_2)Pr(E_2|E_1)Pr(E_1) \\
&= \frac{1}{q_{H_1}}(1 - \frac{1}{q_{H_1}})^{q_{ppk}+q_{Tag}} \cdot \eta
\end{aligned}
\tag{10}
$$

Thus, the probability that $\mathcal{A}_1$ forges a Tag is $\varepsilon' = \frac{1}{2^q} \cdot \varepsilon$.

Due to η is non-negligible, ε is also non-negligible. Thus, $\mathcal{C}_1$ can solve the DL problem with a non-negligible probability. However, it is difficult to solve the DL problem, namely, the proposed CL-PDP scheme is secure against Type I adversary.

Theorem 2. *According to the assumption of the difficulty of the DL problem, the proposed CL-PDP scheme is secure against Type II adversary.*

Proof. Assuming given $P, Q = aP$, where P, Q are two points on elliptic curve E_q, $\mathcal{A}_2$ can forge a one-time-use verification key S^* and a Tag X^* corresponding the challenging identity ID^*. We have built a game between $\mathcal{A}_2$ and a challenger $\mathcal{C}_2$, and $\mathcal{C}_2$ has the ability to run $\mathcal{A}_2$ with a non-negligible probability as a subroutine to solve DL problem.

Setup: The master key s is randomly selected by challenger $\mathcal{C}_2$. And $\mathcal{C}_2$ then calculates the corresponding public key $P_{pub} = sP$. Next, $\mathcal{C}_1$ sends the master key s and system parameters $params = \{q, Z_q^*, P_{pub}, H\}$ to $\mathcal{A}_2$. $\mathcal{C}_2$ chooses a challenging identity ID^* and answers the following queries from $\mathcal{A}_2$.

H_i Queries: Similar to **H_i queries** in the Proof of Theorem 1.

Create Data Owner Query: When $\mathcal{A}_2$ performs a create data owner query on the challenging identity ID^*, $\mathcal{C}_2$ checks whether the form $(ID_i, sk_2, sk_1, PK_i, A_i)$ exists in L_6. If exists, $\mathcal{C}_2$ replies (PK_i, A_i) to $\mathcal{A}_2$. Otherwise, $\mathcal{C}_2$ works as following:

- If $ID_i = ID^*$, $\mathcal{C}_2$ picks three elements $\alpha_i, \tau_{h_1} \in Z_q^*$ randomly and computes $A_i = \alpha_i P$ and $sk_1 = \alpha_i + h_{i,1}s(\bmod q)$. $\mathcal{C}_2$ inserts the tuple (ID_i, A_i, τ_{h_1}) and $(ID_i, \perp, sk_1, PK_i, A_i)$ into L_{h_1} and L_6, respectively.
- Otherwise, $ID_i \neq ID^*$, $\mathcal{C}_2$ picks three elements $\alpha_i, sk_2 \in Z_q^*$ randomly and computes $PK_i = sk_2 P$, $A_i = \alpha_i P$, $\tau_{h_1} = H_1(ID_i\|A_i)$ and $sk_1 = \alpha_i + \tau_{h_1}s(\bmod q)$. $\mathcal{C}_2$ inserts the tuple (ID_i, A_i, τ_{h_1}) and $(ID_i, sk_2, \perp, PK_i, A_i)$ into L_{h_1} and L_6, respectively.

Extract Partial Private Key Query: Upon receiving $\mathcal{A}_2$'s query, $\mathcal{C}_2$ checks whether ID_i already exists in hash list L_{h_2}. If $\mathcal{C}_2$ cannot find the corresponding tuple, $\mathcal{C}_2$ makes H_1 query on ID_i itself to produce τ_{h_1}. Then, $\mathcal{C}_2$ works as following:

- If $ID_i \neq ID^*$, $\mathcal{C}_2$ first checks whether ID_i exists in L_6. If exists, $\mathcal{C}_2$ searches the tuple $(ID_i, sk_2, sk_1, PK_i, A_i)$ and returns (A_i, sk_1) to $\mathcal{A}_2$. Otherwise, $\mathcal{C}_2$ picks two element $sk_1, \tau_{h_1} \in Z_q^*$ and computes $A_i = sk_1 P - \tau_{h_1} P_{pub}$. Then, $\mathcal{C}_2$ returns (A_i, sk_1) to $\mathcal{A}_2$ and stores $(ID_i, sk_2, sk_1, PK_i, A_i)$ to L_6.
- Otherwise, $ID_i = ID^*$, $\mathcal{C}_2$ searches the tuple $(ID_i, sk_2, sk_1, PK_i, A_i)$ and returns (A_i, sk_1) to $\mathcal{A}_2$.

Extract Secret Value Query: Upon receiving $\mathcal{A}_2$'s extract secret value query on ID_i, $\mathcal{C}_2$ answers as following:

- If $ID_i \neq ID^*$, $\mathcal{C}_2$ first checks whether ID_i exists in L_6. If exists, $\mathcal{C}_2$ searches the tuple $(ID_i, sk_2, sk_1, PK_i, A_i)$ and returns sk_2 to $\mathcal{A}_2$. Otherwise, $\mathcal{C}_1$ picks two element $sk_2 \in Z_q^*$ and computes $pk_i = sk_2 P$. Then, $\mathcal{C}_2$ returns sk_2 to $\mathcal{A}_2$ and stores $(ID_i, sk_2, sk_1, PK_i, A_i)$ to L_6.
- Otherwise, $ID_i = ID^*$, $\mathcal{C}_2$ stops the game.

Generate Tag Query: $\mathcal{A}_2$ performs a generate tag query on $(name_l, M_l)$ under (ID_i, PK_i, A_i). $\mathcal{C}_2$ first checks whether ID_i exists in L_6, L_{h_1}, L_{h_4} and L_{h_5}. $\mathcal{C}_2$ answers as following:

- If $ID_i = ID^*$, $\mathcal{C}_2$ stops the game.
- Otherwise, $\mathcal{C}_2$ picks three elements $S_l, \tau_{h_1}, \tau_{h_4}, \tau_{h_5} \in Z_q^*$ randomly and computes $X_l = E_K(M_l)^{-1}(S_l P - \tau_{h_4} PK_i - \tau_{h_5}(A_i + \tau_{h_1} P_{pub}))$. Then, $\mathcal{C}_2$ returns (S_l, X_l) to $\mathcal{A}_2$. Note that if τ_{h_4} or τ_{h_5} already exists in hash list L_{h_4} or L_{h_5}, $\mathcal{C}_2$ picks an element S_l and works again.

Forgery: At last, $\mathcal{A}_2$ outputs a M_l's Tag $\{X_M, S_M^*, Z, Y, X_l, S_l^*, E_K(M_l)\}$ under (ID_i, PK_i, A_i). If $ID_i \neq ID^*$, $\mathcal{C}_2$ aborts the game. Otherwise, on the basis of the forking lemma [10], $\mathcal{C}_2$ has the ability to get two different valid Tags $T_l = (X_l, S_l)$ and $T_l^* = (X_l, S_l^*)$ in polynomial time through $\mathcal{A}_2$, if $\mathcal{C}_2$ repeat the process with a different choice of H_4. We have the following equation:

$$S_l P = E_K(M_l)X_l + h_{i,4}^l PK_i + h_{i,5}^l (A_i + h_{i,1} P_{pub}) \tag{11}$$

$$S_l^* P = E_K(M_l)X_l + h_{i,4}^{l*} PK_i + h_{i,5}^l (A_i + h_{i,1} P_{pub}) \tag{12}$$

Hence, we can get that

$$
\begin{aligned}
(S_l - S_l^*)P &= S_l P - S_l^* P \\
&= E_K(M_l)X_l + h_{i,4}^l PK_i + h_{i,5}^l (A_i + h_{i,1} P_{pub}) \\
&\quad - E_K(M_l)X_l + h_{i,4}^{l*} PK_i + h_{i,5}^l (A_i + h_{i,1} P_{pub}) \\
&= (h_{i,4}^l - h_{i,4}^{l*})PK_i \\
&= sk_2(h_{i,4}^l - h_{i,4}^{l*})P
\end{aligned}
\tag{13}
$$

and

$$sk_2 = \frac{S_l - S_l^*}{(h_{i,4}^l - h_{i,4}^{l*})} \tag{14}$$

Thus, $\mathcal{C}_2$ could solve the DL problem. However, this is in contradiction with the difficulty of DL problem.

Similarly, if $\mathcal{A}_2$ could correctly guess the output of H_2, $\mathcal{C}_2$ also has the ability to get two different valid signatures $\{X_M, S_M, Z, Y\}$ and $\{X_M, S_M^*, Z, Y\}$ based on the forking lemma [10]. $\mathcal{C}_2$ also repeat the process with a different choice of H_1 and we can get $sk_2 = \frac{S_l - S_l^*}{(h_{i,4}^l - h_{i,4}^{l*})}$.

Analysis: The probability that $\mathcal{A}_2$ can correctly guess the output of H_2 is $\frac{1}{2^q}$. Assume $\mathcal{C}_2$ can solve the DL problem with negligible advantage ε. The following three events are used to analyze the probability that $\mathcal{C}_2$ can solve the DL problem.

- Event E_1: $\mathcal{A}_2$ can forge a valid Tag $\{X_M^*, S_M^*, Z^*, Y^*, X_l^*, S_l^*, E_K(M_l)^*\}$ under (ID_i, PK_i, A_i).
- Event E_2: $\mathcal{C}_2$ does not abort when $\mathcal{A}_2$ performs extract secret value query and generate tag query.
- Event E_3: $ID_i = ID^*$.

Under the random oracle model, a probabilistic polynomial-time adversary $\mathcal{A}_2$ forges a Tag in an attack modeled by the forking lemma after making q_{H_i} ($i = 1, 2, 3, 4, 5$) times queries, q_{sev} times extract secret value queries, and q_{tag} times generate tag queries. We can achieve that $Pr(E_1) = \eta$, $Pr(E_2|E_1) = (1 - \frac{1}{q_{H_1}})^{q_{sev}+q_{Tag}}$ and $Pr(E_3|E_1 \wedge E_2) = \frac{1}{q_{H_1}}$. The probability that $\mathcal{C}_2$ can solve the DL problem is

$$\begin{aligned}
\varepsilon &= Pr(E_1 \wedge E_2 \wedge E_3) \\
&= Pr(E_3|E_1 \wedge E_2)Pr(E_2|E_1)Pr(E_1) \\
&= \frac{1}{q_{H_1}}(1 - \frac{1}{q_{H_1}})^{q_{sev}+q_{Tag}} \cdot \eta
\end{aligned} \tag{15}$$

Thus, the probability that $\mathcal{A}_2$ forges a Tag is $\varepsilon' = \frac{1}{2^q} \cdot \varepsilon$.

Due to η is non-negligible, ε is also non-negligible. Thus, $\mathcal{C}_2$ can solve the DL problem with a non-negligible probability. However, it is difficult to solve the DL problem, namely, the proposed CL-PDP scheme is secure against Type II adversary.

4.3   Discussion

Table 1 compares the security and functionality feature analyse of the related schemes [4,5,7,19] and our scheme. The symbol $\sqrt{}$ indicates that the scheme is secure or provides that feature. In contrast, the symbol $\times$ indicates that the scheme is insecure or does not provide that feature. This table indicates that only our proposed scheme can provide better security features than those of existing schemes [4,5,7,19].

Table 1. Comparison of security and functionality features.

Security features	Zhang et al. [19]	Kang et al. [7]	He et al. [5]	Gao et al. [4]	The proposed
Public verifiability	√	√	√	√	√
Storage correctness	√	√	×	×	√
Data privacy preserving	×	×	×	×	√
Tag cannot be forged	√	√	×	√	√
Batch verification	√	√	√	√	√

5 Performance Evaluation

In this section, we discuss comparisons of computation and communication costs of the proposed CL-PDP scheme and other existing related schemes [4,5,7,19]. Because the analyses of the other existing schemes are similar to the analysis of our proposed scheme, we discuss only our proposed scheme in the following subsection.

To compare fairness, bilinear pairing is constructed as follows: bilinear pairing $\bar{e}: G_1 \times G_1 \rightarrow G_2$ are built on the security level of 80-bit. G_1 is an additive group whose order is $\bar{q}$ and the generator is $\bar{p}$, which is a point on the super singular elliptic curve $\bar{E}: y^2 = x^3 + x \mod \bar{p}$ with an embedding degree of 2, where $\bar{p}$ is a 512-bit prime number and $\bar{q}$ is a 160-bit prime number. For elliptic curve-based scheme, we construct an additive group G generated by a point P with order p on a non-singular elliptic curve $E: y^2 = x^3 + ax + b(\mod q)$ to achieve a security level of 80 bits, where p, q are two 160 bit prime numbers.

5.1 Computation Cost

In our experiments, we used a computer that is HP with an Intel(R) Core(TM) i7-6700@ 3.4 GHz processor, 8 GB main memory, and the Ubuntu 14.04 operation system to derive the average execution time of the running 5000 times based on the MiRACL library [16]. To facilitate the analysis of computational cost, we list some notations about execution time, as shown in Table 2.

Table 2. Execution time of different cryptographic operations.

Notations	Definitions	Execution time
T_{bp}	Bilinear pairing operation	5.086 ms
$T_{bp.m}$	The scale multiplication operation based on bilinear pairing	0.694 ms
$T_{bp.a}$	The point addition operation based on bilinear pairing	0.0018 ms
T_H	The hash-to-point operation based on bilinear pairing	0.0992 ms
$T_{e.m}$	The scale multiplication operation based on ECC	0.3218 ms
$T_{e.a}$	Calculating the point addition operation related to ECC	0.0024 ms
T_h	Hash operation	0.001 ms

Table 3 shows the computational overhead of Tag Generation, Generate-Proof and Verify-Proof Algorithms. Note that I represents the size of the subset $I \in \{1, 2, ..., n\}$.

Table 3. Comparison of computation cost.

Schemes	Tag generation	Generate-Proof	Verify-Proof
Zhang et al. [19]	$(3n+3)T_{bp.m} + 4T_{bp.a}$ $+3T_H + nT_h$ $\approx 2.083n + 2.3868$ ms	$(2I)T_{bp.m} + (2I-2)T_{bp.a}$ $\approx 1.3916I - 0.0036$ ms	$4T_{bp} + 5T_{bp.m} + 2T_{bp.a}$ $+5T_H + IT_h$ $\approx 0.001I + 24.3136$ ms
Kang et al. [7]	$(4n)T_{bp.m} + (2n)T_{bp.a}$ $+(n+1)T_H + nT_h$ $\approx 2.8798n + 0.0992$ ms	$(I+1)T_{bp.m} + (I-1)T_{bp.a}$ $+T_h \approx 0.6958I + 0.6932$ ms	$4T_p + (2I+3)T_{bp.m} + IT_H$ $+(2I)T_{bp.a} + (I+1)T_h$ $\approx 1.3926I + 22.5262$ ms
He et al. [5]	$(2n+2)T_{bp.m} + (n)T_{bp.a}$ $+(n+1)T_H + 2T_h$ $\approx 1.488n + 1.4892$ ms	$(I+1)T_{bp.m} + (I-1)T_{bp.a}$ $+T_h \approx 0.6958I + 0.6932$ ms	$2T_p + (I+5)T_{bp.m} + 4T_h$ $+(I+4)T_{bp.a} + (I+1)T_H$ $\approx 0.794I + 13.7524$ ms
Gao et al. [4]	$(2n)T_{bp.m} + (n)T_{bp.a}$ $+T_H + (2n+1)T_h$ $\approx 1.3918n + 0.1002$ ms	$(2I+2)T_{bp.m} + T_H +$ $(2I-1)T_{bp.a} + (I+1)T_h$ $\approx 1.3926I + 1.4864$ ms	$3T_p + (I+3)T_{bp.m} +$ $IT_{bp.a} + T_H + 4T_h$ $\approx 0.6958I + 17.4464$ ms
The proposed	$nT_{e.m} + 2nT_h$ ≈ 0.3266 nms	$(I+1)T_{e.m} + (I-1)T_{e.a}$ $\approx 0.3242I + 0.3194$ ms	$3T_{e.m} + 3T_{e.a} + (2I+1)T_h$ $\approx 0.002I + 0.9736$ ms

For the Tag Generation Algorithm of the proposed scheme, the DO needs to execute one scalar multiplication operation and two hash function operations for each block M_l. Thus, the execution time of n blocks is $nT_{e.m} + 2nT_h \approx 0.3266$ nms. The computation of Generate-Proof Algorithm requires $(I+1)$ scalar multiplication operations and $(I-1)$ point addition operations related to the ECC, thus, the computation time of the phase is $(I+1)T_{e.m} + (I-1)T_{e.a} \approx 0.3242I + 0.3194$ ms. For the Verify-Proof Algorithm of the proposed scheme, verifier executes three scalar multiplication operations, three point addition operations and $(2I+1)$ hash function operations, Therefore, the execution time of the phase is $3T_{e.m} + 3T_{e.a} + (2I+1)T_h \approx 0.002I + 0.9736$ ms.

To make a more significant comparison, Fig. 2 and Fig. 3 are used to show that the computation cost of Generate-Proof Algorithm and Verify-Proof Algorithm increases with an increasing number of blocks, respectively. Based on an analysis and comparison of Table 3, Fig. 2 and Fig. 3, we conclude that the computation cost of the proposed scheme is lower than those of the related schemes [4,5,7,19].

5.2   Communication Cost

As $\bar{p}$ and p are 64 and 20 bytes, the sizes of the elements in G_1 and G are 64 × 2 = 128 bytes and 20 × 2 = 40 bytes, respectively. Set the size of block l be 4 bytes and the length of Z_q^* be 20 bytes. The communication cost of the five scheme are shown in Table 4.

In the proposed scheme, the TPV sends the challenging message $\{j, v_j\}_{j \in I}$ to CS, and the CS generates the response proof $\{S_{cs}, C_{cs}\}$ and returns it to TPV, where $j \in l$, $v_j \in Z_q^*$ and $S_{cs}, C_{cs} \in G$. Therefore, the communication

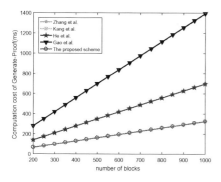

Fig. 2. Cost of Generate-Proof

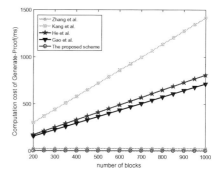

Fig. 3. Cost of Verify-Proof

Table 4. Comparison of communication cost.

Schemes	Communication costs
Zhang et al. [19]	$(|l| + |Z_q^*|)I + 2|G_1| + 2|Z_q^*| = 24I + 196$ bytes
Kang et al. [7]	$(|l| + |Z_q^*|)I + 2|G_1| + 1|Z_q^*| = 24I + 176$ bytes
He et al. [5]	$(|l| + |Z_q^*|)I + 4|G_1| + 1|Z_q^*| = 24I + 532$ bytes
Gao et al. [4]	$(|l| + |Z_q^*|)I + 2|G_1| + 1|Z_q^*| = 24I + 176$ bytes
The proposed	$(|l| + |Z_q^*|)I + 2|G| = 24I + 80$ bytes

cost of the proposed scheme is $(|l| + |Z_q^*|)I + 2|G| = 24I + 80$ bytes. According to Table 4, our proposed scheme expends less communication cost than those of other existing schemes [4,5,7,19].

6   Conclusion

The proposed scheme can realize the confidentiality of outsourcing data and solve the problem of data privacy leakage in the cloud data management system. In order to ensure the integrity of encrypted data, we used not only a third-party verifier to randomly check and verify, but also the public key of cloud services to encrypt random strings, so that the reliability of the stored data can be further ensured. Moreover, the detailed analysis showed that the proposed scheme is secure against the Type-I and Type-II adversaries under the random oracle model. Additionally, we compared and analyzed existing schemes from the perspective of Tag Generation and Generate-Proof and Verify-Proof Algorithms. The results verified that our scheme can effectively reduce delays and improve authentication efficiency.

Acknowledgment. The work was supported by the NSFC grant (No. U1936220, No. 61872001, No. 62011530046), and the Special Fund for Key Program of Science and Technology of Anhui Province, China (Grant No. 202003A05020043).

References

1. Armbrust, M., et al.: A view of cloud computing. Commun. ACM **53**(4), 50–58 (2010)
2. Ateniese, G., et al.: Provable data possession at untrusted stores. In: Proceedings of the 14th ACM Conference on Computer and Communications Security, pp. 598–609 (2007)
3. Fernandes, D.A.B., Soares, L.F.B., Gomes, J.V., Freire, M.M., Inácio, P.R.M.: Security issues in cloud environments: a survey. Int. J. Inf. Secur. **13**(2), 113–170 (2013). https://doi.org/10.1007/s10207-013-0208-7
4. Gao, G., Fei, H., Qin, Z.: An efficient certificateless public auditing scheme in cloud storage. Concurr. Comput. Pract. Exp. **32**(24), e5924 (2020)
5. He, D., Kumar, N., Zeadally, S., Wang, H.: Certificateless provable data possession scheme for cloud-based smart grid data management systems. IEEE Trans. Ind. Inf. **14**(3), 1232–1241 (2018)
6. He, D., Zeadally, S., Wu, L.: Certificateless public auditing scheme for cloud-assisted wireless body area networks. IEEE Syst. J. **12**(1), 64–73 (2015)
7. Kang, B., Wang, J., Shao, D.: Certificateless public auditing with privacy preserving for cloud-assisted wireless body area networks. Mob. Inf. Syst. **2017** (2017)
8. Ming, Y., Shi, W.: Efficient privacy-preserving certificateless provable data possession scheme for cloud storage. IEEE Access **7**, 122091–122105 (2019)
9. Nayak, S.K., Tripathy, S.: SEPDP: secure and efficient privacy preserving provable data possession in cloud storage. IEEE Trans. Serv. Comput. (2018)
10. Pointcheval, D., Stern, J.: Security arguments for digital signatures and blind signatures. J. Cryptol. **13**(3), 361–396 (2000). https://doi.org/10.1007/s001450010003
11. Wang, B., Li, B., Li, H., Li, F.: Certificateless public auditing for data integrity in the cloud. In: 2013 IEEE Conference on Communications and Network Security (CNS), pp. 136–144. IEEE (2013)
12. Wang, C., Wang, Q., Ren, K., Lou, W.: Privacy-preserving public auditing for data storage security in cloud computing. In: 2010 proceedings IEEE INFOCOM, pp. 1–9. IEEE (2010)
13. Wang, F., Xu, L., Gao, W.: Comments on SCLPV: secure certificateless public verification for cloud-based cyber-physical-social systems against malicious auditors. IEEE Trans. Comput. Soc. Syst. **5**(3), 854–857 (2018)
14. Wang, H., He, D., Tang, S.: Identity-based proxy-oriented data uploading and remote data integrity checking in public cloud. IEEE Trans. Inf. Forensics Secur. **11**(6), 1165–1176 (2016)
15. Wang, H., Wu, Q., Qin, B., Domingo-Ferrer, J.: Identity-based remote data possession checking in public clouds. IET Inf. Secur. **8**(2), 114–121 (2013)
16. Wenger, E., Werner, M.: Evaluating 16-bit processors for elliptic curve cryptography. In: Prouff, E. (ed.) CARDIS 2011. LNCS, vol. 7079, pp. 166–181. Springer, Heidelberg (2011). https://doi.org/10.1007/978-3-642-27257-8_11
17. Yang, K., Jia, X.: An efficient and secure dynamic auditing protocol for data storage in cloud computing. IEEE Trans. Parallel Distrib. Syst. **24**(9), 1717–1726 (2012)
18. Yu, Y., et al.: Identity-based remote data integrity checking with perfect data privacy preserving for cloud storage. IEEE Trans. Inf. Forensics Secur. **12**(4), 767–778 (2016)
19. Zhang, Y., Xu, C., Yu, S., Li, H., Zhang, X.: SCLPV: secure certificateless public verification for cloud-based cyber-physical-social systems against malicious auditors. IEEE Trans. Comput. Soc. Syst. **2**(4), 159–170 (2015)

20. Zhang, Y., Yu, J., Hao, R., Wang, C., Ren, K.: Enabling efficient user revocation in identity-based cloud storage auditing for shared big data. IEEE Trans. Dependable Secure Comput. **17**(3), 608–619 (2020)
21. Zhou, C.: Security analysis of a certificateless public provable data possession scheme with privacy preserving for cloud-based smart grid data management system. Int. J. Netw. Secur. **22**(4), 584–588 (2020)
22. Zhu, Y., Hu, H., Ahn, G.J., Yu, M.: Cooperative provable data possession for integrity verification in multicloud storage. IEEE Trans. Parallel Distrib. Syst. **23**(12), 2231–2244 (2012)

Dirty-Data Impacts on Regression Models: An Experimental Evaluation

Zhixin Qi[1] and Hongzhi Wang[1,2]([⊠])

[1] School of Computer Science and Technology,
Harbin Institute of Technology, Harbin, China
`{qizhx,wangzh}@hit.edu.cn`
[2] PengCheng Laboratory, Shenzhen, China

Abstract. Data quality issues have attracted widespread attentions due to the negative impacts of dirty data on regression model results. The relationship between data quality and the accuracy of results could be applied on the selection of appropriate regression model with the consideration of data quality and the determination of data share to clean. However, rare research has focused on exploring such relationship. Motivated by this, we design a generalized framework to evaluate dirty-data impacts on models. Using the framework, we conduct an experimental evaluation for the effects of missing, inconsistent, and conflicting data on regression models. Based on the experimental findings, we provide guidelines for regression model selection and data cleaning.

Keywords: Experimental evaluation · Data quality · Regression model · Model selection · Data cleaning

1 Introduction

Data quality attracts widespread attentions in both database and machine learning communities. The data with data quality problems are called dirty data. For a regression task, dirty data in both training and testing data sets affect the accuracy. Thus, we have to know the relationship between the quality of input data sets and the accuracy of regression model results. Based on such relationship, we could select an appropriate regression model with the consideration of data quality issues and determine the share of data to clean.

Before a regression task, it is usually difficult for users to decide which model should be adopted due to the diversity of regression models. The effects of data quality on regression models are helpful for model selection. Therefore, exploring dirty-data impacts on regression models is in demand.

In addition, data cleaning is necessary to guarantee the data quality of a regression task. Although existing data cleaning methods improve data quality dramatically, the cleaning costs are still expensive [6]. If we know how dirty data affect the accuracy of regression models, we could clean data selectively according to the accuracy requirements instead of cleaning the entire dirty data

© Springer Nature Switzerland AG 2021
C. S. Jensen et al. (Eds.): DASFAA 2021, LNCS 12681, pp. 88–95, 2021.
https://doi.org/10.1007/978-3-030-73194-6_6

with large costs. As a result, the data cleaning costs are reduced. Therefore, the study of relationship between data quality and the accuracy of regression model results is urgently needed. Unfortunately, there is no existing research to explore the impacts of dirty data on regression models in terms of data quality dimensions. Motivated by this, we attempt to fill this gap. Our contributions of this paper are listed as follows.

1. In order to evaluate dirty-data impacts on regression models, we design a multi-functional generalized evaluation framework with the consideration of various data quality dimensions. To the best of our knowledge, this is the first paper that studies this issue.
2. We propose three novel metrics, $SENS(M)$, $KP(M)$, and $CP(M)$, to evaluate the effects of dirty data on regression models. $SENS(M)$, $KP(M)$, and $CP(M)$ are used to measure the sensibility, tolerability, and expected accuracy of a regression model, respectively.
3. Based on the experimental results, we provide guidelines of regression model selection and data cleaning for users.

2 Generalized Evaluation Framework

Existing metrics of regression models, such as RMSD (Root-Mean-Square Deviation), NRMSD (Normalized Root-Mean-Square Deviation), and CV(RMSD) (Coefficient of Variation of the RMSD), are only able to show us the variations of accuracy, but not able to measure the fluctuation degrees quantitatively. Therefore, we propose three novel metrics to evaluate dirty-data impacts on regression models. We first define the sensibility of a regression model as follows.

Definition 1. *Given the values of a metric y of a regression model M with a%, (a+x)%, (a+2x)%, ..., (a+bx)% (a ≥ 0, x > 0, b > 0) error rate, the sensibility of M, denoted by SENS(M), is computed as $\sum_{0 \leq i < b} |y_{a+ix} - y_{a+(i+1)x}|$.*

$SENS(M)$ aims to measure the fluctuation degree of a regression model to dirty data. When the value of $SENS(M)$ is larger, the fluctuation degree is larger. Accordingly, the regression model is more sensitive to dirty data. Therefore, $SENS(M)$ is able to evaluate the dirty-data sensibility of a regression model. Here, we explain the computation of $SENS(M)$ with Fig. 1(a).

Example 1. *In Fig. 1(a), the values of RMSD of the least square regression model (LSRM for brief) with 0%, 10%, ..., 50% missing rate are given. On Iris, $SENS(LSRM) = |RMSD_{0\%} - RMSD_{10\%}| + |RMSD_{10\%} - RMSD_{20\%}| + ... + |RMSD_{40\%} - RMSD_{50\%}| = |0.59 - 0.62| + |0.62 - 0.64| + |0.64 - 0.68| + |0.68 - 0.68| + |0.68 - 0.72| = 0.13. On Servo, SENS(LSRM) = |1.55 - 1.63| + |1.63 - 1.66| + |1.66 - 1.7| + |1.7 - 1.72| + |1.72 - 2.02| = 0.47. On Housing, SENS(LSRM) = |10.53 - 10.59| + |10.59 - 9.92| + |9.92 - 10.04| + |10.04 - 9.96| + |9.96 - 9.79| = 1.1. On Concrete, SENS(LSRM) = |16.21 - 16.57| + |16.57 - 16.49| + |16.49 - 16.67| + |16.67 - 17.46| + |17.46 - 17.35| = 1.52. And on Solar Flare, SENS(LSRM) = |0.3 - 0.32| + |0.32 - 0.31| + |0.31 - 0.31| + |0.31 - 0.33| + |0.33 - 0.29| = 0.09. Thus, the average of SENS(LSRM) is 0.662.*

Though $SENS(M)$ measures the sensibility of a regression model, we could not determine the error rate at which a regression model is unacceptable. Motivated by this, we define the keeping point of a regression model as follows.

Definition 2. *Given the values of a metric y of a regression model with $a\%$, $(a+x)\%$, $(a+2x)\%$, ..., $(a+bx)\%$ ($a \geq 0$, $x > 0$, $b > 0$) error rate, and a number k ($k > 0$). If the larger value of y causes the better accuracy, and $y_{a\%} - y_{(a+ix)\%} > k$ ($0 < i \leq b$), the keeping point of a regression model M, denoted by $KP(M)$, is $min\{(a+(i-1)x)\%\}$. If $y_{a\%} - y_{(a+bx)\%} \leq k$, $KP(M)$ is $min\{(a+bx)\%\}$. If the smaller value of y causes the better accuracy, and $y_{(a+ix)\%} - y_{a\%} > k$ ($0 < i \leq b$), $KP(M)$ is $min\{(a+(i-1)x)\%\}$. If $y_{(a+bx)\%} - y_{a\%} \leq k$, $KP(M)$ is $min\{(a+bx)\%\}$.*

$KP(M)$ is defined to measure the error rate at which a regression model is acceptable. When the value of $KP(M)$ is larger, the error rate at which a regression model is acceptable is larger. Accordingly, the error-tolerability of a regression model is higher. Therefore, $KP(M)$ is useful to evaluate the error-tolerability of a regression model. Here, we take Fig. 1(a) as an example to explain the computation of $KP(M)$.

Example 2. *From Fig. 1(a), we know the values of RMSD of LSRM with 0%, 10%, ..., 50% missing rate, and set 0.1 as the value of k. On Iris, when the missing rate is 50%, $RMSD_{50\%}$-$RMSD_{0\%} = 0.72 - 0.59 = 0.13 > 0.1$, we take 40% as the $KP(LSRM)$. On Servo, when the missing rate is 20%, $RMSD_{20\%} - RMSD_{0\%} = 1.66 - 1.55 = 0.11 > 0.1$, we take 10% as the $KP(LSRM)$. On Housing, when the missing rate is 50%, $RMSD_{50\%} - RMSD_{0\%} = 9.79 - 10.53 = -0.74 \leq 0.1$, we take 50% as the $KP(LSRM)$. On Concrete, when the missing rate is 10%, $RMSD_{10\%} - RMSD_{0\%} = 16.57 - 16.21 = 0.36 > 0.1$, we take 0% as the $KP(LSRM)$. On Solar Flare, when the missing rate is 50%, $RMSD_{50\%} - RMSD_{0\%} = 0.29 - 0.3 = -0.01 \leq 0.1$, we take 50% as the $KP(LSRM)$. Thus, the average of $KP(LSRM)$ is 30%.*

Since the value of k in $KP(M)$ reflects the acceptable fluctuation degree of $SENS(M)$, we could determine the candidate regression models using k and the accuracy requirements of a regression task. To achieve this, we define the critical point of a regression model as follows.

Definition 3. *Given the data set D, accuracy metric accMetric, the number k in $KP(M)$, the critical point of a regression model M on accMetric, denoted by $CP(M)$ is computed as the accuracy of accMetric on D plus k.*

Since $CP(M)$ is defined to measure the expected accuracy of a regression model, its role is to filter the unacceptable regression models whose values of $CP(M)$ are below the required lower bound of accMetric. Here, we explain the function of $CP(M)$ with Fig. 1(a), 1(b), 1(c), 1(d), and 1(e) as an example.

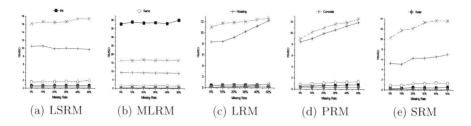

(a) LSRM   (b) MLRM   (c) LRM   (d) PRM   (e) SRM

Fig. 1. RMSD results on regression models varying missing rate

Example 3. *Given the data set Housing with 30% missing rate, the accuracy metric RMSD, and its required lower bound 9.20, we attempt to determine the candidate regression models. From Fig. 1(a), 1(b), 1(c), 1(d), and 1(e), we know the RMSD value of each regression model on Housing. If the value of k in $KP(M)$ is set as 0.1, $CP(LSRM, RMSD) = 10.04 + 0.1 = 10.14 > 9.20$, $CP(MLRM, RMSD) = 9.08 + 0.1 = 9.18 < 9.20$, $CP(LRM, RMSD) = 10.25 + 0.1 = 10.35 > 9.20$, $CP(PRM, RMSD) = 10.57 + 0.1 = 10.67 > 9.20$, and $CP(SRM, RMSD) = 6.32 + 0.1 = 6.42 < 9.20$. Thus, the candidate regression models are LSRM, LRM, and PRM.*

Based on these three metrics, we develop a generalized framework to evaluate dirty-data impacts on models with the consideration of data quality dimensions. The framework is sketched in Fig. 2. It contains five components, (a) dirty data generation, (b) performance testing, (c) metric computation, (d) candidate model selection, and (e) model and cleaning strategy determination.

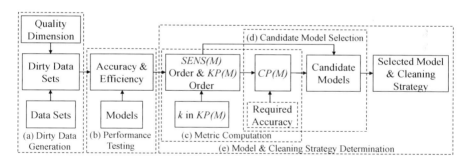

Fig. 2. Generalized evaluation framework

There are three functions in this framework: (i) evaluating dirty-data impacts on models, (ii) selecting models on dirty data, and (iii) generating data cleaning strategies. As Fig. 2 shows, we first generate dirty data sets with different dirty-data types or rates from the given data sets. Dirty-data types correspond to the given data quality dimensions. Then, we test accuracy and efficiency of the given models on the generated dirty data. Based on the model performance results and the given value of k in $KP(M)$, we compute $SENS(M)$ order, $KP(M)$

order, and $CP(M)$ of each model. Using the value of $CP(M)$ and the required accuracy, we are able to select candidate models. Finally, using $SENS(M)$ order and $KP(M)$ order, we determine the selected model and data cleaning strategy.

The multi-functional framework is not only applicable to regression models, but also generalized to use in other kinds of models, such as classification and clustering models.

3 Evaluation Results and Analyses

Based on the evaluation framework, we conduct extensive experiments. In this section, we discuss the evaluation results and analyses.

3.1 Data Sets, Models, and Setup

For the input in generalized framework, we select five typical data sets, that are Iris, Servo, Housing, Concrete, and Solar Flare, from UCI public data sets [1] with various types and sizes. Due to the completeness and correctness of these original data sets, we inject errors of different rates from different data quality dimensions into them, and generate different kinds of dirty data sets. Then, we test the performances of various regression models on them. In experimental evaluation, the original data sets are used as the baselines, and the accuracy of regression models is measured based on the results on original data sets.

We also select five classical regression models, Least Square Regression, Maximum Likelihood Regression, Logistic Regression, Polynomial Regression, and Stepwise Regression, as the input in generalized framework. We choose these models since they are always used as competitive regression models [2–5].

Since there are class labels in the selected original data sets for regression, we use standard RMSD, NRMSD, and CV(RMSD) to evaluate the accuracy of regression models. We vary each error rate from 10% to 50%, and use 10-fold cross validation to generate the training data and the testing data.

All experiments are conducted on a machine powered by two Intel(R) Xeon(R) E5-2609 v3@1.90 GHz CPUs and 32 GB memory, under CentOS7. All the algorithms are implemented in C++ and compiled with g++ 4.8.5.

3.2 Varying Missing Rate

As shown in Table 1, for RMSD, the $SENS(M)$ order is "PRM > MLRM > SRM > LRM > LSRM". For NRMSD, the $SENS(M)$ order is "PRM > SRM > LRM > LSRM > MLRM". For CV(RMSD), the $SENS(M)$ order is "MLRM > SRM > LRM > PRM > LSRM". Thus, for RMSD and CV(RMSD), the least sensitive algorithm is **LSRM**. And for NRMSD, the least sensitive algorithm is **MLRM**. These are due to the fact that the number of parameters in LSRM is small. Hence, there is little chance for the model training to be affected by missing values. For RMSD and NRMSD, the most sensitive algorithm is **PRM**. And for CV(RMSD), the most sensitive algorithm is **MLRM**. These are because that these algorithms perform badly on some original data sets (error rate is 0%).

Table 1. $SENS(M)$ results of regression models (R: RMSD, NR: NRMSD, CV: CV(RMSD))

Model	Missing			Inconsistent			Conflicting		
	R	NR	CV	R	NR	CV	R	NR	CV
Least Square (LSRM)	0.662	0.066	0.192	1.204	0.090	0.278	1.056	0.054	0.284
Maximum Likelihood (MLRM)	1.356	0.034	4.466	2.384	0.046	1.546	1.534	0.060	3.914
Logistic Regression (LRM)	1.268	0.096	0.288	1.296	0.164	1.096	1.030	0.092	0.308
Polynomial Regression (PRM)	1.568	0.106	0.200	2.010	0.174	0.464	1.794	0.116	0.426
Stepwise Regression (SRM)	1.338	0.104	3.466	1.616	0.102	5.996	0.890	0.074	2.748

Table 2. $KP(M)$ Results of Regression Models ($k = 0.1$, Unit: %, R: RMSD, NR: NRMSD, CV: CV(RMSD))

Model	Missing			Inconsistent			Conflicting		
	R	NR	CV	R	NR	CV	R	NR	CV
Least Square (LSRM)	30	50	50	16	50	50	32	50	42
Maximum Likelihood (MLRM)	22	50	40	16	50	34	20	50	40
Logistic Regression (LRM)	18	42	30	12	44	24	14	46	40
Polynomial Regression (PRM)	14	40	20	12	50	22	12	40	26
Stepwise Regression (SRM)	16	40	24	10	42	22	40	50	30

When missing data are injected, the uncertainty of data becomes more, which leads to increasing uncertainty to algorithms. Accordingly, the performances of algorithms become worse.

As shown in Table 2, for RMSD, the $KP(M)$ order is "LSRM > MLRM > LRM > SRM > PRM". For NRMSD, the $KP(M)$ order is "LSRM = MLRM > LRM > SRM = PRM". For CV(RMSD), the $KP(M)$ order is "LSRM > MLRM > LRM > SRM > PRM". Therefore, the most incompleteness-tolerant algorithm is **LSRM**. This is because that the amount of parameters in LSRM is small. Hence, there is little chance to be affected. The least incompleteness-tolerant algorithm is **PRM**. This is due to the fact that there are many parameters in PRM, which makes it susceptible to missing data.

3.3 Varying Inconsistent Rate

As shown in Table 1, for RMSD, the $SENS(M)$ order is "MLRM > PRM > SRM > LRM > LSRM". For NRMSD, the $SENS(M)$ order is "PRM > LRM > SRM > LSRM > MLRM". For CV(RMSD), the $SENS(M)$ order is "SRM > MLRM > LRM > PRM > LSRM". Thus, for RMSD and CV(RMSD), the least sensitive algorithm is **LSRM**. And for NRMSD, the least sensitive algorithm is **MLRM**. The reason is similar as that of the LSRM varying missing rate. For RMSD, the most sensitive algorithm is **MLRM**. For NRMSD, the most sensitive algorithm is **PRM**. And for CV(RMSD), the most sensitive algorithm is **SRM**. These are due to their poor performances on some original data sets (error rate is 0%). When inconsistent data are injected, the uncertainty of data becomes more, which leads to increasing uncertainty to algorithms. Accordingly, algorithms perform worse.

As shown in Table 2, for RMSD, the $KP(M)$ order is "LSRM = MLRM > LRM = PRM > SRM". For NRMSD, the $KP(M)$ order is "LSRM = MLRM = PRM > LRM > SRM". For CV(RMSD), the $KP(M)$ order is "LSRM > MLRM > LRM > PRM = SRM". Therefore, the most inconsistency-tolerant algorithm is **LSRM**. The reason is similar as that of the most incompleteness-tolerant algorithm varying missing rate. The least inconsistency-tolerant algorithm is **SRM**. This is due to the fact that there are many independent variables to be tested in SRM, which makes it easily affected by inconsistent values.

3.4 Varying Conflicting Rate

As shown in Table 1, for RMSD, the $SENS(M)$ order is "PRM > MLRM > LSRM > LRM > SRM". For NRMSD, the $SENS(M)$ order is "PRM > LRM > SRM > MLRM > LSRM". For CV(RMSD), the $SENS(M)$ order is "MLRM > SRM > PRM > LRM > LSRM". Thus, for RMSD, the least sensitive algorithm is **SRM**. This is because that there is a validation step in SRM, which guarantees the regression accuracy. For NRMSD and CV(RMSD), the least sensitive algorithm is **LSRM**. The reason is similar as that of the least sensitive algorithm varying missing rate. For RMSD and NRMSD, the most sensitive algorithm is **PRM**. And for CV(RMSD), the most sensitive algorithm is **MLRM**. The reason is similar as that of the most sensitive algorithms varying missing rate.

As shown in Table 2, for RMSD, the $KP(M)$ order is "SRM > LSRM > MLRM > LRM > PRM". For NRMSD, the $KP(M)$ order is "LSRM = MLRM = SRM > LRM > PRM". For CV(RMSD), the $KP(M)$ order is "LSRM > MLRM = LRM > SRM > PRM". Therefore, the most conflict-tolerant algorithms are **LSRM**, **MLRM**, and **SRM**. This is due to the fact that there are a small amount of parameters in LSRM and MLRM. In SRM, the validation step helps guarantee the regression accuracy. The least conflict-tolerant algorithm is **PRM**. The reason is similar as that of the least incompleteness-tolerant algorithm varying missing rate.

3.5 Lessons Learned

According to the evaluation results, we have the following findings.

- Dirty-data impacts are related to dirty-data type and dirty-data rate. Thus, it is necessary to detect the rate of each dirty-data type in the given data.
- For the regression models whose RMSD is larger than 1, NRMSD is larger than 0.2, or CV(RMSD) is larger than 1 on original data sets, as the data size rises, RMSD, NRMSD, or CV(RMSD) of the models becomes stable.
- When dirty data exist, the regression model with the least $SENS(M)$ is the most stable. For instance, if the $SENS(M)$ order is "PRM > MLRM > SRM > LRM > LSRM", the most stable regression model is LSRM.

- Since the accuracy of the selected regression model becomes unacceptable beyond $KP(M)$, the error rate of each dirty data type needs to be controlled within its $KP(M)$.

Based on the lessons learned from experimental evaluation, we suggest users select regression model and clean dirty data according to the following steps.

- Users are suggested to detect the dirty-data rates (e.g., missing rate, inconsistent rate, conflicting rate) of the given data.
- According to the given task requirements (e.g., lower bound on RMSD, NRMSD, or CV(RMSD)), we suggest users determine the candidate regression models whose $CP(M)$ is better than the required lower bound.
- According to the requirements of a regression task and dirty-data type of the given data, we suggest users find the corresponding $SENS(M)$ order and choose the least sensitive regression model.
- According to the selected model, task requirements, and error rate of the given data, we suggest users find the corresponding $KP(M)$ order and clean each type of dirty data to its $KP(M)$.

4 Conclusion

In this paper, we propose three metrics to measure the sensibility, tolerability, and expected accuracy of a data analysis model. With these metrics, we develop a generalized evaluation framework to evaluate dirty-data impacts on models. Using the framework, we conduct an experimental evaluation to explore the relationship of dirty data and accuracy of regression models. Based on the experimental findings, we provide guidelines for model selection and data cleaning.

Acknowledgment. This paper was partially supported by NSFC grant U1866602, CCF-Huawei Database System Innovation Research Plan CCF-HuaweiDBIR2020007B.

References

1. Data sets: https://archive.ics.uci.edu/ml/index.php
2. Abraham, S., Raisee, M., Ghorbaniasl, G., Contino, F., Lacor, C.: A robust and efficient stepwise regression method for building sparse polynomial chaos expansions. J. Comput. Phys. **332**, 461–474 (2017)
3. Avdis, E., Wachter, J.A.: Maximum likelihood estimation of the equity premium. J. Financ. Econ. **125**(3), 589–609 (2017)
4. Li, L., Zhang, X.: Parsimonious tensor response regression. J. Am. Stat. Assoc. **112**(519), 1131–1146 (2017)
5. Silhavy, R., Silhavy, P., Prokopova, Z.: Analysis and selection of a regression model for the use case points method using a stepwise approach. J. Syst. Softw. **125**, 1–14 (2017)
6. Wang, H., Qi, Z., Shi, R., Li, J., Gao, H.: COSSET+: crowdsourced missing value imputation optimized by knowledge base. JCST **32**(5), 845–857 (2017)

UniTest: A Universal Testing Framework for Database Management Systems

Gengyuan Shi, Chaokun Wang$^{(\boxtimes)}$, Bingyang Huang, Hao Feng, and Binbin Wang

School of Software, Tsinghua University, Beijing 100084, China
{shigy19,hby17,fh20,wbb18}@mails.thu.edu.cn, chaokun@tsinghua.edu.cn

Abstract. With the continuous development of data collection, network transmission, and data storage, Big Data are now rapidly expanding in all science and engineering domains. Considering the characteristics of Big Data including quick generation, large size, and diverse data models, higher requirements are placed on the functionality and performance of database management systems. Therefore, it is essential for users to choose a stable and reliable database management system. However, finding the best way to evaluate the reliability and stability of database management systems is still a huge challenge, and it is difficult for users to design their own test cases for evaluating these systems.

In order to address this problem, we carefully design a universal testing framework, called UniTest, which can perform effective functional testing and performance testing for different types of database management systems. Extensive testing experiments on multiple types of database management systems show the universality and efficiency of our framework.

Keywords: Database management system · Functional testing · Performance testing · UniTest

1 Introduction

In recent years, the fast development of Big Data technologies including cloud computing, the Internet of Things, and social network analysis [3], has greatly changed the way people live. However, massive data not only bring exceptional opportunities to the development of the technologies but also bring huge challenges to data management. Considering the variety of Big Data, efficient multi-model data management has become a fundamental requirement in real-world scenarios with the perspective that "no one size fits all" [16]. How to manage such heterogeneous data effectively and efficiently is still a big challenge for many industries. In order to utilize the existing DBMSs for providing more efficient multi-model data management, there is an urgent need for techniques to evaluate different types of Database Management Systems (DBMSs) [7,8]. Based on

© Springer Nature Switzerland AG 2021
C. S. Jensen et al. (Eds.): DASFAA 2021, LNCS 12681, pp. 96–104, 2021.
https://doi.org/10.1007/978-3-030-73194-6_7

these database evaluation techniques, users can accurately and effectively evaluate the overall abilities of the DBMSs to ensure that, in real-world scenarios, these systems can provide efficient and stable services.

There are many studies on the DBMSs comparison [1, 10, 13]. However, there is little work available for universally evaluating multiple types of DBMSs. For most non-relational databases, as well as for different types of data, there is currently no widely used testing platform [7]. Another problem with most of the existing database testing platforms and benchmarks is that the test cases are designed in advance. As a result, the testers cannot create test cases easily according to their concerns, and hence face great difficulties when testing DBMSs that support different data models.

The challenge behind the problems above is the difficulty in conducting the test for different types of DBMSs universally, both the data models and data management technologies of these systems differ. In order to address this problem, this paper designs UniTest, a universal framework for evaluating a variety of DBMSs in terms of both functionality and performance. UniTest allows users to easily design and execute new test cases for different types of DBMSs.

The main contributions of this paper are summarized as follows:

1. We carefully design UniTest, a universal framework for testing multiple types of DBMSs. Besides the universal pre-defined cases for both functional and performance testing, UniTest allows users to design new statements of different types of query languages as test cases. To the best of our knowledge, UniTest is the first testing framework that can be used to test more than five types of DBMSs universally.
2. We implement the UniTest system with effectiveness and high extensibility. We provide an easy-to-use interface for testers to configure the system environment, select test cases, and check the test results in detail. UniTest can also be easily extended by integrating new DBMSs and designing new test cases. The experimental results show that UniTest can effectively test DBMSs that support different data models.

The rest of this paper is organized as follows. We discuss the related work in Sect. 2. In Sect. 3, we present the architecture of UniTest. We report the experimental results in Sect. 4 and finally give our conclusions in Sect. 5.

2   Related Work

The overall performance of one DBMS is related to many factors such as the system architecture, the scale of data, the hardware environment, and so on. To compare different DBMSs, it is necessary to conduct test cases under the same conditions. Typical performance testing methods include benchmark testing, load testing, and so on. Different test methods perform evaluations of the DBMSs from different perspectives as required [2, 4, 5, 9, 11, 12, 19].

2.1 Benchmark Testing

Benchmark testing refers to the test method used to quantify some particular performance metrics of the systems under test. The benchmark testing process mainly consists of three key steps: test data generation, load type selection, and test indicator selection [15]. Various aspects of the systems under test need to be evaluated, such as reliability, time efficiency, resource consumption, cost performance, and so on.

Some benchmark testing tools are designed for specific typical applications. Facebook's LinkBench is mainly used to test the DBMSs in the social network scenarios [2]. Yahoo's YCSB is proposed for testing NoSQL cloud databases [4]. BigDataBench runs tests using different business models [14,18].

However, there are still some drawbacks to these tools. Firstly, the test methods mainly focus on one or a small number of data models and are difficult to test database management systems of multiple types just using one testing tool. Secondly, the lack of a graphical interface is a big problem for many of them, since testers cannot easily set test configurations and get the test results. For example, LinkBench mainly provides test cases including adding, deleting, and changing for graph data but does not support modification of data of other models. YCSB and BigDataBench do not provide direct supports for testing many types (e.g., graph and message queue) of DBMSs.

2.2 Load Testing

The load testing technologies evaluate the processing limit of the DBMSs by simulating the business scenarios. Specifically, these technologies continuously increase the pressure on the system under test until a certain performance measure (such as the response time) of the system exceeds the expected value or a certain resource is exhausted. By means of the load testing, we can understand the performance capacity of the system, discover possible problems in the system, or provide a reference for system performance tuning.

The data loading model performs load testing on the DBMS under test by generating a large amount of data. It is necessary to consider the representativeness, extensiveness, and data distribution of the test dataset when generating the test data, since these characteristics have a great influence on the performance of the application.

Using the load testing tools, the tester can simulate a series of virtual operations in a real business scenario, thereby testing and evaluating various aspects of the system under test. Currently, many load testing tools help testers conduct performance tests through automated testing. JMeter [6] is one of the typical load testing tools.

JMeter is an open-source software application designed to test the server or client architecture and simulate a massive load to test the stability and performance of the DBMSs. Testers can get test results by creating, configuring, and executing test plans.

3 The UniTest

This section proposes UniTest, a universal framework for testing multiple types of DBMSs. The framework architecture can be divided into three parts: the *User Interface*, the *Test Management* module, and the *Test Execution* module. Firstly, we briefly introduce the architecture of UniTest. Then, the three major components are presented in detail in Sect. 3.1, 3.2 and 3.3, respectively.

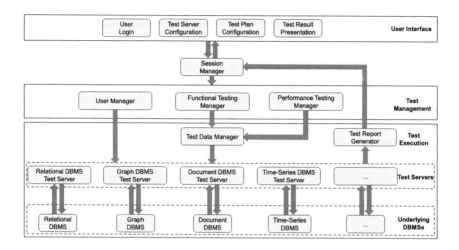

Fig. 1. The architecture of UniTest.

The architecture of UniTest is shown in Fig. 1. The user interface interacts with the tester for test case customization, system configuration, results presentation, and so forth. The test execution module processes the test request sent by the tester, and then forwards it to the corresponding test server and finally obtains the test result from the test server for displaying in the user interface. There is one test server for each of the data models, which is responsible for receiving and executing the specific test cases on the DBMS under test. The test management module receives the test results from the test execution module and returns them to the user interface to display.

With the support of this framework, we also have an evaluation method for testing different types of DBMSs. The tester can log in through the user interface, generate a test plan by selecting the target DBMS and the test cases, and pass the test plan to the test management module. The test server in the test execution module connects to the server where the DBMS under test, namely the target DBMS, is located according to the test plan, and performs corresponding tests. During testing, the test execution module prepares test data by loading the corresponding test dataset or generating test data. The test results are automatically collected and returned to the user interface along with the generated test report.

3.1 User Interface

The user interface includes a user login pane, a test server configuration pane, a test plan configuration pane, and a test result presentation pane. Through the user interface, the tester can complete operations such as logging in, configuring and conducting tests, and viewing test results.

In the test server configuration pane, the tester can configure the address, the user name, and the password for each DBMS test server.

In the test plan configuration pane, the tester can select test cases for each DBMS from the list of test cases given by UniTest, combine them into a test plan, and execute the test. UniTest automatically generates the test plan and passes the plan to the test management module.

In the test result display pane, the log information and the test results returned from the test server are displayed. The tester can download the automatically generated test report.

3.2 Test Management Module

The test management module, consisting of a user manager, a functional testing manager, and a performance testing manager, is responsible for performing all test plans in the framework.

In the user manager, the user permissions of the testers of UniTest can be managed to ensure the security of the DBMSs under test. Roles of the users are divided into two categories: administrator users and ordinary users. The administrator user can add users or delete existing users, and can modify the passwords of other users; ordinary users only have the right to log in and modify their passwords.

The test management module refines the received test plan. There may be precondition or inclusion relationships between a series of given test cases, i.e., test case A must be executed at first, or test case A should contain test case B. UniTest automatically checks these relationships and forms an integrated test plan.

In the functional testing manager, different test cases for different data types are preset for the tester to select. One or more test cases can be selected to be freely combined to form a test plan.

The performance testing manager includes performance test cases for different DBMSs, including data migration efficiency (import/export) and query execution efficiency.

3.3 Test Execution Module

The test execution module receives the test plan from the test management module, dispatches the test cases to the corresponding test servers, and returns the test results and the test report. The test execution module includes a test data manager, a test report generator, and a test server for each type of the DBMSs.

The test data manager prepares test datasets for the target DBMS when the test is performed. For example, this component generates a testing dataset automatically before executing the data migration test which needs a large amount of data to be imported.

The test report generator monitors the test process and collects the test results. It automatically generates a test report document of the current test according to information on the process and results of the test.

For each type of DBMSs, the test server controls the corresponding DBMS server and executes the user-specified test cases. After the test plan is forwarded by the test management module to the specific test server, the test server conducts test cases according to the test plan.

For performance testing, the test execution module conducts the user-specified test plan. The test server controls the target DBMSs to test the performance of data migration and query execution, and returns the results including the response time and data migration speed.

4   Experiments

In this section, we test some typical DBMSs under Ubuntu 16.04 with a 10-core Intel Xeon E5-2630 (2.20 GHz) and 320 GB main memory. This process consists of two parts: functional testing and performance testing, which are the core of our framework.

We choose representative DBMSs of different types in the performance testing experiment to check the performance of the data migration and query execution for different models of data. Specifically, we test the performance of MySQL for relational data management, Neo4j for graph data management, InfluxDB for time series data management, CouchDB for document data management, Redis for key-value data management, MySQL Blob for binary big object data management, and Kafka for message queue data management.

4.1   Functional Testing

We have implemented the testing system according to the UniTest architecture. The pre-defined test cases in UniTest cover seven categories of functionalities, such as the separation of service instances, the authentication of database users, and the ability to trace data sources.

Test cases for functionality testing are conducted for representative DBMSs of different types. For the space limitation, the results of functional testing of all the DBMSs are not presented.

4.2   Performance Testing

We also conduct performance testing for many types of DBMSs. Here, two typical categories of test cases are considered. In the data migration efficiency test, we test the speed of data importing and data exporting. In the query execution efficiency test, we test the speed of query execution under different conditions.

The results of performance testing on Neo4j is shown in Table 1. The graph datasets are generated by FastSGG [17]. The data migration efficiency is tested varying the sizes of nodes and edges, while the query execution efficiency test cases are executed varying the sizes of the graph. For the limit of space, test results of other types of representative DBMSs are not displayed in this paper.

Table 1. Results of performance testing for Neo4j.

Description		Average time of 9 cases
Data migration	Importing node data (56 KB with 1893 nodes)	9333 KB/s, 6 ms
	Importing edge data (162 KB with 4641 edges)	123 KB/s, 1.32 s
	Backuping 1893 nodes with attributes	519 nodes/s
	Restoring 1893 nodes with attributes	2146 nodes/s
Query execution	$|V| = 100$, $|E| = 1000$	35.2 ms
	$|V| = 1000$, $|E| = 10000$	472.5 ms
	$|V| = 10000$, $|E| = 100000$	12802.6 ms

In summary, UniTest can test different types of database management systems universally, and effectively supports testing all of these systems in a specific evaluation method, including functional and performance testing.

5   Conclusion

This paper proposes a universal testing framework called UniTest with a specific evaluation method for conducting evaluation on various types of database management systems. UniTest provides a rich set of test cases with an easy-to-use interface for different types of DBMSs. We carry out extensive experiments of functional and performance testing on some typical DBMSs. The experimental results show that UniTest provides a universal evaluation environment for different types of DBMSs and plays an important role in the application of database products in real-world scenarios.

Acknowledgments. This work is supported in part by the Intelligent Manufacturing Comprehensive Standardization and New Pattern Application Project of MIIT (Experimental validation of key technical standards for trusted services in industrial Internet), and the National Natural Science Foundation of China (No. 61872207).

References

1. Abramova, V., Bernardino, J.: NoSQL databases: MongoDB vs cassandra. In: Proceedings of the International C* Conference on Computer Science and Software Engineering, pp. 14–22. ACM (2013)
2. Armstrong, T.G., Ponnekanti, V., Borthakur, D., Callaghan, M.: LinkBench: a database benchmark based on the Facebook social graph. In: Proceedings of the 2013 ACM SIGMOD International Conference on Management of Data, pp. 1185–1196. ACM (2013)
3. Cai, L., Zhu, Y.: The challenges of data quality and data quality assessment in the big data era. Data Sci. J. **14**, 2 (2015)
4. Cooper, B.F., Silberstein, A., Tam, E., Ramakrishnan, R., Sears, R.: Benchmarking cloud serving systems with YCSB. In: Proceedings of the 1st ACM Symposium on Cloud Computing, pp. 143–154. ACM (2010)
5. Difallah, D.E., Pavlo, A., Curino, C., Cudre-Mauroux, P.: OLTP-bench: an extensible testbed for benchmarking relational databases. Proc. VLDB Endow. **7**(4), 277–288 (2013)
6. Halili, E.H.: Apache JMeter: A Practical Beginner's Guide to Automated Testing and Performance Measurement for Your Websites. Packt Publishing Ltd., Olton (2008)
7. Han, R., John, L.K., Zhan, J.: Benchmarking big data systems: a review. IEEE Trans. Serv. Comput. **11**(3), 580–597 (2017)
8. Han, R., Lu, X., Xu, J.: On big data benchmarking. In: Zhan, J., Han, R., Weng, C. (eds.) BPOE 2014. LNCS, vol. 8807, pp. 3–18. Springer, Cham (2014). https://doi.org/10.1007/978-3-319-13021-7_1
9. Iosup, A., et al.: LDBC Graphalytics: a benchmark for large-scale graph analysis on parallel and distributed platforms. Proc. VLDB Endow. **9**(13), 1317–1328 (2016)
10. Jouili, S., Vansteenberghe, V.: An empirical comparison of graph databases. In: 2013 International Conference on Social Computing, pp. 708–715. IEEE (2013)
11. Kasture, H., Sanchez, D.: Tailbench: a benchmark suite and evaluation methodology for latency-critical applications. In: 2016 IEEE International Symposium on Workload Characterization (IISWC), pp. 1–10. IEEE (2016)
12. Li, M., Tan, J., Wang, Y., Zhang, L., Salapura, V.: SparkBench: a comprehensive benchmarking suite for in memory data analytic platform spark. In: Proceedings of the 12th ACM International Conference on Computing Frontiers, p. 53. ACM (2015)
13. Li, Y., Manoharan, S.: A performance comparison of SQL and NoSQL databases. In: 2013 IEEE Pacific Rim Conference on Communications, Computers and Signal Processing (PACRIM), pp. 15–19. IEEE (2013)
14. Liang, F., Feng, C., Lu, X., Xu, Z.: Performance benefits of DataMPI: a case study with BigDataBench. In: Zhan, J., Han, R., Weng, C. (eds.) BPOE 2014. LNCS, vol. 8807, pp. 111–123. Springer, Cham (2014). https://doi.org/10.1007/978-3-319-13021-7_9
15. Ming, Z., et al.: BDGS: a scalable big data generator suite in big data benchmarking. In: Rabl, T., Jacobsen, H.-A., Raghunath, N., Poess, M., Bhandarkar, M., Baru, C. (eds.) WBDB 2013. LNCS, vol. 8585, pp. 138–154. Springer, Cham (2014). https://doi.org/10.1007/978-3-319-10596-3_11
16. Stonebraker, M., Çetintemel, U.: "one size fits all" an idea whose time has come and gone. In: Making Databases Work: The Pragmatic Wisdom of Michael Stonebraker, pp. 441–462 (2018)

17. Wang, C., Wang, B., Huang, B., Song, S., Li, Z.: FastSGG: efficient social graph generation using a degree distribution generation model. In: Proceedings of the IEEE 37th International Conference on Data Engineering (ICDE), Chania, Greece (2021)
18. Wang, L., et al.: BigDataBench: a big data benchmark suite from internet services. In: 2014 IEEE 20th International Symposium on High Performance Computer Architecture (HPCA), pp. 488–499. IEEE (2014)
19. Wang, M., Wang, C., Yu, J.X., Zhang, J.: Community detection in social networks: an in-depth benchmarking study with a procedure-oriented framework. Proc. VLDB Endow. **8**(10), 998–1009 (2015)

Towards Generating HiFi Databases

Anupam Sanghi$^{(\boxtimes)}$, Rajkumar Santhanam, and Jayant R. Haritsa

Indian Institute of Science, Bengaluru, India
{anupamsanghi,srajkumar,haritsa}@iisc.ac.in

Abstract. Generating synthetic databases that capture essential data characteristics of client databases is a common requirement for database vendors. We recently proposed Hydra, a workload-aware and scale-free data regenerator that provides statistical fidelity on the volumetric similarity metric. A limitation, however, is that it suffers poor accuracy on unseen queries. In this paper, we present HF-Hydra (HiFi-Hydra), which extends Hydra to provide better support to unseen queries through (a) careful choices among the candidate synthetic databases and (b) incorporation of metadata constraints. Our experimental study validates the improved fidelity and efficiency of HF-Hydra.

Keywords: Big data management · Data summarization · Data warehouse · OLAP workload · DBMS testing

1 Introduction

Database vendors often need to generate synthetic databases for a variety of use-cases, including: (a) testing engine components, (b) testing of database applications with embedded SQL, and (c) performance benchmarking. Several approaches to synthetic data generation have been proposed in the literature (reviewed in [7]) – in particular, a declarative approach of workload-aware data regeneration has been advocated over the last decade [1,2,4,5].

Workload-Aware Data Regeneration. Consider the database schema with three relations shown in Fig. 1(a), and a sample SQL query (Fig. 1(b)) on it. In the corresponding query execution plan (Fig. 1(c)), each edge is annotated with the associated cardinality of tuples flowing from one operator to the other. This is called an annotated query plan (AQP). From an AQP, a set of cardinality constraints (CCs) are derived, as enumerated in Fig. 1(d). The goal here is to achieve *volumetric similarity* – that is, on a given query workload, when these queries are executed on the synthetic database, the result should produce similar AQPs. In other words, the database should satisfy all the CCs.

The work of Anupam Sanghi was supported by an IBM PhD Fellowship Award.

C. S. Jensen et al. (Eds.): DASFAA 2021, LNCS 12681, pp. 105–112, 2021.
https://doi.org/10.1007/978-3-030-73194-6_8

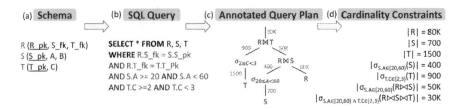

Fig. 1. Example Annotated Query Plan (AQP) and Cardinality Constraints (CC) (Color figure online)

1.1 Hydra

Hydra [5,6] is a workload-aware generator recently developed by our group. For each relation in the database, Hydra first constructs a corresponding denormalized relation (without key columns), called a *view*. To generate a view R, the domain space of R is partitioned into a set of disjoint *regions* determined by the filter predicates in the CCs. Further, a variable is created for each region, representing its row cardinality in the synthetic database. Next, an SMT Problem is constructed, where each CC is expressed as a linear equation in these variables. After solving the problem, the *data generator* picks a unique tuple within the region-boundaries and replicates it as per the region-cardinality obtained from the solution.

To make the above concrete, consider the constraint for relation S (from Fig. 1) shown by the red box (A in [20, 40) and B in [15000, 50000)) in Fig. 2, and having an associated row-count 150 (say). Likewise, a green constraint is also shown – say with row-count 250. Accordingly, the SMT problem constructed is:

$$x_1 + x_2 = 250, \quad x_2 + x_3 = 150, \quad x_1, x_2, x_3, x_4 \geq 0$$

The process of extracting relations from the views, while ensuring referential integrity (RI), forces the addition of some (spurious) tuples in the dimension tables. At the end, the output consists of concise constructors, together called as the *database summary*. An example summary is shown in Fig. 3 – the entries of the type a - b in the PK column (e.g. 101–250 for S_pk), represent rows with

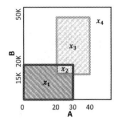

Fig. 2. Domain partitioning (Color figure online)

Fig. 3. Hydra example database summary

R			S			T	
R_pk	S_fk	T_fk	S_pk	A	B	T_pk	C
1-30K	101	601	1-100	0	15K	1-600	0
30001-50K	251	1	101-250	20	15K	601-1500	2
50001-60K	1	601	251-500	30	5K		
60001-80K	501	1	501-700	0	20K		

values $(a, a + 1, ..., b)$ for that column, keeping others unchanged. The summary makes Hydra amenable to handle Big Data volumes because the data can now be generated *dynamically*, i.e., "on-demand" during query execution, thereby obviating the need for materialization. Also, the summary construction time is *data-scale-free*, i.e., independent of the database size.

Limitations. As discussed above, Hydra is capable of efficiently delivering volumetric similarity on *seen* queries. However, the ability to generalize to new queries can be a useful feature for the vendor as part of the ongoing evaluation exercise. This is rendered difficult for Hydra due to the following design choices:

No Preference among Feasible Solutions: There can be several feasible solutions to the SMT problem. However, Hydra does not prefer any particular solution over the others. Moreover, due to the usage of Simplex algorithm internally, the SMT solver returns a sparse solution, i.e., it assigns non-zero cardinality to very few regions. This leads to very different *inter-region distribution* of tuples in the original and synthetic databases.

Artificial Skewed Data: Within a region that gets a non-zero cardinality assignment, Hydra generates a single unique tuple. As a result, a highly skewed data distribution is generated, which leads to an inconsistent *intra-region distribution* of tuples. Furthermore, the artificial skew can cause hindrance in efficient testing of queries, and gives an *unrealistic look* to the data.

Non-compliance with the Metadata: The metadata statistics present at the client site are transferred to the vendor and used to ensure matching plans at both sites. However, these statistics are not used in the data generation process, leading to data that is out of sync with the client meta-data.

1.2   HF-Hydra

In this work, we present **HF-Hydra** (High-Fidelity Hydra), which materially extends Hydra to address the above robustness-related limitations while retaining its desirable data-scale-free and dynamic generation properties.

The end-to-end pipeline of HF-Hydra's data generation is shown in Fig. 4. The client AQPs and metadata stats are given as input to *LP Formulator*. Using the inputs, the module constructs a refined partition, i.e. it gives finer regions. Further, a linear program (LP) is constructed by adding an objective function to pick a *desirable* feasible solution. From the LP solution, which is computed using the popular Z3 solver [8], the *Summary Generator* produces a richer database summary.

A sample summary produced by HF-Hydra on our running example is shown in Fig. 5. We see that the number of regions, characterized by the number of rows in the summary tables, are more in comparison to Hydra. Also, intervals are stored instead of points which support generation of a spread of tuples within each region. Tuples are generated uniformly within the intervals using the *Tuple Generator* module. These stages are discussed in detail in Sects. 2 and 3.

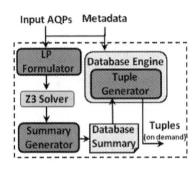

Input AQPs Metadata

LP Formulator
Database Engine
Tuple Generator
Z3 Solver
Summary Generator
Database Summary
Tuples (on demand)

Fig. 4. HF-Hydra pipeline (Color figure online)

Fig. 5. HF-Hydra database summary

R			S		
R_pk	S_fk	T_fk	S_pk	A	B
1-20k	101-180	601-900	1-60	[0-10)	[15k-20k)
20001-30k	181-250	901-1500	61-100	[10-20)	[15k-20k)
30001-45k	251-300	1-400	101-180	[20-30)	[15k-18k)
45001-50k	301-500	401-600	181-250	[20-30)	[18k-20k)
50001-55k	1-60	601-900	251-300	[30-60)	[0,10k) U [50k,60k)
55001-60k	61-100	901-1500			
60001-68k	501-550	1-400	301-500	[30-60)	[10k,15k) U [60k,80k)
68001-80k	551-700	401-600			

T		S (cont.)		
T_pk	C	S_pk	A	B
1-400	[0,1) U [3,7)	501-550	[0-15) U [60,75)	[20k-50k)
401-600	[2,3)		[15-20)	
601-900	[1,2) U [7,10)	551-700	U	[20k-50k)
901-1500	[2,3)		[75,80)	

In a nutshell, the addition of an objective function in the LP improves the inter-region tuple distribution. Further, having refined regions, plus uniform tuple distribution within these finer regions, improves the intra-region tuple distribution. Finally, addition of constraints from the metadata statistics ensures metadata-compliance.

We evaluate the efficacy of HF-Hydra by comparing its volumetric similarity with Hydra on unseen queries. Our results, elaborated in Sect. 4, indicate a substantive improvement – specifically, the volumetric similarity on filter constraints of unseen queries was better by more than **30%**, as measured by the UMBRAE model-comparison metric [3]. Further, we also show that HF-Hydra ensures metadata compliance. A sample table illustrating HF-Hydra delivers more realistic databases in comparison to Hydra is shown in [7].

2   LP Formulation

We now show how the LP is constructed from the AQPs and metadata. The summary steps are the following (complete details in [7]):

1. **Creating Metadata CCs.** Constraints are derived from the metadata statistics. Typically, the statistics include histograms and most common values (MCVs) with the corresponding frequencies. These are encoded as *metadata CCs*:
   i. $|\sigma_{A=a}(R)| = c_a$, for a value a stored in MCVs with frequency c_a (for column A).
   ii. $|\sigma_{A\in[l,h)}(R)| = B$, for a histogram bucket (for column A) with boundary $[l, h)$, having total row-count equal to B.
   Let there be a total of m such metadata CCs.
2. **Region Partitioning.** *Refined* regions are constructed using region-based domain partitioning, leveraging the CCs derived from both AQPs and metadata. Let the total number of resultant regions be n, where the row-cardinality of region i is captured in variable x_i.

$$\text{minimize} \sum_{j=1}^{m} \epsilon_j, \text{ subject to:}$$
1. $-\epsilon_j \leq (\sum_{i:I_{ij}=1} x_i) - k_j \leq \epsilon_j, \forall j \in [m],$
2. $C_1, C_2, ..., C_q,$
3. $x_i \geq 0 \quad \forall i \in [n], \quad \epsilon_j \geq 0 \quad \forall j \in [m]$

$$\text{minimize} \sum_{i=1}^{n} \epsilon_i, \text{ subject to:}$$
1. $-\epsilon_i \leq x_i - \tilde{x}_i \leq \epsilon_i, \forall i \in [n],$
2. $C_1, C_2, ..., C_q,$
3. $x_i, \epsilon_i \geq 0, \quad \forall i \in [n]$

Fig. 6. MDC LP formulation

Fig. 7. OE LP formulation

3. **Formulating LP Constraints.** The CCs from AQPs are added as explicit LP constraints, as in the original Hydra. Let there be q such CCs denoted by $C_1, C_2, ..., C_q$.
4. **Constructing Objective.** An *optimization function* is added to find a feasible solution that is close to the *estimated solution*. We use two notions of estimated solution:
   i. **Metadata Constraints Satisfaction (MDC):** Here the distance between the output cardinalities from metadata CCs and the sum of variables that represent the CCs is minimized. The LP thus obtained is shown in Fig. 6, with I_{ij} being an indicator variable, which takes value 1 if region i satisfies the filter predicate in the jth metadata CC, 0 otherwise.
   ii. **Optimizer Estimates Satisfaction (OE):** Here, instead of directly enforcing metadata CCs, the estimated cardinality for each region is obtained from the database engine using the optimizer's selectivity estimation logic. The objective function minimizes the distance between the solution and these estimates. The estimated cardinality $\tilde{x}_i$ for a region i is computed by constructing an SQL query equivalent for the region and using the query's estimated selectivity obtained from its compile-time plan. The LP produced using OE strategy is shown in Fig. 7.

Our choice of minimizing L1 distance is because query execution performance is linearly dependent on the row count, especially when all joins are PK-FK joins. In picking between the MDC and OE strategies, the following considerations apply: MDC has better metadata compliance due to explicit enforcement of the associated constraints. Further, its solution has higher sparsity because no explicit constraint is applied at a per-region level. However, while sparsity does make summary production more efficient, it adversely affects volumetric accuracy for higher levels of joins, as compared to OE.

3 Data Generation

Post LP-solving, the data generation pipeline proceeds in the following stages (complete details in [7]):

1. **Ensuring Referential Integrity.** Since each view is processed independently, these solutions may have inconsistencies. Specifically, when F, the fact table view, has a tuple whose value combination for the attributes that

it borrows from D, the dimension table view, does not have a matching tuple in D, then it causes a reference violation. To avoid it, for each region f of F, we maintain the populated regions in D that have an interval intersection with f for the borrowed columns. If no such region in D is found, then a new region with the intersection portion is added and assigned a cardinality of 1. This fixes the reference violation but leads to an additive error of 1 in the relation cardinality for the dimension table.

2. **Generating Relation Summary.** Here the borrowed attribute-set in a view is replaced with appropriate FK attributes. In contrast to Hydra's strategy of picking a single value in the FK column for a region, here we indicate a *range* to achieve a good span. To compute the FK column values for a region f, the corresponding matching regions from dimension table are fetched and the union of PK column ranges of these regions is returned.

3. **Tuple Generation.** The aim here is generate tuples uniformly within each region. Based on interval lengths that are contained for an attribute in the region, the ratio of tuples to be generated from each interval is computed. Now, if n values have to be generated within an interval I, then I is split into n equal sub-intervals and the center point within each interval is picked. If the range does not allow splitting into n sub-intervals, then it is split into the maximum possible sub-intervals, followed by a round-robin instantiation. The PK column values are generated consecutively, similar to row-numbers. This deterministic approach is well-suited for dynamic generation. If a materialized output is desired, then random values can be picked within intervals.

4   Experimental Evaluation

We now move on to empirically evaluating the performance of HF-Hydra against Hydra. For our experiments, we used a 1 GB version of the TPC-DS benchmark, hosted on a PostgreSQL v9.6 engine operating on a vanilla workstation. The SMT/LP problems were solved using Z3 [8].

We constructed a workload of 110 representative queries, which was then split randomly into *training* and *testing* sets of 90 and 20 queries. The associated AQPs led to formulation of 225 and 51 CCs, respectively. These CCs were a mix of pure filters on base relations, and CCs that involve filters along with 1 to 3 joins. Further, 2622 metadata CCs were derived from histograms and MCVs data.

4.1   Volumetric Similarity

For evaluating volumetric accuracy, we used the **UMBRAE** (Unscaled Mean Bounded Relative Absolute Error) model-comparison metric [3], with Hydra serving as the reference model. An UMBRAE value U ranges over positive numbers, where $U < 1$ implies $(1 - U) * 100\%$ better performance wrt baseline model, $U > 1$ implies $(U - 1) * 100\%$ worse performance and $U = 1$ shows no improvement.

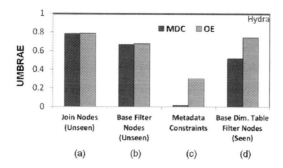

Fig. 8. Accuracy (Color figure online)

The UMBRAE values obtained by the two flavors of HF-Hydra over the 20 test queries are shown in Fig. 8(a)–(b). For clear understanding, the results for base filters and join nodes are shown separately. We see that HF-Hydra delivers more than **30%** better performance on filters, and an improvement of over **20%** with regard to joins. The higher improvement on filters is expected because the accuracy of metadata statistics, being on a per-column basis, is best at the lower levels of the plan tree.

Metadata compliance is evaluated in Fig. 8(c). A substantial improvement over Hydra is seen here – **98%** and **70%** for MDC and OE, respectively.

Interestingly, HF-Hydra outperforms Hydra even on the volumetric accuracy for *seen queries*, as captured in Fig. 8(d). Specifically, an improvement of **48%** and **26%** for MDC and OE, respectively, with regard to the base filter nodes on dimension tables. This benefit is an outcome of better distribution of tuples over regions, reducing the likelihood of mismatch between populated regions in the fact-table and empty regions in the dimension-table.

On the metrics considered thus far, MDC outperformed OE. However, for higher level joins, OE did better than MDC. Specifically, **33%** better for two-join cases and **13%** better for three-join cases. This is primarily because OE adheres to constraints at a per region level while MDC generates a sparse solution.

4.2 Database Summary Overheads

The database summaries generated by HF-Hydra and Hydra differ significantly in their structures. The former has many more regions, and stores intervals instead of points within a region. Due to these changes, a legitimate concern could be the impact on the size of the summary and the time taken to generate data from it at run-time. To quantitatively evaluate this concern, the space and time overheads are enumerated in Table 1. We see here that there is certainly a large increase in summary size, going from kilobytes to megabytes – however, in *absolute* terms, the summary size is still small enough to be easily viable on contemporary computing platforms. When we consider the time aspect, again there is an expected increase in the generation time from a few seconds to several

Table 1. Space and time analysis

	Hydra	MDC	OE
Summary size	40 KB	6 MB	985 MB
Tuple instantiation time	6 s	37 s	51 s

tens of seconds, but here too the absolute values are small enough (sub-minute) to make HF-Hydra usable in practice. Further, it is important to recall that these summary sizes and their construction time are independent of the client database size (experiments validating this claim are described in [7]).

5   Conclusions

Testing database engines efficiently is a critical issue in the industry, and the ability to accurately mimic client databases forms a key challenge in this effort. In contrast to the prior literature which focused solely on capturing database fidelity with respect to a known query workload, in this paper we have looked into the problem of generating databases that are robust to unseen queries. In particular, we presented HF-Hydra, which materially extends the state-of-the-art Hydra generator by bringing the potent power of metadata statistics and optimizer estimates to bear on the generation exercise. The resulting fidelity improvement was quantified through experimentation on benchmark databases, and the UMBRAE outcomes indicate that HF-Hydra successfully delivers high-fidelity databases.

Acknowledgements. We thank Tarun Kumar Patel and Shadab Ahmed for their valuable inputs in the implementation of this work.

References

1. Arasu, A., Kaushik, R., Li, J.: Data generation using declarative constraints. In: ACM SIGMOD Conference, pp. 685–696 (2011)
2. Binnig, C., Kossmann, D., Lo, E., Özsu, M.T.: QAGen: generating query-aware test databases. In: ACM SIGMOD Conference, pp. 341–352 (2007)
3. Chen, C., Twycross, J., Garibaldi, J.M.: A new accuracy measure based on bounded relative error for time series forecasting. PLoS ONE **12**(3), e0174202 (2017)
4. Li, Y., Zhang, R., Yang, X., Zhang, Z., Zhou, A.: Touchstone: generating enormous query-aware test databases. In: USENIX ATC, pp. 575–586 (2018)
5. Sanghi, A., Sood, R., Haritsa, J.R., Tirthapura, S.: Scalable and dynamic regeneration of big data volumes. In: 21st EDBT Conference, pp. 301–312 (2018)
6. Sanghi, A., Sood, R., Singh, D., Haritsa, J.R., Tirthapura, S.: HYDRA: a dynamic big data regenerator. PVLDB **11**(12), 1974–1977 (2018)
7. Sanghi, A., Rajkumar, S., Haritsa, J.R.: High fidelity database generators. Technical report TR-2021-01, DSL/CDS, IISc (2021). dsl.cds.iisc.ac.in/publications/report/TR/TR-2021-01.pdf
8. Z3. https://github.com/Z3Prover/z3

Modelling Entity Integrity
for Semi-structured Big Data

Ilya Litvinenko, Ziheng Wei, and Sebastian Link[(✉)] [iD]

The University of Auckland, Auckland, New Zealand
{ilit874,z.wei,s.link}@auckland.ac.nz

Abstract. We propose a data model for investigating constraints that enforce the entity integrity of semi-structured big data. Particular support is given for the volume, variety, and veracity dimensions of big data.

Keywords: Big data · Functional dependency · JSON · Key · SQL

1 Introduction

Database management systems model some domain of the real-world within a database system. For that purpose, SQL governs data by the rigid structure of relations [5]. Big data must handle potentially large volumes of data that may originate from heterogeneous sources (variety) with different degrees of uncertainty (veracity) [1]. Given the mature and popular technology that SQL provides many organizations use SQL to manage big data, at least when it is semi-structured such as in JSON format. While unstructured data, such as text or images, is not our focus, there is a rich landscape of techniques and tools for converting unstructured into semi-structured or even structured data [1].

We introduce the class of keys and functional dependencies (FDs) over possibilistic SQL data, with the aim to efficiently reason about the entity integrity of semi-structured big data that accommodates the volume, variety and veracity dimensions. Codd stipulated entity integrity as one of the three major integrity principles in databases [5]. Entity integrity refers to the principle of representing each entity of the application domain uniquely within the database. Violations of this principle are common in database practice, resulting in their own fields of research including entity resolution [4] and data cleaning [9].

While keys and FDs have standard definitions in the relational model, simple extensions introduce opportunities to define these concepts differently [18]. SQL, for example, permits occurrences of a so-called *null marker*, denoted by $\perp$, to say that there is no information about the value of this row on this column [10,21]. Moreover, columns can be defined as NOT NULL to prevent null marker occurrences. The interpretation of $\perp$ is deliberately kept simple to uniformly accommodate many types of missing information, including values that do not exist or values that exist but are currently unknown [21]. While such distinction is possible, it would lead to an application logic that is too complex for

C. S. Jensen et al. (Eds.): DASFAA 2021, LNCS 12681, pp. 113–120, 2021.
https://doi.org/10.1007/978-3-030-73194-6_9

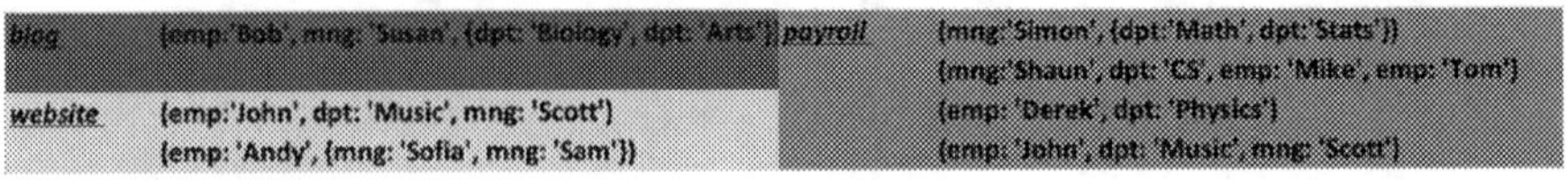

blog {emp:'Bob', mng: 'Susan', {dpt: 'Biology', dpt: 'Arts'}} *payroll* {mng:'Simon', {dpt:'Math', dpt:'Stats'}}
{mng:'Shaun', dpt: 'CS', emp: 'Mike', emp: 'Tom'}
website {emp:'John', dpt: 'Music', mng: 'Scott'}   {emp: 'Derek', dpt: 'Physics'}
{emp: 'Andy', {mng: 'Sofia', mng: 'Sam'}}   {emp: 'John', dpt: 'Music', mng: 'Scott'}

Fig. 1. JSON data from different information sources

database practice [6]. In modern applications, for example data integration, null markers are used frequently by SQL to fit data of heterogeneous structure within a uniform table. This is SQL's answer to the variety dimension of big data. SQL also permits the duplication of rows in support of a multiset semantics, where FDs can no longer express keys [12]. Hence, for SQL we need to study the combined class of keys and FDs. The veracity dimension abandons the view that all data are equal to improve the outcomes of data-driven decision making. Probabilistic and possibilistic databases offer complementary approaches to uncertain data. Essentially, there is a trade-off as probabilistic databases offer continuous degrees of uncertainty and real probability distributions are hard to come by and maintain, while possibilistic databases offer discrete degrees of uncertainty and are simpler to come by and maintain [8, 17].

Contributions and Organization. We introduce our running example in Sect. 2. We propose a framework of data structures capable of handling all combinations of the volume, variety, and veracity dimension of semi-structured big data within possibilistic SQL. Section 3 reviews isolated previous work under this framework. We define our possibilistic SQL model in Sect. 4, and possibilistic SQL constraints in Sect. 5. We conclude in Sect. 6.

2   The Running Example

As a simple running example consider Fig. 1 that shows some JSON data. JSON is the de-facto standard for managing and exchanging semi-structured data, due to its capability to accommodate different information structures [16].

In our example, the information origins from three sources: payroll data, web data, and blog data. As

Table 1. A university employment table

Row	emp	dpt	mng	p-degree	Interpretation	Origin
1	$\perp$	Math	Simon	α_1	Fully possible	Payroll
2	$\perp$	Stats	Simon	α_1	Fully possible	Payroll
3	Mike	CS	Shaun	α_1	Fully possible	Payroll
4	Tom	CS	Shaun	α_1	Fully possible	Payroll
5	Derek	Physics	$\perp$	α_1	Fully possible	Payroll
6	John	Music	Scott	α_1	Fully possible	Payroll
7	John	Music	Scott	α_2	Quite possible	Website
8	Andy	$\perp$	Sofia	α_2	Quite possible	Website
9	Andy	$\perp$	Sam	α_2	Quite possible	Website
10	Bob	Biology	Susan	α_3	Somewhat possible	Blog
11	Bob	Arts	Susan	α_3	Somewhat possible	Blog

the data stewards associate different levels of trust with these sources, they would like to attribute these levels of trust to the data from the sources. We are using an SQL-based DBMS, and the data stewards have transformed the JSON data into SQL-compliant format as shown in Table 1.

The null marker ⊥ indicates that no information is available for a data element of a given attribute, such as information on employees working in Maths. The column *p-degree* represents the levels of trust for the data elements. The highest p-degree α_1 is assigned to payroll data, α_2 to web data, and α_3 to blog data. This application scenario will be used to illustrate concepts.

3   Related Work

As a first main contribution, we introduce a systematic framework for handling the volume, variety, and veracity dimension of big data. We apply the framework to manage and reason about entity integrity in those dimensions. Figure 2 shows all combinations of the three dimensions, ordered as a lattice.

A directed edge means that the target node covers additional dimensions over the source node. For each combination of dimensions, we indicate a data structure for the combination. At the bottom are relations (sets of rows). Arguably, these may already accommodate the volume dimension. The nodes on top of relations are bags, partial relations, and p-relations.

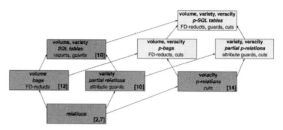

Fig. 2. Framework for semi-structured big data, related work, and new contributions (Color figure online)

Bags accommodate the volume dimension by permitting duplicate rows, partial relations accommodate the variety dimension by permitting null markers (as explained earlier), and p-relations accommodate the veracity dimension. Above these single dimensions we then have any combinations of two dimensions, and the top node combines all three dimensions.

We now use this framework to discuss previous work on entity integrity rules, based on the three dimensions of semi-structured big data. These are marked in cyan in Fig. 2. Well-known are Armstrong's axioms for FDs in the relational model [2] and the corresponding algorithms for deciding implication [7]. For bags, the interaction of keys and FDs was characterized in [12], using the technique of FD-reducts. Over partial relations, the implication problem for keys and FDs in the presence of NOT NULL constraints was solved in [10], using so-called *attribute guards*. The implication problem over SQL tables, which combine bags and partial relations, was solved by [10]. Finally, the implication problem of FDs over p-relations was solved in [14], using so-called β-cuts.

We can view these isolated results under our big data framework. This view motivates us to extend the previous techniques to new combinations of these dimensions. These include the combination of i) volume and veracity, ii) variety and veracity, and iii) volume, variety, and veracity (marked green in Fig. 2).

4 Possibilistic SQL Tables

An *SQL table schema*, denoted by T, is a finite non-empty set of attributes. The domain of each attribute contains the null marker, $\perp$, as a distinguished element. As running example, we use the table schema $\text{WORK} = \{emp, dpt, mng\}$ with information about employees that work in departments under managers.

A *row* (or *tuple*) over table schema T is a function $r : T \to \cup_{A \in T} dom(A)$ assigning to every attribute a value from the attribute's domain. The image $r(A)$ of a row r on attribute A is the *value* of r on A. For $X \subseteq T$, a row r over T is *X-total*, if $r(A) \neq \perp$ for all $A \in X$. A row r is *total*, if it is T-total.

We adopt Zaniolo's interpretation of $\perp$ as "no information" [21]. That is, $r(A) = \perp$ means no information about the value of row r on attribute A is available. It may mean there is no value at all, or that there is a value which is currently unknown. SQL uses this interpretation [10,21].

Our data model handles duplicates as tables are multisets of rows. An *SQL table* over table schema T is a finite multiset t of rows over T. For $X \subseteq T$, the table t is *X-total*, if every row $r \in t$ is X-total. Table t is total, if t is T-total. A total table is called a bag. Table 2 shows an example of a table over WORK. The third row has value *Derek* on *emp*, value *Physics* on *dpt*, and marker $\perp$ on *mng*. The first and second row are total, and the

Table 2. Table over WORK

emp	dpt	mng
John	Music	Scott
John	Music	Scott
Derek	Physics	$\perp$
Derek	Physics	$\perp$

third and fourth row, as well as the table itself are $\{emp, dpt\}$-total. The first and second row, as well as the third and fourth row, respectively, are duplicate tuples, since they have matching values on all attributes.

SQL does not accommodate uncertainty. For example, one cannot say tuple (Derek, Physics, $\perp$) is less likely to occur than tuple (John, Music, Scott). We extend our data model by assigning degrees of possibilities (p-degrees) to

Table 3. Possible worlds of p-SQL table from Table 1

	t_1			t_2			t_3	
emp	dpt	mng	emp	dpt	mng	emp	dpt	mng
$\perp$	Math	Simon	$\perp$	Math	Simon	$\perp$	Math	Simon
$\perp$	Stats	Simon	$\perp$	Stats	Simon	$\perp$	Stats	Simon
Mike	CS	Shaun	Mike	CS	Shaun	Mike	CS	Shaun
Tom	CS	Shaun	Tom	CS	Shaun	Tom	CS	Shaun
Derek	Physics	$\perp$	Derek	Physics	$\perp$	Derek	Physics	$\perp$
John	Music	Scott	John	Music	Scott	John	Music	Scott
			Andy	$\perp$	Sofia	Andy	$\perp$	Sofia
			Andy	$\perp$	Sam	Andy	$\perp$	Sam
						Bob	Biology	Susan
						Bob	Arts	Susan

tuples, thereby also extending the model of [14,15] where no duplicate nor partial information was used. In our example, p-degrees result from the source the tuples originate from. The tuples in Table 1 originate from payroll, website or blog data, respectively. Payroll data is 'fully possible', website data 'quite possible', and blog data 'somewhat possible', while other tuples are 'impossible' to occur in the current table. Since p-degrees can have different meanings, we denote them by abstract symbols $\alpha_1, \ldots, \alpha_k, \alpha_{k+1}$. Table 1 shows an instance with p-degrees assigned to tuples. The table has meta-data columns: 'row' assigns an identifier to each tuple, while 'interpretation' and 'origin' show the interpretation of p-degrees and the source of tuples, respectively.

A *possibility scale* is a strict finite linear order $\mathcal{S}_p = (S_p, >_p)$, denoted by $\alpha_1 >_p \cdots >_p \alpha_k >_p \alpha_{k+1}$, where k is at least one. The elements α_i are *possibility degrees* (p-degrees). In Table 1 we have $k = 3$ for the *possibility scale*. Fully possible rows have p-degree α_1, while the least possible rows have p-degree α_3. The bottom p-degree $\alpha_{k+1} = \alpha_4$ captures rows 'impossible' for the current table. Non-possibilistic tables are a special case of possibilistic ones where $k = 1$.

A *possibilistic SQL table schema* (or p-SQL table schema) is a pair $(T, \mathcal{S}_p)$, where T is a table schema and $\mathcal{S}_p$ is a possibility scale. A *possibilistic SQL table* (or p-SQL table) over $(T, \mathcal{S}_p)$ consists of a table t over T, and a function $Poss_t$ that maps each row $r \in t$ to a p-degree $Poss_t(r) \neq \alpha_{k+1}$ in the p-scale $\mathcal{S}_p$. The p-SQL table of our example is shown in Table 1. It consists of an SQL table over WORK in which every row is assigned a p-degree from α_1, α_2 or α_3. P-SQL tables enjoy a well-founded possible world semantics. The possible worlds form a linear chain of k SQL tables in which the i-th possible world contains tuples with p-degree α_i or higher. Given a p-SQL table t over $(T, \mathcal{S}_p)$, the possible world t_i associated with t is defined by $t_i = \{r \in t \mid Poss_t(r) \geq \alpha_i\}$, that is, t_i is an SQL table of those rows in t that have p-degree α_i or higher. Since t_{k+1} would contain impossible tuples it is not considered a possible world. Table 3 shows the possible worlds of the p-SQL table from Table 1. The possible worlds of t form a linear chain $t_1 \subseteq t_2 \subseteq t_3$.

The linear order of the p-degrees $\alpha_1 > \cdots > \alpha_k$ results in a reversed linear order of possible worlds associated with a p-SQL table t: $t_1 \subseteq \cdots \subseteq t_k$. We point out the distinguished role of the top p-degree α_1. Every row that is fully possible belongs to every possible world. Therefore, every fully possible row is also fully certain. This explains why p-SQL tables subsume SQL tables as a special case.

5   Possibilistic SQL Constraints

We recall the definitions of SQL FDs and NOT NULL constraints [10]. Keys are essential to entity integrity and cannot be expressed by FDs in this context.

Intuitively, a key is an attribute collection that can separate different rows by their values on the key attributes. We adopt the semantics for the SQL constraint UNIQUE by separating different rows whenever they are total on the key attributes. A *key* over an SQL table schema T is an expression $u(X)$ where $X \subseteq T$. An SQL table t over T *satisfies* $u(X)$ over T, denoted by $\models_t u(X)$, if for all $r_1, r_2 \in t$ we have: if $r_1(X) = r_2(X)$ and r_1, r_2 are X-total, then $r_1 = r_2$. The possible world t_1 of Table 3 satisfies $u(emp)$, while t_2 and t_3 violate this key.

The following semantics of FDs goes back to Lien [13]. A *functional dependency* (FD) over an SQL table schema T is an expression $X \to Y$ where $XY \subseteq T$. An SQL table t over T *satisfies* $X \to Y$ over T, denoted by $\models_t X \to Y$, if for all $r_1, r_2 \in t$ we have: if $r_1(X) = r_2(X)$ and r_1, r_2 are X-total, then $r_1(Y) = r_2(Y)$. The possible world t_2 of Table 3 satisfies $emp \to dpt$ and $dpt \to mng$, while t_3 satisfies $dpt \to mng$, but not $emp \to dpt$.

SQL NOT NULL constraints control occurrences of the null marker. They have been studied in combination with FDs and multivalued dependencies [10].

A NOT NULL constraint over an SQL table schema T is an expression $n(X)$ where $X \subseteq T$. An SQL table t over T *satisfies* the NOT NULL constraint $n(X)$ over T, denoted by $\models_t n(X)$, if t is X-total. For a given set Σ of constraints over T we call $T_s = \{A \in T \mid \exists n(X) \in \Sigma \wedge A \in X\}$ the *null-free subschema* (NFS) over T. If $T_s = T$, we call T a *bag schema*, as instances over T are bags. For example, $n(dpt)$ is satisfied by the possible world t_1 in Table 3, but not by t_2 or t_3.

Possibilistic SQL Constraints. We extend our semantics of SQL constraints to possibilistic SQL tables. Following [14], we use the p-degrees of rows to specify with which certainty an SQL constraint holds. Similar to how α_i denotes p-degrees of rows, β_i denotes c-degrees by which constraints hold. Let us inspect some SQL constraints on the possible worlds t_1, t_2, t_3 in Table 3. The constraint $dpt \rightarrow mng$ is satisfied by t_3, and therefore by t_2 and t_1. Since the constraint is satisfied by every possible world, it is 'fully certain' to hold, denoted by β_1. The constraint $emp \rightarrow dpt$ is satisfied by t_2 and therefore by t_1, but it is not satisfied by t_3. Since the constraint is only violated by the 'somewhat possible' world t_3, it is 'quite certain' to hold, denoted by β_2. The constraint $u(emp)$ is satisfied by t_1, but it is not satisfied by t_2 and therefore not by t_3. Since the smallest possible world that violates the constraint is 'quite possible', it is 'somewhat certain' to hold, denoted by β_3. The constraint $n(emp)$ is not even satisfied in the 'fully possible' world t_1. It is 'not certain at all' to hold, denoted by β_4.

The examples illustrate how the p-degrees of rows motivate degrees of certainty (c-degrees) with which constraints hold on p-SQL tables. If the smallest world that violates a constraint has p-degree α_i (this world is impossible only when all possible worlds satisfy the constraint), then the constraint holds with c-degree β_{k+2-i}. For example, the p-key $u(emp)$ holds with c-degree β_3 in the p-SQL table t of Table 1, meaning the smallest possible world that violates $u(emp)$ is t_2, which is 'quite possible', that is $u(emp)$ is 'somewhat certain' to hold in t. We introduce the certainty scale derived from a given possibility scale.

Let $(T, \mathcal{S}_p)$ denote a p-SQL table schema where the bottom p-degree of $\mathcal{S}_p$ is $k+1$. The certainty scale $\mathcal{S}_p^T$ for $(T, \mathcal{S}_p)$ is the strict finite linear order $\beta_1 >_p \cdots >_p \beta_k >_p \beta_{k+1}$. The top c-degree β_1 is for constraints that are 'fully certain', while the bottom c-degree β_{k+1} is for constraints that are 'not certain at all'.

We define by which c-degree an SQL constraint holds on a p-SQL table. Similar to marginal probabilities in probability theory, we call this c-degree the *marginal certainty*. In SQL tables an SQL constraint either holds or does not hold. In a p-SQL table, an SQL constraint always holds with some c-degree.

Definition 1 (Marginal certainty). *Let σ denote an SQL key, FD or NOT NULL constraint over table schema T. The* marginal certainty $c_t(\sigma)$ *by which σ holds in the p-SQL table t over $(T, \mathcal{S}_p)$ is the c-degree β_{k+2-i} that corresponds to the p-degree α_i of the smallest possible world t_i of t in which σ is violated, that is, $c_t(\sigma) = \beta_1$ if $\models_{t_k} \sigma$, and $c_t(\sigma) = \min\{\beta_{k+2-i} \mid \not\models_{t_i} \sigma\}$ otherwise.*

For example, when t denotes the p-SQL table of Table 1, then $c_t(dpt \rightarrow mng) = \beta_1$, $c_t(emp \rightarrow dpt) = \beta_2$, $c_t(u(emp)) = \beta_3$, and $c_t(n(emp)) = \beta_4$.

Constraints specify the semantics of an application domain. They govern which databases are regarded as meaningful for the application. We classify a p-SQL table as meaningful whenever it satisfies a given set of possibilistic constraints (σ, β) (key, FD, NOT NULL constraint), which allow us to stipulate the minimum marginal c-degree β by which the constraint σ must hold in every p-SQL table that is considered to be meaningful in the application domain.

Definition 2 (Possibilistic constraints). *Let* $(T, \mathcal{S}_P)$ *denote a p-SQL table schema. A possibilistic SQL key, possibilistic SQL FD, or possibilistic* NOT NULL *constraint is a pair* (σ, β) *where* σ *denotes an SQL key, FD or* NOT NULL *constraint over* T*, respectively, and* β *denotes a c-degree from* $\mathcal{S}_P^T$*. The p-constraint* (σ, β_i) *is satisfied by a p-SQL table* t *over* $(T, \mathcal{S}_P)$ *iff* $c_t(\sigma) \geq \beta_i$*.*

For example, when t denotes the p-SQL table of Table 1, then the following examples of p-constraints are satisfied by t: $(dpt \rightarrow mng, \beta_3)$ since $c_t(dpt \rightarrow mng) = \beta_1 \geq \beta_3$, $(emp \rightarrow dpt, \beta_2)$ since $c_t(emp \rightarrow dpt) = \beta_2 \geq \beta_2$, and $(u(emp), \beta_4)$ since $c_t(u(emp)) = \beta_3 \geq \beta_4$. In other words, t satisfies these three constraints. On the other hand, t violates (i.e. does not satisfy) any of the following p-constraints: $(emp \rightarrow dpt, \beta_1)$ since $c_t(emp \rightarrow dpt) = \beta_2 < \beta_1$, $(u(emp), \beta_2)$ since $c_t(u(emp)) = \beta_3 < \beta_2$, and $(n(emp), \beta_3)$ since $c_t(n(emp)) = \beta_4 < \beta_3$.

6   Conclusion and Future Work

We aim at a comprehensive toolbox for reasoning about the integrity of real-world entities in semi-structured big data. As underlying data model we chose a possibilistic extension of SQL. We showed how previous work captures some of the big data dimensions as special cases. Our definition of possibilistic keys, FDs, and NOT NULL constraints lays the foundation for investigating fundamental reasoning tasks for them in the future.

Indeed, different approaches should be applied to the big data dimensions, such as probabilistic approaches to the veracity dimension [3], different approaches of handling missing information to the variety dimension such as embedded keys and FDs [20], and different approaches to entity integrity such as key sets [19]. In a different direction, we may want to add further big data dimensions. For example, temporal extensions [11] may support the velocity dimension (Fig. 3).

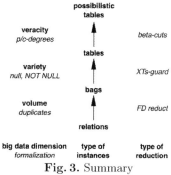

Fig. 3. Summary

References

1. Amalina, F., et al.: Blending big data analytics: review on challenges and a recent study. IEEE Access **8**, 3629–3645 (2020)
2. Armstrong, W.W.: Dependency structures of data base relationships. In: Proceedings of IFIP World Computer Congress, pp. 580–583 (1974)
3. Brown, P., Link, S.: Probabilistic keys. IEEE Trans. Knowl. Data Eng. **29**(3), 670–682 (2017)
4. Christophides, V., Efthymiou, V., Stefanidis, K.: Entity Resolution in the Web of Data, Synthesis Lectures on the Semantic Web. Morgan & Claypool Publishers (2015)
5. Codd, E.F.: A relational model of data for large shared data banks. Commun. ACM **13**(6), 377–387 (1970)
6. Date, C.J.: A critique of the SQL database language. SIGMOD Rec. **14**(3), 8–54 (1984)
7. Diederich, J., Milton, J.: New methods and fast algorithms for database normalization. ACM Trans. Database Syst. **13**(3), 339–365 (1988)
8. Dubois, D., Prade, H., Schockaert, S.: Generalized possibilistic logic: foundations and applications to qualitative reasoning about uncertainty. Artif. Intell. **252**, 139–174 (2017)
9. Ganti, V., Sarma, A.D.: Data Cleaning: A Practical Perspective, Synthesis Lectures on Data Management. Morgan & Claypool Publishers (2013)
10. Hartmann, S., Link, S.: The implication problem of data dependencies over SQL table definitions: axiomatic, algorithmic and logical characterizations. ACM Trans. Database Syst. **37**(2), 13:1–13:40 (2012)
11. Jensen, C.S., Snodgrass, R.T., Soo, M.D.: Extending existing dependency theory to temporal databases. IEEE Trans. Knowl. Data Eng. **8**(4), 563–582 (1996)
12. Köhler, H., Link, S.: Armstrong axioms and Boyce-Codd-Heath normal form under bag semantics. Inf. Process. Lett. **110**(16), 717–724 (2010)
13. Lien, Y.E.: On the equivalence of database models. J. ACM **29**(2), 333–362 (1982)
14. Link, S., Prade, H.: Possibilistic functional dependencies and their relationship to possibility theory. IEEE Trans. Fuzzy Syst. **24**(3), 757–763 (2016)
15. Link, S., Prade, H.: Relational database schema design for uncertain data. Inf. Syst. **84**, 88–110 (2019)
16. Liu, Z.H., Hammerschmidt, B.C., McMahon, D.: JSON data management: supporting schema-less development in RDBMS. In: International Conference on Management of Data, SIGMOD 2014, Snowbird, UT, USA, 22–27 June 2014, pp. 1247–1258 (2014)
17. Suciu, D., Olteanu, D., Ré, C., Koch, C.: Probabilistic Databases, Synthesis Lectures on Data Management. Morgan & Claypool Publishers (2011)
18. Thalheim, B.: Dependencies in relational databases. Teubner (1991)
19. Thalheim, B.: On semantic issues connected with keys in relational databases permitting null values. Elektronische Informationsverarbeitung und Kybernetik **25**(1/2), 11–20 (1989)
20. Wei, Z., Link, S.: Embedded functional dependencies and data-completeness tailored database design. PVLDB **12**(11), 1458–1470 (2019)
21. Zaniolo, C.: Database relations with null values. J. Comput. Syst. Sci. **28**(1), 142–166 (1984)

Graph Data

Label Contrastive Coding Based Graph Neural Network for Graph Classification

Yuxiang Ren[1(✉)], Jiyang Bai[2], and Jiawei Zhang[1]

[1] IFM Lab, Department of Computer Science, Florida State University,
Tallahassee, FL, USA
{yuxiang,jiawei}@ifmlab.org
[2] Department of Computer Science, Florida State University, Tallahassee, FL, USA
bai@cs.fsu.edu

Abstract. Graph classification is a critical research problem in many applications from different domains. In order to learn a graph classification model, the most widely used supervision component is an output layer together with classification loss (e.g., cross-entropy loss together with softmax or margin loss). In fact, the discriminative information among instances are more fine-grained, which can benefit graph classification tasks. In this paper, we propose the novel **L**abel **C**ontrastive Coding based **G**raph **N**eural **N**etwork (LCGNN) to utilize label information more effectively and comprehensively. LCGNN still uses the classification loss to ensure the discriminability of classes. Meanwhile, LCGNN leverages the proposed *Label Contrastive Loss* derived from self-supervised learning to encourage instance-level intra-class compactness and inter-class separability. To power the contrastive learning, LCGNN introduces a dynamic label memory bank and a momentum updated encoder. Our extensive evaluations with eight benchmark graph datasets demonstrate that LCGNN can outperform state-of-the-art graph classification models. Experimental results also verify that LCGNN can achieve competitive performance with less training data because LCGNN exploits label information comprehensively.

1 Introduction

Applications in many domains in the real world exhibit the favorable property of graph data structure, such as social networks [15], financial platforms [20] and bioinformatics [5]. Graph classification aims to identify the class labels of graphs in the dataset, which is an important problem for numerous applications. For instance, in biology, a protein can be represented with a graph where each amino acid residue is a node, and the spatial relationships between residues (distances, angles) are the edges of a graph. Classification of graphs representing proteins can help predict protein interfaces [5].

Recently, graph neural networks (GNNs) have achieved outstanding performance on graph classification tasks [29,33]. GNNs aims to transform nodes to

---

Y. Ren and J. Bai—Contributed equally to this work.

© Springer Nature Switzerland AG 2021
C. S. Jensen et al. (Eds.): DASFAA 2021, LNCS 12681, pp. 123–140, 2021.
https://doi.org/10.1007/978-3-030-73194-6_10

low-dimensional dense embeddings that preserve graph structural information and attributes [34]. When applying GNNs to graph classification, the standard method is to generate embeddings for all nodes in the graph and then summarize all these node embeddings to a representation of the entire graph, such as using a simple summation or neural network running on the set of node embeddings [31]. For the representation of the entire graph, a supervision component is usually utilized to achieve the purpose of graph classification. A final output layer together with classification loss (e.g., cross-entropy loss together with softmax or margin loss) is the most commonly used supervision component in many existing GNNs [6,28,29,32]. This supervision component focuses on the discriminability of class but ignores the instance-level discriminative representations. A recent trend towards learning stronger representations to serve classification tasks is to reinforce the model with discriminative information as more as possible [4]. To be explicit, graph representations, which consider both intra-class compactness and inter-class separability [14], are more potent on the graph classification tasks.

Inspired by the idea of recent self-supervised learning [3] and contrastive learning [7,18], the contrastive loss [17] is able to extract extra discriminative information to improve the model's performance. The recent works [8,18,35] using contrast loss for representation learning are mainly carried out under the setting of unsupervised learning. These contrastive learning models treat each instance as a distinct class of its own. Meanwhile, discriminating these instances is their learning objective [7]. The series of contrastive learning have been verified effective in learning more fine-grained instance-level features in the computer vision [26] domain. Thus we plan to utilize contrastive learning on graph classification tasks to make up for the shortcomings of supervision components, that is, ignoring the discriminative information on the instance-level. However, when applying contrastive learning, the inherent large intra-class variations may import noise to graph classification tasks [14]. Besides, existing contrastive learning-based GNNs (e.g., GCC [18]) detach the model pre-training and fine-tuning steps. Compared with end-to-end GNNs, the learned graph representations via contrastive learning can hardly be used in the downstream application tasks directly, like graph classification.

To cope with the task of graph classification, we propose the label contrastive coding based graph neural network (LCGNN), which employs *Label Contrastive Loss* to encourage instance-level intra-class compactness and inter-class separability simultaneously. Unlike existing contrastive learning using a single positive instance, the label contrastive coding imports label information and treats instances with the same label as multiple positive instances. In this way, the instances with the same label can be pulled closer, while the instances with different labels will be pushed away from each other. Intra-class compactness and inter-class separability are taken into consideration simultaneously. The label contrastive coding can be regarded as training an encoder for a dictionary look-up task [7]. In order to build an extensive and consistent dictionary, we propose a dynamic label memory bank and a momentum-updated graph encoder inspired

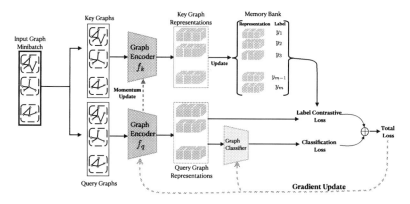

Fig. 1. The high-level structure of LCGNN. LCGNN trains the graph encoder f_q and the graph classifier using a mixed loss. *Label Contrastive Loss* and *Classification Loss* constitute the mixed loss. *Classification Loss* used in LCGNN is cross-entropy loss. *Label Contrastive Loss* is calculated by a dictionary look-up task. The query is each graph of the input graph minibatch, and the dictionary is a memory bank that can continuously update the label-known graph representations. The graph representation in the memory bank is updated by the graph encoder f_k, which is momentum-updated. After training, the learned graph encoder f_q, and the graph classifier can serve for graph classification tasks.

by the mechanism [7]. At the same time, LCGNN also uses *Classification Loss* to ensure the discriminability of classes. LCGNN can utilize label information more effectively and comprehensively from instance-level and class-level, allowing using fewer label data to achieve comparative performance, which can be considered as a kind of label augmentation in essence. We validate the performance of LCGNN on graph classification tasks over eight benchmark graph datasets. LCGNN achieves SOTA performance in seven of the graph datasets. What is more, LCGNN outperforms the baseline methods when using less training data, which verifies its ability to learn from label information more comprehensively.

The contributions of our work are summarized as follows:

- We propose a novel label contrastive coding based graph neural network (LCGNN) to reinforce supervised GNNs with more discriminative information.
- The *Label Contrastive Loss* extends the contrastive learning to the supervised setting, where the label information can be imported to ensure intra-class compactness and inter-class separability.
- The momentum-updated graph encoder and the dynamic label memory bank are proposed to support our supervised contrastive learning.
- We conduct extensive experiments on eight benchmark graph datasets. LCGNN not only achieves SOTA performance on multiple datasets but also can offer comparable results with fewer labeled training data.

2   Related Works

Graph Classification. Several different techniques have been proposed to solve the graph classification problem. One important category is the kernel-based method, which learns a graph kernel to measure similarity among graphs to differentiate graph labels [25]. The Weisfeiler-Lehman subtree kernel (WL) [21], Multiscale Laplacian graph kernels (MLG) [13], and Graphlets kernel(GK) [22] are all representative graph kernels. Another critical category is the deep-learning-based method. Deep Graph Kernel (DGK) [30], Anonymous Walk Embeddings (AWE), and Graph2vec [16] all employ the deep-learning framework to extract the graph embeddings for graph classification tasks. With the rise of graph neural networks (GNNs), many GNNs are also used for graph classification tasks by learning the representation of graphs, which will be introduced below.

Graph Neural Network. The graph neural network learns the low-dimensional graph representations through a recursive neighborhood aggregation scheme [29]. The derived graph representations can be used to serve various downstream tasks, such as graph classification and top-k similarity search. According to the learning method, the current GNN serving graph classification can be divided into end-to-end models and pre-train models. The end-to-end models are usually under supervised or semi-supervised settings, with the goal of optimizing classification loss or mutual information, mainly including GIN [29], CapsGNN [28], DGCNN [32] and InfoGraph [23]. The pre-trained GNNs use certain pre-training tasks [9] to learn the graph's general representation under the unsupervised setting. In order to perform graph classification tasks, a part of label data will be used to fine-tuning the models [18].

Contrastive Learning. Contrastive learning has been widely used for unsupervised learning by training an encoder that can capture similarity from data. The contrastive loss is normally a scoring function that increases the score on the single matched instance and decreases the score on multiple unmatched instances [17,26]. In the graph domain, DGI [24] is the first GNN model utilizing the idea of contrastive learning, where the mutual information between nodes and the graph representation is defined as the contrastive metric. HDGI [19] extends the mechanism to heterogeneous graphs. InfoGraph [23] performs contrastive learning in semi-supervised graph-level representation learning. When faced with the task of supervised learning, such as graph classification, we also need to use the advantage of contrastive learning to capture similarity. GCC [18] utilizes contrastive learning to pre-train a model that can serve for the downstream graph classification task by fine-tuning. Compared to them, our method is an end-to-end model and performs label contrastive coding to encourage instance-level intra-class compactness and inter-class separability.

3   Proposed Method

In this section, we introduce the label contrastive coding based graph neural network (LCGNN). Before introducing LCGNN, we provide the preliminaries about graph classification first.

3.1   Preliminaries

The goal of graph classification is to predict class labels of graphs based on the graph structural information and node contents. Formally, we denote it as follows:

Graph Classification. Given a set of labeled graphs $\mathbb{G}_L = \{(\mathcal{G}_1, y_1),$ $(\mathcal{G}_2, y_2), \dots\}$ and $y_i \in \mathbb{Y}$ is the corresponding label of $\mathcal{G}_i$. The task is to learn a classification function $f : \mathcal{G} \longrightarrow \mathbb{Y}$ to make predictions for unseen graphs $\mathbb{G}_U$.

3.2   LCGNN Architecture Overview

A learning process illustration of the proposed LCGNN is shown in Fig. 1. Usually, for the input graph, we need to extract the latent features that can serve the graph classification through a high-performance graph encoder. In order to cooperate with the proposed mixed loss (*Label Contrastive Loss* & *Classification Loss*), LCGNN contains two graph encoder f_k and f_q, which serve for encoding input key graphs and query graphs respectively. *Label Contrastive Loss* encourages instance-level intra-class compactness and inter-class separability simultaneously by keeping intermediate discriminative representations, while *Classification Loss* ensures the class-level discriminability. A dynamic memory bank containing key graph representations and corresponding labels works for label contrastive loss calculation. A graph classifier takes the representations from the graph encoder f_q as its input to predict the graph labels. In the following parts, we will elaborate on each component and the learning process of LCGNN in detail.

3.3   Label Contrastive Coding

Existing contrastive learning has been proved a success in training an encoder that can capture the universal structural information behind graph data [18]. In the graph classification task, we focus on classification-related structural patterns compared with the universal structural patterns. Therefore, our proposed label contrastive coding learns to discriminate between instances with different class labels instead of treating each instance as a distinct class of itself and contrasting to other distinct classes.

Contrastive learning can be considered as learning an encoder for a dictionary look-up task [7]. We can describe the contrastive learning as follows. Given an encoded query $\mathbf{q}$ and a dictionary containing m encoded keys $\{\mathbf{k}_1, \mathbf{k}_2, \ldots, \mathbf{k}_m\}$, there is only a single positive key $\mathbf{k}_+$ (normally encoded from the same instance as $\mathbf{q}$). The loss of this contrastive learning is low when $\mathbf{q}$ is similar to the positive key $\mathbf{k}_+$ while dissimilar to negative keys for $\mathbf{q}$ (all other keys in the dictionary). A widely used contrastive loss is InfoNCE [17] like:

$$\mathcal{L} = -\log \frac{\exp(\mathbf{q} \cdot \mathbf{k}_+ / \tau)}{\sum_{i=1}^{m} \exp(\mathbf{q} \cdot \mathbf{k}_i / \tau)} \tag{1}$$

Here, τ is the temperature hyper-parameter [26]. Essentially, the loss of InfoNCE is a classification loss aiming to classify $\mathbf{q}$ from $m = 1$ classes to the same class as $\mathbf{k}_+$.

However, when facing graph classification tasks, the class labels have been determined. We hope to import known label information in the training data to assist contrastive learning in serving the graph classification task. In this way, we design the label contrastive coding.

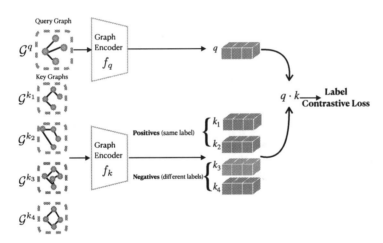

Fig. 2. *Label Contrastive Loss.* The query graph $\mathcal{G}^q$ and key graphs $\mathcal{G}^k$ are encoded by f_q and f_k to low-dimensional representations $\mathbf{q}$ and $\mathbf{k}$ respectively. $\mathbf{k}_1$ and $\mathbf{k}_2$ having the same label as $\mathbf{q}$ are denoted as positive keys. $\mathbf{k}_3$ and $\mathbf{k}_4$ are negative keys due to different labels. The label contrastive loss encourage the model to distinguish the similar pair $(\mathcal{G}^q, \mathcal{G}^{k_1})$ and $(\mathcal{G}^q, \mathcal{G}^{k_2})$ from dissimilar instance pairs, e.g., $(\mathcal{G}^q, \mathcal{G}^{k_3})$.

Define Similar and Dissimilar. In the graph classification task, we seek that instances with the same label can be pulled closer, while instances with different labels will be pushed away from each other. Therefore, in the label contrastive coding, we consider two instances with the same label as a similar pair while treating the pair consisting of different label instances as dissimilar.

Label Contrastive Loss. Still from a dictionary look-up perspective, given an labeled encoded query $(\mathbf{q}, y)$, and a dictionary of m encoded labeled keys $\{(\mathbf{k}_1, y_1), (\mathbf{k}_2, y_2), \ldots, (\mathbf{k}_m, y_m)\}$, the positive keys $\mathbf{k}_+$ in label contrastive coding are the keys $\mathbf{k}_i$ where $y_i = y$. The label contrastive coding looks up the positive keys $\mathbf{k}_i$ that the query $\mathbf{q}$ matches in the dictionary. For the encoded query $(\mathbf{q}, y)$, its label contrastive loss $\mathcal{L}_{LC}$ is calculated by

$$\mathcal{L}_{LC}(\mathbf{q}, y) = -\log \frac{\sum_{i=1}^{m} \mathbb{1}_{y_i=y} \cdot \exp(\mathbf{q} \cdot \mathbf{k}_i / \tau)}{\sum_{i=1}^{m} \exp(\mathbf{q} \cdot \mathbf{k}_i / \tau)} \tag{2}$$

Here, $\mathbb{1}_{statement} \in \{0, 1\}$ is a binary indicator that returns 1 if the statement is true. We illustrate the label contrastive loss in Fig. 2 for reference. In LCGNN, key graph representations are stored in a dynamic memory bank. For the sake of brevity, we have not shown in Fig. 2. We introduce the dynamic memory bank and the updating process next.

The Dynamic Memory Bank. In label contrastive coding, the m-size dictionary is necessary. We use a dynamic memory bank to work as a dictionary. In order to fully utilize label information, the size of the memory bank is equal to the size of the set of labeled graphs $\mathbb{G}_L$, i.e., $m = |\mathbb{G}_L|$. The memory bank contains both the encoded low-dimensional key graph representations along with the corresponding labels, i.e., $\{(\mathbf{k}_1, y_1), (\mathbf{k}_2, y_2), \ldots, (\mathbf{k}_{|\mathbb{G}_L|}, y_{|\mathbb{G}_L|})\}$. Based on the conclusions in MoCo [7], the key graph representations should be kept as consistent as possible when the graph encoder f_k encoder evolves during training. Therefore, in each training epoch, newly encoded key graphs will dynamically replace the old version in the memory bank.

3.4 Graph Encoder Design

For given graphs $\mathcal{G}^q$ and $\mathcal{G}^k$, LCGNN empolys two graph encoders f_q and f_k to encode them to low-dimensional representations.

$$\begin{aligned} \mathbf{q} &= f_q(\mathcal{G}^q) \\ \mathbf{k} &= f_k(\mathcal{G}^k) \end{aligned} \tag{3}$$

In LCGNN, f_q and f_k have the same structure. Graph neural network has proven its powerful ability to encode graph structure data [27]. Many potential graph neural networks can work as the graph encoder in LCGNN.

Two kinds of encoders are considered in LCGNN. The first is Graph Isomorphism Network (GIN) [29]. GIN uses multi-layer perceptrons (MLPs) to conceive aggregation scheme and updates node representations as:

$$h_v^k = \mathrm{MLP}^{(k)}\left((1 + \epsilon^{(k)}) + \sum_{u \in \mathcal{N}(v)} h_u^{(k-1)}\right) \tag{4}$$

where ϵ is a learnable parameter or a fixed scalar, and k represents k-th layer. Given embeddings of individual nodes, the readout function is proposed by GIN to produce the representation g of the entire graph $\mathcal{G}$ for graph classification tasks:

$$g = \overset{K}{\underset{k=1}{\|}} \left(\text{SUM}(\{h_v^k | v \in \mathcal{G}\}) \right) \tag{5}$$

Here, $\|$ is the concatenation operator.

The second encoder we consider is Hierarchical Graph Pooling with Structure Learning (HGP-SL) [33]. HGP-SL incorporates graph pooling and structure learning into a unified module to generate hierarchical representations of graphs. HGP-SL proposes a graph pooling operation to identify a subset of informative nodes to form a new but smaller graph. The details about the Manhattan distance-based pooling operation can be referenced to [33]. For graph $\mathcal{G}$, HGP-SL repeats the graph convolution and pooling operations for several times and achieves multiple subgraphs in different layers: $\mathbf{H}^1, \mathbf{H}^2, \ldots, \mathbf{H}^K$. HGP-SL uses the concatenation of mean-pooling and max-pooling to aggregate all the node representations in the subgraph as follows:

$$\mathbf{r}^k = \mathcal{R}(\mathbf{H}^k) = \sigma \left(\frac{1}{n^k} \sum_{p=1}^{n^k} \mathbf{H}^k(p,:) \| \max_{q=1}^{d} \mathbf{H}^k(:,1) \right) \tag{6}$$

where σ is a nonlinear activation function. n^k is the node number in the k-th layer subgraph. In order to achieve the final representation g of the entire graph $\mathcal{G}$, another readout function is utilized to combine subgraphs in different layers.

$$g = \text{SUM}(\mathbf{r}^k | k = 1, 2, \ldots, K) \tag{7}$$

In the experiment section, we will show the performance along with the analysis of using GIN and HGP-SL as graph encoders in LCGNN.

3.5   LCGNN Learning

The training process illustration is provided in Fig. 1. During the training process, the input of LCGNN is a batch of labeled graphs $\mathbb{G}_b \subset \mathbb{G}_L$. For each mini-batch iteration, the set of key graphs and the set of query graphs are the same as $\mathbb{G}_b$. The graph encoder f_q and f_k will be initialized with the same parameters ($\theta_q = \theta_k$). The memory bank's size is equal to the size of the set of labeled graphs $\mathbb{G}_L$. The labeled graph $\mathcal{G}^i$ with the label y_i is assigned with a random representation to initialize the memory bank. The set of key graphs will be encoded by f_k to low-dimensional key graph representations $\mathbb{K}$, which will replace the corresponding representations in the memory bank. The set of query graphs are encoded by f_q to query graph representations $\mathbb{Q}$, whereas $\mathbb{Q}$ is also the input of the graph classifier. In LCGNN, a logistic regression layer serves as the graph classifier. Based on the output of the graph classifier, *Classification Loss* can be calculated by:

$$\mathcal{L}_{Cla} = -\frac{1}{|\mathbb{Q}|} \sum_{\mathbf{q}_i \in \mathbb{Q}} \sum_{j \in \mathbb{Y}} \mathbb{1}_{\mathbf{q}_i, j} \log(p_{\mathbf{q}_i, j}) \tag{8}$$

where $\mathbb{1}$ is a binary indicator (0 or 1) that indicates whether label j is the correct classification for the encoded query graph $\mathbf{q}_i$. Besides, $p_{\mathbf{q}_i,j}$ is the predicted probability.

$\mathbb{Q}$ and the memory bank work together to implement the label contrastive coding described in previous parts. Based on the Eq. 2, *Label Contrastive Loss* of the mini-batch $\mathbb{G}_b$ is:

$$\mathcal{L}_{LC} = -\frac{1}{|\mathbb{Q}|} \sum_{\mathbf{q}_i \in \mathbb{Q}} \mathcal{L}_{LC}(\mathbf{q}_i, y_{\mathbf{q}_i}) \qquad (9)$$

In order to train the model by utilizing label information more effectively and comprehensively, we try to minimize the following mixed loss combining both the *Label Contrastive Loss* and the *Classification Loss*:

$$\mathcal{L}_{total} = \mathcal{L}_{Cla} + \beta\,\mathcal{L}_{LC} \qquad (10)$$

Here, the hyper-parameter β controls the relative weight between the label contrastive loss and the classification loss. The motivation behind $\mathcal{L}_{total}$ is that $\mathcal{L}_{LC}$ encourages instance-level intra-class compactness and inter-class separability while $\mathcal{L}_{Cla}$ ensures the discriminability of classes. The graph encoder f_q, and the graph classifier can be updated end-to-end by back-propagation according to the loss $\mathcal{L}_{total}$. The parameters θ_k of f_k follows a momentum-based update mechanism as MoCo [7] instead of the back-propagation way. Specifically, the momentum updating process is:

$$\theta_k \longleftarrow \alpha\theta_k + (1 - \alpha)\theta_q \qquad (11)$$

where $\alpha \in [0, 1)$ is the momentum weight to control the speed of f_k evolving. We use this momentum-based update mechanism not only to reduce the overhead of backpropagation but also to keep the key graph representations in the memory bank as consistent as possible despite the encoder's evolution.

After completing the model training, the learned graph encoder f_q along with the graph classifier can be used to perform graph classification tasks for the unlabeled graphs $\mathbb{G}_U$.

4   Experiments

4.1   Experiment Settings

Datasets. We test our algorithms on 8 widely used datasets. Three of them are social networks benchmark datasets: IMDB-B, IMDB-M, and COLLAB; the rest five datasets: MUTAG, PROTEINS, PTC, NCI1, and D&D, belong to biological graphs datasets [28–30]. Each dataset contains multiple graphs, and each graph is assigned with a label. The statistics of these datasets are summarized in Table 1. What should be mentioned is that the biological graphs have categorical node attributes, while social graphs do not come with node attributes. In this paper, for the encoders requiring node attributes as input, we follow [29] to use one-hot encodings of node degrees as the node attributes on datasets without node features.

Table 1. Datasets in the experiments

Datasets	# graphs	Avg # nodes	Avg # edges	# classes
IMDB-B	1000	19.77	96.53	2
IMDB-M	1500	13.00	65.94	3
COLLAB	5000	74.49	2457.78	3
MUTAG	188	17.93	19.79	2
PROTEINS	1113	39.06	72.82	2
PTC	344	25.56	25.56	2
NCI1	4110	29.87	32.30	2
D&D	1178	284.32	715.66	2

Methods Compared. We select 3 categories of models as comparison methods:

- Kernel-based method: Weisfeiler-Lehman subtree kernel (**WL**) [21], **AWE** [10], and Deep Graph Kernel (**DGK**) [30]: They first decompose graphs into sub-components based on the kernel definition, then learn graph embeddings in a feature-based manner. For graph classification tasks, a machine learning model (i.e., SVM) will be used to perform the classification with learned graph embeddings.
- Graph embedding-based methods: **Sub2vec** [1], **Graph2vec** [16]: They extend document embedding neural networks to learn representations of entire graphs. A machine learning model (i.e., SVM) work on the classification tasks with learned graph representations.
- Graph neural network methods: **GraphSAGE** [6], **GCN** [12], **DCNN** [2]: They are designed to learn meaningful node level representations. A readout function is empolyed to summarize the node representations to the graph representation for graph-level classification tasks; **DGCNN** [32], **Caps-GNN** [28], **HGP-SL** [33], **GIN** [29], **InfoGraph** [23]: They are GNN-based algorithms with the pooling operator for graph representation learning. Then a classification layer will work as the last layer to implement graph classification; **GCC** [18]: It follows pre-training and fine-tuning paradigm for graph representation learning. A linear classifier is used to support the fine-tuning targeing graph classification; **LCGNN$_{GIN}$**, **LCGNN$_{HGP-SL}$**: They are two variants of the proposed LCGNN. LCGNN$_{GIN}$ uses GIN [29] as the graph encoders, and LCGNN$_{HGP-SL}$ sets the graph encoders as HGP-SL [33].

Experiment Configurations. We adopt two graph model structures: GIN [29] and HGP-SL [33] as the graph encoders. For LCGNN with different encoders, we follow the model configurations from the initial papers as the default settings. For the LCGNN structure, we choose the hidden representation dimension as 64 and 128 for two respective encoders; the contrastive loss weight

$\beta \in \{0.1, 0.6, \ldots, 1.0\}$; the momentum term $\alpha \in [0.0, 1.0)$; the temperature $\tau = 0.07$. For the graph classification tasks, to evaluate the proposed LCGNN we adopt the procedure in [28,29] to perform 10-fold cross-validation on the aforementioned datasets. For the training process of LCGNN, we select the Adam [11] as the optimizer, and tune the hyperparameters for each dataset as: (1) the batch size $\in \{32, 128, 512\}$; (2) the learning rate $\in \{0.01, 0.001\}$; (3) the dropout rate $\in \{0.0, 0.5\}$; (4) number of training epochs 1000 and select the epoch as the same with [29]. We run the experiments on the Server with 3 GTX-1080 ti GPUs, and all codes are implemented in Python3. Code and supplementary materials are available at: $LCGNN^1$.

4.2 Experimental Results and Analysis

Overall Evaluation. We present the main experimental results in Table 2. For the graph datasets that comparison methods do not have the results in the original papers, we denote it as "−". From the table, we can observe that LCGNN outperforms all comparison methods on 7 out of the total 8 datasets. The improvement is especially evident on datasets such as IMBD-B and D&D, which can be up to about 1.0%. At the same time, we can find that LCGNN using different graph encoders have achieved SOTA performance on different datasets ($LCGNN_{GIN}$ in 3 datasets; $LCGNN_{HGP-SL}$ in 4 datasets). The results also show that for different datasets, the selection of graph encoders has a critical impact on performance. Nonetheless, LCGNN generally outperforms all other baselines methods.

We also note that, compared to the baseline methods GIN and HGP-SL, $LCGNN_{GIN}$ and $LCGNN_{HGP-SL}$ can acquire better results when adopting them as corresponding encoders. For the COLLAB dataset results, LCGNN actually achieves much higher performance compared with the result we get when running GIN source code (71.7 ± 3.5). However, the result reported by the original paper [29] is 80.1 ± 1.9, which we also report in Table 2. To further evaluate the advantages of LCGNN and highlight the effectiveness of *Label Contrastive Loss*, we compare the classification loss during the training processes and show the curves of GIN and LCGNN in Fig. 3. From the figure, we can see that not only LCGNN has a faster convergence rate, but also can finally converge to lower classification loss. The classification loss comparison results on other datasets are also consistent, but we did not show them all due to space limitation. Thus we can conclude that with the support of label contrastive coding, LCGNN has better potential on graph classification tasks.

Besides, through the comparison between GCC and LCGNN, we can find that for the task of graph classification, The proposed label contrastive coding shows more advantages than the contrastive coding in GCC. We believe that the contrastive coding in GCC mainly focuses on learning universal representations. The label contrastive coding in LCGNN has a stronger orientation for

---

[1] https://github.com/YuxiangRen/Label-Contrastive-Coding-based-Graph-Neural-Network-for-Graph-Classification-.

Table 2. Test sets classification accuracy on all datasets. We use **bold** to denote the best result on each dataset.

Categories	Methods	IMDB-B	IMDB-M	COLLAB	MUTAG	PROTEINS	PTC	NCI1	D&D
Kernels	WL	73.4 ± 4.6	49.3 ± 4.8	79.0 ± 1.8	82.1 ± 0.4	76.2 ± 4.0	–	76.7 ± 2.0	76.4 ± 2.4
	AWE	74.5 ± 5.9	51.5 ± 3.6	73.9 ± 1.9	87.9 ± 9.8	–	–	–	71.5 ± 4.0
	DGK	67.0 ± 0.6	44.6 ± 0.5	73.1 ± 0.3	87.4 ± 2.7	75.7 ± 0.5	60.1 ± 2.5	80.3 ± 0.5	73.5 ± 1.0
Graph Embedding	Graph2vec	71.1 ± 0.5	50.4 ± 0.9	–	83.2 ± 9.3	73.3 ± 1.8	60.2 ± 6.9	73.2 ± 1.8	–
	Sub2vec	55.2 ± 1.5	36.7 ± 0.8	–	61.0 ± 15.8	–	60.0 ± 6.4	–	–
GNNs	DCNN	72.4 ± 3.6	49.9 ± 5.0	79.7 ± 1.7	79.8 ± 13.9	65.9 ± 2.7	–	74.7 ± 1.3	–
	GCN	73.3 ± 5.3	51.2 ± 5.1	**80.1 ± 1.9**	87.2 ± 5.1	75.2 ± 3.6	–	76.3 ± 1.8	73.3 ± 4.5
	GraphSAGE	72.4 ± 3.6	49.9 ± 5.0	79.7 ± 1.7	79.8 ± 13.9	65.9 ± 2.7	–	74.7 ± 1.3	–
	DGCNN	70.0 ± 0.9	47.8 ± 0.9	73.8 ± 0.5	85.8 ± 1.7	75.5 ± 0.9	58.6 ± 2.5	74.4 ± 0.5	79.4 ± 0.9
	CapsGNN	73.1 ± 4.8	50.3 ± 2.7	79.6 ± 0.9	86.7 ± 6.9	76.3 ± 3.6	–	78.4 ± 1.6	75.4 ± 4.2
	HGP-SL	–	–	–	82.2 ± 0.6	84.9 ± 1.6	–	78.5 ± 0.8	81.0 ± 1.3
	GIN	75.1 ± 5.1	52.3 ± 2.8	80.2 ± 1.9	89.4 ± 5.6	76.2 ± 2.8	64.6 ± 7.0	82.7 ± 1.7	–
	InfoGraph	73.0 ± 0.9	49.7 ± 0.5	–	89.0 ± 1.1	–	61.7 ± 1.4	–	–
	GCC	73.8	50.3	81.1	–	–	–	–	–
Proposed	LCGNN$_{GIN}$	**76.1 ± 6.9**	**52.4 ± 6.7**	72.3 ± 6.3	89.9 ± 4.8	76.9 ± 6.8	64.7 ± 2.0	**82.9 ± 3.6**	77.4 ± 1.2
	LCGNN$_{HGP-SL}$	75.4 ± 1.5	46.5 ± 1.3	77.5 ± 1.2	**90.5 ± 2.3**	**85.2 ± 2.4**	**65.9 ± 2.8**	78.8 ± 4.4	**81.8 ± 3.6**

representation learning, that is, extracting features that significantly affect the intra-class compactness and inter-class separability.

Table 3. Experiments with less labeled training data

Datasets	Methods	Training ratio				
		60%	70%	80%	90%	100%
IMDB-B	GIN	61.8	65.4	69.2	70.5	75.1
	LCGNN$_{GIN}$	**66.3**	**70.8**	**71.3**	**72.2**	**76.1**
IMDB-M	GIN	40.5	41.4	41.8	46.0	52.3
	LCGNN$_{GIN}$	**43.4**	**42.8**	**43.6**	**48.1**	**52.4**

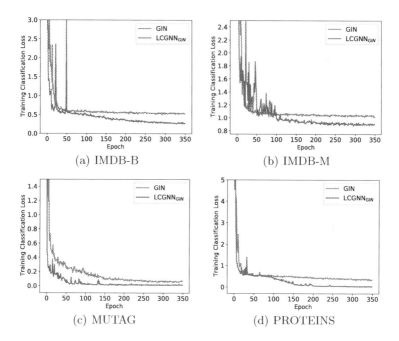

(a) IMDB-B   (b) IMDB-M

(c) MUTAG   (d) PROTEINS

Fig. 3. Training *Classification Loss* versus training epoch

Performance with Less Labeled Data. To validate our claim that LCGNN can utilize label information more comprehensively and use fewer label data to achieve comparative performance, we conduct experiments with less training data. For each fold of cross-validation, we extract only part of the training set (e.g., 60% of the data in the training set) as the training data and maintain the test set as the same. We present the results in Table 3. In Table 3, the training ratio denotes how much data in the training set is extracted as the training data. When the training ratio is 100%, it means using the full training set in each fold. From the results, it is obvious that LCGNN$_{GIN}$ can always outperform the baseline GIN when using less training data. What's more, in many cases when LCGNN$_{GIN}$ with less training data (e.g., 70% training data for LCGNN$_{GIN}$ while 80% for GIN; 60% for LCGNN$_{GIN}$ while 70% for GIN), LCGNN$_{GIN}$ still obtains more competitive results than GIN. The experimental results demonstrate that LCGNN can utilize the same amount of training data more comprehensively and efficiently. The capability also makes LCGNN possible to learn with less training data to obtain a better performance than comparison methods when they need more training data.

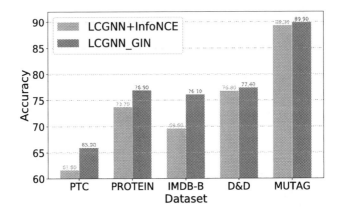

Fig. 4. The effectiveness of *Label Contrastive Loss*

The Effectiveness of the Label Contrastive Coding. In order to further verify the effectiveness of the proposed label contrastive coding on the task of graph classification, we conduct comparison experiments between LCGNN$_{GIN}$ and LCGNN+InfoNCE. Here, LCGNN+InfoNCE replaces the label contrastive loss in LCGNN$_{GIN}$ with InfoNCE loss [17] but keeps other parts the same. We present the results in Fig. 4. The experimental results show that the performance of LCGNN$_{GIN}$ on all data sets exceeds LCGNN+InfoNCE, which also demonstrates that the label contrastive coding can effectively utilize label information to improve model performance. In addition, we observe that the performance of LCGNN+InfoNCE is even worse than GIN. It verifies that the inherent large intra-class variations may import noise to graph classification tasks if we treat the intra-class instances as distinct classes like the existing comparative learning.

Hyper-parameter β Analysis. We consider the influence of label contrastive loss weight term β and conduct experiments with different values. The results is exhibited in Table 4. We select β from $\{0.1, 0.2, \ldots, 1.0\}$, and find the trend of using a relatively larger β inducing better results. Thus in the experiment, we

Table 4. LCGNN$_{GIN}$ with different contrastive loss weight β

Datasets	Contrastive loss weight β							
	0.3	0.4	0.5	0.6	0.7	0.8	0.9	1.0
IMDB-B	73.8	75.1	**76.1**	75.5	76.0	75.4	75.7	75.7
IMDB-M	50.5	51.2	**52.4**	51.9	51.7	51.5	51.5	51.6

empirically select from $\beta \in \{0.5, 0.6, \ldots, 1.0\}$ to achieve the best performance. Nevertheless, we also observed that when β gradually increases, the performance does not continue to increase. Our analysis is that when the label contrastive loss weight is too high, the learning of the model places too much emphasis on instance-level contrast. More fine-grained discriminative features on the instance-level will reduce the generalization performance of the model on the test set.

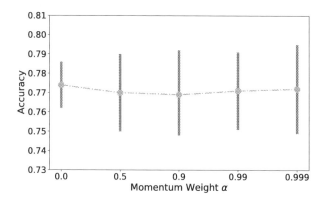

Fig. 5. LCGNN with different momentum weight

Momentum Ablation. The momentum term plays an important role in contrastive learning problems. In our experiments, we also try different momentum weight α when running LCGNN$_{GIN}$ on D&D and show the results in Fig. 5. Unlike [7], LCGNN$_{GIN}$ also achieves good performance when $\alpha = 0$. The main reason should be that the D&D is not extremely large, which makes it easy for representations to ensure consistency during encoder evolving. Furthermore, in this set of experiments, the momentum term did not show much impact on Accuracy, that is, the model performance is relatively stable, which should be caused by the moderate-sized dataset as well.

5   Conclusion

In this paper, we have introduced a novel label contrastive coding based graph neural network, LCGNN, which works on graph classification tasks. We extend the existing contrastive learning to the supervised setting and define the label contrastive coding. The label contrastive coding treats instances with the same label as multiple positive instances, which is different from the single positive instance in unsupervised contrastive learning. The label contrastive coding can pull the same label instances closer and push the instances with different labels away from each other. We demonstrate the effectiveness of LCGNN on graph classification tasks over eight benchmark graph datasets. The experimental results show that LCGNN achieves SOTA performance in 7 datasets. Besides, LCGNN can take advantage of label information more comprehensively. LCGNN outperforms the baseline method when using less training data, which verifies this advantage.

Acknowledgement. This work is also partially supported by NSF through grant IIS-1763365.

References

1. Adhikari, B., Zhang, Y., Ramakrishnan, N., Prakash, B.A.: Sub2Vec: feature learning for subgraphs. In: Phung, D., Tseng, V.S., Webb, G.I., Ho, B., Ganji, M., Rashidi, L. (eds.) PAKDD 2018. LNCS (LNAI), vol. 10938, pp. 170–182. Springer, Cham (2018). https://doi.org/10.1007/978-3-319-93037-4_14
2. Atwood, J., Towsley, D.: Diffusion-convolutional neural networks. In: Advances in Neural Information Processing Systems, pp. 1993–2001 (2016)
3. Chen, T., Kornblith, S., Norouzi, M., Hinton, G.: A simple framework for contrastive learning of visual representations. arXiv preprint arXiv:2002.05709 (2020)
4. Elsayed, G., Krishnan, D., Mobahi, H., Regan, K., Bengio, S.: Large margin deep networks for classification. In: Advances in Neural Information Processing Systems, pp. 842–852 (2018)
5. Fout, A., Byrd, J., Shariat, B., Ben-Hur, A.: Protein interface prediction using graph convolutional networks. In: Advances in Neural Information Processing Systems, pp. 6530–6539 (2017)
6. Hamilton, W., Ying, Z., Leskovec, J.: Inductive representation learning on large graphs. In: Advances in Neural Information Processing Systems, pp. 1024–1034 (2017)
7. He, K., Fan, H., Wu, Y., Xie, S., Girshick, R.: Momentum contrast for unsupervised visual representation learning. arXiv-1911 (2019)
8. Hjelm, R.D., et al.: Learning deep representations by mutual information estimation and maximization. arXiv preprint arXiv:1808.06670 (2018)
9. Hu, W., et al.: Strategies for pre-training graph neural networks. arXiv preprint arXiv:1905.12265 (2019)
10. Ivanov, S., Burnaev, E.: Anonymous walk embeddings. In: International Conference on Machine Learning, pp. 2186–2195 (2018)
11. Kingma, D.P., Ba, J.L.: Adam: a method for stochastic optimization. In: International Conference on Learning Representaion (2015)

12. Kipf, T.N., Welling, M.: Semi-supervised classification with graph convolutional networks. In: International Conference on Learning Representaion (2017)
13. Kondor, R., Pan, H.: The multiscale Laplacian graph kernel. In: Advances in Neural Information Processing Systems, pp. 2990–2998 (2016)
14. Liu, W., Wen, Y., Yu, Z., Yang, M.: Large-margin Softmax loss for convolutional neural networks. In: ICML, vol. 2, p. 7 (2016)
15. Meng, L., Ren, Y., Zhang, J., Ye, F., Philip, S.Y.: Deep heterogeneous social network alignment. In: 2019 IEEE First International Conference on Cognitive Machine Intelligence (CogMI), pp. 43–52. IEEE (2019)
16. Narayanan, A., Chandramohan, M., Venkatesan, R., Chen, L., Liu, Y., Jaiswal, S.: graph2vec: learning distributed representations of graphs. arXiv preprint arXiv:1707.05005 (2017)
17. van den Oord, A, Li, Y., Vinyals, O.: Representation learning with contrastive predictive coding. arXiv preprint arXiv:1807.03748 (2018)
18. Qiu, J., et al..: Gcc: Graph contrastive coding for graph neural network pre-training. In: Proceedings of the 26th ACM SIGKDD International Conference on Knowledge Discovery & Data Mining, pp. 1150–1160 (2020)
19. Ren, Y., Liu, B., Huang, C., Dai, P., Bo, L., Zhang, J.: Heterogeneous deep graph infomax. arXiv preprint arXiv:1911.08538 (2019)
20. Ren, Y., Zhu, H., Zhang, J., Dai, P., Bo, L.: Ensemfdet: An ensemble approach to fraud detection based on bipartite graph. arXiv preprint arXiv:1912.11113 (2019)
21. Shervashidze, N., Schweitzer, P., Van Leeuwen, E.J., Mehlhorn, K., Borgwardt, K.M.: Weisfeiler-Lehman graph kernels. J. Mach. Learn. Res. 12(9), 2539–2561 (2011)
22. Shervashidze, N., Vishwanathan, S., Petri, T., Mehlhorn, K., Borgwardt, K.: Efficient graphlet kernels for large graph comparison. In: Artificial Intelligence and Statistics, pp. 488–495 (2009)
23. Sun, F.Y., Hoffmann, J., Verma, V., Tang, J.: InfoGraph: unsupervised and semi-supervised graph-level representation learning via mutual information maximization. arXiv preprint arXiv:1908.01000 (2019)
24. Veličković, P., Fedus, W., Hamilton, W.L., Liò, P., Bengio, Y., Hjelm, R.D.: Deep graph infomax. arXiv preprint arXiv:1809.10341 (2018)
25. Vishwanathan, S., Schraudolph, N.N., Kondor, R., Borgwardt, K.M.: Graph kernels. J. Mach. Learn. Res. 11, 1201–1242 (2010)
26. Wu, Z., Xiong, Y., Yu, S., Lin, D.: Unsupervised feature learning via non-parametric instance-level discrimination. arXiv preprint arXiv:1805.01978 (2018)
27. Wu, Z., Pan, S., Chen, F., Long, G., Zhang, C., Philip, S.Y.: A comprehensive survey on graph neural networks. IEEE Trans. Neural Netw. Learn. Syst. 32(1), 4–24 (2020)
28. Xinyi, Z., Chen, L.: Capsule graph neural network. In: International Conference on Learning Representations (2018)
29. Xu, K., Hu, W., Leskovec, J., Jegelka, S.: How powerful are graph neural networks? arXiv preprint arXiv:1810.00826 (2018)
30. Yanardag, P., Vishwanathan, S.: Deep graph kernels. In: Proceedings of the 21th ACM SIGKDD International Conference on Knowledge Discovery and Data Mining, pp. 1365–1374 (2015)
31. Ying, Z., You, J., Morris, C., Ren, X., Hamilton, W., Leskovec, J.: Hierarchical graph representation learning with differentiable pooling. In: Advances in Neural Information Processing Systems, pp. 4800–4810 (2018)

32. Zhang, M., Cui, Z., Neumann, M., Chen, Y.: An end-to-end deep learning architecture for graph classification. In: Thirty-Second AAAI Conference on Artificial Intelligence (2018)
33. Zhang, Z., et al.: Hierarchical graph pooling with structure learning. arXiv preprint arXiv:1911.05954 (2019)
34. Zhu, Y., Xu, Y., Yu, F., Liu, Q., Wu, S., Wang, L.: Deep graph contrastive representation learning. arXiv preprint arXiv:2006.04131 (2020)
35. Zhuang, C., Zhai, A.L., Yamins, D.: Local aggregation for unsupervised learning of visual embeddings. In: Proceedings of the IEEE International Conference on Computer Vision, pp. 6002–6012 (2019)

Which Node Pair and What Status? Asking Expert for Better Network Embedding

Longcan Wu[1], Daling Wang[1(✉)], Shi Feng[1], Kaisong Song[2], Yifei Zhang[1], and Ge Yu[1]

[1] Northeastern University, Shenyang, China
{wangdaling,fengshi,zhangyifei,yuge}@cse.neu.edu.cn
[2] Alibaba Group, Hangzhou, China
kaisong.sks@alibaba-inc.com

Abstract. In network data, the connection between a small number of node pair are observed, but for most remaining situations, the link status (i.e., connected or disconnected) of node pair can not be observed. If we can get more useful information hidden in node pairs with unknown link status, it will help improve the performance of network embedding. Therefore, how to model the network with unknown link status actively and effectively remains an area for exploration. In this paper, we formulate a new network embedding problem, which is how to select valuable node pair (**which node pair**) to ask expert about their link status (**what status**) information for improving network embedding. To tackle this problem, we propose a novel active learning method called **ALNE**, which includes a proposed network embedding model **AGCN**, three active node pair selection strategies and an information evaluation module. In this way, we can obtain the real valuable link statuses information between node pairs and generate better node embeddings. Extensive experiments are conducted to show the effectiveness of ALNE.

Keywords: Network embedding · Active learning · Graph convolutional network

1 Introduction

Recently, much effort in the literature has been invested in network embedding (NE) methods, which aims to assign nodes in a network to low-dimensional representations and preserve the network structure [9]. Despite the effectiveness of the existing NE models, they tend to rely on rich network information, including attribute or label of node, and network structure [16].

In terms of network structure, real-world networks are usually partially observed, i.e., the connection between a small number of node pair are observed and there are a large number of node pairs with unknown link status (i.e., connected or disconnected), which leads to the lack of crucial information about

C. S. Jensen et al. (Eds.): DASFAA 2021, LNCS 12681, pp. 141–157, 2021.
https://doi.org/10.1007/978-3-030-73194-6_11

network and brings in difficulties for NE and corresponding downstream tasks. Take online social networks as an example. We can observe there is a close connection between user pair by comments interaction or thumb up. For most user pairs, we can not observe these interactions between them, and we need some experts to judge whether there are close connections between them in other ways.

In order to achieve more information about the original network, active learning (AL) based methods have been adopted [2]. Currently, most existing studies focus on how to obtain label information for specific nodes in the network through AL [3,10]. Only few literature has proposed to apply AL in obtaining link status information in network [6,27]. The aforementioned researches design AL strategies only for specific downstream tasks such as node classification or link prediction. Limited literature has been reported for improving the performance of network embedding through AL, which eventually makes better performance on common downstream tasks. Moreover, in some case, it is more important and easy to get the link status information between node pair than to get the label of a specific node.

In this paper, we formulate a new NE problem, namely how to select the valuable node pairs (**which node pair**) to ask expert about their link statuses (**what status**) information for improving the performance of NE. Several new challenges emerge when addressing this problem. (i) This problem involves two part: NE and AL. How to make AL query and NE interact and reinforce with each other for better NE? (ii) Based on NE model and characteristics of network data, how to design effective AL query strategies for selecting valuable node pairs? (iii) According to the result of link status given by experts, how to judge whether the link status information has positive effect on the NE model?

To tackle these challenges, we propose a method called **ALNE** in the active learning framework, i.e. **A**sking **L**ink status information for better **N**etwork **E**mbedding, as shown in Fig. 1. Firstly, in order to use the latest link status information obtained from AL in the next epoch of network embedding process, we propose a network embedding model **AGCN**, **A**ctive **G**raph **C**onvolutional **N**eural, inspired by [15]. Secondly, we design three effective AL query strategies, namely gradient-based, representativeness-based and uncertainty-based strategies. With the three AL query strategies evaluating node pairs from different perspectives, we can rate the node pairs with unknown link statuses, and further choose a batch of node pairs to ask the expert about their link statuses. Thirdly, we need to make a secondary selection of node pair in the batch according to their effectiveness for network embedding model AGCN. Thus we design an evaluation module to evaluate the information about node pairs. After each embedding iteration, the AL query and information evaluation are conducted for obtaining more and more useful node pairs, which make AGCN yield better network representation. Meanwhile, in AL query process, we can find more valuable node pairs with the help of gradient-based AL criterion generated based on AGCN model.

According to the above description, in ALNE method, the network embedding and active learning query can interact and reinforce with each other.

Therefore, the proposed model ALNE is a good solution to the three challenges raised above. To sum up, the main contributions of this paper are as follows:

- We formulate a new network embedding problem, which is how to select the valuable node pairs to ask their link statuses information for improving the performance of network embedding.
- We propose a novel method ALNE in active learning framework, in which the proposed network embedding model AGCN and AL query strategies can interact and reinforce with each other to generate better node embedding results.
- We propose three AL query strategies from different perspectives and design an information evaluation module to elicit the real valuable node pairs for the network embedding model AGCN.
- We have carried out detailed experiments on five public available datasets and two classic downstream tasks: node classification and link prediction. Experimental results show the validity of ALNE we proposed.

2   Related Work

Recently, a series of network embedding methods have been proposed because of its effectiveness in network data analysis [9, 20, 24]. In these methods, the graph convolutional network (GCN) models have shown obvious advantages in modeling the network data [15]. The GCN model propagated information from each node in the network to other nodes through layer-wise propagation rule, thus allowing node information to spread over the network. Based on the GCN model, the researchers have put forward GAT [21] and DAGCN [4]. A detailed review of the GCN can be found in the papers [22].

In the actual application scenarios, there is very little data with labels. In order to take advantage of a large number of unlabeled data, semi-supervised learning and active learning are used [1]. As a human-computer interaction framework, active learning has attracted great attentions from academic researchers. At present, the applications of AL in network mainly focus on network node classification [2, 13, 25]. In addition, some literature has applied AL for link related task in the network data [8]. Zhu et al. [27] proposed a novel measure to apply AL in anchor link prediction. Chen et al. [6] modeled the link prediction problem as a classification problem, and the most uncertain node pairs were selected to give the network a high information gain.

Based on the success of GCN related models, it is necessary to explore how to integrate the network embedding models into the active learning framework. Based on GCN model, Cai et al. [3] obtained the label of informative node according to AL to enhance the performance of node classification. Gao et al. [10] used a multi-armed bandit mechanism to further fuse AL query strategies. Similarly, Chen et al. [7] adapted the same idea to heterogeneous network embedding. Xiao et al. [14] proposed to transform experts' cognition into a weight link to advance network embedding.

Different from above models, we aim to improve the performance of network embedding by obtaining the link status information of disconnected node pairs in the network through AL queries, and thus improve the performance of downstream tasks with better embedding results.

3   Problem Definition

Given an undirected network $G = (V, E, Y)$, V is the set of nodes with size of N, and E denotes the set of all node pairs with size of $N(N-1)/2$. The elements in E can be divided into three groups: E_p, E_n, E_u. In E_p (E_n), the node pairs have (do not have) an edge. In E_u, the link status between node pair is unknown. Based on E, we can obtain positive adjacency matrix A_p and negative adjacency matrix A_n, where $A_p, A_n \in \mathbb{R}^{(N \times N)}$. If node pair (i, j) belongs to E_p (or E_n), $A_{p_{ij}} = 1$ and $A_{n_{ij}} = 0$ (or $A_{p_{ij}} = 0$ and $A_{n_{ij}} = 1$); if node pair (i, j) belongs to E_u, $A_{p_{ij}} = 0$ and $A_{n_{ij}} = 0$. $Y \in \mathbb{R}^{(L \times C)}$ denotes label matrix, where L is the number of nodes with label, and C is the number of classes.

In real-world network data, we usually get a sparse matrix A_p and a zero matrix A_n, which means we only know the connected status between some node pairs. Moreover, A_p is so sparse that directly implementing network embedding on A_p can not achieve a satisfactory result. Besides, there is a lot of useful information hidden in E_u and we can obtain them through asking expert in the way of active learning. According to the information obtained from E_u, we can enrich A_p and A_n, which is useful to network embedding. Therefore, the problem we study can be formalized as below.

Problem 1 (**Selecting Node Pairs and Asking Link Status Information for Better Network Embedding**). Given the network G, how to design the AL query strategy to select a batch of node pairs from E_u and ask the expert about the node pair's link status information in each iteration; these node pairs are evaluated and then added to A_p or A_n for the next round of embedding, which ultimately improves the performance of the network embedding and common downstream tasks.

4   Our Solution: ALNE

In this section, we first give the overall framework of the ALNE model, then describe the three components of ALNE in detail.

4.1   ALNE Framework

The overview of proposed ALNE is shown in Fig. 1. In general, ALNE follows the active learning framework, which mainly consists of three parts: network embedding, AL query strategies and information evaluation. The ALNE model runs as follows:

(1) We feed network G into our proposed network embedding model AGCN, optimize the loss function in one epoch, and obtain the gradient matrix G_p and G_n.

(2) At each epoch of selecting node pairs, we first obtain candidate node pairs set E_u^{cand} based on E_u according to gradient matrix G_p.

(3) With the help of AGCN model and network structure, we design three different types of AL query strategies to select valuable node pairs.

(4) We use AL query strategies to rate node pair in E_u^{cand} and choose a batch of node pairs to ask the expert about their link statuses based on utility score.

(5) After link status acquisition, we further evaluate the information (gradient and link status) about node pairs to pick out what is really useful to the model.

(6) For the node pairs that do not meet the criteria, we store them in the information pool first.

(7) As the training going on, we re-evaluate the node pairs in the pool in subsequent iterations.

(8) Finally, the really useful node pairs can be added into network G.

(9) The above process repeats until the maximum number of query q exhausts or the AGCN model converges.

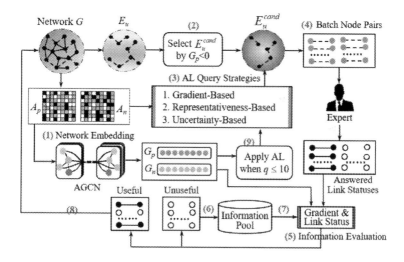

Fig. 1. Framework of proposed model ALNE.

4.2 Network Embedding: AGCN

Given $G = (V, E, Y)$, adjacency matrix A_p and A_n, we propose a network representation method based on GCN [15]. Firstly, we review the layer-wise propagation rule of GCN:

$$H^{(l+1)} = \sigma \left(\widetilde{D_p}^{-\frac{1}{2}} \widetilde{A_p} \widetilde{D_p}^{-\frac{1}{2}} H^{(l)} W_p^{(l)} \right) \tag{1}$$

According to above propagation rule, we can only model the network data on A_p. However, in the setting of AL, we not only have A_p, but also have negative adjacency matrix A_n. So, we propose a new model active graph convolutional network (dubbed as AGCN), which can simultaneously employ A_p and A_n to model the network data. The propagation rule of AGCN is as follow:

$$\begin{aligned} H_p^{(l+1)} &= \widetilde{D_p}^{-\frac{1}{2}} \widetilde{A_p} \widetilde{D_p}^{-\frac{1}{2}} H^{(l)} W_p^{(l)} \\ H_n^{(l+1)} &= \widetilde{D_n}^{-\frac{1}{2}} \widetilde{A_n} \widetilde{D_n}^{-\frac{1}{2}} H^{(l)} W_n^{(l)} \\ H^{(l+1)} &= \sigma \left(H_p^{(l+1)} + H_n^{(l+1)} \lambda_n \right) \end{aligned} \tag{2}$$

where $\widetilde{A_n} = A_n + I_N$ is similar to $\widetilde{A_p}$ in Eq. (1), $\widetilde{D}_{n_{ii}} = \sum_j \widetilde{A}_{n_{ij}}$ is similar to $\widetilde{D_p}^{-\frac{1}{2}}$, $W_p^{(l)}$ and $W_n^{(l)}$ are trainable weight matrix, $H^{(l)}$ is an activation matrix of l_{th} hidden layer, σ denotes the ReLU activation function. Considering the influence of A_n on $H^{(l+1)}$, we use a real number λ_n to denote its contribution on $H^{(l+1)}$.

In our method, we regard adjacent matrix A_p and A_n as two types of matrices, which represent two relationship between nodes. Although these two relationship are in opposition, there exists complex correlation between them. So, our idea is similar to some method using GCN in multi-relationship network [17].

After obtaining node representation matrix H from Eq. (2), we apply softmax function to each row of H to obtain labels of node. Finally, the cross-entropy loss is used for multi-class classification over all the labeled nodes. The above process can be expressed in the following:

$$Z = softmax(H) \tag{3}$$

$$J_{sup} = -\sum_l \sum_{c=1}^{C} Y_{lc} Z_{lc} \tag{4}$$

In order to save the structural proximities between nodes in latent space, we make use of graph structure to further constrain node embedding. Because AUC depicts the relationship between node pairs [11], AUC is considered as a part of the loss function. In order to optimize AUC, we need to select the node pairs with links as positive data, and node pairs without links as negative data. If we only consider positive and negative data in the network, lots of node pairs with unknown link status are not fully utilized. Previous studies have shown that PU-AUC risk R_{PU} is equivalent to supervised AUC risk R_{PN} with a linear transformation [23]. Based on above conclusion, we define AUC-based loss function J_{auc} as the sum of R_{PU} and R_{PN}:

$$J_{auc} = R_{PU} + R_{PN} \tag{5}$$

$$R_{PU} = \sum_{(i,j,k) \in T} \max\left(0, \delta_{PU} + S\left(h_i, h_j\right) - S\left(h_i, h_k\right)\right) \tag{6}$$

$$R_{PN} = \sum_{\substack{(i,j) \in E_p \\ (m,n) \in E_n}} \max\left(0, \delta_{PN} + S\left(h_i, h_j\right) - S\left(h_m, h_n\right)\right) \tag{7}$$

where h_i is node i representation, $S(h_i, h_j)$ denotes the distance between node i and node j and we use L_2 norm to calculate the distance, $\max(0,)$ is the hinge function and δ_{PU} and δ_{PN} are threshold. There are many node pairs with unknown link status. To reduce the computational cost, we construct a set of triplets T by negative sampling when we calculate R_{PU}. For the every triplet (i,j,k) in T, node pair (i,j) has link and the link status of node pair (i,k) is unknown. For the loss function R_{PN}, we directly use positive data from E_p and negative data from E_n. The final loss function for optimization is the sum of the cross-entropy loss J_{sup} and AUC-based loss function J_{auc}:

$$J = J_{sup} + J_{auc} \tag{8}$$

4.3 AL Query Strategy

In this section, we first introduce our proposed three types of AL query criteria, and then explain how to combine different AL query criteria to obtain the final utility score of node pair.

Gradient-Based Query Strategy. We hope that the selected node pairs can influence model significantly. In other words, the link status information of selected node pairs can let the model's loss function decrease the most. As we all know, gradient information is important in optimizing model parameters, through which the loss of model can drop rapidly. Inspired by [5], we proposed a gradient-based query strategy.

Specifically, if we treat A_p and A_n of layer-wise propagation rule in Eq. (2) as a group of variables like $W_p^{(l)}$ and $W_n^{(l)}$, we can extract the gradient of A_p and A_n using loss function J. Take A_p as an example, we can obtain the gradient $g_{p(i,j)}$ of $A_{p(i,j)}$ as shown in Eq. (9). Considering we focus on the undirected network, we denote the gradient of node pair (i,j) as the average of $g_{p(i,j)}$ and $g_{p(j,i)}$ as shown in Eq. (10). Finally, we denote $S^g(i,j) = G_{p(i,j)}$ as gradient-based score for node pair (i,j) based on A_p.

$$g_{p(i,j)} = \frac{\partial J}{\partial A_{p(i,j)}} \tag{9}$$

$$G_{p(i,j)} = G_{p(j,i)} = \left(g_{p(i,j)} + g_{p(j,i)}\right)/2 \tag{10}$$

According to above equations, we can obtain the gradient matrix G_p and G_n. The element in G_p (or G_n) may be positive or negative, which means we need to decrease or increase the corresponding element in A_p (or A_n) for minimizing loss function J.

For any node pair $(i, j) \in E_u$, $G_{p(i,j)}$, $G_{n(i,j)}$ and link status have eight possible combinations as shown in Table 1. We leave out the subscript (i, j) to make the expressions more concise in Table 1 and the following sections where no confusion occurs. We take $G_p < 0$ and $G_n > 0$ as an example, and analyze the effect of link status of (i, j) on the loss function. When $G_p < 0$ and $G_n > 0$, we can improve the value of $A_{p(i,j)}$ or reduce the value of $A_{n(i,j)}$ to minimize the loss function according to the analysis in the previous paragraph. If there is link between node pair (i, j), i.e. nodes are connected, the value of $A_{p(i,j)}$ improves from 0 to 1 and $A_{n(i,j)}$ stays at 0, so the message of link between node pair (i, j) is beneficial to model and we use the symbol $\sqrt{}$ to represent advantage. If the node pair (i, j) is disconnected, the value of $A_{p(i,j)}$ stays at 0 and $A_{n(i,j)}$ improves from 0 to 1. Because of the $G_{n(i,j)} > 0$, the link status information of (i, j) is bad for model and we use the symbol $\times$ to represent disadvantage. We can also analyze other situations in the Table 1 in the same way.

From Table 1 we can see that when $G_p < 0$, the link status information is good for the model in most cases. So, in our proposed model ALNE, we first pick out node pair set from E_u to construct candidate set E_u^{cand} based on the condition of $G_p < 0$, and then choose a batch of node pairs from E_u^{cand} using AL query strategy as shown in step (2) in Fig. 1. It is also important to note that when selecting the node pair (i, j), we prefer ones with higher magnitude of negative gradient $|-S^g(i,j)|$, which means the link status have higher influence on the loss function.

Table 1. Combinations between gradient and link status.

Gradient	Link status	
	Connected	Disconnected
$G_p > 0, G_n > 0$	$\times$	$\times$
$G_p > 0, G_n < 0$	$\times$	$\sqrt{}$
$G_p < 0, G_n > 0$	$\sqrt{}$	$\times$
$G_p < 0, G_n < 0$	$\sqrt{}$	$\sqrt{}$

Representativeness-Based Strategy. If we only rely on the gradient-based query strategy proposed above, we might choose the noisy and unrepresentative node pairs, because the gradient-based query strategy only considers how to make the loss function descend the fastest, not considers how to make the loss function descend to a reasonable area. Therefore, if we can choose those representative node pairs, we may let the loss function down to a reasonable area in the right direction.

In the network, the importance of nodes can be measured by graph centrality [18]. In this paper, we utilize graph centrality of node to measure the representativeness of node pair, which is the sum of graph centrality of nodes. Considering that A_n has fewer nonzero elements, in order to better calculate representativeness of node pairs, we use A_p to calculate the centrality of node. Finally, we denote $S^r(i,j)$ as representativeness score of node pair (i,j).

$$S^r(i,j) = centrality(i) + centrality(j) \tag{11}$$

where $centrality(i)$ represents the graph centrality of node i and we use degree centrality here. Obviously, we prefer to elect node pairs with large representativeness score.

Uncertainty-Based Strategy. In AL query strategies, uncertainty-based strategy is the most common strategy. In order to select the most uncertain node pairs, we propose the uncertainty-based query strategy. In the link prediction task, researchers often use similarity based method [18]. Take the method Shortest Path as an example, if the length of shortest path of node pair $(i,j) \in E_u$ is small, that means there is a high probability of existing a link for (i,j); if there is no edge between (i,j), the node pair (i,j) is uncertain for the graph. Based on above description, we can use similarity of node pair as uncertainty score. In this paper, we define $S^u(i,j)$ as uncertainty score of node pair (i,j) using negative length of Shortest Path as shown in Eq. (12).

$$S^u(i,j) = -|ShortestPath_{i,j}| \tag{12}$$

Similar to representativeness-based query strategy, we use A_p to calculate the uncertainty score of node pair. Obviously, we prefer to elect the node pairs with higher uncertainty score.

Combination of Different AL Query Strategy. The scores obtained based on above AL strategies are not in the same order of magnitude, so we convert them into percentiles as in [3,26] and then implement weighted sum to get the final utility score. We use $P^{strategy}(i,j)$ to represent the percentile score of node pair (i,j) in terms of strategy $\in \{g,r,u\}$ in set E_u^{cand}. Finally, we take the weighted sum of the three types of percentiles score to obtain the utility score $U(i,j)$ of node pair (i,j) in the following equation, where a,b,c are hyperparameter and range in $[0,1]$.

$$U(i,j) = a * P^g(i,j) + b * P^r(i,j) + c * P^u(i,j). \tag{13}$$

4.4 Information Evaluating

When we get link statuses of a batch of node pairs, we need to decide whether to use the link status information. The reason is that the node pair with $G_p < 0$ is not necessarily good for the network embedding model AGCN as shown

44

444444

4 44

in Table 1. When a node pair satisfies the condition, i.e. $G_p < 0$, $G_n > 0$ and disconnected, it is unuseful to the model in this epoch. Therefore, for node pairs that meet the aforementioned criterion, we just store them in the information pool as shown in (6) in Fig. 1. For node pairs that do not meet above criterion, we directly add them into network G, i.e. updating A_p or A_n. In subsequent iterations, if a node pair satisfies $G_p < 0$ and $G_n < 0$, it is valuable for AGCN model, and we will remove it from information pool and use it in the next network embedding learning iteration.

5 Experiments

In this section, we conduct experiments on five real-world datasets to evaluate the proposed ALNE model with respect to two common downstream tasks. We first compare ALNE with other active learning baselines. Then we empirically analyze the effect of proposed three AL query strategies. Finally, we also study how various hyper-parameters, e.g., the weight of A_n (λ_n in Eq.(2)), affect the performance of ALNE.

5.1 Experimental Settings

Datasets. We conduct experiments on five common used real-world datasets: Caltech, Reed, Flickr, Cora and Citeseer. We summarize the statistics of five processed datasets in Table 2. As illustrated in the table, Caltech, Reed, Flickr contain more edges; Cora and Citeseer have more nodes. **Caltech** and **Reed** [12] are two university Facebook network. Node represents student, edge represents "friendship" on Facebook, and label means resident. In **Flickr** [19] dataset, we treat users as nodes, following relationships between them as edges, and groups as labels. **Cora, Citeseer** [21] are two public citation network datasets.

Table 2. Datasets statistics.

Dataset	Caltech	Reed	Flickr	Cora	Citeseer
#Nodes	762	962	1120	2708	3327
#Edges	16651	18812	15939	5429	4732
#Density	0.0574	0.0407	0.0254	0.0015	0.0009
#Labels	9	18	6	7	6

Baselines. Specific information about baselines is described below:

- **GCN** [15]. When A_n is zero matrix, AGCN is approximately equivalent to GCN. We leverage GCN to learn on the initial network data without involving AL.

- **ALNE-R**. It randomly selects a batch of node pairs in each round of AL to ask their link statuses.
- **ALNE-E**. This model is a variant of our model, which dose not have information evaluation process.
- **AGE** [3]. This model used GCN to obtain a node classifier, then used AL to select a batch of node pairs in each round of AL to ask their link statuses.
- **HALLP** [6]. Based on the edge classifier and network data, this model chose a batch of node pairs in each round of AL according to uncertainty and representativeness of them.
- **NEEC** [14]. NEEC firstly selected prototype node using k-medodis. Considering we only use network structure in network data, then NEEC selected a batch of node pairs according to the uncertainty and representativeness of node pair used in our proposed method.

Evaluation Protocols. In order to simulate experts in active learning process, we set the following with reference to related papers [14]: we randomly select a fraction of all edges in the original network as the initial network, and then the remaining edge collection serve as the expert's judgment to answer the link status between the node pair.

We use PyTorch to implement our algorithm. For ALNE and its variants, a two-layer model is used in the network embedding model AGCN, where the first layer has 64 dimensions and the second layer has 16 dimensions. We set the size of node embedding as 16, the number of node pairs asked at every query as 8, the maximum number of queries q as 10, and λ_n as 0.1. For the hyperparameter a, b, c in Eq. (13), we conduct grid-search on numerical interval $[0, 1]$. We utilize Adam for model optimization, and if the evaluation metrics of the task do not change, we early stop the training process. For other baselines, we set the parameters according to the corresponding paper and optimize them to obtain the best performance.

Table 3. Performance on Cora and Citeseer datasets.

Methods	Node classification		Link prediction	
	Cora	Citeseer	Cora	Citeseer
GCN	58.75	37.44	69.76	79.77
ALNE-R	59.72	37.97	79.09	85.97
AGE	59.65	38.32	73.30	84.27
HALLP	59.81	38.14	77.75	84.22
NEEC	59.57	38.15	77.23	85.55
ALNE	59.99*	38.35*	79.95*	86.83*

5.2   Node Classification

In node classification task, for Caltech, Reed and Flick dataset we first remove 90% of the links and ensure that the remaining network data are connected. We randomly select $p\%$ of the node as the training set, the remaining nodes as the testing set. For node classification tasks, we use Micro-F1 as evaluation criteria. The above experimental process repeats ten times and the average Micro-F1 values are reported. The final results of experiment are shown in the Table 4 with the best result highlighted in bold and the second best results are underlined. From the table, we have draw the following conclusions:

Table 4. Node classification performance on Caltech, Reed and Flick datasets.

Dataset	p%	GCN	ALNE-R	ALNE-E	AGE	HALLP	NEEC	ALNE
Caltech	10%	37.11	37.7	37.95	38.01	37.71	37.46	**38.64***
	20%	41.66	41.82	42.26	42.52	42.34	42.04	**42.63***
	30%	43.79	46.99	47.33	47.46	47.52	47.46	**47.7***
	40%	49.78	50.3	50.87	50.31	50.33	50.13	**51.29***
	50%	50.00	50.28	50.76	50.92	51.21	50.00	**51.42***
Reed	10%	36.98	37.17	37.73	37.38	37.30	37.58	**37.82***
	20%	42.14	43.15	43.65	43.52	43.23	43.23	**43.9***
	30%	45.67	46.68	47.21	46.74	47.02	47.29	**47.77***
	40%	48.69	49.13	49.65	49.35	49.51	49.24	**49.68***
	50%	50.15	51.36	51.81	51.28	51.52	51.73	**52.23***
Flickr	10%	66.73	66.39	67.80	66.72	66.50	66.64	**68.06***
	20%	68.61	69.27	69.65	68.83	69.17	69.05	**69.86***
	30%	73.05	73.28	73.99	72.82	73.68	73.51	**74.96***
	40%	75.88	75.82	76.00	75.66	76.14	**76.65**	76.48*
	50%	77.51	77.51	77.79	77.41	77.78	77.77	**78.06***

- Compared with other models, the proposed model ALNE achieves the best performance in most cases. Compared with GCN, ALNE further improves the Micro-F1 value by $3.91, 2.1, 1.91$ in Caltech, Reed, Flickr datasets. We use "$*$" to indicate the improvement of ALNE over GCN is significant based on paired t-test at the significance level of 0.01.
- In ALNE, we can select node pairs really useful to network embedding model AGCN based on information evaluation module. Thus in Table 4, the ALNE model can achieve better performance than ALNE-E.
- Compared with other three active learning models, we can find that our proposed ALNE achieves the best performance except in one case. The reason is that ALNE uses gradient-based query strategy and this strategy helps to find the most influential node pair.

– Besides that, as shown in Table 3, we also conduct node classification on Cora and Citeseer datasets, which contain more nodes than Caltech, Reed and Flickr. We remove 50% of the links and randomly select 10% of the nodes as the training set. From the table, we can see that our proposed model ALNE outperforms all baselines.

5.3 Link Prediction

For the link prediction task, we randomly remove the $p\%$ link and ensure that the remaining networks are connected. For link prediction task, we choose AUC as the evaluation measurements. Similarly, the above procedure repeats ten times and the value of average AUC is reported. The final results of the experiment are shown in Table 5 with the best result highlighted in bold and the second best results are underlined. From the table, we can find:

Table 5. Link prediction performance on Caltech, Reed and Flick datasets.

Dataset	p%	GCN	ALNE-N	ALNE-E	AGE	HALLP	NEEC	ALNE
Caltech	90%	59.26	64.35	65.06	66.06	62.69	62.42	**66.87***
	85%	68.82	70.67	72.14	**73.01**	72.25	70.43	72.59*
	80%	71.73	75.57	76.04	75.99	74.26	74.91	**76.31***
	75%	74.61	77.11	77.3	76.73	76.74	77.08	**78.11***
	70%	76.5	78.65	77.95	79.08	79.53	76.17	**79.97***
Reed	90%	60.79	65.3	66.8	66.53	65.30	65.05	**67.02***
	85%	65.9	70.65	71.36	**71.80**	67.55	69.41	71.68*
	80%	70.06	73.13	73.32	73.57	70.95	71.77	**73.90***
	75%	72.93	74.89	75.52	76.22	72.91	75.11	**76.38***
	70%	74.46	76.66	77.49	74.75	74.37	75.27	**77.6***
Flickr	90%	78.2	83.3	86.92	86.78	85.33	81.79	**87.15***
	85%	87.2	89.33	90.87	**91.29**	89.18	88.90	91.04*
	80%	89.83	90.69	92.2	92.42	91.27	90.03	**92.50***
	75%	90.93	91.5	92.74	**93.27**	91.92	90.77	93.26*
	70%	91.55	91.85	92.85	**94.28**	92.26	91.60	93.43*

– Generally, ALNE achieves the best performance in most cases. We use "∗" to indicate the improvement of ALNE over GCN is significant based on paired t-test at the significance level of 0.01. In addition, we can observe that when the network has fewer edges, ALNE has better performance improvements over GCN. This shows the effectiveness of idea of improving the quality of network embedding by actively querying link status of node pair.

- Compared with other three active learning models, we can find that our proposed ALNE achieves the best performance in most cases. The reason is that ALNE use gradient-based query strategy and information evaluating, which help to find the most influential node pairs.
- Besides that, we also conduct link prediction on Cora and Citeseer datasets as shown in Table 3. From the table, we can see that our proposed model ALNE outperforms all baselines.

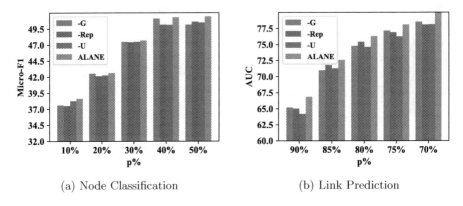

(a) Node Classification   (b) Link Prediction

Fig. 2. Ablation studies on different AL query strategy.

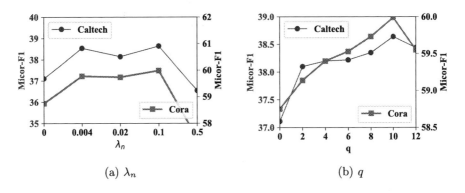

(a) λ_n (b) q

Fig. 3. Parameter sensitivity w.r.t the weight of A_n: λ_n, the number of queries: q.

5.4   Ablation Study

In order to verify the effectiveness of three AL query strategies, we conduct the ablation study based on the model ALNE. Firstly, we remove Q-based query strategy, and denote the models as $-Q$ ($Q \in$ {Gradient, Representation, Uncertainty} and we use {G, Rep, U} for brevity sake). Then node classification and link prediction tasks are carried out on Caltech dataset. Figure 2 reports the ablation study results. We can see that for two tasks, each type of AL query strategy contributes to the final result. Using three strategies together can maximize the model performance.

5.5   Parameter Sensitivity

We investigate the effect hyperparameters: the weight of A_n in propagation rule (λ_n in Eq. (2)), and the number of queries q on the model's performance. We use Caltech and Cora datasets and conduct node classification for sensitivity analysis. From the Fig. 3(a) we can see that the performance of our model improves when we consider the A_n. After λ_n is greater than 0.1, the performance decreases dramatically. This is because A_n has less information compared with A_p and big value of λ_n introduces noises to the model. So, we set the weight of A_n as 0.1. For the number of queries q in Fig. 3(b), one can see that the performance improves with the increase of the number of queries, reaching the highest value at 10.

6   Conclusion

In this paper, we formulate a novel problem about improving the performance of network embedding by asking link statuses of node pairs in AL framework and propose an effective method ALNE to solve this problem. ALNE includes a novel network embedding model AGCN, three types of valid AL query strategy and a information evaluation module, based on which the valuable node pairs can be selected. We have conducted extensive experiments on five open datasets and two classic downstream tasks, and the results demonstrate the effectiveness of our proposed model. In the future, we will extend our model to more domains, such as recommendation and anti-fraud.

Acknowledgement. The work was supported by the National Key R&D Program of China under grant 2018YFB1004700, and National Natural Science Foundation of China (61772122, 61872074)

References

1. Aggarwal, C.C., Kong, X., Gu, Q., Han, J., Yu, P.S.: Active learning: a survey. In: Data Classification: Algorithms and Applications, pp. 571–606 (2014)
2. Bilgic, M., Mihalkova, L., Getoor, L.: Active learning for networked data. In: Proceedings of the 27th International Conference on Machine Learning, pp. 79–86 (2010)
3. Cai, H., Zheng, V.W., Chen-Chuan Chang, K.: Active learning for graph embedding. arXiv e-prints arXiv:1705.05085, May 2017
4. Chen, F., Pan, S., Jiang, J., Huo, H., Long, G.: DAGCN: dual attention graph convolutional networks. In: International Joint Conference on Neural Networks, pp. 1–8. IEEE (2019)

5. Chen, J., Lin, X., Shi, Z., Liu, Y.: Link prediction adversarial attack via iterative gradient attack. IEEE Trans. Comput. Soc. Syst. **7**(4), 1081–1094 (2020)
6. Chen, K., Han, J., Li, Y.: HALLP: a hybrid active learning approach to link prediction task. JCP **9**(3), 551–556 (2014)
7. Chen, X., Yu, G., Wang, J., Domeniconi, C., Li, Z., Zhang, X.: ActiveHNE: active heterogeneous network embedding. In: Proceedings of the Twenty-Eighth International Joint Conference on Artificial Intelligence, pp. 2123–2129 (2019)
8. Cheng, A., et al.: Deep active learning for anchor user prediction. In: IJCAI 2019, Macao, China, 10–16 August 2019, pp. 2151–2157 (2019)
9. Cui, P., Wang, X., Pei, J., Zhu, W.: A survey on network embedding. IEEE Trans. Knowl. Data Eng. **31**(5), 833–852 (2019)
10. Gao, L., Yang, H., Zhou, C., Wu, J., Pan, S., Hu, Y.: Active discriminative network representation learning. In: Proceedings of the Twenty-Seventh International Joint Conference on Artificial Intelligence, IJCAI 2018, pp. 2142–2148 (2018)
11. Grover, A., Leskovec, J.: node2vec: scalable feature learning for networks. In: Proceedings of the 22nd ACM SIGKDD International Conference on Knowledge Discovery and Data Mining, pp. 855–864 (2016)
12. He, K., Li, Y., Soundarajan, S., Hopcroft, J.E.: Hidden community detection in social networks. Inf. Sci. **425**, 92–106 (2018)
13. Hu, X., Tang, J., Gao, H., Liu, H.: ActNet: active learning for networked texts in microblogging. In: Proceedings of the 13th SIAM International Conference on Data Mining, pp. 306–314 (2013)
14. Huang, X., Song, Q., Li, J., Hu, X.: Exploring expert cognition for attributed network embedding. In: Proceedings of the Eleventh ACM International Conference on Web Search and Data Mining, WSDM 2018, pp. 270–278 (2018)
15. Kipf, T.N., Welling, M.: Semi-supervised classification with graph convolutional networks. In: 5th International Conference on Learning Representations, ICLR 2017 (2017)
16. Leroy, V., Cambazoglu, B.B., Bonchi, F.: Cold start link prediction. In: Proceedings of the 16th ACM SIGKDD International Conference on Knowledge Discovery and Data Mining, pp. 393–402 (2010)
17. Ma, Y., Wang, S., Aggarwal, C.C., Yin, D., Tang, J.: Multi-dimensional graph convolutional networks. In: Proceedings of the 2019 SIAM International Conference on Data Mining, SDM 2019, pp. 657–665 (2019)
18. Mutlu, E.C., Oghaz, T.A.: Review on graph feature learning and feature extraction techniques for link prediction. arXiv preprint arXiv:1901.03425 (2019)
19. Perozzi, B., Al-Rfou, R., Skiena, S.: DeepWalk: online learning of social representations. In: The 20th ACM SIGKDD International Conference on Knowledge Discovery and Data Mining, KDD 2014, pp. 701–710 (2014)
20. Tu, C., Zhang, W., Liu, Z., Sun, M.: Max-margin: discriminative learning of network representation. In: Proceedings of the Twenty-Fifth International Joint Conference on Artificial Intelligence, IJCAI, pp. 3889–3895 (2016)
21. Velickovic, P., Cucurull, G., Casanova, A., Romero, A., Liò, P., Bengio, Y.: Graph attention networks. In: 6th International Conference on Learning Representations, ICLR 2018 (2018)
22. Wu, Z., Pan, S., Chen, F., Long, G., Zhang, C., Yu, P.S.: A comprehensive survey on graph neural networks. arXiv preprint arXiv:1901.00596 (2019)
23. Xie, Z., Li, M.: Semi-supervised AUC optimization without guessing labels of unlabeled data. In: Proceedings of the Thirty-Second AAAI Conference on Artificial Intelligence, pp. 4310–4317 (2018)

24. Yang, C., Xiao, Y., Zhang, Y., Sun, Y., Han, J.: Heterogeneous network representation learning: survey, benchmark, evaluation, and beyond. arXiv preprint arXiv:2004.00216 (2020)
25. Yang, Z., Tang, J., Zhang, Y.: Active learning for streaming networked data. In: Proceedings of the 23rd ACM International Conference on Conference on Information and Knowledge Management, pp. 1129–1138 (2014)
26. Zhang, Y., Lease, M., Wallace, B.C.: Active discriminative text representation learning. In: Proceedings of the Thirty-First AAAI Conference on Artificial Intelligence, pp. 3386–3392 (2017)
27. Zhu, J., et al.: Constrained active learning for anchor link prediction across multiple heterogeneous social networks. Sensors **17**(8), 1786 (2017)

Keyword-Centric Community Search over Large Heterogeneous Information Networks

Lianpeng Qiao[1]([✉]), Zhiwei Zhang[2], Ye Yuan[2], Chen Chen[2], and Guoren Wang[2]

[1] Northeastern University, Shenyang, China
qiaolp@stumail.neu.edu.cn
[2] Beijing Institute of Technology, Beijing, China
Yuan-ye@bit.edu.cn

Abstract. Community search in heterogeneous information networks (HINs) has attracted much attention in recent years and has been widely used for graph analysis works. However, existing community search studies over heterogeneous information networks ignore the importance of keywords and cannot be directly applied to the keyword-centric community search problem. To deal with these problems, we propose $k\mathcal{KP}$-core, which is defined based on a densely-connected subgraph with respect to the given keywords set. A $k\mathcal{KP}$-core is a maximal set of $\mathcal{P}$-connected vertices in which every vertex has at least one $\mathcal{KP}$-neighbor and k path instances. We further propose three algorithms to solve the keyword-centric community search problem based on $k\mathcal{KP}$-core. When searching for answers, the basic algorithm Basic-$k\mathcal{KP}$-core will enumerate all paths rather than only the path instances of the given meta-path $\mathcal{P}$. To improve efficiency, we design an advanced algorithm $Advk\mathcal{KP}$-core using a new method of traversing the search space based on trees to accelerate the searching procedure. For online queries, we optimize the approach with a new index to handle the online queries of community search over HINs. Extensive experiments on HINs are conducted to evaluate both the effectiveness and efficiency of our proposed methods.

Keywords: Keyword-centric · Community · Heterogeneous information networks

1 Introduction

Heterogeneous information networks (HINs) [11,17] are the networks involving multiple objects and multiple links denoting different types and relations, and has been widely used to model bibliographic networks, social media networks, and knowledge networks. Figure 1(a) depicts an HIN of the bibliographic network, which describes the relationships between different types of entities. In this network, vertices with labels A, P, V, and T represent authors, papers, venues, and time. It consists of four types of entities. The directed lines between the

© Springer Nature Switzerland AG 2021
C. S. Jensen et al. (Eds.): DASFAA 2021, LNCS 12681, pp. 158–173, 2021.
https://doi.org/10.1007/978-3-030-73194-6_12

vertices denote their semantic relationships. For example, the links between a_1, a_2, p_1, v_1 indicate that the author a_1 and a_2 have written a paper p_1 accepted by CIKM.

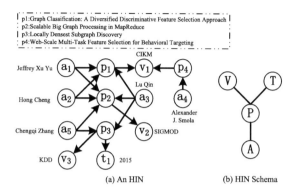

Fig. 1. An example HIN of DBLP network

On the other hand, many real-world networks have a significant community structure, in which vertices in a community are densely connected. Community search has essentially studies in graph analysis, and has attracted much attention in the literature. For HINs, there are also some works focusing on community search problem in them [11,27–29,33]. However, these studies focus on either structure density or keyword cohesiveness, and few of them consider all these at the same time.

Existing works on network community search can be classified into community detection [12,24,27–29,33] and community search [4,6,7]. Community detection algorithms aim to detect all communities for a graph, which means they are not designed for online query. Different from community detection, query-based community search has been studied [7,10,15], in which the vertices and keywords are given, and they aim to find the most cohesive subgrahs related the given vertices and keywords. The vertices and edges in HINs carry different semantic meanings. In this scenario, existing approaches for community search on homogeneous networks cannot solve the community search problem in HINs.

In this paper, we focus on searching communities concerning given keywords set in HINs, in which vertices are with a specific type (e.g., a community of authors in the bibliographical network, as shown in Fig. 1). For the keyword-centric community, we need to deal with three questions. (1) How to combine the keywords with the community? (2) How to measure the connectivity of two vertices of the same type? (3) How to measure the cohesiveness of a community?

For the first two questions, we adopt the meta-path concept [23] to connect two vertices since the vertices with the same type may not be connected directly in the HIN. For the third question, existing solutions adopt a minimum degree [7,9], k-truss [15], or k-clique [6,30] to measure the community cohesiveness.

The minimum degree is the most common metric to ensure every vertex is well engaged in the community. In this paper, we extend such metric for HINs.

For the problem of community search in HINs, the community returned by the queries should be the subgraph in which the distance between vertices is small. Also, the community need to be cohesive considering all the keywords given in the queries. Therefore, we propose a new model called $k\mathcal{KP}$-core in HINs. There are three requirements for a $k\mathcal{KP}$-core S: (1) every vertex v has at least k path instances of the given meta path $\mathcal{P}$ in S starting with v; (2) every vertex v has at least one instance contains the given set of keywords; (3) for the connected graph S, any two vertex u and v could be connected by a path p, and any two adjacent vertices in p should be connected by a path instance. Given an HIN $G(V, E)$, a query $q \in V$, keywords set K, and an integer k, our goal is to find a maximum $k\mathcal{KP}$-core containing q, in which all the vertices are with the same type of q and contain the keywords set K.

In summary, we make the following contributions.

- We propose a keyword-centric community model called $k\mathcal{KP}$-core and formulate the problem of the keyword-centric community search;
- We propose a baseline algorithm to search the community in HINs;
- We design a new method to traversal the search space based on trees to accelerate the community search algorithm, as shown in Algorithm 2 and Algorithm 3. We further propose the optimization for the approach as shown in Sect. 3.3.
- We conduct a series of experiments on real-world HINs to evaluate the effectiveness and efficiency of our algorithms.

The rest of this paper is organized as follows. In Sect. 2, we define the keyword-centric community search problem in HINs. In Sect. 3, we proposed several Algorithms to solve the problem. In Sect. 4, we conduct extensive experiments on real-world HINs to show the effectiveness and efficiency of our methods. We review the related work in Sect. 5 and conclude in Sect. 6.

2   Problem Definition

In this section, we introduce several definitions used in this paper. Furthermore, we define the problem of keyword-centric community search in HINs.

Definition 1 (HIN) [11,17]. *An HIN is a directed graph $G(V, E)$ with a vertex type mapping function $\psi : V \to \mathcal{A}$ and an edge type mapping function $\phi : E \to \mathcal{R}$, where each vertex $v \in V$ belongs to a vertex type $\psi(v) \in \mathcal{A}$, and each edge $e \in E$ belongs to an edge type (also called relation) $\phi(e) \in \mathcal{R}$.*

Definition 2 (HIN schema). *Given an HIN $G(V, E)$ with mappings $\psi : V \to \mathcal{A}$ and $\phi : E \to \mathcal{R}$, its schema G_s is an undirected graph defined over vertex types $\mathcal{A}$ and edge types (as relation) $\mathcal{R}$, i.e., $G_s(\mathcal{A}, \mathcal{R})$.*

Figure 1(b) is the schema of the HIN, in which the vertices labeled A, P, V, and T denote author, paper, venue, time. The schema describes all the edge types between all the vertex types that exist in HIN. In [11,17], HIN schema is defined on directed graph, our approach can also be applied to the HIN schema of directed graph.

In this paper, we try to find a community concerning given keywords set in an HIN, in which all the vertices have the same type, such type is called target type. Since the vertices in a community should be connected cohesively, we define the connection between vertices in an HIN using the symmetric meta-path $\mathcal{P}$, whose source object and target object are with the target type. To describe the cohesiveness between the vertices with the target type in a community, we extend the classic k-core as $k\mathcal{KP}$-core with a symmetric meta-path $\mathcal{P}$.

Definition 3 (Meta-path) [11]. *A meta-path $\mathcal{P}$ is a path defined on an HIN schema $G_s = (\mathcal{A}, \mathcal{R})$, and is denoted in the form $A_1 A_2 ... A_l$, where $A_i \in \mathcal{A}(1 \leq i \leq l)$ and $\phi(A_j, A_{j+1}) \in \mathcal{R}(1 \leq j \leq l - 1)$.*

Definition 4 (Path instance). *Given an HIN $G = \{V, E\}$, a meta-path $\mathcal{P} = (A_1 A_2 ... A_l)$, the path $a_1 \rightarrow a_2 \rightarrow ... \rightarrow a_l$ between vertices a_1 and a_l is a path instance of $\mathcal{P}$, if it satisfies $\psi(a_i) = A_i (1 \leq i \leq l)$, and $\phi(a_j, a_{j+1}) = \phi(A_j, A_{j+1})(1 \leq j \leq l - 1)$.*

According to Definition 4, we say that a vertex v is a $\mathcal{P}$-neighbor of vertex u if an instance of $\mathcal{P}$ can connect them. For example, in Fig. 1, $\mathcal{P} = (APVPA)$ is a meta-path of the given HIN, vertex a_4 is a $\mathcal{P}$-neighbor of vertex a_1, since an instance of $\mathcal{P} = (APVPA)$ can connect them. We say that two vertices u and v are $\mathcal{P}$-connected if a chain of vertices can connect them, such that every vertex is a $\mathcal{P}$-neighbor of its adjacent vertex in the chain. The definition of $\mathcal{P}$-connected and $\mathcal{P}$-neighbor denotes the connectivity between vertices in HIN and meta paths. Furthermore, we propose K-instance to denote the correlations between keywords in the nodes, which is introduced as follows.

Definition 5 ($\mathcal{K}$-instance). *Given an HIN $G(V, E)$, a meta-path $\mathcal{P}$ and a set of input keywords $K = \{k_1, k_2, ..., k_w\}$, a $\mathcal{K}$-instance of $\mathcal{P}$ is a path instance p whose labels set contains all the keywords in K.*

Example 1. Given an HIN as shown in Fig. 1 and $K = \{\text{"Selection"}, \text{"CIKM"}\}$, the path $p = a_1 \rightarrow p_1 \rightarrow v_1 \rightarrow p_4 \rightarrow a_4$ is an $\mathcal{K}$-instance of $\mathcal{P} = (APVPA)$, since p_1 contains keyword "Selection" and v_1 contains keyword "CIKM".

Definition 6 ($\mathcal{KP}$-neighbor). *Given a meta-path $\mathcal{P}$ and a set of input keywords $K = \{k_1, k_2, ..., k_w\}$, we say that a vertex u is a $\mathcal{KP}$-neighbor of a vertex v, if they can be connected by a $\mathcal{K}$-instance of $\mathcal{P}$.*

Definition 7 ($deg(v, S)$). *Given an HIN, a set of vertices S and a meta-path $\mathcal{P}$, we define $deg(v, S)$ as the number of path instances of $\mathcal{P}$ between v and all the other vertices in $S \backslash \{v\}$.*

162   L. Qiao et al.

To ensure the cohesiveness of the keyword-centric community, existing works often use k-core to characterize the cohesiveness of a community. In this paper, we aim to find a community in an HIN containing a query vertex q, in which all the vertices have the same vertex type as $\psi(q)$. We use a symmetric meta-path $\mathcal{P}$ to connect vertices with the target type. Then we can extend the k-core model for HINs as follow.

Definition 8 ($k\mathcal{KP}$-core). *Given an HIN G, a keywords set K, a symmetric meta path $\mathcal{P}$ and an integer k, a $k\mathcal{KP}$-core is a maximal set S of $\mathcal{P}$-connected vertices, s.t. $\forall v \in S$, $deg(v, S) \geq k$ and v has at least one $\mathcal{KP}$-neighbor.*

Example 2. We use $degK(a, S)$ to present the number of $\mathcal{KP}$-neighbors of vertex a in $S\backslash\{a\}$. Consider the HIN $G(V, E)$ in Fig. 1, a meta-path $\mathcal{P} = (APVPA)$ and a keywords set $K = \{\text{"Selection"}, \text{"CIKM"}\}$. Let $k = 3$. Then we can see that the induced subgraph of $\{a_1, a_2, a_3, a_4\}$ denoted as S is a $k\mathcal{KP}$-core. It has $deg(a_1, S) = 5, deg(a_2, S) = 5, deg(a_3, S) = 5, deg(a_4, S) = 3$ and $degK(a_1, S) = 3 > 1, degK(a_2, S) = 3 > 1, degK(a_3, S) = 3 > 1, degK(a_4, S) = 3 > 1$.

Based on Definition 8, we can find different types of communities. As shown in Example 2, we get a community of authors using the meta-path $\mathcal{P} = (APVPA)$, which represents a potential collaborative community of papers. Besides, we could get different communities of vertices with different types by using different meta-paths. Note that all the meta-paths we used in the rest of the paper are symmetric. Now we introduce the keyword-centric community search problem in HINs as follow:

Problem. Given an HIN G, a query vertex q, a keywords set K, a meta-path $\mathcal{P}$ and an integer k, the keyword-centric community we search in G is the corresponding $k\mathcal{KP}$-core containing q.

As shown in Example 2, let $q = a_1$, $K = \{\text{"Selection"}, \text{"CIKM"}\}$, meta-path $\mathcal{P} = (APVPA)$ and $k = 3$. We could get the corresponding keyword-centric community $C(V, E)$, in which $V = \{a_1, a_2, a_3, a_4\}$ and the labels of the vertices include "Jeffrey Xu Yu", "Hong Cheng", "Lu Qin", and "Alexander J. Smola", since C is a maximum $k\mathcal{KP}$-core containing q, which means that there could be more collaborations between the authors in C. Note that Jeffrey Xu Yu and Alexander J. Smola have published papers with the same keyword "Selection" in the same venue "CIKM". According to Definition 8, we can know that the keyword-centric community satisfies the structural maximality and connectivity.

Theorem 1. *Given an HIN G, a query vertex q, keywords set K, a meta-path $\mathcal{P}$ and an integer k, the $k\mathcal{KP}$-core containing q is unique.*

Proof. Suppose that X and Y are two different $k\mathcal{KP}$-cores containing q, we could get that $X \cup Y$ could be a new $k\mathcal{KP}$-core. For each vertex $v \in Y$, we can get that p and v are $\mathcal{P}$-connected. Then we can get that all the vertices in X and Y are $\mathcal{P}$-connected. Since all the vertices in X and Y have k or more than k $\mathcal{P}$-neighbors, we can get that $X \cup Y$ is a new $k\mathcal{KP}$-core containing q, which is against the initial assumption.

3 Search Algorithm

We adopt $k\mathcal{KP}$-core as the model to search the keyword-centric communities in HINs. In this section, we present efficient solutions for the community search problem in HINs.

Algorithm 1: Basic-$k\mathcal{KP}$-core

Input: the HIN graph $G(V, E)$; query vertex q; the keywords set $K = \{k_1, k_2, ..., k_w\}$; the meta-path $\mathcal{P}$ and k

Output: the set of all $k\mathcal{KP} - core$.

1 collect the set S of vertices with the same vertex type as q;

2 $deg(v, S) \leftarrow$ the number of all path instances of $\mathcal{P}$ starting with v and end with u for each $u \in S\backslash\{v\}$;

3 $\mathcal{KP}\_neighbor[v] \leftarrow \mathcal{KP}\_neighbor[v] \cup \{u\}$, if the path instance of $\mathcal{P}$ starting with v and end with u cover all the keywords in K, $\forall u \in S$;

4 **foreach** $v \in S$ **do**

5   **if** $deg(v, S) < k$ *or* $\mathcal{KP}\_neighbor[v] = \emptyset$ **then**

6   remove v from S;

7   **foreach** $u \in S$ **do**

8   **if** u *is the* $\mathcal{KP}$-*neighbor of* v **then** remove v from $\mathcal{KP}\_neighbor[u]$;

9   update $deg(u, S)$;

10 remove all the vertices which are not $\mathcal{P}$-connected with q from S;

11 return S;

3.1 The Basic Algorithm

Based on the concept of $k\mathcal{KP}$-core, we present a basic algorithm, as shown in Algorithm 1 to find the maximum $k\mathcal{KP}$-core containing the query vertex q in an HIN. In general, it consists of four steps: (1) collect the set S of all vertices with target type (line 1); (2) for each vertex $v \in S$, enumerate all path instances of $\mathcal{P}$ starting with v and end with the nodes in $S\backslash\{v\}$, and find the set of $\mathcal{KP}$-neighbors of v in S. Incidentally, we use i-th$(\mathcal{P})$ to present the i-th vertex type of $\mathcal{P}$ (lines 2–3); (3) remove the vertex v which has less than k path instances starting with v or has no $\mathcal{KP}$-neighbor from S and update the remaining vertices iteratively until there is no vertex in S can be removed (lines 4–9); (4) the remaining S is the keyword-centric community we are looking for (lines 10–11).

Example 3. Consider the HIN in Fig. 1 and the meta-path $\mathcal{P} = (APVPA)$. Let the query vertex be a_1, $k = 4$ and keywords $K = \{$"*Selection*", "*CIKM*"$\}$. First, we compute the number of the path instances starting with the vertex in $S = \{a_1, a_2, a_3, a_4, a_5\}$ and the set of $\mathcal{KP}$-neighbors of the vertex in S. By enumerate all the path instances of meta path $\mathcal{P}$, we can get $deg(a_1, S) = 6$, $deg(a_2, S) = 6$, $deg(a_3, S) = 7$, $deg(a_4, S) = 3$, $deg(a_5, S) = 4$, $degK(a_1, S) = 3$, $degK(a_2, S) = 3$, $degK(a_3, S) = 3$, $degK(a_4, S) = 3$, $degK(a_5, S) = 0$. Since we get $deg(a_4, S) = 3 < 4$ and $degK(a_5, S) = 0 < 1$. We have to remove a_4 and a_5 from S. Because a_4 is the $\mathcal{KP}$-neighbor of $\forall v \in \{a_1, a_2, a_3\}$, we can update the $degK(v, S)$ $\forall v \in \{a_1, a_2, a_3\}$ which is 2, 2, 2 respectively. Since we remove a_4 and a_5 from S, we have to recompute the path instance starting with the vertex in $\{a_1, a_2, a_3\}$. Finally, we get $deg(a_1, S) = 4$, $deg(a_2, S) = 4$,

$deg(a_3, S) = 4$, $degK(a_1, S) = 2$, $degK(a_2, S) = 2$, $degK(a_3, S) = 2$ which means that $\{a_1, a_2, a_3\}$ is a keyword-centric community.

In Algorithm 1, we enumerate all the instances starting with each $v \in V$, and the complexity of this process can be bounded by $|S| * |V|^l$, where S is the set of vertices with the target type, and l is the length of the meta-path $\mathcal{P}$.

Algorithm 2: TraversalTreeBuild

Input: the HIN graph $G(V, E)$; the set S of vertices with the target type; the keywords set $K = \{k_1, k_2, ..., k_w\}$; the meta-path $\mathcal{P}$; the parameter of length h

1   **foreach** $v \in S$ **do**
2   push the vertex v into an empty queue C, $count(u, v) \leftarrow 1$ for each $u \in S$ if $(u, v) \in E$;
3     $i \leftarrow 1$;
4   **while** $i \leq h$ *AND* $C \neq \emptyset$ **do**
5       $u \leftarrow C.pop()$;
6   **foreach** z which have $(z, u) \in E$ and $\psi(z) = (i+1) - th(\mathcal{P})$ **do**
7         $C.push(z)$;
8         $root[z] \leftarrow v$;
9         $count(z, v) \leftarrow count(z, v) + count(u, v)$;
10   insert $X \cup \{key[z]\}$ into $key(z, v)$ for each $X \in key(u, v)$;
11     $i++$;

Theorem 2. *Given an HIN $G(V, E)$, a keywords set K, a query vertex q and the meta-path $\mathcal{P}$. Let l be the length of $\mathcal{P}$. The complexity of Algorithm 1 is $O(|S| * n^l)$, where $n = |V|$.*

3.2 Advanced Algorithm

Algorithm 1 is very costly to enumerate all instances, starting with each vertex in S in step (2) and step (3). To speed up step (2) and step (3), we propose Algorithm 2 and Algorithm 3 using a new method of traversing the search space based on trees. Note that we use $K.has(v)$ to judge whether there is a keyword in K which is contained by v, $key[v]$ to express the keyword in K which is contained by v and $key(u, v)$ to express the set of keywords contained by the path from u to v in the trees. The general idea of the approach is to maintain the number of path instances and keywords set between u and v whose vertex types are the same as the target type and the middle vertex type of the given meta path $\mathcal{P}$ respectively. Based on these trees, we can save a lot of time when calculating the number of path instances of $\mathcal{P}$ and judging whether there is a K-instance between two nodes with target type.

The process of building the trees is shown in Algorithm 2. First, we push vertex v into queue C and initialize $count(u, v)$ (lines 1–2). Then, we iteratively push the vertices with the corresponding vertex type as $\mathcal{P}$ into C, until we get the number of the path instances and the group of the keywords sets between v and those vertices with the vertex type h-th($\mathcal{P}$) (lines 3–11).

The next operation is shown in Algorithm 3. First of all, we collect the set S of all vertices with target type and get the set of all index trees rooted at the vertex in S (lines 1–3). Next, we move the vertices which are not $\mathcal{P}$-connected

Algorithm 3: Adv$k\mathcal{KP}$-core

Input: the HIN graph $G(V, E)$; query vertex q; the keywords set $K = \{k_1, k_2, ..., k_w\}$; the meta-path $\mathcal{P}$ and k

Output: the maximum set of $k\mathcal{KP} - core$ containing q.

1   collect the set S of vertices with the same vertex type as q;
2   queue $\mathcal{Q} \leftarrow \emptyset$;

3   $TraversalTreeBuild(G, S, \mathcal{P}, \lceil \frac{sizeof(\mathcal{P})}{2} \rceil)$;
4   **foreach** $v \in S$ **do**
5 | **if** v and q are not $\mathcal{P}$-connected **then** $S \leftarrow S \backslash \{v\}$; $\mathcal{Q}.push(v)$; ;

6   **foreach** $v \in S$ **do**
7 | $deg(v, S) \leftarrow 0$; $k\_neighbor[v] \leftarrow \emptyset$;
8 | **foreach** $u \in S \backslash \{v\}$ **do**
9 | | **foreach** leaf vertex z of v **do**
10 | | | $deg(v, S) \leftarrow deg(v, S) + count(z, v) * count(z, u)$;
11 | | | **if** $X \cup Y = K$ where $X \in key(z, v)$ and $Y \in key(z, u)$ **then**
12 | | | | $k\_neighbor[v].push(u)$;

13 | **if** $deg(v, S) < k$ OR $k\_neighbor[v] = \emptyset$ **then** $S \leftarrow S \backslash \{v\}$, $\mathcal{Q}.push(v)$; ;

14   **foreach** $u \in \mathcal{Q}$ **do**
15 | **foreach** $v \in S$ **do**
16 | | $C \leftarrow$ the commen leaf nodes between u and v;
17 | | **foreach** $z \in C$ **do**
18 | | | $deg(v, S) \leftarrow deg(v, S) - count(z, v) * count(z, u)$;
19 | | | $k\_neighbor[v].remove(z)$;
20 | | **if** $deg(v, S) < k$ OR $k\_neighbor[v] = \emptyset$ OR v and q are not $\mathcal{P}$-connected **then**
     | | | $S \leftarrow S \backslash \{v\}$, $\mathcal{Q}.push(v)$;

21   **return** S;

with q from S to $\mathcal{Q}$ (lines 4–5). Then we get $deg(v, S)$ the number of the path instances starting with the v in S and $k\_neighbor[v]$ the set of $\mathcal{KP}$-neighbors of v. If we have $deg(v, S) < k$ or $k\_neighbor[v] = \emptyset$, we can move all vertices v from S to $\mathcal{Q}$ (lines 6–13). Afterwards, we remove all vertex $u \in \mathcal{Q}$ from S and update $deg(v, S)$ and $k\_neighbor[v]$ for all $v \in S \backslash \mathcal{Q}$ iteratively until there is no vertex can be removed from S (lines 14–20). Finally, we return S as the keyword-centric community we are searching for (line 21).

Example 4. Consider the HIN in Fig. 1(a) and the meta-path $\mathcal{P} = (APVPA)$. Let query vertex be a_1, $k = 4$ and $K = \{$ "Selection", "CIKM"$\}$. First of all, we build the traversal trees rooted at each vertex in $S = \{a_1, a_2, a_3, a_4, a_5,\}$ as shown in Fig. 2. Consider the tree rooted at a_1 as a example, we can get $deg(a_1, S) = count(v_1, a_1) * (count(v_1, a_2) + count(v_1, a_3) + count(v_1, a_4)) + count(v_2, a_1) * (count(v_2, a_2) + count(v_2, a_3) + count(v_2, a_5)) = 1 * (1 + 1 + 1) + 1 * (1 + 1 + 1) = 6$. Secondly, we can get $deg(a_2, S) = 6$, $deg(a_3) = 7$, $deg(a_4, S) = 3$, $deg(a_5, S) = 4$. Next, we can get $k\_neighbor[a_1] = \{a_2, a_3, a_4\}$, $k\_neighbor[a_2] = \{a_1, a_3, a_4\}$, $k\_neighbor[a_3] = \{a_1, a_2, a_4\}$, $k\_neighbor[a_4] = \{a_1, a_2, a_3\}$, $k\_neighbor[a_5] = \emptyset$. According to the above calculation, we can move a_4 and a_5 from S to $\mathcal{Q}$. Then we can update the $deg(v, S)$ and $k\_neighbor[v]$, $\forall v \in \{a_1, a_2, a_3\}$. According to lines 14–20 in Algorithm 3, we can get that $deg(a_1, S) = 6 - 1 * 1 - 1 * 1 = 4$, $deg(a_2, S) = 6 - 1 * 1 - 1 * 1 = 4$, $deg(a_2, S) = 6 - 1 * 1 - 1 * 1 = 4$, $k\_neighbor[a_1] = \{a_2, a_3\}$, $k\_neighbor[a_2] = \{a_1, a_3\}$, and $k\_neighbor[a_3] = \{a_1, a_2\}$. Finally we return $\{a_1, a_2, a_3\}$ as the keyword-centric community.

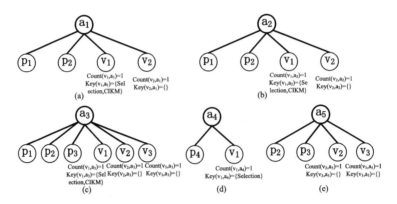

Fig. 2. An example for traversal trees

According to the above example, we can see that when we remove some vertices from S, we can quickly update $deg(v, S)$ and $k\_neighbor[v]$ $\forall v \in S$ using trees.

Theorem 3. *Given an HIN $G(V, E)$, a keywords set K, a query vertex q and the meta-path $\mathcal{P}$. Let l be the length of $\mathcal{P}$. The complexity of Algorithm 3 is $O(|S| * n^{\frac{l}{2}})$, where $n = |V|$.*

3.3 Optimization for the Approaches

In this section, We propose a new index to immediately get the numbers of the path instances of all meta paths, which can save much time when we handle the community search problem over HINs.

Algorithm 4: PreIndexTree

Input: The HIN graph $G(V, E)$; the HIN schema $G_s = (\mathcal{A}, \mathcal{R})$
1 insert all the meta paths of G_s into the empty set Q;
2 **foreach** $v \in V$ **do**
3   construct a spanning tree rooted at v;
4   **foreach** $u \neq v$ *AND* $\psi(u) = \psi(v)$ **do**
5          $ins\_count(v, \mathcal{P}, u) \leftarrow$ the number of the path instances of $\mathcal{P}$ between v and u, for each $\mathcal{P} \in Q$;

The process of building the index is shown in Algorithm 4. Based on the HIN schema, we maintain the number of the path instances of all the meta paths between v and the other vertices (lines 1–5). We can save a lot of time when calculating and updating the number of path instances as shown in lines 10 and 18 in Algorithm 3. The detail of the optimization approach is described in the following example.

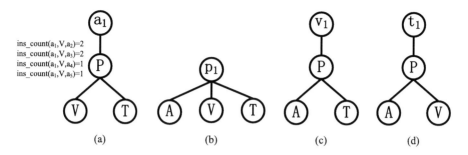

Fig. 3. A part of the new index trees

Example 5. Consider the HIN G in Fig. 1(a), the meta-path $\mathcal{P} = (APVPA)$ and the HIN schema $G_s = (\mathcal{A}, \mathcal{R})$. Let query vertex be a_1, $k = 4$ and $K = \{\text{"Selection"}, \text{"CIKM"}\}$. First of all, we build the index trees rooted at each vertex in V respectively as shown in Fig. 3 which is the part of the index trees. Consider the index tree rooted at a_1 as a example as shown in Fig. 3(a), we can immediately get $deg(a_1, S) = 6$. Secondly we can get $deg(a_1, S) = ins\_count(a_1, V, a_2) + ins\_count(a_1, V, a_3) + ins\_count(a_1, V, a_4) + ins\_count(a_1, V, a_5) = 2 + 2 + 1 + 1 = 6$, $deg(a_2) = 6$, $deg(a_3) = 7$, $deg(a_4, S) = 3$, $deg(a_5, S) = 4$. Next, we can get $k\_neighbor[a_1] = \{a_2, a_3, a_4\}$, $k\_neighbor[a_2] = \{a_1, a_3, a_4\}$, $k\_neighbor[a_3] = \{a_1, a_2, a_4\}$, $k\_neighbor[a_4] = \{a_1, a_2, a_3\}$, $k\_neighbor[a_5] = \emptyset$. According to the above calculation, we can remove a_4 and a_5 from S. Then we can update the $deg(v, S)$ and $k\_neighbor[v]$, $\forall v \in \{a_1, a_2, a_3\}$ and get that $deg(a_1, S) = 6 - 1 * 1 - 1 * 1 = 4$, $deg(a_2, S) = 6 - 1 * 1 - 1 * 1 = 4$, $deg(a_3, S) = 7 - 1 * 1 - 2 * 1 = 4$, $k\_neighbor[a_1] = \{a_2, a_3\}$, $k\_neighbor[a_2] = \{a_1, a_3\}$, $k\_neighbor[a_3] = \{a_1, a_2\}$. Finally we return $\{a_1, a_2, a_3\}$ as the keyword-centric community.

4   Experiments

We now present the experimental results. We first discuss the experimental setup in Sect. 4.1.

4.1   Experimental Setup

To search the keyword-centric community in HINs, we implement three approaches called Baseline, AdvCore, and OptCore, respectively. Baseline is based on the basic algorithm Basic-$k\mathcal{KP}$-core, which is shown in Algorithm 1. AdvCore is the advanced approach with the $Advk\mathcal{KP}$-core algorithm as shown in Algorithm 2 and Algorithm 3. OptCore is the optimized approach as shown in Sect. 3.3. All algorithms are implemented in C++. All the experiments are conducted on a computer with Intel(R) Core(TM) i5-9500 CPU @ 3.00 GHz and 16G main memory. Windows 10 X64 operating system with kernel 18362.1139.

Table 1. Datasets used in the following experiments

Dataset	Vertices	Edges	Vertex types	Edge types
Foursquare	43,199	405,476	5	4
DBLP	682,819	1,951,209	4	3
IMDB	4,467,806	7,597,591	4	3
DBpedia	5,900,558	17,961,887	4	3

Datasets. We use four real datasets: Foursquare[1], DBLP[2], IMDB[3], and DBpedia[4]. Their detailed information is shown in Table 1. Foursquare contains the users' check-in records in the US, and there are five types of vertices in the dataset. DBLP contains the publication information in computer science areas, which has four types of vertices. IMDB contains the movie rating information since 2000, and there are four types of vertices in the dataset (actors, directors, writers, and movies). DBpedia is the data set extracted from Wikipedia.

Queries. For each dataset, we collect a set of meta-paths, and the size of the set is presented in Table 1. Based on the current works, we get that the default lengths of all meta-path we used in this paper do not exceed four unless otherwise specified. We collect all the possible meta-paths of the first two datasets, because the relationships in these two datasets are relatively small. For the other two datasets, there are a lot of relationships in these two datasets. Then we choose 50 meta-paths with the highest frequencies from the sets of the possible meta-paths of these two datasets, respectively. For each dataset, we generate 100 queries. To generate a query, we first randomly choose a meta-path. Then we first choose a vertex that has 50 instances or more starting with it and then get several keywords from a random instance of the meta-path starting with the chosen vertex. By default, we set k as 50, and for the results mentioned in the following, each value in the chart is the average result for these 100 queries.

4.2 Effectiveness Testing

4.2.1 Core Analysis
To analyze the proposed $k\mathcal{KP}$-core, we examine the size distribution of $k\mathcal{KP}$-core, where k ranges from 20 to 120. In this part, we only show results contain two different queries, which are $\mathcal{P}_1 = (TPAPT)$, $K_1 = \{$"20"$\}$ and $\mathcal{P}_2 = (APVPA)$, $K_2 = \{$"report"$\}$ respectively. According to the result shown in Fig. 4, we can get that the proposed $k\mathcal{KP}$-core can achieve strong cohesiveness.

---
[1] https://sites.google.com/site/yangdingqi/home/foursquare-dataset.
[2] http://dblp.uni-trier.de/xml/.
[3] https://www.imdb.com/interfaces/.
[4] https://wiki.dbpedia.org/Datasets.

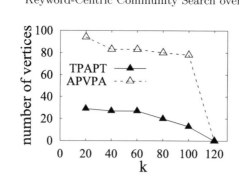

Fig. 4. Number of the vertices in $k\mathcal{KP}$-core

Table 2. Result of a case study on DBLP network.

$\mathcal{P}_1 = APVPA$	$\mathcal{P}_2 = APTPA$
$K_1 = \{Attack, Meltdown\}$	$K_2 = \{Attack, Meltdown\}$
Daniel Genkin, Daniel Gruss, Mickael Schwarz, Mike Hambury, Moritz Lipp, Paul Kocher, Stefan Mangard, Thomas Prescher, Werner Haas, Yuval Yarom	Daniel Genkin, Daniel Gruss, Diego Gragnaniello, Francesco Marra, Giovanni Poggi, Limin Zhang, Lu Feng, Luisa Verdoliva, Michael Schwarz, Mike Hamburg, Moritz Lipp, Paul Kocher, Pengyuan Lu, Stefan Mangard, Thomas Prescher, Werner Haas, Yuval Yarom

4.2.2   Case Study

We perform two queries on DBLP. In the first query, we set q = Prof. Paul Kocher, $\mathcal{P} = (APVPA)$, $K = \{$ "Attack", "meltdown" $\}$, and $k = 10$. Note that we regard the types of conference and journal as V. As shown in Table 2, the first community contains ten researchers who collaborated intensively. On the other hand, some researchers have published papers containing keyword "Attack" in a journal containing "meltdown". This community includes those researchers who can cooperate in a specific field, which will help researchers find new collaborators to expand their research field. In the second query, we set q =Prof. Paul Kocher, $\mathcal{P} = (APTPA)$, $K = \{$ "Attack", "Meltdown" $\}$, and $k = 10$. We get the second community contains seventeen researchers, as shown in Table 2. We can see that the second community has seven more people than the first community because we use "T" to constraint the community instead of "V". Compared with the first community, the second community realizes the discovery of potential collaborators without considering conferences or journals because the keywords "meltdown" contained by the vertices with vertex type "V" in DBLP are all lowercase. Therefore, the second community can help researchers find more potential partners than the first community.

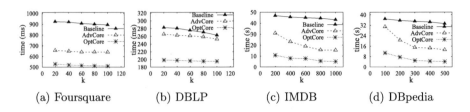

(a) Foursquare   (b) DBLP   (c) IMDB   (d) DBpedia

Fig. 5. Runtime of different algorithms

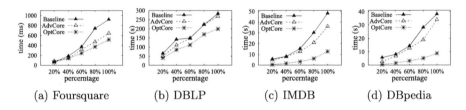

(a) Foursquare   (b) DBLP   (c) IMDB   (d) DBpedia

Fig. 6. Scalability test of different algorithms

4.3   Efficiency Testing

Runtime of Baseline, AdvCore, and OptCore. We evaluate the runtime of Baseline, AdvCore, and OptCore for the keyword-centric community query in HINs. As shown in Fig. 5(a), OptCore is consistently faster than AdvCore and Baseline. Since Foursquare is sparse and small in scale, we can see that all these three algorithms have a short response time. As shown in Fig. 5(c), we can see that only OptCore can respond to the query within ten seconds. Unlike Foursquare, IMDB has many vertices and edges, and there are few vertex types in this data set, which means that for a meta path $\mathcal{P}$, there are many path instances of $\mathcal{P}$ in IMDB. Since OptCore is implemented based on the optimization approach, the response time of OptCore is short than the other two approaches.

Scalability Test. For each dataset, we randomly select 20%, 40%, 60%, 80%, and 100% vertices and get four subgraphs induced by these vertices, respectively. We run Baseline, AdvCore, and OptCore on all datasets. According to the results, as shown in Fig. 6, we can see that these three algorithms scale well with the number of vertices.

5   Related Work

Keyword Search. The keyword search over graphs mainly focuses on the connection between the vertices and the keywords in the query. The semantics used in the existing works can be roughly divided into two categories, one is tree semantics [8,14,18,22] and the other is subgraph semantics [19,21]. Among the tree semantics, Steiner trees are used in [2] to present a new backward search algorithm. In [8], a dynamic programming approach for finding all Steiner trees in graphs. The dynamic programming approach is feasible for input queries with a small number of keywords. The algorithm proposed in [13] follows Lawler's procedure [20] produces Steiner trees with polynomial delay. For the subgraph semantics, Kargar and An [19] find the subgraph containing all keywords in K, which is the set of the keywords. The authors use the sum of the shortest distance between all vertex pairs to measure the weight. Lei et al. [21] study the problem of clustering based on keywords. However, these semantics could be used to solve our problems.

Community Search. Community search aims to find connected subgraphs containing a query vertex. People use some metrics to ensure the cohesiveness of the community found in a graph. The minimum degree metric is the most frequent one used in the problem of the community search. It requires that the degree of each vertex in the community is at least k, which is similar to the constraint of the k-core [1,3,25]. For example, Sozio et al. proposed to find a community as the connected k-core containing the query vertex in [26]. Zhang et al. solve the keyword-centric community search problem over attribute graphs based on k-core in [32]. The other metrics used in the problem of community search are k-truss [5,15,31], k-clique [6,30] and K-ECC respectively. For example, Huang et al. and Chen et al. used k-truss as a metric to search the community in [4,16]; Yuan et al. proposed a k-clique percolation community model based on k-clique to solve the densest clique percolation community search problem in [30]. However, all these works focus on homogeneous graphs. We cannot use them to solve the keyword-centric community search problem over HINs.

6   Conclusion

In this paper, we study the problem of keyword-centric community search over HINs. We propose a basic algorithm, as shown in Algorithm 1 to find the community. However, the basic algorithm is very costly. Then we propose an advanced algorithm using a new method of traversing the search space based on trees. Since the trees are built based on the query vertex and the given meta-path $\mathcal{P}$, the advanced algorithm is not suitable for online query. According to that, we propose an optimization algorithm based on index trees to solve the problem. Extensive experiments on large real-world networks demonstrate the effectiveness and efficiency of our solution.

Acknowledgement. Zhiwei Zhang is supported by National Key R&D Program of China (Grant No. 2020YFB1707902), NSFC (Grant No. 62072035), Hong Kong GRF (Grant No. 12201518) and Zhejiang Lab (Grant No. 2020KE0AB04). Ye Yuan is supported by the NSFC (Grant No. 61932004) and the Fundamental Research Funds for the Central Universities (Grant No. N181605012). Guoren Wang is supported by the NSFC (Grant No. 61732003 and 61729201).

References

1. Batagelj, V., Zaversnik, M.: An O(m) algorithm for cores decomposition of networks. arXiv preprint cs/0310049 (2003)
2. Bhalotia, G., Hulgeri, A., Nakhe, C., Chakrabarti, S., Sudarshan, S.: Keyword searching and browsing in databases using banks. In: ICDE, pp. 431–440. IEEE (2002)
3. Bonchi, F., Khan, A., Severini, L.: Distance-generalized core decomposition. In: ICDM, pp. 1006–1023 (2019)
4. Chen, L., Liu, C., Zhou, R., Li, J., Yang, X., Wang, B.: Maximum co-located community search in large scale social networks. Proc. VLDB Endow. $11(10)$, 1233–1246 (2018)
5. Cohen, J.: Trusses: cohesive subgraphs for social network analysis. National Security Agency Technical Report 16, pp. 3–29 (2008)
6. Cui, W., Xiao, Y., Wang, H., Lu, Y., Wang, W.: Online search of overlapping communities. In: SIGMOD, pp. 277–288 (2013)
7. Cui, W., Xiao, Y., Wang, H., Wang, W.: Local search of communities in large graphs. In: SIGMOD, pp. 991–1002 (2014)
8. Ding, B., Yu, J.X., Wang, S., Qin, L., Zhang, X., Lin, X.: Finding top-k min-cost connected trees in databases. In: ICDE, pp. 836–845. IEEE (2007)
9. Fang, Y., Cheng, R., Luo, S., Hu, J.: Effective community search for large attributed graphs. Proc. VLDB Endow. $9(12)$, 1233–1244 (2016)
10. Fang, Y., et al.: A survey of community search over big graphs. VLDB J. $29(1)$, 353–392 (2020)
11. Fang, Y., Yang, Y., Zhang, W., Lin, X., Cao, X.: Effective and efficient community search over large heterogeneous information networks. Proc. VLDB Endow. $13(6)$, 854–867 (2020)
12. Fortunato, S.: Community detection in graphs. Phys. Rep. $486(3–5)$, 75–174 (2010)
13. Golenberg, K., Kimelfeld, B., Sagiv, Y.: Keyword proximity search in complex data graphs. In: SIGMOD, pp. 927–940 (2008)
14. Hristidis, V., Papakonstantinou, Y.: DISCOVER: keyword search in relational databases. In: VLDB, pp. 670–681. Elsevier (2002)
15. Huang, X., Cheng, H., Qin, L., Tian, W., Yu, J.X.: Querying k-truss community in large and dynamic graphs. In: SIGMOD, pp. 1311–1322 (2014)
16. Huang, X., Lakshmanan, L.V.: Attribute-driven community search. Proc. VLDB Endow. $10(9)$, 949–960 (2017)
17. Huang, Z., Zheng, Y., Cheng, R., Sun, Y., Mamoulis, N., Li, X.: Meta structure: computing relevance in large heterogeneous information networks. In: KDD, pp. 1595–1604 (2016)
18. Kacholia, V., Pandit, S., Chakrabarti, S., Sudarshan, S., Desai, R., Karambelkar, H.: Bidirectional expansion for keyword search on graph databases. In: VLDB, pp. 505–516 (2005)
19. Kargar, M., An, A.: Keyword search in graphs: finding r-cliques. Proc. VLDB Endow. $4(10)$, 681–692 (2011)
20. Lawler, E.L.: A procedure for computing the k best solutions to discrete optimization problems and its application to the shortest path problem. Manage. Sci. $18(7)$, 401–405 (1972)
21. Li, G., Ooi, B.C., Feng, J., Wang, J., Zhou, L.: EASE: an effective 3-in-1 keyword search method for unstructured, semi-structured and structured data. In: SIGMOD, pp. 903–914 (2008)

22. Liu, F., Yu, C., Meng, W., Chowdhury, A.: Effective keyword search in relational databases. In: SIGMOD, pp. 563–574 (2006)
23. Meng, C., Cheng, R., Maniu, S., Senellart, P., Zhang, W.: Discovering meta-paths in large heterogeneous information networks. In: WWW, pp. 754–764 (2015)
24. Newman, M.E., Girvan, M.: Finding and evaluating community structure in networks. Phys. Rev. E **69**(2), 026113 (2004)
25. Seidman, S.B.: Network structure and minimum degree. Soc. Netw. **5**(3), 269–287 (1983)
26. Sozio, M., Gionis, A.: The community-search problem and how to plan a successful cocktail party. In: KDD, pp. 939–948 (2010)
27. Sun, Y., Han, J., Zhao, P., Yin, Z., Cheng, H., Wu, T.: RankClus: integrating clustering with ranking for heterogeneous information network analysis. In: EDBT, pp. 565–576 (2009)
28. Sun, Y., Norick, B., Han, J., Yan, X., Yu, P.S., Yu, X.: PathSelClus: integrating meta-path selection with user-guided object clustering in heterogeneous information networks. TKDD **7**(3), 1–23 (2013)
29. Sun, Y., Yu, Y., Han, J.: Ranking-based clustering of heterogeneous information networks with star network schema. In: KDD, pp. 797–806 (2009)
30. Yuan, L., Qin, L., Zhang, W., Chang, L., Yang, J.: Index-based densest clique percolation community search in networks. IEEE Trans. Knowl. Data Eng. **30**(5), 922–935 (2017)
31. Zhang, Y., Yu, J.X.: Unboundedness and efficiency of truss maintenance in evolving graphs. In: SIGMOD, pp. 1024–1041 (2019)
32. Zhang, Z., Huang, X., Xu, J., Choi, B., Shang, Z.: Keyword-centric community search. In: ICDE, pp. 422–433. IEEE (2019)
33. Zhou, Y., Liu, L.: Social influence based clustering of heterogeneous information networks. In: KDD, pp. 338–346 (2013)

KGSynNet: A Novel Entity Synonyms Discovery Framework with Knowledge Graph

Yiying Yang[1], Xi Yin[1], Haiqin Yang[1(✉)], Xingjian Fei[1], Hao Peng[2(✉)],
Kaijie Zhou[1], Kunfeng Lai[1], and Jianping Shen[1]

[1] Ping An Life Insurance Company of China, Ltd., Shenzhen, China
{yangyiying283,yinxi445,feixingjian568,zhoukaijie002,laikunfeng597,
shenjianping324}@pingan.com.cn, hqyang@ieee.org
[2] BDBC, Beihang University, Beijing, China
penghao@act.buaa.edu.cn

Abstract. Entity synonyms discovery is crucial for entity-leveraging applications. However, existing studies suffer from several critical issues: (1) the input mentions may be out-of-vocabulary (OOV) and may come from a different semantic space of the entities; (2) the connection between mentions and entities may be hidden and cannot be established by surface matching; and (3) some entities rarely appear due to the long-tail effect. To tackle these challenges, we facilitate knowledge graphs and propose a novel entity synonyms discovery framework, named *KGSynNet*. Specifically, we pre-train subword embeddings for mentions and entities using a large-scale domain-specific corpus while learning the knowledge embeddings of entities via a joint TransC-TransE model. More importantly, to obtain a comprehensive representation of entities, we employ a specifically designed *fusion gate* to adaptively absorb the entities' knowledge information into their semantic features. We conduct extensive experiments to demonstrate the effectiveness of our *KGSynNet* in leveraging the knowledge graph. The experimental results show that the *KGSynNet* improves the state-of-the-art methods by 14.7% in terms of hits@3 in the offline evaluation and outperforms the BERT model by 8.3% in the positive feedback rate of an online A/B test on the entity linking module of a question answering system.

Keywords: Entity synonyms discovery · Knowledge graph

1 Introduction

Entity synonyms discovery is crucial for many entity-leveraging downstream applications such as entity linking, information retrieval, and question answering (QA) [19,28]. For example, in a QA system, a user may interact with a chatbot as follows:

Y. Yang and X. Yin—Equal contribution.

© Springer Nature Switzerland AG 2021
C. S. Jensen et al. (Eds.): DASFAA 2021, LNCS 12681, pp. 174–190, 2021.
https://doi.org/10.1007/978-3-030-73194-6_13

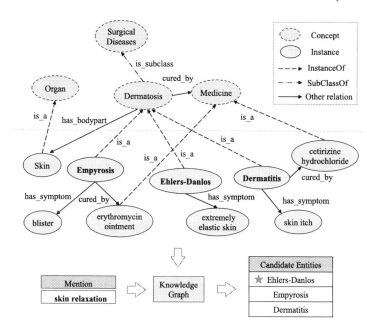

Fig. 1. An illustration of linking the synonymous entity of the mention "skin relaxation" to "Ehlers-Danlos" with the help of an external knowledge graph.

User query: Am I qualified for the new insurance policy as I suffer from **skin relaxation** recently?

System reply: Unfortunately, based on the policy, you may fall into the terms of **Ehlers-Danlos**, which may exclude your protection. Please contact our agents for more details.

In this case, we can correctly answer the user's query only linking the mention of "skin relaxation" to the entity, "Ehlers-Danlos". This is equivent to the entity synonyms discovery task, i.e., automatically identifying the synonymous entities for a given mention or normalizing an informal mention of an entity to its standard form [8,26].

In the literature, various methods, such as DNorm [15], JACCARD-based methods [27], and embedding-based methods [6,11], have been proposed to solve this task. They usually rely on matching of syntactic string [8,27] or lexical embeddings [6,11,25] to build the connections. Existing methods suffer from the following critical issues: (1) the input mentions and the entities are often out-of-vocabulary (OOV) and lie in different semantic spaces since they may come from different sources; (2) the connection between mentions and entities may be hidden and cannot be established by surface matching because they scarcely appear together; and (3) some entities rarely appear in the training data due to the long-tail effect.

To tackle these challenges, we facilitate knowledge graphs and propose a novel entity synonyms discovery framework, named *KGSynNet*. Our *KGSynNet* resolves the OOV issue by pre-training the subword embeddings of mentions and entities using a domain-specific corpus. Moreover, we develop a novel TransC-TransE model to jointly learn the knowledge embeddings of entities by exploiting the advantages of both TransC [17] in distinguishing concepts from instances and TransE [4] in robustly modeling various relations between entities. Moreover, a *fusion gate* is specifically-designed to adaptively absorb the knowledge embeddings of entities into their semantic features. As illustrated in Fig. 1, our *KGSynNet* can discover the symptom of "extremely elastic skin" in the entity of "Ehler-Danlos" and link the mention of "skin relaxation" to it.

In summary, our work consists of the following contributions:

- We study the task of automatic entity synonyms discovery, a significant task for entity-leveraging applications, and propose a novel neural network architecture, namely *KGSynNet*, to tackle it.
- Our proposed *KGSynNet* learns the pre-trained embeddings of mentions and entities from a domain-specific corpus to resolve the OOV issue. Moreover, our model harnesses the external knowledge graph by first encoding the knowledge representations of entities via a newly proposed TransC-TransE model. Further, we adaptively incorporate the knowledge embeddings of entities into their semantic counterparts by a specifically-designed *fusion gate*.
- We conduct extensive experiments to demonstrate the effectiveness of our proposed *KGSynNet* framework while providing detailed case studies and errors analysis. Our model significantly improves the state-of-the-art methods by 14.7% in terms of the offline hits@3 and outperforms the BERT model by 8.3% in the online positive feedback rate.

2   Related Work

Based on how the information is employed, existing methods can be divided into the following three lines:

- The first line of research focuses on capturing the surface morphological features of sub-words in mentions and entities [8,9,27]. They usually utilize lexical similarity patterns and the synonym rules to find the synonymous entities of mentions. Although these methods are able to achieve high performance when the given mentions and entities come from the same semantic space, they fail to handle terms with semantic similarity but morphological difference.
- The second line of research tries to learn semantic embeddings of words or sub-words to discover the synonymous entities of mentions [6,10,11,16,19]. For example, the term-term synonymous relation has been included to train the word embeddings [11]. More heuristic rule-based string features are expanded to learn word embeddings to extract medical synonyms [26]. These methods employ semantic embeddings pretrained from massive text corpora and improve the discovery task in a large margin compared to the direct string matching methods. However, they perform poorly when the terms rarely appear in the corpora but reside in external knowledge bases.

- The third line of research aims to incorporate external knowledge from either the unstructured term-term co-occurrence graph or the structured knowledge graph. For example, Wang et al. [29] utilizes both semantic word embeddings and a term-term co-occurrence graph extracted from unstructured text corpora to discover synonyms on privacy-aware clinical data. More powerful methods, such as SynSetMine [23] and SA-ESF [13], have been proposed to leverages the synonym of entities in knowledge graphs or the knowledge representations. They ignore other relations among entities, e.g., the hypernym-hyponym relations, and lack a unified way to absorb the information. This motivates our further exploration in this work.

3   Methodology

Here, we present the task and the main modules of our *KGSynNet* accordingly.

3.1   Task Definition

The task of entity synonyms discovery is to train a model to map the mention to synonymous entities as accurate as possible given a set of annotated mention-entity pairs $\mathcal{Q}$, a knowledge graph $\mathcal{KG}$, and a domain-specific corpus, $\mathcal{D}$. The mention-entity pairs, $\mathcal{Q} = \{(q_i, t_i)\}_{i=1}^{N}$, record the mentions from queries and their corresponding synonymous entities, where N is the number of annotated pairs, $q_i = q_{i1} \ldots q_{i|q_i|}$ denotes the i-th mention with $|q_i|$ subwords and $t_i = t_{i1} \ldots t_{i|t_i|} \in \mathcal{E}$ denotes the i-th entity in $\mathcal{KG}$ with $|t_i|$ subwords. The knowledge graph is formalized as $\mathcal{KG} = \{\mathcal{C}, \mathcal{I}, \mathcal{R}, \mathcal{S}\}$, where $\mathcal{C}$ and $\mathcal{I}$ denote the sets of concepts and instances, respectively, $\mathcal{R}$ is the relation set and $\mathcal{S}$ is the triple set. Based on the above definition, we have $\mathcal{E} = \mathcal{C} \cup \mathcal{I}$. After we train the model, for a given mention, we can recommend a list of synonymous entities from the knowledge graph. The domain-specific corpus, $\mathcal{D}$, is used for learning the embeddings of mentions and entities.

As illustrated in Fig. 2, our proposed *KGSynNet* consists of four main modules: (1) a semantic encoder module to represent mentions and entities; (2) a knowledge encoder module to represent the knowledge of entities by a jointly-learned TransC-TransE model; (3) a feature fusion module to adaptively incorporate knowledge information via a specifically designed *fusion gate*; (4) a classifier with a similarity matching metric to train the entire model.

3.2   Semantic Encoder

Given a mention-entity pair, (q, t), we may directly apply existing embeddings, e.g., Word2Vec [18], or BERT [7], on q and t to represent the semantic information of mentions and entities. However, it is not effective because many subwords are out-of-vocabulary (OOV), since the pre-trained embeddings are trained from corpora in general domains.

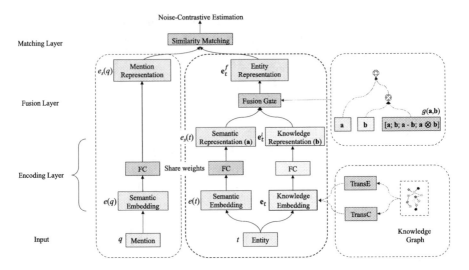

Fig. 2. The architecture of our *KGSynNet*.

To leverage the contextualized information of each mention and entity from $\mathcal{D}$, we train a set of subword-level Word2Vec embeddings from scratch on $\mathcal{D}$, and apply them to initialize the semantic representations of the subwords of the mentions and the entities in $\mathcal{Q}$. Then, similar to the fastText approach [2], we obtain the initialized semantic representations of mentions and entities by averaging their subword representations:

$$e(q) = \frac{1}{|q|} \sum_{k=1}^{|q|} e(q_k), \quad e(t) = \frac{1}{|t|} \sum_{k=1}^{|t|} e(t_k). \tag{1}$$

After that, the semantic embeddings of the mentions and the entities are further fed into a two-layer fully-connected (FC) network to extract deeper semantic features. Here, we adopt shared weights as in [5] to transform the learned embedding $e(v)$ into a semantic space of k-dimension:

$$e_s(v) = \tanh(\mathbf{W}_2 \tanh(\mathbf{W}_1 e(v) + b_1) + b_2) \in \mathbb{R}^k, \tag{2}$$

where v can be a mention or an entity. The parameters, $\mathbf{W}_1 \in \mathbb{R}^{k \times d}$ and $\mathbf{W}_2 \in \mathbb{R}^{k \times k}$, are the weights on the corresponding layers of the FC network. $b_1 \in \mathbb{R}^k$ and $b_2 \in \mathbb{R}^k$ are the biases at the corresponding layers.

3.3   Knowledge Encoder

Though entities can be encoded in the semantic space as detailed above, their representations are not precise enough due to lack of the complementary information included in the knowledge graph.

In the knowledge graph $\mathcal{KG}$, the relation set $\mathcal{R}$ is defined by $\mathcal{R} = \{r_e, r_c\} \cup \mathcal{R}_l \cup \mathcal{R}_{\mathcal{IC}} \cup \mathcal{R}_{cc}$, where r_e is an instanceOf relation, r_c is a subClassOf relation, $\mathcal{R}_l$ is the instance-instance relation set, $\mathcal{R}_{\mathcal{IC}}$ is the Non-Hyponym-Hypernym (NHH) instance-concept relation set, and $\mathcal{R}_{cc}$ is the NHH concept-concept relation set. It is noted that different from the three kinds of relations defined in TransC [17], we specifically categorize the relations into five types to differentiate the NHH relations of the instance-concept pairs from the concept-concept pairs. Therefore, the triple set $\mathcal{S}$ can be divided into the following five disjoint subsets:

1. The instanceOf triple set: $\mathcal{S}_e = \left\{ \left(i, r_e, c \right)_k \right\}_{k=1}^{|\mathcal{S}_e|}$, where $i \in \mathcal{I}$ is an instance, $c \in \mathcal{C}$ is a concept, and r_e is the instanceOf relation.
2. The subClassOf triple set: $\mathcal{S}_c = \left\{ \left(c_i, r_c, c_j \right)_k \right\}_{k=1}^{|\mathcal{S}_c|}$, where $c_i, c_j \in \mathcal{C}$ are concepts, c_i is a sub-concept of c_j, and r_c is the subClassOf relation.
3. The instance-instance triple set: $\mathcal{S}_l = \left\{ \left(i, r_{ij}, j \right)_k \right\}_{k=1}^{|\mathcal{S}_l|}$, where $r_{ij} \in \mathcal{R}_l$ defines the instance-instance relation from the head instance i to the tail instance j.
4. The NHH instance-concept triple set: $\mathcal{S}_{\mathcal{IC}} = \left\{ \left(i, r_{ic}, c \right)_k \right\}_{k=1}^{|\mathcal{S}_{\mathcal{IC}}|}$, where i and c are defined similarly as $\mathcal{S}_e$. $r_{ic} \in \mathcal{R}_{\mathcal{IC}}$ is an NHH instance-concept relation.
5. The NHH concept-concept triple set: $\mathcal{S}_{cc} = \left\{ \left(c_i, r_{c_i c_j}, c_j \right)_k \right\}_{k=1}^{|\mathcal{S}_{cc}|}$, where $c_i, c_j \in \mathcal{C}$ denote two concepts, $r_{c_i c_j} \in \mathcal{R}_{cc}$ is an NHH concept-concept relation.

We now learn the knowledge embeddings of entities. Since TransE [4] is good at modeling general relations between entities while TransC [17] excelling in exploiting the hierarchical relations in the knowledge graph, we propose a unified model, the TransC-TransE model, to facilitate the advantage of both models.

Specifically, TransE represents an entity by $\mathbf{v} \in \mathbb{R}^n$, where n is the size of the knowledge embedding, and defines the loss for the instance-instance triples [4]:

$$f_l(i, r_{ij}, j) = \|\mathbf{v}_i + \mathbf{v}_{r_{ij}} - \mathbf{v}_j\|_2^2, \tag{3}$$

where $(i, r_{ij}, j) \in \mathcal{S}_l$ denotes a triple in the instance-instance relation set, $\mathbf{v}_i$, $\mathbf{v}_{r_{ij}}$, and $\mathbf{v}_j$ denote the corresponding TransE representations.

In TransC, an instance i is represented by a vector, $\mathbf{v}_i \in \mathbb{R}^n$, same as that of an entity in TransE. A concept c is represented by a sphere, denoted by $(\mathbf{p}_c, m_c)$, where $\mathbf{p}_c \in \mathbb{R}^n$ and $m_c \in \mathbb{R}_+$ define the corresponding center and radius for the concept, respectively. The corresponding losses can then be defined as follows:

– The loss for the instanceOf triples [17]:

$$f_e(i, c) = \|\mathbf{v}_i - \mathbf{p}_c\|_2 - m_c, \quad \forall i \in c. \tag{4}$$

– The loss for the subClassOf triples [17]:

$$f_c(c_i, c_j) = \begin{cases} m_{c_i} - m_{c_j}, & c_j \text{ is a subclass of } c_i, \text{ or } c_j \subseteq c_i \\ \|\mathbf{p}_{c_i} - \mathbf{p}_{c_j}\|_2 + m_{c_i} - m_{c_j}, & \text{otherwise} \end{cases}. \tag{5}$$

However, the spherical representation is not precise enough to model the NHH relations. We therefore denote the concept of c by an additional node embedding, $\mathbf{v}_c \in \mathbb{R}^n$, and define the following additional loss functions:

– The loss for the NHH instance-concept triples [4]:

$$f_{IC}(i, r_{ic}, c) = \|\mathbf{v}_i + \mathbf{v}_{r_{ic}} - \mathbf{v}_c\|_2^2, \tag{6}$$

where the triplet $(i, r_{ic}, c) \in \mathcal{S}_{IC}$ denotes the NHH instance-concept relation r_{ic} connecting the instance i to the concept c.
– The loss for the NHH concept-concept triples [4]:

$$f_{cc}(c_i, r_{c_i c_j}, c_j) = \|\mathbf{v}_{c_i} + \mathbf{v}_{r_{c_i c_j}} - \mathbf{v}_{c_j}\|_2^2, \tag{7}$$

where the triplet $(c_i, r_{c_i c_j}, c_j) \in \mathcal{S}_{cc}$ denotes the NHH concept-concept relation $r_{c_i c_j}$ connecting the concept c_i to the concept c_j.

Therefore, the knowledge embeddings of entities are learned by minimizing the following objective function:

$$\mathcal{L}_k = \sum_{(i,r_e,c) \in \mathcal{S}_e} f_e(i,c) + \sum_{(c_i,r_c,c_j) \in \mathcal{S}_c} f_c(c_i, c_j) + \sum_{(i,r_{ij},j) \in \mathcal{S}_l} f_l(i, r_{ij}, j)$$
$$+ \sum_{(i,r_{ic},c) \in \mathcal{S}_{IC}} f_{IC}(i, r_{ic}, c) + \sum_{(c_i,r_{c_i c_j},c_j) \in \mathcal{S}_{CC}} f_{cc}(c_i, r_{c_i c_j}, c_j). \tag{8}$$

It is noted that our objective differs from TransC by explicitly including both the NHH instance-concept relations and the NHH concept-concept relations. Similarly, we apply the negative sampling strategy and the margin-based ranking loss to train the model as in [17].

After training the unified TransC-TrainsE model in Eq. (8), we obtain the knowledge embeddings for both instances and concepts, e.g., $\mathbf{v}_i$ for an instance i, and the representation of $(\mathbf{p}_c, m_c)$ and $\mathbf{v}_c$ for a concept c. For simplicity and effectiveness, we average the center and the node embedding of a concept to yield its final knowledge embedding $\mathbf{e}_t$:

$$\mathbf{e}_t = \begin{cases} \mathbf{v}_t & \forall t \in \mathcal{I} \\ (\mathbf{p}_t + \mathbf{v}_t)/2 & \forall t \in \mathcal{C} \end{cases}. \tag{9}$$

Similar to the semantic embeddings, the learned knowledge embeddings of entities obtained in Eq. (9) are transformed into the same k-dimensional semantic space by a two-layer fully connected network to yield $\mathbf{e}_t^l$:

$$\mathbf{e}_t^l = \tanh(\mathbf{W}_4(\tanh(\mathbf{W}_3 \mathbf{e}_t + b_3)) + b_4) \in \mathbb{R}^k, \tag{10}$$

where $\mathbf{W}_3 \in \mathbb{R}^{k \times q}$ and $\mathbf{W}_4 \in \mathbb{R}^{k \times k}$ are the weights on the corresponding layers of the FC network. $b_3 \in \mathbb{R}^k$ and $b_4 \in \mathbb{R}^k$ are the biases at the layers.

3.4   Fusion Gate

A critical issue in the task is that the semantic features and the knowledge embeddings are learned separately. To effectively integrate these two types of information, we design a fusion network, named *Fusion Gate*, to adaptively absorb the transformed knowledge information $\mathbf{e}_t^l$ into the semantic informa-tion $e_s(t)$ for an entity t. As illustrated in the upper right grid box of Fig. 2, the final representation of an entity t is computed by

$$\mathbf{e}_t^f = e_s(t) + e_t^l \otimes g(e_s(t), \mathbf{e}_t^l). \tag{11}$$

Here, the implementation is motivated by the highway network [24], but is differ-ent on the specific information carrying. Here, we directly feed all the semantic information of the entities to the next level without filtering to guarantee the consistency of the semantic representations between mentions and entities. The interaction of the semantic embeddings and knowledge embeddings of the enti-ties is then fulfilled by the transform gate to determine the amount of knowledge incorporated into the semantic feature, defined by $g(\mathbf{a}, \mathbf{b})$:

$$g(\mathbf{a}, \mathbf{b}) = \text{Softmax}(\mathbf{W}_g[\mathbf{a}; \mathbf{b}; \mathbf{a} - \mathbf{b}; \mathbf{a} \otimes \mathbf{b}]), \tag{12}$$

where $\mathbf{W}_g \in \mathbb{R}^{k \times 4k}$ is the weight of a fully-connected network to reduce the dimension of the concatenated features. The first two features maintain the orig-inal form while the latter two measuring the "similarity" or "closeness" of the two features. This allows to compute the high-order interactions between two input vectors [5,20]. Finally, the Softmax operator is applied to determine the proportion of the flow-in knowledge.

3.5   Similarity Matching and Classification

As the training data only consist of the positive pairs, for each pair (q_i, t_i), we additionally sample some negative pairs $\{(q_i, t_{i_j})\}_{j=1}^{N_i}$, where t_{i_j} is sampled from other mention-to-entity pairs and N_i is the number of sampled negative pairs. Hence, we derive the objective function for the final matching:

$$\mathcal{L}_m = \sum_{i=1}^{N} -\log \left(\frac{\exp\left(e_s(q_i)^T \mathbf{e}_{t_i}^f\right)}{\exp\left(e_s(q_i)^T \mathbf{e}_{t_i}^f\right) + \sum_{j=1}^{N_i} \exp\left(e_s(q_i)^T \mathbf{e}_{t_{i_j}}^f\right)} \right). \tag{13}$$

It is noted that each term in Eq. (13) defines the Noise-Contrastive Estimation (NCE) [12], which is the cross-entropy of classifying the positive pair (q_i, t_i). After training, given a new mention q, we can determine the list of the candidate entities by the rank of $e_s(q_i)^T \mathbf{e}_{t_i}^f$.

4   Experiments

In the following, we present the curated dataset along with the associated knowl-edge graph, as well as the experimental details.

Table 1. Data statistics.

Knowledge Graph	All	Insurance	Occupation	Medicine	Cross Domain
# Entities	75,153	1,409	2,587	71,157	0
# Entity_type	17	2	2	13	0
# Relations	1,120,792	2,827	2,580	1,098,280	17,105
# Relation_type	20	2	2	13	4
# Mention-entity pairs in Train/Dev/Test					45,500/5,896/5,743
# Regular cases/# Difficult cases					5,303/440

4.1 Datasets

Knowledge Graph. The existing open-source knowledge graphs [1,3] cannot be used for this task, because they do not provide sufficient disease entities and relations required by the task. Therefore, we construct a specific knowledge graph ($\mathcal{KG}$) to verify this task. Table 1 records the statistics of the constructed $\mathcal{KG}$, a heterogeneous $\mathcal{KG}$ with entities collected from three categories: *Insurance Products*, *Occupation*, and *Medicine*. In *Insurance Products*, there are 1,393 insurance products and 16 concepts; while in *Occupation*, there are 1,863 instances and 724 concepts obtained from the nation's professional standards[1]. Both *Insurance Products* and *Occupation* contain only two types of relations, i.e., the instanceOf relation and the subClassOf relation. In *Medicine*, 45K disease entities and 9,124 medical concepts are extracted from three different resources: (1) raw text of insurance products' clauses; (2) users' query logs in the app; (3) the diagnostic codes of International Classification of Diseases (ICD-10). Furthermore, 18K other types of medical entities, such as symptom, body part, therapy, and treatment material, are extracted from some open-source medical knowledge graphs[2]. The relation types include not only instanceOf and subClassOf, but also the instance-instance relations, the NHH concept-instance relations, and the NHH concept-concept relations, 13 types in total.

Data. We collect a large-scale Chinese medical corpus from 14 medical textbooks[3], 3 frequently used online medical QA forums, and some QA forums[4]. We also deploy a self-developed BERT-based NER tool to extract 100K disease mentions from users' query logs in the professional app. From the extracted disease mentions and $\mathcal{KG}$ entities, we generate 300K candidate synonymous mention-entity pairs based on the similarity score computed by BERT. The extracted mention-entity candidates are double-blindly labeled to obtain 57,139 high-quality disease mention-entity synonym pairs. After that, the dataset is

---

[1] http://www.jiangmen.gov.cn/attachment/0/131/131007/2015732.pdf.

[2] http://openkg.cn/dataset/symptom-in-chinese; http://openkg.cn/dataset/omaha-data.

[3] https://github.com/scienceasdf/medical-books.

[4] https://github.com/lrs1353281004/Chinese_medical_NLP.

randomly split into the sets of training, development, and test, respectively, approximately at a ratio of 8:1:1. We further divide the test set (the All case group) into two groups based on the surface form similarity. That is, a Regular case means that there is at least one identical subword between the mention and the entity, while the rest pairs belong to the Difficult case group.

4.2   Compared Methods

We compare *KGSynNet* with the following strong baselines:

(1) JACCARD [21]: a frequently used similarity method based on the surface matching of mentions and entities;
(2) Word2Vec [6]: a new subword embedding is trained on the medical corpus to learn representations. Cosine similarity is then applied to the average of subword embeddings of each mention-entity pair to rank their closeness;
(3) CNN [19]: a CNN-based Siamese network is trained using the triplet loss with the newly trained word2vec embeddings for the mentions and entities.
(4) BERT [7]: the [CLS] representations of mentions and entities are extracted from the fine-tuned BERT to compute their cosine similarity;
(5) DNorm [15]: one of the most popular methods that utilizes the TF-IDF embedding and a matching matrix, trained by the margin ranking loss, to determine the similarity score between mentions and entities.
(6) SurfCon [29]: one of the most popular methods that constructs a term-term co-occurrence graph from the raw corpus to capture both the surface information and the global context information for entity synonym discovery.

4.3   Experimental Setup and Evaluation Metrics

The number of sampled negative mention-entity pairs is tuned from $\{10, 50, 100, 200, 300\}$ and set to 200 as it attains the best performance in the development set. ADAM is adopted as the optimizer with an initial learning rate of 0.001. The training batch size is 32, and the dimension of the knowledge graph embedding is 200. Besides, the dimension of the semantic embeddings of both mentions and entities are set to 500, and the dimensions of the first and the second FC networks are set to 300. These parameters are set by a general value and tuned in a reasonable range. Dropout is applied in the FC networks and selected as 0.5 from $\{0.3, 0.5, 0.7\}$. The knowledge embedding is trained by an open-source package[5]. Early stopping is implemented when the performance in the development set does not improve in the last 10 epochs.

To provide fair comparisons, we set the same batch size, embedding sizes, and dropout ratio to all baseline models. For SurfCon, we construct a co-occurrence graph of 24,315 nodes from our collected Chinese medical corpus, and obtain the graph embedding according to [29][6].

---

[5] https://github.com/davidlvxin/TransC.
[6] https://github.com/yzabc007/SurfCon.

Filtered hits@k, the proportion of correct entities ranked in the top k predictions by filtering out the synonymous entities to the given mention in our constructed $\mathcal{KG}$, because it is an effective metric to determine the accuracy of entity synonyms discovery [17]. We follow the standard evaluation procedure [4,17] and set $k = 3, 5, 10$ to report the model performance.

4.4 Experimental Results

Rows three to nine of Table 2 report the experimental results of the baselines and our *KGSynNet*. It clearly shows that

- JACCARD yields no hit on the difficult case because it cannot build connections on mentions and entities when they do not contain a common sub-word.
- Word2Vec yields the worst performance on the *All* case and the *Regular* case since the representations of mentions and entities are simply obtained by their mean subword embeddings, which blur the effect of each subword.
- CNN improves Word2Vec significantly because of the Siamese network, but cannot even beat JACCARD due to the poor semantic representation learned from Word2Vec.
- BERT gains further improvement over JACCARD, Word2Vec, and CNN by utilizing the pre-trained embeddings. The improvement is not significant enough especially in the Difficult case because the representation of the token [CLS] does not fully capture the relations between mentions and entities.
- DNorm further improves the performance by directly modeling the interaction between mentions and entities. SurfCon yields the best performance among all baselines because it utilizes external knowledge bases via the term-term co-occurrence graph.
- Our *KGSynNet* beats all baselines in all three cases. Especially, we beat the best baseline, SurfCon, by 14.7%, 10.3%, and 5.6% for the *All* case, 14.2%, 10.0%, and 5.4% for the *Regular* case, and 45.7%, 24.4%, and 10.2% for

Table 2. Experimental results: − means that *KGSynNet* removes the component while → means that *KGSynNet* replaces the fusion method.

Methods	hits@3			hits@5			hits@10		
	All	Regular	Difficult	All	Regular	Difficult	All	Regular	Difficult
JACCARD [21]	52.28%	56.61%	0.00%	58.03%	62.83%	0.00%	63.76%	69.04%	0.00%
Word2Vec [6]	47.00%	50.88%	0.00%	52.28%	56.59%	2.30%	58.31%	63.10%	4.60%
CNN [19]	51.76%	55.69%	4.33%	57.75%	61.98%	6.38%	65.13%	69.72%	9.34%
BERT [7]	54.60%	58.87%	2.96%	60.41%	65.02%	4.78%	66.50%	71.39%	7.52%
DNorm [15]	56.23%	59.78%	12.76%	63.79%	67.58%	17.77%	71.89%	75.64%	26.42%
SurfCon [29]	58.29%	62.02%	12.98%	66.27%	70.11%	19.59%	75.20%	79.03%	28.93%
KGSynNet	**66.84%**	**70.81%**	18.91%	**73.09%**	**77.13%**	24.37%	**79.41%**	**83.35%**	31.89%
−KE	64.91%	69.07%	14.58%	71.56%	75.77%	20.73%	79.12%	83.14%	30.52%
−TransC	65.80%	69.92%	15.95%	71.44%	75.79%	18.91%	78.94%	83.18%	27.80%
→DA	63.51%	67.19%	**19.13%**	70.85%	74.47%	**27.10%**	78.13%	81.77%	**34.17%**
→EF	61.98%	65.85%	15.26%	68.63%	72.54%	21.41%	76.28%	80.29%	27.79%

the *Difficult* case with respect to Hits@3, Hits@5, and Hits@10, respectively. We have also conducted the statistical significance tests, and observe that for the *All* case group, $p << 0.01$ under the paired t-tests. The significant improvement clearly shows that our *KGSynNet* is effective in integrating the knowledge information with the semantic features.

4.5   Ablation Study

To better understand why our *KGSynNet* works well, we compare it with four variants: (1) $-$KE: removing the knowledge embedding and the Fusion Gate; (2) $-$TransC: removing losses of Eq. (4) and Eq. (5) from Eq. (8) of TransC, to learn the knowledge embedding by utilizing only TransE; (3) $\rightarrow$DA: directly adding the learned semantic features and knowledge features of entities together; and (4) $\rightarrow$EF: fusing the learned semantic features and knowledge information via a FC network [30].

Table 2 reports the results of the variants in the last four rows and clearly shows three main findings:

- By excluding the knowledge embedding (see the last fourth row in Table 2), our *KGSynNet* drops significantly for the All case, i.e., 1.93 for hits@3, 1.53 for hits@3, and 0.29 for hits@10, respectively. Similar trends appear for the Regular case and the Difficult case. The performance decay is more serious than those in other variants, $-$TransC and $\rightarrow$DA. This implies the effectiveness of our *KGSynNet* in utilizing the knowledge information.
- By removing TransC, we can see that the performance decays accordingly in all cases. The results make sense because learning the knowledge representation by TransE alone does not specifically model the *InstanceOf* relation and the *SubclassOf* relation. This again demonstrates the effectiveness of our proposed TransC-TransE framework.
- In terms of the fusion mechanism, the performance exhibits similarly under the three metrics. Here, we only detail the results of hits@3. It shows that the performance by *Fusion Gate* beats "DA" and "EF" 3.3 to 5.0 in both the All and Regular cases. However, "DA" improves the performance significantly on the Difficult case, i.e., no common sub-word appearing in the mention-entity pairs. The results make sense because in the Difficult case, the model depends heavily on the external knowledge. Setting the weight to 1, i.e., the largest weight, on the learned knowledge features can gain more knowledge information. On the contrary, "EF" yields the worst performance on the All and Regular cases, but gains slightly better performance than $-$KE on the Difficult case. We conjecture one reason is that the available data is not sufficient to trained a more complicated network in "EF".

4.6   Online Evaluation

Our *KGSynNet* has been deployed in the entity linking module, a key module of the KBQA system of a professional insurance service app, served more than

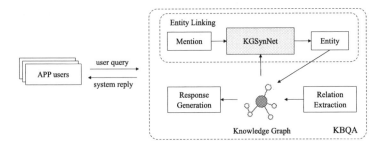

Fig. 3. The architecture of online system.

one million insurance agents. The architecture of the online system is shown in Fig. 3. On average, the requests of the KBQA service of the app are 700K per day with more than 50 requests per second at the peak.

We conducted an A/B test to compare the original BERT model and our *KGSynNet* on the entity linking module of the KBQA system for two weeks. The traffic was evenly split into two groups. Approximately 10% of users' queries involve disease mentions, within which the proportion of queries with user experience feedback is around 5%. Eventually, BERT and *KGSynNet* received about 25K and 26K user feedbacks, respectively. The positive rate of the feedback for BERT is about 34.9%, while the positive rate of *KGSynNet* is about 37.8%, significantly better with $p < 0.05$ under the paired t-test.

Moreover, we randomly selected and labeled 1000 disease related queries from each of the two groups. The proportion of queries involving difficult cases was around 3% in both groups. Results in Table 3 show that *KGSynNet* consistently outperforms BERT in terms of hits@3, hits@5, and hits@10, respectively.

Table 3. Online evaluation results

Metric	BERT			KGSynNet		
	All	Regular	Difficult	All	Regular	Difficult
hits@3	58.2%	59.9%	3.3%	**68.4%**	70.0%	18.8%
hits@5	63.2%	64.9%	6.7%	**75.4%**	77.1%	25.0%
hits@10	70.0%	71.9%	10.0%	**81.7%**	83.4%	31.3%

4.7 Case Studies

We provide several typical examples to show the effectiveness of our *KGSynNet*. In Table 4, four query mentions are selected with the top-5 discovered synonymous entities. The results show that:

– Our *KGSynNet* can successfully detect at least one annotated synonym for each mention. For example, for the mention, "hyperelastic skin", our found top-5 synonymous entities are all correct.

– For the mention of "facial paralysis", other than its synonym "facioplegia", our *KGSynNet* can discover "prosopoplegia" through the semantic equivalence. Other top predicted terms, e.g., "neonatal facial paralysis", "peripheral facial paralysis", and "idiopathic facial paralysis", are all hyponyms of the mention with specific clinical manifestations.

Table 4. Query mentions and the corresponding top 5 synonymous entities: the correct synonyms are underlined.

Mention	Top 5 Found Entities
弹力过度性皮肤 hyperelastic skin	埃莱尔-当洛综合症, 埃勒斯-当洛斯综合症, 皮肤松垂, 埃莱尔-当洛, 皮肤松弛, Ehlers-Danlos syndrome, Ehlers-Danlos syndrome, dermatolysis, Ehlers-Danlos, cutis laxa
肚子痛 stomachache	急性腹泻, 疼痛, 下腹痛, 全身疼痛, 疼痛性脂肪过多症 collywobbles, pain, hypogastralgia, generalized pain, lipomatosis dolorosa
歪嘴风 facial paralysis	面瘫, 面神经痹, 新生儿面部神经麻痹, 周围性面瘫, prosopoplegia, facioplegia, neonatal facial paralysis, peripheral facial paralysis, 特发性面神经瘫痪 idiopathic facial paralysis
倦怠 exhaustion	虚弱,乏力, 张力失常, 失眠症, 弱精 debility, asthenia, dystonia, insomnia, asthenozoospermia

4.8 Error Analysis

We provide a concrete error analysis by sampling 10% of the incorrectly predicted mention-entity pairs in our *KGSynNet*. Table 5 lists the main error types:

– More than half of the errors (54%) occur due to the lack of knowledge in the knowledge graph. For example, since the entity "bow legs" is not in the $\mathcal{KG}$, the mention "knee varus" mistakenly found "knee valgus" and "congenital knee valgus" through surface matching.

– The second largest error comes from hypernyms distraction, which accounts for 29% of the total errors. For example, the mention "pituitary gland cancer" is distracted to its hypernym "brain cancer" and "cerebral cancer", and failed to identify the true entity "pituitary gland malignant tumor".

– Another 12% of the errors are due to the keyword extraction error. For example, the golden entity for the mention, "lung calcification", is "lung mineralization". Our *KGSynNet* makes a wrong extraction on the keyword "calcification" and discovers a wrong entity, "bronchial calcification", for this mention. It seems that this problem may be alleviated by adding an fine-grained feature interaction between mentions and entities in our *KGSynNet*.

Table 5. Error analysis. The "Golden Entity" is the correct entity for the corresponding mention.

Error Type	Proportion	Mention	Golden Entity	Top 2 Predicted Entities
Lack of Knowledge	54%	膝内翻, knee varus	O型腿, bow legs	膝外翻, knee valgus 先天性膝外翻, congenital knee valgus
Hypernym Distraction	29%	脑垂腺癌, pituitary gland cancer	垂体恶性肿瘤, pituitary gland malignant tumor	脑癌, brain cancer 癌性脑病, cerebral cancer
Keyword Extraction Error	12%	肺部钙化, lung calcification	肺矿化, lung mineralization	支气管钙化, bronchial calcification 肺转移瘤, pulmonary metastasis
Others	5%	奥尔布赖特综合症, Albright's syndrome	麦凯恩-奥尔布赖特综合症, McCune-Albright's syndrome	莱特尔综合症, Leiter's syndrome 吉尔伯特综合症, Gilbert's syndrome

5   Conclusion

In this paper, we tackle the task of entity synonyms discovery and propose *KGSynNet* to exploit external knowledge graph and domain-specific corpus. We resolve the OOV issue and semantic discrepancy in mention-entity pairs. Moreover, a jointly learned TransC-TransE model is proposed to effectively represent knowledge information while the knowledge information is adaptively absorbed into the semantic features through *fusion gate* mechanism. Extensive experiments and detailed analysis conducted on the dataset show that our model significantly improves the state-of-the-art methods by 14.7% in terms of the offline hits@3 and outperforms the BERT model by 8.3% in the online positive feedback rate. Regarding future work, we can extend our *KGSynNet* to other domains, e.g., education or justice, to verify its generalization ability.

Acknowledgement. The authors of this paper were supported by NSFC under grants 62002007 and U20B2053.

References

1. Bizer, C., et al.: DBpedia - a crystallization point for the web of data. J. Web Semant. **7**(3), 154–165 (2009)
2. Bojanowski, P., Grave, E., Joulin, A., Mikolov, T.: Enriching word vectors with subword information. Trans. Assoc. Comput. Linguistics **5**, 135–146 (2017)
3. Bollacker, K.D., Evans, C., Paritosh, P., Sturge, T., Taylor, J.: Freebase: a collaboratively created graph database for structuring human knowledge. In: SIGMOD, pp. 1247–1250. ACM (2008)
4. Bordes, A., Usunier, N., García-Durán, A., Weston, J., Yakhnenko, O.: Translating embeddings for modeling multi-relational data. In: NIPS, pp. 2787–2795 (2013)
5. Chen, Q., Zhu, X., Ling, Z., Wei, S., Jiang, H., Inkpen, D.: Enhanced LSTM for natural language inference. In: ACL, pp. 1657–1668 (2017)
6. Cho, H., Choi, W., Lee, H.: A method for named entity normalization in biomedical articles: application to diseases and plants. BMC Bioinform. **18**(1), 1–12, 451 (2017)
7. Devlin, J., Chang, M., Lee, K., Toutanova, K.: BERT: pre-training of deep bidirectional transformers for language understanding. In: NAACL, pp. 4171–4186 (2019)

8. Dogan, R.I., Lu, Z.: An inference method for disease name normalization. In: AAAI (2012)
9. D'Souza, J., Ng, V.: Sieve-based entity linking for the biomedical domain. In: ACL and IJCNLP, pp. 297–302 (2015)
10. Faruqui, M., Dodge, J., Jauhar, S.K., Dyer, C., Hovy, E.H., Smith, N.A.: Retrofitting word vectors to semantic lexicons. In: NAACL, pp. 1606–1615 (2015)
11. Fei, H., Tan, S., Li, P.: Hierarchical multi-task word embedding learning for synonym prediction. In: ACM SIGKDD, pp. 834–842 (2019)
12. Gutmann, M., Hyvärinen, A.: Noise-contrastive estimation: a new estimation principle for unnormalized statistical models. AISTATS **9**, 297–304 (2010)
13. Hu, S., Tan, Z., Zeng, W., Ge, B., Xiao, W.: Entity linking via symmetrical attention-based neural network and entity structural features. Symmetry **11**(4), 453 (2019)
14. Jiang, L., et al.: GRIAS: an entity-relation graph based framework for discovering entity aliases. In: IEEE ICDM, pp. 310–319 (2013)
15. Leaman, R., Dogan, R.I., Lu, Z.: DNorm: disease name normalization with pairwise learning to rank. Bioinformatics **29**(22), 2909–2917 (2013)
16. Li, H., et al.: CNN-based ranking for biomedical entity normalization. BMC Bioinform. **18**(S-11), 79–86 (2017)
17. Lv, X., Hou, L., Li, J., Liu, Z.: Differentiating concepts and instances for knowledge graph embedding. In: EMNLP, pp. 1971–1979 (2018)
18. Mikolov, T., Sutskever, I., Chen, K., Corrado, G.S., Dean, J.: Distributed representations of words and phrases and their compositionality. In: NIPS, pp. 3111–3119 (2013)
19. Mondal, I., et al.: Medical entity linking using triplet network. In: Clinical NLP (2019)
20. Mou, L., et al.: Natural language inference by tree-based convolution and heuristic matching. In: ACL (2016)
21. Niwattanakul, S., Singthongchai, J., Naenudorn, E., Wanapu, S.: Using of Jaccard coefficient for keywords similarity. In: IMECS (2013)
22. Schumacher, E., Dredze, M.: Learning unsupervised contextual representations for medical synonym discovery. JAMIA Open **2**, 538–546 (2019)
23. Shen, J., Lyu, R., Ren, X., Vanni, M., Sadler, B.M., Han, J.: Mining entity synonyms with efficient neural set generation. In: AAAI, pp. 249–256 (2019)
24. Srivastava, R.K., Greff, K., Schmidhuber, J.: Training very deep networks. In: NIPS, pp. 2377–2385 (2015)
25. Sung, M., Jeon, H., Lee, J., Kang, J.: Biomedical entity representations with synonym marginalization. In: ACL, pp. 3641–3650 (2020)
26. Wang, C., Cao, L., Zhou, B.: Medical synonym extraction with concept space models. In: IJCAI, pp. 989–995 (2015)
27. Wang, J., Lin, C., Li, M., Zaniolo, C.: An efficient sliding window approach for approximate entity extraction with synonyms. In: EDBT, pp. 109–120 (2019)

28. Wang, X., et al.: Improving natural language inference using external knowledge in the science questions domain. In: AAAI, pp. 7208–7215 (2019)
29. Wang, Z., Yue, X., Moosavinasab, S., Huang, Y., Lin, S.M., Sun, H.: SurfCon: synonym discovery on privacy-aware clinical data. In: ACM SIGKDD, pp. 1578–1586 (2019)
30. Zhang, Z., Han, X., Liu, Z., Jiang, X., Sun, M., Liu, Q.: ERNIE: enhanced language representation with informative entities. In: ACL, pp. 1441–1451 (2019)

Iterative Reasoning over Knowledge Graph

Liang Xu and Junjie Yao[✉]

East China Normal University, Shanghai, China
junjie.yao@cs.ecnu.edu.cn

Abstract. The concept reasoning is an essential task in text data management and understanding. Recent methods usually capture shallow semantic features and cannot extend to multi-hop reasoning. Knowledge graphs have rich text information and connections. We use a knowledge graph to encode complex semantic relation between evidence and question. The nodes represent valuable information as clue entities and candidate answers in evidence and question, and the edges represent the reasoning rules between nodes.

In this paper, we propose a graph-based reasoning framework with iterative steps. The model obtains the completed evidence chain through iterative reasoning. The new approach iteratively infers the clue entities and candidate answers from the question and clue paragraphs to as new nodes to expand the semantic relation graph. Then we update the semantic representation of the questions and context via memory network and apply the graph attention network to encode the reasoning paths in the knowledge graph. Extensive experiments on commonsense reasoning and multi-hop question answering verified the advantage and improvements of the proposed approach.

Keywords: Iterative reasoning · Knowledge graph · Clue entities

1 Introduction

Recently, some neural network models have outperformed humans on answering questions among several public datasets. Does it indicate that the ability of neural networks to comprehend and reason about text has reached human standards? The answer is obviously unsure.

Much of the existing work has focused on the shallow semantic interaction of the question and evidence, resulting in the inability to capture semantic information at a deeper level and being unable to complete complex reasoning tasks. In addition, the reasoning process of neural network models is implicit, which leads to many unreliable and counterintuitive results. Especially for multi-hop question answering (QA), the model needs to understand complex semantics and have a powerful reasoning engine.

This work was supported by NSFC grant 61972151.

C. S. Jensen et al. (Eds.): DASFAA 2021, LNCS 12681, pp. 191–206, 2021.
https://doi.org/10.1007/978-3-030-73194-6_14

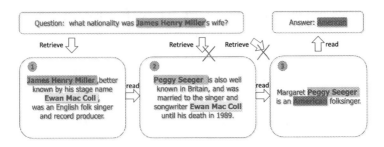

Fig. 1. An example of the reasoning process of the multi-hop question answering. The second and third sentences cannot be obtained directly from the question, and need to be obtained step by step through iterative reasoning.

As shown in Fig. 1, the question is "what nationality was James Henry Miller's wife?". According to the way of reasoning like humans, we must first find the valuable clue information of "James Henry Miller" from the evidence source. Then, we try to find the information about his wife according to the clue information of "James Henry Miller". At last, we get his wife's nationality as answer according to the clue information of his wife. This is a very typical case of multi-hop reasoning. To get the final answer, it must go through a multi-stage iterative reasoning process. If directly based on the text of the question, the second and third step documents cannot be retrieved directly by the method of pattern matching, which does not conform to the reasoning way. The rigorous reasoning must be to iteratively deal with the semantic relationship between the question and the evidence and finally make a prediction. Many current models are similar to black boxes to input directly questions and evidence into deep neural network models to infer answers. This approach is not only inaccurate but also not interpretable.

The inference engine needs to understand the complex semantics in the text. For an open-domain question answering, if all the relevant evidence is retrieved based only on the question at the beginning, then the evidence is related to the question, but the evidence is independent of each other. In addition, the evidence obtained through simple retrieval methods such as pattern matching not only has low accuracy but also creates a lot of noise, making it difficult to get the correct final answer. Because the pattern matching method can only retrieve directly related evidence, and multi-hop reasoning often requires a lot of intermediate clues to form a complete evidence chain, and these intermediate clues are often not directly retrieved by pattern matching. Therefore, a complex multi-hop reasoning model not only needs to be able to handle the semantic relationship between the question and the retrieved evidence, but also the complex semantic relationship between different evidence. Only through iterative reasoning can the complete chain of evidence be accurately retrieved. Most of the previous work is based on the question to retrieve documents in the external information source as the external knowledge of reasoning, but direct retrieval often loses a lot of important indirect related evidence, and at the same time adds a lot of worthless redundant information. In addition, there is still a lot of work that involves step-by-step reasoning through multiple rounds of interaction, and the next round of reasoning is based on the results of the previous step. This is closer to the way people think. However, the reasoning engine proposed by many works is very simple and the effect is not ideal. In addition, the reasoning model based on the attention mechanism used by these methods has poor reasoning ability. With the development of graph neural network models, a lot of work began to do text reasoning based on graph neural networks. However, these works use simple link relationships between question and retrieved evidence to construct graphs, thereby losing a lot of semantic information in the text, resulting in low accuracy of inference results.

In order to solve the above problem, we propose a novel framework for graph-based iterative reasoning. Similar to people's way of thinking, it is not to let the question directly interact with all the information, but to find related clues first, infer the deeper clues of the next hop based on the existing clues, until the answer is found.

We initialize the raw question text as clues, and then construct a semantic relationship graph by iterating between retrieval and reasoning. At each iteration, the semantic relation graph will expand with new nodes and edges. The node contains the representation of the context and question which update iteratively, and the edge encodes the inference path information. Paths of reasoning will be diverse because the process of reasoning is uncertain.

In the current hop, we use the key entities called *clue entities* in the question and context to retrieve information called *clue paragraphs*. Based on the previous hop *clue paragraphs*, we then reasoned to obtain the next hop *clue entities* and *candidate answers* until we find the correct answer. This kind of iteration process of retrieval-reasoning can provide complete evidence chain.

Complex multi-hop reading comprehension QA requires complete evidence chain to describe the rigor of its reasoning logic. Previous deep learning based models cannot provide a sufficient explanation of the reasoning process and lacking persuasiveness. HotpotQA [23] asks the model for supporting facts, which means interpretability at the sentence level. Our approach extracts several entities at each step, and finally provides entity-level interpretability through the explicit inference paths. To improve the reasoning accuracy, we use *clue entities* and *candidate answers* to iteratively construct graph, then utilize a graph attention network [19] to aggregate the information of the inference paths. With the deepening of reasoning, the expression of the question has different from the original question. Therefore, We use a dynamic memory network [10,17,18] to update iteratively the semantic information of the question at each hop. The contribution of this work can be summarized as follows:

1. We propose a novel framework of graph-based reasoning for multi-hop question answering;
2. Our model uses iterative reasoning to continuously retrieve clue evidence to form a complete evidence chain.
3. Experiments on two datasets verify the improvements, compared with the baselines;

2 Related Works

This work is related to several areas. We briefly review them in this section.

Machine Reading Comprehension: R-Net [21] proposes a self-matching attention mechanism to refine the representation by matching the passage against itself, which effectively encodes information from the whole passage. BiDAF [16] network, a multi-stage hierarchical process that represents the context at different levels of granularity and uses bidirectional attention flow mechanism to obtain a query-aware context representation without early summarization. DrQA [4] leverages a neural model to extract the accurate answer from retrieved paragraphs, usually called retrieval-extraction framework. Multi-step Retriever-Reader [5] uses gated recurrent unit to update the query at each step conditioned on the state of the reader and the reformulated query is used to re-rank the paragraphs by the retriever.

Knowledge Graph: Graph modeling is increasingly used in reasoning. In BAG [3], relationships are modeled between nodes in an entity graph and attention information is utilized between a query and the entity graph. Entity-GCN [2] considers three different types of edges that connect different entities in the entity graph. HGN [8] provides multi-level fine-grained graphs with a hierarchical structure for joint answer and evidence prediction. construct a hierarchical graph for each question to capture clues from sources on different levels of granularity: question, paragraphs, sentences, and entities. DFGN [22] constructs a dynamic entity graph, wherein each reasoning step irrelevant entities are softly masked out, and a fusion module is designed to improve the interaction between the entity graph and the documents.

Multi-hop Reasoning [1] introduces a new graphbased recurrent retrieval approach that learns to retrieve reasoning paths over the Wikipedia graph to answer multi-hop open-domain questions. [14] answers Complex Open-domain Questions Through Iterative Query Generation. [13] utilizes evidence extracted from both structured knowledge base ConceptNet and Wikipedia to construct graph. ORQA [11] treats evidence retrieved from open corpus as a latent variable. jointly learn simultaneously the retriever and reader from question-answer string pairs and without any IR system. QANet [24] consists exclusively of convolution and self-attention, where convolution models local interactions and self-attention models global interactions. Cognitive Graph QA [7] employs an machine reading comprehension model to predict answer spans and possible next-hop spans, and then organizes them into a cognitive graph. Coref-GRN [6] extracts and aggregates entity information in different references from scattered paragraphs. It utilizes co-reference resolution to detect different mentions of the same entity. These mentions are combined with a graph recurrent neural network to produce aggregated entity representations. KagNet [12] match tokens in questions and answers to sets of mentioned concepts from the knowledge graph ConceptNet, then construct sub-graph via path finding.

3   Framework Overview

For complex multi-hop reasoning question, the key clues are found from the original question as the frontier clues, and new evidence is found based on the frontier clues in each iteration. This iterative reasoning method makes the evidence found more accurate and comprehensive. While retrieving evidence, it is necessary to deal with the complex semantic relationship between the context and the problem. For many jobs, the complex relationship between them is handled through a simple link method, which leads to the loss of semantics. Therefore, we propose a semantic relationship diagram to indicate complex semantic relationships in text. The nodes in the graph are the key information in each document, and there is a logical deduction between different information. This push-to-relation mining is also achieved through iterative reasoning. At each step of reasoning, new clues will be found to continue to the next round. In the process of iterative reasoning, the representation of the problem also changes. We can see complex semantic relations, from question to answer, there is a clear reasoning path.

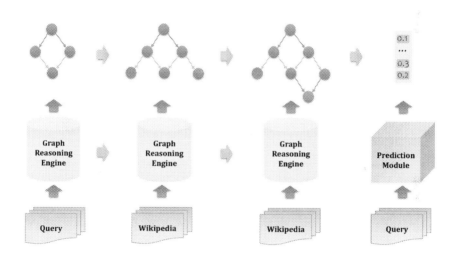

Fig. 2. The framework of iterative reasoning with knowledge graph. We iteratively infer clue entities and candidate answers through the inference engine as new nodes to expand the knowledge graph.

The model of text reasoning usually consists of two parts, one is a powerful reasoning engine, and the other is a knowledge graph. Open field multi-hop question and answer, the model retrieves evidence from external information sources according to the question, and uses the reasoning engine to iteratively reason on the new evidence. In complex situations, only one-time retrieval of evidence based on the problem description often fails to retrieve the evidence in the intermediate process of the reasoning evidence chain, and can only retrieve the evidence that contains partial information. Just like human thinking, it must be a step-by-step search on the basis of existing evidence to obtain new evidence, and iterative reasoning to unearth new clues, and then the next step of search and reasoning. This retrieval method cannot be through pattern matching, but the inference relationship between clues and evidence. The model needs to solve two. First, how to retrieve information that allows us to retrieve comprehensive and accurate information, does not contain redundant information, and has less noise? Second, the answer to the question is contained in several documents retrieved. How to deal with the semantic relationship between documents and between documents and questions? Only by correctly and effectively representing the semantic relationship between them, can complex reasoning be carried out. There are complex semantic relationships between documents, and what we need most is to encode the inference relationships among them. The nodes of the knowledge graph are fragments extracted from the document, and the edges of the graph are the push-to-relationships between nodes. Let the model learn to derive new clues based on the question text. The semantic relationship graph is used to encode the semantic relationship of the text, and the graph neural network is used as the reasoning engine for path information aggregation and answer prediction.

As shown in the Fig. 2, we construct the knowledge graph by generating new clue entities and candidate answers through iterative reasoning. On this basis, the graph neural network is used as the reasoning engine for answer prediction. Our approach encodes the question memory iteratively. We first take the raw question text as the initialization of clue paragraphs, and extract the key information in the clue paragraphs as the clue entities, then retrieve paragraphs related to clue entities as the next hop clue paragraphs. The clue entities and candidate answers are used as nodes, and an knowledge graph is established according to the progressive relationship in the reasoning process. Each node has two vector embeddings representation, one is the question memory and the other is the context semantic embedding. The context semantic embedding requires the graph attention neural network model aggregate neighbor information of the inference paths, and the question memory update is based on the memory of the previous hop and the question semantic embedding of the current hop via iterative memory network.

4   Approach

This chapter mainly introduces the realization of the reasoning model proposed in this paper. The proposed framework in Fig. 3. It consists of three core modules: Knowledge Graph Constructor, Iterative Memory Network, Graph Attention Reasoning. The Semantic Relation Graph Constructor iteratively extracts *clue entities* and *candidate answers*, the Graph Attention Reasoning module encodes the information of the inference paths, and the iterative question memory is updated accordingly at each hop.

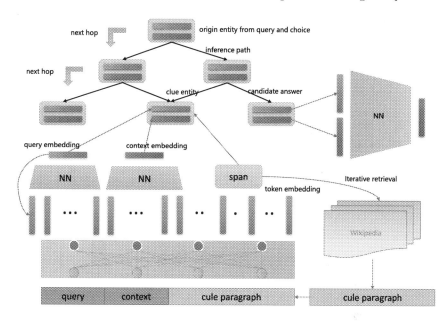

Fig. 3. The framework consists of three core modules: Knowledge Graph Constructor, Iterative Memory Network and Graph Attention Reasoning.

4.1 Knowledge Graph Constructor

In this section, we will introduce the implementation of Semantic Relation Graph Constructor. In order to construct explicit and logical reasoning paths, we propose a semantic relation graph that can encode complex entities-level semantic logical relationships between different text. The nodes of semantic relation graphs encode the representation of query and context, and the directed edges of different nodes directly indicate inference paths. We take questions and choices, and related documents as input to the module to obtain multi-level features that include question embedding, context semantic embedding, and clue entities and candidate answers. Given a question q consisting of l tokens $\{q_1, ..., q_l\}$, several choices are initialized as candidate answers $a\{a_1, ..., a_k\}$ for multiple-choice questions. If no choices are given with question, model need to extract text span from paragraphs as answer for span-extraction QA. Span-extraction QA initializes entities from raw question text by fuzzy matching as first hop clue entities x. Multiple-choices QA initializes entities from raw question text by fuzzy matching and all choices as first hop clue entities$x$. For the task of machine reading comprehension for open-domain QA, model needs to use clue entities retrieve a document or a small set of documents of n paragraphs where a single paragraph p consists of m tokens $\{p_1, ..., p_m\}$ from external knowledge source.

We input Wikipedia, a collection of article, as external source, and using clue entities x to retrieve related document $para[x]$ from it. Then we transform these document mentioned clue entities x in previous hop as $clue[x]$ in current hop to extract candidate answers a and useful next-hop clue entities y from the $para[x]$.

$$(\hat{a}, y, E(x), C(x)) = Bert(q, a, clue[x], para[x]) \tag{1}$$

where $\hat{a}$ is the expansion of candidate answers, $E(x)$ is the question vector representation of the current hop clue entities x, the $C(x)$ is the context semantic vector representation for the combination of question and clue paragraphs.

Pointer Network. We use Pointer Network [20] variants to train the model. Pointer Networks directly takes the probability obtained after softmax as an output, allowing the probability to assume the role of a pointer to a specific element of the input sequence as follows:

$$u_j^i = v^T tanh(W_1 e_j + W_2 d_i) \tag{2}$$

$$P_j^i = softmax(u_j^i) \tag{3}$$

e is pointer vector, d is output vector of BERT, v^T, W_1 and W_2 are learnable parameters. we also utilize S_{hop}, E_{hop}, S_{ans}, E_{ans} as additional learnable pointer vectors to predict targeted spans. The probability of the ith input token to be the start of an candidate answer span $P_{ans}^{start}[i]$ is calculated as follows:

$$P_{ans}^{start}[i] = \frac{e^{v^T tanh(W_1 S_{ans} + W_2 d_i)}}{\sum_j e^{v^T tanh(W_1 S_{ans} + W_2 d_j)}} \tag{4}$$

Let $P_{ans}^{end}[i]$ be the probability of the i th input token to be the end of an candidate answer span, which can be calculated following the same formula. We only focus on the positions with top k start probabilities $start_k$. For each k, the end position end_k is given by:

$$end_k = \underset{start_k \leq j \leq start_k + max\,L}{\arg\max} P_{ans}^{end}[j] \tag{5}$$

Where $maxL$ is the maximum possible length of spans. The process extract the candidate answers based on S_{ans} and E_{ans}, and extracting the next hop clue entities based on S_{hop} and E_{hop}. We use the next hop clue entities y to extract relevant documents $para[y]$ from Wikipedia, and use the predecessor documents $para[x]$ which extracted clue entities y as current hop clue paragraphs $clue[y]$. Both are put into the model to extract the further hop candidate answers and clue entities.

Semantic Encoding: Outputs of BERT at position 0 have the ability to summarize the input sequence. Thus the most straightforward method is to use T_0 as $sem[q, clue[x], para[x]]$. In our experiment, the summary of the four-to-last layer output at position 0 as $sem[q, clue[x], para[x]]$ performs the best. The node of knowledge graph initialize the vector representation by using $h_0 = sem[q, clue[x], para[x]]$.

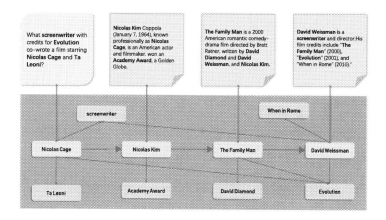

Fig. 4. The example of graph reasoning for multi-hop question answering, which iteratively obtains clue entities and candidate answers as nodes of semantic relation graph.

4.2   Iterative Memory Network

Figure 4 illustrates an example where our framework handles multi-hop reading comprehension.

As the reasoning deepens step by step, the characterization of the question should change iteratively. The question memory m iterates over the question vector representations outputted by the semantic relation graph constructor. In its general form, the question memory module is comprised of an attention mechanism as well as a recurrent network with which it updates its memory [10]. During each iteration,the attention mechanism attends over the question vector representations E and the previous memory m^{i-1}. The scoring function S takes as input the feature set $z(c, m, q)$ and produces a scalar score. We first define a large feature vector that captures a variety of similarities between input, memory and question vectors:

$$z(E, m) = [E, m, E \circ m, E^T W m] \tag{6}$$

In our work, we use a gating function as our attention mechanism. For each iteration i, the mechanism takes the vector representation E of current question and a previous memory m^{i-1} as inputs to compute gating.

$$g^i = S(E, M^{i-1}) = \sigma\left(W_2 tanh\left(W_1 z(E, m^{i-1}) + b_1\right) + b_2\right) \tag{7}$$

Different inference paths have different effects on question memory. According to the inference path, each entity obtained from current $clue[x]$ has different memories. Based on the previous memory m^{i-1} ,we update the question memory for each time of the multi source knowledge extraction. The initial state of this GRU is initialized to the question vector itself: $m^0 = E^0$.

$$h_t^i = g^i GRU(E, h_{t-1}^i) + (1 - g^i)h_{t-1}^i \tag{8}$$

$$m^i = GRU(h_t^i, m^{i-1}) \tag{9}$$

4.3   Graph Attention Reasoning

Algorithm 1 describes the procedure of iterative graph attention reasoning. The nodes of semantic relation graph are iteratively added based on the reasoning of each hop. At each reasoning hop, we will obtain new candidate answers and clue entities from semantic relation graph constructor as new nodes to expand the knowledge graph. At the same time, we use the context semantic vector outputed as the initialization vector representation of the knowledge graph nodes and the current hop question memory vector as the question memory of the candidate answers nodes. Then using the latest clues to retrieve clue paragraphs in Wikipedia, and re-place them into the semantic relation graph constructor to extract candidate answers and clue entities for the next hop to continuously expand the knowledge graph and fill node properties. The knowledge graph can effectively aggregate the information of the inference path through the encoding of multiple neural attention layers. There may be multiple paths passing through the same node. Therefore, using the attention mechanism for different paths of this node will capture the information differences of different inference paths more finely.

Algorithm 1. Iterative Answer Generation.

Require: Question q, Entity Sequence S, Graph G,Preductor F,Full Wiki W, Semantic Relation Graph Constructor SE,Question Encoder QE, Question Embedding qe,Context Semantic Embedding sem,Entities $hop[x]$

1: Initialize S and G with clue entities from question and choices
2: **repeat**
3:     pop a entity x from S
4:     fetch $clue[x]$ and $para[x]$ from W
5:     **if** x is clue entity **then**
6:        generate multi-level features $qe, sem, hop[x] = SE(q, clue[x], para[x])$
7:        **for** y in $hop[x]$ **do**
8:           initialize the representation $G[y] = sem$
9:           update question memory $m_y = QE(m_x, qe])$
10:          add y and $edge(x, y)$ to G
11:          push y in S
12:       **end for**
13:    **end if**
14:    update node representation $G[x]$ via aggregating neighbors information
15: **until** S is null
16: return answer with $arg\ max\ F(M[x], G[x])$

We use graph attention networks (GAT) [19] to encode the information of different inference paths. Specifically, GAT takes all the nodes as input, and updates node feature through its neighbors in the graph. The input to our layer is a set of node features, $h = \{h_1, h_2, ..., h_N\}$, $h_i \in R^F$, where N is the number of nodes, and F is the number of features in each node. The layer produces a new set of node features $h' = \{h'_1, h'_2, ..., h'_N\}$ as its output. Then performing a shared attentional mechanism a to computes attention coefficients

$$e_{ij} = a(Wh_i, Wh_j) \tag{10}$$

That indicate the importance of node j's features to node i. To make coefficients easily comparable across different nodes, we normalize them across all choices of j using the softmax function: In the implementation, the attention mechanism a is a single-layer feedforward neural network, and applying the LeakyReLU nonlinearity. Fully expanded out, the coefficients computed by the attention mechanism may then be expressed as:

$$a_{ij} = \frac{exp\left(LeakyRelu\left(W_{e_{ij}}[h_i||h_j]\right)\right)}{\sum_{j\in\mathcal{N}_i} exp\left(LeakyRelu\left(W_{e_{ik}}[h_i||h_k]\right)\right)} \tag{11}$$

Where $W_{e_{ij}}$ is the weight matrix corresponding to the edge type e_{ij} between the i-th and j-th nodes. Where $W \in R^{d\times d}$ is a weight matrix to be learned, $\sigma(\cdot)$ denotes an activation function, and a_{ij} is the attention coefficients, which can be calculated by:

$$h'_i = \sigma\left(\sum_{j\in\mathcal{N}_i} a_{ij}Wh_j\right) \tag{12}$$

5   Experiment

In this chapter, we will verify the validity and interpretability of our model through experiments. Experiments were conducted from common-sense reasoning and multi-hop machine reading comprehension tasks. In order to further analyze the contribution of each component of the model to the whole, we compare and analyze the independent value of each module through ablation experiments. In addition, we also conducted a case study to analyze how our model uses knowledge to iteratively infer the correct answer step by step.

5.1   Datasets

We use CommonSenseQA[1] and HotpotQA[2] for the evaluation. Common-SenseQA is a dataset for multi-choices commonsense question answering which inferences correct answer with prior knowledge. It collected 12,247 commonsense questions which each question has only one correct answers and four distractors. The full-wiki dump of HotpotQA contains training set (90,564 questions), a development set (7,405 questions).

---

[1] https://www.tau-nlp.org/commonsenseqa.
[2] https://hotpotqa.github.io.

5.2   Baselines

In order to verify the reasoning ability of our proposed model, we conducted comparative experiments with other models.

- Bert [23] follows the retrieval-extraction framework of DrQA and subsumes the advanced techniques in QA, such as self-attention, character-level model, bi-attention.
- MUPPET [9] uses a bidirectional GRU to process the paragraph and obtain the contextualized word representation.
- CogQA [7] framework for multi-hop reading comprehension QA at scale.
- ESIM+ELMO [23] is a strong NLI model.
- CoS-E [15]is used to train language models to automatically generate explanations.
- KagNet [12] effectively utilizes external structured commonsense knowledge graphs to perform explainable inferences.

In order to verify the independent contributions of the two component modules to the model framework and the overall contribution of the combination, we conducted the following four sets of comparative analysis experiments:

- **Baseline** performs self-attention interaction between the context and query to produce the final answer.
- **Baseline + memory** interacts with the context in multi-hop to update the query.
- **Baseline + graph** adds the semantic relation graph constructor on the basis of the baseline model.
- **Baseline + memory + graph** is the complete framework of our proposed model, including memory network and semantic relation graph constructor.

5.3   Quantitative Study of Commonsense Reason

In CommonSenseQA, multiple options are given along with the question, and the model needs to pick the one option as correct answer. The accuracy of the answer is the main indicator of the evaluation model. The results on the CommonSenseQA dataset showing in Fig. 5 illustrates that our model outperform all comparison models.

From the comparison chart of accuracy, our approach makes the accuracy higher. Compared to other models that only use the raw question text as context material, we use external knowledge sources like Wikipedia as supporting facts for inference,and make full use of the semantic information of the relevant documents retrieved. Although there is often a complex reasoning relationship between external knowledge and question, our model can handle them well and thus exhibit higher accuracy. In addition, since we iteratively use external sources as supporting facts, the path of inference is explicit and interpretable.

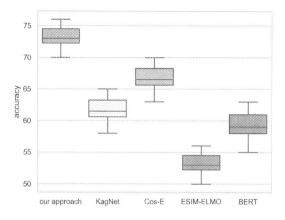

Fig. 5. The results of different models on commonsenseQA, our model achieved the highest accuracy.

5.4 Quantitative Study on Question Answering

For the evaluation of HotpotQA, Exact Match (EM), precision, recall and F1 score of not only answers but also sentence-level supporting facts to verify the model's reasoning ability and explainability.

Accuracy Results: The results on HotpotQA dataset are listed in Table. 1. Not only Exact Match (EM), precision, recall and F1 score of our proposed approach performs much better than the baseline model and MUPPET, but also strong competition with the latest models CogQA.

Table 1. Results on HotpotQA. The evaluation of answer and supporting facts consists of two metrics: Exact Match (EM) and F1. Joint EM is 1 only if answer and supporting facts are both strictly correct.

Methods	Answer				Supporting facts				Joint			
	EM	F1	Pre	Recall	EM	F1	Pre	Recall	EM	F1	Pre	Recall
Baseline	17.70	26.40	27.56	27.71	2.538	27.76	34.88	26.12	0.985	10.21	13.02	10.22
MUPPET	31.09	39.22	41.20	42.76	17.94	50.31	58.02	55.45	10.04	28.89	29.44	31.94
CogQA	36.56	48.49	51.29	49.06	22.44	58.15	63.58	59.90	11.64	34.42	39.22	35.88
Our approach	**37.03**	**50.53**	**51.60**	**49.51**	**22.64**	**62.14**	**64.02**	**60.37**	**12.52**	**38.15**	**40.66**	**35.90**

Explainability Study: For multi-hop reading comprehension QA, not only need to gain the final answer, often the internal process of multi-hop reasoning is very critical. The higher the comprehensive evaluation metrics of supporting facts, the stronger the model reasoning ability. Our proposed model achives the highest value of supporting facts in Table 1, indicating that our model has strong multi-hop reasoning ability and interpretability.

Query memory mechanism provides iterative memory for our reasoning at each hop, and updates simultaneously clue entities and documents, which make it more accurate to get supporting facts. The graph attention neural network has greater advantages in aggregating inference paths information, and obtaining accurate reasoning answers.

5.5   Ablation Study

In order to verify the effectiveness of each module in the framework proposed in the paper, we constructed four sets of comparative experiments. On the basis of the baseline model, we continue to add components to verify the independent contribution of each component to the model. The Table 2 shows the experimental results between different comparison models.

Table 2. Ablation study results of HotpotQA and CommonSenseQA.

Model	CommonSenseQA	HotpotQA							
	Answer	Answer				Supporting facts			
	Accuracy	EM	F1	Pre	Recall	EM	F1	Pre	Recall
Baseline	57.32	27.32	38.97	40.01	38.45	24.37	39.50	38.99	40.45
Baseline + Memory	69.84	34.84	48.98	50.02	47.87	28.84	51.38	49.16	54.27
Baseline + Graph	63.22	33.22	47.45	48.98	46.33	37.11	55.98	57.63	55.64
Baseline + Memory + Graph	74.53	37.03	50.53	51.60	49.51	42.01	61.98	63.16	61.55

The baseline model uses self-attention mechanism. Although it has a certain effect, the score is the lowest in the comparison. Models based on the self-attention mechanism are often the extraction and matching of shallow semantic features, making it difficult to handle complex multi-hop reasoning. It illustrates that the memory network interacts with the context multiple times through external memory modules to successfully capture complex semantic information. Adding semantic relation graph achieves higher accuracy than joining memory network, indicating that the graph structure has a more powerful representation capability and can summarize the deep semantic information of context and query. The semantic relation graph has explicit inference paths, so the neighbor nodes and paths information can be encoded through the graph neural network, which has a more powerful inference ability. Combining the memory network and the semantic relationship graph together to obtain the best results on the two data sets.

5.6   Case Study

Through case studies, we analyze the explicit reasoning process of commonsenseQA and multi-hop reading comprehension tasks. We show how the knowledge graph reason model clearly explains complex reasoning processes in our experiments in Fig. 6. For each question, the model iteratively extracts relevant paragraphs from Wikipedia and constructs a graph based on the semantic relationship between the text. The model not only obtains accurate answers, but also clearly shows the path of reasoning. The result of reasoning is reliable and rigorous, and the process of reasoning is also clear and interpretable.

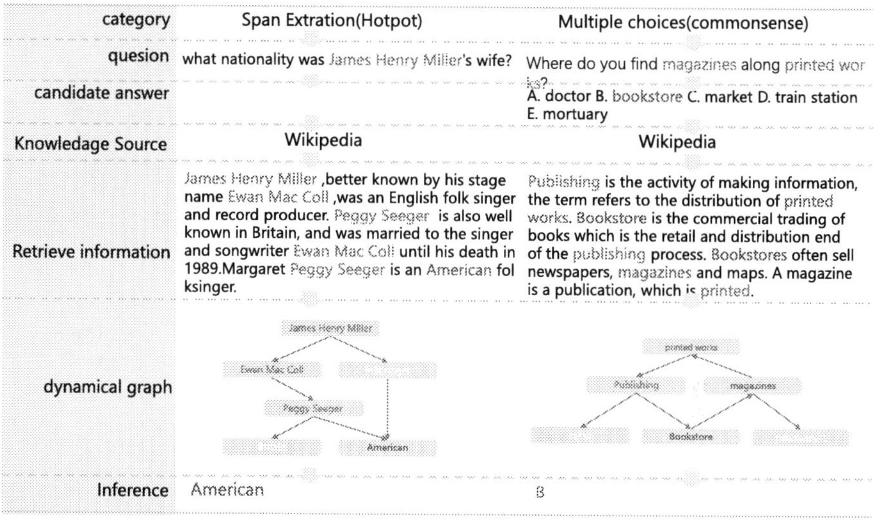

category	Span Extration(Hotpot)	Multiple choices(commonsense)
quesion	what nationality was James Henry Miller's wife?	Where do you find magazines along printed works?
candidate answer		A. doctor B. bookstore C. market D. train station E. mortuary
Knowledage Source	Wikipedia	Wikipedia
Retrieve information	James Henry Miller ,better known by his stage name Ewan Mac Coll ,was an English folk singer and record producer. Peggy Seeger is also well known in Britain, and was married to the singer and songwriter Ewan Mac Coll until his death in 1989.Margaret Peggy Seeger is an American folksinger.	Publishing is the activity of making information, the term refers to the distribution of printed works. Bookstore is the commercial trading of books which is the retail and distribution end of the publishing process. Bookstores often sell newspapers, magazines and maps. A magazine is a publication, which is printed.
dynamical graph		
Inference	American	B

Fig. 6. Examples of knowledge graph construction and iterative reasoning of commonsenseQA and HotpotQA. The model iteratively finds clue entities and candidate answers to construct a knowledge graph.

HotpotQA for general questions in HotpotQA, model need to extract text span from related paragraph as final correct answer. Span extraction module extracts clue entities and candidate answers to construct iteratively directed graph where $edge <x, y>$ is that entity y is extracted from $para[x]$ based on question and $clue[x]$. The construction of iterative directed graph are main three categories: tree, directed acyclic graph and directed cycle graph. We show an example of a tree structure. Because the HotpotQA is multi-hop reading comprehension question answer, the inference path is more inclined to progressive relationship, and more tree structure. CommonsenseQA the structure categories of graph in CommonsenseQA are same with HotpotQA. However, Not only the entity extracted from the question, but also the answer options of the question are initialized as the 1-hop entity, the inference paths tend to establish connection between them. Therefore, it often shows the structure of the directed cycle graph.

References

1. Asai, A., Hashimoto, K., Hajishirzi, H., Socher, R., Xiong, C.: Learning to retrieve reasoning paths over Wikipedia graph for question answering. In: ICLR (2020)

2. Cao, N.D., Aziz, W., Titov, I.: Question answering by reasoning across documents with graph convolutional networks. In: NAACL, pp. 2306–2317 (2019)
3. Cao, Y., Fang, M., Tao, D.: BAG: bi-directional attention entity graph convolutional network for multi-hop reasoning question answering. In: NAACL-HLT, pp. 357–362 (2019)
4. Chen, D., Fisch, A., Weston, J., Bordes, A.: Reading Wikipedia to answer open-domain questions. In: ACL, pp. 1870–1879 (2017)
5. Das, R., Dhuliawala, S., Zaheer, M., McCallum, A.: Multi-step retriever-reader interaction for scalable open-domain question answering. In: ICLR (2019)
6. Dhingra, B., Jin, Q., Yang, Z., Cohen, W., Salakhutdinov, R.: Neural models for reasoning over multiple mentions using coreference. In: NAACL, pp. 42–48 (2018)
7. Ding, M., Zhou, C., Chen, Q., Yang, H., Tang, J.: Cognitive graph for multi-hop reading comprehension at scale. In: ACL 2019, pp. 2694–2703 (2019)
8. Fang, Y., Sun, S., Gan, Z., Pillai, R., Wang, S., Liu, J.: Hierarchical graph network for multi-hop question answering. CoRR abs/1911.03631 (2019)
9. Feldman, Y., El-Yaniv, R.: Multi-hop paragraph retrieval for open-domain question answering. In: ACL, pp. 2296–2309 (Jul 2019)
10. Kumar, A., et al.: Ask me anything: dynamic memory networks for natural language processing. In: ICML, pp. 1378–1387 (2016)
11. Lee, K., Chang, M., Toutanova, K.: Latent retrieval for weakly supervised open domain question answering. In: ACL, pp. 6086–6096 (2019)
12. Lin, B.Y., Chen, X., Chen, J., Ren, X.: KagNet: knowledge-aware graph networks for commonsense reasoning. In: Proceedings of EMNLP-IJCNLP (2019)
13. Lv, S., et al.: Graph-based reasoning over heterogeneous external knowledge for commonsense question answering. CoRR abs/1909.05311 (2019)
14. Qi, P., Lin, X., Mehr, L., Wang, Z., Manning, C.D.: Answering complex open-domain questions through iterative query generation. In: EMNLP-IJCNLP, pp. 2590–2602 (2019)
15. Rajani, N.F., McCann, B., Xiong, C., Socher, R.: Explain yourself! Leveraging language models for commonsense reasoning. In: ACL (2019)
16. Seo, M.J., Kembhavi, A., Farhadi, A., Hajishirzi, H.: Bidirectional attention flow for machine comprehension. In: ICLR (2017)
17. Shen, Y., Huang, P., Gao, J., Chen, W.: ReasoNet: learning to stop reading in machine comprehension. CoRR abs/1609.05284 (2016)
18. Sukhbaatar, S., Szlam, A., Weston, J., Fergus, R.: Weakly supervised memory networks. CoRR abs/1503.08895 (2015)
19. Veličković, P., Cucurull, G., Casanova, A., Romero, A., Liò, P., Bengio, Y.: Graph attention networks. In: ICLR, pp. 1–12 (2018)
20. Vinyals, O., Fortunato, M., Jaitly, N.: Pointer networks. In: Cortes, C., Lawrence, N.D., Lee, D.D., Sugiyama, M., Garnett, R. (eds.) Advances in Neural Information Processing Systems, vol. 28, pp. 2692–2700. Curran Associates, Inc. (2015). http://papers.nips.cc/paper/5866-pointer-networks.pdf
21. Wang, W., Yang, N., Wei, F., Chang, B., Zhou, M.: Gated self-matching networks for reading comprehension and question answering. In: ACL, pp. 189–198 (2017)
22. Xiao, Y., et al.: Dynamically fused graph network for multi-hop reasoning (2019). arxiv:1905.06933Comment. Accepted by ACL 19
23. Yang, Z., Qi, P., Zhang, S., Bengio, Y., Cohen, W.W., Salakhutdinov, R., Manning, C.D.: HotpotQA: a dataset for diverse, explainable multi-hop question answering. In: EMNLP, pp. 2369–2380 (2018)
24. Yu, A.W., et al.: QANet: combining local convolution with global self-attention for reading comprehension. In: ICLR (2018)

Spatial-Temporal Attention Network for Temporal Knowledge Graph Completion

Jiasheng Zhang[1,2], Shuang Liang[1], Zhiyi Deng[1], and Jie Shao[1,2(✉)]

[1] University of Electronic Science and Technology of China, Chengdu 611731, China
{zjss12358,shuangliang,zhiyideng}@std.uestc.edu.cn, shaojie@uestc.edu.cn
[2] Sichuan Artificial Intelligence Research Institute, Yibin 644000, China

Abstract. Temporal knowledge graph completion, which aims to predict missing links in temporal knowledge graph (TKG), is an important research task due to the incompleteness of TKG. Recently, TKG embedding methods have proved to be effective for this task. However, most of existing methods regard TKG as a set of independent facts and consequently ignore the implicit relevance among facts. Actually, as a kind of dynamic heterogeneous graph, the evolving graph structure of TKG is able to reflect a wealth of information. To this end, in this paper we regard temporal knowledge graph as heterogeneous and discrete spatial-temporal resource, and propose a novel spatial-temporal attention network to learn TKG embeddings by modeling spatial-temporal property of TKG while considering its special characteristics. Specifically, our model employs a **M**ulti-**F**aceted **G**raph **At**tention Network (MFGAT) to extract rich structural information from the egocentric network of each entity. Additionally, an **Ad**aptive **T**emporal **At**tention Mechanism (ADTAT) is utilized to flexibly model the correlation of entity representations in the time dimension. Finally, by combing our obtained representations with existing static KG completion methods, they can be extended to spatial-temporal versions to predict missing links in TKG while considering its inherent graph structure and time-evolving property. Experimental results on three real-world datasets demonstrate the superiority of our model over the state-of-the-art methods.

Keywords: Temporal knowledge graph completion · Temporal knowledge graph embedding learning · Spatial-temporal data mining

1 Introduction

Temporal knowledge graph (TKG) is a knowledge base system which contains facts happened in real-world with the corresponding happened times. As shown in Fig. 1, TKG can be represented as a dynamic heterogeneous graph in which nodes denote entities in real-world and labeled edges represent relations among entities. Moreover, nodes and edges in the graph will appear or disappear with the development of time which leads to that the structure of the graph evolves over time and the static graph in each timestamp is called a snapshot.

© Springer Nature Switzerland AG 2021
C. S. Jensen et al. (Eds.): DASFAA 2021, LNCS 12681, pp. 207–223, 2021.
https://doi.org/10.1007/978-3-030-73194-6_15

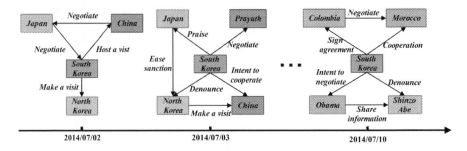

Fig. 1. An example of temporal knowledge graph. In each snapshot we give an example of the egocentric network of *South Korea*.

Compared with static knowledge graph (KG) which ignores the time annotations of facts, TKG is more adequate for real-world scenarios and thus receives a surge of interest in recent years. However, same as static knowledge graph, temporal knowledge graph is also far from complete. Therefore, the task of predicting missing links in TKG, which is known as temporal knowledge graph completion (TKGC) becomes an increasingly important research task in this field.

KG embedding methods, which aim to map each element of KG to a hidden vector representation, is a powerful technique for static knowledge graph completion. However, such methods fail to consider the time annotations of facts. Therefore, some researchers turn to temporal knowledge graph embedding methods for the TKGC task in recent years, several methods have been proposed such as TAE [10] and HyTE [3]. Although these methods outperform KG embedding methods on the TKGC task, they mostly regard TKG as a set of independent facts and thus ignore the graph structure of TKG, which fails to capture the implicit relevance among facts. Furthermore, most of them treat facts in each snapshot separately and thus ignore the time-evolving property of TKG, which fails to obtain more accurate representations based on the information of history snapshots. Therefore, the performance of TKGC is still far from satisfactory and it is necessary to develop a model that can consider graph structure and time-evolving property of TKG simultaneously.

Actually, we notice that temporal knowledge graph can be viewed as a kind of spatial-temporal resource where graph structure in each snapshot reflects its spatial property and the correlation of different snapshots in the time dimension reflects its temporal property. Recently, deep spatial-temporal models [24] have achieved successes in many fields due to their effectiveness in modeling spatial-temporal correlation of data, so we argue that learning TKG embeddings via deep spatial-temporal models can effectively consider its graph structure and time-evolving property. However, there are still no studies applying such models to TKG because TKG has two characteristics: 1) *heterogeneity*, as shown in Fig. 1, nodes in the graph correspond to entities in the real world, which leads to that different nodes have different semantics and thus play different roles in the graph; 2) *discreteness*, facts in TKG are discretely distributed in the time dimension, which leads to that data quantities of different snapshots are inhomogeneous. For a particular entity, some snapshots contain more related facts while others contain fewer or even no related facts.

Based on above considerations, in this paper, we propose a novel spatial-temporal attention network to learn TKG embeddings by modeling its spatial-temporal property. First, in order to model the spatial property and heterogeneity of TKG, we focus on the egocentric network [8], which is defined as the induced graph of a node with its immediate neighbors. It is considered as the basic structure that dominates the attributes and behaviors of nodes in the field of social network analysis [1]. As shown in Fig. 1, we give an example of egocentric networks of *South Korea* in different snapshots. Compared with the star-like structure considered by previous graph neural network (GNN) models, such as GAT [23] and R-GCN [20], which can only consider the binary relationships between nodes, egocentric network can capture the multiple relationships among a node and its neighbors, and thus is able to describe the role of a node in the graph more accurately. In this way, we develop a novel **M**ulti-**F**aceted **G**raph **At**tention Network (MFGAT) based on the egocentric network. Specifically, for each snapshot, it firstly constructs rich structural features from the egocentric network of each entity, and then an attention mechanism is applied for each feature independently. Finally, by fusing different kinds of features, our MFGAT can effectively learn TKG embeddings of each snapshot while considering the graph structure and heterogeneity of TKG.

Additionally, in order to model the time-evolving property of TKG while addressing the inhomogeneity problem brought by discrete distribution, we propose a novel **Ad**aptive **T**emporal **At**tention Mechanism (ADTAT). The core component of ADTAT is a mask function which is able to dynamically select attention position for each entity to focus on the information of active snapshots. Furthermore, it can adaptively model the time span information based on the fact distribution of each entity in the time dimension. By employing an attention mechanism with our mask function, ADTAT is able to flexibly model the temporal correlation of entity representations in different snapshots.

Combining the above two parts, our spatial-temporal attention network can learn TKG embeddings while considering the graph structure and time-evolving property of TKG simultaneously. Furthermore, existing static knowledge graph embedding methods can be extended to a spatial-temporal version for the TKGC task by applying our obtained representations in the score function. Main contributions of our work are summarized as follow:

- We propose a novel spatial-temporal attention network for TKG completion. To the best of our knowledge, this is the first work that learns TKG embeddings from the perspective of spatial-temporal data modeling.
- We introduce egocentric network to the field of TKG, and propose a novel multi-faceted graph attention network based on egocentric network of each entity to capture the structural information of TKG more effectively.

- Experimental results on three real-world datasets demonstrate the superiority of our model. Our source code and datasets are publicly available at https://github.com/zjs123/ST-ConvKB.

2   Related Work

In this section, we first provide an overview of the typical methods for static knowledge graph embedding learning and temporal knowledge graph embedding learning respectively, and then briefly review deep spatial-temporal model and its recent advances in several fields.

2.1   Static Knowledge Graph Embedding Methods

Static knowledge graph embedding methods aim to represent each element of knowledge graph as a low-dimensional vector while preserving its inherent semantic. There exist two kinds of typical methods, namely translation methods and semantic matching methods. TransE [2] is a typical translation method, which maps each entity to a vector and regards relation as the translation from subject entity to object entity. Based on TransE, a number of improved methods have been proposed, such as TransH [25], TransR [15], and TransD [9]. RESCAL [19] is the first semantic matching method that utilizes restricted Tucker decomposition for static knowledge graph embedding learning. Due to too many parameters of RESCAL, DistMult [27] simplifies RESCAL by using diagonal matrix. Other semantic matching methods have been further proposed, such as HoIE [18] and ComplEx [22]. Besides the above two kinds of methods, in recent years, some researchers attempt to learn KG representations based on convolution, such as ConvE [4] and ConvKB [17]. Furthermore, there are also some works attempt to learn KG representations based on graph neural networks, such as R-GCN [20] and KBAT [16].

2.2   Temporal Knowledge Graph Embedding Methods

Temporal knowledge graph embedding methods aim to learn representations for each element of TKG while considering the happened times of facts. TAE [10] is the first work that attempts to incorporate temporal order information between relations into TKG embeddings. Based on this, TKGFrame [28] formally defines the relation chain of TKG and incorporates it into TKG embeddings. Inspired by the objective of TransH, HyTE [3] projects the embeddings of entity and relation to a time-specific hyperplane and applies TransE score function for the embeddings in each hyperplane. TTransE [14] is an extension of TransE by considering time embeddings in the score function. TA-DistMult [6] constructs temporal relation embeddings for each fact by encoding corresponding time annotation with an LSTM model. Recently, DE-DistMult [7] provides a diachronic entity embedding function to distinguish entities in different time stamps. Inspired by the canonical decomposition of tensors of order 4, TNTComplEX [13] proposes a new regularization scheme and presents a temporal extension of ComplEX. Although these methods have achieved significant performance on the TKGC task, all of them ignore graph structure of TKG and they mostly are unable to capture the correlation of facts in the time dimension. RE-NET [11] is the only work that considers both of them, but this model is designed for extrapolation problem rather than learning embeddings for TKGC.

2.3 Deep Spatial-Temporal Models

Deep spatial-temporal models are a kind of spatial-temporal data mining model based on deep learning techniques. These models mostly contain a spatial part to model the spatial property of data, and the most used deep learning models are convolutional neural network and graph convolutional network (GCN) [12]. A temporal part is used to capture the temporal correlation of data, in which recurrent neural network (RNN) is widely used. Based on the above architecture, several models have been proposed in different fields to model data with spatial-temporal property. For example, GMAN [29] combines a spatial attention model and a temporal attention model with a gated mechanism to predict future traffic conditions, ConvLSTM [21] integrates the structure of CNN and LSTM to predict the spatial-temporal sequences and ST-GCN [26] combines spatial and temporal convolutions for action recognition. These successful attempts demonstrate the universality of deep spatial-temporal models and inspire us to design a spatial-temporal model for temporal knowledge graph embedding learning. More detailed introduction of deep spatial-temporal models can be viewed in [24].

3 Preliminaries

Definition 1 (Temporal Knowledge Graph). *Temporal knowledge graph can be denoted as a sequence of static snapshots $G = \{G^1, G^2, ..., G^{|T|}\}$, where each snapshot contains facts happened in the same time. $G^t = \{(s_i, r_i, o_i, t)\}$ in which $s_i \in \mathcal{E}$ and $o_i \in \mathcal{E}$ are subject entity and object entity respectively, $r_i \in \mathcal{R}$ is the relation and $t \in \mathcal{T}$ denotes the happened time of these facts.*

Definition 2 (Temporal Knowledge Graph Completion). *Temporal knowledge graph completion (as known as link prediction) aims to predict fact (s, r, o, t) when s or o is missing. It can be divided into two subtasks, one is subject entity prediction to predict s given r and o in time t, and the other is object entity prediction to predict o given r and s in time t.*

Definition 3 (Egocentric Network). *Given a node u in network G, the egocentric network of u is a subgraph which is composed of u, its neighbors $\mathcal{N}(u)$, and edges between them, which can be denoted as $G_u = (V_u, E_u)$, where $V_u = u \cup \mathcal{N}(u)$ and E_u are node set and edge set of G_u respectively. Particularly, in this paper we use G_e^t to denote the egocentric network of entity e in the snapshot G^t.*

4   Proposed Model

In this section, we give an introduction of our model in detail. As shown in Fig. 2, our model takes a sequence of snapshots $\{G^1, G^2, ..., G^{|T|}\}$ as input (part (b)), the multi-faceted graph attention network (part (a)) is first used to obtain entity and relation representations in each snapshot, and then adaptive temporal attention mechanism is utilized to model the temporal correlation of entity representations in different snapshots. After obtaining final entity and relation representations, they can be used to predict missing links via a score function (part (c)).

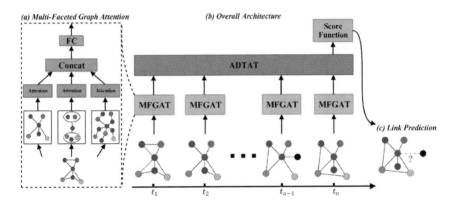

Fig. 2. We give an overview of the architecture of our proposed model in part (b), the detailed illustration of MFGAT is shown in part (a), and after obtaining the final embeddings, they will be used to predict missing links as shown in part (c).

4.1   Multi-faceted Graph Attention Network

As shown in Fig. 2(a), first, due to the complex structure of egocentric network, our MFGAT constructs three kinds of structural features called triple feature, group feature, and path feature based on egocentric network of each entity to adequately describe its structure. Then, the attention mechanism is applied for each feature independently to screen out important information. Finally, the representation of each entity is obtained via a fully connected layer. In this part, we take entity e in snapshot G^t as an example to introduce the detailed process of our MFGAT to obtain its representation and representations of other entities can be obtained in the same way.

Triple Feature. Triple is the basic structure in temporal knowledge graph which can describe the binary relation among entities. In the egocentric network G_e^t, triples that involve e are able to illustrate the direct relevance between e and its neighbors, therefore it is important to integrate the information of such basic structure. We construct triple feature for each fact (e, r_i, e_i, t) in the egocentric network G_e^t as follows:

$$\mathbf{u}_i^t = \mathbf{r}_i \circledast \mathbf{e}_i, \tag{1}$$

in which $\mathbf{r}_i \in \mathbb{R}^d$ is the initial embedding of relation r_i and $\mathbf{e}_i \in \mathbb{R}^d$ is the initial embedding of entity e_i. We obtain the triple feature $\mathbf{u}_i^t \in \mathbb{R}^d$ via circular-correlation operation $\circledast$ which is employed in HoIE [18] due to its high expressivity. Finally, by constructing triple feature for each fact that involves e, we can obtain a set of triple features $\{\mathbf{u}_1^t, \mathbf{u}_2^t, ..., \mathbf{u}_{|\mathcal{N}^t(e)|}^t\}$ where $|\mathcal{N}^t(e)|$ is the number of neighbors of entity e in snapshot G^t.

Group Feature. Neighbors in egocentric network can be divided into several independent groups based on their connectivity and the connected neighbors in each group are generally a set of entities that have similar characteristics to the central entity. Specifically, we regard each group in the egocentric network as a set of nodes that can be connected through paths that do not go through the central entity. As shown in Fig. 1, there are two groups in the egocentric network of *South Korea* in 2014/07/10, one contains *Obama* and *Shinzo Abe* which are the presidents of partner countries of *South Korea* while the other contains *Colombia* and *Morocco* which are cooperation countries. Groups in the egocentric network can reflect the multiple relations among neighbor entities and provide an abstract perspective for the relevance between an entity and its neighbors. Therefore, in order to consider the information of such structure, we define the graph feature of each graph in the egocentric network G_e^t as follows:

$$\mathbf{v}_i^t = \mathbf{MAXPOOL}\{\mathbf{e}_1, \mathbf{e}_2, ..., \mathbf{e}_n\}, \tag{2}$$

where $\mathbf{e}_k \in \mathbb{R}^d$ is the initial embedding of each entity in the group and n is the total number of entities in the group. The group feature $\mathbf{v}_i^t \in \mathbb{R}^d$ is obtained by applying max-pooling operation for entities in the group to screen out the most prominent features of them. Finally, we can obtain a set of group features $\{\mathbf{v}_1^t, \mathbf{v}_2^t, ..., \mathbf{v}_{|\mathcal{G}^t(e)|}^t\}$ where $|\mathcal{G}^t(e)|$ is the number of groups in G_t^e.

Path Feature. Relational path is widely used to model complex graph structure of knowledge graph because it can reflect multi-hop relations between entities. In the egocentric network G_e^t, relational path between e and each of its neighbors is able to illustrate indirect relevance between them. In this part, for each neighbor entity e_i in G_e^t, we randomly find a relational path of length 2 from e to e_i in the egocentric network, which can be denoted as (e, r_{i1}, r_{i2}, e_i). The corresponding path feature is obtained as follows:

$$\mathbf{o}_i^t = \mathbf{W}_o[\mathbf{r}_{i1} : \mathbf{r}_{i2} : \mathbf{e}_i], \tag{3}$$

in which $\mathbf{r}_{i1} \in \mathbb{R}^d$ and $\mathbf{r}_{i2} \in \mathbb{R}^d$ are initial embeddings of relations involved in the path and $\mathbf{e}_i \in \mathbb{R}^d$ is the initial embedding of neighbor entity e_i, $\mathbf{W}_o \in \mathbb{R}^{d \times 3d}$ denotes the linear transform matrix and $[:]$ is concatenation operation. By constructing path feature for each neighbor, we can obtain a set of path features $\{\mathbf{o}_1^t, \mathbf{o}_2^t, ..., \mathbf{o}_{|\mathcal{N}^t(e)|}^t\}$ where $|\mathcal{N}^t(e)|$ is the number of neighbors of entity e in snapshot G^t.

Feature Fusion. After obtaining the above three kinds features $\{\mathbf{u}_i^t\}$, $\{\mathbf{v}_i^t\}$ and $\{\mathbf{o}_i^t\}$, we then apply the attention mechanism to each of them independently, and for each kind of feature we can obtain a set of attention weights $\{\alpha_1^t, \alpha_2^t, ..., \alpha_{N_c}^t\}$ which quantify the importance of feature $\{\mathbf{c}_1^t, \mathbf{c}_2^t, ..., \mathbf{c}_{N_c}^t\}$ for entity e. $\mathbf{c}$ can be $\mathbf{u}$, $\mathbf{v}$ and $\mathbf{o}$, and N_c is the length of each feature sequence.

$$\alpha_i^t = \frac{exp(\mathbf{e}^\top \mathbf{U} \mathbf{c}_i^t)}{\sum_{j=1}^{N_c} exp(\mathbf{e}^\top \mathbf{U} \mathbf{c}_j^t)}, \tag{4}$$

$$\tilde{\mathbf{c}}^t = \sum_{i=1}^{N_c} \alpha_i^t \mathbf{c}_i^t, \tag{5}$$

in which $\mathbf{U} \in \mathbb{R}^{d \times d}$ is the transfer matrix to be learned and $\mathbf{e}$ is the initial embedding of entity e. As shown in Eq. 5, we obtain the corresponding output vector $\tilde{\mathbf{c}}^t \in \mathbb{R}^d$ of each kind of feature as the weighted average. Finally, we concatenate the obtained three kinds of feature vectors with the initial embedding of e and employ a fully connected layer to obtain the output representation of entity e in snapshot G^t as follows:

$$\tilde{\mathbf{e}}^t = \sigma(\mathbf{W}[\mathbf{e} : \tilde{\mathbf{u}}^t : \tilde{\mathbf{v}}^t : \tilde{\mathbf{o}}^t] + \mathbf{b}). \tag{6}$$

Unseen Entity Transform. If there are no related facts of entity e in snapshot G^t, our MFGAT obtains the corresponding representation via another fully connected layer as follows:

$$\tilde{\mathbf{e}}^t = \sigma(\mathbf{W}_{ent}\mathbf{e} + \mathbf{b}_{ent}). \tag{7}$$

Finally, by applying MFGAT for entity e in different snapshots, we can obtain a sequence of output representation vectors for different snapshots, which can be denoted as $\{\tilde{\mathbf{e}}^1, \tilde{\mathbf{e}}^2, ..., \tilde{\mathbf{e}}^{|\mathcal{T}|}\}$, where $|\mathcal{T}|$ is the total number of snapshots.

Relation Transform. Further, after obtaining entity representations via multi-faceted graph attention network, the relation representations are also transformed as follows:

$$\tilde{\mathbf{r}} = \mathbf{r} \cdot \mathbf{W}_{rel}, \tag{8}$$

where $\mathbf{r} \in \mathbb{R}^d$ is the initial relation embedding and $\tilde{\mathbf{r}} \in \mathbb{R}^d$ is the transformed relation embedding. $\mathbf{W}_{rel} \in \mathbb{R}^{d \times d}$ is the learnable transform matrix used to project relation embeddings to the same vector space as entity embeddings.

4.2 Adaptive Temporal Attention Mechanism

In temporal knowledge graph, the temporal correlation of entity representations in different snapshots mainly relies on two parts. First, it is affected by the inherent semantic correlation of entity representations. As shown in Fig. 1, representations of *South Korea* in 2014/07/02 and 2014/07/03 tend to have high correlation because *South Korea* interact with *Japan*, *China*, and *North Korea* in both two snapshots. Second, it is also affected by the time span between snapshots, and entity representations with long time span tend to have low correlation because the effects of facts will attenuate over time. Our MFGAT can effectively learn entity representations in each snapshot, but it fails to model the correlation of entity representations in different snapshots. Furthermore, as we mentioned, data quantities of different snapshots are inhomogeneous in temporal knowledge graph which leads to the complexity of modeling temporal correlation of entity representations. To this end, we develop a novel adaptive temporal attention mechanism (ADTAT) to flexibly capture the correlation of entity representations in different snapshots. For each entity e, our ADTAT takes the output representation sequence $\{\tilde{\mathbf{e}}^1, \tilde{\mathbf{e}}^2, ..., \tilde{\mathbf{e}}^{|\mathcal{T}|}\}$ of our MFGAT as input and the correlation of its representations in time t and t_j ($t_j \leq t$) is measured as follows:

$$\beta^{t,t_j} = \frac{m_e(t, t_j)exp(\sigma(\mathbf{a}^\top \cdot [\mathbf{W}_1\tilde{\mathbf{e}}^t : \mathbf{W}_2\tilde{\mathbf{e}}^{t_j}]))}{\sum_{t_k \leq t} m_e(t, t_k)exp(\sigma(\mathbf{a}^\top \cdot [\mathbf{W}_1\tilde{\mathbf{e}}^t : \mathbf{W}_2\tilde{\mathbf{e}}^{t_k}]))}, \tag{9}$$

where $\tilde{\mathbf{e}}^t \in \mathbb{R}^d$ and $\tilde{\mathbf{e}}^{t_j} \in \mathbb{R}^d$ are representations of entity e in time t and t_j respectively, $\mathbf{W}_1 \in \mathbb{R}^{d \times d}$ and $\mathbf{W}_2 \in \mathbb{R}^{d \times d}$ are two learned transform matrices, and $\mathbf{a} \in \mathbb{R}^{2d}$ is the attention vector. $m()$ is a mask function, in which firstly, in order to avoid attention smooth problem brought by inhomogeneous data distribution, for each entity e, if there are no facts that involve e in the snapshot G^{t_j}, $m_e(t, t_j)$ will be set as 0, which forces our attention mechanism to focus on the active snapshots of entity e. In addition, in order to capture time span information of TKG, we employ a temporal attenuation function with a dynamic attenuation coefficient γ_e^t, since facts of each entity are distributed inhomogeneously in the time dimension, too large attenuation coefficient will lead to local sparse entities fails to capture sufficient history information, but too small attenuation coefficient will make our model unable to adequately consider the effect of time span. Therefore, we define the dynamic attenuation coefficient as follows:

$$\gamma_e^t = \frac{\sum_{|t_i - t| \leq \frac{\sqrt{|\mathcal{T}|}}{2}} |\mathcal{N}^{t_i}(e)|}{\sqrt{|\mathcal{T}|} - 1} \cdot \lambda, \tag{10}$$

in which $|\mathcal{N}^{t_i}(e)|$ is the number of neighbors of entity e in the snapshot G^{t_i}, and λ is the basic attenuation coefficient. For each entity e, the size of γ_e^t is related to the distribution of facts around snapshot G^t, and the sparser distribution will lead to the smaller attenuation coefficient. Combining above two parts, the mask function of our ADTAT can be defined as follows:

$$m_e(t, t_j) = \begin{cases} exp(-\gamma_e^t(|t - t_j|)), & e \in G^{t_j} \\ 0, & otherwise \end{cases}. \tag{11}$$

Based on the mask function, our ADTAT is able to model the temporal correlation of entity representations while effectively tackle the inhomogeneity problem of TKG. The output representation of each entity e in time t is obtained as follows:

$$\mathbf{h}_e^t = \sum_{t_j \leq t} \beta^{t,t_j} \tilde{\mathbf{e}}^{t_j}. \tag{12}$$

Finally, our ADTAT can obtain the final representations $\{\mathbf{h}_e^1, \mathbf{h}_e^2, ..., \mathbf{h}_e^{|\mathcal{T}|}\}$ of each entity e in different snapshots while considering the graph structure and temporal correlation of TKG.

4.3   Training

After obtaining the final representations of entity and relation, they can be used in the score function of existing static knowledge graph embedding methods such as TransE [2] and DistMult [27] to obtain the spatial-temporal version of these methods for TKGC. Here, we give the illustration of using ConvKB [17] score function because it achieves the best performance in our experiment and the performances of different score functions will be presented in Sect. 5. The score function of each fact (s, r, o, t) can be defined as follows:

$$f(s, r, o, t) = \mathbf{contact}(g([\mathbf{h}_s^t : \tilde{\mathbf{r}} : \mathbf{h}_o^t] * \Omega)) \cdot \mathbf{w}, \tag{13}$$

where $\mathbf{h}_s^t \in \mathbb{R}^d$ and $\mathbf{h}_o^t \in \mathbb{R}^d$ are obtained representations of s and o in time t respectively, and $\tilde{\mathbf{r}} \in \mathbb{R}^d$ is the obtained relation representation for r. After obtaining the score of each fact, the model is then trained using soft-margin loss as follows:

$$L = \sum_{x \in \{S \cup S'\}} log(1 + exp(l_x \cdot f(x))) + \frac{\lambda}{2}||\mathbf{w}||_2^2, \tag{14}$$

where S is the set of positive facts, and S' is a set of negative facts obtained by randomly replacing subject or object entity of each positive fact. l_x is the indicator variable which is set as 1 when $x \in S$ and -1 when $x \in S'$.

5   Experiments

In this section, we first provide an overview of the detailed settings in our experiment, and then we report extensive experimental evaluations and provide the analysis of the experimental results.

5.1   Experimental Settings

Datasets. We evaluate our model and baselines on three public datasets released by TA-DistMult [6], which are derived from two popular temporal knowledge graph resources, namely ICEWS and Wikidata [5]. Simple statistics of three datasets are summarized in Table 1, and we detail each dataset as follows:

- **ICEWS14:** This is a short-range version subset of ICEWS recourse by collecting all facts from 2014/1/1 to 2014/12/31 with the granularity of daily, and there are 7,128 distinct entities and 230 types of relations in this dataset.
- **ICEWS05-15:** This is a long-range version subset of ICEWS recourse which is almost 5 times larger than ICEWS14. It contains facts from 2005/1/1 to 2015/12/31 with the granularity of daily and there are 10,488 distinct entities and 251 types of relations in this dataset.

– **WIKIDATA11k:** This is a subset of Wikidata which contains 11,134 distinct entities, 95 types of time-sensitive relations, and in total of 28.5k facts with the granularity of year.

Table 1. Statistics of datasets.

Datasets	Entity	Relation	Fact			Time
			Train	Valid	Test	
ICEWS14	6,869	230	72.8k	8.9k	8.9k	365
ICEWS05-15	10,094	251	368k	46.2k	46k	4017
WIKIDATA11k	11,134	95	121k	14.3k	14.2k	306

Since our model is designed for the TKGC task rather than extrapolation, we utilize random-split and sample roughly 80% of instances as training, 10% as validation, and 10% for testing on each dataset.

Baselines. We compare our model with a suite of state-of-the-art baselines which have been introduced in Sect. 2, such as TAE [10], HyTE [3], and DE-DistMult [7]. Note that, we did not compare our model with RE-NET [11] because RE-Net is designed for extrapolation task rather than TKGC. Furthermore, in order to compare the performance of our model using different score functions, we refer to the resulting models as ST-X, such as ST-TransE and ST-DistMult, where ST is short for **S**patial-**T**emporal.

Metrics. For each test fact (s, r, o, t), we corrupt it by replacing the subject or object entity by all possible entities in turn and obtain a list of candidate facts, and then these candidate facts and original fact are ranked in descending order of their plausibility score. The rank of original fact denoted as $rank(s, r, o, t)$ is the basic metric of the TKGC task, and then we use two kinds of refined metrics based on this to evaluate the performance of each model. One is mean reciprocal rank (MRR) defined as $MRR = \frac{1}{|Test|} \sum_{(s,r,o,t) \in Test} \frac{1}{rank(s,r,o,t)}$, which is the average of the reciprocal of the rank of each test fact, and the higher MRR denotes the better model performance. The other is Hits@N which is defined as $Hits@N = \frac{1}{|Test|} \sum_{(s,r,o,t) \in Test} ind(rank(s, r, o, t) \leq N)$, where $ind()$ is 1 if the inequality holds and 0 otherwise.

Implementation. We implement our model in PyTorch, and all the experiments are performed on an Intel Xeon CPU E5-2640(v4) with 128 GB main memory, and Nvidia TITAN RTX. We initialize all the baselines with the parameter settings in the corresponding papers and then turn them on our datasets for the best performance for a fair comparison. For our model, we create 100 mini-batches for each epoch during training. The dimension of embedding representations $d \in \{50, 100, 200\}$, learning rate $l \in \{10^{-2}, 10^{-3}, 10^{-4}\}$, negative sampling ratio $n \in \{1, 3, 5, 10\}$, basic attenuation coefficient $\lambda \in \{1, 3, 5\}$. The best configuration is chosen based on MRR on the validation dataset. The final parameters are $d = 100$, $l = 10^{-2}$, $n = 5$, $\lambda = 1$ for the ICEWS14 dataset. For the WIKIDATA11k and and ICEWS05-15 datasets, the best configuration is $d = 100$, $l = 10^{-2}$, $n = 3$, $\lambda = 3$.

Table 2. Comparison of different methods on three datasets for link prediction. The best and second best results in each column are boldfaced and underlined respectively (the higher is better for each metric).

Dataset	ICEWS14				ICEWS05-15				WIKIDATA11k			
Models	MRR	Hit@1	Hit@3	Hit@10	MRR	Hit@1	Hit@3	Hit@10	MRR	Hit@1	Hit@3	Hit@10
TransE	0.280	9.4	–	63.7	0.294	9.0	–	66.3	0.316	18.1	–	65.9
DistMult	0.439	32.3	–	67.2	0.456	33.7	–	69.1	0.316	18.1	–	66.1
ConvKB	0.335	22.4	38.7	56.6	–	–	–	–	0.267	12.2	29.6	63.1
TAE	0.263	10.1	49.7	66.2	0.295	10.4	49.0	71.4	0.319	18.3	39.2	65.7
TA-DistMult	0.435	31.5	49.1	68.3	0.468	35.2	51.8	72.8	0.557	40.6	58.6	<u>78.4</u>
TTransE	0.227	7.2	30.1	58.2	0.243	7.6	26.5	57.8	0.294	18.3	35.2	60.9
HyTE	0.297	10.8	41.6	65.5	0.316	11.6	44.5	68.1	0.371	21.5	45.9	75.1
DE-DistMult	0.501	39.2	56.9	70.8	0.484	36.6	54.6	71.8	0.396	24.1	45.7	74.5
TNTComplEX	<u>0.616</u>	**51.8**	65.7	75.8	0.665	<u>59.0</u>	70.5	80.7	0.408	23.9	47.8	75.6
ST-TransE	0.396	9.1	66.8	**86.4**	0.457	12.4	<u>76.2</u>	**93.2**	<u>0.647</u>	<u>56.3</u>	<u>70.4</u>	**78.8**
ST-DistMult	0.603	48.3	<u>67.2</u>	83.0	<u>0.673</u>	55.1	75.0	91.6	0.625	54.9	67.0	75.8
ST-ConvKB	**0.629**	<u>51.0</u>	**71.5**	<u>85.1</u>	**0.704**	**59.3**	**79.6**	<u>91.9</u>	**0.649**	**57.3**	**73.4**	77.9

5.2 Performance Comparison

Table 2 illustrates the results of baselines and our proposed models using different score functions in the link prediction task. According to the results, firstly, our proposed model outperforms all the baselines by a significant improvement, which demonstrates the superiority of our model to obtain more accurate representation for temporal knowledge graph. The improvement of Hits@10 on the ICEWS05-15 dataset is the highest, which may be because that ICEWS05-15 is relatively larger and hence the subgraph in each snapshot is denser, so that our MFGAT can capture richer structural information. TNTComplEX [13] fails to achieve good performance on the WIKIDATA11k dataset because its model is sensitive to data sparsity. Furthermore, the spatial-temporal version of each static method outperforms original counterpart on all metrics, which gives evidence of the merit of considering graph structure and temporal correlation of TKG. DE-DistMult [7] outperforms static KG method DistMult [27] on all datasets, which demonstrates the importance of integrating temporal information for the TKGC task. However, DE-DistMult fails to consider structural information of TKG, therefore, our ST-DistMult consistently outperforms DE-DistMult, which shows the necessity of considering graph structure in the TKGC task. DistMult-based models consistently outperform TransE-based models [2] due to the higher expressivity of DistMult score function. ConvKB [17] has the highest expressivity and thus achieves the best performance. What is more, ST-TransE gets low Hit@1 on ICEWS14 and ICEWS05-15 but high on WIKI-DATA11k because the number of relations in ICEWS14 and ICEWS05-15 is much larger than that of WIKIDATA11k which leads to higher complexity.

5.3   Model Variants and Ablation Study

We run experiments on the ICEWS14 dataset with several variants of our proposed model to provide a better understanding of the effectiveness of each part in our model. The results are shown in Table 3, which includes ST-ConvKB and its variants.

Table 3. Performance of different variants of our model for link prediction.

Variants	MRR	Hit@1	Hit@10
Replacing MFGAT with GAT [23]	0.480	29.7	81.7
Replacing MFGAT with KBAT [16]	0.582	45.8	82.4
Replacing MFGAT with R-GCN [20]	0.531	34.2	83.1
MFGAT without triple feature	0.568	42.6	83.6
MFGAT without group feature	0.583	46.3	81.5
MFGAT without path feature	0.581	46.0	81.1
ADTAT without temporal attenuation	0.598	47.4	85.4
ADTAT with static temporal attenuation coefficient ($\lambda = 0.1$)	0.605	48.9	84.8
ADTAT with static temporal attenuation coefficient ($\lambda = 1$)	0.610	49.6	84.3
ST-ConvKB	0.629	51.0	85.1

Effect of Different Spatial Models. First, as shown in Table 3, the performance of variants with different graph neural network models outperform most of baselines, which indicates the importance of integrating structure information of temporal knowledge graph. Hit@1 of the variant with GAT is lower than other variants because GAT only considers neighbor entities but ignores the information of relations. Hit@10 of all variants are at the same level because all of them are able to capture the co-occurrence relationship among entities. Furthermore, ST-ConvKB outperforms all these variants, which illustrates the superiority of egocentric network considered in our model.

Effect of Each Feature in MFGAT. As shown in Table 3, we compare our model with three variants without triple feature, group feature, and path feature respectively. First, all of these variants are unable to outperform our original model which illustrates that all three kinds of features are effective and contribute to the final performance of our model. Furthermore, the performance of the variant without triple feature drops most because triple feature provides the most intuitive relevance of an entity with its neighbors.

Effect of Adaptive Temporal Attenuation Function. We first compare our model with a simple attention version without temporal attenuation. The Hit@10 result of this variant is at the same level as ST-ConvKB, which indicates that both our original model and this variant are able to capture adequate history information for each snapshot. However, the MRR and Hit@1 results of this variant are lower because it is unable to consider time span and thus the information of long-range snapshots will confuse the model to obtain more accurate predictions. Furthermore, we compare our model with variants using different static temporal attenuation coefficients ($\lambda = 0.1$ and $\lambda = 1$). They are unable to outperform our original model because large attenuation coefficient will let the model fail to capture sufficient history information for locally sparse entities, and small attenuation coefficient will let the model fail to consider the effect of time span adequately.

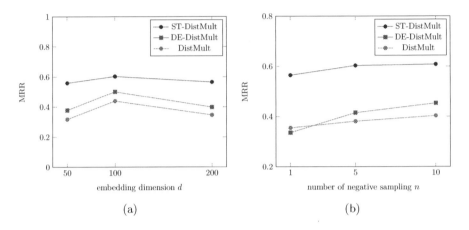

Fig. 3. Influence of the embedding dimension and negative sampling number.

5.4   Parameter Analysis

We study the impact of the training parameters of our model in this part, including the dimension of embedding representations d and the number of negative samples n.

Dimension of Embedding Representation. Here, we analyze the performance of ST-DistMult which considers both the graph structure and time-evolving property of TKG, DE-DistMult which only considers the time-evolving property and static KG method DistMult on changing the dimension of embedding representations. As shown in Fig. 3(a), with the increase of dimension d, the performance of each model increases firstly and then decreases. This is because when d is too small, representations have insufficient capacity to capture rich information from temporal knowledge graph, and when d is large, the model will be trapped in overfitting problem. Furthermore, we notice that with the representation dimension d changes, the performance of our ST-DistMult changes less compared with the other two models, which is because ST-DistMult can extract more effective information from TKG and thus it is more stable.

Number of Negative Sampling. As shown in Fig. 3(b), by comparing the performance of ST-DistMult, DE-DistMult and DistMult with different negative sampling numbers, we observe that with the increase of negative sampling number n, the performance of each model increases consistently. This is because a larger negative sampling number can provide more positive-negative pairs for each model to learn, and thus provide more information. However, we notice that when n is large, keep on increasing n leads to small performance improvement of ST-DistMult, which is because obtaining negative facts by random sampling can only provide coarse-grained information. Furthermore, compared with the other two models, ST-DistMult can still achieve significant performance when n is small, which demonstrates our ST-DistMult is able to obtain richer representations and thus each positive-negative pair can provide more information for model to learn.

6   Conclusion

In this work, we study the temporal knowledge graph completion task. We take temporal knowledge graph as a kind of spatial-temporal resource, and develop a spatial-temporal attention network which is able to obtain representation for each element of TKG while considering the graph structure and time-evolving property of TKG simultaneously. Our model contains a multi-faceted graph attention network used to capture structural information of each snapshot, and an adaptive temporal attention mechanism to model the temporal correlation of different snapshots. The representations obtained by our model can be used in the score function of existing static knowledge graph methods and result in the spatial-temporal version of these methods for the TKGC task. We test our proposed model on the link prediction task on three benchmark datasets. The experimental results show the superiority of our model and the effectiveness of each component in our model. In the future work, we aim to model the temporal correlation of TKG based on the structure evolution of egocentric network of each entity.

Acknowledgments. This work is supported by the National Nature Science Foundation of China (No. 61832001) and Sichuan Science and Technology Program (No. 2021JDRC0067 and No. 2019YFG0535).

References

1. Arnaboldi, V., Conti, M., Gala, M.L., Passarella, A., Pezzoni, F.: Ego network structure in online social networks and its impact on information diffusion. Comput. Commun. **76**, 26–41 (2016)
2. Bordes, A., Usunier, N., García-Durán, A., Weston, J., Yakhnenko, O.: Translating embeddings for modeling multi-relational data. In: NIPS, pp. 2787–2795 (2013)
3. Dasgupta, S.S., Ray, S.N., Talukdar, P.P.: HyTE: hyperplane-based temporally aware knowledge graph embedding. In: EMNLP, pp. 2001–2011 (2018)
4. Dettmers, T., Minervini, P., Stenetorp, P., Riedel, S.: Convolutional 2D knowledge graph embeddings. In: AAAI, pp. 1811–1818 (2018)

5. Erxleben, F., Günther, M., Krötzsch, M., Mendez, J., Vrandečić, D.: Introducing Wikidata to the linked data Web. In: Mika, P., et al. (eds.) ISWC 2014. LNCS, vol. 8796, pp. 50–65. Springer, Cham (2014). https://doi.org/10.1007/978-3-319-11964-9_4

6. García-Durán, A., Dumancic, S., Niepert, M.: Learning sequence encoders for temporal knowledge graph completion. In: EMNLP, pp. 4816–4821 (2018)

7. Goel, R., Kazemi, S.M., Brubaker, M., Poupart, P.: Diachronic embedding for temporal knowledge graph completion. In: AAAI, pp. 3988–3995 (2020)

8. Gupta, S., Yan, X., Lerman, K.: Structural properties of ego networks. In: Agarwal, N., Xu, K., Osgood, N. (eds.) SBP 2015. LNCS, vol. 9021, pp. 55–64. Springer, Cham (2015). https://doi.org/10.1007/978-3-319-16268-3_6

9. Ji, G., He, S., Xu, L., Liu, K., Zhao, J.: Knowledge graph embedding via dynamic mapping matrix. In: ACL, pp. 687–696 (2015)

10. Jiang, T., et al.: Encoding temporal information for time-aware link prediction. In: EMNLP, pp. 2350–2354 (2016)

11. Jin, W., Qu, M., Jin, X., Ren, X.: Recurrent event network: autoregressive structure inference over temporal knowledge graphs. In: EMNLP (2020)

12. Kipf, T.N., Welling, M.: Semi-supervised classification with graph convolutional networks. In: ICLR (2017)

13. Lacroix, T., Obozinski, G., Usunier, N.: Tensor decompositions for temporal knowledge base completion. In: ICLR (2020)

14. Leblay, J., Chekol, M.W.: Deriving validity time in knowledge graph. In: Champin, P., Gandon, F.L., Lalmas, M., Ipeirotis, P.G. (eds.) WWW, pp. 1771–1776 (2018)

15. Lin, Y., Liu, Z., Sun, M., Liu, Y., Zhu, X.: Learning entity and relation embeddings for knowledge graph completion. In: AAAI, pp. 2181–2187 (2015)

16. Nathani, D., Chauhan, J., Sharma, C., Kaul, M.: Learning attention-based embeddings for relation prediction in knowledge graphs. In: ACL, pp. 4710–4723 (2019)

17. Nguyen, D.Q., Nguyen, T.D., Nguyen, D.Q., Phung, D.Q.: A novel embedding model for knowledge base completion based on convolutional neural network. In: NAACL-HLT, pp. 327–333 (2018)

18. Nickel, M., Rosasco, L., Poggio, T.A.: Holographic embeddings of knowledge graphs. In: AAAI, pp. 1955–1961 (2016)

19. Nickel, M., Tresp, V., Kriegel, H.: A three-way model for collective learning on multi-relational data. In: ICML, pp. 809–816 (2011)

20. Schlichtkrull, M., Kipf, T.N., Bloem, P., van den Berg, R., Titov, I., Welling, M.: Modeling relational data with graph convolutional networks. In: Gangemi, A., et al. (eds.) ESWC 2018. LNCS, vol. 10843, pp. 593–607. Springer, Cham (2018). https://doi.org/10.1007/978-3-319-93417-4_38

21. Shi, X., Chen, Z., Wang, H., Yeung, D., Wong, W., Woo, W.: Convolutional LSTM network: a machine learning approach for precipitation nowcasting. In: NIPS, pp. 802–810 (2015)

22. Trouillon, T., Welbl, J., Riedel, S., Gaussier, É., Bouchard, G.: Complex embeddings for simple link prediction. In: ICML, pp. 2071–2080 (2016)

23. Velickovic, P., Cucurull, G., Casanova, A., Romero, A., Liò, P., Bengio, Y.: Graph attention networks. In: ICLR (2018)

24. Wang, S., Cao, J., Yu, P.S.: Deep learning for spatio-temporal data mining: a survey. IEEE Trans. Knowl. Data Eng. (2020). https://doi.org/10.1109/TKDE.2020.3025580

25. Wang, Z., Zhang, J., Feng, J., Chen, Z.: Knowledge graph embedding by translating on hyperplanes. In: AAAI, pp. 1112–1119 (2014)

26. Yan, S., Xiong, Y., Lin, D.: Spatial temporal graph convolutional networks for skeleton-based action recognition. In: AAAI, pp. 7444–7452 (2018)
27. Yang, B., Yih, W., He, X., Gao, J., Deng, L.: Embedding entities and relations for learning and inference in knowledge bases. In: ICLR (2015)
28. Zhang, J., Sheng, Y., Wang, Z., Shao, J.: TKGFrame: a two-phase framework for temporal-aware knowledge graph completion. In: Wang, X., Zhang, R., Lee, Y.-K., Sun, L., Moon, Y.-S. (eds.) APWeb-WAIM 2020. LNCS, vol. 12317, pp. 196–211. Springer, Cham (2020). https://doi.org/10.1007/978-3-030-60259-8_16
29. Zheng, C., Fan, X., Wang, C., Qi, J.: GMAN: a graph multi-attention network for traffic prediction. In: AAAI, pp. 1234–1241 (2020)

Ranking Associative Entities in Knowledge Graph by Graphical Modeling of Frequent Patterns

Jie Li, Kun Yue$^{(\boxtimes)}$, Liang Duan, and Jianyu Li

School of Information Science and Engineering, Yunnan University, Kunming, China
{jiel,jylee}@mail.ynu.edu.cn, {kyue,duanl}@ynu.edu.cn

Abstract. Ranking associative entities in Knowledge Graph (KG) is critical for entity-oriented tasks like entity recommendation and associative inference. Existing methods benefit from explicit linkages in KG w.r.t. exactly two query entities via the closely appearing co-occurrences. Given a query including one or more entities in KG, it is necessary to obtain the implicit associative entities and uncover the strength of associations from data. To this end, we leverage KG with Web resources and propose an approach to ranking associative entities based on frequent pattern mining and graph embedding. First, we construct an entity dependency graph from the frequent patterns of entities generated from both KG and Web resources. Thus, the existence and strength of associations between entities could be depicted effectively in a holistic way. Second, we embed the dependency graph into a lower-dimensional space and consequently fulfill entity ranking on the embedding. Finally, we conduct an extensive experimental study on real-life datasets, and verify the effectiveness of our proposed approach compared to competitive baselines.

Keywords: Knowledge graph · Associative entity · Association ranking · Frequent entity · Graph embedding

1 Introduction

Many entity-oriented applications, like entity alignment [4], entity recommendation [20] and entity associations inference [16], benefit from the results of top-ranked associative entities in knowledge graph (KG). The task of ranking associative entities (a.k.a. association ranking) is to sort candidate entities w.r.t. a query including one or more given entities in KG. For example, {*1. Microsoft*; *2. COVID-19*; *3. Windows*} is a ranking list of candidate entities w.r.t. the query entity *Bill Gates* sorted by their association strength.

It is straightforward to represent associative entities based on the triple-structured data of KG. One feasible solution to ranking associative entities is

ⓒ Springer Nature Switzerland AG 2021
C. S. Jensen et al. (Eds.): DASFAA 2021, LNCS 12681, pp. 224–239, 2021.
https://doi.org/10.1007/978-3-030-73194-6_16

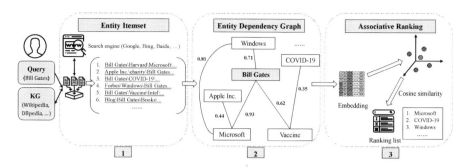

Fig. 1. Overview of EDGM.

based on the semantic associations (a.k.a. relatedness) between words or entities [19] upon the prerequisite that frequently occurring entities are regarded to be highly associated. However, only the frequencies of two closely appearing entities are considered, while the highly associated entities do not necessarily co-occur significantly in the neighboring context. The underlying local co-occurrence principle leads to limited coverage and precision. Recently, multiple association features between words, concepts, and entities are combined to construct an association network [9,15] to improve the relatedness measurement. However, these models could not be well learned in an unsupervised manner. Meanwhile, these methods focus on measuring the semantic association between exactly two entities within KG. It will be more scalable if any number of query entities are allowed and multiple Web resources could be introduced.

By using linked Web resources, explicit associations could be found easily to enhance Web applications like search engines [21]. However, implicit associations between entities show usefulness in many domains including national security and biomedical research [5]. For example, it is necessary to identify the importance of implicit associations such as common preferences and similar behaviors in social networks. Potential connections between a group of users may contribute to suspect search. Thus, additional Web resources outside KG are incorporated to improve the ranking results [14,26]. Figure 1 illustrates an example of ranking associative entities based on both KG and Web resources. As shown in part 2, *Bill Gates* and *Vaccine* are associative with the strength of 0.62, which corresponds to the news topic of *"Bill gates pledges $1.6 billion to vaccine research against COVID-19"*. That is, even though *Bill Gates* and *Vaccine* are not directly linked in KG, there is still a strong association between them. By introducing retrieved results of Web pages w.r.t. *Bill Gates* in part 1, it is available to uncover these kinds of implicit associations.

Thus, we consider refining the association features by incorporating KG and Web resources to fulfill the task of ranking associative entities, in which we will have to solve the following 2 questions:

(1) How to find the associative entities w.r.t. one or more query entities in KG using association features from both KG and Web resources?

(2) How to measure the strength of associations between entities by holistically aggregating multiple co-occurrences?

In this paper, as shown in Fig. 1, we propose an Entity Dependency Graph Model (EDGM) to rank associative entities by graph embedding. In our EDGM, we use association features from both KG and Web resources based on associative Wikipedia articles and contents of Web pages w.r.t. the query entities. The associations especially for up-to-date situations could be determined by the rapidly changing or generated Web resources or user behavioral records that we regard as the *transactions* in frequent pattern mining [1]. By this way, we propose a method to bridge the gap between frequent pattern mining and graphical model. Upon the graph structure of frequent entities, we aggregate frequencies of both single entities and the co-occurrences of entities to evaluate the associations quantitatively. To obtain highly represented embedding of associative entities and fulfill effective ranking, we adopt a BFS-biased random walk sampling mechanism based on node2vec [10]. This enables our EDGM to better measure the strength of associations by capturing neighboring and co-occurring features accurately. The contributions of this paper are as follows:

First, as illustrated in part 1 of Fig. 1, we generate an entity itemset containing sequences of candidate entities to the query from both KG and Web resources. By incorporating the extracted Web resources, it is practical to integrate various statistics of entities. Then, we adopt the frequent pattern mining algorithm on the entity itemset to build an undirected weighted graph, where each node represents an entity, and each edge represents the associations between entities. By an unsupervised manner, co-occurrence associations w.r.t. one or more query entities could be discovered.

Second, to improve the effectiveness of ranking associative entities, we measure the weight of each edge on EDG, which could present the strength of associations by refining both the informativeness and specificness of co-occurrences simultaneously. To fulfill entity ranking for each associative candidate to the given query, we use graph embedding to transform the nodes on the weighted graph into a low-dimensional space and then rank the candidate entities based on the similarity between node embeddings.

Finally, we conduct extensive experiments on two real-life datasets to evaluate the effectiveness of our EDGM. Experimental results illustrate that our approach outperforms some state-of-the-art competitors in ranking associative entities.

The rest of this paper is organized as follows: Sect. 2 introduces related work and preliminaries. Section 3 presents our methods for learning EDG and ranking associative entities. Section 4 shows experiments and performance studies. Section 5 concludes and discusses future work.

2   Related Work and Preliminaries

In this section, we review related work, followed by giving necessary definitions and formulating the problem.

2.1 Related Work

Most research efforts for ranking associative entities could be divided into 3 categories: entity relatedness ranking, association ranking of KG, and entity ranking by graph embedding.

Entity Relatedness Ranking. Entity relatedness ranking optimizes the partial order of the associative entities into desired positions upon semantic relatedness [20]. For measuring relatedness between exactly two entities, text-based methods [2,8] build high-dimensional weighted vectors to represent words and Wikipedia concepts. Other graph-based approaches [27] adopt the link structure of Wikipedia to obtain the distance of entities. These methods are insufficient to uncover more profound co-occurrences with only text semantics or graphical structural relatedness. Better results could be achieved by integrating existing methods through designing comprehensive frameworks [25]. To further leverage more types of co-occurrences in KG, network-based methods [9,15] specify associations among words and concepts in a supervised manner upon well-generated datasets from psychological studies.

Association Ranking of KG. Techniques for ranking associations between two or more entities are developed with the emergence of graph-structured Web resources, which could be divided into data-centric and user-centric. Data-centric techniques mainly use various statistical information of entities, and user-centric techniques focus on user preference. Typically, the associations are regarded as paths connecting two or more entities in KG [7]. Simple associations could be obtained directly by triple-linked data from KG, but implicit associations are more preferred in some domains [5]. To search and rank implicit associations, the frequent pattern mining algorithm has been proved to be efficient and effective [6]. By counting the frequency of canonical codes uniquely representing entity patterns, associations could be ranked upon the edit distance between graph structures.

Entity Ranking by Graph Embedding. Graph embedding techniques like DeepWalk [23] are effective for association analysis in graphical structures [3], in which low-dimensional representations of the nodes with neighboring and co-occurrence relations are learned. Zhang et al. [29] propose a graph embedding-based neural ranking framework to overcome the query-entity sparsity problem by integrating features in click-graph data. On heterogeneous information networks, recent studies for proximity search [18] learn graph embedding models to rank associative nodes by given semantic relations. These techniques are based on user intent with a certain amount of behavior preference labels. Differently, we choose node2vec [10] to embed the associations between entities, since the co-occurrences on EDG, together with their strength, could be expressed by using the biased and dynamic random walk.

2.2 Definitions and Problem Formulation

Firstly, the symbols and notations are given in Table 1. Then, we define several concepts as the basis of later discussions.

Table 1. Notations.

Notation	Description
$D_{\vec{q}}$	Associative datasets w.r.t. query $\vec{q}$
$\Psi(\vec{q})$	Entity itemset w.r.t. $\vec{q}$
Υ	Set of all 1-frequent entities
A_x	A maximal set of frequent entities
$G_E = (V, E, W)$	Entity dependency graph with nodes V, edges E and weights W
$\overline{v_i v_j}$	Edge between 1-frequent entities v_i and v_j
$H^{|V| \times d}$	Representation space of EDG with the dimension of $|V| \times d$
$L(\vec{q})$	Ranking list of associative entities w.r.t. $\vec{q}$

To obtain $D_{\vec{q}}$ from both the KG and the Web, associative data like Web pages and Wikipedia articles w.r.t. the query entities $\vec{q}$ could be retrieved and collected by search engines.

Definition 1. *A knowledge graph is denoted as $\mathcal{G} = (\mathcal{E}, \mathcal{R})$, where $\mathcal{E}$ represents a finite set of nodes indicating entities, and $\mathcal{R}$ is a set of directed edges representing relations between entities.*

Sequences of associative entities could be generated based on items in $D_{\vec{q}}$ and named entities of $\mathcal{G}$. The definition of $\Psi(\vec{q})$ is as following:

Definition 2. *Let $\psi = \{e_1, e_2, ..., e_M\}$ be a sequence of entities, where $e_i \in \mathcal{E}$ and $\psi \in \Psi(\vec{q})$. Each ψ is corresponding to an item in $D_{\vec{q}}$.*

Based on the idea of frequent pattern mining [1], $\Psi(\vec{q})$ could be regarded as the *transactions* of $D_{\vec{q}}$. Next, we define the *set of frequent entities*.

Definition 3. *$v(v \in \mathcal{E})$ is called a 1-frequent entity if $p(v) \geq \sigma$, where $p(v)$ is the support of v (i.e., the proportion of sequences in $\Psi(\vec{q})$ containing v) and σ is the threshold of minimal-support. The set of all 1-frequent entities is denoted as Υ.*

Definition 4. *A set of frequent entities $A_x \subset \Upsilon$ is called* maximal, *if there are no other super-sets A_y in $\mathbb{A}$ satisfying $A_x \subset A_y$, where $\mathbb{A} = \{A_1, ..., A_m\}$ includes all the sets of frequent entities.*

Following, we define the entity dependency graph (EDG) to describe the existence and strength of associations between entities.

Definition 5. *An EDG is an undirected weighted graph, denoted as $G_E = (V, E, W)$. V is the set of nodes, and $V \subset \Upsilon$. Each edge $\overline{v_i v_j} \in E$ ($v_i, v_j \in V, i \neq j$) indicates the co-occurrence association between v_i and v_j. Each $w_{ij} \in W$ represents the weight of $\overline{v_i v_j}$.*

Problem Formulation. Given the query $\vec{q}$, we first extract its itemset $\Psi(\vec{q})$ from $D_{\vec{q}}$ as the input to construct EDG. For each node in EDG, the representation space H is learned as:

$$f : V \longrightarrow H^{|V| \times d} \tag{1}$$

Upon the matrix $H^{|V| \times d}$, we measure the strength of associations between each candidate entity and $\vec{q}$ from a global perspective, and output the ranking list of candidate entities $L(\vec{q})$ w.r.t. $\vec{q}$.

3 Methodology

In this section, we introduce the approach to ranking associative entities by our EDGM. First, the structure of EDG is learned by mining frequent patterns from the *transactions* of both KG and Web resources, and then the weights of edges on EDG are measured based on an extension principle of co-occurrences. Finally, the ranking process is implemented by graph embedding.

For the given KG $\mathcal{G}$ and query $\vec{q}$, the sequences of entities recognized from KG are *transactions* of $D_{\vec{q}}$, for which the entity itemset $\Psi(\vec{q})$ is generated from $D_{\vec{q}}$ (e.g., Wikipedia articles and Web pages retrieved w.r.t. $\vec{q}$) by entity linking. Then, by learning the graphical structure and measuring the weights of edges, the EDG $G_E = (V, E, W)$ is constructed to depict the associations between frequently co-occurring entities in a holistic way.

3.1 Structure Learning

Learning the structure of G_E aims to determine the set of nodes V and the set of edges E. The nodes in V are generated by mining frequent entities in $\Psi(\vec{q})$, and the edges in E depend on the test of conditional independence [17] between frequent entities.

To achieve a high recall in line with the inherence of co-occurrence between entities, the node set V should contain the candidate entities related to the query as many as possible. Given $\Upsilon = \{v_1, v_2, ..., v_n\}$ as a set of 1-frequent entities in $D_{\vec{q}}$, we generate V from Υ by neglecting the entities whose support values are less than the threshold σ according to the probability cut defined as follows:

$$p_{\sigma}(I) = \begin{cases} 0 & p(I) < \sigma \\ p(I) & p(I) \geq \sigma \end{cases} \tag{2}$$

As is known that only frequent entities are concerned when computing $p(I)$ by the classic Apriori algorithm [11]. If I is a set of frequent entities, then all the non-empty subsets of I must also be frequent. If there is no set of frequent entities J in such Υ that $I \subset J$, we call I is the maximal. To include the entities concerning all co-occurrences, we adopt the entities in the maximal set of frequent entities as nodes in V.

To determine the edges among the nodes in V, we first generate completely connected subgraphs over each maximal set of frequent entities. According to the conclusion in [17], the associations between frequent entities imply probabilistic conditional independences. Thus, two entities in the set of frequent entities are not connected in G_E by an edge if they are conditionally independent. By testing conditional independence, the graphical topology of frequent entities could be obtained.

The conditional independence between entities is closely related to the frequent set, to which the entities belong. Let I, J and K be three disjoint subsets of Υ. We use $< I|K|J >$ to denote that "I is independent of J given K", namely $p(I \cap J|K) = p(I|K)p(J|K)$. By focusing on the conditional independence relations between frequent entities, we analyze possible associations between them.

Specifically, let I and J ($I, J \subseteq \Upsilon$) be two different sets of maximal frequent entities and $I \cap J = K$. We consider the following three cases. For two entities in different sets with intersections, an edge is added between these two entities to reflect their mutual dependency. Two entities in different sets without overlap are unconnected. Two entities are also unconnected if they are already in one set and do not co-occur in any other sets.

Case 1. If there exist 1-frequent entities $v_a \in I$ and $v_b \in J$ such that the entity set $\{v_a, v_b\}$ is non-frequent, then $< v_a|K|v_b >$ is false, denoted as $\overline{< v_a|K|v_b >}$. That is, v_a and v_b are associative and there is an edge $\overline{v_a v_b}$.

Case 2. If $\{v_a, v_b\}$ is non-frequent for all $v_a \in I$ and $v_b \in J$ when $K = \varnothing$, then $< v_a|\Upsilon - v_a - v_b|v_b >$ is true. In other words, if there is no any frequent entity v_c such that $\{v_a, v_c\}$ and $\{v_b, v_c\}$ are frequent when $\{v_a, v_b\}$ is non-frequent, v_a and v_b are independent and there is no edge between them.

Case 3. Suppose $< v_a|I - v_a - v_c|v_c >$ is true, where $v_a, v_c \in I$. If there is no such J that $v_a, v_c \in J$, then $< v_a|\Upsilon - v_a - v_c|v_c >$ is true. That is, two conditionally independent entities v_a and v_c do not share an edge if they co-occur only in one maximal set of frequent entities.

Algorithm 1. Structure learning of EDG

Input: Υ; $\mathbb{A} = \{A_1, ..., A_m\}$, where each $A_x \in \mathbb{A}$, $A_x = \{v_{xy}|v_{xy} \in \Upsilon, 1 \leq y \leq n\}(x \in [1, m])$
Output: V, the set of nodes in G_E; E, the set of edges in G_E
1: $V \leftarrow \Upsilon$, $E \leftarrow \{\}$, $G_{\mathbb{A}} \leftarrow \{\}$
2: **for** each $A_x \in \mathbb{A}$ **do**
3:     Generate $G_{A_x}(V_{A_x}, E_{A_x})$ // Join each pair of distinct entities in A_x
4:     $G_{\mathbb{A}} \leftarrow G_{\mathbb{A}} \cup G_{A_x}$ // G_{A_x} is the complete graph of A_x and $G_{\mathbb{A}}$ is the set of G_{A_x}
5: **end for**
6: **for** each pair $(A_x, A_y) \in \mathbb{A} \times \mathbb{A}$ **do** // Case 1
7:     **if** $A_x \cap A_y \neq \varnothing$ **then**
8:         **for** each edge $\overline{v_{xs} v_{yt}}$ **do** // $v_{xs} \in A_x - A_y$ and $v_{yt} \in A_y - A_x$
9:             $E \leftarrow E \cup \overline{v_{xs} v_{yt}}$ // Add $\overline{v_{xs} v_{yt}}$ to the set of edges
10:         **end for**
11:     **end if**
12: **end for**
13: **for** each $G_{A_x} \in G_{\mathbb{A}}$ **do**
14:     **for** each edge $\overline{v_{xs} v_{xt}} \in G_{A_x}$ **do**
15:         **if** $< v_{xs}|A_x - v_{xs} - v_{xt}|v_{xt} >$ **then** // Case 2
16:             $E \leftarrow E - \overline{v_{xs} v_{xt}}$
17:         **end if**
18:         **for** each $A_y \in \mathbb{A} - A_x$ **do** // Case 3
19:             **if** $< v_{xs}|A_y - v_{xs} - v_{xt}|v_{xt} >$ or $v_{xs} \notin A_y$ or $v_{xt} \notin A_y$ **then**
20:                 $E \leftarrow E - \overline{v_{xs} v_{xt}}$
21:             **end if**
22:         **end for**
23:     **end for**
24: **end for**
25: **return** V, E

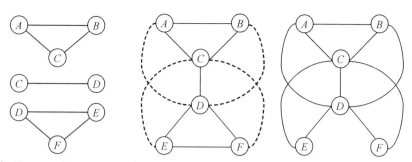

(a) Three undirected complete subgraphs
(b) Undirected graph generated by Algorithm 1
(c) Results EDG over Υ

Fig. 2. A running example of Algorithm 1.

Next, we illustrate the execution of Algorithm 1 by the following example. Given $\Upsilon = \{A, B, C, D, E, F\}$ as the set of 1-frequent entities. $\{A, B, C\}$, $\{C, D\}$, and $\{D, E, F\}$ are three maximal sets of frequent entities over Υ.

Firstly, we add edges for entities within one maximal set of frequent entities respectively in Fig. 2(a) to generate three undirected complete subgraphs according to Case 1. Secondly, we add undirected edges $\overline{AD}, \overline{CE}, \overline{BD}$ and $\overline{CF}$ shown by dotted lines in Fig. 2(b) to represent the possible associations. According to Case 2, following edges should not exist: $\overline{AE}, \overline{AF}, \overline{BE}, \overline{BF}$. Finally, suppose that the conditional independence tests show $< E|D|F >$ and $< E|C|F >$ are true. Then, according to Case 3, $\overline{EF}$ will be deleted. The actual structure of EDG is shown in Fig. 3(c).

Step 2 in Algorithm 1 could be done in $O(|A_1|^2 + \dots + |A_m|^2)$ time, and does not exceed $O(m \times n^2)$, where $|A_x|$ is the number of entities in A_x and $|A_x| \leq n (1 \leq x \leq m)$. Step 6 could be done in $O(m \times n^2)$ time at most. Step 13 could be achieved in $O(|A_1|^2 + \dots + |A_m|^2)$ time and no larger than $O(m \times n^2)$. The overall time complexity of Algorithm 1 is $O(m \times n^2)$. Besides, the Apriori algorithm directly provides all probability values for the construction of EDG.

3.2   Calculation of Weights

Given the structure of graph G_E, it is necessary to accurately quantify the weights of edges by further exploring the co-occurrences statistics from data. Thus, we introduce *coefficient of association* $coa(v_i, v_j)$ as the weight $w_{ij} \in W$ of each edge $\overline{v_i v_j}$. According to the intuition of $coa(v_i, v_j)$, the following properties should be satisfied:

- *Symmetry*: $coa(v_i, v_j) = coa(v_j, v_i)$.
- *Non-negativity*: $coa(v_i, v_j) > 0$.
- *Identical boundedness*: $coa(v_i, v_j) \leq 1$, $coa(v_i, v_j) = 1$ only if $v_i = v_j$.
- *Informativeness of co-occurrence*: The fewer occurrences of ψ in $\Psi(\vec{q})$ containing an entity pair (v_i, v_j), the more informative the (v_i, v_j) is, corresponding to a higher $coa(v_i, v_j)$.

- *Specificness of entity frequency*: Entity frequency (EF) denotes the proportion of an entity to the total number of entities in $\Psi(\vec{q})$. The greater the difference in frequency between v_i and v_j, the smaller the $coa(v_i, v_j)$.

To compute $coa(v_i, v_j)$, we first consider the *informativeness* of co-occurrence by describing the ratio of the number of co-occurrence entries for entity pairs in $\Psi(\vec{q})$:

$$\ln \frac{SN\left(\Psi(\vec{q})\right)}{TN\left(v_i, v_j\right)} \tag{3}$$

where $SN\left(\Psi(\vec{q})\right)$ denotes the total number of *transactions* in $\Psi(\vec{q})$, and $TN\left(v_i, v_j\right)$ represents the number of entity sequences containing both v_i and v_j.

Actually, we aim to distinguish the importance of different entity pairs by *informativeness* of co-occurrence. If (v_i, v_j) appears frequently and dispersedly in multiple entity sequences, we consider the co-occurrence of (v_i, v_j) is trivial and less informative. In contrast, if v_i and v_j co-occur in a smaller number of entity sequences, the associations between them are more representative and informative, which leads to a larger strength. Next, we consider the difference of frequency at the single entity level:

$$\exp\left|EF\left(v_i\right) - EF\left(v_j\right)\right| \tag{4}$$

where $EF\left(v_i\right)$ and $EF\left(v_j\right)$ means entity frequency of v_i and v_j respectively.

Equation (4) takes the *specificness* of entity frequency into account. The smaller the difference between $EF(v_i)$ and $EF(v_j)$ the closer of v_i and v_j. We choose exponential function to ensure that the overall value of Eq. (4) is a number greater than or equal to 1. At the same time, the trend of Eq. (4) is positively correlated with the frequency difference between v_i and v_j.

To combine Eq. (3) and Eq. (4) to jointly measure the weights of edges, we form Eq. (5) to reasonably reflect both trends. The unnormalized weight of $\overline{v_i v_j}$ is defined as follows:

$$\xi(v_i, v_j) = \frac{\ln \frac{SN(\Psi(\vec{q}))}{TN(v_i, v_j)}}{\exp\left|EF\left(v_i\right) - EF\left(v_j\right)\right|} \tag{5}$$

Here, the upper bound of $\xi(v_i, v_j)$ is not constrained, which does not facilitate our specific comparison between the weights of any two edges. The sum of all $\xi(v_i, v_j)$ in G_E is specified as follows:

$$Sum(G_E) = \sum_{v_i, v_j \in V, i \neq j} \xi(v_i, v_j) \tag{6}$$

Then, $\xi(v_i, v_j)$ could be normalized by combining Eq. (5) and Eq. (6).

$$coa(v_i, v_j) = \frac{\xi(v_i, v_j)}{Sum(G_E)} \tag{7}$$

We measure the $coa(v_i, v_j)$ individually to get the actual weights w_{ij} of each edge $\overline{v_i v_j} \in E$. Finally, the set of weights W of EDG could be obtained.

3.3 Ranking Associative Entities

To measure the association strength of any two entities in the EDG, we transform the nodes of G_E into low-dimensional vector space by graph embedding.

Specifically, given an EDG $G_E = (V, E, W)$, we learn the co-occurrence features and the neighboring relations among nodes in two steps: random walk sampling and skip-gram.

We use a tunable bias random walk mechanism [10] in the procedure of neighborhood sequences sampling. Let $v_s \in V$ be a source node, and c_l be the lth node in the walk, $c_0 = v_s$. The unnormalized transition probability $\pi(c_l, c_{l+1})$ is:

$$\pi(c_l, c_{l+1}) = \eta_{mn}(c_{l-1}, c_{l+1}) \times coa(c_l, c_{l+1}) \tag{8}$$

where $\eta_{mn}(c_{l-1}, c_{l+1})$ is a hyper parameter determined by the shortest path distance between c_{l-1} and c_{l+1}. m and n are user-defined parameters to control the bias of random walk. Then, the actual transition probability from v_i to v_j is κ_t, defined as follows:

$$\kappa_t = (c_l = v_j | c_{l-1} = v_i) = \begin{cases} \frac{\pi(v_i, v_j)}{Z} & \overline{v_i v_j} \in E \\ 0 & \text{otherwise} \end{cases} \tag{9}$$

where Z is the normalizing constant.

Upon the sample sequences, we aim to map each $v_i \in V$ into the same space: $f : v_i \longrightarrow \mathbb{R}^d$ (equivalent to Eq. (1)) by maximizing the log-probability function:

$$\max_f \sum_{v_i \in V} log[p(N_b(v_i)|f(v_i))] \tag{10}$$

where $N_b(v_i) \subset V$ is the network neighboring [10] of v_i generated by the random walk sampling strategy controlling by Eq. (8).

A matrix $H^{|V| \times d}$ could be obtained by Eq. (10). Each entry of $H^{|V| \times d}$ represents the vector of a specific entity in EDG. The association strength $ad(v_i, \vec{q})$ of v_i to $\vec{q}$ in EDG could be measured by the cosine similarity of vectors in $H^{|V| \times d}$:

$$ad(v_i, \vec{q}) = \frac{\sum_{j=1}^d H_{ij} \times H_{\vec{q}}}{\sqrt{\sum_{j=1}^d H_{ij}^2 \times \sum_{j=1}^d H_{\vec{q}}^2}} \tag{11}$$

where $H_{\vec{q}}$ denotes the vector representation of query $\vec{q}$. Note that if there are more than one entities in $\vec{q}$, the final $ad(v_i, \vec{q})$ is the average of similarities between the vector of v_i to the vectors of different entities in $\vec{q}$.

Finally, we could obtain a top-k ranked list $L(\vec{q}) = \{ad(v_1, \vec{q}), ..., ad(v_k, \vec{q})\}$, where $ad(v_i, \vec{q})(1 \leq i \leq n)$ is the ith maximal value in $L(\vec{q})$.

4 Experiments

In this section, we present experimental results on two real-life datasets to evaluate our proposed method. We first introduce the experimental settings, and then

234 J. Li et al.

conduct three sets of experiments: (1) ranking associative entities, (2) entity relatedness ranking, and (3) impacts of parameters to evaluate our method compared with existing methods.

4.1 Experiment Settings

Datasets. We perform experiments on two widely used datasets for evaluating entity relatedness, KORE [13] and ERT [12], and extract the datasets containing associative Wikipedia articles and Web pages from search engines.

Table 2. Statistics of datasets.

Dataset	Query entities	Candidate entities	Wikipedia articles	Google & Bing URLs
KORE	21	420	4,200	12,600
ERT	40	937	8,000	24,000

- **KORE**, extracts entities from YAGO2 covering four popular domains: IT companies, Hollywood celebrities, video games, and television series. For each query entity, 20 candidate entities linked to the query's Wikipedia article are ranked in descending order of human rating association scores and regarded as the ground-truth of the most relevant entities to the query.
- **ERT**, consists of query entity pairs within two topics: the first 20 groups are from the music Website last.fm, and the last 20 groups originate in the movie dataset IMDb. Several to dozens of candidate entities with association scores are given for each entity pair. The scores are computed by considering multiple properties of entities from DBpedia

To generate $D_{\vec{q}}$ of each query, we extract the associative texts of all query entities in the two datasets from search engines and Wikipedia. Specifically, we first crawl the Web pages of the top 300 URLs from Google and Bing by using the query entities as queries. We then collect the top-ranked 200 Wikipedia articles by inputting the query entities into the Wikipedia dump. We finally combine these text contents as $D_{\vec{q}}$. Note that we also pre-process these texts by removing redundancy and building indices. The important statistics about these datasets are summarized in Table 2.

Evaluation Metrics. Three groups of metrics are chosen in our experiments: (1) Normalized discounted cumulative gain (NDCG) [20] to evaluate the accuracy of each entity ranking method, (2) Pearson, Spearman correlation coefficients and their harmonic mean [25] to evaluate the consistency between ranking results and ground truth, and (3) precision, recall and F1-score to evaluate the effectiveness of EDG with varying the user-defined threshold of minimal-support σ.

Comparison Methods. We choose six methods for comparison with our EDGM.

- Milne-Witten (MW) [27] is a typical graph-based approach to measure associations between entities using hyperlink structures of Wikipedia.
- ESA [8] is a representative text-based method by using entity co-occurrence information and TF-IDF weights.
- Entity2vec (E2V) [28] jointly learns vector representations of words and entities from Wikipedia by the skip-gram model.
- TSF [25] is a two-stage entity relevance computing framework for Wikipedia by first generating a weighted subgraph for co-occurrence information and then computing the relatedness on the subgraph.
- E-PR is the EDG with PageRank [22] to rank the associative entities.
- E-DW is the EDG with the DeepWalk for graph embedding.

Implementation. To generate *transactions* in our EDGM, we use the tool WAT [24] to link entities in $D_{\vec{q}}$ to their corresponding entity-IDs in Wikipedia, and filter the Top-300 matching candidates with the highest similarity to the given query based on KG embedding by OpenKE.[1] For ESA and MW, we take the current query and candidate entities in the corresponding EDG as input, and generate the ranking list based on the relatedness between the candidate entities and the query. For E2V, we obtain the representations of words and entities based on the same version of Wikipedia chosen for EDGM. For TSF, we adopt the recommended configurations [25] to achieve the optimal results. We transform the undirected edges of EDG to bidirectional directed edges for E-PR and perform E-DW on the unweighted graph structure of EDG[2].

To balance the effectiveness and efficiency of EDG construction, we fix the threshold σ to 11 on KORE and 6 on ERT. For EDGs constructed by each query, they contain an average of 37 entity nodes on KORE and 49 entity nodes on ERT. We also set the node2vec parameters $dimensions, walklength, numberofwalks$ to 128, 30 and 200 on KORE and 128, 30 and 100 on ERT for better graph embedding. Besides, we find that the BFS random walk strategy ($m = 1, n = 2$) is more conducive to achieving the best results for our model.

4.2 Experimental Results

Exp-1: Ranking Associative Entities. To test the accuracy of associative entities ranking by our EDGM, we record NDCG of the top-k ranked lists found by all methods when k is fixed to 5, 10, 15 and 20 on KORE and 3, 5 and 10 on ERT. The results are shown in Fig. 3(a) and Fig. 3(b) respectively. All methods rank the entities that exist in the current EDG, and missing entities are ignored and skipped.

The results tell us that (1) our method EDGM achieves the highest NDCG scores and outperforms other methods on all datasets by taking the advantages of weighted associations between entity nodes in EDG, (2) our EDGM performs consistently better than other methods on all datasets by presenting the frequency characteristics including *informativeness* of co-occurrence and *specificness* of entity frequency, while some methods perform unstably on different

---

[1] http://139.129.163.161/index/toolkits.
[2] https://github.com/opp8888/ConstructionofEDG.

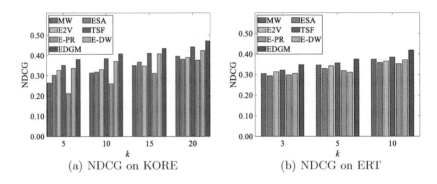

Fig. 3. Results of associative entities ranking.

datasets. For example, ESA performs better than MW in KORE but works worse than MW in ERT, and (3) E-DW is better than E-PR, which indicates that graph embedding is effective for our entity ranking model, and EDGM outperforms E-DW, which also suggests that node2vec is more suitable for embedding the EDG than DeepWalk. In fact, our EDGM improves NDCG by 6.7% and 7.5% over the second-highest method TSF on KORE and ERT, respectively. This verifies the effectiveness of our proposed method.

Table 3. Comparison of entity relatedness ranking on KORE.

Domain	Metrics	MW	ESA	E2V	TSF	E-PR	E-DW	EDGM
IT companies	Pearson	0.496	0.489	0.579	**0.753**	0.192	0.652	0.749
	Spearman	0.537	0.664	0.613	0.741	0.425	0.688	**0.767**
	Harmonic	0.516	0.563	0.596	0.747	0.265	0.670	**0.758**
Hollywood celebrities	Pearson	0.515	0.577	0.675	0.727	0.216	0.613	**0.811**
	Spearman	0.634	0.692	0.589	0.792	0.372	0.582	**0.805**
	Harmonic	0.568	0.629	0.629	0.758	0.273	0.597	**0.808**
Video games	Pearson	0.607	0.552	0.616	0.781	0.18	0.587	**0.793**
	Spearman	0.592	0.621	0.542	**0.810**	0.489	0.675	0.791
	Harmonic	0.599	0.584	0.577	**0.795**	0.263	0.628	0.792
Television series	Pearson	0.671	0.521	0.637	**0.833**	0.261	0.712	0.691
	Spearman	0.735	0.585	0.671	0.732	0.491	0.716	**0.754**
	Harmonic	0.702	0.551	0.654	**0.779**	0.341	0.714	0.721

Exp-2: Entity Relatedness Ranking. Exp-2 aims to test whether our EDGM could generate the ranking lists having a high degree of consistency compared with the ground truth. The results on KORE and ERT are shown in Table 3 and Table 4, respectively. Since the number of entities of EDG are not fixed, the top-5 candidate entities in the current EDG are selected for discussion.

The results tell us that (1) EDGM performs better than the traditional text-based, graph-based methods (MW and ESA) and the pure entity representation approach (E2V) in all domains of the two datasets, (2) EDGM outperforms TSF in most domains of ERT and performs as well as TSF on KORE, and (3) EDGM achieves the highest harmonic mean in most domains of the two datasets. Our

Table 4. Comparison of entity relatedness ranking on ERT.

Domain	Metrics	MW	ESA	E2V	TSF	E-PR	E-DW	EDGM
Music	Pearson	0.677	0.531	0.652	**0.795**	0.257	0.694	0.781
	Spearman	0.589	0.663	0.598	0.732	0.386	0.660	**0.787**
	Harmonic	0.630	0.590	0.624	0.762	0.309	0.677	**0.784**
Movie	Pearson	0.615	0.466	0.681	**0.828**	0.190	0.785	0.825
	Spearman	0.463	0.569	0.626	0.764	0.429	0.682	**0.771**
	Harmonic	0.528	0.512	0.652	0.795	0.263	0.730	**0.797**

EDGM gives better results in total rank, which verifies the effectiveness of EDG that generates a powerful presentation of the associations upon neighboring and co-occurrence features of entities.

Exp-3: Impacts of Parameters. To evaluate the impacts of the threshold σ, we vary σ from 9 to 12 on KORE and from 3 to 7 on ERT. The results are reported in Fig. 4(a)–Fig. 4(f), respectively.

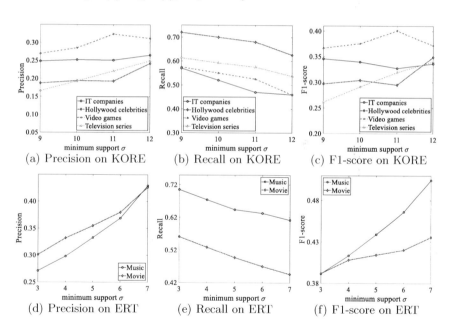

(a) Precision on KORE   (b) Recall on KORE   (c) F1-score on KORE

(d) Precision on ERT   (e) Recall on ERT   (f) F1-score on ERT

Fig. 4. Results of impacts of parameters.

The results tell us that (1) the precision increases (recall decreases) with the increase of σ, which is consistent with the theoretical expectation that the number of entity nodes in EDG decrease when σ increases, and (2) the F1-score remains relatively stable when varying σ, which demonstrates that our method could efficiently recall candidate entities in the ground truth. Note that our model achieves better recall than precision, which is suitable for the application scenarios of ranking problems requiring a higher recall. Hence, we fix σ to 11 on KORE and 6 on ERT to balance the size of EDG and guarantee high recall.

5   Conclusions and Future Work

In this paper, we propose the entity dependency graph model (EDGM) to rank associative entities in KG by graph embedding upon frequent entities. By incorporating multiple features of the association from both KG and Web resources effectively, one or more entities are allowed as a query to achieve better scalability. EDGM facilitates the discovery of associative entities with high recall, since the co-occurrence of entities in KG and the behavioral associations could be represented by a global model in an unsupervised manner.

However, the path and label information in KG, as well as the impacts of neighbors in a random walk on EDGM have not been well considered, which needs further exploration. Moreover, open world KG completion is worthwhile to study further by incorporating with semantic/implicit associations between entities achieved by our method.

Acknowledgements. This paper is supported by National Natural Science Foundation of China (U1802271, 62002311), Science Foundation for Distinguished Young Scholars of Yunnan Province (2019FJ011), China Postdoctoral Science Foundation (2020M673310), Program of Donglu Scholars of Yunnan University. We thank Prof. Weiyi Liu from Yunnan University for his insightful advice.

References

1. Aggarwal, C.C., Bhuiyan, M.A., Hasan, M.A.: Frequent pattern mining algorithms: a survey. In: Aggarwal, C.C., Han, J. (eds.) Frequent Pattern Mining, pp. 19–64. Springer, Cham (2014). https://doi.org/10.1007/978-3-319-07821-2_2
2. Aggarwal, N., Buitelaar, P.: Wikipedia-based distributional semantics for entity relatedness. In: AAAI (2014)
3. Cai, H., Zheng, V.W., Chang, K.C.C.: A comprehensive survey of graph embedding: problems, techniques, and applications. IEEE Trans. Knowl. Data Eng. **30**(9), 1616–1637 (2018)
4. Chen, J., Gu, B., Li, Z., Zhao, P., Liu, A., Zhao, L.: SAEA: self-attentive heterogeneous sequence learning model for entity alignment. In: DASFAA, pp. 452–467 (2020)
5. Cheng, G.: Relationship search over knowledge graphs. SIGWEB Newsl., 8 p. (2020). https://doi.org/10.1145/3409481.3409484. Article ID 3
6. Cheng, G., Liu, D., Qu, Y.: Fast algorithms for semantic association search and pattern mining. IEEE Trans. Knowl. Data Eng. **33**(04), 1490–1502 (2019). https://doi.org/10.1109/TKDE.2019.2942031. ISSN 1558–2191
7. Cheng, G., Shao, F., Qu, Y.: An empirical evaluation of techniques for ranking semantic associations. IEEE Trans. Knowl. Data Eng. **29**(11), 2388–2401 (2017)

8. Gabrilovich, E., Markovitch, S.: Wikipedia-based semantic interpretation for natural language processing. J. Artif. Intell. Res. **34**, 443–498 (2009)
9. Gong, X., Xu, H., Huang, L.: Han: Hierarchical association network for computing semantic relatedness. In: AAAI (2018)
10. Grover, A., Leskovec, J.: node2vec: Scalable feature learning for networks. In: SIGKDD, pp. 855–864 (2016)
11. Han, J., Kamber, M., Pei, J.: Data Mining (Third Edition). Morgan Kaufmann, Burlington (2012)
12. Herrera, J.E.T., Casanova, M.A., Nunes, B.P., Leme, L.A.P.P., Lopes, G.R.: An entity relatedness test dataset. In: ISWC, pp. 193–201 (2017)
13. Hoffart, J., Seufert, S., Nguyen, D.B., Theobald, M., Weikum, G.: KORE: keyphrase overlap relatedness for entity disambiguation. In: CIKM, pp. 545–554 (2012)
14. Imrattanatrai, W., Kato, M.P., Tanaka, K., Yoshikawa, M.: Entity ranking for queries with modifiers based on knowledge bases and web search results. IEICE Trans. Inf. Syst. **101**(9), 2279–2290 (2018)
15. Li, J., Chen, W., Gu, B., Fang, J., Li, Z., Zhao, L.: Measuring semantic relatedness with knowledge association network. In: DASFAA, pp. 676–691 (2019)
16. Li, L., Yue, K., Zhang, B., Sun, Z.: A probabilistic approach for inferring latent entity associations in textual web contents. In: DASFAA, pp. 3–18 (2019)
17. Liu, W., Yue, K., Wu, H., Fu, X., Zhang, Z., Huang, W.: Markov-network based latent link analysis for community detection in social behavioral interactions. Appl. Intell. **48**(8), 2081–2096 (2017). https://doi.org/10.1007/s10489-017-1040-y
18. Liu, Z., et al.: Distance-aware dag embedding for proximity search on heterogeneous graphs. In: AAAI, pp. 2355–2362 (2018)
19. Navigli, R., Martelli, F.: An overview of word and sense similarity. Nat. Lang. Eng. **25**(6), 693–714 (2019)
20. Nguyen, T., Tran, T., Nejdl, W.: A trio neural model for dynamic entity relatedness ranking. In: CoNLL, pp. 31–41 (2018)
21. Noy, N., Gao, Y., Jain, A., Narayanan, A., Patterson, A., Taylor, J.: Industry-scale knowledge graphs: lessons and challenges. Commun. ACM **62**(8), 36–43 (2019)
22. Page, L., Brin, S., Motwani, R., Winograd, T.: The pagerank citation ranking: bringing order to the web. Technical report, Stanford InfoLab (1999)
23. Perozzi, B., Al-Rfou, R., Skiena, S.: DeepWalk: online learning of social representations. In: SIGKDD, pp. 701–710 (2014)
24. Piccinno, F., Ferragina, P.: From TagME to WAT: a new entity annotator. In: The First International Workshop on Entity Recognition and Disambiguation, pp. 55–62 (2014)
25. Ponza, M., Ferragina, P., Chakrabarti, S.: On computing entity relatedness in wikipedia, with applications. Knowl.-Based Syst. **188**, 105051 (2020)
26. Schuhmacher, M., Dietz, L., Ponzetto, S.P.: Ranking entities for web queries through text and knowledge. In: CIKM, pp. 1461–1470 (2015)
27. Witten, I.H., Milne, D.N.: An effective, low-cost measure of semantic relatedness obtained from wikipedia links. In: AAAI (2008)
28. Yamada, I., Shindo, H., Takeda, H., Takefuji, Y.: Joint learning of the embedding of words and entities for named entity disambiguation. In: SIGNLL, pp. 250–259 (2016)
29. Zhang, Y., Wang, D., Zhang, Y.: Neural IR meets graph embedding: a ranking model for product search. In: WWW, pp. 2390–2400 (2019)

A Novel Embedding Model for Knowledge Graph Completion Based on Multi-Task Learning

Jiaheng Dou, Bing Tian, Yong Zhang$^{(\boxtimes)}$, and Chunxiao Xing

BNRist, Department of Computer Science and Technology, RIIT,
Institute of Internet Industry, Tsinghua University, Beijing, China
{djh19,tb17}@mails.tsinghua.edu.cn, {zhangyong05,xingcx}@tsinghua.edu.cn

Abstract. Knowledge graph completion is the task of predicting missing relationships between entities in knowledge graphs. State-of-the-art knowledge graph completion methods are known to be primarily knowledge embedding based models, which are broadly classified as translational models and neural network models. However, both kinds of models are single-task based models and hence fail to capture the underlying inter-structural relationships that are inherently presented in different knowledge graphs. To this end, in this paper we combine the translational and neural network methods and propose a novel multi-task learning embedding framework (TransMTL) that can jointly learn multiple knowledge graph embeddings simultaneously. Specifically, in order to transfer structural knowledge between different KGs, we devise a global relational graph attention network which is shared by all knowledge graphs to obtain the global representation of each triple element. Such global representations are then integrated into task-specific translational embedding models of each knowledge graph to preserve its transition property. We conduct an extensive empirical evaluation of multi-version TransMTL based on different translational models on two benchmark datasets WN18RR and FB15k-237. Experiments show that TransMTL outperforms the corresponding single-task based models by an obvious margin and obtains the comparable performance to state-of-the-art embedding models.

1 Introduction

Knowledge Graphs (KGs) such as WordNet [16] and Freebase [1] are graph-structured knowledge bases whose facts are represented in the form of relations (edges) between entities (nodes). This can be represented as a collection of triples (*head entity*, *relation*, *tail entity*) denoted as (h, r, t), for example (*Beijing*, *CapitalOf*, *China*) is represented as two entities: *Beijing* and *China* along with a relation *CapitalOf* linking them. KGs are important sources in many applications such as question answering [2], dialogue generation [10] and

---

J. Dou and B. Tian – contribute equally to this work.

© Springer Nature Switzerland AG 2021
C. S. Jensen et al. (Eds.): DASFAA 2021, LNCS 12681, pp. 240–255, 2021.
https://doi.org/10.1007/978-3-030-73194-6_17

recommender systems [34]. Containing billions of triples though, KGs still suffer from incompleteness, that is, missing a lot of valid triples [24,31]. Therefore, many research efforts have concentrated on the Knowledge Graph Completion (KGC) or link prediction task which entails predicting whether a given triple is valid or not [4,24]. Recent state-of-the-art KGC methods are known to be primarily knowledge embedding based models, which are broadly classified as translational models [3,21,32] and neural network models [8,20,23]. Translational models aim to learn embeddings by representing relations as translations from head to tail entities. For example, the pioneering work TransE [3] assumes that if (h, r, t) is a valid fact, the embedding of head entity h plus the embedding of relation r should be close to the embedding of tail entity t, i.e. $v_h + v_r \approx v_t$ (here, v_h, v_r and v_t are embeddings of h, r and t respectively). In order to learn more deep expressive features, recent embedding models have raised interests in applying deep neural networks for KGC such as Convolutional Neural Network (CNN) [8] and capsule network [20]. Recently, some studies explored a new research direction of adopting Graph Neural Network (GNN) [23] for knowledge graph completion, which demonstrates superior effectiveness and advantages than traditional translational methods since it takes the relationship of different triples into consideration. Among the GNN models, Graph Attention Network (GAT) [29] is an effective and widely used model which utilizes attentive nodes aggregation to learn neighborhood information. Although the effectiveness of these models, they are all single-task based models and ignore the inter-structural relations that are inherently presented in different knowledge graphs. To that end, such methods need to train different models for each knowledge graph, which involves substantial extra efforts and resources.

Fig. 1. An example of shared structure pattern

Nevertheless, we find that different knowledge graphs are structurally inter-related and one knowledge graph can benefit from others. On the one hand, since different knowledge graphs have different data characteristics, they can complement each other by simultaneously learning the representations. For example, WordNet provides semantic knowledge of words. It contains few types of relations but each relation corresponds to a large number of triples. In comparsion, Freebase provides more specific facts of the world and contains a lagre number of relations with fewer entities. Therefore, knowledge representation model based on WordNet would be good at modeling and inferring the patterns of (or between) each relation such as symmetry/antisymmetry, inversion and composition [25] whereas the model based on Freebase enables to model more complex relations. As such, simultaneously learning the representations of these knowledge graphs can definitely promote and benefit each other. On the other hand, we observe that one knowledge graph may contain some common structural patterns that are beneficial for other knowledge graphs. An example is shown in Fig. 1 where the dotted line is the link needs to be predicted. For the missing link (*Dauphin Country, ?, United Stated dollar*) in Freebase, it is essential to understand the structural pattern that two entities connected by a symmetric relation usually exist in some triples linked by the same relations. However, it is hard for Freebase based embedding model to capture this kind of pattern since it is rare in this knowledge graph. As this kind of struture pattern is very common in WordNet which is shown in the left of the figure, the knowledge graph completion task based on Freebase can definitely benefit from them.

Motivated by such observations, in this paper we propose a novel embedding model for knowledge graph completion based on multi-task learning (TransMTL) where multiple knowledge graphs can be trained and represented simultaneously and benefit from each other. Specifically, in order to preserve the transition property of knowledge graphs, we first adopt the widely used translational models such as TransE, TransH and TransR to represent the entities and relations of each single knowledge graph. And then, we devise a global Translation preserved Relational-Graph Attention Network (TR-GAT) which is shared by all knowledge graphs to capture the inter-structural information between different knowledge graphs and obtain the global representation of each triple element. Such global representations are then integrated intotask-specific translational embedding models of each knowledge graph to enhance its transition property. In this way, each single knowledge graph can benefit from the common inter-structural information from other knowledge graphs through the global shared layer. Recall the example in Fig. 1, with the help of MTL, the information learned from WordNet can be transferred to Freebase representation task by means of the global sharing mechanism. Specifically, in WordNet, there exists a triple (*austronesia, similar_to, oceania*) containing the symmetry realtion *similar_to*. Then the head entity *austronesia* and tail entity *oceania* would exist in some triples linked by the same relation such as *instance_hypernym* and *has_part*. Once recognizing this kind of pattern in WordNet, the multi-task learning model could take advantage of such knowledge for link prediction in Freebase dataset. As there exists a triple (*Dauphin Country, adjoins, Cumberland Country*) with the symmetric relation *adjoins* and the head entity *Dauphin Country* and tail entity *Cumberland Country* exist in triples linked by the same relation *time_zones*, we can assume that the entity *Dauphin Country* may also linked by the relation *currency* since the *Cumberland Country* and *United Stated dollar* are linked by *currency*. We conduct an extensive empirical evaluation TransMTL based on different translational models on two benchmark datasets WN18RR and FB15k-237. Experiments show that our TransMTL outperforms the corresponding single-task based models by an obvious margin and obtains the comparable performance to state-of-the-art embedding models.

Contributions of this paper are summarized as follows:

- We propose a novel embedding model for knowledge graph completion based on multi-task learning (TransMTL) that can learn embeddings of multiple knowledge graphs simultaneously. To the best of our knowledge, this is the first attempt of multi-task learning in the field of knowledge representation for knowledge graph completion.
- We devise a translation preserved relational-graph attention network (TR-GAT) to utilize the shared information from multiple knowledge graphs, capturing inter-structural information in different knowledge graphs.
- We conduct extensive experiments on WN18RR and FB15k-237. Experimental results show the effectiveness of our model TransMTL.

2   Related Work

2.1   Knowledge Graph Completion (KGC)

Representation learning has been widely adopted in a variety of applications [15, 35, 36]. Recently, several variants of KG embeddings have been proposed following the paradigm of representation learning. These methods can be broadly classified as: semantic matching, translational and neural network based models. Firstly, semantic matching models such as DistMult [32], ComplEx [28] and Holographic Embeddings model (HolE) [22] use similarity-based functions to infer relation facts. Differently, translational models aim to learn embeddings by representing relations as translations from head entities to tail entities. For example, Bordes et al. [3] proposed TransE by assuming that the added embedding of $h + r$ should be close to the embedding of t with the scoring function defined under $L1$ or $L2$ constraints. Starting with it, many variants and extensions of TransE have been proposed to additionally use projection vectors or matrices to translate embeddings into the vector space, such as TransH [30], TransR [13] and TransD [11]. In recent studies, neural network models that exploit deep learning techniques have yielded remarkable predictive performance for KG embeddings. Dettmers et al. [8] introduced ConvE that used 2D convolution over embeddings and multiple layers of non-linear features to model knowledge graphs. To preserve the transitional characteristics, Nguyen et al. [19] proposed ConvKB that applied the convolutional neural network to explore the global relationships among same dimensional entries of the entity and relation embeddings. To capture long-term relational dependency in knowledge graphs, recurrent networks are utilized. Gardner et al. [9] and Neelakantan et al. [18] proposed Recurrent Neural Network (RNN)-based models over relation path to learn vector representation without and with entity information, respectively. To cover the complex and hidden information that is inherently implicit in the local neighborhood surrounding a triple, some studies used Graph Neural Networks (GNNs) for knowledge embeddings such as R-GCN [23] and KBGAT [17] etc.

Though the effectiveness of these models, they are all single-task based models and hence fail to capture the underlying inter-structural relationships that are inherently present in different knowledge graphs.

2.2   Multi-Task Learning

Multi-Task Learning (MTL) [5] is a learning paradigm in machine learning aiming at leveraging potential correlations and common features contained in multiple related tasks to help improve the generalization performance of all the tasks. It has been widely adopted in many machine learning applications from various areas including web applications, computer vision, bioinformatics, health informatics, natural language processing and so on. For example, Chapelle et al. [6] introduced a multi-task learning algorithm based on gradient boosted decision trees that is specifically designed with web search ranking in mind. Yim et al. [33] proposed a multi-task deep model to rotate facial images to a target pose and the auxiliary task aimed to use the generated image to reconstruct the original image. Chowdhury et al. [7] provided an end-to-end multi-task encoder-decoder framework for three adverse drug reactions detection and extraction tasks by leveraging the interactions between different tasks. Tian et al. [26] devised a multi-task hierarchical inter-attention network model to improve the task-specific document representation in nature language processing for document classification. In this paper, we utilize the idea of multi-task learning to transfer structural knowledge between different KGs by jointly learning multiple knowledge graph embeddings simultaneously.

3   Method

We begin this section by introducing the notations and definitions used in the rest of the paper, followed by a brief background on GAT [29]. Immediately afterwards, we introduce the details of our TransMTL framework as displayed in Fig. 2. It consists of two components: the task-specific knowledge embedding layer and the global shared layer. The task specific knowledge embedding layer is a translational model in which each KG learns low-dimensional embeddings of entities and relations. The global shared layer enables multi-task learning: we devise a Translation preserved Relational-Graph Attention Network (TR-GAT) to acquire the entity and relation embeddings simultaneously consisting of common structural information from all the knowledge graphs. Then such information is shared and integrated into the task-specific knowledge embedding layer to further enhance the entity and relation representations.

3.1   Background and Definition

A knowledge graph G is donated by $\mathscr{G} = (E, R, T)$ where E, R and T represent the set of entities (nodes), relations (edges) and triplets, respectively. It contains a collection of valid factual triples in the form of (head entity, relation, tail entity) denoted as (h, r, t) such that $h, t \in E$ and $r \in R$, representing the specific relation r linking from the head entity h to tail entity t. Knowledge embedding models aim to learn an effective representation of entities, relations, and a scoring function f which gives an implausibility score for each triple (h, r, t) such that

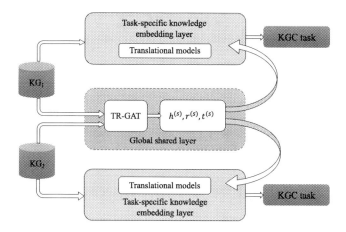

Fig. 2. The overall architecture of TransMTL

valid triples receive lower scores than invalid triples. With the learned entity and relation embeddings, the knowledge graph completion is to predict the missing head entity h given query $(?, r, t)$ or tail entity t given query $(h, r, ?)$.

3.2 Graph Attention Networks (GAT)

The concept of Graph Convolutional Networks (GCN) was first proposed in [12], which extended existing neural networks for processing the graph structured data. It gathers information from the entity's neighborhood and all neighbors contribute equally in the information passing. To resolve the shortcomings the GCNs, Velickovic et al. [29] proposed Graph Attention Networks (GAT). The advantage of GAT lies in the aspect that it leverages attention mechanism to assign varying levels of importance to nodes in every node's neighborhood, which enables the model to filter out noises and concentrate on important adjacent nodes. Specifically, the convolution layer attentionally aggregates features of each node in the graph as well as its one-hop neighbors as new features. The convolution process on the tth layer for node v is formalized as Eq. (1)–(2).

$$h_v^{(t)} = \sigma\left(\sum_{u \in \mathcal{N}(v) \cup v} \alpha_{vu} W^{(t)} h_u^{(t-1)} \right) \tag{1}$$

$$\alpha_{vu} = softmax(f(a_t^T [W^{(t)} h_v^{(t-1)} || W^{(t)} h_u^{(t-1)}]))$$
$$= \frac{exp(f(a_t^T [W^{(t)} h_v^{(t-1)} || W^{(t)} h_u^{(t-1)}]))}{\sum\limits_{j \in \mathcal{N}(v) \cup v} exp(f(a_t^T [W^{(t)} h_v^{(t-1)} || W^{(t)} h_j^{(t-1)}]))} \tag{2}$$

where $W^{(t)}$ is the weight matrix, α_{vu} is the attention coefficient of node u to v, $\mathcal{N}(v)$ presents the neighborhoods of node v in the graph, f donates the *LeakyReLU* function and a_t is the weight vector.

3.3 Task-Specific Knowledge Embedding Layer

In order to integrate the strength of translational property in knowledge graphs, we adopt the widely-used translation-based methods for each involved KG in task specific knowledge embedding layer, which benefit the multi-task learning tasks by representing embeddings uniformly in different contexts of relations. Here, we take the basic translational model TransE as an example to describe the embedding model. TransE [3] projects both relations and entities into the same continuous low-dimension vector space, in which the relations are considered as translating vectors from head entities to tail entities. Following the energy-based framework in TransE, the energy of a triplet is equal to $d(h+r,t)$ for some dissimilarity measure d. Specifically, the energy function is defined as:

$$E(h,r,t) = \|h + r - t\| \tag{3}$$

To learn such embeddings, we minimize the margin-based objective function over the training set, defined as:

$$K = \Sigma_{(h,r,t)\in T} L(h,r,t), \tag{4}$$

where L(h, r, t) is a margin-based loss function with respect to the triple (h, r, t):

$$L(h,r,t) = \Sigma_{(h',r',t')\in T'}[\gamma + E(h,r,t) - E(h',r',t')]_+, \tag{5}$$

where $[x]_+ = max(0,x)$ represents the maximum between 0 and x. T' stands for the negative sample set of T, donated as follows:

$$T' = \{(h',r,t)|h' \in E\} \cup \{(h,r,t')|t' \in E\}$$
$$\cup \{(h,r',t)|r' \in R\}, (h,r,t) \in T. \tag{6}$$

The set of corrupted triplets, constructed according to Eq. (6), is composed of training triplets with either the head or tail replaced by a random entity (but not both at the same time). The objective function is optimized by stochastic gradient descent (SGD) with mini-batch strategy.

Note that in this paper, we aim at providing a general multi-task leaning solution to take advantage of the inter-structural knowledge between different KGs and not limited to any knowledge representation learning method. In other words, this task specific knowledge embedding layer can also be implemented through any other knowledge representation learning methods, including translational models and neural network models. In order to illustrate the effectiveness of multi-task learning in knowledge graph representation and KGC task, we implemented our TransMTL model based on TransE, TransH [30] and TransR [13] in this paper. The energy functions of TransH and TransR are defined as in Eq. (7) and Eq. (8) respectively.

$$E(h_\perp, r, t_\perp) = \|h + d_r - t\|$$
$$= \left\|(h - w_r^\top h w_r) + d_r - (t - w_r^\top t w_r)\right\|, \|w_r\|_2 = 1 \qquad (7)$$
$$E(h_r, r, t_r) = \|h_r + r - t_r\| = \|h M_r + r - t M_r\| \qquad (8)$$

3.4 Global Shared Layer for Multi-task Learning

On the basis of task-specific model, we then utilize MTL techniques to improve the entity and relation representations. The intuition is that different knowledge graphs share some common structural knowledge, which can help improve the entity and relation representations of each knowledge graph and contribute to a better knowledge graph completion performance. The key factor of multi-task learning is the sharing scheme among different tasks. Considering the observations that existing KG embedding models treat triples independently and thus fail to cover the complex and hidden information that is inherently implicit in the local neighborhood surrounding a triple, we propose TR-GAT in Fig. 3 to acquire the entity and relation embeddings simultaneously by capturing both entity and relation features in any given entity's neighborhood.

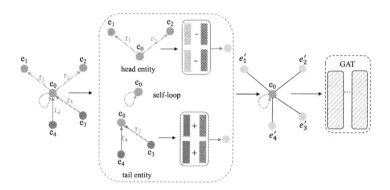

Fig. 3. Embedding processing in TR-GAT. Orange represents the center entity, brown represents relations connected with it, and green and blue represent its neighboring entities. If the entity has the head role, accumulating its neighboring tail nodes and relations with $t - r$. If it has the tail role, accumulating its neighboring head nodes and relations with $h + r$. Then the role discrimination representations passed through a GAT network during which the embeddings of entities and relations are updated.(Color figure online)

TR-GAT integrates the strength of GAT and the translational property in knowledge graphs ($h + r \approx t$) to design the new propagation model. As such, we modify the update rule in GAT for the entity and relation embeddings and the convolution process on the tth layer for node v is formalized as Eq. 9:

$$h_v^l = \sigma(W^l(\sum_{r \in N_r} \sum_{t \in N_t^r} \alpha_{vt} c(h_t^{l-1}, h_r^{l-1}) + \sum_{r \in N_r} \sum_{h \in N_h^r} \alpha_{vh} \tilde{c}(h_h^{l-1}, h_r^{l-1}) + \alpha_v h_v^{l-1}))$$

$$(9)$$

where N_r denotes the set of relations connecting the entity i, N_t^r represents the set of tail entities connected with the entity i by the relation r, N_h^r is the set of head entities connected with the entity i by the relation r. $h_h^l \in R^{d(l)}$, $h_r^l \in R^{d(l)}$ and $h_t^l \in R^{d(l)}$ denote the lth layer embedding of the head entity, relation and tail entity respectively in the neural network and $d(l)$ is the dimension of this layer. σ is the activation function. $c(\cdot, \cdot)$ is the function to describe the relationship between h_t^l and h_r^l, and $\tilde{c}(\cdot, \cdot)$ describes the relationship between h_h^l and h_r^l. W^l is the weight matrix of the lth layer.

Equation 9 features the role discrimination criterion to identify if entity v in the knowledge graph takes the role of head or tail entity regarding a specific relation r. It performs different convolution operations for them: if v has the head entity role, its embedding is calculated by combining the related tail entity $h_t^{(l-1)}$ and relation $h_r^{(l-1)}$. Otherwise, its embedding is calculated with the related head entity $h_h^{(l-1)}$ and relation $h_r^{(l-1)}$. Thereafter, all occurrences of head roles and tail roles of v are added, together with a single self-connection representation h_v^{l-1}, to infer the l representation of the entity v.

The design of function c and $\tilde{c}$ features the translation adoption criterion which is $h+r \approx t$ for a triplet (h, r, t) in the graph. Alternatively, the translational property can be transformed into $h \approx t - r$ and $t \approx h + r$. Therefore,

$$c(h_t^l, h_r^l) = h_t^l - h_r^l \tag{10}$$

$$\tilde{c}(h_h^l, h_r^l) = h_h^l + h_r^l \tag{11}$$

Based on the TR-GAT, the overall embedding process of our TransMTL is as follows: for each entity and relation in a knowledge graph k ($G^{[k]}$), we first compute the global output based on the global shared TR-GAT with Eq. (12) to utilize the global shared inter-structural information of all KGs.

$$(\boldsymbol{h}^{(s)}, \boldsymbol{r}^{(s)}, \boldsymbol{t}^{(s)}) = \text{TR-GAT}((E, R, T), \Theta^{(s)}) \tag{12}$$

Here, we use a TR-GAT$(\cdot, \cdot)$ as a shorthand for convolution process (Eq. (9)–(11)) in the global shared layer and Θ represents the parameters of global shared layer which are shared by all KGs. Then such global information is integrated into each task to enhance the embeddings of each KG. Formally, for a specific task k, the energy function in Eq. (3) is modified as follows:

$$(h^{(k)} + M_h h^{(s)}) + (r^{(k)} + M_r r^{(s)}) \approx (t^{(k)} + M_t t^{(s)}) \tag{13}$$

$$E(h, r, t)^{[k]} = \left\| (h^{(k)} + M_h h^{(s)}) + (r^{(k)} + M_r r^{(s)}) - (t^{(k)} + M_t t^{(s)}) \right\| \tag{14}$$

where $h^{(k)}, r^{(k)}, t^{(k)}$ are the task specific embeddings of knowledge graph k, $h^{(s)}, h^{(s)}, h^{(s)}$ are global embeddings obtained through the shared TR-GAT layer, and M_h, M_r, M_t are transform matrix to guarantee the consistency of the vector spaces between the embeddings of task-specific and global shared layers. In this way, every entity and relation of each single knowledge graph can benefit from common knowledge and extra information from all other knowledge graphs.

3.5   Training

Following the translational models, to learn such embeddings, we adopt a margin-based loss function respect to the triples for all tasks:

$$L = \sum_{k \in K} \sum_{(h,r,t) \in S} \sum_{(h',r',t') \in S} [\gamma + d(h + r, t) - d(h' + r, t')]_+ \qquad (15)$$

$$h = h^{(k)} + M_h h^{(s)}; r = r^{(k)} + M_r r^{(s)}; t = t^{(k)} + M_t t^{(s)} \qquad (16)$$

where K is the number of knowledge graphs.

4   Experiments

In this section, we evaluate the performance of our framework. We conduct an extensive empirical evaluation of multi-version TransMTL based on TransE, TransH and TransR respectively to verify the effectiveness of the proposed model. We further vary the training set size to illustrate that our proposed multi-task learning framework can still perform well on low resource settings.

4.1   Experiment Setup

Data Sets. Our model is evaluated on two widely used knowledge graphs: WordNet [16] and Freebase [1]. WordNet provides semantic knowledge of words. Entities in WordNet are synonyms which express distinct concepts. Relations in WordNet are conceptual-semantic and lexical relations. In this paper, we employ the dataset WN18RR [8] from WordNet. Freebase provides general facts of the world. In this paper, we employ data set FB15k-237 [27] from Freebase.

Notably, WN18RR and FB15k-237 are correspondingly subsets of two common data sets WN18 and FB15k. It is firstly discussed by [27] that WN18 and FB15k suffer from test leakage through inverse relations, i.e. many test triplets can be obtained simply by inverting triplets in the training set. To address this issue, Toutanova and Chen et al. [27] generated FB15k-237 by removing redundant relations in FB15k and greatly reducing the number of relations. Likewise, Dettmers et al. [8] removed reversing relations in WN18. As a consequence, the difficulty of reasoning on these two data sets is increased dramatically. The statistics of the two datasets are described in Table 1.

Table 1. Statistics of the datasets

Dataset	#Relation	#Entity	#Train	#Valid	#Test
WN18RR	11	40,943	86,835	3,034	3,134
FB15k-237	237	14,541	272,115	17,535	20,466

Baselines. We first compared our TransMTL with the corresponding single-task models, namely TransE [3], TransH [30] and TransR [13] respectively. To further illustrate the effectiveness of multi-task learning, we then compared our model with recent knowledge embedding models, including both non-neural models and neural models. Specifically, we compared our TransMTL with Dist-Mult [32], ConvE [8], ComplEx [28], KBGAT [17], ConvKB [19], RotatE [25] and DensE [14] for comparison.

Evaluation Protocol. Link prediction aims to predict the missing h or t for a triplet (h, r, t). In this task, the model is asked to rank a set of candidate entities from the KG, instead of giving one best result. For each test triplet (h, r, t), we replace the head/tail entity by all possible candidates in the KG, and rank these entities in ascending order of scores calculated by score function showed in Eq. 14. We follow the evaluation protocol in [3] to report filtered results. Because a corrupted triplet, generated in the aforementioned process of removal and replacement, may also exist in KG, and should be considered as correct. In other words, while evaluating on test triples, we filter out all the valid triples from the candidate set, which is generated by either corrupting the head or tail entity of a triple. We report three common measures as our evaluation metrics: the average rank of all correct entities (Mean Rank), the mean reciprocal rank of all correct entities (MRR), and the proportion of correct entities ranked in top 10 (Hits@10). We report average results across 5 runs. We note that the variance is substantially low on all the metrics and hence omit it. A good link predictor should achieve lower Mean Rank, higher MRR,

Training Protocol. We use the common Bernoulli strategy [13,30] when sampling invalid triples. We select 500 as the batch size, which is not too big or too small for both the two datasets. There are two learning rates in our multi-task model: one for the global shared layer and the other for the task specific knowledge embedding layer. We use grid search method to find the appropriate learning rate for the two parts. And finally in our experiments, we use learning rate 0.5 for task-specific knowledge embedding layer and 0.01 for the global shared TR-GAT model. We use the Stochastic Gradient Descent (SGD) optimizer for training. In our model, the embedding size of entities and relations from the two knowledge graphs should be equal and we set it to 200. We use a two-layer GAT for the global shared TransMTL model that allows message passing among nodes that are two hops away from each other. As a result, although for some entity pairs, there are no direct edges in the knowledge graph, the two-layer

GAT is still capable to learn the inter-entity relations and enables the information exchange between pairs of entities. In our preliminary experiment, we found that a two-layer GAT performs better than a one-layer GAT, while more layers do not improve the performances. We set the dropout rate as 0.1 in order to release overfitting.

For multi-task learning, the training data come from completely different datasets, so our training process is conducted by looping over the tasks as follow:

1. Select a random task.
2. Select a mini-batch of examples from this task.
3. Backward the model and update the parameters of both task-specific layer and global shared layer with respect to this mini-batch.
4. Go to 1.

4.2   Results and Analysis

Table 2. Link prediction results of WN18RR and FB15k-237 compared with translational models. [*]: Results are taken from [19]. Best scores are highlighted in bold.

Models	WN18RR			FB15k-237		
	MR	MRR	Hit@10 (%)	MR	MRR	Hit@10 (%)
TransE[*]	3384	0.226	50.1	347	0.294	46.5
TransH	3048	0.286	50.3	348	0.284	48.8
TransR	3348	0.303	51.3	310	0.310	50.6
TransMTL-E	3065	0.363	54.1	116	0.336	52.6
TransMTL-H	**2521**	**0.498**	**57.0**	**111**	**0.349**	**53.7**
TransMTL-R	3154	0.465	54.6	133	0.333	52.2

Table 2 compares the experimental results of our TransMTL with different task specific knowledge embedding models to the corresponding single-task based models, using the same evaluation protocol. Here, TransMTL-E, TransMTL-H and TransMTL-R are models with TransE, TransH and TransR as their task specific knowledge embedding models respectively. From the table, we can see that our multi-task learning models outperform the corresponding single-task based models by an obvious margin. Specifically, TransMTL-E shows an improvement of Hit@10 4%, 6.1% to TransE on WN18RR and FB15k-237 respectively. TransMTL-H shows an improvement of Hit@10 6.7%, 4.9% to TransH and such numbers are 3.3% and 1.6% for the pair of TransMTL-R and TransR on dataset WN18RR and FB15k-237. Moreover, our TransMTL also obtains better MR and MRR scores than single-task models on both datasets. We argue that it is because with the global shared TR-GAT layer, the entities and relations of each single task can benefit from extra information from other tasks for better representations.

To further illustrate the effectiveness of multi-task learning, we then compared our model with recent knowledge embedding models, including both non-neural models and neural models. The experimental results are shown in Table 3. Since the datasets are same, we directly copy the experiment results of several baselines from [14,17]. From the table, we can see that even with the basic translational models, our TransMTL can obtain comparable performance to these recent models that integrate much additional information and new technologies into their models. Moreover, our TransMTL performs better on FB15k-237 than on WN18RR. The reason may be that there are rich conceptual-semantic and lexical relations in WN18RR and the entities and relations in FB15k-237 can benefit from these information through multi-task learning.

Table 3. Link prediction results for WN18RR and FB15k-237. Best scores are highlighted in bold.

Models	WN18RR			FB15k-237		
	MR	MRR	Hit@10 (%)	MR	MRR	Hit@10 (%)
DistMult	5110	0.430	49.0	512	0.281	44.6
ConvE	4187	0.43	52.0	244	0.325	50.1
ComplEx	7882	0.449	53.0	546	0.278	45.0
KBGAT	1921	0.412	55.4	270	0.157	33.1
ConvKB	**1295**	0.265	55.8	216	0.289	47.1
RotatE	3340	0.476	57.1	177	0.338	53.3
DensE	3052	0.491	**57.9**	169	0.349	53.5
TransMTL-E	3065	0.363	54.1	116	0.336	52.6
TransMTL-H	2521	**0.498**	57.0	**111**	**0.349**	**53.7**
TransMTL-R	3154	0.465	54.6	133	0.333	52.2

Varying the Data Size. In order to illustrate the robustness of our proposed multi-task learning framework, we vary the data sizes by randomly sampling different ratios of the training data for training and test them on the whole test sets of two datasets. Figure 4 shows the experimental results of Hit@10 scores of our TransMTL and the corresponding single task translational models: TransE, TransH, and TransR on two datasets respectively. Here, for the performance of single task translational models in different training data sizes, we use the same settings to the task-specific models of our TransMTL. From the figure, we can readily see that TransMTL consistently outperforms all single-task models across all datasets. Besides, we can see that with the decrease of training triples, the Hit@10 metrics decrease with different degrees. More specifically, the performance gap between our TransMTL models and the baselines are larger in small dataset settings than in big dataset settings. For example, with only the 60% of the training data, the performance of our TransMTL is still competitive, which shows an improvement of Hit@10 8.5% and 8% to TransE on WN18RR and FB15k-237 respectively. We argue that this is because our multi-task learning framework can exploit the underlying inter-structural relationships that are inherently presented in different knowledge graphs, thus it can alleviate the data insufficiency problem and achieve good results with less data.

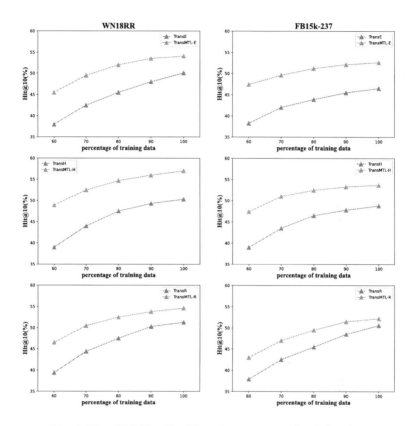

Fig. 4. The Hit@10 with different percentage of training data

5   Conclusions

In this paper, we propose a novel embedding model based on multi-task learning that can jointly learn multiple knowledge graph embeddings simultaneously for knowledge graph completion. We devise a global translation preserved relational graph attention network which is shared by all knowledge graphs to capture and transfer structural knowledge between different KGs. To preserve the transition property of each KG, we then integrate the global information learned by the global shared layer into the translational models for each KG. Experimental results on two benchmark datasets WN18RR and FB15k-237 show that our proposed model outperforms the corresponding single-task based models by an obvious margin and obtains the comparable performance to state-of-the-art embedding models, indicating the effectiveness of multi-task learning on knowledge graph representations.

Acknowledgements. This work was supported by NSFC (91646202), National Key R&D Program of China (2018YFB1404401, 2018YFB1402701).

References

1. Bollacker, K.D., Evans, C., Paritosh, P., Sturge, T., Taylor, J.: Freebase: a collaboratively created graph database for structuring human knowledge. In: SIGMOD, pp. 1247–1250 (2008)
2. Bordes, A., Chopra, S., Weston, J.: Question answering with subgraph embeddings. In: EMNLP, pp. 615–620 (2014)
3. Bordes, A., Usunier, N., García-Durán, A., Weston, J., Yakhnenko, O.: Translating embeddings for modeling multi-relational data. In: NIPS, pp. 2787–2795 (2013)
4. Bordes, A., Weston, J., Collobert, R., Bengio, Y.: Learning structured embeddings of knowledge bases. In: AAAI (2011)
5. Caruana, R.: Multitask learning. Mach. Learn. **28**(1), 41–75 (1997)
6. Chapelle, O., Shivaswamy, P.K., Vadrevu, S., Weinberger, K.Q., Zhang, Y., Tseng, B.L.: Multi-task learning for boosting with application to web search ranking. In: ACM SIGKDD, pp. 1189–1198 (2010)
7. Chowdhury, S., Zhang, C., Yu, P.S.: Multi-task pharmacovigilance mining from social media posts. In: WWW, pp. 117–126 (2018)
8. Dettmers, T., Minervini, P., Stenetorp, P., Riedel, S.: Convolutional 2D knowledge graph embeddings. In: AAAI, pp. 1811–1818 (2018)
9. Gardner, M., Talukdar, P.P., Krishnamurthy, J., Mitchell, T.M.: Incorporating vector space similarity in random walk inference over knowledge bases. In: EMNLP, pp. 397–406 (2014)
10. He, H., Balakrishnan, A., Eric, M., Liang, P.: Learning symmetric collaborative dialogue agents with dynamic knowledge graph embeddings. In: ACL, pp. 1766–1776 (2017)
11. Ji, G., He, S., Xu, L., Liu, K., Zhao, J.: Knowledge graph embedding via dynamic mapping matrix. In: ACL, pp. 687–696 (2015)
12. Kipf, T.N., Welling, M.: Semi-supervised classification with graph convolutional networks. In: ICLR (2017)
13. Lin, Y., Liu, Z., Sun, M., Liu, Y., Zhu, X.: Learning entity and relation embeddings for knowledge graph completion. In: Bonet, B., Koenig, S. (eds.) AAAI, pp. 2181–2187. AAAI Press (2015)
14. Lu, H., Hu, H.: Dense: An enhanced Non-Abelian group representation for knowledge graph embedding. CoRR abs/2008.04548 (2020)
15. Luo, L., et al.: Beyond polarity: interpretable financial sentiment analysis with hierarchical query-driven attention. In: IJCAI, pp. 4244–4250 (2018)
16. Miller, G.A.: WordNet: a lexical database for English. Commun. ACM **38**(11), 39–41 (1995)
17. Nathani, D., Chauhan, J., Sharma, C., Kaul, M.: Learning attention-based embeddings for relation prediction in knowledge graphs. In: ACL, pp. 4710–4723 (2019)
18. Neelakantan, A., Roth, B., McCallum, A.: Compositional vector space models for knowledge base completion. In: ACL, pp. 156–166 (2015)

19. Nguyen, D.Q., Nguyen, T.D., Nguyen, D.Q., Phung, D.Q.: A novel embedding model for knowledge base completion based on convolutional neural network. In: NAACL-HLT, pp. 327–333 (2018)
20. Nguyen, D.Q., Vu, T., Nguyen, T.D., Nguyen, D.Q., Phung, D.Q.: A capsule network-based embedding model for knowledge graph completion and search personalization. In: NAACL-HLT, pp. 2180–2189 (2019)
21. Nguyen, D.Q., Sirts, K., Qu, L., Johnson, M.: Neighborhood mixture model for knowledge base completion. In: CoNLL, pp. 40–50 (2016)
22. Nickel, M., Rosasco, L., Poggio, T.A.: Holographic embeddings of knowledge graphs. In: AAAI, pp. 1955–1961 (2016)
23. Schlichtkrull, M.S., Kipf, T.N., Bloem, P., van den Berg, R., Titov, I., Welling, M.: Modeling relational data with graph convolutional networks. In: ESWC, pp. 593–607 (2018)
24. Socher, R., Chen, D., Manning, C.D., Ng, A.Y.: Reasoning with neural tensor networks for knowledge base completion. In: NIPS, pp. 926–934 (2013)
25. Sun, Z., Deng, Z., Nie, J., Tang, J.: Rotate: Knowledge graph embedding by relational rotation in complex space. In: ICLR (2019)
26. Tian, B., Zhang, Y., Wang, J., Xing, C.: Hierarchical inter-attention network for document classification with multi-task learning. In: IJCAI, pp. 3569–3575 (2019)
27. Toutanova, K., Chen, D.: Observed versus latent features for knowledge base and text inference. In: CVSM (2015)
28. Trouillon, T., Welbl, J., Riedel, S., Gaussier, É., Bouchard, G.: Complex embeddings for simple link prediction. In: ICML, pp. 2071–2080 (2016)
29. Velickovic, P., Cucurull, G., Casanova, A., Romero, A., Liò, P., Bengio, Y.: Graph attention networks. In: ICLR (2018)
30. Wang, Z., Zhang, J., Feng, J., Chen, Z.: Knowledge graph embedding by translating on hyperplanes. In: AAAI, pp. 1112–1119 (2014)
31. West, R., Gabrilovich, E., Murphy, K., Sun, S., Gupta, R., Lin, D.: Knowledge base completion via search-based question answering. In: WWW, pp. 515–526 (2014)
32. Yang, B., Yih, W., He, X., Gao, J., Deng, L.: Embedding entities and relations for learning and inference in knowledge bases. In: ICLR (2015)
33. Yim, J., Jung, H., Yoo, B., Choi, C., Park, D., Kim, J.: Rotating your face using multi-task deep neural network. In: IEEE CVPR, pp. 676–684 (2015)
34. Zhang, F., Yuan, N.J., Lian, D., Xie, X., Ma, W.: Collaborative knowledge base embedding for recommender systems. In: ACM SIGKDD, pp. 353–362 (2016)
35. Zhao, K., et al.: Modeling patient visit using electronic medical records for cost profile estimation. In: DASFAA, pp. 20–36 (2018)
36. Zhao, K., et al.: Discovering subsequence patterns for next POI recommendation. In: IJCAI, pp. 3216–3222 (2020)

Gaussian Metric Learning for Few-Shot Uncertain Knowledge Graph Completion

Jiatao Zhang, Tianxing Wu, and Guilin Qi$^{(\boxtimes)}$

School of Computer Science and Engineering, Southeast University, Nanjing, China
{zjt,tianxingwu,gqi}@seu.edu.cn

Abstract. Recent advances in relational information extraction have allowed to automatically construct large-scale knowledge graphs (KGs). Nevertheless, an automatic process entails that a significant amount of uncertain facts are introduced into KGs. Uncertain knowledge graphs (UKGs) such as NELL and Probase model this kind of uncertainty as confidence scores associated to facts for providing more precise knowledge descriptions. Existing UKG completion methods require sufficient training examples for each relation. However, most relations only have few facts in real-world UKGs. To solve the above problem, in this paper, we propose a novel method to complete few-shot UKGs based on Gaussian metric learning (GMUC) which could complete missing facts and confidence scores with few examples available. By employing a Gaussian-based encoder and metric function, GMUC could effectively capture uncertain semantic information. Extensive experiments conducted over various datasets with different uncertainty levels demonstrate that our method consistently outperforms baselines.

1 Introduction

Knowledge graphs (KGs) describe structured information of entities and relations, which have been widely used in many intelligent applications such as question-answering and semantic search. Despite large scales of KGs, they are still far from complete to describe infinite real-world facts. In order to complete KGs automatically, many efforts [3,7,18,22,25,31] have been studied to infer missing facts.

Most KGs such as Freebase [2], DBpedia [1], and Wikidata [28] consist of deterministic facts, referred to as Deterministic KGs (DKGs). Due to the automatic process has been widely applied in the construction of large-scale KGs, there are many uncertain facts that make it hard to guarantee the determination of knowledge. Besides, lots of knowledge in some fields such as medicine and finance cannot be represented as deterministic facts. Therefore, Uncertain KGs (UKGs) such as NELL [4] and ConceptNet [24] represent the uncertainty as confidence scores associated to facts. Since such scores could provide more precise information, UKGs benefit many knowledge-driven applications, especially for highly risk-sensitive applications such as drug discovery [23] and investment decisions [19].

C. S. Jensen et al. (Eds.): DASFAA 2021, LNCS 12681, pp. 256–271, 2021.
https://doi.org/10.1007/978-3-030-73194-6_18

Inspired by the completion methods for DKGs, some research efforts [6,13] have been made to complete UKGs. Existing research on UKG completion usually assumes the availability of sufficient training examples for all relations. However, due to the long-tail distribution of relations, most relations only have few facts in real-world UKGs. It is crucial and challenging to deal with such cases.

Table 1. Example facts of relation "synonymfor" and entity "redhat" in NELL.

Relation: synonymfor	Entity: redhat
<(macos, synonymfor, linux), 0.94>	<(redhat, categories, software), 1.00>
<(adobe, synonymfor, flash), 0.94>	<(redhat, categories, enterprise), 1.00>
<(america, synonymfor, us), 1.00>	<(redhat, synonymfor, linux), 1.00>
<(ford, synonymfor, ibm), 1.00>	<(redhat, synonymfor, fedora), 1.00>

In UKGs, entities and relations usually have significant uncertainty of its semantic meaning. For example, in Table 1, the fact *(america, synonymfor, us)* reflects the semantic meaning of *synonymfor* precisely, while other facts such as *(adobe, synonymfor, flash)* and *(macos, synonymfor, linux)* are obviously not precise for the original semantic meaning of *synonymfor*. Another example is entity *redhat*, which has different meaning in facts *(redhat, categories, software)* and *(redhat, categories, enterprise)*. Such a condition is very common in UKGs. We refer this uncertainty as internal uncertainty of entities and relations.

Completing UKGs in few-shot settings is a non-trivial problem for the following reasons: (1) The internal uncertainty of entities and relations is essential to complete UKGs in few-shot scenarios but ignored by previous works. Existing UKG relational learning methods [6,12,13] interpret entities and relations as "points" in low-dimensional spaces. Since there are nothing different about these "points" except their positions, different internal uncertainty of entities and relations cannot be expressed. The ignorance of internal uncertainty leads to insufficient modeling of entities and relations, especially under settings with few and noisy facts. (2) Existing methods of few-shot DKGs completion [30,33] could not be used to complete UKGs directly. These models assume that all facts in KGs are entirely correct without any noise and ignore different qualities of facts. This assumption is obviously not reasonable for UKGs and leads to poor performance in a completion process, which can also be validated in our experiments. Besides, these methods could only complete missing facts but could not estimate confidence scores of completion results.

To address the above issues, we propose a novel method to complete few-shot UKGs based on Gaussian metric learning (GMUC). Given a set of few-shot facts for each relation, our model aims at learning a metric of similarity that can be used to complete missing facts and their confidence scores. Specifically, we first propose a Gaussian neighbor encoder to represent the facts of a relation as multi-dimensional Gaussian distributions, in which the semantic feature (mean) and internal uncertainty (variance) can be learned simultaneously. Gaussian-based representation could innately express internal uncertainty of entities and relations, and enable more expressive parameterization of decision boundaries [26]. Next, a Gaussian matching function considering fact qualities is designed to discover new facts and predict their confidence scores.

In experiments, we newly construct a four datasets under few-shot settings. In order to examine completion performance in real-world UKGs, these datasets have different amounts of noisy facts (i.e., uncertainty levels) to simulate an automatic construction process. We then evaluate our model on two tasks, including link prediction and confidence prediction. The results demonstrate that our model could achieve the best performance on all tasks.

Our main contributions are summarized as follows:

- We are the first to consider the long-tail distribution of relations in UKG completion tasks and formulate the problem as few-shot UKG completion.
- We propose a novel method to complete few-shot UKGs based on Gaussian metric learning. Our model could predict missing facts and their confidence scores by considering fact qualities and internal uncertainty of entities and relations.
- We newly construct a set of datasets for few-shot UKG completion, which contains four datasets with different uncertainty levels.
- We evaluate our model on two tasks, including link prediction and confidence prediction, and our model could achieve promising performances.

2   Related Works

2.1   Completion Methods for DKGs

Various works have been proposed to automatically complete DKGs by learning relation representation. RESCAL [18] represents inherent structures of relational data as tensor factorization. TransE [3] regards the relation between entities as a translation operation on low-dimensional embeddings. More advanced approaches have been invested, such as DistMult [31] and ComplEx [25]. Recently, methods utilizing deep neural networks, such as ConvE [7], have also been proposed.

2.2   Completion Methods for UKGs

Inspired by completion methods for DKGs, some UKG completion methods are have also been invested. UKGE [6] is the first UKG embedding model that is able to capture both semantic and uncertain information in embedding space. GTransE [13] uses confidence-margin-based loss function to deal with uncertainty on UKGs. PKGE in [12] employs Markov Logic Network (MLN) to learn first-order logic and encodes uncertainty.

2.3 Few-Shot Learning

Recent few-shot learning methods can be divided into two categories: (1) Metric-based methods [15,21,27,32], trying to learn a similarity metric between new instances and instances in the training set. Most of the methods in this category use the general matching framework proposed in the deep siamese networks given in [15]. An example is the matching networks [27], which make predictions by comparing input examples with small support set with labels. (2) Methods based on meta-learners [9,16,17,20], aiming to directly predict or update parameters of the model according to training data.

Recently, few-shot learning has been applied in DKG completion. Gmatching [30] designs a matching metric by considering both learned embeddings and local-subgraph structures. FSRL [33] proposes a more effective neighbor encoder module to capture heterogeneous graph structure of knowledge. Unlike metric-based methods, MetaR [5] focuses on transferring meta information of a relation to learn models effectively and efficiently. CogKR [8] solves one-shot DKG completion problem by combining summary and reasoning modules. However, these methods are intractable for UKG completion since they ignore the fact qualities and cannot predict confidence scores.

As far as we know, this is the first work to study the few-shot UKG completion problem.

3 Problem Definition

Definition 1. Uncertain Knowledge Graph
An Uncertain Knowledge Graph (UKG) is denoted as $G = \{< (h,r,t), s >\}$, where $(h,r,t) \in E \times R \times E$ represents a fact as a triple, E and R are the sets of all entities and relations, $s \in [0,1]$ is the confidence score which means the confidence of this triple to be true.

Definition 2. Few-Shot Uncertain Knowledge Graph Completion
For a relation r and one of its head entities h_j in an UKG G, few-shot UKG problem is to predict corresponding tail entities and confidence scores based on a few-shot support set $\mathcal{S}_r = \{< (h_i, t_i), s_i > |(< (h_i, r, t_i), s_i >\in G\}$. The problem can be formally represented as $r :< (h_j, ?), ? >$.

Table 2. Examples of a training task and a testing task in 3-shot UKG completion problem

Phase	Training	Testing
Task	Relation: productby	Relation: synonymfor
Support set	<(word, microsoft), 1.00> <(alphago, google), 0.50> <(ps4, nintendo), 0.37>	<(us, america), 1.00> <(adobe, flash), 0.94> <(linux, microsoft), 0.50>
Query	<(iphone, apple), 0.94>	(Hewlett-Packard, HP)

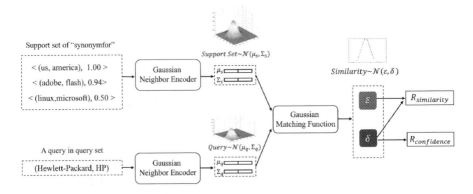

Fig. 1. The framework of GMUC: it first encodes support set and queries into multi-dimensional Gaussian distributions by Gaussian neighbor encoders. Then a Gaussian matching function is employed to construct similarity distribution between queries and the support set which is denoted as *Similarity*. Based on *Similarity*, we define two matching results $R_{similarity}$ and $R_{confidence}$ to complete missing triples and confidence scores.

As above definition, a few-shot UKG completion task can be always defined for a specific relation. During a testing process, there usually is more than one triple to be completed. We denote such triples as a query set $\mathcal{Q}_r = \{r :< (h_j, ?), ? >\}$ (Table 2).

A few-shot UKG completion method aims to gain the capability to predict new triples and their confidence scores about a relation r with only observing a few triples about r. Therefore, its training process is based on a set of tasks $\mathcal{T}_{train} = \{\mathcal{T}_i\}_{i=1}^{M}$, where $\mathcal{T}_i = \{\mathcal{S}_i, \mathcal{Q}_i\}$ indicates an individual few-shot UKG completion task. Every task has its own support set and query set. In the testing process, a set of new tasks $\mathcal{T}_{test} = \{\mathcal{T}_j\}_{j=1}^{N}$ ($\mathcal{T}_{test} \cap \mathcal{T}_{train} = \emptyset$), which can be constructed similarity.

4   Methodology

In this section, we present the detail of our proposed model GMUC (Fig. 1). First, we give the architecture of Gaussian neighbor encoder to represent queries and support set by capturing the semantic information and uncertainty. Then, we focus on Gaussian matching to measure the similarity between the queries and the support set. Finally, we describe the learning process of GMUC.

4.1   Gaussian Neighbor Encoder

Many point-based UKG relational learning methods [6,12,13] have desirable performances with sufficient training data, but these models ignore internal uncertainty which is essential in few-shot settings. Inspired by [10,33], we design the

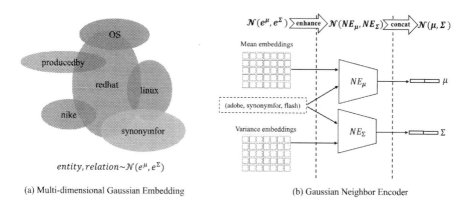

(a) Multi-dimensional Gaussian Embedding   (b) Gaussian Neighbor Encoder

Fig. 2. (a) Multi-dimensional Gaussian embedding for entities and relations. The mean embeddings e^μ of such multi-dimensional Gaussian distribution indicates its semantic feature, and the variance embeddings e^Σ indicates the corresponding internal uncertainty; (b) A diagram of Gaussian Neighbor Encoder. Two neighbor encoders NE_μ and NE_Σ are employed to learn enhanced mean embeddings μ and variance embeddings Σ of triples respectively.

Gaussian neighbor encoder to encode a support set and queries, which could naturally capture internal uncertainty by employing multi-dimensional distributions.

As Fig. 2(a) shows, we first represent each entity and relation as a multi-dimensional Gaussian distribution $\mathcal{N}(e^\mu, e^\Sigma)$, where $e^\mu \in \mathbb{R}^{d\times 1}$ is the mean embedding and $e^\Sigma \in \mathbb{R}^{d\times 1}$ is the variance embedding of entity or relation, d is embedding dimension. The mean embedding indicates its semantic feature, and the variance embedding indicates the corresponding internal uncertainty.

Based on the Gaussian-based representation of entities and relations, we then use heterogeneous neighbor encoders [33] to enhance the representation of each entity with its local structure in a knowledge graph (Fig. 2(b)). Specifically, for a entity h, we denote the enhanced representation as $\mathcal{N}(NE_\mu(h), NE_\Sigma(h))$, where NE_μ and NE_Σ are two heterogeneous neighbor encoders. The set of neighbors of h denoted as $N_h = \{<(r_i, t_i), s_i> \mid <(h, r_i, t_i), s_i> \in G\}$, where r_i and t_i represent the i-th relation and corresponding tail entity of h, s_i is confidence score of this triple. Besides, an attention module is introduced to consider different impacts of neighbors $<(r_i, t_i), s_i> \in N_h$. The calculation process of a heterogeneous neighbor encoder is defined as follows:

$$NE_*(h) = Tanh(\sum_i s_i \alpha_i e_{t_i}^*) \tag{1}$$

$$\alpha_i = \frac{\exp\left\{u_{rt}^T\left(W_{rt}\left(e_{r_i}^* \oplus e_{t_i}^*\right) + b_{rt}\right)\right\}}{\sum_j \exp\left\{u_{rt}^T\left(W_{rt}\left(e_{r_j}^* \oplus e_{t_j}^*\right) + b_{rt}\right)\right\}} \tag{2}$$

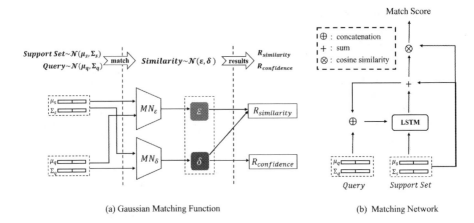

(a) Gaussian Matching Function   (b) Matching Network

Fig. 3. (a) A diagram of Gaussian matching function. Two LSTM-based matching networks MN_ε and MN_δ are used to calculate mean values ε and variance values δ of *Similarity*, respectively. (b) The structure of a LSTM-based matching network.

where $*$ could be μ or Σ, $e_{t_i}^\mu$ and $e_{r_i}^\mu$ are mean embeddings of t_i and r_i, $e_{t_i}^\Sigma$ and $e_{r_i}^\Sigma$ are variance embeddings of t_i and r_i. Moreover, $u_{rt} \in \mathbb{R}^{d \times 1}$, $W_{rt} \in \mathbb{R}^{d \times 2d}$ and $b_{rt} \in \mathbb{R}^{d \times 1}$ are learnable parameters, $\oplus$ is a concatenation operator.

Each triple in a support set and queries is interpreted as $\mathcal{N}(\mu, \Sigma)$, where mean embedding μ and variance embedding Σ are defined as follows:

$$\mu = [NE_\mu(h_k) \oplus NE_\mu(t_k)] \tag{3}$$

$$\Sigma = [NE_\Sigma(h_k) \oplus NE_\Sigma(t_k)] \tag{4}$$

By above approach, we can get the representation of a query $\mathcal{N}(\mu_q, \Sigma_q)$. For the representations of support set $\{\mathcal{N}(\mu_i, \Sigma_i) | <(h_i, t_i), s_i> \in \mathcal{S}_r\}$, we use max-pooling to aggregate these distributions into one multi-Gaussian distribution $\mathcal{N}(\mu_s, \Sigma_s)$, where μ_s and Σ_s are defined as follows:

$$\mu_s = pool_{max}(s_i \cdot \mu_i) \tag{5}$$

$$\Sigma_s = pool_{max}(s_i \cdot \Sigma_i) \tag{6}$$

4.2 Gaussian Matching Function

Given the Gaussian neighbor encoder module, we now present the Gaussian matching function to measure the similarity of queries and the support set (Fig. 3(a)). Most existing metric-based functions complete missing triples by a single value similarity, but they cannot give confidence scores to the completion results, which is inadequate for UKG completion. To address this issue, we propose Gaussian matching function to complete missing triples and their confidence scores simultaneously.

We first define the matching similarity $Similarity$ as a one-dimensional Gaussian distribution $\mathcal{N}(\varepsilon, \delta)$, where mean value $\varepsilon \in \mathbb{R}$ can be regarded as the most likely similarity value and the variance value $\delta \in [0, 1]$ refers to the uncertainty of such similarity value.

Then, we employ LSTM-based matching networks [27] to calculate $Similarity$. Compared with a simple cosine similarity, the matching networks perform a multi-step matching process, which could effectively improve the matching capability [30]. Figure 3(b) shows the structure of a matching network. The calculation process of a matching network MN is defined as follows:

$$MN(x, y) = g_t \cdot x$$
$$g_t = g_t' + y$$
$$g_t', c_t = LSTM(y, [g_{t-1} \oplus x, c_{t-1}]) \tag{7}$$

where x and y are embeddings to be matched, $LSTM(z, [g_t, c_t])$ is a LSTM cell [11] with input z, hidden state g_t and cell state c_t. After T processing steps, we use the inner product between g_t and x as the matching score of x and y.

Two matching networks MN_ε and MN_δ are used to get the mean value ε and variance value δ of $Similarity$ by the following formulas:

$$\varepsilon = MN_\varepsilon(\mu_s, \mu_q) \tag{8}$$

$$\delta = sigmoid(W \cdot MN_\delta(\Sigma_s, \Sigma_q) + b) \tag{9}$$

where $sigmoid(x) = 1/(1 + exp(-x))$, W and b are learnable parameters.

To complete missing triples and their confidence scores, we define two matching results $R_{similarity}$ and $R_{confidence}$ based on the $Similarity$ as follows:

$$R_{similarity} = \varepsilon + \lambda(1 - \delta) \tag{10}$$
$$R_{confidence} = 1 - \delta \tag{11}$$

where λ is hyper-parameter. Finally, we use the $R_{similarity}$ as ranking scores to complete missing triples and the $R_{confidence}$ to predict confidence scores.

4.3   The Learning Process

For a relation r, we randomly sample a set of few positive entity pairs $\{< (h_k, t_k), s_k > \,|\, < (h_k, r, t_k), s_k > \in G\}$ and regard them as the support set $\mathcal{S}_r$. The remaining positive entity pairs $\mathcal{Q}_r = \{< (h_l, t_l), s_l > \,|\, < (h_l, r, t_l), s_l > \in G \cap < (h_l, t_l), s_l > \notin \mathcal{S}_r\}$ are utilized as positive queries.

$\mathcal{L}_{mse}$ is designed to minimize the mean squared error (MSE) between the ground truth confidence score s and our predicting confidence score $R_{confidence}$ for each triple $< (h, t), s > \in \mathcal{Q}_r$. Specifically, $\mathcal{L}_{mse}$ is defined as:

$$\mathcal{L}_{mse} = \sum_{<(h,t),s>\in\mathcal{Q}_r} |R_{confidence} - s|^2 \tag{12}$$

Following TransE [3], we design a margin-based ranking loss $\mathcal{L}_{rank}$ to make the mean value ε of positive entity pairs to be higher than those of negative entity pairs. In order to reduce the impact of poor quality queries, we filter queries by threshold thr and get $\mathcal{Q}_r^{thr} = \{< (h_l, t_l), s_l > \mid < (h_l, r, t_l), s_l >\in \mathcal{Q}_r \text{ and } s_l \geq thr\}$. Then we construct a group of negative entity pairs $\mathcal{Q}_r^{thr-} = \{(h_l, t_l^-) \mid < (h_l, r, t_l^-), * >\notin G\}$ by polluting the tail entities. The ranking loss is formulated as:

$$\mathcal{L}_{rank} = \sum_{<(h,t),s>\in \mathcal{Q}_r^{thr}} \sum_{(h,t')\in \mathcal{Q}_r^{thr-}} s \cdot [\gamma + \varepsilon_{(h,t)} - \varepsilon_{(h,t')}]_+ \quad (13)$$

where $[x]_+ = max[0, x]$ and γ is a safety margin distance, $\varepsilon_{(h,t)}$ and $\varepsilon_{(h,t')}$ are mean value of $Similarity$ between query $(h, t_l/t_l')$ and support set $\mathcal{S}_r$. Here the triple confidence score s instructs our model to pay more attention on those more convincing queries.

Finally, we define the final objective function as:

$$\mathcal{L}_{joint} = \mathcal{L}_{rank} + \mathcal{L}_{mse} \quad (14)$$

Our objective is to minimize $\mathcal{L}_{joint}$ in the training process for all query tasks. The detail of this process can be summarized in Algorithm 1.

Algorithm 1: GMUC Training Procedure

Input:
  a) Meta-training task (relation) set $\mathcal{T}_{train}$;
  b) Embeddings of entities and relations φ;
  c) Initial parameters θ of the metric model

1 **for** $epoch:=0$ to MAX_{epoch} **do**
2   **for** $\mathcal{T}_r$ in $\mathcal{T}_{train}$ **do**
3   Sample few entity pairs as support set $\mathcal{S}_r$
4   Sample a batch of positive queries $\mathcal{Q}_r$ and filtered queries $\mathcal{Q}_r^{thr}$
5   Pollute the tail entity of queries to get $\mathcal{Q}_r^{thr-}$
6   Calculate the loss by Eq. (14)
7   Update parameters θ and φ

8 **return** Optimal model parameters θ and φ

---

5   Experiments

In this section, we present the detail of experiments. First, we introduce newly constructed datasets under few-shot settings. Then, we describe baseline models and the experimental setup. Finally, we evaluate our model on two tasks, including link prediction and confidence prediction.

5.1   Datasets

In this paper, we evaluated our model based on NL27K [6], which is a typical UKG dataset extracted from NELL [4]. However, the triples in NL27K are high quality (confidence scores ≥ 0.95) which rarely has noises or uncertain data. Therefore, similar to the work given in CRKL [29], we generated new datasets with different amounts of noisy triples (i.e., uncertainty levels) based on NL27K to simulate the real-world UKGs constructed by an automatic process with less human supervision. Specifically, based on NL27K, we constructed four datasets: NL27K-N0, NL27K-N1, NL27K-N2 and NL27K-N3 which include 0%, 10%, 20% and 40% negative triples of positive triples. Then we utilized CKRL [29] to assign confidence scores to the triples in datasets. The confidence scores are calculated by the following function:

$$C(h,r,t) = \omega_1 \cdot LT(h,r,t) + \omega_2 \cdot PP(h,r,t) + \omega_3 \cdot AP(h,r,t) \qquad (15)$$

where LT is the local triple confidence which concentrates on the inside of a triple. PP is the prior path confidence which utilizes the co-occurrence of a relation and a path to represent their dissimilarity. AP is the adaptive path confidence which could flexibly learn relation-path qualities. ω_1, ω_2 and ω_2 are hyper-parameters. Following [29], we selected $\omega_1 = 0.75$, $\omega_2 = 0.05$ and $\omega_3 = 0.2$ to create the datasets.

After assigning confidence scores to triples of datasets, following [30], we selected the relations with less than 500 but more than 50 triples to construct few-shot tasks. We referred to the rest of the relations as background relations since their triples provide important background knowledge to match entity pairs. Table 3 shows the datasets statistics. We used $101/13/20$ tasks for training/validation/testing separately.

Table 3. Statistics of the Datasets. #Entities denotes the number of unique entities and #Relations denotes the number of all relations. #Triples denotes the number of all triples. #Tasks denotes the number of relations we used as few-shot tasks. #Neg_Triples denotes the number of negative examples. Avg(s) and Std(s) are the average and standard deviation of the confidence scores.

Dataset	#Entities	#Relations	#Triples	#Tasks
NL27K-N0	27,221	404	175,412	134

Datasets	NL27K-N0	NL27K-N1	NL27K-N2	NL27K-N3
#Neg_Triples	0	17,541	35,082	70,164
Avg(s)	0.863	0.821	0.787	0.732
Std(s)	0.111	0.176	0.210	0.244

5.2   Baseline Methods

Three categories of baseline methods are considered.

Embedding Models for UKG Completion. UKGE [6] is a recently proposed UKG embedding model. UKGE preserves the semantic and uncertain information by matching the representation of entities and relations in embedding space.

Metric-based Models for Few-Shot DKG Completion. GMatching [30] and FSRL [33] are metric-based few-shot DKG completion models. However, these models cannot deal with the confidence scores of triples, which may suffer poor-quality triples. Besides, these models can only complete missing triples but cannot predict their confidence scores.

Variant Models of GMUC. We proposed two variant models of GMUC, called GMUC-noconf and GMUC-point. GMUC-noconf removes all the processes considering triple qualities. GMUC-point only uses the mean embedding μ of queries and a support set and the mean value ε of *Similarity* to calculate ranking scores and confidence scores, which is a point-based model.

5.3   Experimental Setup

Adam optimizer [14] is used for training. For baseline models, we reported results based on their best hyper-parameter. We identified each model based on the validation set performance. For hyper-parameter tuning, we searched the best hyper-parameter as follows: learning rate $lr \in \{0.001, 0.005, 0.01\}$, dimension $d \in \{64, 128, 256, 512\}$, batch size $b \in \{128, 256, 512, 1024\}$, margin $\gamma \in \{1.0, 5.0\}$, threshold $thr \in \{0.2, 0.3, 0.4, 0.5\}$, trade-off factor $\lambda \in \{0.1, 0.5, 1.0\}$. The training was stopped using early stopping based on Hit@10 on the validation set, computed every 10 epochs. The maximum number of local neighbors in Gaussian neighbor encoder is set to 30 for all datasets. For the LSTM module in Gaussian matching function, the hidden state is set to 128 and the number of recurrent steps equals 2. Specifically, for each dataset, the optimal configuration is $\{lr = 0.01, d = 128, b = 128, \gamma = 5.0\}$. For NL27K-N0, NL27K-N1 and NL27K-N2, we set $\{thr = 0.3, \lambda = 0.1\}$, while $\{thr = 0.5, \lambda = 1.0\}$ for NL27K-N3. Besides, the few-shot size $|\mathcal{S}_r|$ is set to 3 for the following experiments.

5.4   Link Prediction

This task is to complete missing tail entities for a given relation r and a head entity h,denoted as $(h, r, ?)$.

Evaluation Protocol. We followed the same protocol as in FSRL [33]: In the testing phase, for each positive query $<(h, r, t), s>$, we replaced the tail entity by candidate entities in UKG and ranked these entities in descending order of ranking scores. The maximum candidate entities size is set to 1000 for all datasets. Based on these entity ranking lists, we used three evaluation metrics by aggregation over all the queries: first, the mean reciprocal rank of correct entities (denoted as MRR); then, the proportion of correct entities in the top1 and top10 in entity ranking lists (denoted as Hits@1, Hits@10). A good method should obtain higher MRR, Hits@1 and Hits@10. Considering some corrupted triples for (h, r, t) also exists in datasets, such a prediction should also be regarded correct. To eliminate this factor, we removed those corrupted triples that already appear in training, validation and testing sets before obtaining the ranking entity list of each query. We termed the setting as "Filter" and used it for our evaluation.

Table 4. Result of link prediction

Dataset	NL27K-N0			NL27K-N1			NL27K-N2			NL27K-N3		
Metrics	MRR	Hit@1	Hit@10	MRR	Hit@1	Hit@10	MRR	Hit@1	Hit@10	MRR	Hit@1	Hit@10
GMatching	0.361	0.272	0.531	0.193	0.123	0.315	0.125	0.066	0.253	0.025	0.005	0.051
FSRL	0.397	0.304	0.589	0.188	0.101	0.333	0.123	0.052	0.264	0.027	0.007	0.045
UKGE	0.053	0.058	0.138	0.071	0.107	0.153	0.057	0.066	0.153	0.092	0.091	0.144
GMUC-noconf	0.420	0.324	0.611	0.179	0.113	0.310	0.127	0.071	0.271	0.092	0.048	0.155
GMUC-point	0.413	0.316	0.603	0.215	0.130	**0.344**	0.131	**0.113**	0.272	0.065	0.006	0.156
GMUC	**0.433**	**0.342**	**0.644**	**0.219**	**0.148**	0.332	**0.143**	0.110	**0.292**	**0.148**	**0.107**	**0.194**

Results. Table 4 shows the results of link prediction in datasets with different uncertainty levels, from which we could observe that:

(1) Our model outperforms baselines on all datasets. Compared with UKGE, GMUC has consistent improvements, demonstrating that GMUC could better complete a UKG in few-shot settings. Additionally, GMUC outperforms few-shot DKG completion methods (i.e., GMatching and FSRL), especially for NL27K-N3 GMUC achieves 0.194 of Hit@10 while GMatching only has 0.051, which indicates the promising effectiveness of GMUC for KG completion in uncertain scenarios.

(2) Comparing evaluation results between different datasets, we found that FSRL and Gmatching achieve good performance for NL27K-N0 but have a great descent when the uncertainty level goes up. Taking Hit@10 of FSRL as an example, it achieves 0.589 for NL27K-N0, but it only has 0.045 for NL27K-N3. It demonstrates the few-shot DKG completion methods could not be used to complete UKGs directly. Conversely, the performance of UKGE is worse than FSRL and GMatching for the datasets with lower uncertainty level, including NL27K-N0, NL27K-N1 and NL27K-N3, but keeps stable from NL27K-N0 to NL27K-N3. GMUC consistently outperforms FSRL, GMatching and UKGE. A possible reason is that FSRL and GMatching are based on ranking loss which is sensitive for noisy data, while UKGE is based on MSE-loss which is better suitable for noisy data in UKGs. GMUC based on the similarity distribution which can be regarded as a combination of these two methods.

(3) For a more detailed analysis of the component effectiveness, we compared GMUC and its variant models to do ablation studies. First, to investigate the design of the multi-dimensional Gaussian representation of Gaussian neighbor encoder, we compared GMUC with GMUC-point which can be seen as a point-based function. The results of GMUC-point are worse than GMUC, demonstrating the benefit of Gaussian representation. It can also suggest that confidence scores of triples are essential information that could be used to enhance the performance of link prediction. Then, GMUC outperforms GMUC-noconf, demonstrating our strategy considering the triple qualities is useful.

5.5   Confidence Prediction

The objective of this task is to predict confidence scores of triples, formulated as $< (h, r, t), ? >$.

Evaluation Protocol. For each triple (h, r, t) in the query set, we predicted its confidence score and reported the Mean Squared Error (MSE) and Mean Absolute Error (MAE).

Table 5. Result of confidence prediction

Dataset	NL27K-N0		NL27K-N1		NL27K-N2		NL27K-N3	
Metrics	MSE	MAE	MSE	MAE	MSE	MAE	MSE	MAE
UKGE	0.070	0.198	0.061	0.177	0.063	0.184	0.072	0.199
GMUC-noconf	0.019	0.106	0.022	0.111	0.028	0.126	0.029	0.130
GMUC-point	0.038	0.154	0.035	0.143	0.042	0.156	0.046	0.157
GMUC	**0.013**	**0.086**	**0.018**	**0.096**	**0.022**	**0.104**	**0.027**	**0.113**

Results. Table 5 shows the results of confidence prediction. We could find that:

(1) UKGE has larger MAE and MSE for NL27K-N0 (few-shot settings dataset) than the original NL27K (non-few-shot dataset) in [33], which validates that UKGE could not complete UKGs well in few-shot scenarios. Our model consistently outperforms UKGE, demonstrating the effectiveness of GMUC for UKG completion in few-shot settings. Comparing with evaluation results between different datasets, we found that our model and UKGE keep stable. It is the reason why these methods can get stable results on the link prediction task with different uncertainty levels.

(2) To investigate the effect of using Gaussian-based representation, we compared GMUC and its variant model GMUC-point which could be regarded as a point-based method. The results of GMUC-point are worse than

GMUC, demonstrating the benefit of Gaussian representation. A possible reason why the Gaussian representation could enhance the completion performance is that the point-based UKG completion methods try to capture the semantic and uncertain information in one embedding space simultaneously, while Gaussian representation uses mean embedding and variance embedding to learn such information respectively in two embedding spaces with different learning targets.

(3) By comparing GMUC and GMUC-noconf, we could find that our strategy considering triple qualities can also improve the performance of confidence prediction.

6   Conclusion and Future Work

In this paper, we proposed a novel method to complete few-shot UKGs based on Gaussian metric learning (GMUC), which could complete missing triples and confidence scores with few examples available. Compared with the state-of-the-art UKG completion model and few-shot DKG completion models on few-shot UKG datasets, our model has comparable effectiveness of capturing uncertain and semantic information. Experimental results also show our method consistently outperforms baselines in datasets with different uncertain levels. The source code and datasets of this paper can be obtained from https://github.com/zhangjiatao/GMUC.

In the future, we will explore the following research directions:

(1) The meta-information of relation could provide the common knowledge which could help model learn more efficiently in few-shot setting. We will explore to combine the metric-based method and meta-based method to better complete UKG with few examples.

(2) We observe that the uncertainty levels of datasets could highly adverse the completion result in few-shot settings. In the future, we may design a metric to measure the uncertainty of data, which is set by manual in this work.

(3) External knowledge such as logic rules could enrich KGs. Our future work will introduce external knowledge to further enhance the precision of completion.

Acknowledgements. This work was partially supported by the National Key Research and Development Program of China under grants (2018YFC0830200, 2017YFB1002801), the National Natural Science Foundation of China grants (U1736204, 62006040), and the Judicial Big Data Research Centre, School of Law at Southeast University. In addition, we wish to thank Prof. Qiu Ji for her valuable suggestion.

References

1. Auer, S., Bizer, C., Kobilarov, G., Lehmann, J., Cyganiak, R., Ives, Z.: DBpedia: a nucleus for a web of open data. In: Aberer, K., et al. (eds.) ASWC/ISWC-2007. LNCS, vol. 4825, pp. 722–735. Springer, Heidelberg (2007). https://doi.org/10.1007/978-3-540-76298-0_52
2. Bollacker, K.D., Evans, C., Paritosh, P., Sturge, T., Taylor, J.: Freebase: a collaboratively created graph database for structuring human knowledge. In: Proceedings of the 2008 ACM SIGMOD International Conference on Management of Data, SIGMOD 2008, pp. 1247–1250 (2008)
3. Bordes, A., Usunier, N., García-Durán, A., Weston, J., Yakhnenko, O.: Translating embeddings for modeling multi-relational data. In: Proceedings of the 2013 Annual Conference on Neural Information Processing Systems, NeurIPS 2013, pp. 2787–2795 (2013)
4. Carlson, A., Betteridge, J., Kisiel, B., Settles, B., Jr., E.R.H., Mitchell, T.M.: Toward an architecture for never-ending language learning. In: Proceedings of the 2010 AAAI Conference on Artificial Intelligence, AAAI 2010 (2010)
5. Chen, M., Zhang, W., Zhang, W., Chen, Q., Chen, H.: Meta relational learning for few-shot link prediction in knowledge graphs. In: Proceedings of the 2019 Conference on Empirical Methods in Natural Language Processing, EMNLP 2019, pp. 4216–4225 (2019)
6. Chen, X., Chen, M., Shi, W., Sun, Y., Zaniolo, C.: Embedding uncertain knowledge graphs. In: Proceedings of the 2019 AAAI Conference on Artificial Intelligence, AAAI 2019, pp. 3363–3370 (2019)
7. Dettmers, T., Minervini, P., Stenetorp, P., Riedel, S.: Convolutional 2d knowledge graph embeddings. In: Proceedings of the 2018 AAAI Conference on Artificial Intelligence, AAAI 2018, pp. 1811–1818 (2018)
8. Du, Z., Zhou, C., Ding, M., Yang, H., Tang, J.: Cognitive knowledge graph reasoning for one-shot relational learning. CoRR (2019)
9. Finn, C., Abbeel, P., Levine, S.: Model-agnostic meta-learning for fast adaptation of deep networks. In: Proceedings of the 2017 International Conference on Machine Learning, ICML 2017, vol. 70, pp. 1126–1135 (2017)
10. He, S., Liu, K., Ji, G., Zhao, J.: Learning to represent knowledge graphs with Gaussian embedding. In: Proceedings of the 2015 ACM International Conference on Information and Knowledge Management, CIKM 2015, pp. 623–632 (2015)
11. Hochreiter, S., Schmidhuber, J.: Long short-term memory. Neural Comput. **9**, 1735–1780 (1997)
12. Huang, Z., Iyer, R.G., Xiao, Z.: Uncertain knowledge graph embedding using probabilistic logic neural networks (2017)
13. Kertkeidkachorn, N., Liu, X., Ichise, R.: GTransE: generalizing translation-based model on uncertain knowledge graph embedding. In: Ohsawa, Y., et al. (eds.) JSAI 2019. AISC, vol. 1128, pp. 170–178. Springer, Cham (2020). https://doi.org/10.1007/978-3-030-39878-1_16
14. Kingma, D.P., Ba, J.: Adam: a method for stochastic optimization. In: Bengio, Y., LeCun, Y. (eds.) Proceedings of the 2015 International Conference on Learning Representations, ICLR 2015 (2015)
15. Koch, G., Zemel, R., Salakhutdinov, R.: Siamese neural networks for one-shot image recognition. In: Proceedings of the 2015 Workshop on International Conference on Machine Learning, ICML 2015, vol. 2 (2015)

16. Li, Z., Zhou, F., Chen, F., Li, H.: Meta-SGD: Learning to learn quickly for few shot learning. CoRR abs/1707.09835 (2017)
17. Munkhdalai, T., Yu, H.: Meta networks. Proc. Mach. Learn. Res. **70**, 2554 (2017)
18. Nickel, M., Tresp, V., Kriegel, H.: A three-way model for collective learning on multi-relational data. In: Proceedings of the 2011 International Conference on Machine Learning, ICML 2011, pp. 809–816 (2011)
19. Qi, G., Gao, H., Wu, T.: The research advances of knowledge graph. Technol. Intell. Eng. **3**, 4–25 (2017)
20. Ravi, S., Larochelle, H.: Optimization as a model for few-shot learning. In: Proceedings of the 2017 International Conference on Learning Representations, ICLR 2017 (2017)
21. Snell, J., Swersky, K., Zemel, R.S.: Prototypical networks for few-shot learning. In: Proceedings of the 2017 Annual Conference on Neural Information Processing Systems, NeurIPS 2017, pp. 4077–4087 (2017)
22. Socher, R., Chen, D., Manning, C.D., Ng, A.Y.: Reasoning with neural tensor networks for knowledge base completion. In: Proceedings of the 2013 Annual Conference on Neural Information Processing Systems, NeurIPS 2013, pp. 926–934 (2013)
23. Sosa, D.N., Derry, A., Guo, M., Wei, E., Brinton, C., Altman, R.B.: A literature-based knowledge graph embedding method for identifying drug repurposing opportunities in rare diseases. bioRxiv, p. 727925 (2019)
24. Speer, R., Chin, J., Havasi, C.: ConceptNet 5.5: an open multilingual graph of general knowledge. In: Proceedings of the 2017 AAAI Conference on Artificial Intelligence, AAAI 2017, pp. 4444–4451 (2017)
25. Trouillon, T., Welbl, J., Riedel, S., Gaussier, É., Bouchard, G.: Complex embeddings for simple link prediction. In: Proceedings of the 2016 International Conference on Machine Learning, ICML 2016, pp. 2071–2080 (2016)
26. Vilnis, L., McCallum, A.: Word representations via gaussian embedding. In: Bengio, Y., LeCun, Y. (eds.) Proceedings of the 2015 International Conference on Learning Representations, ICLR 2015 (2015)
27. Vinyals, O., Blundell, C., Lillicrap, T., Kavukcuoglu, K., Wierstra, D.: Matching networks for one shot learning. In: Proceedings of the 2016 Annual Conference on Neural Information Processing Systems, NeurIPS 2016, pp. 3630–3638 (2016)
28. Vrandečić, D., Krötzsch, M.: Wikidata: a free collaborative knowledgebase. Commun. ACM **57**, 78–85 (2014)
29. Xie, R., Liu, Z., Lin, F., Lin, L.: Does William Shakespeare REALLY Write Hamlet? knowledge representation learning with confidence. In: Proceedings of the 2018 AAAI Conference on Artificial Intelligence, AAAI 2018, pp. 4954–4961 (2018)
30. Xiong, W., Yu, M., Chang, S., Wang, W.Y., Guo, X.: One-shot relational learning for knowledge graphs. In: Proceedings of the Conference on Empirical Methods in Natural Language Processing, EMNLP 2018, pp. 1980–1990 (2018)
31. Yang, B., Yih, W., He, X., Gao, J., Deng, L.: Embedding entities and relations for learning and inference in knowledge bases. In: Proceedings of the 2015 International Conference on Learning Representations, ICLR 2015 (2015)
32. Yu, M., et al.: Diverse few-shot text classification with multiple metrics. In: Proceedings of the 2018 Conference of the North American Chapter of the Association for Computational Linguistics: Human Language Technologies, NAACL-HLT 2018, pp. 1206–1215 (2018)
33. Zhang, C., Yao, H., Huang, C., Jiang, M., Li, Z., Chawla, N.V.: Few-shot knowledge graph completion. In: Proceedings of the 2020 AAAI Conference on Artificial Intelligence, AAAI 2020, pp. 3041–3048 (2020)

Towards Entity Alignment in the Open World: An Unsupervised Approach

Weixin Zeng[1], Xiang Zhao[1(✉)], Jiuyang Tang[1], Xinyi Li[1], Minnan Luo[2], and Qinghua Zheng[2]

[1] Science and Technology on Information Systems Engineering Laboratory, National University of Defense Technology, Changsha, China
{zengweixin13,xiangzhao,jiuyang_tang}@nudt.edu.cn
[2] Department of Computer Science, Xi'an Jiaotong University, Xi'an, China
{minnluo,qhzheng}@mail.xjtu.edu.cn

Abstract. Entity alignment (EA) aims to discover the equivalent entities in different knowledge graphs (KGs). It is a pivotal step for integrating KGs to increase knowledge coverage and quality. Recent years have witnessed a rapid increase of EA frameworks. However, state-of-the-art solutions tend to rely on labeled data for model training. Additionally, they work under the closed-domain setting and cannot deal with entities that are unmatchable.

To address these deficiencies, we offer an unsupervised framework that performs entity alignment in the open world. Specifically, we first mine useful features from the side information of KGs. Then, we devise an unmatchable entity prediction module to filter out unmatchable entities and produce preliminary alignment results. These preliminary results are regarded as the pseudo-labeled data and forwarded to the progressive learning framework to generate structural representations, which are integrated with the side information to provide a more comprehensive view for alignment. Finally, the progressive learning framework gradually improves the quality of structural embeddings and enhances the alignment performance by enriching the pseudo-labeled data with alignment results from the previous round. Our solution does not require labeled data and can effectively filter out unmatchable entities. Comprehensive experimental evaluations validate its superiority.

Keywords: Entity alignment · Unsupervised learning · Knowledge graph

1 Introduction

Knowledge graphs (KGs) have been applied to various fields such as natural language processing and information retrieval. To improve the quality of KGs, many efforts have been dedicated to the alignment of KGs, since different KGs usually contain complementary information. Particularly, entity alignment (EA), which aims to identify equivalent entities in different KGs, is a crucial step of

© Springer Nature Switzerland AG 2021
C. S. Jensen et al. (Eds.): DASFAA 2021, LNCS 12681, pp. 272–289, 2021.
https://doi.org/10.1007/978-3-030-73194-6_19

KG alignment and has been intensively studied over the last few years [1–8]. We use Example 1 to illustrate this task.

Example 1. *In Fig. 1 are a partial English KG and a partial Spanish KG concerning the director* Hirokazu Koreeda, *where the dashed lines indicate known alignments (i.e., seeds). The task of EA aims to identify equivalent entity pairs between two KGs, e.g.,* (Shoplifters, Manbiki Kazoku).

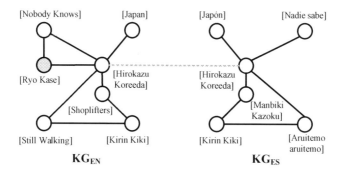

Fig. 1. An example of EA.

State-of-the-art EA solutions [9–12] assume that equivalent entities usually possess similar neighboring information. Consequently, they utilize KG embedding models, e.g., TransE [13], or graph neural network (GNN) models, e.g., GCN [14], to generate structural embeddings of entities in individual KGs. Then, these separated embeddings are projected into a unified embedding space by using the seed entity pairs as connections, so that the entities from different KGs are directly comparable. Finally, to determine the alignment results, the majority of current works [3,15–17] formalize the alignment process as a ranking problem; that is, for each entity in the source KG, they rank all the entities in the target KG according to some distance metric, and the closest entity is considered as the equivalent target entity.

Nevertheless, we still observe several issues from current EA works:

– **Reliance on labeled data.** Most of the approaches rely on pre-aligned seed entity pairs to connect two KGs and use the unified KG structural embeddings to align entities. These labeled data, however, might not exist in real-life settings. For instance, in Example 1, the equivalence between Hirokazu Koreeda in KG_{EN} and Hirokazu Koreeda in KG_{ES} might not be known in advance. In this case, state-of-the-art methods that solely rely on the structural information would fall short, as there are no seeds to connect these individual KGs.

– **Closed-domain setting.** All of current EA solutions work under the closed-domain setting [18]; that is, they assume every entity in the source KG has an equivalent entity in the target KG. Nevertheless, in practical settings, there always exist unmatchable entities. For instance, in Example 1, for the source entity Ryo Kase, there is no equivalent entity in the target KG. Therefore, an ideal EA system should be capable of predicting the unmatchable entities.

In response to these issues, we put forward an unsupervised EA solution UEA that is capable of addressing the unmatchable problem. Specifically, to mitigate the reliance on labeled data, we mine useful features from the KG side information and use them to produce preliminary pseudo-labeled data. These preliminary seeds are forwarded to our devised **progressive learning framework** to generate unified KG structural representations, which are integrated with the side information to provide a more comprehensive view for alignment. This framework also progressively augments the training data and improves the alignment results in a self-training fashion. Besides, to tackle the unmatchable issue, we design an **unmatchable entity prediction** module, which leverages thresholded bi-directional nearest neighbor search (TBNNS) to filter out the unmatchable entities and excludes them from the alignment results. We embed the unmatchable entity prediction module into the progressive learning framework to control the pace of progressive learning by dynamically adjusting the thresholds in TBNNS.

Contribution. The main contributions of the article can be summarized as follows:

- We identify the deficiencies of existing EA methods, namely, requiring labeled data and working under the closed-domain setting, and propose an unsupervised EA framework UEA that is able to deal with unmatchable entities. This is done by (1) exploiting the side information of KGs to generate preliminary pseudo-labeled data; and (2) devising an unmatchable entity prediction module that leverages the thresholded bi-directional nearest neighbor search strategy to produce alignment results, which can effectively exclude unmatchable entities; and (3) offering a progressive learning algorithm to improve the quality of KG embeddings and enhance the alignment performance.
- We empirically evaluate our proposal against state-of-the-art methods, and the comparative results demonstrate the superiority of UEA.

Organization. In Sect. 2, we formally define the task of EA and introduce related work. Section 3 elaborates the framework of UEA. In Sect. 4, we introduce experimental results and conduct detailed analysis. Section 5 concludes this article.

2   Task Definition and Related Work

In this section, we formally define the task of EA, and then introduce the related work.

Task Definition. The inputs to EA are a source KG G_1 and a target KG G_2. The task of EA is defined as finding the equivalent entities between the KGs, i.e., $\Psi = \{(u,v)|u \in E_1, v \in E_2, u \leftrightarrow v\}$, where E_1 and E_2 refer to the entity sets in G_1 and G_2, respectively, $u \leftrightarrow v$ represents the source entity u and the target entity v are *equivalent*, i.e., u and v refer to the same real-world object.

Most of current EA solutions assume that there exist a set of seed entity pairs $\Psi_s = \{(u_s, v_s)|u_s \in E_1, v_s \in E_2, u_s \leftrightarrow v_s\}$. Nevertheless, in this work, we focus on unsupervised EA and do not assume the availability of such labeled data.

Entity Alignment. The majority of state-of-the-art methods are supervised or semi-supervised, which can be roughly divided into three categories, i.e., methods merely using the structural information, methods that utilize the iterative training strategy, and methods using information in addition to the structural information [19].

The approaches in the first category aim to mine useful structural signals for alignment, and devise structure learning models such as recurrent skipping networks [20] and multi-channel GNN [17], or exploit existing models such as TransE [3,9,21–23] and graph attention networks [3]. The embedding spaces of different KGs are connected by seed entity pairs. In accordance to the distance in the unified embedding space, the alignment results can hence be predicted.

Methods in the second category iteratively label likely EA pairs as the training set and gradually improve alignment results [15,21–24]. A more detailed discussion about these methods and the difference from our framework is provided in Sect. 3.3. Methods in the third category incorporate the side information to offer a complementing view to the KG structure, including the attributes [10,25–29], entity descriptions [16,30], and entity names [12,24,31–34]. These methods devise various models to encode the side information and consider them as features parallel to the structural information. In comparison, the side information in this work has an additional role, i.e., generating pseudo-labeled data for learning unified structural representations.

Unsupervised Entity Alignment. A few methods have investigated the alignment without labeled data. Qu et al. [35] propose an unsupervised approach towards knowledge graph alignment with the adversarial training framework. Nevertheless, the experimental results are extremely poor. He et al. [36] utilize the shared attributes between heterogeneous KGs to generate aligned entity pairs, which are used to detect more equivalent attributes. They perform entity alignment and attribute alignment alternately, leading to more high-quality aligned entity pairs, which are used to train a relation embedding model. Finally, they combine the alignment results generated by attribute and relation triples using a bivariate regression model. The overall procedure of this work might seem similar to our proposed model. However, there are many notable differences; for instance, the KG embeddings in our work are updated progressively, which can lead to more accurate alignment results, and our model can deal with unmatchable entities. We empirically demonstrate the superiority of our model in Sect. 4.

We notice that there are some entity resolution (ER) approaches established in a setting similar to EA, represented by PARIS [37]. They adopt collective alignment algorithms such as similarity propagation so as to model the relations among entities. We include them in the experimental study for the comprehensiveness of the article.

3   Methodology

In this section, we first introduce the outline of UEA. Then, we elaborate its components.

As shown in Fig. 2, given two KGs, UEA first mines useful features from the *side information*. These features are forwarded to the *unmatchable entity prediction* module to generate initial alignment results, which are regarded as pseudo-labeled data. Then, the *progressive learning framework* uses these pseudo seeds to connect two KGs and learn unified entity structural embeddings. It further combines the alignment signals from the side information and *structural information* to provide a more comprehensive view for alignment. Finally, it progressively improves the quality of structural embeddings and augments the alignment results by iteratively updating the pseudo-labeled data with results from the previous round, which also leads to increasingly better alignment.

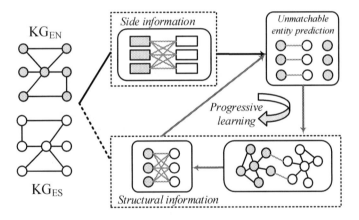

Fig. 2. Outline of UEA. Colored lines represent the progressive learning process.

3.1   Side Information

There is abundant side information in KGs, such as the attributes, descriptions and classes. In this work, we use a particular form of the attributes—the entity name, as it exists in the majority of KGs. To make the most of the entity name information, inspired by [12], we exploit it from the semantic level and string-level and generate the textual distance matrix between entities in two KGs.

More specifically, we use the averaged word embeddings to represent the semantic meanings of entity names. Given the semantic embeddings of a source and a target entity, we obtain the semantic distance score by subtracting their cosine similarity score from 1. We denote the semantic distance matrix between the entities in two KGs as $\mathbf{M^n}$, where rows represent source entities, columns denote target entities and each element in the matrix denotes the distance score between a pair of source and target entities. As for the string-level feature, we adopt the Levenshtein distance [38] to measure the difference between two sequences. We denote the string distance matrix as $\mathbf{M^l}$.

To obtain a more comprehensive view for alignment, we combine these two distance matrices and generate the textual distance matrix as $\mathbf{M^t} = \alpha \mathbf{M^n} + (1-\alpha)\mathbf{M^l}$, where α is a hyper-parameter that balances the weights. Then, we forward the textual distance matrix $\mathbf{M^t}$ into the unmatchable entity module to produce alignment results, which are considered as the pseudo-labeled data for training KG structural embeddings. The details are introduced in the next subsection.

Remark. The goal of this step is to exploit available side information to generate useful features for alignment. Other types of side information, e.g., attributes and entity descriptions, can also be leveraged. Besides, more advanced textual encoders, such as misspelling oblivious word embeddings [39] and convolutional embedding for edit distance [40], can be utilized. We will investigate them in the future.

3.2   Unmatchable Entity Prediction

State-of-the-art EA solutions generate for each source entity a corresponding target entity and fail to consider the potential unmatchable issue. Nevertheless, as mentioned in [19], in real-life settings, KGs contain entities that other KGs do not contain. For instance, when aligning YAGO 4 and IMDB, only 1% of entities in YAGO 4 are related to movies, while the other 99% of entities in YAGO 4 necessarily have no match in IMDB. These unmatchable entities would increase the difficulty of EA. Therefore, in this work, we devise an unmatchable entity prediction module to predict the unmatchable entities and filter them out from the alignment results.

More specifically, we put forward a novel strategy, i.e., thresholded bi-directional nearest neighbor search (TBNNS), to generate the alignment results, and the resulting unaligned entities are predicted to be unmatchable. As can be observed from Algorithm 1, given a source entity u and a target entity v, if u and v are the nearest neighbor of each other, and the distance between them is below a given threshold θ, we consider (u, v) as an aligned entity pair. Note that $\mathbf{M}(u, v)$ represents the element in the u-th row and v-th column of the distance matrix $\mathbf{M}$.

Algorithm 1: TBNNS in the unmatchable entity prediction module

---

Input   : G_1 and G_2: the two KGs to be aligned; E_1 and E_2: the entity sets in
              G_1 and G_2; θ: a given threshold; $\mathbf{M}$: a distance matrix.
Output : S: Alignment results.

1 **foreach** $u \in E_1$ **do**
2      $\quad v \leftarrow \underset{\hat{v} \in E_2}{\arg \min} \mathbf{M}(u, \hat{v})$;
3      $\quad$ **if** $\underset{\hat{u} \in E_1}{\arg \min} \mathbf{M}(v, \hat{u}) = u$ **and** $\mathbf{M}(u, v) < \theta$ **then**
4      $\quad \quad \lfloor \ S \leftarrow S + \{(u, v)\}$

5 **return** S.

The TBNNS strategy exerts strong constraints on alignment, since it requires that the matched entities should both prefer each other the most, and the distance between their embeddings should be below a certain value. Therefore, it can effectively predict unmatchable entities and prevent them from being aligned. Notably, the threshold θ plays a significant role in this strategy. A larger threshold would lead to more matches, whereas it would also increase the risk of including erroneous matches or unmatchable entities. In contrast, a small threshold would only lead to a few aligned entity pairs, and almost all of them would be correct. This is further discussed and verified in Sect. 4.4. Therefore, our progressive learning framework dynamically adjusts the threshold value to produce more accurate alignment results (to be discussed in the next subsection).

3.3   The Progressive Learning Framework

To exploit the rich structural patterns in KGs that could provide useful signals for alignment, we design a progressive learning framework to combine structural and textual features for alignment and improve the quality of both structural embeddings and alignment results in a self-training fashion.

Structural Information. As mentioned above, we forward the textual distance matrix $\mathbf{M^t}$ generated by using the side information to the unmatchable entity prediction module to produce the preliminary alignment results, which are considered as pseudo-labeled data for learning unified KG embeddings. Concretely, following [25], we adopt GCN[1] to capture the neighboring information of entities. We leave out the implementation details since this is not the focus of this paper, which can be found in [25].

Given the learned structural embedding matrix $\mathbf{Z}$, we calculate the structural distance score between a source and a target entity by subtracting the cosine similarity score between their embeddings from 1. We denote the resultant structural distance matrix as $\mathbf{M^s}$. Then, we combine the textual and structural information to generate more accurate signals for alignment: $\mathbf{M} = \beta \mathbf{M^t} + (1 - \beta) \mathbf{M^s}$, where

---

[1] More advanced structural learning models, such as recurrent skipping networks [20], could also be used here. We will explore these alternative options in the future.

β is a hyper-parameter that balances the weights. The fused distance matrix $\mathbf{M}$ can be used to generate more accurate matches.

The Progressive Learning Algorithm. The amount of training data has an impact on the quality of the unified KG embeddings, which in turn affects the alignment performance [10,41]. As thus, we devise an algorithm (Algorithm 2) to progressively augment the pseudo training data, so as to improve the quality of KG embeddings and enhance the alignment performance. The algorithm starts with learning unified structural embeddings and generating the fused distance matrix $\mathbf{M}$ by using the preliminary pseudo-labeled data S_0 (line 1). Then, the fused distance matrix is used to produce the new alignment results ΔS using TBNNS (line 2). These newly generated entity pairs ΔS are added to the alignment results (which are considered as pseudo-labeled data for the next round), and the entities in the alignment results S are removed from the entity sets (line 3–6). In order to progressively improve the quality of KG embeddings and detect more alignment results, we perform the aforementioned process recursively until the number of newly generated entity pairs is below a given threshold γ (line 7–13).

Notably, in the learning process, once a pair of entities is considered as a match, the entities will be removed from the entity sets (line 5–6 and line 12–13). This could gradually reduce the alignment search space and lower the difficulty for aligning the rest entities. Obviously, this strategy suffers from the error propagation issue, which, however, could be effectively mitigated by the progressive learning process that dynamically adjusts the threshold. We will verify the effectiveness of this setting in Sect. 4.3.

Algorithm 2: Progressive learning.

Input   : G_1 and G_2: the two KGs to be aligned; E_1 and E_2: the entity sets in G_1 and G_2; $\mathbf{M^t}$: textual distance matrix; S_0: preliminary labeled data; θ_0: the initial threshold.

Output : S: Alignment results.

1 Use S_0 to learn structural embeddings, generate $\mathbf{M^s}$ and $\mathbf{M}$;
2 $\Delta S \leftarrow$TBNNS$(G_1, G_2, E_1, E_2, \theta_0, \mathbf{M})$;
3 $S \leftarrow S_0 + \Delta S$;
4 $\theta \leftarrow \theta_0 + \eta$;
5 $E_1 \leftarrow \{e | e \in E_1, e \notin S\}$;
6 $E_2 \leftarrow \{e | e \in E_2, e \notin S\}$;
7 **while** the number of the newly generated alignment results is above γ **do**
8 $\quad$ Use S to learn structural embeddings, generate $\mathbf{M^s}$ and $\mathbf{M}$;
9 $\quad$ $\Delta S \leftarrow$TBNNS$(G_1, G_2, E_1, E_2, \theta, \mathbf{M})$;
10 $\quad$ $S \leftarrow S + \Delta S$;
11 $\quad$ $\theta \leftarrow \theta + \eta$;
12 $\quad$ $E_1 \leftarrow \{e | e \in E_1, e \notin S\}$;
13 $\quad$ $E_2 \leftarrow \{e | e \in E_2, e \notin S\}$;

14 **return** S.

Dynamic Threshold Adjustment. It can be observed from Algorithm 2 that, the matches generated by the unmatchable entity prediction module are not only part of the eventual alignment results, but also the pseudo training data for learning subsequent structural embeddings. Therefore, to enhance the overall alignment performance, the alignment results generated in each round should, ideally, have both large *quantity* and high *quality*. Unfortunately, these two goals cannot be achieved at the same time. This is because, as stated in Sect. 3.2, a larger threshold in TBNNS can generate more alignment results (large quantity), whereas some of them might be erroneous (low quality). These wrongly aligned entity pairs can cause the error propagation problem and result in more erroneous matches in the following rounds. In contrast, a smaller threshold leads to fewer alignment results (small quantity), while almost all of them are correct (high quality).

To address this issue, we aim to balance between the quantity and the quality of the matches generated in each round. An intuitive idea is to set the threshold to a moderate value. However, this fails to take into account the characteristics of the progressive learning process. That is, in the beginning, the quality of the matches should be prioritized, as these alignment results will have a long-term impact on the subsequent rounds. In comparison, in the later stages where most of the entities have been aligned, the quantity is more important, as we need to include more possible matches that might not have a small distance score. In this connection, we set the initial threshold θ_0 to a very small value so as to reduce potential errors. Then, in the following rounds, we gradually increase the threshold by η, so that more possible matches could be detected. We will empirically validate the superiority of this strategy over the fixed weight in Sect. 4.3.

Remark. As mentioned in the related work, there are some existing EA approaches that exploit the iterative learning (bootstrapping) strategy to improve EA performance. Particularly, BootEA calculates for each source entity the alignment likelihood to every target entity, and includes those with likelihood above a given threshold in a maximum likelihood matching process under the 1-to-1 mapping constraint, producing a solution containing confident EA pairs [22]. This strategy is also adopted by [15,23]. Zhu et al. use a threshold to select the entity pairs with very close distances as the pseudo-labeled data [21]. DAT employs a bi-directional margin-based constraint to select the confident EA pairs as labels [24]. Our progressive learning strategy differs from these existing solutions in three aspects: (1) we exclude the entities in the confident EA pairs from the test sets; and (2) we use the dynamic threshold adjustment strategy to control the pace of learning process; and (3) our strategy can deal with unmatchable entities. The superiority of our strategy is validated in Sect. 4.3.

4   Experiment

This section reports the experiment results with in-depth analysis. The source code is available at https://github.com/DexterZeng/UEA.

4.1 Experiment Settings

Datasets. Following existing works, we adopt the DBP15K dataset [10] for evaluation. This dataset consists of three multilingual KG pairs extracted from DBpedia. Each KG pair contains 15 thousand inter-language links as gold standards. The statistics can be found in Table 1. We note that state-of-the-art studies merely consider the labeled entities and divide them into training and testing sets. Nevertheless, as can be observed from Table 1, there exist unlabeled entities, e.g., 4,388 and 4,572 entities in the Chinese and English KG of DBP15K$_{ZH-EN}$, respectively. In this connection, we adapt the dataset by including the unmatchable entities. Specifically, for each KG pair, we keep 30% of the labeled entity pairs as the training set (for training the supervised or semi-supervised methods). Then, to construct the test set, we include the rest of the entities in the first KG and the rest of the labeled entities in the second KG, so that the unlabeled entities in the first KG become unmatchable. The statistics of the test sets can be found in the *Test set* column in Table 1.

Table 1. The statistics of the evaluation benchmarks.

Dataset	KG pairs	#Triples	#Entities	#Labeled ents	#Relations	#Test set
DBP15K$_{ZH-EN}$	DBpedia (Chinese)	70,414	19,388	15,000	1,701	14,888
	DBpedia (English)	95,142	19,572	15,000	1,323	10,500
DBP15K$_{JA-EN}$	DBpedia (Japanese)	77,214	19,814	15,000	1,299	15,314
	DBpedia (English)	93,484	19,780	15,000	1,153	10,500
DBP15K$_{FR-EN}$	DBpedia (French)	105,998	19,661	15,000	903	15,161
	DBpedia (English)	115,722	19,993	15,000	1,208	10,500

Parameter Settings. For the *side information* module, we utilize the fastText embeddings [42] as word embeddings. To deal with cross-lingual KG pairs, following [32], we use Google translate to translate the entity names from one language to another, i.e., translating Chinese, Japanese and French to English. α is set to 0.5. For the *structural information learning*, we set β to 0.5. Noteworthily, since there are no training set or validation set for parameter tuning, we set α and β to the default value (0.5). We will further verify that the hyper-parameters do not have a large influence on the final results in Sect. 4.4. For *progressive learning*, we set the initial threshold θ_0 to 0.05, the incremental parameter η to 0.1, the termination threshold γ to 30. Note that if the threshold θ is over 0.45, we reset it to 0.45. These hyper-parameters are default values since there is no extra validation set for hyper-parameter tuning.

Evaluation Metrics. We use *precision* (P), *recall* (R) and *F1 score* as evaluation metrics. The *precision* is computed as the number of correct matches divided by the number of matches found by a method. The *recall* is computed as the number of correct matches found by a method divided by the number of gold matches. The *F1 score* is the harmonic mean between *precision* and *recall*.

Competitors. We select the most performant state-of-the-art solutions for comparison. Within the group that solely utilizes structural information, we compare with BootEA [22], TransEdge [15], MRAEA [41] and SSP [43]. Among the methods incorporating other sources of information, we compare with GCN-Align [25], HMAN [16], HGCN [11], RE-GCN [44], DAT [24] and RREA [45]. We also include the unsupervised approaches, i.e., IMUSE [36] and PARIS [37]. To make a fair comparison, we only use entity name labels as the side information.

Table 2. Alignment results.

	ZH-EN			JA-EN			FR-EN		
	P	R	F1	P	R	F1	P	R	F1
BootEA	0.444	0.629	0.520	0.426	0.622	0.506	0.452	0.653	0.534
TransEdge	0.518	0.735	0.608	0.493	0.719	0.585	0.492	0.710	0.581
MRAEA	0.534	0.757	0.626	0.520	0.758	0.617	0.540	0.780	0.638
SSP	0.521	0.739	0.611	0.494	0.721	0.587	0.512	0.739	0.605
GCN-Align	0.291	0.413	0.342	0.274	0.399	0.325	0.258	0.373	0.305
HMAN	0.614	0.871	0.720	0.641	**0.935**	0.761	0.674	**0.973**	0.796
HGCN	0.508	0.720	0.596	0.525	0.766	0.623	0.618	0.892	0.730
RE-GCN	0.518	0.735	0.608	0.548	0.799	0.650	0.646	0.933	0.764
DAT	0.556	0.788	0.652	0.573	0.835	0.679	0.639	0.922	0.755
RREA	0.580	0.822	0.680	0.629	0.918	0.747	0.667	0.963	0.788
IMUSE	0.608	0.862	0.713	0.625	0.911	0.741	0.618	0.892	0.730
PARIS	**0.976**	0.777	0.865	**0.981**	0.785	0.872	**0.972**	0.793	0.873
UEA	0.913	**0.902**	**0.907**	0.940	0.932	**0.936**	0.953	0.950	**0.951**

4.2 Results

Table 2 reports the alignment results, which shows that state-of-the-art supervised or semi-supervised methods have rather low precision values. This is because these approaches cannot predict the unmatchable source entities and generate a target entity for each source entity (including the unmatchable ones). Particularly, methods incorporating additional information attain relatively better performance than the methods in the first group, demonstrating the benefit of leveraging such additional information.

Regarding the unsupervised methods, although IMUSE cannot deal with the unmatchable entities and achieves a low precision score, it outperforms most of the supervised or semi-supervised methods in terms of recall and F1 score. This indicates that, for the EA task, the KG side information is useful for mitigating the reliance on labeled data. In contrast to the abovementioned methods, PARIS attains very high precision, since it only generates matches that it believes to be highly possible, which can effectively filter out the unmatchable entities. It also achieves the second best F1 score among all approaches, showcasing its effectiveness when the unmatchable entities are involved. Our proposal, UEA, achieves the best balance between precision and recall and attains the best F1 score, which outperforms the second-best method by a large margin, validating its effectiveness. Notably, although UEA does not require labeled data, it achieves even better performance than the most performant supervised method HMAN (except for the recall values on DBP15K$_{JA-EN}$ and DBP15K$_{FR-EN}$).

4.3   Ablation Study

In this subsection, we examine the usefulness of proposed modules by conducting the ablation study. More specifically, in Table 3, we report the results of UEA w/o Unm, which excludes the unmatchable entity prediction module, and UEA w/o Prg, which excludes the progressive learning process. It shows that, removing the unmatchable entity prediction module (UEA w/o Unm) brings down the performance on all metrics and datasets, validating its effectiveness of detecting the unmatchable entities and enhancing the overall alignment performance. Besides, without the progressive learning (UEA w/o Prg), the precision increases, while the recall and F1 score values drop significantly. This shows that the progressive learning framework can discover more correct aligned entity pairs and is crucial to the alignment progress.

To provide insights into the progressive learning framework, we report the results of UEA w/o Adj, which does not adjust the threshold, and UEA w/o Excl, which does not exclude the entities in the alignment results from the entity sets during the progressive learning. Table 3 shows that setting the threshold to a fixed value (UEA w/o Adj) leads to worse F1 results, verifying that the progressive learning process depends on the choice of the threshold and the quality of the alignment results. We will further discuss the setting of the threshold in the next subsection. Besides, the performance also decreases if we do not exclude the matched entities from the entity sets (UEA w/o Excl), validating that this strategy indeed can reduce the difficulty of aligning entities.

Moreover, we replace our progressive learning framework with other state-of-the-art iterative learning strategies (i.e., MWGM [22], TH [21] and DAT-I [24]) and report the results in Table 3. It shows that using our progressive learning framework (UEA) can attain the best F1 score, verifying its superiority.

Table 3. Ablation results.

	ZH-EN			JA-EN			FR-EN		
	P	R	F1	P	R	F1	P	R	F1
UEA	0.913	0.902	**0.907**	0.940	0.932	**0.936**	0.953	0.950	**0.951**
w/o Unm	0.553	0.784	0.648	0.578	0.843	0.686	0.603	0.871	0.713
w/o Prg	0.942	0.674	0.786	0.966	0.764	0.853	0.972	0.804	0.880
w/o Adj	0.889	0.873	0.881	0.927	0.915	0.921	0.941	0.936	0.939
w/o Excl	**0.974**	0.799	0.878	0.982	0.862	0.918	0.985	0.887	0.933
MWGM	0.930	0.789	0.853	0.954	0.858	0.903	0.959	0.909	0.934
TH	0.743	**0.914**	0.820	0.795	**0.942**	0.862	0.807	**0.953**	0.874
DAT-I	**0.974**	0.805	0.881	**0.985**	0.866	0.922	**0.988**	0.875	0.928
UEA-M^l	0.908	0.902	0.905	0.926	0.924	0.925	0.937	0.931	0.934
M^l	0.935	0.721	0.814	0.960	0.803	0.875	0.948	0.750	0.838
UEA-M^n	0.758	0.727	0.742	0.840	0.807	0.823	0.906	0.899	0.903
M^n	0.891	0.497	0.638	0.918	0.562	0.697	0.959	0.752	0.843

4.4 Quantitative Analysis

In this subsection, we perform quantitative analysis of the modules in UEA.

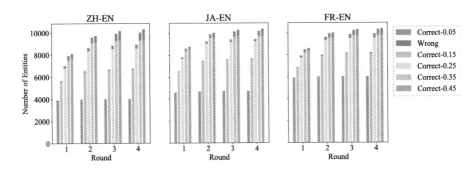

Fig. 3. Alignment results given different threshold values. Correct-θ refers to the number of correct matches generated by the progressive learning framework at each round given the threshold value θ. Wrong refers to the number of erroneous matches generated in each round.

The Threshold θ in TBNNS. We discuss the setting of θ to reveal the trade-off between the risk and gain from generating the alignment results in the progressive learning. Identifying a match leads to the integration of additional structural information, which benefits the subsequent learning. However, for the same reason, the identification of a false positive, i.e., an incorrect match, potentially leads to mistakenly modifying the connections between KGs, with the risk of amplifying the error in successive rounds. As shown in Fig. 3, a smaller θ (e.g., 0.05) brings low risk and low gain; that is, it merely generates a small number of matches, among which almost all are correct. In contrast, a higher θ (e.g., 0.45) increases the risk, and brings relatively higher gain; that is, it results in much more aligned entity pairs, while a certain portion of them are erroneous. Additionally, using a higher threshold leads to increasingly more alignment results, while for a lower threshold, the progressive learning process barely increases the number of matches. This is in consistency with our theoretical analysis in Sect. 3.2.

Unmatchable Entity Prediction. Zhao et al. [19] propose an intuitive strategy (U-TH) to predict the unmatchable entities. They set an NIL threshold, and if the distance value between a source entity and its closest target entity is above this threshold, they consider the source entity to be unmatchable. We compare our unmatchable entity prediction strategy with it in terms of the percentage of unmatchable entities that are included in the final alignment results and the F1 score. On DBP15K$_{\text{ZH-EN}}$, replacing our unmatchable entity prediction strategy with U-TH attains the F1 score at 0.837, which is 8.4% lower than that of UEA. Besides, in the alignment results generated by using U-TH, 18.9% are unmatchable entities, while this figure for UEA is merely 3.9%. This demonstrates the superiority of our unmatchable entity prediction strategy.

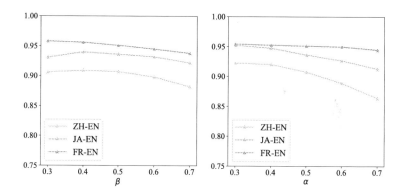

Fig. 4. The F1 scores by setting α and β to different values.

Influence of Parameters. As mentioned in Sect. 4.1, we set α and β to 0.5 since there are no training/validation data. Here, we aim to prove that different values of the parameters do not have a large influence on the final results. More specifically, we keep α at 0.5, and choose β from [0.3, 0.4, 0.5, 0.6, 0.7]; then we keep β at 0.5, and choose α from [0.3, 0.4, 0.5, 0.6, 0.7]. It can be observed from Fig. 4 that, although smaller α and β lead to better results, the performance does not change significantly.

Influence of Input Side Information. We adopt different side information as input to examine the performance of UEA. More specifically, we report the results of UEA-M^l, which merely uses the string-level feature of entity names as input, UEA-M^n, which only uses the semantic embeddings of entity names as input. We also provide the results of M^l and M^n, which use the string-level and semantic information to directly generate alignment results (without progressive learning), respectively.

As shown in Table 3, the performance of solely using the input side information is not very promising (M^l and M^n). Nevertheless, by forwarding the side information into our model, the results of UEA-M^l and UEA-M^n become much better. This unveils that UEA can work with different types of side information and consistently improve the alignment results. Additionally, by comparing UEA-M^l with UEA-M^n, it is evident that the input side information does affect the final results, and the quality of the side information is of significance to the overall alignment performance.

Pseudo-Labeled Data. We further examine the usefulness of the preliminary alignment results generated by the side information, i.e., the pseudo-labeled data. Concretely, we replace the training data in HGCN with these pseudo-labeled data, resulting in HGCN-U, and then compare its alignment results with the original performance. Regarding the F1 score, HGCN-U is 4% lower than HGCN on DBP15K$_{ZH-EN}$, 2.9% lower on DBP15K$_{JA-EN}$, 2.8% lower on DBP15K$_{FR-EN}$. The minor difference validates the effectiveness of the pseudo-labeled data generated by the side information. It also demonstrates that this strategy can be applied to other supervised or semi-supervised frameworks to reduce their reliance on labeled data.

5 Conclusion

In this article, we propose an unsupervised EA solution that is capable of dealing with unmatchable entities. We first exploit the side information of KGs to generate preliminary alignment results, which are considered as pseudo-labeled data and forwarded to the progressive learning framework to produce better KG embeddings and alignment results in a self-training fashion. We also devise an unmatchable entity prediction module to detect the unmatchable entities. The experimental results validate the usefulness of our proposed model and its superiority over state-of-the-art approaches.

Acknowledgments. This work was partially supported by Ministry of Science and Technology of China under grant No. 2020AAA0108800, NSFC under grants Nos. 61872446 and 71971212, NSF of Hunan Province under grant No. 2019JJ20024, Postgraduate Scientific Research Innovation Project of Hunan Province under grant No. CX20190033.

References

1. Hao, Y., Zhang, Y., He, S., Liu, K., Zhao, J.: A joint embedding method for entity alignment of knowledge bases. In: Chen, H., Ji, H., Sun, L., Wang, H., Qian, T., Ruan, T. (eds.) CCKS 2016. CCIS, vol. 650, pp. 3–14. Springer, Singapore (2016). https://doi.org/10.1007/978-981-10-3168-7_1

2. Shi, X., Xiao, Y.: Modeling multi-mapping relations for precise cross-lingual entity alignment. In: EMNLP, pp. 813–822 (2019)

3. Li, C., Cao, Y., Hou, L., Shi, J., Li, J., Chua, T.S.: Semi-supervised entity alignment via joint knowledge embedding model and cross-graph model. In: EMNLP, pp. 2723–2732 (2019)

4. Sun, Z., et al.: Knowledge graph alignment network with gated multi-hop neighborhood aggregation. In: AAAI, pp. 222–229 (2020)

5. Xu, K., Song, L., Feng, Y., Song, Y., Yu, D.: Coordinated reasoning for cross-lingual knowledge graph alignment. In: AAAI, pp. 9354–9361 (2020)

6. Chen, J., Gu, B., Li, Z., Zhao, P., Liu, A., Zhao, L.: SAEA: self-attentive heterogeneous sequence learning model for entity alignment. In: Nah, Y., Cui, B., Lee, S.-W., Yu, J.X., Moon, Y.-S., Whang, S.E. (eds.) DASFAA 2020, Part I. LNCS, vol. 12112, pp. 452–467. Springer, Cham (2020). https://doi.org/10.1007/978-3-030-59410-7_31

7. Wu, Y., Liu, X., Feng, Y., Wang, Z., Zhao, D.: Neighborhood matching network for entity alignment. In: ACL, pp. 6477–6487 (2020)

8. Sun, Z., et al.: A benchmarking study of embedding-based entity alignment for knowledge graphs. Proc. VLDB Endow. 13(11), 2326–2340 (2020)

9. Chen, M., Tian, Y., Yang, M., Zaniolo, C.: Multilingual knowledge graph embeddings for cross-lingual knowledge alignment. In: IJCAI, pp. 1511–1517 (2017)

10. Sun, Z., Hu, W., Li, C.: Cross-lingual entity alignment via joint attribute-preserving embedding. In: d'Amato, C., et al. (eds.) ISWC 2017, Part I. LNCS, vol. 10587, pp. 628–644. Springer, Cham (2017). https://doi.org/10.1007/978-3-319-68288-4_37

11. Wu, Y., Liu, X., Feng, Y., Wang, Z., Zhao, D.: Jointly learning entity and relation representations for entity alignment. In: EMNLP, pp. 240–249 (2019)

12. Zeng, W., Zhao, X., Tang, J., Lin, X.: Collective entity alignment via adaptive features. In: ICDE, pp. 1870–1873 (2020)

13. Bordes, A., Usunier, N., García-Durán, A., Weston, J., Yakhnenko, O.: Translating embeddings for modeling multi-relational data. In: NIPS, pp. 2787–2795 (2013)

14. Kipf, T.N., Welling, M.: Semi-supervised classification with graph convolutional networks. CoRR, abs/1609.02907 (2016)

15. Sun, Z., Huang, J., Hu, W., Chen, M., Guo, L., Qu, Y.: TransEdge: translating relation-contextualized embeddings for knowledge graphs. In: Ghidini, C., et al. (eds.) ISWC 2019, Part I. LNCS, vol. 11778, pp. 612–629. Springer, Cham (2019). https://doi.org/10.1007/978-3-030-30793-6_35

16. Yang, H.W., Zou, Y., Shi, P., Lu, W., Lin, J., Sun, X.: Aligning cross-lingual entities with multi-aspect information. In: EMNLP, pp. 4430–4440 (2019)

17. Cao, Y., Liu, Z., Li, C., Liu, Z., Li, J., Chua, T.S.: Multi-channel graph neural network for entity alignment. In: ACL, pp. 1452–1461 (2019)

18. Hertling, S., Paulheim, H.: The knowledge graph track at OAEI. In: Harth, A., et al. (eds.) ESWC 2020. LNCS, vol. 12123, pp. 343–359. Springer, Cham (2020). https://doi.org/10.1007/978-3-030-49461-2_20

19. Zhao, X., Zeng, W., Tang, J., Wang, W., Suchanek, F.: An experimental study of state-of-the-art entity alignment approaches. IEEE Trans. Knowl. Data Eng. 01, 1 (2020)

20. Guo, L., Sun, Z., Hu, W.: Learning to exploit long-term relational dependencies in knowledge graphs. In: ICML, pp. 2505–2514 (2019)
21. Zhu, H., Xie, R., Liu, Z., Sun, M.: Iterative entity alignment via joint knowledge embeddings. In: IJCAI, pp. 4258–4264 (2017)
22. Sun, Z., Hu, W., Zhang, Q., Qu, Y.: Bootstrapping entity alignment with knowledge graph embedding. In: IJCAI, pp. 4396–4402 (2018)
23. Zhu, Q., Zhou, X., Wu, J., Tan, J., Guo, L.: Neighborhood-aware attentional representation for multilingual knowledge graphs. In: IJCAI, pp. 1943–1949 (2019)
24. Zeng, W., Zhao, X., Wang, W., Tang, J., Tan, Z.: Degree-aware alignment for entities in tail. In: SIGIR, pp. 811–820 (2020)
25. Wang, Z., Lv, Q., Lan, X., Zhang, Y.: Cross-lingual knowledge graph alignment via graph convolutional networks. In: EMNLP, pp. 349–357 (2018)
26. Trisedya, B.D., Qi, J., Zhang, R.: Entity alignment between knowledge graphs using attribute embeddings. In: AAAI, pp. 297–304 (2019)
27. Yang, K., Liu, S., Zhao, J., Wang, Y., Xie, B.: COTSAE: co-training of structure and attribute embeddings for entity alignment. In: AAAI, pp. 3025–3032 (2020)
28. Chen, B., Zhang, J., Tang, X., Chen, H., Li, C.: JarKA: modeling attribute interactions for cross-lingual knowledge alignment. In: Lauw, H.W., Wong, R.C.-W., Ntoulas, A., Lim, E.-P., Ng, S.-K., Pan, S.J. (eds.) PAKDD 2020, Part I. LNCS (LNAI), vol. 12084, pp. 845–856. Springer, Cham (2020). https://doi.org/10.1007/978-3-030-47426-3_65
29. Tang, X., Zhang, J., Chen, B., Yang, Y., Chen, H., Li, C.: BERT-INT: a BERT-based interaction model for knowledge graph alignment. In: IJCAI, pp. 3174–3180 (2020)
30. Chen, M., Tian, Y., Chang, K.W., Skiena, S., Zaniolo, C.: Co-training embeddings of knowledge graphs and entity descriptions for cross-lingual entity alignment. In: IJCAI, pp. 3998–4004 (2018)
31. Xu, K., et al.: Cross-lingual knowledge graph alignment via graph matching neural network. In: ACL, pp. 3156–3161 (2019)
32. Wu, Y., Liu, X., Feng, Y., Wang, Z., Yan, R., Zhao, D.: Relation-aware entity alignment for heterogeneous knowledge graphs. In: IJCAI, pp. 5278–5284 (2019)
33. Fey, M., Lenssen, J.E., Morris, C., Masci, J., Kriege, N.M.: Deep graph matching consensus. In: ICLR (2020)
34. Zeng, W., Zhao, X., Tang, J., Lin, X., Groth, P.: Reinforcement learning based collective entity alignment with adaptive features. ACM Transactions on Information Systems. to appear (2021)
35. Qu, M., Tang, J., Bengio, Y.: Weakly-supervised knowledge graph alignment with adversarial learning. CoRR, abs/1907.03179 (2019)
36. He, F., et al.: Unsupervised entity alignment using attribute triples and relation triples. In: Li, G., Yang, J., Gama, J., Natwichai, J., Tong, Y. (eds.) DASFAA 2019, Part I. LNCS, vol. 11446, pp. 367–382. Springer, Cham (2019). https://doi.org/10.1007/978-3-030-18576-3_22
37. Suchanek, F.M., Abiteboul, S., Senellart, P.: PARIS: probabilistic alignment of relations, instances, and schema. PVLDB 5(3), 157–168 (2011)
38. Levenshtein, V.I.: Binary codes capable of correcting deletions, insertions, and reversals. Soviet Phy. Doklady 10, 707–710 (1966)
39. Edizel, K., Piktus, A., Bojanowski, P., Ferreira, R., Grave, E., Silvestri, F.: Misspelling oblivious word embeddings. In: NAACL-HLT, pp. 3226–3234 (2019)
40. Dai, X., Yan, X., Zhou, K., Wang, Y., Yang, H., Cheng, J.: Convolutional embedding for edit distance. In: SIGIR, pp. 599–608 (2020)

41. Mao, X., Wang, W., Xu, H., Lan, M., Wu, Y.: MRAEA: an efficient and robust entity alignment approach for cross-lingual knowledge graph. In: WSDM, pp. 420–428 (2020)
42. Bojanowski, P., Grave, E., Joulin, A., Mikolov, T.: Enriching word vectors with subword information. Trans. Assoc. Comput. Linguist. **5**, 135–146 (2017)
43. Nie, H., et al.: Global structure and local semantics-preserved embeddings for entity alignment. In: IJCAI, pp. 3658–3664 (2020)
44. Yang, J., Zhou, W., Wei, L., Lin, J., Han, J., Hu, S.: RE-GCN: relation enhanced graph convolutional network for entity alignment in heterogeneous knowledge graphs. In: Nah, Y., Cui, B., Lee, S.-W., Yu, J.X., Moon, Y.-S., Whang, S.E. (eds.) DASFAA 2020, Part II. LNCS, vol. 12113, pp. 432–447. Springer, Cham (2020). https://doi.org/10.1007/978-3-030-59416-9_26
45. Mao, X., Wang, W., Xu, H., Wu, Y., Lan, M.: Relational reflection entity alignment. In: CIKM, pp. 1095–1104 (2020)

Sequence Embedding for Zero or Low Resource Knowledge Graph Completion

Zhijuan Du[1,2]([✉]) [iD]

[1] Inner Mongolia University, Hohhot 010021, China
[2] Inner Mongolia Discipline Inspection and Supervision Big Data Laboratory,
Hohhot 010015, China

Abstract. Knowledge graph completion (KGC) has been proposed to improve KGs by filling in missing links. Previous KGC approaches require a large number of training instances (entity and relation) and hold a closed-world assumption. The real case is that very few instances are available and KG evolve quickly with new entities and relations being added by the minute. The newly added cases are zero resource in training. In this work, we propose a **S**equence **E**mbedding **w**ith **A**dversarial learning approach (SEwA) for zero or low resource KGC. It transform the KGC into a sequence prediction problem by making full use of inherently link structure of knowledge graph and resource-easy-to-transfer feature of adversarial contextual embedding. Specifically, the triples ($<h, r, t>$) and higher-order triples ($<h, p, t>$) containing the paths ($p = r_1 \rightarrow \cdots \rightarrow r_n$) are represented as word sequences and are encoded by pre-training model with multi head self-attention. The path is obtained by a non-parametric learning based on the one-class classification of the relation trees. The zero and low resources issues are further optimizes by adversarial learning. At last, our SEwA is evaluated by low resource datasets and open world datasets.

Keywords: Knowledge graph · Zero/low resource · Structure sequence · Multi head attention · Non-parameter · Adversarial learning

1 Introduction

Knowledge Graphs (KGs) organize facts in a structured way as triples in the form of $<head\ entity,\ relation,\ tail\ entity>$, abridged as $<h,\ r,\ t>$, where r builds relations between entities h and t. In this formalism a statement like *"Beijing is the capital of China"* can be represented as $<Beijing,\ capitalOf,\ China>$. Usually KG is not a complete graph, which has many potential links or new facts to discover. This problem is called KG completion (KGC). Its equivalent

---

Supported by the Natural Science Foundation of Inner Mongolia in China (2020BS06005, 2018BS06001), the Inner Mongolia Discipline Inspection and Supervision Big Data Laboratory Open Project (IMDBD2020010), and the High-level Talents Scientific Research Foundation of Inner Mongolia University (21500-5195118).

C. S. Jensen et al. (Eds.): DASFAA 2021, LNCS 12681, pp. 290–306, 2021.
https://doi.org/10.1007/978-3-030-73194-6_20

task is link prediction or triples classification. For example, predict the missing part ? in $<?, capitalOf, China>$, $<Beijing,?, China>$, $<Beijing, capitalOf,?>$ or assess plausibility of $<Beijing, capitalOf, China>$. Previous KGC approaches (TransE [3], ComplEx [21], ConvE [4], RotatE [20], TuckER [1]) require a large number of training instances (An entity with only a small training samples is limited to update opportunities due to fewer occurrences, resulting in it being represented by an approximately random vector.) and are restricted by the closed world assumption [17] (Untrained entities cannot be updated from any inference function and can only be represented by an initial random vector.). As shown in Fig. 1(b), the fewer entity occurrences, the larger MRR (mean reciprocal rank) (larger MRR means less accurate)[1].

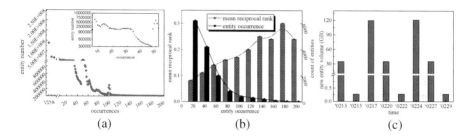

Fig. 1. (a) occurrences ≤ 200; (b) entity frequency vs MRR; (c) new entity growth.

However, the real case is that very few instances are available and evolve quickly with new entities and relations being added by the minute. As shown in Fig. 1(a), 53% entities appear only once and 78% entities appear up to 4 times in Freebase[2], as shown in Fig. 1(c), Wikidata grows 12 GB new entities per day in February this year. We call this case with only a small or zero training sample as a low-zero-resource issue. The small and zero training sample are also called few-shot and zero-shot (or new entity or open-world). Currently, few methods such as wRAN [31] and LAN [24] are working on low resource issue. The former incorporates text information beyond KG triples and the latter designs a neighbor-assisted strategy within KG triples. Besides, OWE [18], ConMask [19], KG-BERT [29] incorporates text information to solve zero-shot issue. Those works yet still have some limitations. First, they can only solve one type of issue in low or zero resource. Second, the text information are entity descriptions or Web corpus. But not all entities have text information, and not all text information is valid and available, e.g. only 10, 159, 119 out of 184, 346, 843 entities in Freebase have descriptions. More importantly, text is extra information beyond KG triples. Third, newly added or few shot cases often have few neighboring

---

[1] This results are measured by the classic TransE model on the baseline dataset FB15K. FB15K is a subset of Freebase, Freebase is a subset of Wikidata.

[2] The results come from 2016 release data in Freebase. The total number of entities in Freebase is 507, 480, 694, of which 271, 330, 531 entities occur only once and 124, 378, 009 entities occur 2 to 4 times. 15449 entities occur from 1257 to 97922175.

entities (in and out edges). For example, let's take the one-shot as an example, as shown in Fig. 2. It's a special case of few-shot.

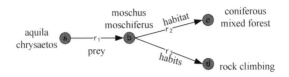

Fig. 2. An example of KG structure

The entity a, c and d appear once and b 3 times in Fig. 2. Hence, b is not a one-shot entity. The basic idea of neighbor-assisted strategy is to use the sum of neighbors to represent entities. In Fig. 2, a has a neighbor b (or (r_1, b)), b has two neighbors c (or (r_2, c)) and d (or (r_3, d)), and there are no neighbors for c and d. So the neighbor-assisted strategy is not very effective for entities with fewer neighbors. But given an entity a, if a path is added, there will be 3 triples containing entity a, such as $<a, r_1, b>$, $<a, r_1 \rightarrow r_2, c>$, $<a, r_1 \rightarrow r_3, d>$. So that the entity a can train twice more than before. However, there are 3 issues that need to be addressed: (1) There are many paths in KG for a given entity pair. E.g. 100 relations can form billions 4-step paths. So it's better to learn relational features rather than hand-crafted. (2) Each relation expresses different semantics in different paths. [13] shares the view that words might exhibit different meanings when they appear in different textual contexts. (3) Usually, only its related high-resource entities and relations can really help low-resource objects. This is particularly important for low and zero-resource objects.

To solve those problems, we borrow the idea of relational one-class classification (relOCC) [9]) and first-order logical decision trees (TILDE) [2] to learn relational features, introduce pre-trained language model [5] to realize knowledge share and keep language patterns, use multi-head self-attention mechanism [23] to capture the diverse meaning and importance of each word in different contexts, adversarial learning [22] to further optimized against zero-resources issues. So far, we propose a novel KGC approach with low and zero resource references, namely SEwA (**S**equence **E**mbedding **w**ith **A**dversarial learning). The key contributions as follows: (1) We propose a pre-training model SEwA, which modeling (higher-order) triples as word sequences and turn KGC into a sequence prediction. (2) We obtain the path via non-parametric relational one-class classification to improve path learning ability. (3) We use adversarial learning to further optimized against low and zero-resources issues. (4) We evaluate our SEwA from different aspects on low or zero resource datasets.

The rest of the paper is organized as follows. Section 2 describes related works. Section 3 gives a problem formulation, elaborates the methodology, which details the pre-training sequence embedding model SEwA, non-parametric path acquisition and adversarial learning. Section 4 evaluates the performance of various methods in multi-type datasets. Section 5 concludes the paper.

2   Related Work

The relevant work is introduced from knowledge graph embedding and pre-trained language model.

2.1   Knowledge Graph Embedding

Embedding is the mainstream method of knowledge graph completion. The earlier approach focused on four types of relations (1-to-1, 1-to-n, n-to-1, n-to-n [3]), especially latter three. There are many classic methods, *Trans* family such as Trans (E, H, R, D, Sparse) [3,7,8,12,25]. PTransE [11], Bilinear family such as DistMult [28], ComplEx [21] HolE [15] and deep learning approaches such as ConvE [4], ConvKG [14]. Later methods target more complex relation types, such as (anti)symmetry, inversion relation and composition relation. The most famous RotatE [20] and TuckER [1]) methods fall into this category. Recently, research has focused on zero-shot and few-shot issues. OWE [18], ConMask [19], KG-BERT [29] incorporates text information. LAN [24] designs a neighbor-assisted strategy within KG triples to solve zero-shot issue. wRAN [31] claims to solve low-resource problems, but essentially it solves few-shot issue via incorporates text information beyond KG triples. Neural networks are basically used in these methods. But they all treat the three elements of a triple group as separate symbols. Moreover, newly added or few shot cases often have few neighboring entities. So far, the zero-low resources issue still has a long way to go.

2.2   Pre-trained Language Model

The language model has the ability to capture syntactic and semantic information about words from large-scale unlabeled text. So word vectors are standard components of most of the latest natural language processing (NLP) architectures. A literature survey of pre-trained language models has been conducted by [16]. It can be seen that the attention mechanism is a milestone in the language model. Since the introduction of the attention mechanism, the Seq2seq model that joined attention has improved on all tasks, so the current seq2seq model refers to the model that combines RNN and attention. Then Google put forward a Transformer model [23] to solve the problem of to sequence, replacing LSTM with the full structure of self-attention mechanism. To that end, he became another watershed. Then a multilayer bidirectional Transformer encoder BERT [5] is built. It is a state-of-the-art pre-trained contextual language representation model. There are two steps in BERT framework: pre-training and fine-tuning. During pre-training, BERT is trained on large-scale unlabeled general domain corpus over two self-supervised tasks: masked language modeling and next sentence prediction. In masked language modeling, BERT predicts randomly masked input tokens. In next sentence prediction, BERT predicts whether two input sentences are consecutive. For fine-tuning, BERT is initialized with the pre-trained parameter weights, and all of the parameters are fine-tuned using labeled data from downstream tasks such as sentence pair.

3   Methodology

This section will introduce our pre-training model based on sequence embedding with adversarial learning for zero or low resource knowledge graph completion.

3.1   Problem Formulation

We denote triple by $<h, r, t>$, entity by $e \in \{h, t\}$, relation by r, path by p, column (matrix) vector by bold lower (upper) case letter, KG by $G = \{E, R, T\} = \{\langle h, r, t \rangle \mid h, t \in E, r \in R\}$. The formal definitions are as follows:

Definition 1 (Entity). *Given KG G, it is divided into train set Δ_{train} and test set Δ_{test}, and it satisfies $G = \{\{\Delta_{train} \cup \Delta_{test}\} \mid \Delta_x = \{E^x, R^x, T^x\}, x \in \{train, test\} \left(O^{\Delta_{test}} \cap O^{\Delta_{train}} \neq \emptyset\right) \cap \left(T^{\Delta_{test}} \cap T^{\Delta_{train}} = \emptyset\right), O \in \{E, R\}$. The regular entity is denoted as $e_{com} = \{e \mid N_e > \delta, e \in E\}$, the few-shot entity is denoted as $e_{few} = \{e \mid N_e \leq \delta, e \in E\}$, the one-shot entity is denoted as $e_{one} = \{e \mid N_e = 1, e \in E\}$, the zero-shot entity is denoted as $e_{zero} = \{e \mid N_e = 0, e \in E^{\Delta_{test}} \cap e \notin E^{\Delta_{train}}\}$, N_e is e occurrences in Δ_{train}.*

Definition 2 (Low-Zero-Resource KG, LZKG). *Low-zero-resource KG satisfies $G_{lz} = \{G \mid |e_{few}| \geq \theta, |e_{zero}| \geq \sigma\}$. Here, e_{few} and e_{zero} is few-shot and zero-shot entity. $|\bullet|$ is the number $\bullet$.*

Definition 3 (Higher-order Triples, HT). *Given two entities $h, t \in E$, the direct relation $r \in R$ and the indirect relation set $p_i \in P_{h,t}$ between them, the high-order triple is represented as $<h, p_i, t>$. Here, $P_{h,t} = \{p_1, p_1, \cdots p_N\}$, $p_i = r_1 \rightarrow r_2 \rightarrow \cdots \rightarrow r_n$ is a path linking entity h, t.*

The paper use word sequence representation, e.g. given a triple $<h, r, t>$ and higher-order triple $<h, p, t>$, then they are represented as follows:

$<h, r, t>$: $h \rightarrow r \rightarrow t$

$<h, p, t>$: $h \rightarrow r_1 \rightarrow r_i \rightarrow \cdots \rightarrow r_n \rightarrow t$

here h, r, p, t is the name (words sequences, e.g. aquila chrysaetos) rather than a symbolic mark (e.g. a). $\rightarrow$ denote sequence formed by a list of two elements.

Definition 4 (LZKG completion, LZC). *With a low-zero-resource KG G_{lz}, given a support set $S \in \{<h, r, t>, <h, p, t>\}$ about entity h and t, predicting a missing entity ($<?, r, t>$ or $<?, p, t>$ or $<h, r, ?>$ or $<h, p, ?>$) in the input sequence s_i ($s_i \in S$), is called low-zero resource KG completion.*

3.2   Framework Overview

Our goal is to use paths to help complete zero and low-resource KGs. The key idea is to transform it into a sequence prediction problem. Specifically, the triples ($<h, r, t>$) and higher-order triples ($<h, p, t>$) are represented as word sequences. Then encode the word sequence with a stack of Transformer [23] blocks. So architecture of the our SEwA is shown in Fig. 3.

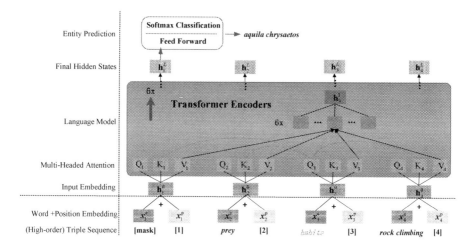

Fig. 3. Overview of our approach.

Input Encoder. In Fig. 3, we first represent entities, relations, triples, and higher-order triples as names lists, i.e., natural language sentences. For example, various objects in Fig. 2 can be represented as follows:

> **Entities:** *aquila chrysaetos, moschus moschiferus,*et al.
> **Relations:** *prey, habitat, habits.*
> **Triple:** *aquila chrysaetos prey moschus moschiferus,* et al.
> **Higher-order triples:** *aquila chrysaetos prey habits rock climbing,*et al.

The above word sequence serves as input into a stack of L successive Transformer encoders [23], and each entity and relation serves as each element in the word sequence, separated by [SEP]. Here, as usual, each element is represented by an element vector and a position vector, as shown in Eq. (1).

$$\mathbf{X} = \{x_1, \cdots, x_n\} = \{[\mathbf{x}_1^e + \mathbf{x}_1^p], \cdots, [\mathbf{x}_n^e + \mathbf{x}_n^p]\} \tag{1}$$

$x_1, x_n \in E, x_{i \in [2, n-1]} \in R$. $\mathbf{x}_i^e$ is element embedding used to identify the current element. $\mathbf{x}_i^p$ is the position embeddings used to represent its position in the sequence. $\mathbf{x}_i^p$ is calculated using Eq. (2).

$$\mathbf{x}_i^p = \begin{cases} sin(i/10000^{k/d}), k\%2 = 0, k \leq d \\ cos(i/10000^{k/d}), k\%2 = 1, k \leq d \end{cases} \tag{2}$$

d is embedding dimension, $k\%2 = 0(or1)$ represents the even (or odd) dimension. This not only eliminates parameter training, but also adapts to the length of sentences that have never been met. So input sentence is denoted as $\mathbf{h}_i^0 = \mathbf{x}_i^e + \mathbf{x}_i^p$.

Non-parametric Path Acquisition. We have adopted a path-assisted strategy. So the path information needs to be obtained. Usually, the path can be expressed in the form of predicate logic. For example, $\exists x r_1(h, x) \wedge r_2(x, t)$ formulas capture 2-step paths between h and t, $\exists x, y r_1(h, x) \wedge r_2(x, y) \wedge r_3(y, t)$ formulas capture 3-step paths between h and t. Then these structure learning is done with one-class classification problem for relations (relOCC) [9]. Specifically, top-down induction of first-order logical decision trees (TILDE) [2] is used to learn relation trees. The trees represents a decision list of relation rules. The relational distance between a pair of instances u and v is calculated by lowest common ancestors (LCA), as shown in Eq. (3). Here, u, v is the relational representation of triples, such as the $Prey(aquila\_chrysaetos, moschus\_moschiferus)$.

$$dis(u, v) = \begin{cases} 0 & , LCA(u, v) \ is \ leaf \\ e^{-\lambda depth(LCA(u,v))} & , otherwise \end{cases} \quad (3)$$

Here, $depth(LCA(u, v))$ is the depth of instances u and v to lowest common ancestors. E.g. $depth(LCA(r_1(a, b), r_2(b, c))) = 1$, $depth(LCA(r_1(a, b), r_3(b, d))) = 0$ in Fig. 4 (A first-order logic decision tree of Fig. 2).

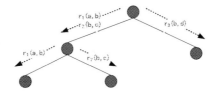

Fig. 4. A case of lowest common ancestors.

There is usually more than one tree. The total distance function can be represented as the weighted sum of the individual tree-level distances, as shown in Eq. (4).

$$Dis(u, v) = \sum_i \omega_i dis_i(u, v), \sum_i \omega_i = 1, \omega_i \geq 0 \quad (4)$$

$Dis(\cdot, \cdot)$ can then be used to compute the density expectation for a new relational instance o as a weighted sum of the distance of o from all training instances $x \in T$, as shown in Eq. (5).

$$Ed(o \notin class) = \sum_{x \in T} \varepsilon_x Dis(x, o), \sum_{x \in T} \varepsilon_x = 1, \varepsilon_x \geq 0 \quad (5)$$

We learn a tree-based distance iteratively [9] to introduce new relational features that perform one-class classification. The left-most path in each relational tree is a conjunction of predicates (clause), which can be used as a relational feature. The splitting criteria is the squared error over the instances and the goal is to minimize squared error in each node as shown in Eq. (6) and Eq. (7).

$$L = \min \left(\sum_{y \in x_r} \left(I\left(o\right) - Ed\left(o \notin class\right) - \sum_{j:x_j \in x_l} \varepsilon_j \omega_i dis_i\left(x_j, o\right) \right)^2 \right.$$
$$\left. + \sum_{y \in x_l} \left(I\left(o\right) - Ed\left(o \notin class\right) - \sum_{j:x_j \in x_r} \varepsilon_j \omega_i dis_i\left(x_j, o\right) \right)^2 \right) \qquad (6)$$

$$I\left(o\right) = \begin{cases} 0, o\,is\,an\,labeled\,instance \\ 1, otherwise \end{cases} \qquad (7)$$

Here, x_l and x_r are the examples that take the left and right branch respectively. A greedy search approach is employed for tree learning, thereby it is a non-parametric approach.

Sequence Encoder. The input representations $\mathbf{h}_i^0 = \mathbf{x}_i^e + \mathbf{x}_i^p$ synthesize a matrix $\mathbf{H} = \left[\mathbf{h}_1^0, \mathbf{h}_2^0, \cdots \mathbf{h}_n^0\right]^T$ are fed into a block of L successive Transformer encoders [23]. $\mathbf{h}_i^0$ is the row vector, n is the number of elements. Then self-attention use 3 linear transformations to get query (Q), key (K) and value (V) as shown in Eq. (8).

$$\begin{cases} \mathbf{Q} = linear_q(\mathbf{H}) = \mathbf{H} \times \mathbf{W}^Q \\ \mathbf{K} = linear_k(\mathbf{H}) = \mathbf{H} \times \mathbf{W}^K \\ \mathbf{V} = linear_v(\mathbf{H}) = \mathbf{H} \times \mathbf{W}^V \end{cases} \qquad (8)$$

The $linear_k$, $linear_q$ and $linear_v$ are independent of each other, weights $\mathbf{W}^Q, \mathbf{W}^K, \mathbf{W}^V$ are different and can be obtained by training. The scaled dot-product attention (Q, K, V) is calculated by Eq. (9).

$$Attention\left(Q, K, V\right) = soft\max\left(\frac{\mathbf{QK}^T}{\sqrt{d_k}}\right)\mathbf{V} \qquad (9)$$

In Eq. (9), d_k is the column numbers of the Q, K matrix, which is the vector dimension. Equation (9) allows each element to attend to all elements in the sequence. It can not only increase the training times of the low-shot objects, but also transfer high resource knowledge to low or zero resource objects.

Multi-head attention allows the model to jointly attend to information from different representation subspaces at different positions. With a single attention head, averaging inhibits this, as shown in Eq. (10) and Eq. (11).

$$MultiHeadAttention\left(Q, K, V\right) = Concat\left(head_1, \cdots, head_h\right)\mathbf{W}^O \qquad (10)$$

$$head_i = Attention_i\left(QW_i^Q, KW_i^K, VW_i^V\right) \qquad (11)$$

Where the projections are parameter matrices $W_i^Q, W_i^K \in \mathbb{R}^{d \times d_k}$, $W_i^V \in \mathbb{R}^{d \times d_v}$ and $W^O \in \mathbb{R}^{hd_v \times d}$. h is parallel attention layers, $d_k = d_v = d/h$. d is embedding dimension. Due to the reduced dimension of each head, the total computational cost is similar to that of single-head attention with full dimensionality.

In addition to attention sub-layers, each block contains a *add-norm* layer and a fully connected position-wise feed-forward network, which is applied to each position separately and identically. This consists of two linear transformations with a ReLU activation in between, as shown in Eq. (12)–(14).

$$\mathbf{h}_i^l = LayerNorm\left(\mathbf{h}_i^{l-1} + \left(SubLayer\left(\mathbf{h}_i^{l-1}\right)\right)\right) \tag{12}$$

$$SubLayer\left(\mathbf{h}_i^l\right) = \begin{cases} MultiHeadAttention\left(\mathbf{h}_i^l\right) \\ FeedForward\left(\mathbf{h}_i^l\right) \end{cases} \tag{13}$$

$$FeedForward\left(\mathbf{h}_i^l\right) = \max(0, \mathbf{h}_i^l W_1 + b_1)W_2 + b_2 \tag{14}$$

Here, $l \in [1, L], i \in [1, n]$, $\mathbf{h}_i^l$ is the hidden state of x_i after the l-th layer. When L-th block is finished, we will get the final hidden states $\mathbf{h}_i^L$.

Entity Prediction. According to Definition 4, our task is to predict ? in $<$ $?, r, t>$ or $<?, p, t>$ or $<h, r, ?>$ or $<h, p, ?>$. According to the masked language model(MLM) [29], the corresponding position of ? in the sequence is replaced by a special token [MASK]. E.g. if predict h, x_1 in the sequence needs to be replaced with [MASK], as shown in Fig. 3. Here the masked and predicted entities are only selected in the given triple or high-order triple. The final hidden state $\mathbf{h}_1^L$ or $\mathbf{h}_n^L$ corresponding to [MASK] is obtain by Transformer [23] encoding. $\mathbf{h}_1^L$ or $\mathbf{h}_n^L$ cannot yet be used directly for prediction. It also needs to go through feed forward and standard softmax classification, as shown in Eq. (15).

$$\begin{cases} \mathbf{z}_i^L = Feedforward\left(\mathbf{h}_i^L\right), i \in \{1, n\} \\ \mathbf{p}_i = soft\max\left(\mathbf{E}^e \mathbf{z}_i^L\right), i \in \{1, n\} \end{cases} \tag{15}$$

Here, $\mathbf{E}^e \in R^{V \times H 3}$ is classification weight that shared with the input element embedding matrix, V is entity vocabulary size, H is hidden state size and $\mathbf{p}_i$ is predicted distribution of x_i over all entities. $\mathbf{p}_i$ is equivalent to the scoring function in the traditional KGC models. We compute the cross-entropy loss with $\mathbf{p}_i$ and label $y_i, i \in \{1, n\}$ as shown in Eq. (16).

$$\mathcal{L} = -\sum_k y_i^k log p_i^k, y_i^k = \begin{cases} \varepsilon, \ target \ entity \\ 1 - \varepsilon, \ others \end{cases} \tag{16}$$

Generally, rather than requiring one best answer, entity prediction emphasizes more on ranking a set of candidate entities [3,29]. So, y_i^k is not a one-hot label. y_i^k and p_i^k are the k-th component of y_i and $\mathbf{p}_i, i \in \{1, n\}$. Where $1 - N$ scoring in ConvE [4] may be a solution to improve the efficiency. It can fast evaluation for link prediction tasks because it take one (e, r) pair and score it against all entities $o \in E$ simultaneously.

---

[3] It is the only new parameters introduced during entity prediction fine-tuning.

Adversarial Learning. We introduce adversarial learning to further enhance the high-resource objects transfer knowledge to related low-zero-resource objects. The overview of adversarial procedure in Fig. 5. Wherein, the feature extractor F, which encode the instance semantics into a vector to learn common features adaptable from source to target object. The adversarial discriminator D is trained to distinguish the source object from the target object, and adaptation discriminator D_a is trained to identify the unrelated source objects.

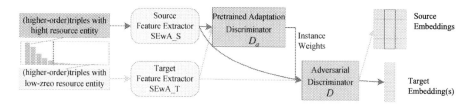

Fig. 5. The adversarial procedure for SEwA.

The feature extractor F adopt the method of Fig. 3. The hight and low-zero resource objects adopt unshared feature extractors [22].

The essence of adversarial procedure is to minimize the loss of the label predictor of the source data while maximize the loss of the adaptation discriminator, as shown in Eq. (17).

$$\min_{\substack{SEwA\_S \\ SEwA\_T}} \max_{D} \mathcal{L}_{st} = \sum p_{s(x)} log D\left(SEwA\_S\left(x\right)\right) + \sum p_{t(x)}\left(1 - \log D\left(SEwA\_T\left(x\right)\right)\right) \tag{17}$$

As mentioned before, only its related high-resource entities and relations can really help low-resource objects. In other words, the unrelated high-resource objects are nontransferable. So the source object weight function should be inversely related to the adaptation discriminator D_a [30], which learning by the output of the optimal parameters of D_a. For example Eq. (18).

$$w = \frac{D_a\left(SEwA\_T\right)}{D_a\left(SEwA\_T\right) + D_a\left(SEwA\_S\right)} = \tag{18}$$

Where the optimal parameters of D_a come from Eq. (19) and Eq. (20).

$$\min_{SEwA\_T} \max_{D_a} L_a = \sum p_{s(x)} log D_a\left(SEwA\_S\left(x\right)\right) + \sum p_{t(x)}\left(1 - \log D_a\left(SEwA\_T\left(x\right)\right)\right) \tag{19}$$

$$D_a\left(\cdot\right) = p\left(y = \varepsilon \left|x\right.\right) \tag{20}$$

Where x is the input from the source and the target object.

After adding the relevance weights to the source objects for the adversarial discriminator D_w, the loss of D_w is is shouwn in Eq. (21) and Eq. (22).

$$\min_{SEwA\_T} \max_{D_w} L_w = \sum p_{s(x)} w \log D_w \left(SEwA\_S \left(x \right) \right) + \sum p_{t(x)} \left(1 - \log D_w \left(SEwA\_T \left(x \right) \right) \right) \tag{21}$$

$$D_w \left(x \right) = \frac{w p_{s(x)}}{w p_{s(x)} + p_{t(x)}} \tag{22}$$

4   Performance Evaluation

To demonstrate the effectiveness of SEwA, we will conduct experiments on 2 tasks: low resource link prediction and open-world KG completion. We use 3 evaluation metrics: (1) *MRR* mean reciprocal rank of correct entities or relations, (2) *Hit@k* the proportion of valid entities or relations ranked in top-k predictions and (3) *MR* mean rank of correct entities or relations. A higher *MRR* or *Hit@k* or low *MR* is better. Low resource datasets WN11, FB13, Wiki, NELL [6,19, 27] and open-world datasets DB50k,DB500k [19], FB12K and FB20K [26] are used as our datascts. FB12K is constructed according to [19] from FB15K. The statistics of these data sets are listed in Table 1. Here, #Rel and #Ent represents the number of relations and entities, respectively. #Train, #Valid and #Test represents training, validation, and test datasets, respectively.

Table 1. Datasets used in our experiments

Dataset	Low resource datasets				Open-world datasets			
	WN11	FB13	Wiki	NELL	DB50k	DB500k	FB12K	FB20K
#Rel	11	13	822	358	654	654	1,192	1,345
#Ent	38,551	75,043	4,838,244	68,545	49,900	517,475	13,456	19,923
#Train	110,361	316,232	4,859,240	360,239	32,388	3,102,677	360,239	494,328
#Valid	2,602	5,908	5,000,000	10,000	399	10,000	40,000	50,000
#Test	10,462	23,733	5,000,000	10,000	10,969	1,155,937	40,822	50,000

We also use a filtered setting, removing all corrupted triples appearing in training, validation, or test dataset before getting the rank of each testing triple[4]. Unlike rank of TransE, SEwA replace $x_{i,i\in\{1,n\}}$ with [MASK] and feed the sequence into itself when given a test triple $<h, r, t>$. x_1 and x_n are sequence representations of h and t, respectively. Then obtain the predicted distribution of s over all entities. We sort the distribution probabilities in descending order and get the rank of $x_{i,i\in\{1,n\}}$. We use the following configuration for our SEwA:

---

[4] Note: If a corrupted triple exists in the knowledge graph, it is also correct. It may be ranked above the test triple, but this should not be counted as an error because both triples are true.

the number of Transformer blocks $L = 6$, number of self-attention heads $A = 4$, hidden size $H = 256$. The maximum input sequence length is $K = 3$ in triple and $K = 4$ in path. All layers have dropout with the rate $\rho \in \{0.1, 0.5\}$. The label value of target entity $\varepsilon \in (0, 1]$, $steps = 0.05$. Learning rate of Adam [10] $\eta \in \{3e^{-4}, 5e^{-4}\}$. Batch size $B \in \{512, 4096\}$ for at most 1000 epochs.

4.1 Low Resource Link Prediction

This section conduct 2 experiments. One is re-evaluating the link prediction effects of various models on the low-resource datasets. Another is analyze the impact of few-shot size K. The results are shown in Table 2 and Fig. 6.

Table 2. The results of link prediction on low resource datasets (%).

Model	WN11		FB13		Wiki		NELL	
	MRR	Hit@10	MRR	Hit@10	MRR	Hit@10	MRR	Hit@10
TransE	9.7	16.9	25.3	33.4	30.5	46.4	13.7	22.3
ComplEx	10.8	17.3	25.7	33.6	31.6	43,2	11.2	15.3
ConvE	12.0	18.7	26.3	35.4	33.2	48.7	17.3	29.2
RotatE	10.1	17.1	25.5	33.5	30.7	46.6	13.9	22.6
TuckER	12.3	19.0	26.6	35.5	31.4	47.3	14.6	26.7
ConMask	12.4	18.8	26.5	35.7	36.3	49.2	35.7	43.5
KG-BERT	13.6	20.1	27.9	37.0	42.0	51.5	45.2	59.8
TransN	11.6	17.9	27.0	35.4	38.7	51.9	37.3	57.6
GMatching	12.9	18.8	27.3	36.3	38.7	51.9	37.3	57.6
LAN	13.2	19.0	27.6	36.9	40.1	52.3	39.6	58.7
PTransE	15.3	24.9	28.3	36.9	43.9	52.6	46.8	61.2
PaSKoGE	15.6	25.6	29.1	37.4	45.1	54.0	47.0	61.4
SEwA_T	14.6	20.6	28.8	38.1	43.6	52.7	46.5	60.7
SEwA_P	20.8	30.0	31.7	41.3	47.1	56.3	49.3	64.1

As we can see Table 2, SEwA_P performs extremely well on 4 datasets, which illustrates the correctness of our motivation and the effectiveness of our methods. Path-aided methods PTransE and PaSKoGE and our SEwA_P are clearly very dominant. This shows that the path has a greater effect on low resource KGC than other features. In path-aided models, our SEwA_P performs best, and in purely triple-oriented models, our SEwA_T is optimal. This shows that the word sequence is more suitable for encoding KGs inherently link structure and capturing words syntax and semantics characteristics. Looking closely, we also find that almost all models have the highest accuracy on the NELL dataset, followed by Wiki and FB13, and the worst on WN11. This is mainly due to the different degree of few-shot of the datasets.

We further verify that the few-shot size K has an impact on KG completion. To do so, we consider a link prediction task on the 4 datasets. Figure 6 reports

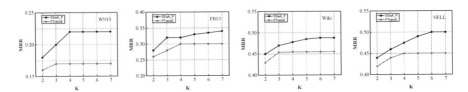

Fig. 6. Impact of few-shot size K.

the performances of our model and path-aided models in 4 test datasets with different settings of K. We reach the following conclusions according to the Fig. 6: (1) With the increment of K, performances of all models increase. It indicates that larger path set may produce better path embedding for the low resource objects. (2) Our model consistently outperforms other path-aided models in different K, demonstrating the stability of the proposed model for low resource KG completion.

4.2 Open-World KG Completion

Zero-shot learning [19] and open-world knowledge graph completion [19] are essentially the same. The tasks of open-world knowledge graph completion focus on the situation when at least one entity in a test triple is out of KGs [19]. Datasets DB50k, DB500k [19], FB12K and FB20K [26] with new entities are used here. The results as shown in Fig. 7.

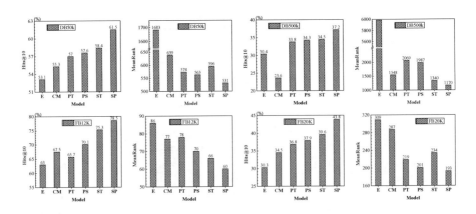

Fig. 7. The results of link prediction on Open-world datasets.

In Fig. 7, the model names are abbreviated. E.g. E, CM, PT, PS, ST and SP denotes TransE, ConMask, PTransE, PaSKoGE, SEwA_T and SEwA_P. From the results shown in Fig. 7, we observe the following conclusions: (1) All models have better predictions on dataset DB50k than on DB500k. This is probably because the random sampling procedure used to create DB500k generates a sparse graph. All knowledge graph completion models, which rely exclusively on structural features, have a more difficult time with sub-sampled KGs. (2) our SEwA_T and SEwA_P are significantly better than others, which shows that the sequence embedding with multi-head attention and adversarial learning are very beneficial to zero resource knowledge graph completion. (3) All models have better predictions on dataset FB12K and FB20K than on DB50k and DB500k. Mainly because of the following five aspects: a) They have more relation types than DB50k and DB500k. There are also significant differences in the number of attribute and relational entities. b) Link structures among entities are different. c) Different relation types cover different number of triples, such as 1-to-1, 1-to-n,n-to-1, n-to-n. d) Semantic differences between head and tail entities linked by a relation. e) The occurrences of entities and relations are different.

5   Conclusions

The paper proposes a novel completion or embedding approach for low and zero resource knowledge graph. It is a **S**equence **E**mbedding **w**ith **A**dversarial learning approach (SEwA). The key differences between SEwA and previous approaches are as follows: (1) It is a pre-training model with multi head self-attention. (2) It introduces path and models triples and higher-order triples as word sequences. (3) It turns knowledge graph completion into a sequence prediction problem. (4) The path acquisition adopts a non-parametric learning based on the one-class classification of the relation trees. (5) It optimizes zero and low resources issues via adversarial learning. Finally, SEwA is evaluated by low resource datasets and open world datasets, which shows our model can both deal with zero resource and low resource problem well. We also found some phenomena in the experiments. For example, although neighbors are not as efficient as path for low-resource knowledge graph completion, they can still play a role. Therefore, we will investigate the effectiveness of link structure beyond edges and paths in the future.

References

1. Balazevic, I., Allen, C., Hospedales, T.M.: Tucker: tensor factorization for knowledge graph completion. In: Proceedings of the 2019 Conference on Empirical Methods in Natural Language Processing and the 9th International Joint Conference on Natural Language Processing, EMNLP-IJCNLP 2019, Hong Kong, China, 3–7 November 2019, pp. 5184–5193 (2019)
2. Blockeel, H., Raedt, L.D.: Top-down induction of first-order logical decision trees. Artif. Intell. **101**(1–2), 285–297 (1998)
3. Bordes, A., Usunier, N., García-Durán, A., Weston, J., Yakhnenko, O.: Translating embeddings for modeling multi-relational data. In: Advances in Neural Information Processing Systems 26: 27th Annual Conference on Neural Information Processing Systems 2013. Proceedings of a meeting held 5–8 December 2013, Lake Tahoe, Nevada, USA, pp. 2787–2795 (2013)

4. Dettmers, T., Minervini, P., Stenetorp, P., Riedel, S.: Convolutional 2D knowledge graph embeddings. In: Proceedings of the Thirty-Second AAAI Conference on Artificial Intelligence, New Orleans, Louisiana, USA (2018)
5. Devlin, J., Chang, M., Lee, K., Toutanova, K.: BERT: pre-training of deep bidirectional transformers for language understanding. In: Burstein, J., Doran, C., Solorio, T. (eds.) Proceedings of the 2019 Conference of the North American Chapter of the Association for Computational Linguistics: Human Language Technologies, NAACL-HLT 2019, Minneapolis, MN, USA, 2–7 June 2019, Volume 1 (Long and Short Papers), pp. 4171–4186. Association for Computational Linguistics (2019)
6. Guu, K., Miller, J., Liang, P.: Traversing knowledge graphs in vector space. In: Proceedings of the 2015 Conference on Empirical Methods in Natural Language Processing, EMNLP 2015, Lisbon, Portugal, pp. 318–327 (2015)
7. Ji, G., He, S., Xu, L., Liu, K., Zhao, J.: Knowledge graph embedding via dynamic mapping matrix. In: Proceedings of the 53rd Annual Meeting of the Association for Computational Linguistics and the 7th International Joint Conference on Natural Language Processing of the Asian Federation of Natural Language Processing, ACL 2015, 26–31 July 2015, Beijing, China, Volume 1: Long Papers, pp. 687–696 (2015)
8. Ji, G., Liu, K., He, S., Zhao, J.: Knowledge graph completion with adaptive sparse transfer matrix. In: Proceedings of the Thirtieth AAAI Conference on Artificial Intelligence, Phoenix, Arizona, USA, pp. 985–991 (2016)
9. Khot, T., Natarajan, S., Shavlik, J.W.: Relational one-class classification: A nonparametric approach. In: Brodley, C.E., Stone, P. (eds.) Proceedings of the Twenty-Eighth AAAI Conference on Artificial Intelligence, 27–31 July 2014, Québec City, Québec, Canada, pp. 2453–2459. AAAI Press (2014)
10. Kingma, D.P., Ba, J.: Adam: a method for stochastic optimization. In: 3rd International Conference on Learning Representations, ICLR 2015, San Diego, CA, USA, 7–9 May 2015, Conference Track Proceedings (2015)
11. Lin, Y., Liu, Z., Luan, H., Sun, M., Rao, S., Liu, S.: Modeling relation paths for representation learning of knowledge bases. In: Proceedings of the 2015 Conference on Empirical Methods in Natural Language Processing, EMNLP 2015, Lisbon, Portugal, pp. 705–714 (2015)
12. Lin, Y., Liu, Z., Sun, M., Liu, Y., Zhu, X.: Learning entity and relation embeddings for knowledge graph completion. In: Proceedings of the Twenty-Ninth AAAI Conference on Artificial Intelligence, 25–30 January 2015, Austin, TX, USA, pp. 2181–2187 (2015)
13. Mehta, S., Rangwala, H., Ramakrishnan, N.: Low rank factorization for compact multi-head self-attention. CoRR abs/1912.00835 (2019)
14. Nguyen, D.Q., Nguyen, T.D., Nguyen, D.Q., Phung, D.Q.: A novel embedding model for knowledge base completion based on convolutional neural network. In: Proceedings of the 2018 Conference of the North American Chapter of the Association for Computational Linguistics: Human Language Technologies, NAACL-HLT, New Orleans, Louisiana, USA, 1–6 June 2018, Volume 2 (Short Papers), pp. 327–333 (2018)
15. Nickel, M., Rosasco, L., Poggio, T.A.: Holographic embeddings of knowledge graphs. In: Proceedings of the Thirtieth AAAI Conference on Artificial Intelligence, 12–17 February, 2016, Phoenix, Arizona, USA, pp. 1955–1961 (2016)
16. Qiu, X., Sun, T., Xu, Y., Shao, Y., Dai, N., Huang, X.: Pre-trained models for natural language processing: A survey. CoRR abs/2003.08271 (2020)

17. Reiter, R.: On closed world data bases. Logic and Data Bases. In: 1977 Symposium on Logic and Data Bases, Centre d'études et de recherches de Toulouse, France, pp. 55–76 (1977)
18. Shah, H., Villmow, J., Ulges, A., Schwanecke, U., Shafait, F.: An open-world extension to knowledge graph completion models. In: The Thirty-Third AAAI Conference on Artificial Intelligence, AAAI 2019, The Thirty-First Innovative Applications of Artificial Intelligence Conference, IAAI 2019, The Ninth AAAI Symposium on Educational Advances in Artificial Intelligence, EAAI 2019, Honolulu, Hawaii, USA, 27 January – 1 February 2019, pp. 3044–3051 (2019)
19. Shi, B., Weninger, T.: Open-world knowledge graph completion. In: McIlraith, S.A., Weinberger, K.Q. (eds.) Proceedings of the Thirty-Second AAAI Conference on Artificial Intelligence, (AAAI-18), the 30th innovative Applications of Artificial Intelligence (IAAI-18), and the 8th AAAI Symposium on Educational Advances in Artificial Intelligence (EAAI-18), New Orleans, Louisiana, USA, 2–7 February 2018, pp. 1957–1964. AAAI Press (2018)
20. Sun, Z., Deng, Z., Nie, J., Tang, J.: Rotate: knowledge graph embedding by relational rotation in complex space. In: 7th International Conference on Learning Representations, ICLR 2019, New Orleans, LA, USA (2019)
21. Trouillon, T., Welbl, J., Riedel, S., Gaussier, É., Bouchard, G.: Complex embeddings for simple link prediction. In: Proceedings of the 33nd International Conference on Machine Learning, ICML 2016, New York City, NY, USA, 19–24 June 2016, pp. 2071–2080 (2016)
22. Tzeng, E., Hoffman, J., Saenko, K., Darrell, T.: Adversarial discriminative domain adaptation. In: 2017 IEEE Conference on Computer Vision and Pattern Recognition, CVPR 2017, Honolulu, HI, USA, 21–26 July 2017, pp. 2962–2971. IEEE Computer Society (2017)
23. Vaswani, A., Shazeer, N., Parmar, N., Uszkoreit, J., Jones, L., Gomez, A.N., Kaiser, L., Polosukhin, I.: Attention is all you need. In: Advances in Neural Information Processing Systems 30: Annual Conference on Neural Information Processing Systems 2017, pp. 5998–6008 (2017)
24. Wang, P., Han, J., Li, C., Pan, R.: Logic attention based neighborhood aggregation for inductive knowledge graph embedding. In: The Thirty-Third AAAI Conference on Artificial Intelligence, AAAI 2019, The Thirty-First Innovative Applications of Artificial Intelligence Conference, IAAI 2019, The Ninth AAAI Symposium on Educational Advances in Artificial Intelligence, EAAI 2019, Honolulu, Hawaii, USA, 27 January – 1 February 2019, pp. 7152–7159 (2019)
25. Wang, Z., Zhang, J., Feng, J., Chen, Z.: Knowledge graph embedding by translating on hyperplanes. In: Proceedings of the Twenty-Eighth AAAI Conference on Artificial Intelligence, Québec City, Québec, Canada, pp. 1112–1119 (2014)
26. Xie, R., Liu, Z., Jia, J., Luan, H., Sun, M.: Representation learning of knowledge graphs with entity descriptions. In: Proceedings of the Thirtieth AAAI Conference on Artificial Intelligence, Phoenix, USA, pp. 2659–2665 (2016)
27. Xiong, W., Yu, M., Chang, S., Guo, X., Wang, W.Y.: One-shot relational learning for knowledge graphs. In: Proceedings of the 2018 Conference on Empirical Methods in Natural Language Processing, Brussels, Belgium, 31 October – 4 November 2018, pp. 1980–1990 (2018)
28. Yang, B., Yih, W., He, X., Gao, J., Deng, L.: Embedding entities and relations for learning and inference in knowledge bases. CoRR abs/1412.6575 (2014)
29. Yao, L., Mao, C., Luo, Y.: KG-BERT: BERT for knowledge graph completion. CoRR abs/1909.03193 (2019)

30. Zhang, J., Ding, Z., Li, W., Ogunbona, P.: Importance weighted adversarial nets for partial domain adaptation. In: 2018 IEEE Conference on Computer Vision and Pattern Recognition, CVPR 2018, Salt Lake City, UT, USA, 18–22 June 2018, pp. 8156–8164. IEEE Computer Society (2018)
31. Zhang, N., Deng, S., Sun, Z., Chen, J., Zhang, W., Chen, H.: Relation adversarial network for low resource knowledge graph completion. In: WWW 2020: The Web Conference 2020, Taipei, Taiwan, pp. 1–12 (2020)

HMNet: Hybrid Matching Network for Few-Shot Link Prediction

Shan Xiao[1], Lei Duan[1(✉)], Guicai Xie[1], Renhao Li[1], Zihao Chen[1],
Geng Deng[1], and Jyrki Nummenmaa[2]

[1] School of Computer Science, Sichuan University, Chengdu, China
{shanxiao,guicaixie,lirenhao,chenzihao}@stu.scu.edu.cn,
leiduan@scu.edu.cn
[2] Tampere University, Tampere, Finland
jyrki.nummenmaa@tuni.fi

Abstract. Knowledge graphs (KGs) are widely used in many real-world applications, such as information retrieval, question answering system, and personal recommendation. However, most KGs are suffering from the incompleteness problem. To deal with the task of link prediction, previous knowledge graph embedding methods require numerous reference instances for each relation. It is worth noting that most relations in KGs have only a few reference instances available. Existing works for few-shot link prediction evaluate the authenticity of triplets from a single relation perspective. In this paper, we propose Hybrid Matching Network (HMNet) for few-shot link prediction, evaluating triplets from entity and relation two perspectives. At the entity-aware matching network, HMNet uses attentive inductive embedding layer to aggregate entity features and relation-aware topology, and then provides entity-aware score to implement first perspective evaluation. At the relation-aware matching network, HMNet integrates feature attention mechanism to implement relation perspective evaluation. Experiments on two public datasets indicate that HMNet achieves promising performance in few-shot link prediction.

Keywords: Few-shot link prediction · Hybrid matching network · Feature attention mechanism

1 Introduction

Knowledge graphs (KGs), collection of triplets (e.g., <*head entity, relation, tail entity*>), have been widely used in a range of applications, such as question answering [3], recommender system [29], and information retrieval [5]. A typical large-scale KG, such as Freebase [1] or YAGO [18], contains billions of triplets. However, they are suffering from the incompleteness problem [26]. For instance, 75% of person entities have no nationality information in Freebase [7]. As a

This work was supported in part by the National Natural Science Foundation of China (61972268), the Sichuan Science and Technology Program (2020YFG0034), and the Academy of Finland (327352).

ⓒ Springer Nature Switzerland AG 2021
C. S. Jensen et al. (Eds.): DASFAA 2021, LNCS 12681, pp. 307–322, 2021.
https://doi.org/10.1007/978-3-030-73194-6_21

308   S. Xiao et al.

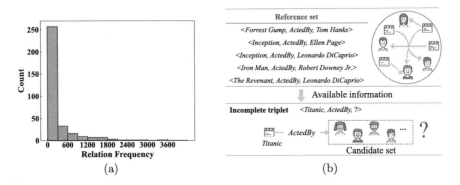

(a)   (b)

Fig. 1. (a) The histogram of relation frequencies in NELL-One dataset; and (b) Illustration of an example of few-shot link prediction (Each edge denotes a reference instance).

result, it is hard to accurately answer questions like "How many users are from the same country as *Leonardo DiCaprio?*".

Clearly, it is time-consuming and labor intensive to deal with all incomplete triplets manually. Thus automatically completing a knowledge graph, which is also called *link prediction*, has become an important research task. There are some studies [2,6,15,16,20,27] that have been proposed to predict missing values in incomplete triplets based on existing knowledge. The main idea of these methods is consummating incomplete triplets via low-dimensional representations of entities and relations. However, the precondition of these methods solving the incompleteness problem is that each relation contains numerous reference instances.

It is worth noting that the relation frequency in many KGs always shows a long-tail distribution, as shown in Fig. 1(a). That is, a large portion of relations have only a few reference instances [26]. The automatic completion of knowledge graph under long-tail distribution is called few-shot link prediction task. In this task, only a few reference instances are available for each relation. To better understand the task, an example is shown in Fig. 1(b).

Example 1. In an existing KG, we need to perform link prediction on relation "*ActedBy*", that is, predicting tail entity from the candidate set given head entity "*Titanic*" and relation "*ActedBy*". Different from other relations, "*ActedBy*" has only 5 reference instances. Therefore, there is a little information available for it. Few-shot link prediction task focuses on these less informativeness relations.

There are some few-shot learning studies [4,26,28] for the above task. Specifically, these methods first construct query triplets by splicing all candidate tail entities with the given head entity, and then learn embeddings of query triplets and representation of the relation of reference set. Finally, they match each query triplet with the reference set to get a score. The goal is to make true triplets rank high. Obviously, for each query triplet, these methods evaluate its authenticity from a single relation perspective. In addition, most of existing works calculate the score for each triplet by applying dot product under the assumption that all features of the relation contribute equally. Although these methods achieve encouraging improvements, the performance remains unsatisfactory.

In this paper, we propose a novel model <u>H</u>ybrid <u>M</u>atching <u>Net</u>work (HMNet) for few-shot link prediction. It consists of an entity-aware matching network and a relation-aware matching network. HMNet can evaluate the authenticity of triplets from two perspectives:

- *Entity perspective*: The entity-aware matching network obtains the entity-aware scores between different candidates and reference instances as the first perspective evaluation.
- *Relation perspective*: The relation-aware matching network obtains the relation-aware scores as the second perspective. It simultaneously weights different features of relation with unequal contributions when calculating the score.

The final prediction result of each triplet is acquired by combining these two matching scores.

The contributions of this work are summarized as follows:

- designing a novel model HMNet, which employs hybrid matching and integrates attention mechanism for few-shot link prediction.
- pointing out the importance of entity-aware matching, and providing an extra perspective evaluation.
- evaluating HMNet model on two public datasets. Empirical results prove the effectiveness of our proposed model HMNet over many competitive baselines.

The rest of the paper is organized as follows. We review related work in Sect. 2, and formulate the problem of few-shot link prediction in Sect. 3. In Sect. 4, we discuss the critical techniques of the proposed model HMNet. We report a systematic empirical evaluation in Sect. 5, and conclude the paper in Sect. 6.

2 Related Work

Our work is related to the existing research on knowledge graph embedding and few-shot learning. We introduce the related work briefly below.

2.1 Knowledge Graph Embedding

To consummate incomplete triplets in KGs, it is vital to obtain embeddings of entities and relations in the continuous low-dimensional space. Existing knowledge graph embedding models can be divided into two main categories: distance based models and bilinear based models.

Aiming to translate distance between entity pairs, Bordes *et al.* [2] first proposed a translational distance based method TransE. It can obtain low-dimensional embeddings by optimizing the distance function between triplets of the relational semantic. After that, in order to break through the limitation of TransE in dealing with complex relations, several models have been proposed, such as TransH (Wang et al. [24]) and TransR (Lin et al. [13]). Focused on tensor decomposition, Nickel et al. [15] firstly designed a bilinear model RESCAL, which can obtain relation embeddings by modeling the potential structure of KGs. Later, DistMult proposed by Yang et al. [27] simplifies RESCAL by limiting the relational matrix to the diagonal matrix. ComplEx introduced by Trouillon et al. [20] extends DistMult into the complex space to better model reversible relations in KGs.

The performance of above methods strongly relies on numerous reference instances. In practical applications, these methods fail to achieve their expected performance, due to the relation frequency in real datasets often has a long-tail distribution.

2.2 Few-Shot Learning

Few-shot learning enables models to achieve impressive results with insufficient data. Existing approaches include learning a metric space over input features [12,17,21,26], such that similar instances are close together while dissimilar can be more easily differentiated. Recently, meta-learning is proposed to solve few-shot learning problem. Specially, the meta-learner gradually learns generic information (meta-knowledge) across tasks, and task-learner generalizes to the new task based on meta-knowledge and specific information of the new task [8,14,23]. Although few-shot learning has developed fast in recent years, it mainly focus on computer vision applications and text classification.

To the best of our knowledge, the work proposed by Xiong et al. [26] is the first research on few-shot link prediction. It's a metric based model called GMatching, which includes two components: neighbor encoder and matching processor. The neighbor encoder uses entities' one-hop neighbors to obtain their embeddings. And then each relation representation is obtained by concatenating the embeddings of the head entity and tail entity. The matching processor matches each query instance with the reference set. Following the work of GMatching [26], Zhang et al. [28] proposed a relation-aware heterogeneous neighbor encoder based on the attention mechanism to learn entity embedding, and used recurrent auto-encoder to aggregate information from reference instances. Chen et al. [4] employed the relation-specific meta information transferring from the reference set to query set and proposed the MetaR model.

Previous methods solve the few-shot link prediction task only considering the relation perspective evaluation. This work is attempting to design a new framework that can evaluate the authenticity of triplets from two perspectives by leveraging valuable semantic information provided by the reference set.

3 Problem Definition

We start with some preliminaries. Let $\mathcal{E}$ and $\mathcal{R}$ be the sets of entities and relations, respectively. A knowledge graph is viewed as a graph $\mathcal{G} = \{(h, r, t)\} \subseteq$

$\mathcal{E} \times \mathcal{R} \times \mathcal{E}$, where $h \in \mathcal{E}$ and $t \in \mathcal{E}$ represent the head entity and tail entity, respectively, and $r \in \mathcal{R}$ denotes a specific relation connecting h and t. The goal of link prediction is to predict the missing values in incomplete triplets when two elements are given. In this study, we focus on predicting the tail entity given the head entity and query relation.

Under the few-shot learning setting, the model can be optimized on the set of training tasks $\mathcal{T}_{train} = \{\mathcal{T}_i\}_{i=1}^{M}$ and its generalization can be evaluated on the set of test tasks $\mathcal{T}_{test} = \{\mathcal{T}_j\}_{j=1}^{N}$. Each task $\mathcal{T}_i = \{\mathcal{D}^{ref}, \mathcal{D}^{query}\}$ corresponds to a few-shot learning task with reference set $\mathcal{D}^{ref}$ and query set $\mathcal{D}^{query}$. Each task $\mathcal{T}_j \in \mathcal{T}_{test}$ is similar to $\mathcal{T}_i$. According to the reference instances, the model needs to make prediction for instances in the query set. It should be noted that all tasks in testing are invisible in training, that is, $\mathcal{T}_{train} \cap \mathcal{T}_{test} = \varnothing$.

Definition 1 (Few-shot link prediction). *Few-shot link prediction is defined as a task to predict the true tail entity t_j of the missing triplet $(h_j, r, ?)$, given the reference set $\mathcal{D}_r^{ref} = \{(h_i, t_i) \mid (h_i, r, t_i) \in \mathcal{G}\}$ of relation r. $K = |\mathcal{D}_r^{ref}|$ represents the number of triplets in reference set, which is a small number. The set of all instances to be predicted of relation r is the query set $\mathcal{D}_r^{query} = \{(h_j, c_j) | c_j \in C_{h_j, r}\}$, where $C_{h_j, r}$ is candidate tail entities set for a given head entity h_j and relation r ($C_{h_j, r}$ including the true tail entity t_j).*

In the few-shot link prediction task, $\mathcal{R}_1$ and $\mathcal{R}_2$ are sets of relations involved in training and testing, respectively, and $\mathcal{R}_1 \cap \mathcal{R}_2 = \varnothing$. Each task corresponds to a relation $r \in \mathcal{R}_1 \cup \mathcal{R}_2$. Following the standard problem definition of work [26], we assume that the method to solve the task can access a background graph $\mathcal{G}'$, where $\mathcal{G}' = \{(h, r, t) | (h, r, t) \in \mathcal{G} \wedge r \in \mathcal{R} \setminus (\mathcal{R}_1 \cup \mathcal{R}_2)\}$.

4   The Design of HMNet

In this section, we present the details of HMNet. Figure 2 shows the framework of HMNet, which includes two components: entity-aware matching network and relation-aware matching network. Different from previous studies that focus on single relation perspective evaluation, HMNet can evaluate the authenticity of triplets from two perspectives. In Sect. 4.1, we describe the mechanism of entity-aware matching network. Relation-aware matching network for evaluation is described in Sect. 4.2.

4.1   Entity-Aware Matching Network

The objective of entity-aware matching network is to evaluate triplets from entity perspective. Specifically, it assigns high scores for true tail entities of triplets with tail entities' information in the reference set. This component consists of following two modules: attentive inductive embedding layer and entity-aware score.

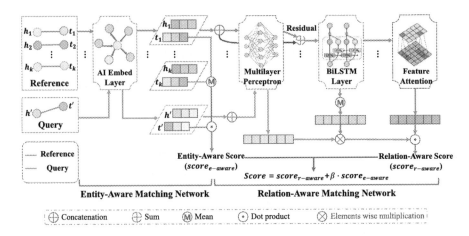

Fig. 2. Illustration of the proposed HMNet model.

Attentive Inductive Embedding Layer (AI Embed Layer). Low dimensional representations of nodes in the network have been proved useful in a variety of graph analysis tasks [10]. Existing works show that it is beneficial to use the relation-aware topology of an entity for link prediction task [26,30]. In addition, attention mechanism is widely used in recent deep learning studies [25,28]. Different from existing few-shot link prediction works [26,28] only modeling the relation-aware topology explicitly, HMNet employs AI Embed Layer to obtain entity embedding. Since it simultaneously captures the relation-aware topology and entity features, AI Embed Layer retains the advantages of previous methods and fully aggregates the information provided by reference set.

Specifically, for any entity e, the set of link information with head entity e in $\mathcal{G}'$ denotes as $\mathcal{I}_e = \{(r,t)|(e,r,t) \in \mathcal{G}'\}$. Hence, entity e is assigned relation-aware topology embedding as follows,

$$\mathcal{F}_e = \sigma(\sum\nolimits_{(r,t)\in\mathcal{I}_e} a_{(r,t)}(\mathbf{W_1}[\boldsymbol{v}_r \oplus \boldsymbol{v}_t]))$$
$$a_{(r,t)} = \frac{exp(\mathbf{P}(\mathbf{W_1}[\boldsymbol{v}_r \oplus \boldsymbol{v}_t]))}{\sum_{(r',t')\in\mathcal{I}_e} exp(\mathbf{P}(\mathbf{W_1}[\boldsymbol{v}_{r'} \oplus \boldsymbol{v}_{t'}]))} \quad (1)$$

where $\oplus$ represents concatenation operation, $\boldsymbol{v}_r \in \mathbb{R}^d$ and $\boldsymbol{v}_t \in \mathbb{R}^d$ are pretrained embeddings of the relation r and tail entity t, respectively. d is the embedding size. σ represents the $Tanh$ activation function, and $a_{(r,t)}$ indicates the weight of link information (r,t) when representing the entity e. $\mathbf{P} \in \mathbb{R}^{1\times d}$ and $\mathbf{W_1} \in \mathbb{R}^{d\times 2d}$ are trainable weight matrices.

Aggregating features of entity e with its relation-aware topology representation has been widely used in many tasks and achieves good performance [22,30]. In order to make full use of the information of reference set, AI Embed Layer further combines them to get the entity embedding of e: $\omega_e = \mathbf{W_2}\boldsymbol{v}_e + \mathcal{F}_e$, where v_e is the pre-trained embedding of e and $\mathbf{W_2} \in \mathbb{R}^{d \times d}$ is a trainable weight matrix.

Entity-Aware Score. The problem we tackle is that given a *head entity* and a *relation*, we need to predict the *tail entity*. According to the information of tail entities in the reference set, for each query instance (h', t'), we can calculate the entity-aware score to implement the first perspective evaluation. First, by applying the AI Embed Layer to each tail entity t from $\mathcal{D}_r^{ref}$ and (h', t'), HMNet gets the representation $\boldsymbol{\omega}_t$ of t. Then, HMNet summarizes output features of tail entities in the reference set $\mathcal{D}_r^{ref}$ as follows,

$$\mathcal{E}_{ref} = \frac{1}{K} \sum_{i=1}^{K} \boldsymbol{\omega}_{t_i} \tag{2}$$

where $t_i \in \left\{ t \mid (h, t) \in \mathcal{D}_r^{ref} \right\}$. Finally, HMNet calculates the entity-aware score for (h', t'). Without loss of generality, HMNet employs the following way to calculate entity-aware score,

$$score_{e-aware} = \mathcal{E}_{ref} \odot \boldsymbol{\omega}_{t'} \tag{3}$$

where $\odot$ represents dot product.

4.2 Relation-Aware Matching Network

To implement the relation perspective evaluation, we design a relation-aware matching network. In this section, we start by describing how to get the embedding of corresponding relation based on reference set. And then we discuss how to select more discriminative features to achieve more appropriate relation perspective evaluation.

Relation Encoder. HMNet assumes that each query instance expresses a special relation, and then measures whether this relation is similar to the relation expressed by reference set. HMNet employs the multilayer perceptron to encoder entity pairs. It can obtain the relation embedding represented by entity pair (h, t) as follows,

$$e_{r \leftarrow (h,t)} = \mathbf{W}_r(\mathbf{W}[\boldsymbol{\omega}_h \oplus \boldsymbol{\omega}_t]) + [\boldsymbol{\omega}_h \oplus \boldsymbol{\omega}_t] \tag{4}$$

where $\boldsymbol{\omega}_h$ and $\boldsymbol{\omega}_t$ are embeddings of the head entity and tail entity, respectively, obtained by applying AI Embed Layer to h and t. $\mathbf{W}_r \in \mathbb{R}^{2d \times 4d}$ and $\mathbf{W} \in \mathbb{R}^{4d \times 2d}$ are trainable weight matrices.

If the relation representation of each instance in $\mathcal{D}_r^{ref}$ is far away from each other, the resulting prototype vector of relation cannot capture common and representative features. Here, we employ a network to perform information propagation between reference instances, so that reference instances are closer in the metric space. LSTM network [11] has achieved good performance in the NLP field based on the long-distance information memory characteristic. But it can

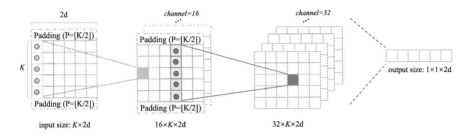

Fig. 3. The architecture of feature attention module.

only achieve unidirectional information propagation. Therefore, HMNet uses the BiLSTM network to implement bidirectional information propagation.

Given relation representations of reference instances, calculated by Eq. 4, BiLSTM performs information propagation on them to obtain the new relation representation $s_r(h_i, t_i)$ for each reference instance (h_i, t_i): $s_r(h_i, t_i) =$ BiLSTM$(e_{r \leftarrow (h_i, t_i)})$.

Relation-Aware Score. The works by Xiong *et al.* [26] and Zhang *et al.* [28] use dot product to calculate the score for each query instance. These methods believe that all features of relation contribute equally. However, when few-shot reference instances are used to represent the corresponding relation, the obtained information is limited and may contain noise information. It is hard to accurately capture all unique features of the relation. Therefore, we should pay more attention to discriminative features. HMNet is required to measure the importance of captured features when calculating score under the few-shot setting.

Since there are only a few reference instances for each relation, it is difficult to extract important features using feature engineering algorithms. Inspired by the work of [9] on text classification, HMNet uses a feature attention module to measure the importance of relation features, which enhances generality applicability of the model.

The feature attention module uses the convolution operation to iteratively update feature weights. The relation representation of each instance in the reference set is combined to form a matrix $s_r^K \in \mathbb{R}^{K \times 2d}$. Figure 3 shows the module framework.

In order to aggregate the information of reference set when measuring the importance of each feature, the size of all convolution kernels of this module is set to $K \times 1$. More specific steps are as follows:

Step 1: HMNet uses 16 convolution kernels to perform convolution operation on s_r^K, and sets stride to 1×1. It pads the bias to participate in the calculation. The output $H_r^K \in \mathbb{R}^{16 \times K \times 2d}$ can be obtained.

Step 2: HMNet uses 32 convolution kernels to perform convolution operation again. Channels are 16, other settings are the same as previous step. The output is $H_r^K \in \mathbb{R}^{32 \times K \times 2d}$.

Step 3: A convolution kernel with 32 channels is used, and stride is set to $K \times 1$ to obtain feature attention weight $O_r^K \in \mathbb{R}^{1 \times 1 \times 2d}$ for relation matching.

Algorithm 1. HMNet

Input: Training task set $\mathcal{T}_{train}$; Background graph $\mathcal{G}'$; The number of training steps N_{step}; Learning rate α; Margin distance γ; Hyperparameter β
Output: θ: Learning parameters of HMNet
1: Load the pre-trained embeddings;
2: Initialize θ;
3: **for** $i = 1 \to N_{step}$ **do**
4:    Shuffle the tasks in $\mathcal{T}_{train}$;
5:    **for** each task $\mathcal{T}_r$ in $\mathcal{T}_{train}$ **do**
6:       Sample reference set $\mathcal{D}_r^{ref}$ and positive query instances set $\mathcal{Q}_r$ from $\mathcal{T}_r$;
7:       Construct negative query instances set $\mathcal{Q}_r^-$ by replacing tail entities of $\mathcal{Q}_r$;
8:       Compute the matching score for each triplet in $\mathcal{Q}_r \cup \mathcal{Q}_r^-$ using Eq.6;
9:       Compute the loss $\mathcal{L}$ using Eq. 7;
10:      $\theta \leftarrow \text{Adam}(\nabla_\theta \mathcal{L}, \theta, \alpha, \beta)$;
11:   **end for**
12: **end for**
13: **return** θ;

After applying the relation encoder module, we obtain the relation representation of each reference instance in a metric space. Taking these representations as input of the feature attention module, HMNet gets feature attention weights. Like the tail entity information aggregation, HMNet uses the average of representations of reference instances to get the prototype representation of relation r: $c_r = \frac{1}{K} \sum_{i=1}^{K} s_r(h_i, t_i)$. Then HMNet reduces the dimensionality of the feature attention weights, so that $O_r^K \in \mathbb{R}^{1 \times 1 \times 2d} \to O_r^K \in \mathbb{R}^{2d}$. For each query instance (h', t'), HMNet uses O_r^K to calculate the final relation-aware score:

$$score_{r-aware} = O_r^K \odot (c_r \otimes e_{r \leftarrow (h',t')}) \tag{5}$$

where $\otimes$ denotes elements wise multiplication, and $e_{r \leftarrow (h',t')}$ represents the relation embedding between h' and t' which is calculated by Eq. 4.

For a query instance (h', t'), the final score is:

$$Score = score_{r-aware} + \beta \cdot score_{e-aware} \tag{6}$$

where β is the hyperparameter indicating the weight of entity perspective evaluation.

4.3 Learning Objective and Algorithm

Given the background graph $\mathcal{G}'$, and reference set $\mathcal{D}_r^{ref}$ of relation r, we aim to select the true tail entity t_j from the candidate set for each h_j. Based on this learning objective, we rewrite the loss function following the definition of [28],

$$\mathcal{L}_\theta = \sum_{r \in \mathcal{R}_1} \sum_{(h_j, t_j) \in \mathcal{Q}_r} \sum_{(h_j, t_j^-) \in \mathcal{Q}_r^-} [\gamma - Score_{(h_j, t_j)} + Score_{(h_j, t_j^-)}]_+ \tag{7}$$

where $\mathcal{Q}_r = \{(h_j, t_j)|(h_j, r, t_j) \in \mathcal{G}\}$ is the set of positive query instances of relation r, $\mathcal{Q}_r^- = \{(h_j, t_j^-)|(h_j, r, t_j^-) \notin \mathcal{G}\}$ is the set of negative query instances of relation r, which is constructed by replacing tail entities of positive instances. $\mathcal{L}_\theta$ is standard hinge loss, and γ is margin distance.

Based on the discussions above, we present the pseudo-code of HMNet in Algorithm 1.

5   Experiments

We evaluate the performance of HMNet on two public datasets. All experiments are conducted on a server with an RTX2080 Ti and 11 GB memory. The model HMNet is implemented by Python 3.6 based on Pytorch 1.5.1.

5.1   Experimental Setup

Datasets: 1) NELL-One[1] consists of 181,109 triplets, 68,545 entities, and 358 relations. 2) Wiki-One, which is a subset of Wikidata[2], consists of 5,829,240 triplets, 4,838,244 entities, and 822 relations. Following the experimental settings of work [26], we select relations with less than 500 but more than 50 triplets as few-shot link prediction tasks. Table 1 shows the statistics of two datasets (#Training/Validation/Test denotes the number of relations for training/validation/testing).

Table 1. Statistics of the datasets.

Dataset	#Entities	#Relations	#Triplets	#Training/Validation/Test
NELL-One	68,545	358	181,109	51/5/11
Wiki-One	4,838,244	822	5,829,240	133/16/34

Baselines: In our experiments, several related methods are selected as baselines.

- **Knowledge graph embedding methods.** Knowledge graph embedding methods map relations and entities into continuous low-dimensional space. **TransE** [2] is a translational distance based method which defines the score function as $f_r(h, t) = -\|h + r - t\|_{1/2}$. **RESCAL** [15] is a bilinear based method. This method represents each relation as a full rank matrix M_r. **DistMult** [27] uses a bilinear score function to compute scores of knowledge triplets. **ComplEx** [20] extends DistMult to the complex space instead of real-valued ones.

---

[1] http://rtw.ml.cmu.edu/rtw/.
[2] https://test.wikidata.org.

– **Few-shot learning methods.** These models use a background graph $\mathcal{G}'$ to get the pre-trained embeddings of entities and learn a representation of relation. Then they adopt different score functions to get the ranking. **GMatching** [26] tackles the problem by enhancing the representation of entity and learning a relation metric space. **MetaR** [4] proposes relation-meta and gradient-meta two kinds of relation-specific meta information to solve this problem. **FSRL** [28] extends GMatching [26], from one-shot link prediction to few-shot link prediction.

Evaluation Metrics: Two metrics Hits@k and MRR are applied to evaluate the performance of the proposed model. Hits@k is the proportion of the correct tail entities in the top-k of all candidate entities. MRR (Mean Reciprocal Rank) is the average of all correct tail entities reciprocal ranking.

Implementation Details: For TransE, ComplEx, and DistMult, the implementation[3] released by Sun *et al.* [19] is adopted in our experiments. For RESCAL, we implement it by ourselves. For the above knowledge graph embedding methods, all the triplets from $\mathcal{G}'$ and training set are utilized for training. In addition, for each relation, K triplets from validation and test sets are chosen for training. In iterative training, only one negative sample is constructed for each true triplet in the batch task by replacing tail entity. Following GMatching [26], the embedding dimension is set to 100 and 50 for NELL-One and Wiki-One datasets, respectively. The maximum number of neighbors is set to 50 and margin distance is set to 5 for two datasets. The pre-trained embedding is set to ComplEx for all models. During the training procedure, HMNet uses Adam with the initial learning rate as 0.0001 to update parameters. The size of the hidden layer in the BiLSTM structure is set to $2d$, where d is the embedding dimension of datasets. The β is set to 0.5 for both datasets. All learning parameters are randomly initialized. For GMatching, we employ max/mean pooling (denoted as MaxP/MeanP) to obtain the prototype vector of the relation in reference set. Following FSRL [28], the maximum score between a query instance and K instances in the reference set is also considered as the final ranking score of this query instance (denoted as Max). For MetaR, we use pre-trained mode to maintain a consistent experimental environment. The results reported in the paper [28] are under the setting where the maximum size of the candidate set is 1000. Here, entities that satisfy type constraints [26] are added to the candidate set, where all candidates are considered in our work. In the absence of specific knowledge to choose otherwise, K is set to 5.

5.2 Results

The performance comparison results on two datasets are presented in Table 2, where the best results are shown in bold. We have following observations:

---

[3] https://github.com/DeepGraphLearning/KnowledgeGraphEmbedding.

Table 2. Link prediction results on two datasets. Results with * are reported in [26].

Model	NELL-One				Wiki-One			
	MRR	Hits@10	Hits@5	Hits@1	MRR	Hits@10	Hits@5	Hits@1
TransE [2]	.131	.220	.182	.081	.144	.204	.172	.108
RESCAL [15]	.033	.055	.038	.019	.060	.112	.081	.029
DistMult [27]	.051	.134	.081	.010	.027	.058	.035	.010
ComplEx* [20]	.200	.325	.269	.133	.033	.066	.046	.015
GMatching (MaxP) [26]	.189	.301	.225	.136	.134	.287	.181	.066
GMatching* (MeanP) [26]	.201	.311	.264	**.143**	.242	.419	.318	.163
GMatching (Max) [26]	.190	.305	.247	.123	.125	.251	.167	.065
MetaR [4]	.164	.320	.252	.083	.220	.347	.287	.158
FSRL [28]	.184	.341	.248	.105	.126	.242	.154	.068
HMNet	**.209**	**.364**	**.296**	.129	**.294**	**.423**	**.353**	**.230**

- HMNet yields the best performance under most evaluation metrics. Taking Hits@10 and MRR as examples, HMNet improves over the strongest baselines w.r.t. Hits@10 by 6.74% in NELL-One and w.r.t. MRR by 21.48% in Wiki-One, respectively. In particular, HMNet yields 41.1% higher performance w.r.t. Hits@1 than GMatching on Wiki-One. This verifies the significance of entity perspective evaluation. Moreover, compared with the fixed weights used in the other three few-shot learning methods, HMNet verifies the effectiveness of the feature attention mechanism.
- It can be seen that graph embedding methods work poorly on relations that have only a few triplets to train. It demonstrates the limitations of previous graph embedding methods for few-shot link prediction.

5.3   Further Analysis

Impact of Few-Shot Size: We conduct experiments to analyze the impact of few-shot size settings with $K \in \{3, 4, 5, 6\}$. The test results on NELL-One of different methods measured using Hits@k and MRR are shown in Fig. 4. HMNet outperforms other baseline methods on most evaluation metrics in different few-shot size settings, indicating its stability on few-shot link prediction. In addition, the performance of most methods does not improve with the few-shot size increasing in this experimental setting. The reason may be that unrepresentative instances are added to the reference set, which are far away from the prototype representation of the relation.

Impact of Embedding Methods: To observe the impact of different embedding methods for relation representation, we compare the performance between our model HMNet and the latest method FSRL [28]. Figure 5 shows the results of HMNet and FSRL on NELL-One dataset. We can see that FSRL obtains the best performance when using ComplEx as the embedding method. Compared with FSRL, HMNet achieves better performance in four different embedding method settings. It further indicates the superior performance of our model in terms of few-shot link prediction in KGs.

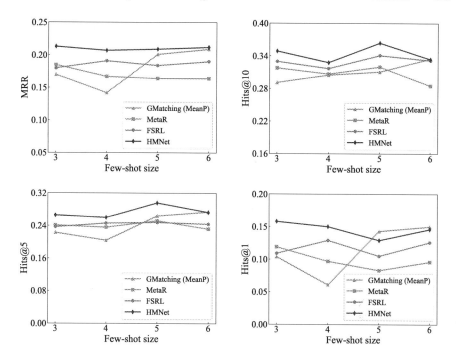

Fig. 4. Impact of few-shot size.

Table 3. Impact of hyperparameter β.

Hyperparameter β	MRR	Hits@10	Hits@5	Hits@1
0.2	.184	.325	.242	.112
0.5	**.209**	**.364**	**.296**	**.129**
1.0	.184	.356	.233	.111

On Parameter Selection for HMNet: We investigate the impact of different entity-aware score weights β on the few-shot link prediction performance. We conduct the experiment with hyperparameter $\beta \in \{0.2, 0.5, 1.0\}$ while other factors are fixed. The results on NELL-One are reported in Table 3 with the best results bold. HMNet reaches the best performance with $\beta = 0.5$. Moreover, the performance of our model first improves and then declines when β increases. The reason is that the entities matching can provide a valuable evaluation indicator. However, when β is larger than 0.5, it greatly reduces the influence of relation perspective evaluation which harms the hybrid matching performance.

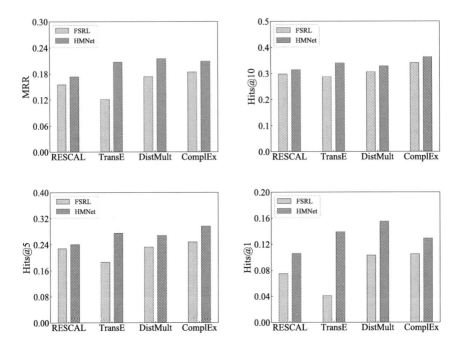

Fig. 5. Impact of embedding methods.

5.4   Ablation Study

HMNet consists of two components, and each component contains different modules. To get deep insight into HMNet, we analyze the contribution of each module. Specifically, we remove the entity-aware score module and only keep the AI Embed Layer (denoted as HMNetw/oEntityMatching). For the relation encoder module, we remove the BiLSTM network and only keep the multi-layer perceptron to get the representation of relation of each instance (denoted as HMNetw/oBiLSTM). To explore the impact of score function selection, we remove the feature attention module (denoted as HMNetw/oCNN). The parameters follow the above settings, and the results on NELL-One dataset are reported in Table 4 with the best results bold. Several observations from these results are worth noting:

– The best results of most evaluation metrics on the NELL-One dataset are obtained by complete HMNet.
– Removing the entity-aware score or CNN from the complete model causes the most significant performance drop on all evaluation metrics, showing the crucial role of entity perspective evaluation and the feature attention module in general.
– Removing the BiLSTM causes performance drop on some evaluation metrics but not all. All the components of HMNet together lead to the robust performance of our approach.

Table 4. Results of ablation study on NELL-One.

Model	MRR	Hits@10	Hits@5	Hits@1
HMNetw/oEntityMatching	.189	.326	.256	.119
HMNetw/oBiLSTM	.205	.352	.277	**.134**
HMNetw/oCNN	.193	.348	.263	.120
HMNet	**.209**	**.364**	**.296**	.129

6 Conclusion

In this paper, we propose a novel few-shot link prediction model, named HMNet. HMNet with entity-aware matching network and relation-aware matching network can evaluate the authenticity of triplets from two different perspectives. The comprehensive results on two public datasets indicate that HMNet can obtain more superior performance than state-of-the-art baseline methods. With in-depth analysis and ablation empirical evidence, we show the effectiveness and importance of each module of the HMNet model.

In the future, we will study the impact of different entity feature aggregation methods on experimental performance. Furthermore, we plan to integrate extra information (e.g., text information) to improve performance.

References

1. Bollacker, K.D., Evans, C., Paritosh, P., Sturge, T., Taylor, J.: Freebase: a collaboratively created graph database for structuring human knowledge. In: SIGMOD, pp. 1247–1250 (2008)
2. Bordes, A., Usunier, N., García-Durán, A., Weston, J., Yakhnenko, O.: Translating embeddings for modeling multi-relational data. In: NIPS, pp. 2787–2795 (2013)
3. Bordes, A., Weston, J., Usunier, N.: Open question answering with weakly supervised embedding models. In: ECML-PKDD, pp. 165–180 (2014)
4. Chen, M., Zhang, W., Zhang, W., Chen, Q., Chen, H.: Meta relational learning for few-shot link prediction in knowledge graphs. In: EMNLP-IJCNLP, pp. 4216–4225 (2019)
5. Dalton, J., Dietz, L., Allan, J.: Entity query feature expansion using knowledge base links. In: SIGIR, pp. 365–374 (2014)
6. Dettmers, T., Minervini, P., Stenetorp, P., Riedel, S.: Convolutional 2D knowledge graph embeddings. In: AAAI, pp. 1811–1818 (2018)
7. Dong, X., et al.: Knowledge vault: a web-scale approach to probabilistic knowledge fusion. In: SIGKDD, pp. 601–610 (2014)
8. Finn, C., Abbeel, P., Levine, S.: Model-agnostic meta-learning for fast adaptation of deep networks. In: ICML, pp. 1126–1135 (2017)
9. Gao, T., Han, X., Liu, Z., Sun, M.: Hybrid attention-based prototypical networks for noisy few-shot relation classification. In: AAAI, pp. 6407–6414 (2019)
10. Hamilton, W.L., Ying, Z., Leskovec, J.: Inductive representation learning on large graphs. In: NIPS, pp. 1024–1034 (2017)

11. Hochreiter, S., Schmidhuber, J.: Long short-term memory. Neural Comput. **9**(8), 1735–1780 (1997)
12. Koch, G., Zemel, R., Salakhutdinov, R.: Siamese neural networks for one-shot image recognition. In: ICML (2015)
13. Lin, Y., Liu, Z., Sun, M., Liu, Y., Zhu, X.: Learning entity and relation embeddings for knowledge graph completion. In: AAAI, pp. 2181–2187 (2015)
14. Munkhdalai, T., Yu, H.: Meta networks. In: ICML, pp. 2554–2563 (2017)
15. Nickel, M., Tresp, V., Kriegel, H.: A three-way model for collective learning on multi-relational data. In: ICML, pp. 809–816 (2011)
16. Schlichtkrull, M., Kipf, T.N., Bloem, P., van den Berg, R., Titov, I., Welling, M.: Modeling relational data with graph convolutional networks. In: Gangemi, A., et al. (eds.) ESWC 2018. LNCS, vol. 10843, pp. 593–607. Springer, Cham (2018). https://doi.org/10.1007/978-3-319-93417-4_38
17. Snell, J., Swersky, K., Zemel, R.S.: Prototypical networks for few-shot learning. In: NIPS, pp. 4077–4087 (2017)
18. Suchanek, F.M., Kasneci, G., Weikum, G.: YAGO: a core of semantic knowledge. In: WWW, pp. 697–706 (2007)
19. Sun, Z., Deng, Z., Nie, J., Tang, J.: Rotate: Knowledge graph embedding by relational rotation in complex space. In: ICLR (2019)
20. Trouillon, T., Welbl, J., Riedel, S., Gaussier, É., Bouchard, G.: Complex embeddings for simple link prediction. In: ICML, pp. 2071–2080 (2016)
21. Vinyals, O., Blundell, C., Lillicrap, T., Kavukcuoglu, K., Wierstra, D.: Matching networks for one shot learning. In: NIPS, pp. 3630–3638 (2016)
22. Wang, X., He, X., Cao, Y., Liu, M., Chua, T.: KGAT: knowledge graph attention network for recommendation. In: KDD, pp. 950–958 (2019)
23. Wang, Y., Yao, Q., Kwok, J.T., Ni, L.M.: Generalizing from a few examples: a survey on few-shot learning. ACM Comput. Surv. **53**(3), 63:1–63:34 (2020)
24. Wang, Z., Zhang, J., Feng, J., Chen, Z.: Knowledge graph embedding by translating on hyperplanes. In: AAAI, pp. 1112–1119 (2014)
25. Xie, Y., Xiong, Y., Zhu, Y.: SAST-GNN: a self-attention based spatio-temporal graph neural network for traffic prediction. In: DASFAA, pp. 707–714 (2020)
26. Xiong, W., Yu, M., Chang, S., Guo, X., Wang, W.Y.: One-shot relational learning for knowledge graphs. In: EMNLP, pp. 1980–1990 (2018)
27. Yang, B., Yih, W., He, X., Gao, J., Deng, L.: Embedding entities and relations for learning and inference in knowledge bases. In: ICLR (2015)
28. Zhang, C., Yao, H., Huang, C., Jiang, M., Li, Z., Chawla, N.V.: Few-shot knowledge graph completion. In: AAAI, pp. 3041–3048 (2020)
29. Zhang, F., Yuan, N.J., Lian, D., Xie, X., Ma, W.: Collaborative knowledge base embedding for recommender systems. In: SIGKDD, pp. 353–362 (2016)
30. Zhang, Z., Zhuang, F., Zhu, H., Shi, Z., Xiong, H., He, Q.: Relational graph neural network with hierarchical attention for knowledge graph completion. In: AAAI, pp. 9612–9619 (2020)

OntoCSM: Ontology-Aware Characteristic Set Merging for RDF Type Discovery

Pengkai Liu, Shunting Cai, Baozhu Liu, and Xin Wang[✉]

College of Intelligence and Computing, Tianjin University, Tianjin, China
{liupengkai,caishunting,liubaozhu,wangx}@tju.edu.cn

Abstract. With the growing popularity and application of knowledge-based artificial intelligence, the scale of knowledge graph data is dramatically increasing. The RDF, as one of the mainstream models of knowledge graphs, is widely used to describe the characteristics of Web resources due to its simplicity and flexibility. However, RDF datasets are usually incomplete (without `rdf:type` information) and noisy, which hinders downstream tasks. RDF entities can be characterized by their *characteristic sets* that is the sets of predicates of the RDF entities. Since untyped entities can be assigned to closest types by merging characteristic sets, optimally merging characteristic sets has become a crucial issue. In this paper, aiming at the Optimal Characteristic Set Merge Problem (OCSMP), we propose an <u>Onto</u>logy-Aware <u>C</u>haracteristic <u>S</u>et <u>M</u>erging algorithm, called OntoCSM, which extracts an ontology hierarchy using RDF characteristic sets and guides the merging process by optimizing the objective function. Extensive experiments on various datasets show that the efficiency of OntoCSM is generally higher than that of the state-of-the-art algorithms and can be improved by orders of magnitude in the best case. The accuracy and scalability of our method have been verified, which shows that OntoCSM can reach competitive results to the existing algorithms while being ontology-aware.

Keywords: RDF data · Ontology-aware · Type discovery

1 Introduction

With the rapid development of artificial intelligence, knowlege graphs have been widely used in many fields. In the Semantic Web community, the *Resource Description Framework* (RDF) [1] is a model for representing Web resources, which has become a standard format for knowledge graphs and has been extensively applied. However, as the data volume increasing, due to the flexible structure of RDF and fewer constraints on instances, some RDF datasets often contain incomplete or noisy data, especially the RDF type information, which makes reasoning tasks more difficult and inconvenience for type-based storage and query

© Springer Nature Switzerland AG 2021
C. S. Jensen et al. (Eds.): DASFAA 2021, LNCS 12681, pp. 323–339, 2021.
https://doi.org/10.1007/978-3-030-73194-6_22

processing. Thus, specifying types for untyped entities in the dataset has become an essential problem for effective RDF knowledge graph data management.

To group entities with identical or similar types and solve the various problems caused by untyped entities, various methods have been proposed. Among the existing RDF type discovery methods, statistical methods [2–5] divide entities into different clusters according to the similarity between them, however, these methods suffer from scalability issues and rarely consider the hierarchical relationships between the entities in RDF datasets [6]. Some reasoning-based approaches [7,8] apply inference rules to make implicit facts explicit to obtain the type information of the entity. However, only those facts strictly abided by logic rules can be inferred, which makes reasoning-based approaches not suitable for type discovery task on datasets with erroneous or conflicting statements. Additionally, some works [9,10] use machine learning methods to classify data whose type information is partially available, but they require labeled datasets and cannot deal with completely untyped data. Related to the work of this paper, the predicates and objects of RDF entities have been utilized in some research works to obtain the schema information of the knowledge graph. In [11], the proposed approach processes the RDF dataset by grouping the properties of the entities according to the similarity of their subjects or/and objects.

The set of predicates of each entity, refered to as the characteristic set [12], can be used to represent the features of the entity, as shown in Fig. 1. By constructing the hierarchical structure between characteristic sets, the type of untyped entities can be specified by defining the distance between the entity characteristic sets and merging them into the closest entity set of the known-type. Entities can be classified into the same cluster when they have the same characteristic set or the distance between their characteristic sets is less than a threshold. To this end, we propose an Ontology-aware Characteristic Set Merging algorithm, named OntoCSM. Moreover, OntoCSM relaxes the constraints for prior type information on RDF datasets, significantly broadening the applicability of the proposed method.

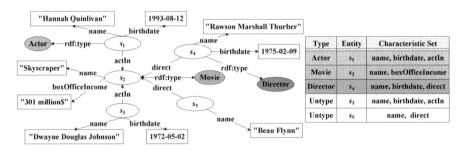

Fig. 1. An example of RDF graph and its corresponding characteristic sets

Our contributions in this paper can be summarized as follows:

(1) We propose a novel ontology-aware characteristic set merging algorithm, OntoCSM, which merges the entity characteristic sets using an optimization method to solve the problem of RDF type discovery.

(2) In order to obtain the hierarchical structure between untyped entities, we devise an algorithm that can extract ontology information from the characteristic sets of entities.

(3) Extensive experiments have been conducted to verify the efficiency and accuracy of the proposed method on various datasets. The experimental results show that the proposed method can significantly improve the efficiency in most cases and is suitable for processing large-scale datasets while ensuring accuracy and scalability.

The rest of this paper is organized as follows. Section 2 reviews related works. In Sect. 3, we introduce preliminary definitions of OCSMP. In Sect. 4, we describe the OntoCSM algorithm for RDF type discovery in detail. Section 5 shows experimental results, and we conclude in Sect. 6.

2   Related Work

As most large open knowledge bases lack type information, which is an essential component, in order to deal with untyped entities, researchers have raised several approaches to solve RDF type discovery problem in recent years. The existing methods can be divided into the following three categories:

(1) Statistical approaches. Among the state-of-the-art methods, statistical approaches are more robust to noisy data, therefore more suitable for the type prediction task, making them become the mainstream algorithm to solve the type discovery problem in RDF. The key challenge of RDF type discovery is the noisy data in the knowledge bases, which hampers the collection of statistics over RDF graphs. Statistical approaches divide a dataset into different clusters according to dataset statistics, increasing intra-cluster similarity while reducing inter-cluster similarity.

An entity can be assigned to a type by data mining algorithms such as KNN [13], as mentioned in [2]. However, this algorithm requires some parameters given in advance, e.g., the number of types in the dataset. DBSCAN [3] is a density-based clustering algorithm, which does not need to declare the number of clusters in advance. SDA [4] proposes the concept of probabilistic type profiles based on DBSCAN, in which, however, the similarity threshold parameter must be specified. SDA++ [5], an extension of SDA, proposes an incremental algorithm that can effectively solve the problem of dataset expansion and assign multiple types to an entity, at the same time automatically detecting the similarity threshold. In [6], hierarchical clustering is used to support structure inference from RDF resources which only contain instance-level data.

(2) Reasoning-based approaches. Specific solutions based on logical reasoning to the RDF type inference problem have also been proposed recently. Logical reasoning can be used to solve the type inference problem by applying RDF schema (RDFS) or OWL entailment regimes. Logical methods only allow reasoning information from datasets that strictly obey the rules, which cannot be applied to the knowledge base which contains erroneous or conflicting

statements. Although it is identified in [7] that the problems of inference on noisy data in the Semantic Web, and since then a technique is given in [8] to process noisy semantic data, the inference-based approach is still not applicable to type inference scenario, where most `rdf:type` values are missing.

(3) Machine learning-based approaches. Machine learning approaches are also used to solve the problem of formulating RDF types with labeled training sets. The method proposed in [9] discusses an approach of iterative classification to train the machine learning models with relational data and iteratively utilizes the models to process untyped instances. DL-Learner system [10] conducts inductive learning on Semantic Web data, provides an OWL-based machine learning tool to solve supervised learning tasks, and infers types in knowledge graphs through induction process. Machine learning-based approaches do not make available for situations when there is no prior RDF types information.

Approaches that more closely related to ours build a hierarchy of types from the characteristics of entities. The similarity of entities contained in the data is discussed in [14], which regards building categories as a learning process. Dynamic generation of concepts hierarchies [15] preprocesses linked data and then formalizes the hierarchy. Statistical schema induction is proposed in [16] to mine association rules from RDF data, which can help obtain schema level ontology knowledge. Standard ascending hierarchical clustering is presented in [6] to build structured abstraction of linked data.

Our method differs from all the aforementioned algorithms in terms of objectives and methods. To the best of our knowledge, OntoCSM is the first method to classify entities based on the distance between the characteristic sets, using the idea of optimal merging to make the RDF type discovery algorithm be able to deal with two cases, in which the type information of the dataset is partially known or completely unknown. Moreover, OntoCSM can guide the classification process by integrating the structure information of ontology, which makes it ontology-aware.

3   Preliminaries

In this section, we introduce the definitions of relevant background knowledge. Table 1 gives the main notations used throughout this paper.

Table 1. List of notations.

Notation	Description
$S_C(s)$	The characteristic set of entity s
$\mathscr{P}(V)$	The power set of V
α_i	The membership of property p_i to cluster C
$sem(C)$	The semantic feature of cluster C
$dist(C_i, C_j)$	Distance between cluster C_i and C_j
$\sigma(\mathcal{CS})$	The evaluation function of $\mathcal{CS}$ (the set of clusters)
$avg(C_i)$	Average distance between characteristic sets in cluster C_i

Definition 1 (RDF Graph). *Consider three disjoint infinite sets U, B, and L representing Uniform Resource Identifiers (URI), blank nodes, and literals, respectively. RDF graph is a finite set of RDF triples $(s, p, o) \in (U \cup B) \times U \times (U \cup B \cup L)$, in which s is the subject, p is the predicate, and o is the object. A triple (s, p, o) is a statement of a fact, which means there is a connection p between s and o or the value of property p for s is o.*

In most RDF graphs, entities can be uniquely identified by a proper subset of their emitting edges. While we might not be able to clearly identify an entity as one type (due to the lack of `rdf:type` information), an entity can be characterized by its emitting edges [12], i.e., characteristic set.

Definition 2 (Characteristic Set Partition). *Formally, for each entity s that appears in an RDF dataset R, its characteristic set is defined as follows:*

$$S_C(s) := \{p \mid \exists o : (s, p, o) \in R\} \tag{1}$$

Let V be the entity set of R, for an entity set $\mathcal{V} \subseteq V$, the set of all S_C for $\mathcal{V}$ is: $S_C(\mathcal{V}) := \{S_C(s) \mid s \in \mathcal{V}\}$. Characteristic Set Partition $\mathcal{CSP} = \{SP_1, \cdots, SP_n\}$ is a subset of $\mathscr{P}(V)$, where $n = |S_C(R)|$. $\mathcal{CSP}$ needs to satisfy the following conditions:

(1) $\forall SP_i \in \mathcal{CSP}, \ SP_i \neq \emptyset$;
(2) $\forall SP_i, SP_j \in \mathcal{CSP}, \ SP_i \cap SP_j = \emptyset$ if $i \neq j$;
(3) $\bigcup_{i=1}^{n} SP_i = V$;
(4) $\forall SP_i \in \mathcal{CSP}, \ S_C(s_p) = S_C(s_q)$ if $s_p \in SP_i \wedge s_q \in SP_i$;
(5) $\forall SP_i, SP_j \in \mathcal{CSP}, \ S_C(SP_i) \neq S_C(SP_j)$ if $i \neq j$.

Among them, (4) ensures there is only one characteristic set in each SP_i, and (5) prevents different elements of $\mathcal{CSP}$ from holding the same characteristic.

The characteristic set provides a node-centric division for the entities in a knowledge graph, based on the node structure [17]. Each entity only belongs to one single characteristic set. According to the definition of characteristic set, entities with slightly different properties should also be divided into different characteristic sets, but in fact, they may belong to the same type. In our work, the entities will be merged into the same cluster if their characteristic sets are similar.

Definition 3 (Cluster). *According to the $\mathcal{CSP}$, the cluster C is composed of entities from several SP_i, and $\mathcal{CS}$ is the set of all clusters, which is defined as $\mathcal{CS} = \{C_1, C_2, \cdots, C_m\}, \forall C_j \in \mathcal{CS}, \text{where } C_j = \bigcup \{SP_1, SP_2, \cdots, SP_k\} (1 \leq k \leq n, 1 \leq j \leq m)$.*

Based on the idea of *word frequency* and *reverse document frequency*, we use α_i to describe the membership of property p_i to cluster C, which can be defined as

$$\alpha_i = \frac{count(C, p_i)}{count(C)} \log \frac{m}{m_i} \tag{2}$$

In Eq. 2, m_i represents the number of clusters containing property p_i in $\mathcal{CS}$, m represents the current total number of clusters, $count(C)$ shows the number of entities in cluster C, $count(C, p_i)$ corresponds to the number of entities in cluster C containing property p_i. The semantic feature of cluster C is defined as $sem(C) = \{(p_1, \alpha_1), (p_2, \alpha_2), \cdots, (p_k, \alpha_k)\}$, in which $k = |\bigcup_{i=1}^{m} C_i|$ represents the number of properties in the C_i of $\mathcal{CS}$.

The process of merging characteristic sets involves two extreme cases. In the first case, the entities in each SP_i form a seperate cluster, which makes entities of the same type in different SP_i divided into different clusters. In the second case, all entities are in one single cluster, which will cause the problem of sparse properties contained in the characteristic set within the cluster. OntoCSM uses the semantic statistical information of characteristic sets which represents the similarity between clusters, combining with the hierarchical relationships between types to help get the most appropriate set of clusters. The main issue discussed in this paper is to merge similar entities into the same cluster according to their characteristic sets. In order to get the best merging result, we consider the issue as an optimization problem, and thus put forward the Optimal Characteristic Set Merge Problem (OCSMP).

Example 1. As shown in Fig. 2, entities $(e_1 - e_{12})$ are partitioned according to characteristic sets, and $\mathcal{CSP}$ is obtained after the partitioning. SP_1 and SP_3 are suppoesd to be merged into the same cluster C_1 because of the similar characteristic sets they have. SP_2 and SP_4 are supposed to be merged into the same cluster C_2 for the same reason.

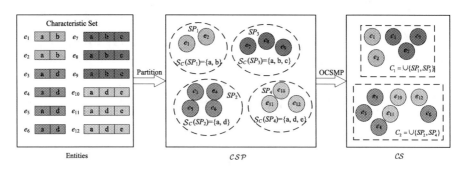

Fig. 2. An example of OCSMP

Definition 4 (Optimal Characteristic Set Merge Problem, OCSMP).
To estimate the outcome of merging process, the evaluation function is defined as

$$\sigma(\mathcal{CS}) = \frac{1}{k} \sum_{i=1}^{k} \max_{j \neq i} \left(\frac{avg(C_i) + avg(C_j)}{dist(C_i, C_j)} \right) \tag{3}$$

The numerator measures the average distance of characteristic sets in each cluster, where $avg(C)$ represents for the average value of the distance between each characteristic set in cluster C, where $sem(C) = \{(p_1, \alpha_1), \cdots, (p_k, \alpha_k)\}$, and *avg(C) is calculated by the difference of the predicates contained in each characteristic set, which is defined as*

$$avg(C) = \frac{\sum_{C_i \in C} \sum_{p_j \in \{p_1, \dots, p_k\}} \alpha_j b_j}{|C|} \qquad (4)$$

if $p_j \in C_i, b_j = 0$, else $b_j = 1$. The denominator measures the distance between clusters, and $dist(C_i, C_j)$ can be calculated with the semantic features from cluster C_i and C_j. OCSMP is the process to obtain the optimal solution of $\mathcal{CS}$ from $\mathcal{CSP}$ by evaluation function $\sigma(\mathcal{CS})$.

4   Ontology-Aware Characteristic Set Merging Approach

In this section, we present the ontology-aware characteristic set merging algorithm, i.e., OntoCSM, aiming to solve the problem of RDF type discovery, which exploits the characteristic set partitioning introduced in Sect. 3. First, we illustrate the workflow of our approach, then we describe the extraction process of the ontology hierarchy from the characteristic sets and the implementation of the ontology-aware optimization algorithm. Finally, we analyze the complexity results of the proposed algorithm.

4.1   Workflow

The entities in the RDF graph can be divided according to their characteristic sets, and entities with the same characteristic set are initialized into the same cluster after entity partitioning. Through the recursive merging of clusters, similar characteristic sets are merged into one cluster. Since there is a hierarchical relationship between the types of entities, the ontology hierarchy that represents the hierarchical relationship between clusters can help each entity to be merged in an optimal way. After the merging process, we can assign the type of the entity to the cluster where the entity resides.

Example 2. Figure 3 briefly depicts the workflow of OntoCSM, which can be divided into three stages. (1) In the first phase (Fig. 3(a)–(b)), after the RDF data is loaded, entities (representing s_1–s_4) are given different colors according to their types, and white nodes (e.g. s_5) represent entities without types. Entities are divided into $SP_1 \dots SP_n$ based on their characteristic sets; (2) In the second phase (Fig. 3(b)–(c)), the ontology hierarchy for clusters is constructed according to the similarity between the characteristic sets of entities. Specifically, each SP_i is initialized as a cluster, and the undirected graph $\mathbb{G}$ is formed according to the distance between the clusters, where each cluster corresponds to a node in the graph. The minimum spanning tree T is composed of nodes whose shortest distance to neighboring nodes is between e_1 and e_2 (e.g., SP_3), and is selected as the ontology hierarchy; (3) In the third phase (Fig. 3(c)–(d)), the remaining clusters $(C_1, ..., C_i)$ are merged into the generated ontology hierarchy according to their distance from the nodes on the tree T. Based on the extracted ontology hierarchy, the untyped entities are merged and their types are assigned as the type of the cluster in which it is located.

4.2   Characteristic Set-Based Ontology Extraction

As an extension of the basic RDF vocabulary, RDFS provides mechanisms for describing groups of related resources and the relationships between these resources, using `rdf:type` (stating that a resource is an instance of a type) and `rdfs:subClassOf` (stating that one class is a subclass of another). In order to identify the relationships among characteristic sets and guide the merging process of characteristic sets, it is essential to extract the type hierarchy from an RDF graph using RDFS vocabulary. Though the ontology information may not be directly provided by RDFS, the ontology hierarchy can be extracted and constructed based on the distance among the characteristic sets.

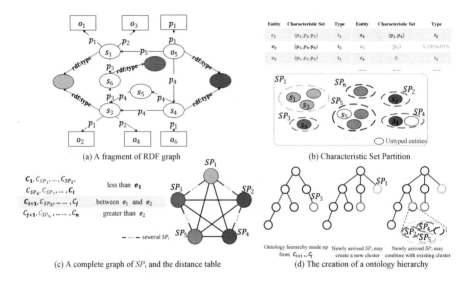

(a) A fragment of RDF graph

(b) Characteristic Set Partition

(c) A complete graph of SP_i and the distance table

(d) The creation of a ontology hierarchy

Fig. 3. The overview of OntoCSM

In the initial phase, Algorithm 1 divides entities according to their characteristic sets by hash function, maps entities with the same characteristic set to the same index value, and returns a hash table $HT = (CS, Index, S, Type)$, including the characteristic sets CS, index value $Index$, corresponding entity set S, and the type set $Type$. Each characteristic set is initialized as a cluster (line 2–3), containing all the corresponding entities. For newly arrived entities, if their characteristic sets have already been added in the hash table and their types information is given, the rows of corresponding CS will be updated (line 6–7). The type of a given cluster is the union of entity types within this cluster. Clusters are sorted in descending order according to the characteristic set with the largest number of properties in the cluster.

In order to solve the problem of missing type information, we design an ontology information extraction approach based on characteristic sets. By calculating the distance between clusters, an undirected graph $\mathbb{G}$ can be constructed regarding each cluster as a node. Given two clusters C_i and C_j,

---

Algorithm 1: Characteristic Set Hashing

Input: Entity set S and characteristic set $S_C(s)$ for each entity s in S
Output: Hash Table $HT = (CS, Index, S, Type)$

1 **for** *each* $s_i \in S$ **do**
2   **if** $HT.get(S_C(s_i)) = \text{NULL}$ **then**
       // the CS appears for the first time
3        $HT.insert(S_C(s_i), s_i)$;
       // insert characteristic set and its corresponding entities
         into the CS and S columns of the hash table

4   **else**
5        $HT.get(S_C(s_i)).S \cup \{s_i\}$;
       // insert a new entity into the corresponding row of the hash
         table
6   **if** $s_i.type \neq \text{NULL}$ **then**
         // the type of the entity is given
7          $HT.get(S_C(s_i)).Type \cup \{s_i.type\}$;
         // insert a new type into the corresponding row of the
           hash table

8 **return** HT

$sem(C_i) = \{(p_1, \alpha_1), ..., (p_m, \alpha_m)\}, sem(C_j) = \{(p_1, \alpha_1), ..., (p_n, \alpha_n)\}, P_i = \{p_1, ..., p_m\}, P_j = \{p_1, ..., p_n\}, P = \{p_1, ..., p_t\} = P_1 \cup P_2$, then $A_i = [\alpha_{i1}, ..., \alpha_{it}]$, if $p_k \notin P_i, \alpha_{ik} = 0$, else α_{ik} is the α value corresponding to p_k in $sem(C_i)$. The same is true for $A_j = [\alpha_{j1}, ..., \alpha_{jt}]$. The formula for calculating the distance between two clusters is defined as follows:

$$dist(C_i, C_j) = \sum_{k=1}^{t} |A_{ik} - A_{jk}| \tag{5}$$

Algorithm 2 presents an overview of the ontology extraction process. We define e_1 and e_2 as two thresholds to decide whether a node should be merged, used to construct the ontology hierarchy, or regarded as noisy data. For all nodes in graph $\mathbb{G} = \langle \mathbb{V}, \mathbb{E} \rangle$, where the node set $\mathbb{V}$ stands for the clusters (every cluster is regarded as one node) and the edge set $\mathbb{E}$ represents the links between nodes, we calculate the distance between nodes to obtain a distance matrix. Then we divide nodes into three disjoint sets V_1, V_2, and V_3 by the distance from a node to other nodes comparing with e_1 and e_2. For any node v in $\mathbb{G}$:

(1) if $minDis(v) < e_1$, $V_1 = V_1 \cup \{v\}$, where V_1 represents the node set corresponding to the characteristic sets to be merged (line 4);
(2) if $e_1 < minDis(v) < e_2$, $V_2 = V_2 \cup \{v\}$, where V_2 represents the node set corresponding to the characteristic sets used to construct the ontology hierarchy (line 6);

(3) if $e_2 < minDis(v)$, $V_3 = V_3 \cup \{v\}$, where V_3 represents the set of noisy nodes, and the characteristic sets corresponding to these nodes do not participate in the merging process or the construction of the ontology hierarchy (line 8).

The minimum spanning tree T is the ontology hierarchy constructed from the nodes belonging to V_2. Each node in V_2 is considered as an independent tree node. We find the shortest edge $\langle v_1, v_2 \rangle$ from graph $\mathbb{G}$, given that v_1 and v_2 belong to V_2. If v_1 and v_2 are not in the same tree, v_1, v_2 will be connected, and $\langle v_1, v_2 \rangle$ is added to the edge set E' of the spanning tree T (line 12), and the algorithm is repeated until all edges in $\mathbb{G}$ have been traversed. In the end, we can obtain the minimum spanning tree $T = \langle V_2, E' \rangle$ for graph $\mathbb{G}$.

4.3   Ontology-Aware Characteristic Set Merging Algorithm

According to the characteristic set of the entity and the extracted ontology hierarchy, we propose an Ontology-Aware Characteristic Set Merging Algorithm, i.e., OntoCSM. Based on the idea of optimizing the objective function, OntoCSM simulates the process of merging the characteristic set of the untyped entities to obtain the best threshold and then achieves the best classification results.

Algorithm 2: Ontology Hierarchy Extraction

Input: Characteristic sets relationship graph $\mathbb{G} = \langle \mathbb{V}, \mathbb{E} \rangle$;
Similarity threshold e_1 and e_2
Output: Ontology hierarchy $T = \langle V_2, E' \rangle$

1 **for** *each* $v \in \mathbb{V}$ *of* $\mathbb{G}$ **do**
2 $\quad$ $minDis := min(dist(v,u))$;
$\quad$ // distance between node v and the nearest neighbor u
3 $\quad$ **if** $minDis < e_1$ **then**
4 $\quad\quad$ $V_1 := V_1 \cup \{v\}$;
$\quad\quad$ // the set of nodes to be merged
5 $\quad$ **else if** $minDis < e_2$ **then**
6 $\quad\quad$ $V_2 := V_2 \cup \{v\}$;
$\quad\quad$ // the set of nodes to construct the ontology hierarchy
7 $\quad$ **else**
8 $\quad\quad$ $V_3 := V_3 \cup \{v\}$; $\qquad\qquad\qquad$ // the set of noisy nodes
9 $C := V_2$; $\qquad\qquad\qquad\qquad\qquad\qquad$ // nodes to be added to T
10 **while** $C \neq \emptyset$ **do**
11 $\quad$ $(u_0, v_0) := findMin(C)$;
$\quad$ // find the nodes u_0 and v_0 in V_2, which have the minimium
$\quad\quad$ distance, u_0 and v_0 are not in the same connected component
12 $\quad$ $E' := E' \cup \{(u_0, v_0)\}$;
$\quad$ // add (u_0, v_0) as an edge to E'
13 **return** T

Algorithm 3 introduces the details of the characteristic set merging process. After the preprocessing of characteristic set partitioning, the remaining clusters will be merged with an existing cluster in the currently constructed ontology hierarchy if the distance between the newly added cluster and the clusers in the ontology hierarchy is less than e_1 (line 3–5). The $merge(u_0, v_0)$ function merges the clusters corresponding to u_0 and v_0. If the distance is between e_1 and e_2, the cluster will be used to form a hierarchical structure based on its distance from the nearest cluster (line 6–9). The clusters without specified types are also merged in this manner. The remaining clusters are regarded as noisy data.

Algorithm 3: Ontology-Aware Characteristic Set Merging

Input: The Set of nodes to be merged V_1 and ontology hierarchy $T = \langle V_2, E' \rangle$
Output: Characteristic sets merged sturcture tree $T' = \langle V_2, E' \rangle$

1   **for** *each* $v_i \in V_1$ **do**
2     $(u_0, v_0) := findMin(V_2, V_1)$;
     `// find nodes` $u_0 \in V_2$ `and` $v_0 \in V_1$ `which have the minimium distance`
3   **if** $dist(u_0, v_0) < e_1$ **then**
4       $merge(u_0, v_0)$; `// merge the clusters corresponding to` u_0 `and` v_0
5       $V_1 := V_1 \setminus \{v_0\}$;
6   **else if** $e_1 < dist(u_0, v_0) < e_2$ **then**
7       $v_0.head := u_0$; `//` u_0 `is the father of` v_0 `in the ontology hierarchy`
8       $V_2 := V_2 \cup \{v_0\}$;
       `// add` v_0 `to the ontology hierarchy collection` V_2
9       $E' := E' \cup \{(u_0, v_0)\}$;
10      $V_1 := V_1 \setminus \{v_0\}$;
11 **return** T'

Case Study. It is conceivable that in the real-world RDF graph, as shown in Fig. 1, the type of the entities whose characteristic sets are {name, birthdate} would probably be People, which is the superclass of Singer, Actor, etc. The type of entities that with {name} as their characteristic sets could be a higher-level superclass of People and Movie, i.e., the owl:Thing in the RDF graph. For the more general form given in Fig. 3(a), using OntoCSM, the characteristic sets that are relatively far away from each other will construct the hierarchy among type classes, and characteristic sets that are relatively close to each other will be merged, thereby untyped entities will be merged into the closest set of entities of known types. For RDF graphs with little or no entity type information, OntoCSM can efficiently provide a hierarchical structure constructed from the data to guide type discovery.

4.4 Scalability and Complexity

Scalability. The scale of the datasets may expand, and new entities are constantly being added at any time. Therefore, the algorithm needs to be scalable to incrementally cope with the expansion of the dataset. For the newly arrived entity, OntoCSM can remerge or reconstruct the ontology hierarchy according to the relationship between the threshold and the closest distance to the nodes on the ontology hierarchy, which is constructed using the existing entities, according to the process similar to Algorithm 3, without affecting the previous merged results.

Complexity. The time complexity of the OntoCSM algorithm is bounded by $O\left(|V|^2 + |M|^2 \cdot \log|M| + |M| \cdot |N|\right)$, where $|V|$ is the total number of characteristic sets, $|M|$ is the number of characteristic sets used to construct the ontology hierarchy, and $|N|$ is the number of characteristic sets used for merging.

Proof (Sketch). The time complexity of OntoCSM consists of three parts: (1) The algorithm firstly calculates the $|N| \cdot |N|$ distance matrix between the characteristic sets; (2) the ontology hierarchy is generated from $|M|$ nodes that satisfy the threshold condition, with the complexity of $|M|^2 \cdot \log|M|$; (3) finally, the remaining $|N|$ characteristic sets is merged with the $|M|$ nodes in the ontology hierarchy. Hence, the overall time complexity of the proposed algorithm is $O\left(|V|^2 + |M|^2 \cdot \log|M| + |M| \cdot |N|\right)$.   □

5 Experiments

In this section, we implement our algorithm and verify the effectiveness, efficiency, and scalability of OntoCSM, compared with the baseline algorithms on several datasets.

5.1 Experimental Settings

The proposed algorithm was implemented in Python, which is deployed on a single-node server. The server has an 8-core Intel(R) Xeon(R) Platinum 8255C@ 2.5 GHz CPU, with 32 GB of memory, running 64-bit CentOS 7.6 operating system.

Datasets. Our experiments were conducted on four different datasets which are also used in other methods [4,5] solving the relevant problem. (1) The Conference[1] dataset contains data about several Semantic Web conferences, keynotes, and workshops with 1,430 triples; (2) Histmunic[2] dataset with 119,151 triples is an open government dataset; (3) another dataset is extracted from DBpedia[3]

---

[1] http://data.semanticweb.org/dumps/conferences/dc-2010-complete.rdf.
[2] https://opendata.swiss/dataset.
[3] http://dbpedia.org/.

with 19,696 triples and consider the following types: *Politician, Soccer player, Museum, Movie, Book,* and *Country*; (4) BNF[4] dataset, includes data about the French National Library (Bibliothèque Nationale de France) with 381 triples. Table 2 shows the statistics about each dataset, including the number of triples, the number of instances, and the number of types.

Table 2. Statistics of datasets

Datasets	#triples	#instances	#types
BNF	381	31	5
Conference	1,430	403	12
DBpedia	19,696	100	6
Histmunic	119,151	12,132	14

Baselines. We compare OntoCSM against two clustering algorithms, which have already been mentioned in Sect. 2. SDA++ [5] utilizes a method to transform group instances of RDF data into types in a deterministic and automatic way. StaTIX [18] uses the inference technique to leverage a new hierarchical clustering algorithm. The rest of methods are neither directly applicable nor as good as these two algorithms. Therefore, we take these two state-of-the-art methods, SDA++ and StaTIX, as the baselines for our evaluation.

Evaluation Criteria. In order to evaluate the quality of results produced by different algorithms, we use the metrics proposed in [5]. In these metrics, algorithms are run on the datasets without type information. Each cluster C_i is appointed the most frequent type label of its entities. For each type label L_i corresponding to type T_i in the dataset, the precision P_i is defined as $P_i(T_i, C_i) = |T_i \cap C_i| / C_i$ and the recall R_i is defined as $R_i(T_i, C_i) = |T_i \cap C_i| / T_i$. The final precision P and final recall R is defined as follows:

$$P = \sum_{i=1}^{k} \frac{|Ci|}{n} \times P_i(Ti, C_i), \qquad R = \sum_{i=1}^{k} \frac{|Ci|}{n} \times R_i(Ti, C_i)$$

in which k is the number of clusters. Furthermore, $F1$ score is the harmonic average of precision and recall, which represents the robustness of the model and the quality of the result. The formula is defined as $F1 = \frac{2PR}{P+R}$.

5.2   Experimental Results

Exp 1. Effectiveness of the Algorithms in Accuracy. To evaluate the effectiveness of OntoCSM, we appoint the most frequent type shown in the merged cluster as the type of the characteristic set corresponding to the cluster, using the original type information (`rdf:type`) of each entity. While calculating the

---

[4] http://datahub.io/fr/dataset/data-bnf-fr.

relevant evaluation index, the thresholds e_1 and e_2 are set to the value which optimize the objective function. The precision P, recall R, and score $F1$ obtained by different methods on the four datasets are defined by the formula introduced in Sect. 5.1.

Figure 4 shows all the effectiveness results of different algorithms executed on four datasets. It is noteworthy that SDA++ could not finish within limited time on the Histmunic dataset, so the experimental results of SDA++ on Histmunic are eliminated. OntoCSM can achieve a comparable result against the other two algorithms on the Conference and BNF dataset. Since both of datasets have regular data, considering the statistical information of characteristic sets, OntoCSM reaches better effectiveness. However, because these two datasets do not have a strict hierarchical structure, the extraction of the ontology hierarchy may have a certain impact on the results. In addition, OntoCSM can achieve the same effectiveness in handling noisy data compared with SDA++, which manages noisy data deliberately. It should be noticed that some of the precision and $F1$ values can reach 1, since the Histmunic dataset has extremely regular data.

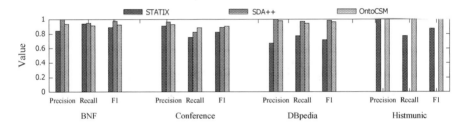

Fig. 4. The experimental results of effectiveness on four datasets

Exp 2. Time Efficiency of the Algorithms. In order to verify the efficiency of OntoCSM, we record the execution time of different algorithms on four datasets. Without type information given (`rdf:type` does not participate in the process of algorithms), entities are merged according to their characteristic sets.

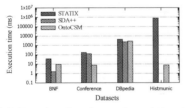 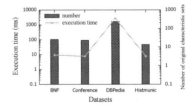

(a) Running time on various datasets   (b) Characteristic sets number effect

Fig. 5. The experimental results of efficiency on four datasets

The execution time of different algorithms is shown in Fig. 5(a). OntoCSM shows the best performance in processing the Histmunic dataset because Histmunic is regular and has a few numbers of characteristic sets, which further demonstrates that the characteristic set-based approach has excellent results in the case of fewer characteristic sets but more entities. Figure 5(b) reveals that the time complexity of OntoCSM depends on the number of original characteristic sets in the dataset.

Exp 3. Scalability of the Algorithms. In order to verify the scalability of our method, we add entities incrementally to the Histmunic dataset. Since STATIX is not incremental and SDA++ is not suitable for handling large-scale data, we fix the number of entities to construct the ontology hierarchy and observe the performance changes of the algorithm when different number of entities are added ($U_1 - U_5$ represents the test set consisting of tuples randomly selected from the original dataset with 50%−85%, and about 9% as the interval).

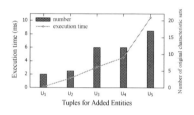
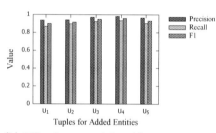

(a) Running time while adding entities

(b) Effectiveness while adding entities

Fig. 6. The experimental results of scalability on the Histmunic dataset

As the entities added, Fig. 6(a) shows the change in the execution time of the algorithm, and the relationship between time and the number of added entities is nearly linear. Figure 6(b) shows the precision, recall, and $F1$ value obtained as entities added. It can be noticed that different numbers of added entities would not affect the effectiveness results, which shows the scalability of the proposed algorithm.

6 Conclusion

In this paper, we present an ontology-aware characteristic set merging algorithm, OntoCSM, a novel method to solve the problem of RDF type discovery from the perspective of the ontology hierarchy. We model the problem as an optimization problem and use the objective function to direct the process of merging entities. Compared with the other state-of-the-art algorithms, OntoCSM exhibits better results on execution time, the scale of datasets, and awareness of ontology.

Besides, OntoCSM can also handle datasets regardless of whether ontology information is provided in advance. The effectiveness, efficiency, and scalability of our method have been verified by extensive experiments. We will consider to extend our algorithm to cope with multiple type hierarchies and multiple inheritance in the future work.

Acknowledgment. This work is supported by the National Key Research and Development Program of China (2019YFE0198600), National Natural Science Foundation of China (61972275), and CCF-Huawei Database Innovation Research Plan (CCF-Huawei DBIR2019004B).

References

1. World Wide Web Consortium: RDF 1.1 concepts and abstract syntax (2014)
2. Rizzo, G., Fanizzi, N., d'Amato, C., Esposito, F.: Prediction of class and property assertions on OWL ontologies through evidence combination. In: Proceedings of the International Conference on Web Intelligence, Mining and Semantics, pp. 1–9 (2011)
3. Ester, M., Kriegel, H.P., Sander, J., Xu, X., et al.: A density-based algorithm for discovering clusters in large spatial databases with noise. In: KDD, vol. 96, pp. 226–231 (1996)
4. Kellou-Menouer, K., Kedad, Z.: Schema discovery in RDF data sources. In: Johannesson, P., Lee, M.L., Liddle, S.W., Opdahl, A.L., López, Ó.P. (eds.) ER 2015. LNCS, vol. 9381, pp. 481–495. Springer, Cham (2015). https://doi.org/10.1007/978-3-319-25264-3_36
5. Kellou-Menouer, K., Kedad, Z.: A self-adaptive and incremental approach for data profiling in the semantic web. In: Hameurlain, A., Küng, J., Wagner, R. (eds.) Transactions on Large-Scale Data- and Knowledge-Centered Systems XXIX. LNCS, vol. 10120, pp. 108–133. Springer, Heidelberg (2016). https://doi.org/10.1007/978-3-662-54037-4_4
6. Christodoulou, K., Paton, N.W., Fernandes, A.A.A.: Structure inference for linked data sources using clustering. In: Hameurlain, A., Küng, J., Wagner, R., Bianchini, D., De Antonellis, V., De Virgilio, R. (eds.) Transactions on Large-Scale Data- and Knowledge-Centered Systems XIX. LNCS, vol. 8990, pp. 1–25. Springer, Heidelberg (2015). https://doi.org/10.1007/978-3-662-46562-2_1
7. Polleres, A., Hogan, A., Harth, A., Decker, S.: Can we ever catch up with the web? Semantic Web **1**(1, 2), 45–52 (2010)
8. Ji, Q., Gao, Z., Huang, Z.: Reasoning with noisy semantic data. In: Antoniou, G., et al. (eds.) ESWC 2011. LNCS, vol. 6644, pp. 497–502. Springer, Heidelberg (2011). https://doi.org/10.1007/978-3-642-21064-8_42
9. Neville, J., Jensen, D.: Iterative classification in relational data. In: Proceedings of the AAAI-2000 Workshop on Learning Statistical Models from Relational Data, pp. 13–20 (2000)
10. Bühmann, L., Lehmann, J., Westphal, P.: DL-learner-a framework for inductive learning on the semantic web. J. Web Semant. **39**, 15–24 (2016)
11. Čebirić, Š., Goasdoué, F., Manolescu, I.: Query-oriented summarization of RDF graphs. In: Maneth, S. (ed.) BICOD 2015. LNCS, vol. 9147, pp. 87–91. Springer, Cham (2015). https://doi.org/10.1007/978-3-319-20424-6_9

12. Neumann, T., Moerkotte, G.: Characteristic sets: accurate cardinality estimation for RDF queries with multiple joins. In: 2011 IEEE 27th International Conference on Data Engineering, pp. 984–994. IEEE (2011)
13. Dasarathy, B.V.: Nearest Neighbor (NN) norms: NN pattern classification techniques. IEEE Computer Society Tutorial (1991)
14. Chen, J.X., Reformat, M.Z.: Learning categories from linked open data. In: Laurent, A., Strauss, O., Bouchon-Meunier, B., Yager, R.R. (eds.) IPMU 2014. CCIS, vol. 444, pp. 396–405. Springer, Cham (2014). https://doi.org/10.1007/978-3-319-08852-5_41
15. Zong, N., Im, D.H., Yang, S., Namgoon, H., Kim, H.G.: Dynamic generation of concepts hierarchies for knowledge discovering in bio-medical linked data sets. In: Proceedings of the 6th International Conference on Ubiquitous Information Management and Communication, pp. 1–5 (2012)
16. Völker, J., Niepert, M.: Statistical schema induction. In: Antoniou, G., et al. (eds.) ESWC 2011. LNCS, vol. 6643, pp. 124–138. Springer, Heidelberg (2011). https://doi.org/10.1007/978-3-642-21034-1_9
17. Meimaris, M., Papastefanatos, G., Mamoulis, N., Anagnostopoulos, I.: Extended characteristic sets: graph indexing for sparql query optimization. In: IEEE 33rd International Conference on Data Engineering (ICDE), pp. 497–508. IEEE (2017)
18. Lutov, A., Roshankish, S., Khayati, M., Cudré-Mauroux, P.: Statix-statistical type inference on linked data. In: 2018 IEEE International Conference on Big Data (Big Data), pp. 2253–2262. IEEE (2018)

EDKT: An Extensible Deep Knowledge Tracing Model for Multiple Learning Factors

Liangliang He[1], Xiao Li[2], Jintao Tang[1(✉)], and Ting Wang[1(✉)]

[1] College of Computer, National University of Defense Technology, Changsha, China
{heliangliang19,tangjintao,tingwang}@nudt.edu.cn
[2] Information Center, National University of Defense Technology, Changsha, China
xiaoli@nudt.edu.cn

Abstract. Knowledge Tracing (KT) refers to the problem of predicting learners' future potential performance given their past learning history in e-learning systems. In order to better trace the learners' knowledge, KT tasks have become increasingly complicated recently, and various factors related to learning (such as skill, exercise, hint, etc.) have been incorporated into the modeling of KS of the learner, which renders it inadequate for the traditional KT definition to formalize these tasks. Therefore, this paper first gives a more general formal definition of KT tasks, and then proposes an Extensible Deep Knowledge Tracing model for multiple learning factors based on this general definition, named EDKT. EDKT can integrate various different learning factors by extending or ablating factors in two plug-ins on the basis of minor modifications. To demonstrate the effectiveness of the proposed model, we conduct extensive experiments on three real-world benchmark datasets, and the results show that EDKT comprehensively outperforms the state-of-the-art KT models on predicting future learner responses.

Keywords: Knowledge Tracing (KT) · Deep Knowledge Tracing · Knowledge state modeling

1 Introduction

Knowledge Tracing (KT) [3] is an important task in e-learning [16,20], whose goal is to model the knowledge state (KS) of the learner, i.e., the level of the learner's mastery of skills [21], based on the history of the learner's interaction with the e-learning platform. On an e-learning platform, learners can learn related skills by completing specific exercises. And the platform traces the learner's KS about the learned skills based on a KT model. Finally, the platform determines whether the learners have mastered these skills by a when-to-stop policy [17].

In a traditional KT task, given a learner's historical interaction sequence $\mathbf{X}_t = (\mathbf{x}_1, \mathbf{x}_2 \ldots \mathbf{x}_t)$ up to the timestamp t, KT models try to predict the probability that the learner will correctly perform a learning action (e.g., responding an exercise) in the next timestamp $t + 1$, i.e., $p(r_{t+1} = 1 | q_{t+1}, \mathbf{X}_t)$, where

© Springer Nature Switzerland AG 2021
C. S. Jensen et al. (Eds.): DASFAA 2021, LNCS 12681, pp. 340–355, 2021.
https://doi.org/10.1007/978-3-030-73194-6_23

$\mathbf{x}_t = (q_t, r_t)$ is an input tuple containing the exercise tag q_t at the timestamp t and the learner's response r_t to q_t [9, 21, 22].

Recent years, Deep Learning based Knowledge Tracing (DLKT) methods [4, 16, 21, 22] have been shown to be significantly better than traditional models [6], such as Bayesian Knowledge Tracing (BKT) [3], Latent Factor Models (LFM) [1, 15] and Item Response Theory (IRT) [20]. Therefore, this paper only focuses on the field of DLKT. Deep Knowledge Tracing (DKT) [16], as the first DLKT model, uses Long Short-Term Memory (LSTM) network [7] to capture the sequential dependency between each skill. Dynamic Key-Value Memory Network (DKVMN) [22] models the relationship between underlying concepts and traces the learner's KS about each concept. The self-attentive knowledge tracing (SAKT) method [13] is the first to apply attention mechanism to the field of KT, to deal with sparse data caused by learners' interaction with few skills.

In order to better trace learners' KSs, KT tasks have become increasingly complicated in recent years, and more and more side factors (such as hints [2], forgetting [12], text [8], etc.) have been incorporated into the modeling work of learner's KS. Nagatani et al. [12] extends the DKT model by modeling the information related to forgetting. Chaudhry et al. [2] extends the DKVMN model by modeling learners' hints data. Ghosh et al. [4] propose a new DLKT method by building context-aware representations of exercises and responses. Liu et al. proposes an Exercise-aware Knowledge Tracing (EKT) [8] framework by incorporating the information of skills and the textual content existed in each exercise.

However, while each of these models shows better performance than models adapted to the definition of the traditional KT task, each model is only suitable for the KT tasks with specific factors. To address this problem, this paper first gives a more general formal definition of KT tasks, and then proposes an Extensible Deep Knowledge Tracing model for multiple learning factors based on this general definition, named EDKT. EDKT is compatible with various KT tasks with different factors by extending or ablating factors in both extended plug-ins on DKVMN. To demonstrate the effectiveness of the proposed model, we conduct extensive experiments on three real-world benchmark datasets, and the results show that EDKT comprehensively outperforms the state-of-the-art KT model on predicting future learner responses.

To summarize, the contributions of this paper are summarized as follows:

(1) Based on the analysis of various existing KT tasks, a general definition of KT tasks is given. And this definition not only applies to all current KT tasks, but the generality allows for more factors to be extended in the future.
(2) Proposing an Extensible Deep Knowledge Tracing model for multiple learning factors based on DKVMN, named EDKT. EDKT is compatible with various KT tasks with different factors by extending or ablating factors in the both extended plug-ins on DKVMN.
(3) Conducting extensive experiments on three real-world benchmark datasets, and the results show that EDKT comprehensively outperforms the state-of-the-art KT model on predicting future learner responses.

The rest of this paper is organized as follows. Section 2 reviews the previous works. Section 3 introduce the general formal definition of KT task and the DKVMN model. Section 4 proposes the EDKT model. Section 5 conducts experiments and result analysis. The last section concludes this paper and presents the future work.

2   Related Works

In this section, we will give a brief overview of the existing works in the field of DLKT according to the number of learning factors extended by the KT model.

2.1   Single-Factor Models

For single-factor KT models, the single factor usually refers to skill or exercise. Since the exercise library is considerably larger than the skill set and many exercises are only learned by few learners in most e-learning patterns [13]. To avoid over-parameterization [4], researchers usually use skills to retrieve exercises, i.e., all exercises covering the same skill are treated as a single exercise $(s_t = q_t)$, unless the learning setting does not provide skill tags for the exercises. Representative methods are as follows.

The earliest single-factor DLKT model is Deep Knowledge Tracing (DKT) [16], which applies Long Short-Term Memory (LSTM) network [7] to the KT task. DKT uses hidden states as a kind of summary of the past learning sequence. Dynamic Key-Value Memory Networks (DKVMN) [22] with the ability to exploit the relationship between underlying concepts is the best single-factor model in recent years. DKVMN can automatically learn the correlation between the input skills and the underlying concepts, so as to trace the KS of the learner about each underlying concept. The self-attentive knowledge tracing (SAKT) method [13] is the first to apply attention mechanism to the field of KT, to deal with sparse data caused by learners' interaction with few exercises. SAKT models a learner's interaction sequence up to the current timestamp and predicts his(or her) performance on the next timestamp by considering the relevant exercises from his(or her) past interactions. In addition, there are some works [10,11,19,21] about the application of the above single-factor model to the specific KT tasks.

2.2   Multi-factor Models

Unlike the single-factor models, the multi-factor KT models require no less than two factors. Because there are many factors involved, the multi-factor models usually show better performance than the single-factor models.

Nagatani et al. [12] extends DKT by modeling the information related to forgetting. They consider both the learning sequence and the forgetting behavior by explicitly modeling the different forgetting behaviors of a learner using multiple features. Chaudhry et al. [2] extends DKVMN by modeling learners' hints data. In the process of updating the learner's KS, they use the triplet of adding hint instead of the doublet of only skill and response as the input to acquire the learner's *knowledge growth* after a response. Pandey and Srivastava propose a Relation-aware self-attention Knowledge Tracing (RKT) model [14] based on the SAKT model. They take into account the relations between skills involved in the interactions and time elapsed since the last interaction to inform the self-attention mechanism. Ghosh et al. propose a new KT method which is completely dependent on attention network Attentive Knowledge Tracing (AKT) [4]. AKT improves upon existing KT methods by building context-aware representations of exercises and responses, using a monotonic attention mechanism to summarize past learner performance in the right time scale. Liu et al. proposes an Exercise-aware Knowledge Tracing (EKT) [8] framework by incorporating the information of skill tags, exercise tags and the textual content of each exercise into a single model. They extend the LSTM model in different ways to extract the content features of the exercises and trace the KSs of learners.

3 Preliminaries

In this section, we introduce a general formal definition of KT tasks, and briefly explain the DKVMN model, which is the basic component of our model.

3.1 Knowledge Tracing Tasks

To formalize the increasingly complex KT tasks, a more general form is defined as follows: given a learner's historical interaction sequence $\mathbf{X}_t = (\mathbf{x}_1, \mathbf{x}_2 \ldots \mathbf{x}_t)$ up to the timestamp t (where $\mathbf{x}_t = (q_t, r_t[, s_t][, \mathbf{o}_t])$ is an input tuple containing the exercise tag q_t, the skill tag s_t and the set of other factor tags $\mathbf{o}_t$ at the timestamp t and the learner's response r_t (correct/incorrect) to q_t.), the exercise tag q_{t+1}, the skill tag s_{t+1} and other factors $\mathbf{o}_{t+1}$ at timestamp $t+1$ on a specific learning scenario, KT models try to predict the probability that the learner will correctly perform a learning action (e.g., responding an exercise) at timestamp $t+1$, i.e., $p(r_{t+1} = 1|q_{t+1}[, s_{t+1}][, \mathbf{o}_{t+1}], \mathbf{X}_t)$. In this formal definition, q and r exist by default, and s and $\mathbf{o}$ are optional to meet different KT tasks.

Compared with the traditional formal definition of KT tasks, the new definition extends $\mathbf{x}_t$ from doublet (q_t, r_t) to quadruplet $(q_t, r_t[, s_t][, \mathbf{o}_t])$, where $\mathbf{o}_t$ is an open set which allows other factors related to learning (such as hints, forgetting, text, etc.) to be extended. For the traditional formal definition, skills usually replace exercises as input to a KT model due to the sparsity of the exercise data. However, the success of more KT models (such as EKT [8] and AKT [4]) has shown that both skills and exercises can be used as inputs to obtain superior KT performance.

In a word, this new formal definition of KT tasks not only applies to all current KT tasks, but the generality allows for more factors to be extended by enriching the set of other factor tags $\mathbf{o}$ in the future.

3.2 Dynamic Key-Value Memory Networks

The knowledge tracing process of DKVMN [22] is shown in Fig. 2(a). At the timestamp t, DKVMN traces the KS of the learner by reading and writing to the *value*-memory matrix $\mathbf{M}_t^v$ using the correlation weight computed from the input skill and the *key*-memory matrix $\mathbf{M}^k$, and predicts the response of the learner to the skill based on the read memory content $\mathbf{r}_t$ and the input skill embedding $\mathbf{k}_t^s$. $\mathbf{M}^k$ and $\mathbf{M}^v$ are used to store the underlying concepts and the mastery levels of each concept, respectively. Details of DKVMN are as follows:

Correlation Weight. The correlation weight vector $\mathbf{w}_t$, which denotes the correlation between the input skill and each underlying concept, involves in two processes: response prediction and memory update.

$$\mathbf{w}_t = \mathrm{Softmax}(\mathbf{M}^{k^T} \mathbf{k}_t) \tag{1}$$

Response Prediction. Firstly, $\mathbf{r}_t$ is retrieved by the weighted sum of all memory slots in $\mathbf{M}_t^v$ based on $\mathbf{w}_t$:

$$\mathbf{r}_t = \mathbf{w}_t \left(\mathbf{M}_t^v\right)^T, \tag{2}$$

Then, $\mathbf{r}_t$ and $\mathbf{k}_t$ are input into the Tanh activation layer after being concatenated to generate $\mathbf{f}_t$:

$$\mathbf{f}_t = \mathrm{Tanh}(\mathbf{W}_f\left[\mathbf{r}_t, \mathbf{k}_t\right] + \mathbf{b}_f), \tag{3}$$

where $\mathbf{f}_t$ denotes a summary vector contained both the learner's mastery level and the difficulty of s_t; $\mathbf{W}_f$ and $\mathbf{b}_f$ are the linear transformation matrix and bias vector, respectively. Finally, $\mathbf{f}_t$ is passed through a fully connected layer to predict the probability p_t:

$$p_t = P\left(s_t\right) = \sigma\left(\mathbf{W}_p \mathbf{f}_t + \mathbf{b}_p\right), \tag{4}$$

where p_t is a scalar that denotes the probability of responding s_t correctly; $\mathbf{W}_p$ and $\mathbf{b}_p$ are the linear transformation matrix and bias vector, respectively.

Memory Update. The *knowledge growth* embedding $\mathbf{v}_t$ is used to update the *value*-memory $\mathbf{M}_t^v$ after working on s_t. Firstly, $\mathbf{M}_t^v$ is erased based on $\mathbf{v}_t$, and the erase process is as follows:

$$\mathbf{e}_t = \sigma\left(\mathbf{W}_e \mathbf{v}_t + \mathbf{b}_e\right). \tag{5}$$

$$\tilde{\mathbf{M}}_{t+1}^v = \mathbf{M}_t^v \otimes \left(\mathbf{1} - \mathbf{w}_t \mathbf{e}_t\right)^T, \tag{6}$$

where $\tilde{\mathbf{M}}_{t+1}^v$ is the intermediate of memory update. Then, $\mathbf{M}_t^v$ is added based on $\mathbf{w}_t$ and $\mathbf{v}_t$, and the added process is as follows:

$$\mathbf{a}_t = \tanh\left(\mathbf{W}_a \mathbf{v}_t + \mathbf{b}_a\right), \tag{7}$$

$$\mathbf{M}_{t+1}^v = \tilde{\mathbf{M}}_{t+1}^v + \mathbf{w}_t \mathbf{a}_t^T, \tag{8}$$

where $\mathbf{e}_t$ and $\mathbf{a}_t$ are the erase vector and the add vector computed from $\mathbf{v}_t$, respectively. And $\mathbf{M}_t^v$ is updated as the *value*-memory of the next timestamp $t+1$, $\mathbf{M}_{t+1}^v$. The *erase*-followed-by-*add* mechanism [5] allows forgetting and strengthening concept states in the learning process of a learner, a step similar to forget gates in LSTMs [7].

4   Model

DKVMN is just a single factor KT model based on skill or exercise. This section will introduce the EDKT model by integrating more factors based on DKVMN. We utilize DKVMN as our basic model because it is the state-of-the-art single-factor KT model, and a deep neural network that can easily incorporate multiple side factors.

4.1   Correlation Factors

In the real world, the completion of an exercise for a learner is affected by various factors, such as the difficulty of the exercise, the difficulty of the related skill, with or without hints, the current KS of the learner, forgetting factors, etc. Usually, specific exercises can only be completed if one or more of these factors are present. For example, when a learner has fully mastered all skills of a field, the learner can theoretically complete any exercise in the field without looking at the hints, and without making mistakes or forgetting; or when learners have the basic knowledge of a field, they can finish the exercise by using the hints, so as to realize the self-learning of the knowledge of the field. In this section, we analyze the four objective factors that may affect the learners' performance by combining real-world e-learning platform ASSISTments[1].

Exercise. As an object in the learning process, the difficulty of exercise greatly affects the learners' correct rate of doing exercises. However, due to the large size of exercise library, many exercises are seldom learned (refer to Table 1). Therefore, just using the exercise in the single-factor KT models may lead to over-parameterization of the related models [4].

Skill. In most cases, skills appear at the same time as the exercise, i.e., each exercise is assigned to a skill or multiple skills. Like the exercise, the skill itself also has the difficulty. Although the difficulty of the skill may determine the difficulty of the exercise to some extent, it is not exactly the same as the difficulty of the exercise. For example, if *Addition* and *Multiplication* are two different skills, the latter is obviously more difficult than the former. However, there is also the difference in the difficulty between their exercises of "$1 + 1$" and "$413 + 926$" or "1×1" and "413×926".

Template. In some cases, the exercise is generated based on the template. Compared with the exercise, the scale of the template set is smaller, but compared with the skill, the scale of the template set is relatively considerable. Therefore, using the template as a bridge between the exercise and the skill can help model the difficulty of exercise together with the skill (the relationship between the template and the skill is usually many-to-many).

---

[1] https://new.assistments.org.

Hint. In e-learning, the way for learners to achieve self-learning is to look at the outcome after attempting. A better way is to break the final outcome into multiple steps and set them as the hints, so that the learner will not look at the outcome directly without working on it, ensuring sufficient time to think. In general, the number of hints assigned by the platform for each exercise can reflect the difficulty of the exercise to a certain extent. We can see from Fig. 1 that the more the number of hints for an exercise, the more difficult the exercise is, given that ACR indicates the difficulty of the exercise.

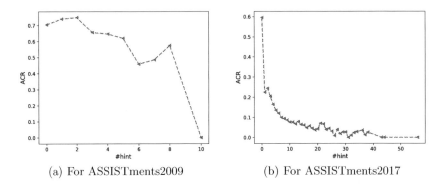

(a) For ASSISTments2009   (b) For ASSISTments2017

Fig. 1. The distributions of ACR on the number of hint for datasets ASSISTments2009 and ASSIST2017 (refer to Table 1), where ACR is the Average Correct Rate of all exercises with #hint hints.

The factors in different e-learning setting are different, so the objective factors that affect the learning effectiveness of learners are far from limited to these four kinds. However, it is obvious that if one or more effective factors can be incorporated into the single model, the effect of tracing learners' knowledge will be greatly improved on the basis of minimal time efficiency.

4.2 Extensible Deep Knowledge Tracing

For DKVMN (refer to Sect. 3.2), the main idea is to a predict first and update later. In other words, the response of the current skill is firstly predicted, and then the model is updated based on the real response of the learner to the skill. Unfortunately, only features related to the skill participate in both processes.

As we described in Sect. 4.1, the completion of an exercise for a learner is affected by various factors. Usually, specific exercises can only be completed if one or more of these factors are present. In addition, the essence of learning is not only the improvement of the level of doing exercises, but also the improvement of other abilities.

Therefore, we propose an Extensible Deep Knowledge Tracing model for multiple learning factors, named EDKT, and the overall model architecture is shown in Fig. 2(b). EDKT extends DKVMN by two plug-ins with similar structure: the Learning Factor Plug-in (LFP) and the Knowledge Growth Plug-in (KGP).

The former is used to provide features of the extended factors in the process of *Response Prediction*. The latter is used to enrich the learner's *knowledge growth* in the process of *Memory Update*. Both of them are indispensable in the whole learning process of learners.

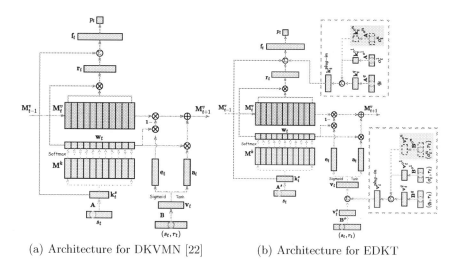

(a) Architecture for DKVMN [22]   (b) Architecture for EDKT

Fig. 2. In both architecture, the model is only drawn at the timestamp t, where the yellow components describe the extraction process of the correlation weight; the purple components describe the read process and the red components describe the write process; the green plug-in and the blue plug-in describe the preparation process of the extended factors and the extended *knowledge growth*, respectively. (Color figure online)

Learning Factor Plug-in. The basic model, DKVMN, only uses the features of the skill in *Response Prediction*, except for the KS extracted in the reading process. In order to model the KS of the learner reasonably, more other factors must be extended. LFP accomplishes this purpose and the structure is shown in Fig. 3(a). Where the green shaded region means that more factors can be extended; $k_t^{plug-in}$ represents the feature vector of the extended *plug-in*, and is formed by concatenating all the embedding vectors ($\mathbf{k}_t^q$, $\mathbf{k}_t^{o^1}$,...,$\mathbf{k}_t^{o^N}$) of the exercise and other factors. Each of these embedding vectors is embedded by the corresponding embedding matrix $\mathbf{A}$.

Knowledge Growth Plug-in. Same as the structure of LFP, the purpose of KGP (refer to in Fig. 3(b)) is to extract learners' *knowledge growth* about the exercise and other factors after one interaction. Where the blue shaded region represents an extensible structure corresponding to LFP (i.e., one doublet per extended factor); $\mathbf{v}_t^{plug-in}$ represents the *knowledge growth* vector of the extended *plug-in*, which is formed by concatenating all the *knowledge growth* vector of the

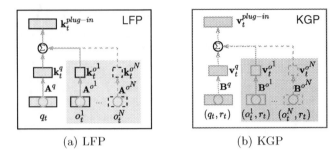

(a) LFP   (b) KGP

Fig. 3. Learning factor plug-in (LFP) and knowledge growth plug-in (KGP) in EDKT.

exercise and other factors $(\mathbf{v}_t^q, \mathbf{v}_t^{o^1}, ..., \mathbf{v}_t^{o^N})$; each of these vectors is embedded by the corresponding embedded matrix $\mathbf{B}$.

Knowledge Tracing Based on EDKT. The KT process of EDKT is similar to that of DKVMN, and the difference is that embedding of the extended *plug-ins* needs to be added before predicting and writing, respectively.

Embedding Layer. All factors and factor-response tuples are input into the EDKT model after embedding. The former contains s, q and $\mathbf{o}$ which denotes the set of side factors need to be extended, i.e. $\mathbf{o} = o^1, o^2, ..., o^N$, and the latter contains $(s, r), (q, r), (o^1, r), (o^2, r), ..., (o^N, r)$. Where N denotes the number of extended factors in EDKT. Each input factor is multiplied by $\mathbf{A}^{factor}$ to get $\mathbf{k}^{factor}$:

$$\mathbf{k}^{factor} = factor \cdot \mathbf{A}^{factor}, \tag{9}$$

where $factor$ represents one of all factors (including s, q and $\mathbf{o}$); $\mathbf{A}^{factor}$ and $\mathbf{k}^{factor}$ represents the continuous embedding matrix and vector of this factor, respectively. Each input tuple is multiplied by $\mathbf{B}^{factor}$ to get $\mathbf{v}^{factor}$:

$$\mathbf{v}^{factor} = tuple \cdot \mathbf{B}^{factor}, \tag{10}$$

where $tuple$ represents the joint tuple of $factor$ and r; $\mathbf{B}^{factor}$ and $\mathbf{v}^{factor}$ represents the continuous embedding matrix and vector of $tuple$, respectively.

Attention Layer. Due to the sparsity of the exercise data, we use the one-hot embedding of the skill tag to compute the correlation weight between the skill and the underlying concepts. Given the continuous embedding vector $\mathbf{k}_t^s$ of the input skill s_t at timestamp t, the correlation weight vector $\mathbf{w}_t$ is computed by taking the softmax activation of the inner product between $\mathbf{k}_t^s$ and key-memory $\mathbf{M}^k$ (refer to Eq. 1).

Response Prediction. When a skill s_t comes, the read vector $\mathbf{r}_t$ is firstly retrieved by the weighted sum of all memory slots in $\mathbf{M}_t^v$ using $\mathbf{w}_t$ (refer to Eq. 2), and the extracted read vector $\mathbf{r}_t$ is treated as a summary of the learner's mastery level of this skill. Secondly, the extended LFP embedding $\mathbf{k}_t^{plug-in}$ is formed by concatenating $\mathbf{k}_t^s$, $\mathbf{k}_t^q$ and $\mathbf{k}_t^{o^1}, ..., \mathbf{k}_t^{o^N}$:

$$\mathbf{k}_t^{plug-in} = [\mathbf{k}_t^q, \mathbf{k}_t^{o^1}, ..., \mathbf{k}_t^{o^N}], \tag{11}$$

where $\mathbf{k}_t^{plug-in}$ is treated as a summary of the difficulty of the exercise and the difficulties of the other side factors. Thirdly, we concatenate the read vector $\mathbf{r}_t$, the input skill embedding $\mathbf{k}_t^s$ and the input LFP embedding $\mathbf{k}_t^{plug-in}$ and then pass it through a fully connected layer with a Tanh activation to get a summary vector $\mathbf{f}_t$, which contains the learner's mastery level, the prior difficulty of the skill, the prior difficulty of the exercise and other prior factor:

$$\mathbf{f}_t = \text{Tanh}(\mathbf{W}_f^T[\mathbf{r}_t, \mathbf{k}_t^s, \mathbf{k}_t^{plug-in}] + \mathbf{b}_f), \tag{12}$$

Finally, $\mathbf{f}_t$ is passed through another fully connected layer with a Sigmoid activation to predict the performance of the learner (refer to Eq. 4).

Memory Update. After the learner complete the skill s_t, the *value*-memory matrix is updated according to the correctness of the learner's response. Specifically, $\mathbf{v}_t$ is written to $\mathbf{M}_t^v$ with the same correlation weight $\mathbf{w}_t$ used in predicting process (refer to Eq. 5−8). Where $\mathbf{v}_t$ is formed by concatenating the skill-response embedding $\mathbf{v}_t^s$ and the extended KGP embedding $\mathbf{v}_t^{plug-in}$ and is treated as the *knowledge growth* of the learner after the interaction at timestamp t. And $\mathbf{v}_t^{plug-in}$ is formed by concatenating the exercise-response joint embedding $\mathbf{v}_t^q$ and the extensible factor-response joint embedding $\mathbf{v}_t^{o^1}, ..., \mathbf{v}_t^{o^N}$:

$$\mathbf{v}_t^{plug-in} = [\mathbf{v}_t^q, \mathbf{v}_t^{o^1}, ..., \mathbf{v}_t^{o^N}]. \tag{13}$$

Model Training. All model parameters in EDKT are jointly learned by minimizing the following cross entropy loss between p_t and r_t during training.

$$\mathcal{L} = -\sum_t (r_t \log(p_t) + (1 - r_t) \log(1 - p_t)) \tag{14}$$

Note that if a skill tag is not set for each exercise in the online learning environment, the correlation weight vector $\mathbf{w}_t$ in *Attention Layer* will be computed by taking the softmax activation of the inner product between $\mathbf{k}_t^q$ and *key*-memory $\mathbf{M}^k$. At the same time, $\mathbf{k}_t^q$ will be removed from $\mathbf{k}_t^{plug-in}$. Similarly, $\mathbf{v}_t^q$ will participate in the process of *Memory Update* instead of $\mathbf{v}_t^s$ and $\mathbf{v}_t^q$ will be removed from $\mathbf{v}_t^{plug-in}$.

Extensibility Specification. For EDKT, $\mathbf{k}_t^{plug-in}$ and $\mathbf{v}_t^{plug-in}$ can be extended to integrate various factors related to learning. The whole extension process is divided into: i) each extended factor and the corresponding extended tuple are respectively embedded by Eq. 9 and Eq. 10 to get the corresponding factor embedding and tuple embedding; ii) the factor embedding and tuple embedding are concatenated in $\mathbf{k}_t^{plug-in}$ and $\mathbf{v}_t^{plug-in}$, respectively.

5   Experiments

In this section, we conduct extensive experiments on three real-world benchmark datasets to demonstrate the effectiveness of our proposed model, including
ablation experiments on two datasets.

5.1   Experimental Setup

Datasets. The performances of EDKT and three baselines are respectively tested
on three benchmark datasets ASSISTments2009[2], ASSISTments2017[3] and Stat-
ics2011[4]. The datasets of ASSISTments are collected from an online tutoring
platform. The dataset of Statics2011 is collected from a college-level engineering
course on statics[5]. For all these datasets, the users with only one interaction are
removed. The main statistics of all datasets are listed in Table 1.

Table 1. Statistical information for all datasets. Where # is the identifier of "the
number of"; "−" means the corresponding item is missing.

Statistics	Datasets		
	ASSISTments2009	ASSISTments2017	Statics2011
#records	525,534	942,816	261,947
#learners	4,217	1,709	333
#skills	124	102	−
#exercises	26,688	3,162	300
#steps	−	−	382
#templates	816	−	−
#maximum_hints	10	56	−

Baselines and Metric. We compare EDKT against several baseline KT methods,
including DKT [16], DKVMN [22], and the recently proposed model AKT [4],
which is a two-factor KT model that integrating both exercise and skill. We use
the Area Under the Curve (AUC) as the metric to evaluate the performances of
all methods on predicting binary-valued future learner responses to exercises. The
role of AUC is to evaluate the prediction accuracy of a model on specific datasets:
AUC of 0.5 is equivalent to the predicted results obtained by random guessing;
the higher the AUC value, the better the prediction performance of a model [22].

---

[2] https://sites.google.com/site/assistmentsdata/home/assistment-2009-2010-data.
[3] https://sites.google.com/view/assistmentsdatamining/dataset.
[4] https://pslcdatashop.web.cmu.edu/DatasetInfo?datasetId=507.
[5] http://oli.stanford.edu/engineering-statics.

Implementation Details. The input factors are presented to neural networks using "one-hot" input vectors. Take skill data for example, if S different skill exist in total, then the skill tag s_t for the *key*-memory part is a length S vector whose entries are all zero except for the s_t-th entry, which is one. Similarly, the tuple input (s_t, r_t) for the *value*-memory matrix is a length $2S$ vector, where the $(s_t + r_t \times S)$-th entry is one.

For evaluation purposes, we perform 5-fold cross-validation for all models and all datasets. Thus, for each fold, 20% learners are used as the test set, 20% are used as the validation set, and 60% are used as the training set. For each fold, we use the validation set to perform early stopping and tune the parameters for every KT method.

5.2 Results and Discussion

Overall Model Performance. Table 2 lists the performance of all KT methods across all datasets on predicting future learner responses, and Fig. 4 presents the results in an intuitive way. For EDKT, the best performing instance of all variants is selected, where EDKT_STH for ASSISTments2009, EDKT_SQH for ASSISTments2017, and EDKT_QS (exercise and step) for Statics2011 (refer to Table 3). we report the averages as well as the standard deviations across five test folds.

Table 2. Performances of all methods on all datasets. Where EDKT comprehensively outperforms the state-of-the-art KT models on each of the datasets. Best models are bold and second best models are '*'.

Datasets	Knowledge Tracing Models			
	DKT	DKVMN	AKT	EDKT
ASSISTments2009	0.8170.0043	0.83720.0120	0.86010.0089*	**0.8638**0.0097
ASSISTments2017	0.72630.0054	0.71830.0038	0.76640.0031*	**0.8690**0.0024
Statics2011	0.82330.0039*	0.80850.0077	0.8180.0068	**0.8607**0.0067

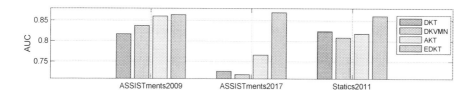

Fig. 4. Performances of all methods on all datasets.

The proposed EDKT (using the parameter optimization method in [18]) exhibits advanced performances on all datasets, and this result suggests that the importance of other factors to model the KS of the learner. Compared with the state-of-the-art methods, EDKT improves the AUC value by 0.43%, 13.39%, and 4.54% on datasets ASSISTments2009, ASSISTments2017 and Statics2011, respectively. Compared with the basic model DKVMN, EDKT improves the AUC value by 3.17%, 20.98%, and 6.46% on datasets ASSISTments2009, ASSISTments2017 and Statics2011, respectively. The reason for the limited performance improvement for ASSISTments2009 is that the exercise is not integrated with other factors in EDKT, due to over-parameterization (refer to **Factor Ablation Study** below for details).

Factor Ablation Study. In order to determine the contribution of different factors in EDKT to the final performance, we conduct a series of ablation experiments on ASSISTments2009 and ASSISTments2017 with more learning factors. Table 3 show the results of the ablation experiments, and Fig. 5 presents the results in an intuitive way.

Table 3. Ablation experiments of EDKT on ASSISTments2009 and ASSISTments2017. Where EDKT_Q and EDKT_S as the single-factor instances of EDKT are equivalent to DKVMN taking exercise and skill as input, respectively; EDKT_ST is the double-factor EDKT integrating skill and template; EDKT_SQ is the double-factor EDKT integrating exercises and skills; EDKT_STH is the triple-factor EDKT integrating skill, template and hint; EDKT_SQH is the triple-factor EDKT model integrating skill, exercise and hint.

ASSISTments2009	EDKT_Q	EDKT_S	EDKT_ST	EDKT_STH
	0.7540	0.8372	0.8577	0.8638
ASSISTments2017	EDKT_Q	EDKT_S	EDKT_SQ	EDKT_SQH
	0.7183	0.6908	0.7734	0.8678

In general, the single-factor model shows the worst performance; the integration of more factors in the model results in better performance; the contribution

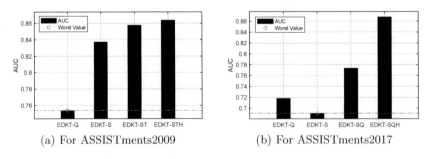

(a) For ASSISTments2009   (b) For ASSISTments2017

Fig. 5. Ablation experiments of EDKT.

of different factors to the final performance of the model is different. The results fully illustrate that one or more effective factors can be incorporated into the single model, the effect of tracing learner's KS will be affirmatively improved.

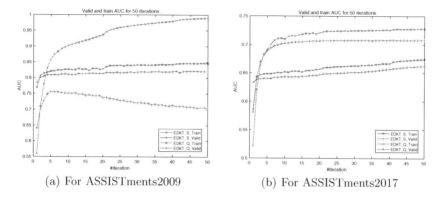

(a) For ASSISTments2009   (b) For ASSISTments2017

Fig. 6. Validation and training AUC of EDKT_S and EDKT_Q on ASSISTments2009 and ASSISTments2017.

For ASSISTments2009, EDKT_Q that only integrates exercises achieves worse performance compared with the single-factor model EDKT_S. The analysis of the results show that, EDKT_Q suffers severe over-fitting. As indicated in Fig. 6, no huge gap exists between the training AUC and the validation AUC of EDKT_S, and the validation AUC of EDKT_S increases smoothly. However, as the epoch proceeds, the training AUC of EDKT_Q increases continuously, and the validation AUC of EDKT only increases in the first several epochs and begins to decrease slowly. As many exercises are only learned by few learners (refer to Fig. 7), the model is over-parameterized in the training process.

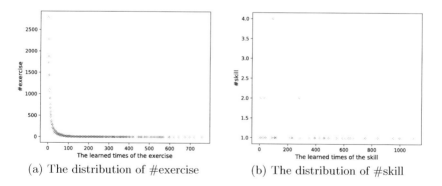

(a) The distribution of #exercise   (b) The distribution of #skill

Fig. 7. The distribution of the number of exercise and skill on the learned times for ASSISTments2009, where many exercises are rarely assigned to learners, while skill data is relatively evenly distributed.

The data analysis also finds that some of the exercises in the validation set and the test set are completely invisible to all learners in the training set. Therefore, exercises are not integrated in EDKT with other factors together. And the over-parameterization problem caused by data sparsity will be alleviated in future work.

For ASSISTments2017, EDKT_S only integrates exercises achieves worse performance compared with EDKT_Q. As indicated in Fig. 6, EDKT_Q does not suffer over-fitting, thanks to the suitable size of the exercise library. However, the size of the skill set is too small to reflect the difficulty of the exercises adequately, resulting in the worst performance. Compared with ASSISTments2009, hints makes a significantly greater contribution on ASSISTments2017, the reason is that the number of hint for an exercise is a better indicator of the difficulty of the exercise in the latter (refer to Fig. 1).

6   Conclusion and Future Work

This paper first gives a more general formal definition of KT tasks, and then proposes an Extensible Deep Knowledge Tracing model for multiple learning factors based on this general definition, named EDKT. EDKT is an extended version of DKVMN and is compatible with various KT tasks with different factors by extending or ablating factors in both extended plug-ins: Learning Factor Plug-in (LFP) and Knowledge Growth Plug-in (KGP). To demonstrate the effectiveness of the proposed model, we conduct extensive experiments on three real-world benchmark datasets, and the results show that EDKT comprehensively outperforms the state-of-the-art KT models on predicting future learner responses, with a maximum improvement of 20%. In the future, our work will focus on the over-fitting of models in the face of large-scale exercises and mining more side factors.

Acknowledgment. We would like to thank the anonymous reviewers for their helpful comments. The research is supported by the National Key Research and Development Program of China (2018YFB1004502) and the National Natural Science Foundation of China (61702532, 61532001, 61690203).

References

1. Cen, H., Koedinger, K., Junker, B.: Learning factors analysis – a general method for cognitive model evaluation and improvement. In: Ikeda, M., Ashley, K.D., Chan, T.-W. (eds.) ITS 2006. LNCS, vol. 4053, pp. 164–175. Springer, Heidelberg (2006). https://doi.org/10.1007/11774303_17
2. Chaudhry, R., Singh, H., Dogga, P., Saini, S.K.: Modeling hint-taking behavior and knowledge state of students with multi-task learning. International Educational Data Mining Society (2018)
3. Corbett, A.T., Anderson, J.R.: Knowledge tracing: modeling the acquisition of procedural knowledge. User Model. User-Adap. Inter. **4**(4), 253–278 (1994)

4. Ghosh, A., Heffernan, N., Lan, A.S.: Context-aware attentive knowledge tracing. In: Proceedings of the 26th ACM SIGKDD International Conference on Knowledge Discovery & Data Mining, pp. 2330–2339 (2020)
5. Graves, A., Wayne, G., Danihelka, I.: Neural turing machines. arXiv preprint arXiv:1410.5401 (2014)
6. He, L., Tang, J., Li, X., Wang, T.: ADKT: adaptive deep knowledge tracing. In: Huang, Z., Beek, W., Wang, H., Zhou, R., Zhang, Y. (eds.) WISE 2020. LNCS, vol. 12342, pp. 302–314. Springer, Cham (2020). https://doi.org/10.1007/978-3-030-62005-9_22
7. Hochreiter, S., Schmidhuber, J.: Long short-term memory. Neural Comput. 9(8), 1735–1780 (1997)
8. Huang, Z., Yin, Y., Chen, E., Xiong, H., Su, Y., Hu, G., et al.: Ekt: exercise-aware knowledge tracing for student performance prediction. IEEE Trans. Knowl. Data Eng. 33(1), 100–115 (2019)
9. Khajah, M., Lindsey, R.V., Mozer, M.C.: How deep is knowledge tracing? arXiv preprint arXiv:1604.02416 (2016)
10. Minn, S., Desmarais, M.C., Zhu, F., Xiao, J., Wang, J.: Dynamic student classification on memory networks for knowledge tracing. In: Yang, Q., Zhou, Z.-H., Gong, Z., Zhang, M.-L., Huang, S.-J. (eds.) PAKDD 2019. LNCS (LNAI), vol. 11440, pp. 163–174. Springer, Cham (2019). https://doi.org/10.1007/978-3-030-16145-3_13
11. Minn, S., Yu, Y., Desmarais, M.C., Zhu, F., Vie, J.J.: Deep knowledge tracing and dynamic student classification for knowledge tracing. In: 2018 IEEE International Conference on Data Mining (ICDM), pp. 1182–1187. IEEE (2018)
12. Nagatani, K., Zhang, Q., Sato, M., Chen, Y.Y., Chen, F., Ohkuma, T.: Augmenting knowledge tracing by considering forgetting behavior. In: The World Wide Web Conference, pp. 3101–3107 (2019)
13. Pandey, S., Karypis, G.: A self-attentive model for knowledge tracing. arXiv preprint arXiv:1907.06837 (2019)
14. Pandey, S., Srivastava, J.: RKT: relation-aware self-attention for knowledge tracing. In: Proceedings of the 29th ACM International Conference on Information & Knowledge Management, pp. 1205–1214 (2020)
15. Pavlik Jr., P.I., Cen, H., Koedinger, K.R.: Performance factors analysis-a new alternative to knowledge tracing. Online Submission (2009)
16. Piech, C., et al.: Deep knowledge tracing. In: Advances in Neural Information Processing Systems, pp. 505–513 (2015)
17. Rollinson, J., Brunskill, E.: From predictive models to instructional policies. International Educational Data Mining Society (2015)
18. Tan, Z., He, L.: An efficient similarity measure for user-based collaborative filtering recommender systems inspired by the physical resonance principle. IEEE Access 5, 27211–27228 (2017)
19. Wang, L., Sy, A., Liu, L., Piech, C.: Learning to represent student knowledge on programming exercises using deep learning (2017)
20. Wilson, K.H., Karklin, Y., Han, B., Ekanadham, C.: Back to the basics: Bayesian extensions of IRT outperform neural networks for proficiency estimation. arXiv preprint arXiv:1604.02336 (2016)
21. Yeung, C.K.: Deep-IRT: make deep learning based knowledge tracing explainable using item response theory. arXiv preprint arXiv:1904.11738 (2019)
22. Zhang, J., Shi, X., King, I., Yeung, D.Y.: Dynamic key-value memory networks for knowledge tracing. In: Proceedings of the 26th International Conference on World Wide Web, pp. 765–774 (2017)

Fine-Grained Entity Typing via Label Noise Reduction and Data Augmentation

Haoyang Li$^{(\boxtimes)}$, Xueling Lin, and Lei Chen

The Hong Kong University of Science and Technology, Hong Kong, China
{hlicg,xlinai,leichen}@cse.ust.hk

Abstract. Fine-grained entity typing aims to assign one or more types for entity mentions in the corpus. Recently, distant supervision has been utilized to generate training data. However, it has two drawbacks. First, the same labels are assigned to every entity mention in a context-agnostic manner, which introduces label noise. Some approaches alleviate this issue by hand-crafted features. However, they require efforts from experts. Second, the entity mentions out of Knowledge Base (KB) are ignored and hence cannot be added to the training data, which decreases the size of the training data. Furthermore, the existing entity typing systems neglect the types of other entity mentions in the same context which provide evidence to infer the types of the target entity mentions. In this paper, we first propose graph-based and sampling-based approaches, to reduce the label noise generated by the distant supervision, and then augment the training data by finding potential entity mentions in the corpus and inferring their types. Moreover, we propose a hierarchical neural network, which involves the types of other mentions in the context and satisfies the type consistency, to predict the types. Experiments on two datasets show that our system outperforms state-of-the-art entity typing systems.

Keywords: Entity typing · Noise reduction · Data augmentation

1 Introduction

Fine-grained entity typing [19,25] is proposed to assign one or more specific fine-grained types to an entity mention, which contributes to many real-world applications, such as question answering [4] and KB population [12]. Specifically, such fine-grained types can be organized in a tree-structured hierarchy. For example, `actor` is a child type of `person`, and `person` is the parent type of `singer`. One of the major challenge of this task is the absence of training dataset. Most traditional entity typing systems [16] generate training data manually, which requires extensive human efforts. To address it, distant supervision [15] is proposed to generate training data [2,17,18,23,27], by annotating the types of an entity mention based on its types recorded in an existing KB.

However, distant supervision will invoke two major problems. First, it may introduce label noise, since it assigns type labels in a context-agnostic manner. Take Fig. 1(a) as an example, under distant supervision, the type labels of

© Springer Nature Switzerland AG 2021
C. S. Jensen et al. (Eds.): DASFAA 2021, LNCS 12681, pp. 356–374, 2021.
https://doi.org/10.1007/978-3-030-73194-6_24

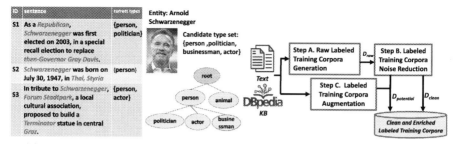

(a) The exmaple of distant supervision.   (b) Framewok overview.

Fig. 1. Example of distant supervision and framework overview

Arnold Schwarzenegger in all the sentences are {person, politician, actor, businessman} by consulting the KB. However, based on their context, the correct labeling should be {person, politician}, {person}, and {person, actor} for sentences s_1, s_2, s_3, respectively. Recently, some works reduce label noise by pruning heuristics such as deleting entity mentions with conflicting types [6]. However, such strategies reduce the size of the training dataset sharply. Moreover, some studies utilize heterogeneous graphs to reduce the label noise with hand-crafted features [18,19], which need the efforts from experts.

Second, the training dataset generated by distant supervision is limited. The reason is that only the entity mentions, that can be linked to the entities in the KB, are labeled. However, the coverage of entities in the KB is always limited. For example, the singer *Jay Chou* has a nickname *President Chou* that does not exist in the KBs such as DBpedia [10]. Hence, utilizing limited entities in the KB for annotation causes the *KB-restriction* problem that the entity mentions in the corpus out of KB will be ignored and not labeled. Thus, distant supervision cannot take full advantage of the corpus and decrease the size of training data.

Besides, there are two additional problems in fine-grained entity typing system. First, when the state-of-the-art methods [23,25] predict types for a target entity mention, they typically ignore the types of other entity mentions. Intuitively, the types of other entity mentions in the same context also provide valuable evidence. As shown in s_3 of Fig. 1(a), the type of *Terminator* is film role, which indicates that *Arnold* tends to be an actor. Secondly, current works [21,23] regards types independently without involving the hierarchy structure information, and cannot satisfy *hierarchical consistency* that the probability of an entity with a parent type should not be less than the probability of this entity with a child type. Hence, they may suffer from the underfitting problem [24]. For example, in s_1, the probability of *Arnold* being person should not be less than politician. The reason is that an entity mention is politician implies that it is also person, but the reverse does not holds.

To address these challenges, we propose a novel framework. First, to reduce the label noise and address the *KB-restriction* problem, we propose the graph-based and sampling-based approaches. (1) Graph-based approach: we assume that the types of entity mentions in the same context should be coherent. Specifically, for each context, we build a coherence graph of the entity mentions, with type coherence scores between each pair of entity mentions as the edge weights. We then reduce label noise by finding the maximum weight subgraph, where the maximal coherent score among types of entity mentions is selected. Finally, we assign types to potential entity mentions based on both the semantic score as well as the type coherence. (2) Sampling-based approach: we assume that two entity mentions share similar feature representation in the latent space if they have similar types. Specifically, we first obtain the feature representations of entity mentions and potential entity mentions based on their context, then reduce the label noise by a sampling-based iterative algorithm. Finally, we assign the potential entity mentions types based on their feature representation. The major difference between these two approaches is that the graph-based approach only considers entity mentions in the context, while the sampling-based approach involves all words, including the non-entity mention words. In addition, to satisfy the hierarchical consistency and involve the type information of other mentions, we propose a novel hierarchical neural network with an attention mechanism to predict types for the target entity mentions in the context.

In this paper, we first introduce the important definitions in Sect. 2 and framework overview in Sect. 3. We then propose graph-based and sampling-based approaches in Sect. 4 and 5, respectively. In Sect. 6, we propose a hierarchical neural network. We present the evaluation results of experiments in Sect. 7, discuss the related work in Sect. 8 and conclude in Sect. 9.

2   Problem Definition

We first briefly introduce the important definitions used in this paper as follows.

Definition 1 (Knowledge Base and Target Type Hierarchy). *A Knowledge Base (KB) Ψ stores a lot of ontological knowledge, including the types of entities. We denote the entities stored in the KB Ψ as $\mathcal{E}_\Psi$, and the type schema of the KB Ψ as $\mathcal{Y}_\Psi$. Specifically, a target type hierarchy $\mathcal{Y} \subseteq \mathcal{Y}_\Psi$ is organized into a tree, where the nodes represent the types. Furthermore, we denote the set of entities with target types in the KB Ψ as $\mathcal{T} = \{(e, y)\} \subset \mathcal{E}_\Psi \times \mathcal{Y}$. Specifically, following [18,19], we assume that the types of each entity consist of one type-path, which indicates one entity has at most one type in each level.*

Definition 2 (Labeled Training Corpus). *A labeled corpus generated by distant supervision consists of three major elements: (1) a set of entity mentions, i.e., $\{m_i\}_{i=1}^N$, where $m = \{w_1, w_2..., w_{|m|}\}$ is a word span in the text representing a real-world entity; (2) a set of the context regrading to each entity mention m_i, i.e., $\{c_i\}_{i=1}^N$, where $c_i = \{w_1, w_2..., w_{|c_i|}\}$ to represent its context (e.g., sentences or paragraphs in the text corpus) (3) a set of candidate type set of each entity mention m_i, i.e., $\{\mathcal{Y}_i\}_{i=1}^N$. Therefore, we denote the raw training dataset as $\mathcal{D}_{raw} = \{d_i\}_{i=1}^N = \{(m_i, c_i, \mathcal{Y}_i)\}_{i=1}^N$, where $d_i = (m_i, c_i, \mathcal{Y}_i)$ is an instance.*

Definition 3 (Problem Definition). *Given a KB Ψ with its hierarchy type schema $\mathcal{Y}_\Psi$, a target type hierarchy $\mathcal{Y}$, the entity-type facts $\mathcal{T} = \{(e, y)\} \subset \mathcal{E}_\Psi \times \mathcal{Y}$ and a text corpus $\mathcal{D}$, our task is to predict correct types $\mathcal{Y}'_i \subset \mathcal{Y}$ for each entity mention m_i based on its context c_i.*

3   Framework Overview

The framework is illustrated in Fig. 1(b), and we present details of each stage.

Stage 1: Labeled Training Corpora Generation. Given corpus $\mathcal{D}$, a KB Ψ with hierarchy type $\mathcal{Y}_\Psi$, and a target type $\mathcal{Y} \subseteq \mathcal{Y}_\Psi$, our target is to automatically generate a clean and large labeled training corpus. There are three major steps:

A. Raw Labeled Training Corpora Generation. Following distant supervision, we apply entity linking tools on $\mathcal{D}$ to obtain the candidate types for each entity mention m_i from the KB Ψ. The output is $\mathcal{D}_{raw} = \{(m_i, c_i, \mathcal{Y}_i)\}_{i=1}^{|\mathcal{D}_{raw}|}$.

B. Labeled Training Corpora Noise Reduction. Given $\mathcal{D}_{raw}$, we propose graph-based and sampling-based approaches to reduce the label noise in Sect. 4.1 and 5.1, respectively. We then obtain a clean training dataset $\mathcal{D}_{clean} = \{(m_i, c_i, \mathcal{Y}'_i)\}_{i=1}^{|\mathcal{D}_{clean}|}$.

C. Labeled Training Corpora Augmentation. We augment the training corpora by adding the entity mentions out of the KB Ψ, i.e., *potential entity mentions*. We first obtain a training dataset $R_{potential} = \{(p_i, c_i)\}_{i=1}^{|R_{potential}|}$ by extracting high-quality entity mentions $\{p_i\}$ from the text corpus $\mathcal{D}$. We then infer types for each p_i based on its context c_i and $\mathcal{D}_{clean}$, and output the potential training dataset $\mathcal{D}_{potential} = \{(p_i, c_i, \mathcal{Y}'_i)\}_{i=1}^{|\mathcal{D}_{potential}|}$. Two approaches are proposed for training data augmentation in Sect. 4.2 and 5.2.

Stage 2: Type Prediction via a Hierarchical Neural Network. We design a hierarchical neural network in Sect. 6 that satisfies the type consistency and is trained on the clean and potential training datasets obtained in Stage 1.

4   Graph Based Improvement

In this section, we propose a novel graph-based approach to reduce the label noise for entity mentions in a *top-to-down* manner, and then augment the training dataset by adding potential entity mentions and their types, based on a *type coherence graph* among mentions and the *semantic score*. The intuition is that the types of entity mentions in the same context should be coherent. Take Fig. 1(a) as an example, in the sentence s_3, the type of *Terminator* is `film role`, which indicates that *Arnold* tends to be an `actor` instead of a `politician`, since `film role` is more coherent with `actor` than `politician`.

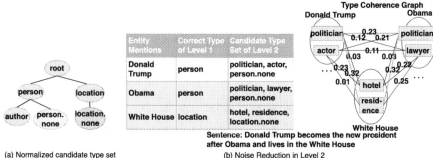

Fig. 2. Normalized candidate type set and the noise reduction in Level 2 (Color figure online)

4.1 Labeled Training Corpora Noise Reduction

In this task, we first conduct data normalization on $\mathcal{D}_{raw}$, and then reduce the label noise in a *top-to-down* manner.

Step 1: Data Normalization. Given $\mathcal{D}_{raw} = \{(m_i, c_i, \mathcal{Y}_i)\}_{i=1}^{|\mathcal{D}_{raw}|}$, we first transform $\mathcal{Y}_i$ into a normalized type set $\mathcal{Y}_i^*$, by adding a none type as the child type for each level (except for the root level). Please note that the none types in different type path are different. Figure 2(a) shows an example. We then transform $\mathcal{D}_{raw}$ into $\mathcal{D}'_{raw} = \{C_i\}_{i=1}^{|\mathcal{D}'_{raw}|}$ where $C_i = \{(m_k, \mathcal{Y}_k^*)\}_{j=1}^{|C_i|}$ and $\{m_k\}_{k=1}^{|C_i|}$ denotes entity mentions in the same context c_i.

Step 2: Label Noise Reduction. As discussed earlier, there is a type-path for each entity mention involving its context [18,19]. Thus, one entity mention can only have at most one type in each level. Following this, we conduct the label noise reduction in a top-to-bottom manner. Specifically, we select at most one type as the correct type for each level, and gather its children type as candidate types for the next level. Finally, these correct types can consist of only one type-path. For simplicity and generality, let $C_i^l = \{(m_1, \mathcal{Y}_1^l), \cdots, (m_n, \mathcal{Y}_n^l)\}$ denote entity mentions in context c_i with their normalized candidate type set in the l-th level. There are two steps to conduct the noise reduction (1) to construct the type coherence graph G_i^l based on C_i^l, and (2) to find the maximum weight subgraph in the graph G_i^l. We define the type coherence graph and maximum weight subgraph as follows.

Definition 4 (Coherence Score). *Given two types y_i and y_j, the coherence score is defined as defined as:* $w(y_i, y_j) = \dfrac{o_{ij}/f_{y_i} + o_{ij}/f_{y_j}}{2}$, *where f_{y_i} and f_{y_j} are the frequency of the type y_i and y_j in D'_{raw}, and o_{ij} is the co-occurrence times between y_i and y_j in D'_{raw}.*

Definition 5 (Type Coherence Graph). *A type coherence graph describes the coherence among all the types of entity mentions in one context. Formally, given a context c_i consisting of n entity mentions with candidate type sets of the l-th level, i.e., $C_i^l = \{(m_1, \mathcal{Y}_1^l), \cdots, (m_n, \mathcal{Y}_n^l)\}$, the type coherence graph of C_i^l is an undirected graph $G_i^l = (Y_i, E_i)$. Specifically, Y_i denotes the types $y \in \mathcal{Y}_i^l$. The edge $e \in E_i$ is connected by a pair of types that are from different entity mentions. The weight of an edge denotes the coherence score between two types.*

Definition 6 (Maximum Weight Subgraph). *Given a type coherence graph G_i^l constructed on $C_i^l = \{(m_k, \mathcal{Y}_k^l)\}_{k=1}^n$, the target is to find a maximum weight subgraph by selecting only one type y_k^l from $\mathcal{Y}_k^l$ for m_k. We denote the total and the average edge weight of the maximum weight subgraph as $w_i^{l_{max}}$ and $w_i^{l_{avg}}$, respectively, i.e., $w_i^{l_{avg}} = \frac{2 \cdot w_i^{l_{max}}}{n(n-1)}$ since there are $\frac{n(n-1)}{2}$ edges in the subgraph.*

Example. Figure 2(b) illustrates a toy example of label noise reduction of the second level. The candidate type sets in the second level of these three entity mentions are shown in the table. We first construct a coherence graph by adding edges among types that are from different entity mentions, and the edge weight is the coherence score. Then we obtain the maximum weight subgraph that is shown by green lines. As a result, the correct types for these three entity mentions are `politician`, `politician` and `official residence`, respectively, $w_i^{2_{max}} = 0.87$ and $w_i^{2_{avg}} = 0.29$.

Algorithm 1 shows the details of graph-based label noise reduction. Specifically, for the l-th level, we obtain the candidate type set $\mathcal{Y}_i^l$ for each entity mention m_i, by involving the children types of the correct type y_i^{l-1} in the last level from the normalized candidate type set $\mathcal{Y}_i^*$ (line 4). We then construct the type coherence graph G_i^l based on C_i^l (line 5), and reduce the label noise by finding a maximum weight subgraph from G_i^l (line 6). Specifically, since finding a maximum weight subgraph is NP-hard [1], we propose a greedy algorithm that consequently select the edge with the maximum weight until we find a subgraph. We repeat the label noise reduction process for C_i for all levels (lines 3–6). Finally, we obtain the result by fusing the clean types $\{y_i^l\}$ from 1 to the H-th level into one type set $\mathcal{Y}_i'$ (line 7). Please note that we can transform $\mathcal{D}'_{clean}$ into another format $\mathcal{D}_{clean}$ by deleting all `none` types of $\mathcal{Y}_i'$ from $\mathcal{D}'_{clean}$.

Time Complexity. We construct H coherence graphs and find their maximum weight subgraphs that traverse all edges in the coherence graphs. Hence, the time complexity is $O(\sum_{l=1}^H E_l \cdot |\mathcal{D}'_{raw}|)$, where $E_l = \frac{n(n-1)}{2} m_l^2$, n is the number of mentions, and m_l is the average number of types in the l-th level.

Algorithm 1: Label Noise Reduction by Type Coherence Graph

Input: $\mathcal{D}'_{raw} = \{C_i\}_{i=1}^{|\mathcal{D}'_{raw}|}$ where $C_i = \{(m_i, \mathcal{Y}_i^*)\}_{i=1}^{|C_i|}$

Output: $\mathcal{D}'_{clean} = \{C_i'\}_{i=1}^{|\mathcal{D}'_{clean}|}$ where $C_i' = \{(m_i, \mathcal{Y}_i')\}_{i=1}^{|C_i'|}$

1 $\mathcal{D}'_{clean} \leftarrow \emptyset$

2 **foreach** $instance\ C_i = \{(m_i, \mathcal{Y}_i^*)_{i=1}^{|C_i|}\} \in \mathcal{D}'_{raw}$ **do**

3   **foreach** l in $range(H)$ **do**

4         $C_i^l = \{(m_i, \mathcal{Y}_i^l)\}_{i=1}^{|C_i|} \leftarrow$ get the l-th level candidate types

5         $G_i^l \leftarrow$ construct the l-th level type coherence graph

6         $C_i^{l'} = \{(m_i, \{y_i^l\})\}_{i=1}^{|C_i|} \leftarrow$ maximum weight subgraph of G_i^l

7     $C_i' \leftarrow Union(C_1^{1'}, \cdots, C_H^{H'})$

8     $\mathcal{D}'_{clean} \leftarrow \mathcal{D}'_{clean} \cup C_i'$

9 **Return** $\mathcal{D}'_{clean} = \{C_i'\}_{i=1}^{|\mathcal{D}'_{clean}|}$

4.2 Labeled Training Corpora Augmentation

In this subsection, we infer the types of the potential entity mention p_i based on its context c_i in a top-to-down manner. Specifically, when we assign the type y_i^l for potential entity mention p_i as the correct type in the *l-th* level, it should satisfy two conditions: (1) p_i should be semantically coherent with y_i^l, and (2) y_i^l should be coherent with the types of other entity mentions in the context c_i.

Semantic Score. Given the potential entity mention p_i and the type y_i^l, the semantic score $Sem(p_i, y_i^l) = max(cosine(\mathbf{r}_{p_i}, \mathbf{r}_{e_j}))$, where $e_j \in \{e|(e, y_i^l)\} \subset \mathcal{T}$, and $\mathbf{r}_{p_i}$ and $\mathbf{r}_{e_j}$ are computed by averaging the word embedding of the words in p_i and e_j, respectively.

Coherence Degree. Given the type y_i^l and $C_i^{l'} = \{(m_k^l, \{y_k^l\})\}_{k=1}^{|C_i^{l'}|}$ obtained in Sect. 4.1, the coherent degree between y_i^l and the correct types of other mentions in C_i' is $W_{(y_i^l, c_i)} = \sum_{k=1}^{|C_i^{l'}|} w(y_i^l, y_k^l)$, where $w(y_i^l, y_k^l)$ is coherence score.

Algorithm 2 shows the detail of labeled training corpora augmentation. The basic idea is to assign types for each instance $(p_i, c_i) \in \mathcal{R}_{potential}$ in a *top-to-down* manner (lines 3–11). More specifically, similar to Algorithm 1, we first construct the candidate type set $\mathcal{Y}_i^l$. We then assign type y_i^l to p_i, if $y_i^l = \arg\max_{y_i^l \in \mathcal{Y}_i^l}(W_{(y_i^l, c_i)} \cdot Sem(p_i, y_i^l))$ and $\frac{W_{(y_i^l, c_i)}}{|C_i'|} \geq w_i^{l_{avg}}$. The type assignment terminates when there is no suitable type for p_i, or we have assigned the type in the H level for p_i. Finally, the output is the augmented training dataset $\mathcal{D}_{potential} = \{(p_i, c_i, \mathcal{Y}_i')\}_{i=1}^{|\mathcal{D}_{potential}|}$.

Algorithm 2: Graph-based Training Dataset Augmentation

Input: $\mathcal{R}_{potential} = \{p_i, c_i\}^{|\mathcal{R}_{potential}|}$ and $\mathcal{D}'_{clean} = \{C'_j\}_{j=1}^{|\mathcal{D}'_{clean}|}$

Output: $\mathcal{D}_{potential} = \{(p_i, c_i, \mathcal{Y}'_i)\}_{i=1}^{|\mathcal{D}_{potential}|}$

1 $\mathcal{D}_{potential} \leftarrow \emptyset$
2 **foreach** *instance* $d_i = (p_i, c_i) \in \mathcal{R}_{potential}$ **do**
3     $\mathcal{Y}'_i \leftarrow \emptyset$
4     $C'_i \leftarrow$ get from $\mathcal{D}'_{clean}$
5   **for** l **in** $range(H)$ **do**
6        $\mathcal{Y}^l_i \leftarrow$ gets the l-th level candidate types
7        $y^l_i \leftarrow \arg\max_{y^l_i \in \mathcal{Y}^l_i} (W_{(y^l_i, c_i)} \cdot Sem(p_i, y^l_i))$
8   **if** $\frac{W_{(y^l_i, c_i)}}{|C'_i|} \geq w^{avg_l}_i$ **then**
9           $\mathcal{Y}'_i \leftarrow \mathcal{Y}'_i \cup \{y^l_i\}$
10   **else**
11   break
12    $\mathcal{D}_{potential} \leftarrow \mathcal{D}_{potential} \cup (p_i, c_i, \mathcal{Y}'_i)$
13 **Return** $\mathcal{D}_{potential} = \{(p_i, c_i, \mathcal{Y}'_i)\}_{i=1}^{|\mathcal{D}_{potential}|}$

Time Complexity. Suppose that on average there are n entity mentions in each instance of $\mathcal{D}'_{clean}$. We compute the semantic score and the type coherence score for each potential instance, which takes $O(|\mathcal{Y}|n + |\mathcal{T}|)$ time. Hence, the time complexity is $O((|\mathcal{Y}| \cdot n + |\mathcal{T}|) \cdot |\mathcal{R}_{potential}|)$.

5   Sampling Based Improvement

The graph-based approach is efficient, but ignores the non-entity mention words in the context, which also indicate the types of the target entity mentions. For example, in the sentence s_2 of Fig. 1(a), *a local cultural association* provides an evidence that *Forum Stadtpark* tends to be an `organization`.

Therefore, we propose a novel sampling-based approach to reduce the label noise and augment the training dataset, by utilizing the distributed feature representation of all the words in context. Our intuition is that if two entity mentions have similar types, their feature representation in the latent space in the context should be similar.

5.1   Labeled Training Corpora Noise Reduction

We propose four major steps to reduce the label noise in $\mathcal{D}_{raw}$.

Step 1. Entity Mention Centered Embedding Method. BERT [5] is trained on unlabeled data and has proved its capability to capture the semantic meaning of words in many tasks. In this step, we utilize the training dataset $\mathcal{D}_{raw}$

to fine tune the pre-trained BERT model to capture the feature representation of the entity mentions based on their context. In our work, the pre-trained BERT model is *bert-base-uncased*, and its output dimension for each word is 768. As illustrated in Fig. 3(a), for each instance $d_i = (m_i, c_i, \mathcal{Y}_i) \in \mathcal{D}_{raw}$, we transform it into $([CLS], m_i, [SEP], c_i, [SEP])$ format, where the [SEP] symbol is to separate the entity mention and its context, and [CLS] is used as a symbol to aggregate the features from other words in the same context. Moreover, we add a fully connected layer with $R^{768 \times |\mathcal{Y}|}$ dimensions. We use the output vector [CLS] as its input to predict entity types, and fine tune the BERT model by minimizing the cross-entropy loss between the predicted types and $\mathcal{Y}_i$. We assume that a mention based on its context should be embedded closer to its relevant types rather than its irrelevant types. Hence, the fune-tuned BERT groups entity mentions with the same types into the same latent space based on their context.

We then generate the feature representation of the entity mentions based on context. Given the raw training dataset $\mathcal{D}_{raw} = \{(m_i, c_i, \mathcal{Y}_i)\}_{i=1}^{|\mathcal{D}_{raw}|}$ and the fine-tuned BERT model, we transform m_i and c_i as the input of BERT, and obtain the output vector [CLS] as its feature representation $\mathbf{r}_i$. Also, for each instance $(p_j, c_j) \in \mathcal{R}_{potential}$, we get the feature representation $\mathbf{r}_j$ using the same method. Please note that obtaining the representation of potential mentions based on the context does not need the label.

Step 2. Similarity Between Instances in the Labeled Training Corpora. Given two instances in $\mathcal{D}_{raw}$, i.e., $d_i = (m_i, c_i, \mathcal{Y}_i)$ and $d_j = (m_j, c_j, \mathcal{Y}_j)$, we define the similarity between d_i and d_j as $sim(d_i, d_j) = \frac{r_i \cdot r_j}{\|r_i\| \cdot \|r_j\|}$, where r_i and r_j are the feature representations of d_i and d_j. $sim(d_i, d_j)$ is larger if the types of entity mention m_i and m_j are similar. Specifically, d_j is a positive sample of d_i if $sim(d_i, d_j) > 0.5$, or else d_j is a negative sample. We use Pos_i and Neg_i to denote the set of positive and negative samples of d_i, respectively.

Step 3. Confidence Score Between Types. The confidence score $cof(y_j|y_i)$ as the probability that an entity mention has the type y_j under the condition that it has the type y_i. Formally, we define $cof(y_j|y_i) = \frac{|\mathcal{T}_{y_i} \cap \mathcal{T}_{y_j}|}{|\mathcal{T}_{y_i}|}$ where $\mathcal{T}_{y_i} = \{e|(e, y_i)\} \subset \mathcal{T}$ and $\mathcal{T}_{y_j} = \{e|(e, y_j)\} \subset \mathcal{T}$.

Step 4. Probability of Type Assignment. For each instance $d_i = (m_i, c_i, \mathcal{Y}_i)$, the probability $p(y_i|m_i, c_i)$ that each $y_i \in \mathcal{Y}_i$ can be assigned for m_i is computed as follows: $p(y_i|m_i, c_i) = \frac{r(y_i|m_i, c_i)}{r(y_i|m_i, c_i) + r(\bar{y}_i|m_i, c_i)}$, where $r(y_i|m_i, c_i)$, $r(\bar{y}_i|m_i, c_i)$ are the reliability of type y_i belonging to, not belong to m_i in terms of c_i, respectively.

Algorithm 3: Label Noise Reduction by Sampling

Input: The raw training dataset $\mathcal{D}_{raw} = \{(m_i, c_i, \mathcal{Y}_i)\}_{i=1}^{|\mathcal{D}_{raw}|}$, and $\{r_i\}_{i=1}^{|\mathcal{D}_{raw}|}$.

Output: The clean training dataset $\mathcal{D}_{clean} = \{(m_i, c_i, \mathcal{Y}_i')\}_{i=1}^{|\mathcal{D}_{clean}|}$

1 $\mathcal{D}_{clean} \leftarrow \emptyset$

2 **while** *unconverge* **do**

3   **foreach** *instance* $d_i = (m_i, c_i, \mathcal{Y}_i) \in \mathcal{D}_{raw}$ **do**

4        $Pos_i, Neg_i \leftarrow RandomSelectFrom(\mathcal{D}_{raw})$

5   **foreach** $y_i \in \mathcal{Y}_i$ **do**

6           $p(y_i|m_i, c_i) \leftarrow$ step 4

7 **foreach** *instance* $d_i = (m_i, c_i, \mathcal{Y}_i) \in \mathcal{D}_{raw}$ **do**

8     $\mathcal{Y}_i' \leftarrow \emptyset$

9   **for** l **in** *range(H)* **do**

10       $\mathcal{Y}_i^l \leftarrow$ get the `1`-th level candidate types

11       $y_i^l = \arg\max_{y_i^l \in \mathcal{Y}_i^l} p(y_i^l|m_i, c_i)$

12   **if** $p(y_i^l|m_i, c_i) \geq 0.5$ **then**

13          $\mathcal{Y}_i' \leftarrow \mathcal{Y}_i' \cup \{y_i^l\}$

14   **else**

15   break

16    $\mathcal{D}_{clean} = \mathcal{D}_{clean} \cup \{(m_i, c_i, \mathcal{Y}_i')\}$

17 **Return** $\mathcal{D}_{clean} = \{(m_i, c_i, \mathcal{Y}_i')\}_{i=1}^{|\mathcal{D}_{clean}|}$

Intuitively, if the representation of two instance are similar, the types of their entity mentions should be similar and conversely, it also holds. Hence, we define the reliability $r(y_i|m_i, c_i) = \frac{1}{|Pos_i|} \sum_{d_j \in Pos_i} sim(d_i, d_j) cof(y_i|y_j) p(y_j|m_j, c_j)$ and $r(\bar{y}_i|m_i, c_i) = \frac{1}{|Neg_i|} \sum_{d_j \in Neg_i} (1 - sim(d_i, d_j)) cof(y_i|y_j) p(y_j|m_j, c_j)$ where $y_j = \arg\max_{y_j \in \mathcal{Y}_j} cof(y_i|y_j)$.

Remark. For each instance $d_i \in \mathcal{D}_{raw}$, we can utilize the rest $|\mathcal{D}_{raw}| - 1$ instances to infer its correct types. However, it is time-consuming especially when $\mathcal{D}_{raw}$ is large. Hence, for efficient purpose, we propose a sampling method that we sample η positive samples and η negative samples respectively for d_i. We assume the similarity distribution for an instance d_i fits the uniform distribution from 0 to 1. Hence, the sample number expectation is less than $\frac{\eta}{0.5} + \frac{\eta}{0.5} = 4\eta$.

Algorithm 3 shows the detail of sampling-based label noise reduction. Specifically, for each $d_i = (m_i, c_i, \mathcal{Y}_i) \in \mathcal{D}_{raw}$, we first randomly select η positive samples and η negative samples (lines 4) to compute the probability $p(y_i|m_i, c_i)$ following step 4, where $y_i \in \mathcal{Y}_i$. After I iterations or when the difference of the average type probability of all instances in $\mathcal{D}_{raw}$ between two adjacent iterations is less than 0.01, the type assignment procedure converges. We then select the clean types in a *top-to-down* manner (lines 14–22). Specifically, we select the type y_i^l from the candidate type set $\mathcal{Y}_i^l$ as the clean type of the l-th level if $y_i^l = \arg\max_{y_i^l \in \mathcal{Y}_i^l} p(y_i^l|m_i, c_i)$ and $p(y_i^l|m_i, c_i) > 0.5$. We then construct the candidate type set $\mathcal{Y}_i^{(l+1)}$ of the $(l+1)$-th level by involving the children types of

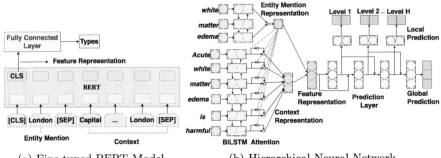

(a) Fine-tuned BERT Model.   (b) Hierarchical Neural Network.

Fig. 3. Fine-tuned BERT model and hierarchical neural network

y_i^l from $\mathcal{Y}_i$. If we cannot find clean type in the *l-th* level, we regard there is no clean type in the *l-th* and deeper level. Finally, we get the clean training dataset $\mathcal{D}_{clean} = \{(m_i, c_i, \mathcal{Y}_i')\}$ as output. The initial value of $p(y_i|m_i, c_i)$ is set as 0.5.

Time Complexity. Suppose on average there are m types for each instance. For each instance, we sample η positive samples and η negative samples to compute the probability of types, which runs I iterations, so this procedure takes $O((4\eta + 2\eta m) \cdot I)$ time. Also, label noise reduction takes $O(m)$ time. Overall, the time complexity of Algorithm 3 is $O((4\eta + 2\eta m) \cdot I|\mathcal{D}_{raw}|)$.

5.2   Labeled Training Corpora Augmentation

The basic idea is to utilize $\mathcal{D}_{clean}$ to infer types for potential mention p_i based on its representation r_i. Specifically, for each $(p_i, c_i) \in \mathcal{R}_{potential}$, we select η positive and η negative samples from $\mathcal{D}_{clean}$ based on r_i. Second, we adopt a *top-to-down* manner to infer a type-path for p_i from the first level to the *H-th* level. Specially, in the *l-th* level, type y_i^l is regarded as the correct type, only if $y_i^l = \arg\max_{y_i^l \in \mathcal{Y}_i^l} p(y_i^l|m_i, c_i)$ and $p(y_i^l|m_i, c_i) > 0.5$. We then construct the *(l+1)-th* candidate type set $\mathcal{Y}_i^{l+1}$ by involving the children types of y_i^l from $\mathcal{Y}$. Finally, we obtain $\mathcal{D}_{potential} = \{(p_i, c_i, \mathcal{Y}_i')\}_{i=1}^{|\mathcal{D}_{potential}|}$. Similar to Algorithm 3, the time complexity is $O((4\eta + 2\eta \cdot |\mathcal{Y}|) \cdot |\mathcal{R}_{potential}|)$.

6   Type Prediction via Hierarchical Neural Network

In this section, we propose a hierarchical neural network, that involves the type of other entity mentions and satisfies hierarchical consistency, to predict types for the target entity mention. The model architecture is shown in Fig. 3(b).

6.1   The Architecture

Embedding Layers. For each word w in m_i, we get the word embedding value $\mathbf{w}^e$ from the pretrained embedding. For each word w in context c_i, we incorporate the word position embedding $\mathbf{w}^p$ to reflect relative distances between w and m_i. Similarly, we incorporate the type of other entity mentions into the embedding. Every type y can be mapped to a type embedding $\mathbf{w}_y$. Then, the type embedding value of w_i can be generated by $\mathbf{w}^t = \frac{1}{T}\sum_{k=1}^{T}\mathbf{w}_{y_k}$, where y_k is the types of the other mention m containing w_i. Finally, the word embedding of w_i is $\mathbf{w}_i^c = [\mathbf{w}^e, \mathbf{w}^p, \mathbf{w}^t]$.

Representation Layers. Given an entity mention $\mathbf{m}_i = \{\mathbf{w}_i^e\}$ and a BiLSTM that consists of two sub-networks for the forward and the backward pass, the mention representation can be obtained as $\mathbf{r}_{m_i} = \overrightarrow{\mathbf{h}_{right}} \oplus \overleftarrow{\mathbf{h}_{left}}$, where $\oplus$ denotes element-wise plus, $\overrightarrow{\mathbf{h}_{right}}$ and $\overleftarrow{\mathbf{h}_{left}}$ denote the outputs of forward and backward pass. Similarly, given context $\mathbf{c}_i = \{\mathbf{w}_i^c\}$ and a BiLSTM, for each word w_j in the context c_i, the feature vector $\mathbf{h}_{i,j}$ can be generated by $\mathbf{h}_{i,j} = \overrightarrow{\mathbf{h}_{i,j}} \oplus \overleftarrow{\mathbf{h}_{i,j}}$. Since the word in context have different impact on inferring the type of the entity mention, we employ the attention mechanism to obtain the context representation $\mathbf{r}_{c_i} = \sum_{j=0}^{l}\beta_{i,j}\mathbf{h}_{i,j}$, where $\beta_{i,j} = \frac{exp(\mathbf{r}_{m_i}tanh(W \cdot \mathbf{h}_{i,j}))}{\sum_{k=0}^{l}exp(\mathbf{r}_{m_i}tanh(W \cdot \mathbf{h}_{i,k}))}$ denotes the context word weight on m_i, and W is the parameter matrix. Then, the final feature representation is $\mathbf{r}_i = \mathbf{r}_{m_i} \odot \mathbf{r}_{c_i}$, where $\odot$ denotes vector concatenation.

Prediction Layers. We then feed $\mathbf{r}_i$ through $H+1$ fully connected layers with residual format input [7], and get the global predicted results of all types in $\mathcal{Y}$, and the local predicted results in each hierarchical level. Formally, let $\mathcal{Y}^l$ be the set of types of $\mathcal{Y}$ in the l-th level. The activation value F_G^l after the l-th global fully connected layer is defined as $\mathbf{F}_G^l = \phi(\mathbf{W}_G^l(\mathbf{F}_G^{l-2} \odot \mathbf{F}_G^{l-1}) + \mathbf{b}_G^l)$, where $\mathbf{W}_G^l \in \mathbb{R}^{|\mathbf{F}_G^l| \times (|\mathbf{F}_G^{l-2}| + |\mathbf{F}_G^{l-1}|)}$ is the weight parameter and $\mathbf{b}_G^l \in \mathbb{R}^{|\mathbf{F}_G^l| \times 1}$ is the bias parameter. Please note that $\mathbf{F}_G^0 = \mathbf{r}_i$. The global predicted results are obtained as: $\mathbf{P}_G = \sigma(\mathbf{W}_G^{H+1}(\mathbf{F}_G^{H-1} \odot \mathbf{F}_G^H) + \mathbf{b}_G^{H+1})$, where $\mathbf{W}_G^{H+1} \in \mathbb{R}^{|\mathcal{Y}| \times (|\mathbf{F}_G^{H-1}| + |\mathbf{F}_G^H|)}$, $\mathbf{b}_G^{H+1} \in \mathbb{R}^{|\mathcal{Y}| \times 1}$, and σ denotes the sigmoid function. For the l-th level, we can get the local predicted results: $\mathbf{P}_L^l = \sigma(\mathbf{W}_L^l\mathbf{F}_G^l + \mathbf{b}_L^l)$, where $\mathbf{W}_L^l \in \mathbb{R}^{|\mathcal{Y}^l| \times |\mathbf{F}_G^l|}$ is a weight parameters matrix and $\mathbf{b}_L^l \in \mathbb{R}^{|\mathcal{Y}^l| \times 1}$ is the bias parameter vector. Then the final predicted result is generated by: $\mathbf{P}_F = \beta\mathbf{P}_L + (1-\beta)\mathbf{P}_G$ where $\beta \in [0,1]$ is to control the proportion between the local and global information.

6.2   Optimization

We optimize the neural network by minimizing the global loss L_G, and the local loss L_L, and the hierarchical inconsistency loss L_H. Formally, the global loss $L_G = -\frac{1}{N}\sum_{i=0}^{N}\sum_{j=1}^{|\mathcal{Y}|}\hat{y_{ij}}logp_i(y_j) + (1-\hat{y_{ij}})log(1-p_i(y_j))$, the local loss $L_L = -\frac{1}{N}\sum_{i=0}^{N}\sum_{l=1}^{H}\sum_{j=|\mathcal{Y}^{l-1}|}^{|\mathcal{Y}^l|}\hat{y_{ij}}logp_i^l(y_j) + (1-\hat{y_{ij}})log(1-p_i^l(y_j))$, where

$\hat{y}_{ij}$ is a binary type indicator denoting whether m_i belongs to y_j type, $p_i(y_j)$ and $p_i^l(y_j)$ are the global and local predicted result, respectively. The inconsistency loss, that can guarantee the hierarchical consistency and keep the type dependency in the type path, is denoted as $L_H = \frac{1}{H}\sum_{l=1}^{H} L_l$ where $L_l = |max(0, p_i^l(y_j) - p_i^{l-1}(y_a))|^2$, y_j is the type in l-th level and y_a is the parent of y_j. Therefore the total loss of this model L_M is $L_M = L_L + L_G + L_H + \lambda\|\Theta\|^2$, where Θ denotes all parameters in the model. Finally, we optimize the neural network by minimizing L_M.

7   Experiments

7.1   Datasets and Preprocessing

Description of Datasets. We use two real-world datasets. (1) BC5CDR[1]: the corpora is from recent BioCreativew V Chemical and Disease Mention Recognition task, which is a medicine-domain dataset that mainly includes entity mentions with `drug`, `chemicalsubstance` and `disease` types. (2) NYT: The training corpus consists of 1000 articles from *New York Times* collected by [11] on general domains, such as `politics`, `food`, `sports` and `movies`.

Table 1. Statistics of the datasets.

	#target types	#all types	#train	#test	max depth H	Average $|\mathcal{Y}_i|$
BC5CDR	36	97	18181	2377	3	1.61
NYT	68	108	19725	2060	3	2.64

Generation of Evaluation Dataset. For each dataset, we follow a 90/10 ratio to separate the data into training corpora and evaluation corpora. To evaluate our model, we annotate the evaluation set semi-automatically. We first employ a sentence tokenization tool NLTK [13] to separate each evaluation corpus into sentences. We then utilize an entity linking tool DBpedia Spotlight [14] to recognize the entity mentions in each sentence and link them to the entities in DBpedia. Specifically, we keep the types in evaluation dataset whose frequency is over 10 as the target type set $\mathcal{Y}$, and regard the other types as side information. Afterwards, we delete the wrong annotations manually. We further randomly select 10% mentions from the evaluation corpora as the validation dataset.

Generation of Training Dataset. We first employ NLTK [13] and DBpedia Spotlight [14] to generate $\mathcal{D}_{raw}$. Specifically, if the words of an entity mention are labeled separately, we regard the union of the types of such words as the candidate type set. We then obtain the training dataset $\mathcal{D}_{raw} = \{(m_i, c_i, \mathcal{Y}_i)\}_{i=1}^{|\mathcal{D}_{raw}|}$. The details are listed in Table 1.

---

[1] https://www.ncbi.nlm.nih.gov/research/bionlp/Data/.

Generation of Potential Training Dataset. Given entity-type facts $\mathcal{T} = \{(e,y)\} \subset \mathcal{E}_\Psi \times \mathcal{Y}$ and corpus $\mathcal{D}$, we use AutoPhrase [20] to extract high quality phrases from $\mathcal{D}$ that are similar to e in $\mathcal{T}$, and we regard these high-quality phrases as potential entity mentions. Finally, we get the $R_{potential} = \{(p_i, c_i)\}$.

Type Confidence Score. We utilize 2016-10 DBpedia, which has $6.6M$ entities, $13B$ triples, and 760 classes, to compute the confidence score between types.

Hyper Parameter. We set the sampling number in Sect. 5 to be 60 and 170 for BC5CDR and NYT, respectively. The parameters of the neural network are searched by Hyperopt [3]. Due to the space limit, we do not report them here.

Baselines. We compare our model with several state-of-the-art approaches. We search hyper parameters for PLE, AFET by grid search and for NFETC by Hyperopt [3] approach. We then run each model three times and use the average value as their results. Duo to space limit, we only report the results.

1. **AFET** [19] proposes an embedding method to model entity mentions in terms of the number of their type path and train the model with partial label loss.
2. **PLE** [18] proposes heterogeneous partial-label embedding for label noise reduction to improve the performance of entity typing systems. We compare the PLE model with FIGER typing system as a baseline.
3. **NFETC** [25] proposes a neural network with hierarchy loss functions to handle the overly-specific labels. We compare the NFETC with partial label and hierarchical loss functions that are denoted as **NFETC** and NFETC$_{hier}$.
4. **CType** is our proposed entity typing model. We report results with five variants. In **CType**, the hierarchical neural network is directly trained on the D_{raw}. In CType$_{gc}$ and CType$_{sc}$, the hierarchical neural network is trained on the D_{clean} obtained by the graph-based and sampling-based noise reduction algorithm in Sect. 4 and 5 respectively. In **CType**$_{gca}$ and **CType**$_{sca}$, the neural network is trained on the both D_{clean} and $D_{potential}$.

Table 2. Performance evaluation in the *BC5CDR* dataset.

Method	Accuracy	Macro-P	Macro-R	Macro-F1	Micro-P	Micro-R	Micro-F1
AFET	0.825	0.885	0.868	0.876	0.875	0.845	0.860
PLE+FIGER	0.657	0.945	0.773	0.850	0.865	0.712	0.781
NFETC	0.660	0.909	0.768	0.833	0.917	0.714	0.803
NFETC$_{hier}$	0.754	0.882	0.796	0.837	0.891	0.749	0.813
CType	0.873	0.926	0.911	0.918	0.958	0.893	0.924
CType$_{gc}$	0.886	0.930	0.921	0.925	0.957	0.903	0.929
CType$_{gca}$	0.913	0.943	0.939	0.941	0.956	0.928	0.942
CType$_{sc}$	0.899	0.939	0.928	0.933	0.958	0.917	0.937
CType$_{sca}$	**0.928**	**0.958**	**0.955**	**0.957**	**0.963**	**0.946**	**0.954**

Table 3. Performance evaluation in the *NYT* dataset.

Method	Accuracy	Macro-P	Macro-R	Macro-F1	Micro-P	Micro-R	Micro-F1
AFET	0.585	0.780	0.731	0.755	0.807	0.755	0.780
PLE+FIGER	0.452	0.918	0.668	0.773	0.698	0.672	0.685
NFETC	0.639	0.842	0.818	0.830	0.869	0.825	0.847
NFETC$_{hier}$	0.668	0.858	0.827	0.842	0.895	0.832	0.862
CType	0.801	0.930	0.938	0.934	0.938	0.942	0.941
CType$_{gc}$	0.824	0.944	0.942	0.943	0.951	0.945	0.948
CType$_{gcu}$	0.755	0.900	0.909	0.904	0.910	0.914	0.912
CType$_{sc}$	**0.860**	**0.954**	**0.949**	**0.952**	**0.963**	**0.952**	**0.957**
CType$_{sca}$	0.822	0.935	0.927	0.931	0.947	0.932	0.939

Evaluation Metrics. Let P denote the evaluation set. For mention $m \in P$, we denote its ground-truth types as t_m and the predicted types as $\hat{t}_m$. Similarly to [18,19,26], we use the following metrics (1) **Accuracy: Accuracy** $= \sum_{m \in P} 1(t_m = \hat{t}_m)/|P|$. (2) **Loose Macro.** The Macro Precision and Macro Recall are computed for each mention: **Ma-P** $= \frac{1}{|P|} \sum_{m \in P} |t_m \cap \hat{t}_m|/|\hat{t}_m|$ and **Ma-R** $= \frac{1}{|P|} \sum_{m \in P} |t_m \cap \hat{t}_m|/|t_m|$. (3) **Loose Micro.** The Micro Precision and Micro Recall are computed by averaging all entity mentions: **Mi-P** $= \sum_{m \in P} |t_m \cap \hat{t}_m|//\sum_{m \in P} |\hat{t}_m|$ and **Mi-R** $= \sum_{m \in P} |t_m \cap \hat{t}_m|/\sum_{m \in P} |t_m|$.

7.2   Comparison and Analysis

Comparing Baselines with Our Model. Tables 2 and 3 summarize the results on BC5CDR and NYT, respectively. All variants of our model outperform other baselines in strict accuracy, Macro-F1 and Micro-F1 metrics. AFET and PLE suffer from low strict accuracy. The reason is that they manually define features to capture the shallow semantic meaning of entity mentions with their context, which may not represent the semantic meaning fully. Moreover, NFETC selects the type with the maximum confidence among candidate types. However, it is a suboptimal in the early stage whose parameters are randomly initialized, and introduces errors. CType$_{gc}$ and CType$_{sc}$ outperform other baselines, which implies that our framework can achieve satisfying performance on noisy datasets.

Comparing Variants of Our Model. The performance of CType$_{gc}$ and CType$_{sc}$ is better than CType on both dataset. This result indicates that the reduction of label noise improves the performance. Moreover, CType$_{sc}$ and CType$_{sca}$ perform better than CType$_{gc}$ and CType$_{gca}$, since the graph-based approach ignores the non-entity mentions. Also, CType$_{gca}$ and CType$_{sca}$ achieve better performance compared with CType$_{gc}$ and CType$_{sc}$ on BC5CDR, while the trend is different at NYT. The reason is that most of the entity mentions in NYT corpora have similar contexts. Therefore, it is easy to introduce errors for the type inference of potential entity mentions, which hurts the performance.

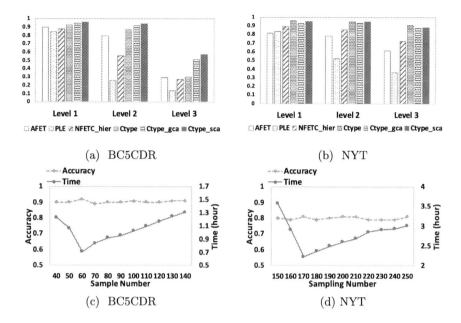

(a) BC5CDR

(b) NYT

(c) BC5CDR

(d) NYT

Fig. 4. Experiment results

Test at Different Type Levels. Figure 4(a)(b) reports the accuracy of six models at different levels. The results show that AFET and PLE obtain satisfying performance at the first level. Nevertheless, they are difficult to detect correct types in deeper levels. For example, they only reach 27.6% and 13.5% in the third level on BC5CDR. The variants of our model outperform all baselines in three levels on both dataset, and also improve the accuracy in the deeper levels significantly. More specifically, $CType_{sca}$ achieves a 27.5% improvement in level 3 compared with the beset baseline AFET on BC5CDR, and CType achieves a 18.3% improvement compared with the best baseline $NFETC_{hier}$ on NYT. These gains prove the effectiveness of our entity typing system.

Efficiency and Effectiveness of Sampling Algorithm. We explore Algorithm 3 in terms of different sampling numbers. As illustrated in Fig. 4(c)(d), when the sampling number is less than 40 (150) in BC5CDR (NYT), the algorithm will not converge within 5 h. The accuracy of both datasets is steady when Algorithm 3 converges. Also, for BC5CDR (NYT), the time cost decreases when the sampling number increases from 40 to 60 (150 to 170), then increases as the sampling number becomes larger. The reason is that smaller sampling number leads to more iterations before convergence, and larger sampling number results in longer time for each iteration.

8 Related Work

Fine-Grained Entity Typing. To reduce label noise, Gillick [6] proposes three pruning heuristics to get clean training dataset. However, these heuristics decrease the size of training dataset sharply. Recently, AFET [19] models `clean` and `noisy` mentions separately with different loss function incorporating type hierarchy information obtained from training dataset. Moreover, PLE [18] proposes a heterogeneous partial-label framework to reduce label noise. However, these two methods rely on hand-crafted features heavily. NFETC [25] proposes a hierarchical loss function to reduce label noise, but it suffers from error aggregation problem since the model at the beginning is sub-optimal.

Some research works utilize neural networks. Shimaoka *et al.* [21] propose an attentive neural network with LSTM to capture the context feature. However, it neglects the label noise. AAA [2] proposes a neural network incorporating label noise information with a hinge loss function. However, it treats types equally without considering the correlations among types. Xin *et al.* [23] utilizes the attention mechanism and incorporates information from KBs and text, but ignores the label noise. Moreover, the above approaches do not consider the side information of other entity mentions in the context and the hierarchical consistency.

Data Augmentation. It deals with the problem that the training dataset is not enough. [22] replaces words in the context with synonyms on the text classification task, but how to choose word is a non-trivial problem. [8] combines a variational auto-encoder and attribute discriminator to generate fake data. [9] proposes a bi-directional language model to replace words based on the context. However, training a variational auto-encoder or bidirectional language model need a lot of work. More importantly, above approaches cannot solve KB restriction problem since they cannot detect new entity mentions.

9 Conclusion

In this paper, we propose a novel framework CType for the fine-grained entity typing task. We first propose two approaches to reduce label noise from training dataset introduced by distant supervision approach and augment the training dataset by finding potential entity mentions from corpora. Then, we design a hierarchical neural network considering the side information of other entity mentions and satisfying the type hierarchical consistency, to predict types for the target entity mentions based on their context. Experiments on two real-world datasets demonstrate that our framework outperforms state-of-the-art models.

Acknowledgment. This work is partially supported by the Hong Kong RGC GRF Project 16202218, CRF Project C6030-18G, C1031-18G, C5026-18G, AOE Project AoE/E-603/18, China NSFC No. 61729201, Guangdong Basic and Applied Basic Research Foundation 2019B151530001, Hong Kong ITC ITF grants ITS/044/18FX and

ITS/470/18FX, Microsoft Research Asia Collaborative Research Grant, Didi-HKUST joint research lab project, and Wechat and Webank Research Grants.

References

1. Althaus, E., Blumenstock, M., Disterhoft, A., Hildebrandt, A., Krupp, M.: Algorithms for the maximum weight connected k-induced subgraph problem. In: Zhang, Z., Wu, L., Xu, W., Du, D.-Z. (eds.) COCOA 2014. LNCS, vol. 8881, pp. 268–282. Springer, Cham (2014). https://doi.org/10.1007/978-3-319-12691-3_21
2. Anand, A., et al.: Fine-grained entity type classification by jointly learning representations and label embeddings. arXiv (2017)
3. Bergstra, J., et al.: Hyperopt: a python library for optimizing the hyperparameters of machine learning algorithms. In: SciPy, pp. 13–20. Citeseer (2013)
4. Cui, W., et al.: KBQA: learning question answering over QA corpora and knowledge bases. PVLDB **10**(5), 565–576 (2017)
5. Devlin, J., et al.: Bert: pre-training of deep bidirectional transformers for language understanding. arXiv preprint arXiv:1810.04805 (2018)
6. Gillick, D., et al.: Context-dependent fine-grained entity type tagging (2014)
7. He, K., et al.: Deep residual learning for image recognition. In: CVPR, pp. 770–778 (2016)
8. Hu, Z., Yang, Z., et al.: Toward controlled generation of text. In: ICML, vol. 70, pp. 1587–1596. JMLR. org (2017)
9. Kobayashi, S.: Contextual augmentation: data augmentation by words with paradigmatic relations. arXiv:1805.06201 (2018)
10. Lehmann, J., et al.: DBpedia-a large-scale, multilingual knowledge base extracted from Wikipedia. Semant. Web **6**(2), 167–195 (2015)
11. Lin, X., Chen, L.: Canonicalization of open knowledge bases with side information from the source text. In: ICDE, pp. 950–961. IEEE (2019)
12. Lin, X., et al.: KBPearl: a knowledge base population system supported by joint entity and relation linking. PVLDB **13**(7), 1035–1049 (2020)
13. Loper, E., et al.: Nltk: the natural language toolkit. arXiv preprint (2002)
14. Mendes, P.N., et al.: DBpedia spotlight: shedding light on the web of documents. In: I-SEMANTICS, pp. 1–8. ACM (2011)
15. Mintz, M., Bills, S., Snow, R., Jurafsky, D.: Distant supervision for relation extraction without labeled data. In: ACL—AFNLP, pp. 1003–1011 (2009)
16. Nadeau, D., et al.: A survey of named entity recognition and classification. Lingvisticae Investigationes **30**(1), 3–26 (2007)
17. Ren, X., El-Kishky, A., et al.: Clustype: effective entity recognition and typing by relation phrase-based clustering. In: SIGKDD. ACM (2015)
18. Ren, X., He, W.O.: Label noise reduction in entity typing by heterogeneous partial-label embedding. In: SIGKDD, pp. 1825–1834. ACM (2016)
19. Ren, X., et al.: AFET: automatic fine-grained entity typing by hierarchical partial-label embedding. In: EMNLP, pp. 1369–1378 (2016)
20. Shang, J., Liu, J., Jiang, M., Ren, X., Voss, C.R., Han, J.: Automated phrase mining from massive text corpora. IEEE TKDE **30**(10), 1825–1837 (2018)
21. Shimaoka, S., et al.: An attentive neural architecture for fine-grained entity type classification. arXiv preprint arXiv:1604.05525 (2016)
22. Wei, J.W., et al.: EDA: easy data augmentation techniques for boosting performance on text classification tasks. arXiv:1901.11196 (2019)

23. Xin, J., Lin, Y., Liu, Z., Sun, M.: Improving neural fine-grained entity typing with knowledge attention. In: AAAI (2018)
24. Xu, D., et al.: A survey on multi-output learning. arXiv (2019)
25. Xu, P., Barbosa, D.: Neural fine-grained entity type classification with hierarchy-aware loss. arXiv:1803.03378 (2018)
26. Yogatama, D., et al.: Embedding methods for fine grained entity type classification. In: ACL—IJCNLP (2015)
27. Zeng, D., et al.: Distant supervision for relation extraction via piecewise convolutional neural networks. In: EMNLP, pp. 1753–1762 (2015)

DMSPool: Dual Multi-Scale Pooling for Graph Representation Learning

Hualei Yu, Chong Luo, Yuntao Du, Hao Cheng, Meng Cao,
and Chongjun Wang[✉]

National Key Laboratory for Novel Software Technology, Department of Computer
Science and Technology, Nanjing University, Nanjing, China
{hlyu,duyuntao,chengh,caomeng}@smail.nju.edu.cn, chjwang@nju.edu.cn

Abstract. Graph neural networks (GNNs) have recently become a powerful graph representation technique for graph-related tasks. However, the existing GNN models mainly focus on generalizing convolution and pooling operations in a pre-defined unified architecture, limiting the model's ability to capture meaningful information of nodes or local structures. Besides, the importance of subgraphs at various levels has not been well-reflected. To address the above challenges, we propose Dual Multi-Scale Pooling (DMSPool), which uses multiple architectures concurrently to integrate graph convolution and pooling modules in an end-to-end fashion. Specifically, these modules adopt multiple GNN architectures to learn node-level embeddings and nodes' importance from different aggregation iterations. Additionally, we employ attention mechanism to adaptively determine the contribution of subgraphs' representations at varying levels to graph classification and integrate them to perform the cross-scale graph level representation. Experiment results show that DMSPool achieves superior graph classification performance over the state-of-the-art graph representation learning methods.

Keywords: Graph neural networks · Graph convolution · Graph pooling · Multiple GNN architectures

1 Introduction

Graph neural networks (GNNs) have recently become a powerful graph representation technique for numerous graph-related tasks in various fields [1,2]. The existing GNN models can be generally classified into two categories: spectral and spatial approaches. The spectral methods focus on defining convolution operation utilizing graph Fourier transform and convolution theorem in the spectral domain [3,4]. For the spatial methods [1,5–7], convolution operations follow the message-passing process, in which the key steps involve transferring, transforming, and aggregating the node feature information from topological neighbors. The aforementioned methods mainly focus on generalizing convolution operations in a pre-defined architecture. [8] devises an appropriate aggregation strategy for each node to obtain useful information. It is worth noting that their goal is

© Springer Nature Switzerland AG 2021
C. S. Jensen et al. (Eds.): DASFAA 2021, LNCS 12681, pp. 375–384, 2021.
https://doi.org/10.1007/978-3-030-73194-6_25

performing node classification tasks. Commonly adopted approach to obtain the corresponding graph-level representations is global pooling. With growing interest in graph pooling, several innovative methods have been proposed to learn hierarchical representations of graphs [9–12]. However, these methods mentioned above have the following problems: (1) Firstly, they generally utilize the convolution operations designed for node classification tasks to extract node features. (2) Additionally, these methods fail to consider the fact that the graphs with different sizes and properties may need different architectures to fully capture useful information. (3) Lastly, the importance of subgraphs at various levels has not been well-reflected.

The main contributions can be summarized as follows:

- A novel graph pooling framework DMSPool is presented to perform hierarchical graph representation learning in an end-to-end fashion. Combined with attention mechanism, the contributions of subgraphs' representations at different levels to graph classification can be adequately fused.
- To the best of our knowledge, we are the first to propose the hypothesis that graph convolution operation in graph classification tasks should be different from that in node classification tasks.
- We evaluate our algorithm DMSPool on five widely used benchmarks. Experimental results clearly show that DMSPool achieves superior performance over other state-of-the-art graph representation learning methods.

2   Related Work

Bruna et al. [13] first design the convolution for graph data from the spectral domain. Then, the ChebNet [14] and AGCN [3] are proposed to further improve the convolution performance. GCN [1] simplifies the convolution operation and aggregates the node features from the one-hop neighbors. GAT [6] introduces the attention mechanism to specify varying weights to different nodes' features. GraphSAGE [5], a general inductive framework, can sample and aggregate features of nodes with mean/max/LSTM pooling. AM-GCN [15] extracts the specific and common embeddings from node features, topological structures and their combination simultaneously.

Graph pooling operation is of vital importance for graph classification tasks and can be grouped into global and hierarchical pooling. The global pooling methods do not learn the structure information of subgraphs, which is essential to the whole graph. DiffPool [9] learns a soft cluster assignment matrix in layer l, which contains the probability values of nodes being assigned to clusters. The gPool [16] designs a novel SortPooling layer that sorts graph vertices consistently so that traditional neural networks can be trained on the graphs. SAGPool [11] introduces self-attention using graph convolution, allows the pooling method to consider both node features and graph topology. Besides these, numerous graph hierarchical pooling methods have emerged recently, including EigenPooling [17], Relational Pooling [18], HGP-SL [12] and StructPool [19].

3   Problem Formulation

We represent a graph $\mathcal{G}$ as (V, E, A, X), where the set $V = (v_1, v_2, \ldots, v_n)$ collects all the n nodes of graph $\mathcal{G}$, and each $e \in E$ denotes an edge between nodes in graph $\mathcal{G}$. $A \in R^{n \times n}$ denotes the adjacency matrix, where the entry $A_{ij} = 1$ if there is an edge between v_i and v_j, and $X \in R^{n \times d}$ is the node feature matrix, and d is the dimension of node features. In the graph classification setting, given a set of graphs $\mathbf{G} = \{\mathcal{G}_1, \mathcal{G}_2, \ldots, \mathcal{G}_n\}$ and each being associated with a label. Formally, we can define our problem as follows:

Input: Given a set of graphs $\mathbf{G}_L = \{\mathcal{G}_1, \mathcal{G}_2, \ldots, \mathcal{G}_L\}$ with corresponding label information $\{y_1, y_2, \ldots, y_L\}$, graph neural network architecture $\mathbf{M}$, pooling rate k, and other hyperparameters.

Output: Predicting the unseen graphs $\mathbf{G}/\mathbf{G}_L$ in the dataset with $\mathbf{M}$ in an end-to-end fashion.

4   Methodology

4.1   The Proposed DMSPool Layer

Our proposed DMSPool is a multi-layer hierarchical GNN model, and shown in Fig. 1. Each layer consists of a convolution module and a pooling module. In the convolution module, DMSPool launches multiple GCNs with different aggregation iterations to learn node embeddings. Then, various node embeddings resulted from multiple GCNs are merged to generate the new node features for the next module. In the pooling module, DMSPool employs multiple GCN architectures, which acquire several nodes' scores based on self-attention by considering both node features and graph topology from multiple scales simultaneously. The scores can be summed to obtain the final scores representing the importance of nodes within the graph. According to the final importance scores, we retain a partion of nodes of the input graph to generate a subgraph as the new input graph for the next layer.

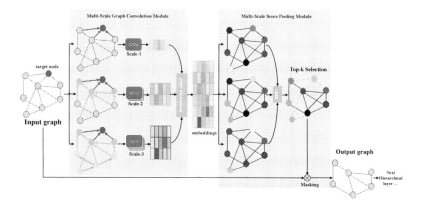

Fig. 1. Framework of the DMSPool layer.

4.2   Multi-scale Graph Convolution Module

We discuss how to 'diversely' model the process of learning a node features aggregation strategy. One of the most representative methods is Graph Convolution
Network (GCN) [1]. And the k-th layer in GCN can be written as:

$$H^{(k+1)} = \sigma(\widetilde{D}^{-\frac{1}{2}} \widetilde{A} \widetilde{D}^{-\frac{1}{2}} H^{(k)} W^{(k)}) \tag{1}$$

where $\sigma(\cdot)$ is a non-linear activation function, $\widetilde{A}$ is the adjacent matrix with
self-connections, $\widetilde{D}$ is the diagonal degree matrix of $\widetilde{A}$, and $H^{(k)}$ represents the
hidden representation matrix of nodes in the k-th layer, where $H^{(0)} = \mathbf{X}$ is the
origin node feature matrix. $W^{(k)} \in R^{d_k \times d_{k+1}}$ is a trainable weight matrix.

We insist that the derived low-dimensional vector may not retain the complete and accurate information if a pre-defined unified convolution architecture
is adopted. In our model, we use multiple GCN architectures with a different
number of aggregation iterations concurrently to learn various node embeddings.
Suppose that there are n_c GCN architectures running concurrently at each layer
l, we will have n_c node hidden representations $H_1^{(l)}, H_2^{(l)}, \ldots, H_{n_c}^{(l)}$ as follows:

$$
\begin{aligned}
H_1^{(l)} &= \sigma(\mathbf{GCN}_1(H^{(l\text{-}1)}, A^{(l)})), \\
H_2^{(l)} &= \sigma(\mathbf{GCN}_2(\sigma(\mathbf{GCN}_1(H^{(l\text{-}1)}, A^{(l)})), A^{(l)})), \\
&\vdots \\
H_{n_c}^{(l)} &= \sigma(\mathbf{GCN}_{n_c}(\sigma(\mathbf{GCN}_{n_c-1}(\sigma(\mathbf{GCN}_{n_c-2}(\ldots)), A^{(l)})), A^{(l)})).
\end{aligned}
\tag{2}
$$

where the $H_2^{(l)}$ denotes the indirect aggregation of two-hop nodes; $H_{n_c}^{(l)}$
denotes the indirect aggregation of n_c-hop nodes. Then we integrate $H_1^{(l)}$,
$H_2^{(l)}, \ldots, H_{n_c}^{(l)}$ to generate complete node embeddings $H^{(l)}$ as follow (concatenation):

$$H^{(l)} = H_1^{(l)} \oplus H_2^{(l)} \oplus \ldots \oplus H_{n_c}^{(l)} \tag{3}$$

4.3   Multi-scale Graph Pooling Module

In this subsection, we introduce the graph pooling operation. Here, we follow
SAGPool [11]. Specifically, if the convolution formula of GCN [1] is used, at
each layer l, the multi-scale nodes' scores $Z_1^{(l)}, Z_2^{(l)}, \ldots, Z_{n_s}^{(l)}$ are calculated
as follows:

$$
\begin{aligned}
Z_1^{(l)} &= \sigma(\mathbf{GCN}_{score}(H^{(l)}, A^{(l)})), \\
Z_2^{(l)} &= \sigma(\mathbf{GCN}_{score}(\sigma(\mathbf{GCN}_{score}(H^{(l)}, A^{(l)})), A^{(l)})), \\
&\vdots \\
Z_{n_s}^{(l)} &= \sigma(\mathbf{GCN}_{score}(\sigma(\mathbf{GCN}_{score}(\sigma(\mathbf{GCN}_{score}(\ldots)), A^{(l)})), A^{(l)})).
\end{aligned}
\tag{4}
$$

The formulation of basic block self-attention $\mathbf{GCN}_{score}$ is:

$$\mathbf{GCN}_{score} = \sigma((\widetilde{D}^{-\frac{1}{2}} \widetilde{A} \widetilde{D}^{-\frac{1}{2}})^{(l)} H^{(l)} \Theta_{att}^{(l)}) \tag{5}$$

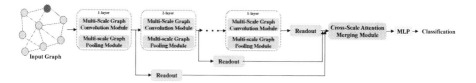

Fig. 2. Multi-layering hierarchical architecture.

where $\Theta_{att}^{(l)} \in R^{F \times 1}$ is a trainable weight matrix. Then we merge the various scores by summing to derive the final node importance scores:

$$Z^{(l)} = Z_1^{(l)} + Z_2^{(l)} + \cdots + Z_{n_s}^{(l)} \qquad (6)$$

Following the node selection proposed by [10], the pooling ratio $k \in (0,1)$ determines the number of nodes to retain. The top $\lceil kN \rceil$ nodes are selected:

$$\begin{aligned} idx &= top\_rank(Z^{(l)}, \lceil kN \rceil), \\ Z_{mask}^{(l)} &= Z_{idx}^{(l)}, \\ H_{out}^{(l)} &= H_{idx,:}^{(l)} \odot Z_{mask}^{(l)}, \\ A_{out}^{(l)} &= A_{idx,idx}^{(l)} \end{aligned} \qquad (7)$$

where top_rank is the function that returns the indices of the top $\lceil kN \rceil$, Z_{idx} is an indexing operation and $Z_{mask}^{(l)}$ is the feature attention mask. $H_{idx,:}^{(l)}$ is the row-wise indexed feature matrix, $\odot$ is the elementwise product. $H_{out}^{(l)}$ and $A_{out}^{(l)}$ are the node feature matrix and adjacent matrix of the new subgraph, respectively.

4.4 Multi-layering

In this subsection, we introduce the residual portion of the DMSPool architecture (Fig. 2), which involves readout layer, attention merging module, and linear layer.

Readout Layer. The outputs of each basic layer are summarized in the **readout layer**, which uses the formulation as: $s = \frac{1}{N_l} \sum_{i=0}^{N_l} h_i \| \max_{i=0}^{N_l} h_i$, where N_l is the number of nodes in l-th layer's input graph, h_i is the node hidden representation of i-th node, and $\|$ denotes concatenation.

Attention Merging Module. Now we have $s_1, s_2, \ldots, s_l$. In our implementation, we set $l = 3$. We use $att(s_1, s_2, s_3)$ to learn their corresponding importance as: $(\alpha_1, \alpha_2, \alpha_3) = att(s_1, s_2, s_3)$, where $\alpha_1, \alpha_2, \alpha_3$ indicate the attention values of different level's graph embeddings. For s_1, we get the w_1 by $w_1 = \mathbf{q}^T \cdot tanh(\mathbf{W} \cdot (s_1)^T + b)$. Similarly, we can get w_2, w_3. Normalizing the attention values with *softmax*. Then we combine these to obtain the final graph level representation emb_G:

$$emb_G = \alpha_1 \cdot s_1 + \alpha_2 \cdot s_2 + \alpha_3 \cdot s_3 \tag{8}$$

Use HSIC Strategy for Enhancing the Disparity. To ensure s_1, s_2, s_3 can capture diverse information from different perspectives, we employ the Hibert-Schmidt Independence Criterion (HSIC) [20] to enhance the disparity of any two hidden representations. Formally, the HSIC constraint is defined as:

$$HSIC(s_i, s_j) = (n-1)^{-2} tr(\mathbf{R}\mathbf{K}_i \mathbf{R}\mathbf{K}_j) \tag{9}$$

where the i, j = 1, 2, 3 and i != j, $\mathbf{K}_1$ and $\mathbf{K}_2$ are the Gram matrices with $k_{1,ij} = k_1(s^i{}_1, s_1{}^j)$, and $k_{2,ij} = k_2(s_2{}^i, s_2{}^j)$. $\mathbf{R} = \mathbf{I} - \frac{1}{n}ee^T$, where I is an identity matrix and e is an all-one column vector. In our implementation, we use the inner product kernel function for $\mathbf{K}_1$ and $\mathbf{K}_2$. So the whole constraints can be discribed as follows:

$$\mathbf{L}_d = HSIC(s_1, s_2) + HSIC(s_1, s_3) + HSIC(s_2, s_3) \tag{10}$$

Finally, we feed the emb_G into MLP layer, and the loss function is defined as follows:

$$\mathbf{L}_{loss} = -\sum_{i \in G_L} \sum_{j=1}^{c} Y_{ij} log(softmax(MLP(emb_G))) \tag{11}$$

where G_L denotes the training set of graphs that have lables, c represents the number of graph labels, Y_{ij} is the ground truth.

Objective Function. Combining the graph classification task and constraints, we have the overall objective function as follows:

$$\mathcal{L} = \mathbf{L}_{loss} + \gamma \mathbf{L}_d \tag{12}$$

where γ is the parameter of the disparity constraint terms.

5   Experiment

5.1   Datasets and Baselines

Our proposed DMSPool is evaluated on five widely used public benchmarks datasets. The statistics and properties are summarized in Table 1. We compare the proposed DMSPool method with several state-of-the-art graph convolution network methods: **GCN** [1], **GraphSAGE** [5], **GAT** [6], **DGCNN** [16], **Diff-Pool** [9], **EigenPool** [17], **gPool** [16], **SAGPool** [11], **HGP-SL** [12].

Table 1. The Statistics of five widely used datasets.

Dataset	#graph	#[min, max]node	#[min, max]edge	# class
PROTEINS	1113	[4, 620]	[5, 1049]	2
D&D	1178	[30, 5748]	[63, 14267]	2
NCI1	4110	[3, 111]	[2, 119]	2
NCI109	4127	[4, 111]	[3, 119]	2
Mutagenicity	4337	[4, 417]	[3, 112]	2

5.2 Parameter Setting

In order to ensure a fair comparison, we randomly split each dataset: 80% as the training set, 10% as the validation set, and the remaining 10% as the test set. This randomly splitting process is repeated 10 times, and the average performance with standard derivation is reported. We implement our proposed DMSPool with PyTorch, and the Adam optimizer is utilized to optimize the model. Early stopping criterion, patience, and weights decay strategy are employed in the training process. In addition, we adopt the widely used evaluation metric, i.e., accuracy, for graph classification to evaluate the performance.

5.3 Performance Comparison on Benchmark Datasets

The performance of DMSPool model and these state-of-the-art baselines are reported in Table 2. From the table, we have the following observations:

- First of all, our proposed DMSPool generally achieves the best performance all the datasets. Especially, DMSPool achieves maximum relative improvements of **1.47%** on PROTEINS and **1.55%** on D&D compared with the best baseline, HGP-SL.

Table 2. The average on graph classification.

Baselines	PROTEINS	D&D	NCI1	NCI109	Mutagenicity
GCN	75.23 ± 3.63	73.26 ± 4.46	76.29 ± 1.79	75.91 ± 1.84	79.81 ± 1.58
GraphSAGE	74.01 ± 4.27	75.78 ± 3.91	74.73 ± 1.34	74.17 ± 2.89	78.75 ± 1.18
GAT	74.72 ± 4.01	77.30 ± 3.68	74.90 ± 1.72	75.81 ± 2.68	78.89 ± 2.05
DGCNN	79.99 ± 0.44	70.07 ± 1.21	74.08 ± 2.19	78.23 ± 1.31	80.41 ± 1.02
DiffPool	79.90 ± 2.95	78.61 ± 1.32	77.73 ± 0.83	77.13 ± 1.49	80.78 ± 1.12
EigenPool	78.84 ± 1.06	78.63 ± 1.36	77.24 ± 0.96	75.99 ± 1.42	80.11 ± 0.73
gPool	80.71 ± 1.75	77.02 ± 1.32	76.25 ± 1.39	76.61 ± 1.39	80.30 ± 1.54
SAGPool	81.72 ± 2.19	78.70 ± 2.29	77.88 ± 1.59	75.74 ± 1.47	79.72 ± 0.79
HGP-SL	84.37 ± 1.71	80.21 ± 1.19	77.89 ± 0.75	79.83 ± 1.38	80.85 ± 0.74
DMSPool	**85.84 ± 0.80**	**81.76 ± 0.65**	**78.23 ± 0.72**	**80.97 ± 0.63**	**82.13 ± 0.89**

- DMSPool consistently outperforms GCN, GraphSAGE, and GAT on all the datasets, indicating the effectiveness of hierarchical structure in DMSPool, because it can extract more useful information than these inherently "flat" graph neural networks.
- The SAGPool model is the most similar one to our DMSPool. Our model outperforms SAGPool by an average of **3.04%**. We can learn that it makes sense to integrate multiple-scale node level representations from different perspectives, which help us learn more complete and richer structure information.

5.4   Variants of DMSPool

In this subsection, we compare DMSPool with its three variants to validate the effectiveness of adopting multiple architectures simultaneously. In our comparative experiments, we still follow the multi-channel setting in convolution and pooling modules. If all channels adopt the unified architectures in convolution/pool module, it means that the iteration number of convolution operations in each layer's will be equal to 3. (a) **DMSPool-mcup**, the settings are multiple architectures in convolution module and unified architecture in pooling module. (b) **DMSPool-ucmp**: The settings are unified architecture in convolution module and multiple architectures in pooling module. (c) **DMSPool-ucup**: The settings are unified architecture in convolution module and unified architecture in pooling module.

Table 3. The results(%) of DMSPool and its variants on five datasets.

Variants	PROTEINS	D&D	NCI1	NCI109	Mutagenicity
DMSPool-ucup	81.16	70.51	75.18	75.85	81.89
DMSPool-mcup	81.43	74.79	73.97	79.83	77.28
DMSPool-ucmp	83.48	76.98	76.72	80.45	78.87
DMSPool	**85.84**	**81.76**	**78.23**	**80.97**	**82.13**

From the results in Table 3, we can draw the following conclusions: (1) The results of DMSPool are consistently better than all the other three variants, indicating the effectiveness of using the multiple architecture concurrently in both convolution module and pooling module. (2) The DMSPool-ucmp is generally better than DMSPool-mcup, DMSPool-ucup, which implies the multiple architectures in pooling module play a more vital role in this framework. (3) The results of DMPool-mcup are usually better than DMSPool-ucup, verifying the usefulness of adopting multiple GNNs architectures in convolution module.

6   Conclusion

In this paper, we proposed a simple but effective model DMSPool for hierachical graph representation learning. DMSPool adopts multiple GNN architectures concurrently to learn node-level embeddings and nodes' importance from different aggregation iterations, then integrate and perform top-ranked nodes selections to generate a subgraph as the input graph for the next layer. Futhermore, we employ attention mechanism to determine the contribution of subgraphs' representations at various levels to graph classification. Our results show that DMSPool has gained performance improvement over the state-of-the-art graph representation learning methods. In the future, we are particularly interested in further providing theoretical proofs to explain the similarities and differences of graph convolution operations in different downstream tasks, thus improving the interpretability of the model.

Acknowledgements. This paper is supported by the National Key Research and Development Program of China (Grant No. 2018YFB1403400), the National Natural Science Foundation of China (Grant No. 61876080), the Collaborative Innovation Center of Novel Software Technology and Industrialization at Nanjing University.

References

1. Kipf, T., Welling, M.: Semi-supervised classification with graph convolutional networks. arXiv, abs/1609.02907 (2017)
2. Fan, W., et al.: Graph neural networks for social recommendation. In: The World Wide Web Conference (2019)
3. Li, R., Wang, S., Zhu, F., Huang, J.: Adaptive graph convolutional neural networks. In: AAAI (2018)
4. Bianchi, F.M., Grattarola, D., Livi, L., Alippi, C.: Graph neural networks with convolutional arma filters. arXiv, abs/1901.01343 (2019)
5. Hamilton, W.L., Ying, Z., Leskovec, J.: Inductive representation learning on large graphs. In: NIPS (2017)
6. Velickovic, P., Cucurull, G., Casanova, A., Romero, A., Liò, P., Bengio, Y.: Graph attention networks. arXiv, abs/1710.10903 (2018)
7. Xu, K., Hu, W., Leskovec, J., Jegelka, S.: How powerful are graph neural networks? arXiv, abs/1810.00826 (2019)
8. Xie, Y., Li, S., Yang, C., Wong, R.C.-W., Han, J.: When do GNNs work: understanding and improving neighborhood aggregation. In: IJCAI (2020)
9. Ying, R., You, J., Morris, C., Ren, X., Hamilton, W.L., Leskovec, J.: Hierarchical graph representation learning with differentiable pooling. arXiv, abs/1806.08804 (2018)
10. Gao, H., Ji, S.: Graph u-nets. In: ICML (2019)
11. Lee, J., Lee, I., Kang, J.: Self-attention graph pooling. In: ICML (2019)
12. Zhang, Z., et al.: Hierarchical graph pooling with structure learning. arXiv, abs/1911.05954 (2019)
13. Bruna, J., Zaremba, W., Szlam, A., LeCun, Y.: Spectral networks and locally connected networks on graphs. CoRR, abs/1312.6203 (2014)
14. Defferrard, M., Bresson, X., Vandergheynst, P.: Convolutional neural networks on graphs with fast localized spectral filtering. In: NIPS (2016)
15. Wang, X., Zhu, M., Bo, D., Cui, P., Shi, C., Pei, J.: AM-GCN: adaptive multi-channel graph convolutional networks. In: Proceedings of the 26th ACM SIGKDD International Conference on Knowledge Discovery & Data Mining (2020)

16. Zhang, M., Cui, Z., Neumann, M., Chen, Y.: An end-to-end deep learning architecture for graph classification. In: AAAI (2018)
17. Ma, Y., Wang, S., Aggarwal, C., Tang, J.: Graph convolutional networks with eigenpooling. In: Proceedings of the 25th ACM SIGKDD International Conference on Knowledge Discovery & Data Mining (2019)
18. Murphy, R., Srinivasan, B., Rao, V., Ribeiro, B.: Relational pooling for graph representations. arXiv, abs/1903.02541 (2019)
19. Yuan, H., Ji, S.: Structpool: structured graph pooling via conditional random fields. In: ICLR (2020)
20. Song, L., Smola, A., Gretton, A., Borgwardt, K., Bedo, J.: Supervised feature selection via dependence estimation. In: ICML 2007 (2007)

A Parameter-Free Approach for Lossless Streaming Graph Summarization

Ziyi Ma[1,2], Jianye Yang[1(✉)], Kenli Li[1], Yuling Liu[1(✉)], Xu Zhou[1], and Yikun Hu[1]

[1] College of Computer Science and Electronic Engineering,
Hunan University, Changsha, China
{maziyi,jyyang,lkl,yuling_liu,yikunhu}@hnu.edu.cn
[2] Academy of Military Sciences PLA China, Beijing, China

Abstract. In rapid and massive graph streams, it is often impractical to store and process the entire graph. Lossless graph summarization as a compression technique can provide a succinct graph representation without losing information. However, the problem of lossless streaming graph summarization is computationally and technically challenging. Although the state-of-the-art method performs well with respect to efficiency, its summarization quality is usually unstable and unsatisfactory. This is because it is a randomized algorithm and depends heavily on the pre-tuned parameters. In this paper, we propose a parameter-free lossless streaming graph summarization algorithm. As the graph changes over time, we incrementally maintain the summarization result, by carefully exploring the influenced subgraph, which is shown to be a bounded neighborhood of the inserted edge. To enhance the performance of our method, we further propose two optimization techniques regarding candidate supernodes refinement and destination supernode selection. The experiment results demonstrate that the proposed methods outperform the state-of-the-art by a large margin in terms of compression quality with comparable running time on the majority of datasets.

Keywords: Incremental algorithms · Lossless graph summarization · Parameter-free · Streaming graph

1 Introduction

Graph model is ubiquitous and has been used to model the relationship between entities in a wide range of applications, such as social networks, online transaction networks, transportation networks, citation networks, to name just a few. Two common properties of these graphs are large in scale and highly dynamic. Take Facebook as an example, there are more than 2.5 billion monthly active users and approximately 6 new users join Facebook each second[1]. Such large dynamic graphs are naturally represented as a streaming graph, i.e., a sequence of time evolving edges $\langle e_1, e_2, ..., e_t \rangle$.

---

[1] https://www.statista.com/topics/751/facebook/.

© Springer Nature Switzerland AG 2021
C. S. Jensen et al. (Eds.): DASFAA 2021, LNCS 12681, pp. 385–393, 2021.
https://doi.org/10.1007/978-3-030-73194-6_26

Motivation. To manage such highly dynamic large graphs, a useful technique is graph summarization, which aims to represent the graph in a *succinct* form [3,4]. For example, Twitter are evolving with user interactions generated rapidly. As a result, an important task is to summarize the connections in real time by grouping users and topics together to reveal the public trend on time. Meanwhile, lossless graph summarization is one of the most effective graph compression techniques, which can be applied to many applications, such as information exchange and data visualization. By lossless, we mean the original graph can be precisely restored from the compressed graph.

In this paper, we study the problem of lossless streaming graph summarization. Formally, given a streaming graph $G = (E, V)$, the goal of lossless streaming graph summarization is to construct a summary graph $\mathcal{G} = (\mathcal{S}, \mathcal{L})$ together with corrections $C = (C^+, C^-)$, such that G can be precisely restored from $\mathcal{G}$ and C, and the edges in $\mathcal{G}$ and C is minimized, where C^+ denotes the set of edges to be inserted and C^- denotes the set of edges to be deleted.

Challenges. The problem of lossless streaming graph summarization is computationally challenging. On one hand, it is shown to be NP-hard to even summarize a static graph optimally [8], which means that frequently re-summarizing the graph from the scratch is computationally unaffordable. On the other hand, in a streaming environment, edges usually arrives rapidly, which implies that updating the summarization result in near constant time complexity is needed. These present great computational and technical challenges to us.

Recently, several studies are devoted to dynamic graph summarization [1,2, 9]. Most of existing solutions are incremental and heuristic. Gou et al. [1] adjust the size of windows to summarize graph streams and thus cannot obtain the whole summary graph. Tsalouchidou et al. [9] consider to incrementally update previous timestamps. Nevertheless, the data structure employed can only store a fixed the number of nodes, and thus is inapplicable for large scale graphs. The state-of-the-art, called MoSSo [2], subjects to incrementally summarize a streaming graph by randomly select movable nodes and destination nodes, which can achieve high efficiency. However, this method tends to provide unstable and unsatisfactory summarization result regarding compression ratio. This is because the node movement strategy is randomly determined. Besides, this method utilizes two important parameters when making decisions on node movement. As a result, the performance of MoSSo in terms of both efficiency and effectiveness is heavily dependent on the parameters tuned. Moreover, for large streaming graphs, the task of tuning the parameter itself is time-consuming.

In this paper, we resort to designing novel and efficient techniques to deal with the problem of lossless streaming graph summarization by considering both computation efficiency and compression quality.

Contributions. (1) A parameter-free lossless streaming graph summarization framework. (2) Two optimization techniques to further improve the performance. (3) Extensive performance studies on real datasets.

2   Preliminaries

Streaming Graph. A streaming graph $G_t = (V_t, E_t)$ consists of an unbounded time evolving sequence of edges, i.e., $E_t = \langle e_1, e_2, ..., e_t \rangle$, where each edge $e_i \in E_t$

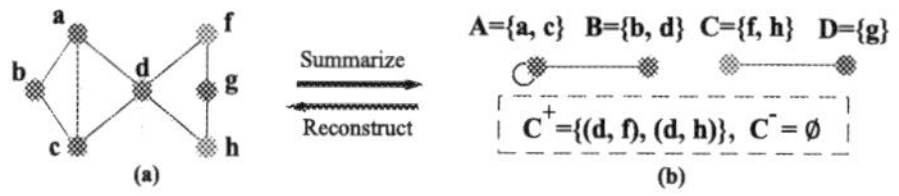

Fig. 1. The example illustrates the processes of summarization and reconstruction. (a) a graph G, (b) a summary graph $\mathcal{G}$ and corrections C.

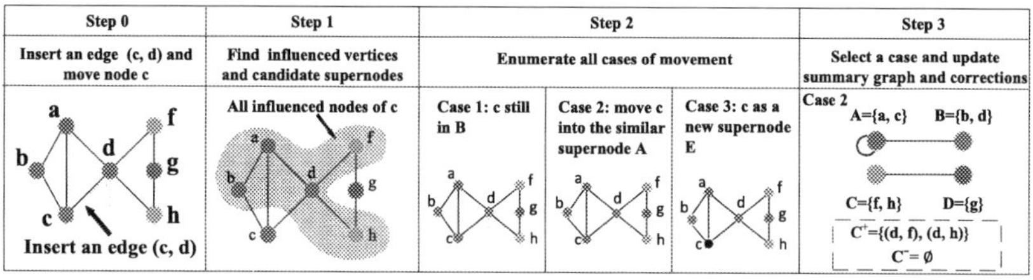

Fig. 2. Framework overview.

is an unweighted and undirected relation between two distinct nodes $u, v \in V_t$. Given a node $v \in V_t$, we use $N(v)$ to denote the neighborhood of v, i.e., the set of nodes adjacent to v. In addition, we use $N_2(v)$ to denote the 2-hop neighborhood of v, i.e., the set of nodes that v can reach in 2-hop. In the following, for presentation simplicity, we omit the time stamp t in the notations by referring to $G_t = (V_t, E_t)$ simply as $G = (V, E)$ if the context is self-evident.

Summary Graph. Given a streaming graph $G = (V, E)$, its summary graph $\mathcal{G} = (\mathcal{S}, \mathcal{L})$ is defined as the concise representation of G, where $\mathcal{S}$ is a partition of V, i.e., each node $S \in \mathcal{S}$ contains a set of nodes in V, and $V = \cup_{S_i \in \mathcal{S}} S_i$ with $S_i \cap S_j = \emptyset$ for any two distinct nodes $S_i, S_j \in \mathcal{S}$. Besides, there is an edge between two nodes $S_i, S_j \in \mathcal{S}$ in $\mathcal{G}$ if there exists an edge $(u, v) \in E$ with $u \in S_i$ and $v \in S_j$, respectively. For presentation clarity, we call nodes and edges in a summary graph supernodes and superedges. By $S(v)$, we denote the supernode in $\mathcal{S}$ that contains node v. Consider the example in Fig. 1. A graph G and its summary graph $\mathcal{G}$ are shown in Fig. 1(a) and Fig. 1(b), respectively. In particular, the 7 nodes in G are merged into 4 supernodes in $\mathcal{G}$. Take supernode A for example. It contains two nodes of G, i.e., a and c. The superedges are constructed accordingly. Note that $\mathcal{G}$ might contain self-loops (e.g., (A, A) in $\mathcal{G}$).

Definition 1 (Lossless Summarization). *Given a graph $G = (V, E)$, a lossless summarization of G consists of a summary graph $\mathcal{G}$ with the corresponding corrections $C = (C+, C-)$, such that G can be precisely restored from $\mathcal{G}$ and C, where C^+ (Resp. C^-) denotes the set of edges to be inserted (Resp. deleted).*

Problem Statement. Given a streaming graph $G = (V, E)$, the goal of this work is to efficiently find for G a lossless summarization instance, namely a summary graph $\mathcal{G} = (\mathcal{S}, \mathcal{L})$ and corrections $C = (C^+, C^-)$, at each time a new edge coming, such that the following object function is minimized, $\Phi = |\mathcal{L}| + |C^+| + |C^-|$. Note that, $|\mathcal{L}|$ counts only non-loop superedges since the self-loops in $\mathcal{L}$ can be encoded concisely using 1 bit per supernode regardless of their count.

Lemma 1. *The problem of graph lossless summarization is NP-Hard [8].*

3   Proposed Method

3.1   Framework Overview

At each time a new edge coming, a straightforward method is to re-summarize the entire graph from the scratch using a static graph summarization method (e.g., SWeG [7]). However, this method is cost-prohibitive since edges tend to arrive rapidly in a streaming graph, which implies that it is necessary to update the summarization result quickly. To avoid such high recomputation cost, we resort to a incremental method. Particularly, we propose a parameter-free incremental method, called *Streaming Graph Summarization* (SGS).

Algorithm 1: Overview of SGS

Input: new edge $e_t = (u, v)$, streaming graph G_{t-1}, summary graph $\mathcal{G}_{t-1}$, edge corrections C_{t-1}
Output: summary graph $\mathcal{G}_t$, corrections C_t
1 Identify the influenced nodes in G_{t-1} and candidate supernodes in $\mathcal{G}_{t-1}$;
   // Section 3.2
2 Select the most promising influenced supernodes as the destination supernode;
   // Section 3.3
3 Update the summarization $\mathcal{G}_{t-1}$ and C_{t-1}; // Section 3.4
4 **return** $\mathcal{G}_t$ and C_t

Overview. The overview of SGS is illustrated in Algorithm 1. When an edge $e_t = (u, v)$ is inserted, SGS greedily moves u and v, while fixing the other nodes, such that the objective function Φ is minimized. We observe that candidate supernodes can be bounded within local neighborhood of u and v. Then, we greedily select the best candidate in the bounded candidates as the destination supernode and update the summarization. Figure 2 shows an example for moving c, where node color indicate the membership to supernodes.

3.2   Identify Candidate Supernodes

We devise effective techniques to identify the candidate supernodes when moving the two end nodes of the inserted edge. For presentation simplicity, we only discuss for one of them (e.g., v) in the following, since we can process the other in exactly the same way.

Key Idea. We merge nodes with similar neighborhood together based aggregation summarization methods [5,7], and can reduce the number of edges in the summary graph substantially [10]. Clearly, it is enough to just keep one of the nodes as a delegate in the summary graph, since the neighborhood of other nodes can be restored by the delegate node. In a streaming graph, when an edge is inserted, the neighborhood of influenced nodes might change, and therefore we need to update the membership of nodes to supernodes to maintain high quality compression result. Fortunately, we observe that the number of influenced nodes can be bounded in a local neighborhood of the two end nodes of the inserted edge. Thus, the number of candidate supernodes can also be bounded.

Definition 2 (Node Similarity). *Given two nodes u and v in streaming graph G, the structural similarity between u and v is defined as the Jaccard similarity of their neighbor sets, i.e., $J(u,v) = \dfrac{|N(u) \cap N(v)|}{|N(u) \cup N(v)|}$.*

Lemma 2. *Given a streaming graph G, when an edge (u,v) is inserted, the number of node pairs that might have changed similarity can be bounded by $|N_2(u)| + |N_2(v)|$.*

With the help of Lemma 2, we can exclude all supernodes that do not contain any nodes in $N_2(u) \cup N_2(v)$ when moving u and v. Formally, by $CS(v)$, we denote the set of supernodes that contain at least one node in $N_2(v)$, which is called the candidate supernodes of v. Clearly, the number of candidate supernodes of v, i.e., $|CS(v)|$, can be bounded by $|N_2(v)|$. Consider the example in Fig. 2. When edge (c,d) is inserted, we only need to recalculate the similarity between c and its 2-hop neighbors, i.e., a, b, d, f, h. We process for node d in a similar way. When moving node c, the candidate supernodes are A, B and C.

3.3 Destination Supernode Selection (DSS)

In this section, we discuss how to select the most promising supernode to move v. Intuitively, node v should move to a candidate supernode that contains many nodes that are similar to v. Next, Definition 3 is extended by Definition 2.

Definition 3 (Supernode Similarity). *Given a node $v \in V$ and a supernode $S \in \mathcal{S}$, the supernode similarity between v and S is defined as follows:*

$$SJ(v,S) = \frac{\sum_{u \in N(v) \cup N(S)} \min(w(u,v), w(S,u))}{\sum_{u \in N(v) \cup N(S)} \max(w(u,v), w(S,u))}, \tag{1}$$

where $N(S) = \cup_{u \in S} N(u)$ is the set of nodes adjacent to any node in S, and $w(S,u) = |p \in S : \{p,u\} \in E|$ is the number of nodes in S adjacent to u.

Generally, $SJ(v,S)$ measures the similarity between v and S in terms of their common neighbors. It is worth noticing that $SJ(v,S)$ is 1 if v and S have same neighbors (i.e., $w(u,v) = w(S,u)$ for every $u \in N(v) \cup N(S)$), and 0 otherwise if they have no common neighbors (i.e., $N(v) \cap N(S) = \emptyset$). We calculate similarity between a node $v \in V$ and each supernode $S \in CS(v)$, and then selects the one with maximum similarity.

Creating New Supernode. In addition to the existing candidate supernodes, we also consider creating a new supernode for v if the current supernode size $|S(v)|$ is large than 1.

Determine Proposal of Movement. From the above discussion, we have the following three choices of movement for v, namely (*i*) staying in its previous supernode, (*ii*) moving to adjacent supernode, (*iii*) forming a new supernode.

Next, we introduce the concept of *moving cost* to evaluate the benefit of each moving choice. Here, we assume the summarization result is available after moving v.

Definition 4. *Given two supernodes A and B, the moving cost of moving node v from A to B is*

$$MC(v, A, B) = 1 - \frac{Cost_{aft}(v, A, B)}{Cost_{bef}(v, A, B)}, \tag{2}$$

where $Cost_{bef}(v, A, B)$ (*Resp.* $Cost_{aft}(v, A, B)$) is the number of edges that adjacent to A and B together with the number of edges that adjacent to v in corrections C before (*Resp.* after) moving v.

A positive value of $MC(v, A, B)$ implies that we reduce the number edges in the summarization by moving v to the corresponding supernode, i.e., one of the above two cases (*ii*) and (*iii*). We therefore accept the proposal of node moving.

3.4   Update of Summarization

After settling down the moving plan, e.g., moving v from A to B, we update the summarization to achieve least size of summary graphs and corrections by MDL techniques [5]. Π_{AB} (*Resp.* Ψ_{AB}) denotes all possible (*Resp.* actual) edges of G between nodes in A and B. For the inter-supernode, if $|\Psi_{AB}| > |\Pi_{AB}|/2$, we create a superedge between A and B, and insert into C^- all edges in $\Pi_{AB} \setminus \Psi_{AB}$. Otherwise, Ψ_{AB} is inserted into C^+. Similarly, for intra-supernode, we create a superloop for A and insert edges in $\Pi_{AA} - \Phi_{AA}$ into C^- if $|\Psi_{AA}| > |\Pi_{AA}|/4$. Otherwise, only edges between nodes in A are inserted into C^+.

4   Optimizations

Motivation. When an edge (u, v) is inserted, SGS needs to consider all 2-hop neighbors of u and v for similarity computation, which is rather time-consuming. To enhance the performance of SGS, we devise advanced techniques towards the following two aspects. First, we observe that some nodes in $N_2(u)$ may have a decreased similarity with u after the insertion (u, v). This implies that we can safely exclude these nodes from the influenced node set since we are only interested in increasing the similarity. Based on this property, we can retrieve a set of refined candidate supernodes (Sect. 4.1). Second, we observe that calculating the similarity between a supernode and a movable node is still computationally expensive. To alleviate this issue, we resort to a utility function based heuristic method to select the most promising supernode (Sect. 4.2).

4.1   Candidate Supernode Refinement

However, we observe that the similarity between v and nodes in $N_2(v)$ might increase, remain the same, or decrease. Since our goal is to find the supernode with high similarity, we can skip processing those with non-increasing similarity to reduce computation cost.

Lemma 3. *Given a streaming graph G, when an edge (u,v) is inserted, for node v, only nodes in $N(u)$ would have an increased similarity with v. We have similar property for node u.*

Clearly, the number of candidate supernodes of v, i.e., $|CS(v)|$, can be bounded by $|N(v)|$. By RSGS, we denote the new version of SGS that is equipped with the advanced method of candidate supernode collecting.

4.2 Advanced Destination Supernode Selection

In Sect. 3.3, we select the most promising destination supernode by calculating the supernode similarity (Eq. 1), which is a rather time-consuming operation under the streaming envoriment where edges may arrive rapidly. To facilitate the computation, a straightforward method is to randomly select a candidate supernode in $CS(v)$. However, this simple method may lead to unacceptable compression ratio. In the following, we tackle this issue by employing an effective and efficient utility function.

Definition 5 (Utility Function). *Given an inserted edge (u,v) in a streaming graph G, let $CS(v)$ be the candidate supernodes of v. Then, for a supernode $S \in CS(v)$, its utility function is defined as .*

$$\tau(v,S) = \frac{|N(u) \cap S|}{|S|}. \tag{3}$$

Intuitively, the utility function evaluates the raio of "useful" nodes in a candidate supernode. By "useful", we mean the nodes with increased similarity to v, i.e., nodes in $N(u)$ according to Lemma 3. Next, guided by this utility function, we present two candidate supernode selection strategies.

Greedy Based Method. In this method, we simply choose the supernode with the largest utility score. We omit the details of algorithm GU-DSS due to space constraint.

Algorithm 2: RU-DSS: Random Utility Function based DSS

Input: a node $v \in V$, candidate supernodes $CS(v)$ of v
Output: a destination supernode Dst

1 $Sum \leftarrow 0$;
2 $Thr \leftarrow 0$;
3 $Dst \leftarrow NULL$;
4 **for** *each $S \in CS(v)$* **do**
5 $\quad \lfloor \ Sum \leftarrow Sum + \tau(v,S)$;
6 sample $X \sim uniform(0,1)$;
7 **for** *each $S \in CS(v)$* **do**
8 $\quad Thr \leftarrow Thr + \tau(v,S)$;
9 $\quad \theta \leftarrow Thr/Sum$;
10 $\quad$ **if** $X \leq \theta$ **then**
11 $\quad \quad \lfloor \ $ **return** S;

12 **return** Dst;

Random Based Method. In some cases, GU-DSS might fall into local optimal. To remedy this issue, we further propose a random selection based method. In specific, a candidate supernode with a larger utility score would be selected as a destination supernode with higher probability. Precisely, for a supernode $S \in CS(v)$, the probability of S being selected is the ratio of its utility score to the overall utility score of supernodes in $CS(v)$, i.e.,

$$Prob(S) = \frac{\tau(v, S)}{\sum_{T \in CS(v)} \tau(v, T)}. \tag{4}$$

In Algorithm 2, we first calculate the overall utility score of nodes in $CS(v)$ (Lines 4–5). Then, we uniformly sample a random number (Line 6). Finally, we find the region of this random number falls and return the corresponding candidate supernode (Lines 7–11). The time complexity of Algorithm 2 is $O(|N(v)|)$.

5   Experiments

Experimental Setting. All experiments are conducted on a machine with a 3.3 GHz Intel i9-7900X CPU and 64 GB memory, running Linux. RSGS-GU-DSS, RSGS-RU-DSS, and MoSSo are implemented by C++. We evaluate the algorithms on 4 real-world graph datasets that can obtained from [6], such that web-wiki, cit-patent, soc-livejournal, and delaunay-n24.

Evaluation Metric. Given a summary graph $\mathcal{G} = (\mathcal{S}, \mathcal{L})$ and edge corrections $C = <C^+, C^->$ of a streaming graph $G = (V, E)$, the compression ratio is $(|\mathcal{L}| + |C^+| + |C^-|)/|E|$.

Performance Evaluation. Based on the comprehensive comparisons of performance tuning, we use RSGS-GU-DSS and RSGS-RU-DSS to compare with MoSSo. We set the same parameters following [2] for MoSSo.

Experimental Results. In compression ratio, as shown in Fig. 3, RSGS-GU-DSS and RSGS-RU-DSS consistently outperform MoSSo in all datasets. In the average processing time for each new edge insertion, Table 1 reports the efficiencies of RSGS-GU-DSS and RSGS-RU-DSS are worse than MoSSo within 1 order of magnitude.

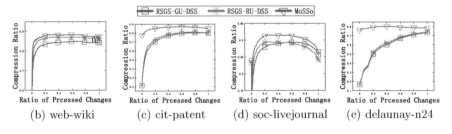

(b) web-wiki   (c) cit-patent   (d) soc-livejournal   (e) delaunay-n24

Fig. 3. Evaluating compression ratio.

Table 1. Processing time in microseconds for each edge update.

Dataset	RSGS-GU-DSS	RSGS-RU-DSS	MoSSo
web-wiki	**6526.47**	8037.26	1792.14
cit-patent	203.22	**189.84**	81.06
soc-livejournal	1190.06	**673.11**	159.42
delaunay-n24	**61.41**	62.80	42.86

6   Conclusion

In this paper, we investigate the problem of lossless streaming graph summarization. To efficiently deal with this problem, we present a novel parameter-free framework to incrementally summarize a streaming graph and two optimization techniques.

Acknowledgment. This research was supported in part by National Key Research and Development Program of China (2018YFB0204302), and NSFC (Grant No. 62002108, 61772182, 61802032, 61872134).

References

1. Gou, X., Zou, L., Zhao, C., Yang, T.: Fast and accurate graph stream summarization. In: ICDE (2019)
2. Ko, J., Kook, Y., Shin, K.: Incremental lossless graph summarization. In: SIGKDD (2020)
3. Koutra, D., Vreeken, J., Bonchi, F.: Summarizing graphs at multiple scales: new trends. In: ICDM (2018)
4. Liu, Y., Safavi, T., Dighe, A., Koutra, D.: Graph summarization methods and applications: a survey. ACM Comput. Surv. (CSUR) **51**(3), 1–34 (2018)
5. Navlakha, S., Rastogi, R., Shrivastava, N.: Graph summarization with bounded error. In: SIGMOD (2008)
6. Rossi, R.A., Ahmed, N.K.: The network data repository with interactive graph analytics and visualization. In: AAAI (2015)
7. Shin, K., Ghoting, A., Kim, M., Raghavan, H.: Sweg: lossless and lossy summarization of web-scale graphs. In: WWW (2019)
8. Tian, Y., Hankins, R.A., Patel, J.M.: Efficient aggregation for graph summarization. In: SIGMOD (2008)
9. Tsalouchidou, I., Bonchi, F., Morales, G.D.F., Baeza-Yates, R.: Scalable dynamic graph summarization. TKDE **32**(2), 360–373 (2020)
10. Yang, J., Zhang, W., Wang, X., Zhang, Y., Lin, X.: Distributed streaming set similarity join. In: ICDE (2020)

Expanding Semantic Knowledge for Zero-Shot Graph Embedding

Zheng Wang[1,2]($\boxtimes$), Ruihang Shao[2], Changping Wang[3], Changjun Hu[2], Chaokun Wang[4], and Zhiguo Gong[1]

[1] State Key Laboratory of Internet of Things for Smart City,
Department of Computer and Information Science,
University of Macau, Macao, China
fstzgg@um.edu.mo
[2] Department of Computer Science and Technology, University of Science
and Technology Beijing, Beijing, China
wangzheng@ustb.edu.cn
[3] Kwai Inc., Beijing, China
[4] School of Software, Tsinghua University, Beijing, China

Abstract. Zero-shot graph embedding is a major challenge for supervised graph learning. Although a recent method RECT has shown promising performance, its working mechanisms are not clear and still needs lots of training data. In this paper, we give deep insights into RECT, and address its fundamental limits. We show that its core part is a GNN prototypical model in which a class prototype is described by its mean feature vector. As such, RECT maps nodes from the raw-input feature space into an intermediate-level semantic space that connects the raw-input features to both seen and unseen classes. This mechanism makes RECT work well on both seen and unseen classes, which however also reduces the discrimination. To realize its full potentials, we propose two label expansion strategies. Specifically, besides expanding the labeled node set of seen classes, we can also expand that of unseen classes. Experiments on real-world datasets validate the superiority of our methods.

Keywords: Graph embedding · Zero-shot learning · Data mining

1 Introduction

Graph embedding is becoming a major trend among various graph processing methods [14,18]. Most recently, there has been an increasing interest in supervised graph embedding [4]. However, little work has considered the *zero-shot graph embedding (ZGE)* problem where some classes have no labeled data at the training time. This problem has practical significance, especially in domains where the graph size is typically large and node class labels can take on many values. Moreover, general supervised methods would deliver very unsatisfying results in this setting.

© Springer Nature Switzerland AG 2021
C. S. Jensen et al. (Eds.): DASFAA 2021, LNCS 12681, pp. 394–402, 2021.
https://doi.org/10.1007/978-3-030-73194-6_27

To fix this problem, RSDNE [16] relaxes the constraints of intra-class similarity and inter-class dissimilarity, so as to avoid the negative influence of missing the labeled data from unseen classes. However, this method cannot model the high non-linear information or the rich information of a graph. A recently proposed graph neural network (GNN) [12] method named RECT [15] overcomes these limits, having shown favorable performance. Nevertheless, its working mechanisms are still not clear, significantly hindering its practicality.

In this paper, we demystify the RECT for ZGE. In particular, we show that its core part (named RECT-L) can be thought as a GNN prototypical model which learns a nearest class mean (NCM) classifier [17]. This explains why RECT works on seen classes. On the other hand, the learned prototypical model maps nodes from the raw-input feature space into a "semantic" space where a class is described by its mean feature vector. This enables transferring knowledge from seen classes to unseen classes, which is the fundamental reason why RECT works well on the nodes coming from unseen classes. However, it also leads to the ineffectiveness of RECT, as semantic knowledge contains much less discriminative information than the original binary labels.

To overcome this limit and realize the full potentials of RECT, we design two label expansion strategies. The first is to expand the labeled node set of seen classes, which will make RECT "see" more labels. This overcomes the localized nature of the used GNN model [6]. The other one is to jointly expand the labeled node sets of both seen and unseen classes. This improves the diversity of labels, which would yield more robust embedding results. Combining these two strategies can substantially improve the performance of RECT, especially when the labeled data is very limited. In addition, we further provide some theoretical analysis for the proposed expansion strategies. Finally, we conduct extensive experiments to demonstrate the effectiveness of our methods.

2 Why RECT Work

2.1 Problem Definition

The problem of zero-shot graph embedding (ZGE) in this paper follows [16]. A graph generally consists of a set of nodes that are possibly connected by edges. We are given a labeled training node set $\mathcal{L}$ whose label set is $\mathcal{C}^s$ (i.e., the seen class set). The rest are testing nodes some of which come from an unseen class set $\mathcal{C}^u$, i.e., $\mathcal{C}^s \cap \mathcal{C}^u = \emptyset$. By using the labeled nodes only from $\mathcal{C}^s$ where no labeled nodes of $\mathcal{C}^u$ is available, we aim to learn low-dimensional node representation vectors, such that the nodes with similar neighbors, features, or labels are close to each other in the learned embedding space.

2.2 Preliminaries: RECT

RECT contains two sub-parts: RECT-N and RECT-L, both of which utilize GNN [12] layers for embedding learning. The first part RECT-N is unsupervised,

aiming to preserve the original graph structure. The other and most noteworthy part is the supervised method RECT-L. Inspired by the success of ZSL,
RECT-L learns with the class-semantic descriptions of seen classes, i.e., semantic knowledge is introduced for transferring supervised knowledge from seen to
unseen classes. Unlike traditional ZSL methods whose semantic knowledge is
human annotated or provided by some third-party resources (like the word2vec
tools [8]), RECT-L obtains this knowledge in a practical domain-dependent manner with a "readout" function. Specifically, for each seen class c, it uses the mean
feature of all corresponding nodes in this class as its class-semantic description
vector $\hat{y}_c$: $\hat{y}_c = \text{MEAN}(\{x_i | \forall_i \, C_i^s = c\})$, where x_i and C_i^s are node i's feature vec-
tor and seen class label, respectively. Finally, RECT-L minimizes the difference
between the predicted and the actual class-semantic description vectors:

$$\mathcal{J} = \sum_{i \in \mathcal{L}} \ell(\hat{y}'_{C_i^s}, \hat{y}_{C_i^s}) \tag{1}$$

where $\hat{y}'_{C_i^s}$ and $\hat{y}_{C_i^s}$ stand for the predicted and actual class-semantic vector of
node i respectively, and $\ell(\cdot, \cdot)$ is a sample-wise loss function.

2.3   RECT-L v.s. ZSL Methods

Theoretically, a typical ZSL method can be thought of a semantic output code
classifier $\mathcal{F} : X^d \rightarrow Y$, such that $\mathcal{F}$ contains two other functions, $\mathcal{S}$ and $\mathcal{Q}$ [9]:

$$\begin{aligned}
\mathcal{F} &= \mathcal{Q}(\mathcal{S}(\cdot)) \\
\mathcal{S} &: X^d \rightarrow Z^p \\
\mathcal{Q} &: Z^p \rightarrow Y
\end{aligned} \tag{2}$$

where $\mathcal{S}$ is a semantic mapping function which maps from a d-dimensional raw-
input space X^d into a p-dimensional semantic space Z^p; and $\mathcal{Q}$ is a semantic
decoding function which maps the obtained semantic encoding to a class label
from a label set Y. The classifier $\mathcal{F}$ is given a knowledge base $\mathcal{K}$ which guides the
learning of $\mathcal{S}$ and $\mathcal{Q}$. Practically, $\mathcal{K}$ is usually simplified as a one-to-one encoding
between class labels and semantic space points. A commonly used encoding is:
a class label and its corresponding class-semantic description vector.

   In RECT-L, a class (prototype) is described by its mean feature vector, indicating the used semantic space is directly constructed from the d-dimensional
raw-input features. As such, the knowledge base $\mathcal{K}$ could only guide the learning of semantic mapping function $\mathcal{S}$ rather than the semantic decoding function
$\mathcal{Q}$. This is because only the one-to-one encoding between seen class labels and
semantic space points (i.e., a seen class and its mean feature vector) is known
in ZGE problem. In other words, $\mathcal{K}$ does not contain any knowledge about the
relationship between semantic space points and unseen classes, since it is impossible to obtain the mean feature vectors of unseen classes when there exists no
labeled nodes from unseen classes. This is the fundamental difference between
RECT-L and ZSL methods.

Remark 1 (The Difference Between RECT-L and ZSL Methods). In the semantic space of ZSL methods, class prototypes are described by human annotation or third-part resources; while in the semantic space of RECT-L, class prototypes are described by their mean feature vectors. In addition, in RECT-L, the knowledge of relationship between unseen classes and semantic space points is unknown.

2.4   The Mechanisms of RECT

We continue with it's core part RECT-L. As analysed above, RECT-L adopts GNN layers and finally ends with a semantic loss (i.e., Eq. 1), where class prototypes are represented by their mean feature vectors. From the viewpoint of classification theory, this is the NCM classifier loss [7].

Remark 2 (The Reasonability of RECT-L). As shown above, RECT-L actually learns a prototypical model with the labeled data of seen classes, reflecting its reasonability on seen classes. On the other hand, as shown in Remark 1, the learned prototypical model maps the data from the raw-input space into a semantic space, like ZSL methods. As validated by lots of ZSL methods, this enables the success of transferring supervised knowledge of seen classes to unseen classes, indicating its reasonability on unseen classes.

3   How to Improve RECT

3.1   The Proposed Method

We overcome the limit of RECT by designing two label expansion strategies. The first is to expand the seen class label set. As directly learning with the binary labels would get unappealing results in ZGE problem [16], we preform label expansion based on the semantic method RECT-L. This naturally leads to a self-training strategy. Specifically, we first train a RECT-L model as described in Sect. 2.2. Then, we use the learned model to get the predicted class-semantic descriptions of unlabeled nodes. After that, for each seen class, we can find top k closest unlabeled nodes to its class-semantic description vector in the semantic space, and finally add them to the labeled node set of this class.

The other is to expand both the seen and unseen class label sets. This would improve the diversity of labels and obtain more robust node embeddings. Although we know little about unseen classes, we can still find some "labeled" data for them. Our idea is quite simple: exploring the discriminative information of both seen and unseen classes via clustering. Specifically, we first train a RECT-L model to get the node embeddings. Then, we apply K-means clustering on the resulted embeddings. After that, for each cluster (class), we can find top k nearby nodes w.r.t. each class center, and finally use them as the labeled data of this class. As K-means clustering is performed on all classes, the newly obtained labeled node set is expected to cover all of them.

3.2   Risk Bounds Analysis

We apply the related learning theories in domain adaptation [1] to our method. Let $D^{train} = \{D^{train}_{original} \cup D^{train}_{expand}\}$ denote the final labeled training node set, where $D^{train}_{original}$ denotes the original labeled node set and D^{train}_{expand} denotes the newly added labeled set via label expansion. Let $D^{test} = D - D^{train}_{original}$ denote the testing node set, where D is the whole node set. The distribution of D^{train} is P_{train} and of D^{test} is P_{test}. The true class-semantic description labeling function is $h(x)$ and the learned prediction function is $f(x)$. We define the prediction error in D^{train} and D^{test} as:

$$\epsilon_{train}(f) = \mathbb{E}_{x \sim P_{train}}[|h(x) - f(x)|]$$
$$\epsilon_{test}(f) = \mathbb{E}_{x \sim P_{test}}[|h(x) - f(x)|] \tag{3}$$

We can consider it as a domain adaptation problem. Suppose the hypothesis space $\mathcal{H}$ containing f is of VC-dimension $\bar{d}$. According to Theorem 1 in [1], with probability at least $1-\delta$, for every $f \in \mathcal{H}$, the expected error $\epsilon_{test}(f)$ is bounded:

$$\epsilon_{test}(f) \le \hat{\epsilon}_{train}(f) + \sqrt{\frac{4}{l}(\bar{d} \log \frac{2el}{\bar{d}} + \log \frac{4}{\delta})}$$
$$+ d_{\mathcal{H}}(D^{train}, D^{test}) + \rho \tag{4}$$

where $\hat{\epsilon}_{train}(f)$ is the empirical error of f in D^{train}, e is the base of natural logarithm, l is the labeled node number after label expansion, $\rho = \inf_{h \in \mathcal{H}}[\epsilon_{train}(f) + \epsilon_{test}(f)]$, and $d_{\mathcal{H}}(D^{train}, D^{test})$ is the distribution distance between D^{train} and D^{test}.

The first term in Eq. 4 is explicitly minimized by training with D^{train} in Eq. 1. If we have high quality D^{train}_{expand}, it is expected that we can learn a model that has a small error on D^{train}. On the other hand, the bad D^{train}_{expand}, e.g., random labels, may lead to a large empirical error. For the second term, we can notice that the final labeled node number l (after label expansion) is definitely larger than the original one. This verifies the reasonability of our label expansion strategy. The third term reflects the relatedness between training and testing data. In the best situation where D^{train} and D^{test} have the same conditional distribution given a class, and suppose all instances are i.i.d., the distribution distance will be small. Besides, introducing more correctly labeled nodes will also reduce this distance [2], as we have $D^{train}_{expand} \subseteq D^{test}$.

4   Experiments

In this section, we conduct extensive experiments to demonstrate the effectiveness of our methods: 1) Ours$_{SL}$: only expanding the labeled node set of seen classes; 2) Ours$_{SUL}$: expanding the labeled node sets of both seen and unseen classes, when the real class number is given; 3) Ours$_{SUL^*}$: expanding the labeled

Table 1. The statistics of datasets.

Dataset	Nodes	Edges	Classes	Features
Citeseer	3,312	4,732	6	3,703
Cora	2,708	5,429	7	1,433
Pubmed	19,717	44,338	3	500

Table 2. Micro-F1 scores on node classification tasks.

	Citeseer			Cora			Pubmed		
	1%	3%	5%	1%	3%	5%	1%	3%	5%
DeepWalk	0.1941	0.2935	0.3713	0.1972	0.3401	0.4916	0.3766	0.5879	0.6350
LSHM	0.1779	0.2143	0.2648	0.1284	0.1295	0.2233	0.3331	0.3591	0.3965
RSDNE	0.2291	0.3066	0.4035	0.2465	0.3869	0.5167	0.4193	0.6219	0.6862
GCN	0.4194	0.5211	0.5478	0.4756	0.5984	0.6266	0.6067	0.6479	0.6664
APPNP	0.4192	0.5397	0.5692	0.4921	0.6380	0.6791	0.6036	0.6287	0.6514
TEA	0.2554	0.3564	0.4010	0.2996	0.4966	0.5770	0.4953	0.5848	0.6431
RECT-L	0.4506	0.5754	0.6204	0.4964	0.6564	0.7325	0.6679	0.7495	0.7668
$Ours_{SL}$	0.5001	0.6004	0.6326	0.5288	0.6748	0.7374	0.7206	0.7622	0.7586
$Ours_{SUL}$	**0.5343**	0.6228	0.6497	0.5125	0.6761	0.7275	0.6641	0.7419	0.7336
$Ours_{SUL*}$	0.5281	0.6226	0.6500	0.4984	0.6636	0.7208	0.6612	0.7406	0.7309
$Ours_{SL-SUL}$	0.5297	**0.6229**	0.6513	0.5450	**0.6963**	**0.7515**	0.7224	0.7704	0.7688
$Ours_{SL-SUL*}$	0.5293	0.6226	**0.6518**	**0.5474**	0.6919	0.7507	**0.7353**	**0.7752**	**0.7730**

node sets of both seen and unseen classes, when the real class number is estimated automatically[1]; 4) $Ours_{SL-SUL}$: concatenating the embeddings obtained by $Ours_{SL}$ and $Ours_{SUL}$; and 5) $Ours_{SL-SUL*}$: concatenating the embeddings obtained by $Ours_{SL}$ and $Ours_{SUL*}$.

4.1   Setup

We conduct our experiments on three widely used citation networks: Citeseer, Cora, and Pubmed [13]. Table 1 shows their statistics. In each dataset, nodes are documents, edges are citations among them, and labels are research topics. Their features are all bag-of-words features. Besides RECT-L, we further compare a famous unsupervised method DeepWalk [10] and some other supervised methods (LSHM [3], RSDNE [16], GCN [4], APPNP [5] and TEA [20]). Following [16], we set the embedding dimension to 200. For all baselines, we adopt their best hyper-parameters. In RECT-L and our methods, we all adopt two GCN layers, PReLU activation, mean squared error loss, and Xavier initialization. We also follow [19] to reduce the node feature dimension to 200 via SVD decomposition, and follow [6] to expand the original labeled node set size to n/ζ^τ, where n is the graph node number, ζ is the average node degree, and τ is the number of the

---

[1] The optimal class number is determined by silhouette coefficient [11].

400 Z. Wang et al.

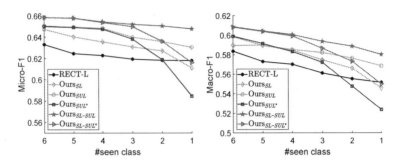

Fig. 1. Classification performance w.r.t. #seen class on Citeseer with 5% label rate.

used GCN layers. In addition, we fix the training epoch number to 100, adopt Adam SGD optimizer, and use the 200-dimensional outputs of the first hidden layer as embedding results, following [15].

4.2 Node Classification

This experiment follows the same procedure as in [16]. Specifically, we first randomly choose two classes as unseen in Citeseer and Cora, and one class as unseen in Pubmed. After that, we remove all the nodes from the unseen classes in the training data, and then apply various graph embedding methods. Finally, an SVM classifier, which is trained based on the resulted embeddings and the original balanced training data, is used to classify the testing nodes.

Table 2 reports the classification performance in terms of Micro-F1. At a glance, we can see the advantage of our label expansion strategies. Generally, our methods outperform the original RECT-L and other baselines by a large margin in most label settings. This improvement would become more significant when the training size is very small. In addition, a very surprising finding is that the performance of $Ours_{SUL}$ is closely related to the performance of $Ours_{SUL^*}$. This indicates that we can always find discrimination information for unseen classes through clustering, even if the true class number is unknown. At last, we can find that combing two label expansion results would get the best performance. This indicates that our two label expansion strategies are complementary for effective embedding learning.

4.3 The Effect of Seen/Unseen Class Number

We continue to use the Citeseer dataset with 5% label rate. As shown in Fig. 1, although all the performance declines smoothly when the seen class number decreases, our two label expansion strategies (especially when combing both of them) steadily improve the performance of RECT-L. This clearly reflects the effectiveness of our methods.

5 Conclusion

In this paper, we give new insights into the mechanisms of RECT and its application in ZGE. In particular, we analyse its relationship with ZSL methods, and the possible limits that it has. To fully realize its potentials, we propose two label expansion strategies. Specifically, we propose to expand the label sets of both seen and unseen classes. In addition, we also study the theoretical properties of our methods. Finally, we conduct extensive experiments to demonstrate the effectiveness of our methods.

Acknowledgment. This work is supported in part by National Key D&R Program of China (2019YFB1600704), National Natural Science Foundation of China (61902020, 61872207), Macao Youth Scholars Program (AM201912), FDCT (FDCT/0045/2019/A1, FDCT/0007/2018/A1), GSTIC (EF005/FST-GZG/2019/GSTIC), University of Macau (MYRG2018-00129-FST), and Baidu Inc.

References

1. Ben-David, S., Blitzer, J., Crammer, K., Pereira, F.: Analysis of representations for domain adaptation. In: NIPS, pp. 137–144 (2007)
2. Hastie, T., Tibshirani, R., Friedman, J.: The Elements of Statistical Learning. SSS. Springer, New York (2009). https://doi.org/10.1007/978-0-387-84858-7
3. Jacob, Y., Denoyer, L., Gallinari, P.: Learning latent representations of nodes for classifying in heterogeneous social networks. In: WSDM, pp. 373–382 (2014)
4. Kipf, T.N., Welling, M.: Semi-supervised classification with graph convolutional networks. In: ICLR (2017)
5. Klicpera, J., Bojchevski, A., Günnemann, S.: Predict then propagate: graph neural networks meet personalized PageRank. In: ICLR (2019)
6. Li, Q., Han, Z., Wu, X.M.: Deeper insights into graph convolutional networks for semi-supervised learning. In: AAAI, pp. 3538–3545 (2018)
7. Mensink, T., Verbeek, J., Perronnin, F., Csurka, G.: Metric learning for large scale image classification: generalizing to new classes at near-zero cost. In: Fitzgibbon, A., Lazebnik, S., Perona, P., Sato, Y., Schmid, C. (eds.) ECCV 2012. LNCS, vol. 7573, pp. 488–501. Springer, Heidelberg (2012). https://doi.org/10.1007/978-3-642-33709-3_35
8. Mikolov, T., Chen, K., Corrado, G., Dean, J.: Efficient estimation of word representations in vector space. arXiv preprint arXiv:1301.3781 (2013)
9. Palatucci, M., Pomerleau, D., Hinton, G.E., Mitchell, T.M.: Zero-shot learning with semantic output codes. In: NIPS, pp. 1410–1418 (2009)
10. Perozzi, B., Al-Rfou, R., Skiena, S.: DeepWalk: Online learning of social representations. In: KDD, pp. 701–710 (2014)
11. Rousseeuw, P.J., Kaufman, L.: Finding Groups in Data. Wiley Online Library, Hoboken (1990)
12. Scarselli, F., Gori, M., Tsoi, A.C., Hagenbuchner, M., Monfardini, G.: The graph neural network model. IEEE Trans. Neural Netw. **20**(1), 61–80 (2008)
13. Sen, P., Namata, G., Bilgic, M., Getoor, L., Galligher, B., Eliassi-Rad, T.: Collective classification in network data. AI Mag. **29**(3), 93–106 (2008)

402 Z. Wang et al.

14. Wang, C., Wang, C., Wang, Z., Ye, X., Yu, J.X., Wang, B.: DeepDirect: learning directions of social ties with edge-based network embedding. TKDE **31**(12), 2277–2291 (2018)
15. Wang, Z., Ye, X., Wang, C., Cui, J., Yu, P.S.: Network embedding with completely-imbalanced labels. TKDE (2020). https://doi.org/10.1109/TKDE.2020.2971490
16. Wang, Z., Ye, X., Wang, C., Wu, Y., Wang, C., Liang, K.: RSDNE: exploring relaxed similarity and dissimilarity from completely-imbalanced labels for network embedding. In: AAAI, pp. 475–482 (2018)
17. Webb, A.R.: Statistical Pattern Recognition. Wiley, Hoboken (2003)
18. Xiao, G., Guo, J., Da Xu, L., Gong, Z.: User interoperability with heterogeneous IoT devices through transformation. TII **10**(2), 1486–1496 (2014)
19. Yang, C., Liu, Z., Zhao, D., Sun, M., Chang, E.Y.: Network representation learning with rich text information. In: IJCAI, pp. 2111–2117 (2015)
20. Yang, Y., Chen, H., Shao, J.: Triplet enhanced autoencoder: model-free discriminative network embedding. In: IJCAI, pp. 5363–5369 (2019)

Spatial and Temporal Data

Online High-Cardinality Flow Detection over Big Network Data Stream

Yang Du[1], He Huang[1(✉)], Yu-E Sun[2], An Liu[1], Guoju Gao[1], and Boyu Zhang[1]

[1] School of Computer Science and Technology, Soochow University, Suzhou, China
{duyang,huangh,anliu,gjgao}@suda.edu.cn
[2] School of Rail Transportation, Soochow University, Suzhou, China
sunye12@suda.edu.cn

Abstract. High-cardinality flow detection over the big network data stream plays an important role in many practical applications. To process large and fast data streams in real-time, most existing work uses compact data structures like sketches to fit themself in high-speed but small on-chip memory. However, this design suffers from expensive computation and thus only supports periodical high-cardinality flow detection. Although NDS can provide online flow cardinality estimation, it is designed to estimate all flows accurately. In contrast, high-cardinality flow detection only concerns whether a flow's cardinality exceeds a certain threshold. This paper complements the prior work by proposing an online high-cardinality flow detection method with high resource efficiency. Based on the on-chip/off-chip design, the proposed method reduces large flows' resource consumption by constructing a virtual bitmap sharing module over the physical bitmap. We evaluate the performance of the proposed method using the real-world Internet traces downloaded from CAIDA. The experimental results show that our method can save up to 65.8% on-chip memory when bounding the same constraints for false-positive rates and false-negative rates.

Keywords: Data stream processing · Network data stream · Online high-cardinality flow detection

1 Introduction

With the proliferation of Internet-connected devices, the big network data stream that flows on the Internet has become the largest data stream in the world [1–5]. For example, Google gets over 3.5 billion searches daily. A high-speed router can forward the network packets at hundreds of Gbps or even multiple Tbps. Such a velocity poses great challenges for real-time network traffic measurement, which has attracted significant attention in recent years [6–10]. This paper focuses on online high-cardinality flow detection, a fundamental problem in network traffic measurement. It has many practical applications in load balancing, access profiling, and attack detection [11–16].

© Springer Nature Switzerland AG 2021
C. S. Jensen et al. (Eds.): DASFAA 2021, LNCS 12681, pp. 405–421, 2021.
https://doi.org/10.1007/978-3-030-73194-6_28

In a general definition, online high-cardinality flow detection reports high-cardinality flows when their estimated cardinalities (number of distinct elements) exceed a certain threshold for the first time. Here a flow is a set of network packets sharing the same flow label (e.g., destination address). The elements in the flow can be source address or any user-defined values. We want to stress that high-cardinality flow detection is highly correlated to the concept named per-flow cardinality estimation [6–8,17] but has a different focus. In particular, it only needs to judge whether a flow's cardinality exceeds a threshold rather than accurately estimating all flows' cardinalities.

The function of network traffic measurement (including per-flow cardinality estimation and high-cardinality flow detection) is often implemented as a module placed on the router/gateway, whose network processor unit can catch up with the ever-higher line rate. However, the actual bottleneck of implementing such network functions is the memory update speed when recording the flow traffic. To solve this problem, most recent work uses high-speed on-chip memory like SRAM to store the flow traffic. Because a large SRAM is expensive and slow, the on-chip memory of a modern router/gateway is usually less than 8.25 MB, which is very limited, considering the numerous flows in the network data stream. It is impossible to assign each flow a separate counter in the limited on-chip memory [18], let alone the larger and more complex data structure required for estimating flow cardinalities.

To address the mismatch between on-chip memory size and flow amount, most existing work uses compact data structures like sketches to store all flow traffic at limited on-chip memory [19–21]. However, adopting this aggressive bit-sharing (or register-sharing) strategy means their estimation accuracy has to give. Moreover, due to the expensive computation for cardinality estimation, they download the flow traffic data for off-chip analysis at the end of each measurement period, thus only providing periodical high-cardinality flow detection, e.g., every 5 or 10 min. Such a lag between the appearance and the identification of a high-cardinality flow will delay the reaction of upstream applications, making the network vulnerable to sudden events like scan attack or DDoS attack.

Given the limitations of sketch-based methods, NDS [16] adopts an on-chip/off-chip model to provide accurate per-flow cardinality estimation. It uses on-chip memory to sample each distinct element with a predefined probability and stores sampled elements in the off-chip memory. The choice of using off-chip memory also make itself support online flow cardinality query. However, when performing high-cardinality flow detection, it is not resource-efficient to provide accurate estimation results for all flows since we only care whether the flow is larger than the threshold.

We aim to fill the gap left by prior arts by addressing their limitations. In particular, we expect a flow's resource consumption grows slower when its cardinality exceeds the threshold. However, it is tricky to tune the resource consumption for different flows since network packets arrive in real-time, and we do not know the exact flow cardinalities. There is recently some work [22,23] trying to predict the flow traffic, assigning different sampling probabilities to different flows. However, predicting flow cardinality is challenging due to the network data stream's dynamicity and uncertainty. Thus, these methods cannot provide any performance guarantees for their prediction results.

This paper presents a novel approach for online high-cardinality flow detection with high resource efficiency. The proposed method adopts an on-chip/off-chip design in [16], maintaining a bitmap at on-chip memory to filter the duplicates. Then we implement a virtual bitmap sharing module to reduce large flows' resource consumption, *i.e.*, mapping a flow's elements to a virtual bitmap which partially overlaps with the physical bitmap. With this design, we limit the number of physical bits a flow is correlated to, reducing large flows' resource consumption while ensuring high-cardinality flow detection performance.

In summary, the contributions of our method are as follows:

- We propose a novel online high-cardinality flow detection method based on the on-chip/off-chip model, reducing the lag between a large flow's appearance and identification. It can provide real-time support for applications like scan detection and DDoS detection.
- We propose a virtual bitmap sharing module to reduce the resource consumptions of large flows by ensuring a flow can only correlate to limited physical bits. We also provide optimal parameter selection for the proposed method with probabilistic performance guarantees.
- We run extensive experiments on real-world Internet traces to evaluate the proposed method's performance. The experimental results show that our method outperforms state-of-the-art solutions by significantly reducing the on-chip memory usage while providing accurate results for online high-cardinality flow detection.

2   Problem Statement

The input to online high-cardinality flow detection is a network data stream $\mathcal{P} = \{\mathbb{P}_1, \mathbb{P}_2, \mathbb{P}_3, \cdots\}$. From each network packet $\mathbb{P}_i \in \mathcal{P}$, we can extract a flow label f and an element label e, where both labels can be flexibly defined to meet the interests of different applications. For example, we may treat the source address as the flow label, use the destination address as the element label, or configure the labels as arbitrary subsets in the packet header. By abstracting the packets with the same flow label as a flow, we model the network data stream as a set of flows $\mathcal{F} = \{f_1, f_2, f_3, \cdots\}$. We use notation n_f to represent the *flow cardinality* of a flow f, representing the number of *distinct* elements in this flow.

The problem of online high-cardinality flow detection is to configure a firewall or defense system that identifies the flows whose cardinalities are no less than a certain threshold T during a measurement interval. We refer to these flows as *high-cardinality flows*. Notice that the high-cardinality flows are reported when their estimated cardinalities exceed the threshold for the first time.

Given the constraints of on-chip memory and packet processing speed, we cannot accurately count the cardinalities of all flows. Instead, we only provide estimations for each flow's cardinality. Let $\widehat{n_f}$ be the estimation of n_f, *i.e.*, the estimated cardinality of flow f. We want our high-cardinality flow detection method satisfying the probabilistic performance constraints as in [13]: Let α and β be two probability values that satisfy $0 < \alpha, \beta < 1$. Supposing we have two integers h and l ($l < T < h$), the objective is to report any flow whose cardinality is h or larger as a high-cardinality flow with a probability no less than α and report any flow whose cardinality is l or lower as a high-cardinality flow with a probability no more than β. Formally, $\widehat{n_f}$, the estimation of n_f, should satisfy the following constraints:

$$\begin{aligned} \Pr(\widehat{n_f} \geq T) &\geq \alpha, \quad n_f \geq h, \\ \Pr(\widehat{n_f} \geq T) &\leq \beta, \quad n_f \leq l \end{aligned} \tag{1}$$

In this case, a false positive refers to misclassifying a normal flow (whose cardinality is no larger than l) as a high-cardinality flow. Besides, a false negative refers to misclassifying a flow (whose cardinality is no less than h) as a normal flow. Thus, the above probabilistic performance objective is equivalent to bound the false-negative rate by $1 - \alpha$ and bound the false-positive rate by β.

This paper focuses on minimizing the on-chip memory requirement of achieving the above objective.

3   Design of Online High-Cardinality Flow Detection

This section presents our design of online high-cardinality flow detection. Unlike traditional sketch-based methods that store all flow traffic in the limited on-chip memory, our solution is based on the on-chip/off-chip model [16], using both on-chip memory and off-chip memory to process the big network data stream.

As shown in Fig. 1, the proposed method is deployed as a module at the router/gateway, identifying the high-cardinality flows from the network data stream passing through it. The proposed method contains two components: *on-chip sampling* and *off-chip recording*, designed to serve different goals.

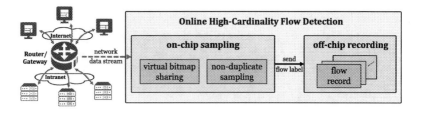

Fig. 1. The system model of online high-cardinality flow detection

The on-chip sampling is designed to process the data stream whenever a packet arrives, meeting real-time processing requirements. Particularly, it selects and sends only a part of flow traffic to the off-chip recording. Unlike [16], which samples all distinct elements with the same probability, we will not treat those elements equally. Instead, we assign each flow a virtual bitmap, which partially overlaps with the physical bitmap. The benefits of virtual bitmap sharing are two-fold: First, we can limit the bits a large flow is mapped to, reducing its resource consumption. Second, we can tune the virtual map size and overlapping ratio to ensure that our estimation results can achieve the desired accuracy for flows whose cardinalities are nearby the threshold T.

The function of the off-chip recording is to store the selected flow statistics. By maintaining a separate counter for each flow, we eliminate the noises introduced by other flows. This design also enables us to answer queries in real-time, while sketch-based methods can only answer queries after the measurement epoch.

3.1   Data Structure

As shown in Fig. 2, our on-chip data structure contains a bitmap B of m bits and a counter c that records the number of ones in B. At the beginning of each measurement epoch, we initialize all bits in the bit array to 0. Thus, counter c will also be set to 0. Our off-chip data structure is composed of a hash table, where we assign each flow f a separate counter, namely c_f.

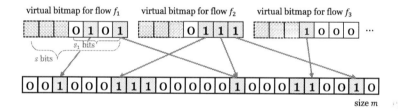

Fig. 2. The on-chip data structure of the proposed method

For each flow f, we assign it a virtual bitmap of s virtual bits. Notice that, if we connect all virtual bits to the physical bitmap, the small flows will consume too much on-chip memory. Therefore, we design two kinds of virtual bits. One is valid virtual bits, each of which is pseudo-randomly mapped to one physical bit in B. The other kind is called void bits, which we will not map them to physical bits in B. In other words, only the valid part of the virtual bitmap overlaps with the physical bitmap.

Given s virtual bits, we use s_1 to represent the number of valid virtual bits and let the remained $s - s_1$ bits be void bits. Thus, the overlapping ratio between virtual bitmap and physical bitmap, *i.e.*, the probability of an element being mapped to a valid virtual bit, is $p_1 = \frac{s_1}{s}$.

3.2   Algorithm Design

In the following, we will demonstrate our method for online high-cardinality flow detection. The proposed method supports two operations: update operation and online query operation. The update operation is performed on each arrival packet, while the online query operation returns the current estimated cardinality for an arbitrary flow. Notice that online high-cardinality flow detection is built on the online query operation. Whenever a flow label is sent to off-chip analysis, the proposed method will compare the flow cardinality with the threshold T. It will report the flow as a high-cardinality flow if the estimated flow cardinality exceeds the threshold T for the first time.

Update Operation: As shown in Algorithm 1, for each packet in the network data stream, the proposed method will perform the update operation when it arrives, determining whether to select the element this packet carries and updating the flow traffic if it is selected.

Algorithm 1: Update Operation

INPUT   : Bit array size m, counter c, network data stream $\mathcal{P}$, non-duplicate sampling probability p_2, virtual bitmap size s, valid virtual bit number s_1

Create an on-chip bitmap of size m and initialize all bits to 0;

Initialize counter c to 0;

for *each arrival packet* $\mathbb{P}_i$ *in packet stream* $\mathcal{P}$ **do**

    Extract flow label f and element label e;

    Map element $< f, e >$ to virtual bit $h_1 = H(f \oplus e) \mod s$;

    if $h_1 < s_1$ **then**

        Map virtual bit h_1 to a physical bit $h_2 = H'(f \oplus h_1) \mod m$;

        if $B[h_2] = 0$ **then**

            Compute a hash value $h_3 = H''(f \oplus e)$;

            Compute $p_3 = p_2 \frac{m}{m-c}$;

            if $h_3 < p_3 X$ **then**

                Send flow label f to off-chip recording;

                if *label* f *appears for the first time* **then**

                    | Initialize counter c_f to 0;

                end

                Set $c_f = c_f + 1$;

            end

            Set $B[h_2] = 0$ to 1;

            Set $c = c + 1$;

        end

    end

end

We want to ensure that each virtual bit in a virtual bitmap has the same probability of being downloaded and stored at off-chip memory, simplifying the computation for cardinality estimation. Let overall sampling probability p represent such probability. Recall that a virtual bit has a probability of $p_1 = \frac{s_1}{s}$ to be a valid virtual bit. Then we adopt the non-duplicate sampling as in [16] to sample each valid virtual bit with a probability p_2. To ensure the overall sampling probability is p, we have $p = p_1 p_2$.

At the beginning of a measurement epoch, we will initialize all bits in B to 0 and set counter c to 0. Then update operation for each arrival packet is performed as follows: Consider an arbitrary packet; we first extract the flow label f and element label e from the information this packet carries. Then, we map this element to one virtual bit h_1 in the virtual bitmap by executing a hash $h_1 = H(f \oplus e) \mod s$, where H is a hash function. Notice that we only map the s_1 valid virtual bits to physical bits. Thus, when $h_1 < s_1$, we say this packet is mapped to a valid virtual bit, and its corresponding bit in B can be obtained by $h_2 = H'(f \oplus h_1) \mod m$, where H' is another hash function.

Given the status of $B[h_2]$, there are two cases to consider: First, when $B[h_2] = 1$, we regard this element as a duplicate that has been seen before and take no further operation. The second case is $B[h_2] = 0$, which means this is the first appearance of $<f, e>$. At this time, we sample this element with a temporal probability $p_3 = p_2 \frac{m}{m-c}$. To perform the above sampling, we run another hash $h_3 = H''(f \oplus h_1)$ and check if $h_3 < p_3 X$, where X is the maximum output of the hash function. When $h_3 < p_3 X$, we will select $<f, e>$ and send flow label f to off-chip recording and increase the flow f's counter, i.e., c_f, by 1. We want to stress that, no matter if a new element $<f, e>$ is selected, we will always set $B[h_2]$ to 1, ensuring all its subsequent appearances will be regarded as duplicates. Therefore, we will not download the same flow element twice.

Online Query Operation: The choice of storing flow traffic in the off-chip memory enables us to perform online query operation during the measurement epoch. Specifically, we can query an arbitrary flow's cardinality at any time. When querying the flow cardinality of flow f, we first retrieve the counter value c_f for this flow. Then we can estimate the number of ones in the valid virtual bits as $\frac{c_f}{p_2}$ and approximate the fraction of ones in the virtual bitmap by $\frac{c_f}{sp}$. According to the probabilistic counting algorithm, we can estimate the flow cardinality as follows:

$$\widehat{n_f} = -s \ln \left(1 - \frac{c_f}{sp}\right) \tag{2}$$

3.3   Online High-Cardinality Flow Detection

Based on the online query operation, our online high-cardinality flow detection is triggered whenever a flow label arrives. It will report a flow as the high-cardinality flow when its estimated cardinality exceeds the threshold for the first time. Based on Eq. 2, we can obtain a counter threshold c_T with respect to threshold T as follows:

$$c_T = sp(1 - e^{-\frac{T}{s}}) \tag{3}$$

Apparently, when a flow's counter value is c_T or larger, its estimated cardinality is no less than T, and we will report it as a high-cardinality flow. To simplify the computation, when performing online detection for a flow f, instead of querying estimated cardinality, we will compare the counter value c_f with c_T. The proposed method will report f as a high-cardinality flow if $c_f = \lceil c_T \rceil$, which ensures a flow will be reported as a high-cardinality flow when this flow's current estimated cardinality exceeds the threshold for the first time.

4   Optimal System Parameters

In this section, we present the parameter selection for the proposed mechanism. As discussed in Sect. 2, our goal is to minimize the on-chip memory requirement while providing the following probabilistic performance guarantees. The proposed method will report any flow whose cardinality is h or larger as a high-cardinality flow with a probability no less than α and report any flow whose cardinality is l or lower as a high-cardinality flow with a probability no more than β.

We need to select four parameters for the proposed model: physical bitmap size m, virtual bitmap size s, valid virtual bit number s_1, and the non-duplicate sampling probability p_2. As discussed above, for each virtual bit, its overall sampling probability is always $p = \frac{s_1 p_2}{s}$.

4.1   Report Probability

Consider a flow f with cardinality n_f (n for short in the following context); when its estimated cardinality $\widehat{n}$ is T or larger, we will report this flow as a high-cardinality flow. Given sampling probability p and threshold T; let $F(n, s, p, T)$ represent the probability of reporting a flow with cardinality n as a high-cardinality flow, namely, *report probability*. As shown in Eq. 7, it can be computed by the sum of probabilities when the estimated cardinality $\widehat{n}$ is T or larger:

$$F(n, s, p, T) = \sum_{i=T}^{\infty} \Pr\{\widehat{n} = i | s, p\} \tag{4}$$

As demonstrated in Sect. 3.2, for each flow f, we assign an s-bit virtual bitmap to it for counting the cardinality. Suppose both valid virtual bits and void ones can record values, and they are set to 0 at the beginning of the measurement. Whenever an element of this flow is pseudo-randomly hashed onto one bit of 0, we set this bit to 1 and use v to count the number of ones in total s virtual bits. According to the probabilistic counting algorithm, v is expected to be $s(1 - e^{-\frac{n}{s}})$. In this paper, we operate a ceiling operation on the expected value and define v as follows:

$$v = \lceil s(1 - e^{-\frac{n}{s}}) \rceil \tag{5}$$

Let c represent the counter value for flow f with cardinality n. Apparently, it follows a Binomial distribution Binomial(v, p), where v is the number of trails,

p is the probability of success, and v is the number of successes. Then we can calculate $\Pr\{c = j|v, p\}$, the probability of c being j, as follows:

$$\Pr\{c = j|v, p\} = C_v^j p^j (1 - p)^{v-j} \tag{6}$$

According to Eq. 2, when c is $\lceil c_T \rceil$ or larger, we will report flow f as a high-cardinality flow. Therefore, $F(n, s, p, T)$ is equivalent to adding up the probabilities when c varies from $\lceil c_T \rceil$ to v, which we denote as $F'(v, p, \lceil c_T \rceil)$. Formally, the following equation always holds:

$$F(n, s, p, T) = F'(v, p, \lceil c_T \rceil) = \sum_{j=\lceil c_T \rceil}^{v} \Pr\{c = j\} \tag{7}$$

4.2   Constraints for System Parameters

The probabilistic performance objective can be stated as two constraints. First, the reporting probability of a flow with cardinality $n \geq h$ must be at least α. That is, $F'(v, p, \lceil c_T \rceil) \geq \alpha, \forall n \geq h$. The second constraint requires that our method report a flow with cardinality $n \leq l$ as a high-cardinality flow with a probability no more than β. Formally, $F'(v, p, \lceil c_T \rceil) \leq \beta, \forall n \leq l$.

Before explaining the selection of system parameters, we start by proving that report probability $F'(v, p, \lceil c_T \rceil)$ is a monotonically increasing function in v.

Theorem 1. *Suppose c is drawn from a Binomial distribution $Binomial(v, p)$. Report probability $F'(v, p, \lceil c_T \rceil) = \sum_{j=\lceil c_T \rceil}^{v} Pr\{c = j\}$ is a monotonically increasing function in v.*

Proof. Let $c_{v,p}$ denote the counter value given binomial distribution parameters v and p. Based on the definition of report probability, we say $F'(v+1, p, \lceil c_T \rceil)$ is equivalent to $\Pr\{c_{v+1,p} \geq \lceil c_T \rceil\}$, the probability when $c_{v+1,p}$ is $\lceil c_T \rceil$ or larger. It can be computed by separately considering the cases when $c_{v,p}$ is at least $\lceil c_T \rceil$ or equal to $\lceil c_T \rceil - 1$.

$$\begin{aligned}
\Pr\{c_{v+1,p} &\geq \lceil c_T \rceil\} \\
&= \Pr\{c_{v+1,p} \geq \lceil c_T \rceil | c_{v,p} \geq \lceil c_T \rceil\} \times \Pr\{c_{v,p} \geq \lceil c_T \rceil\} \\
&\quad + \Pr\{c_{v+1,p} \geq \lceil c_T \rceil | c_{v,p} = \lceil c_T \rceil - 1\} \times \Pr\{c_{v,p} = \lceil c_T \rceil - 1\}
\end{aligned} \tag{8}$$

Apparently, $\Pr\{c_{v+1,p} \geq \lceil c_T \rceil | c_{v,p} \geq \lceil c_T \rceil\}$ equals to 1 since $c_{v+1,p} \geq \lceil c_T \rceil$ always holds when $c_{v,p} \geq \lceil c_T \rceil$. Consider when $c_{v,p} = \lceil c_T \rceil - 1$; inequality $c_{v+1,p} \geq \lceil c_T \rceil$ is true only when the $(v+1)$-th trail is successful, whose probability is p. Therefore, we have the following equation:

$$\Pr\{c_{v+1,p} \geq \lceil c_T \rceil\} = \Pr\{c_{v,p} \geq \lceil c_T \rceil\} + p \times \Pr\{c_{v,p} = \lceil c_T \rceil - 1\} \tag{9}$$

Apparently, $\Pr\{c_{v,p} = \lceil c_T \rceil - 1\}$ is a positive value, which means $\Pr\{c_{v+1,p} \geq \lceil c_T \rceil\}$ is always larger than $\Pr\{c_{v,p} \geq \lceil c_T \rceil\}$. In summary, report probability $F'(v, p, \lceil c_T \rceil)$ is a monotonically increasing function in v. $\square$

As we discussed above, $F(n, s, p, T)$ is equivalent to $F'(v, p, \lceil c_T \rceil)$. From Eq. 5, we know that v increases when n grows. Therefore, $F(n, s, p, T)$ is also a monotonically increasing function in n.

Consider the probabilistic performance guarantees in Eq. 1. To ensure that $F(n, s, p, T) \geq \alpha$ is satisfied for all $n \geq h$, we only have to consider the worst case, i.e., $n = h$. When $F(h, s, p, T) \geq \alpha$ holds, for any flow with cardinality $n \geq h$, its report probability is at least α, according to Theorem 1. Similarly, for the second constraint $F(n, s, p, T) \leq \beta, \forall n \leq l$, it is satisfied if the report probability of worst-case $F(l, s, p, T)$ is β or lower. Therefore, the probabilistic performance guarantee in Eq. 1 can be transformed into the following constraint:

$$(F(h, s, p, T) \geq \alpha) \wedge (F(l, s, p, T) \leq \beta) \tag{10}$$

We want to stress that the above constraint is only related to s and p. In the following, we will explain how to select the optimal value of m, $p_1 = \frac{s_1}{s_2}$, and p_2 when s and p are fixed.

Suppose there are two parameters s and p satisfying the probabilistic performance constraints. We use notation N' to represent the total number '1's in all virtual bitmaps, which can be computed by adding up the number of ones in each flow's virtual bitmap. Formally, N' can be obtained by the following equation:

$$N' = \sum_{f \in \mathcal{F}} s(1 - e^{-\frac{n_f}{s}}) \tag{11}$$

In the update operation, each virtual bit has a probability of p_1 to be a valid virtual bit. Therefore, the number of ones in the valid virtual bits will be $N'p_1$. Recall that the probability of selecting a new valid virtual bit that has been hashed onto one bit of 0 is $p_3 = p_2 \frac{m}{m-c} \leq 1$. According to the probabilistic counter algorithm, we can estimate the counter's maximum value, i.e., $c = m(1 - e^{-\frac{N'p_1}{m}})$, reached when we hashed $N'p_1$ valid bits to B and set the corresponding bits to 1. Gathering the above results, we have $m \geq \frac{N'p_1}{\ln p_2}$. Recall that $p = p_1 p_2$, we can transform this inequality into a function of p_2 as follows:

$$m \geq -\frac{N'p}{p_2 \ln p_2} \tag{12}$$

To ensure that $p_1 = \frac{p}{p_2}$ is a valid probability value (within $(0, 1]$), the range of p_2 is $[p, 1]$. The first derivative of $-\frac{N'p}{p_2 \ln p_2}$ is $\frac{N'p(1+\ln p_2)}{(p_2 \ln p_2)^2}$, which is lower than 0 when $p_2 < \frac{1}{e}$ and larger than 0 when $p_2 > \frac{1}{e}$. Thus, to minimize the value of m, the optimal values of p_1 and p_2 is as follows:

$$p_1 = \min\{ep, 1\}; \quad p_2 = \max\{p, \frac{1}{e}\} \tag{13}$$

Consider that $s_1 = sp_1$ is an integer. We will set s_1 to $\lceil sp_1 \rceil$ to ensure the performance of the proposed method. Therefore, the parameter selection can be expressed as the following problem.

$$\min m$$

$$\text{s.t.} \begin{cases} m \geq -\dfrac{N'\lceil s\min\{ep,1\}\rceil}{s\ln(\max\{p,\frac{1}{e}\})} \\ F(h,s,p,T) \geq \alpha \\ F(l,s,p,T) \leq \beta \end{cases} \tag{14}$$

In the above problem, the first constraint ensures that the selected m is sufficient so that the sampling probability p_3 is within $(0,1]$. The second and third constraints ensure that the optimal parameters s and p can bound the false-negative rate by $1 - \alpha$ and bound the false-negative rate by β.

5 Experimental Evaluation

5.1 Experiment Setup

This section evaluates the performance of the proposed algorithm through extensive experimental evaluations on 5-min Internet trace downloaded from CAIDA [24]. This dataset has 513889 distinct per-destination flows, 3150740 distinct elements. In the following table, we present the distributions of flow cardinalities. Our goal is to identify the high-cardinality per-destination flows in this dataset when setting different threshold T (Table 1).

Table 1. The distribution of per-destination flows in different cardinality ranges

Cardinality	1∼10	11∼20	21∼50	51∼100	101∼200	201∼500	501∼1000	1001∼
#Flows	490998	7734	6917	3932	2039	1383	594	292

We run our evaluation on a server equipped with two six-core Intel Xeon E5-2643 v4 3.40 GHz CPU and 256 GB RAM. We have implemented our solution in C++. For comparison purposes, we also implemented NDS [16] in C++. The hash functions used in our experiments are MURMUR3 hash with different initial seeds.

In the following, we first compare our method and NDS in the on-chip memory they require to satisfy different constraints. Then we compare the proposed method and NDS for their performance on high-cardinality flow detection in terms of estimation accuracy, false-positive rate (FPR), and false-negative rate (FNR). FPR refers to the fraction of normal flows (cardinalities are less than l) that are falsely identified as high-cardinality flows. FNR is defined as the fraction of high-cardinality flows that are not identified as high-cardinality flows.

5.2 Comparison in Terms of Memory Requirements

We first compare our method and NDS for the amount of memory needed to satisfy the constraints given in Eq. 14. Table 2 shows the memory requirements of our method and NDS with respect to h and l, which were computed by the parameter selection methods proposed in this work and [16]. Notice that, given the values of h and l, the threshold T is set to $(h+l)/2$. Besides, the values of α and β are set to 0.95 and 0.05, which means we want to bound the false-negative rate by $1 - \alpha = 0.05$ and bound the false-positive rate by $\beta = 0.05$ at the same time.

Table 2. On-chip memory requirements of NDS and our method (MB)

h	l									
	0.5 h		0.6 h		0.7 h		0.8 h		0.9 h	
	NDS	Ours	NDS	Ours	NDS	Ours	NDS	Ours	NDS	Ours
100	0.303	0.223	0.410	0.317	0.610	0.477	1.097	0.924	4.505	3.779
200	0.219	0.152	0.271	0.220	0.378	0.332	0.690	0.593	2.311	2.121
300	0.177	0.115	0.226	0.173	0.303	0.269	0.497	0.460	1.683	1.476
500	0.149	0.077	0.170	0.110	0.226	0.186	0.348	0.342	0.974	0.964
1000	0.117	0.040	0.134	0.060	0.170	0.108	0.241	0.213	0.574	0.568

From Table 2, we found that when setting different h and l, NDS always requires more on-chip memory than the proposed method, which indicates the resource-efficiency of the proposed method. For example, we observe that our solution save 26.4% on-chip memory than NDS when $h = 100, l = 50$ and save 35.2% on-chip memory than NDS when $h = 500, l = 300$. This is because the proposed method can reduce the resource consumption of large flows, *i.e.*, utilizing the on-chip memory with higher efficiency.

5.3 Comparison in Terms of High-Cardinality Flow Detection

In this part of the evaluation, we compare the proposed method and NDS for high-cardinality flow detection performance under different probabilistic performance constraints.

In the first set of experiments, we compare our method and NDS when $h = 200, l = 120, \alpha = 0.95, \beta = 0.05$. The system parameters of NDS and the proposed method are configured according to the parameter selection methods proposed in [16] and this work. At this time, NDS requires 0.271 MB on-chip memory to satisfy the above constraints, while our method only requires 0.229 MB, saving 15.4% on-chip memory. This is because our method can reduce the resource consumption of large flows and show better space efficiency. Figure 3(a) and Fig. 3(b) show the estimation accuracy of NDS and our method. The x-axis represents the

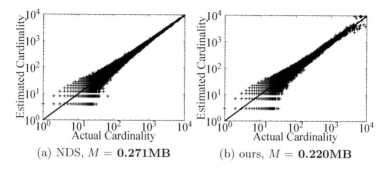

(a) NDS, $M = \mathbf{0.271MB}$ (b) ours, $M = \mathbf{0.220MB}$

Fig. 3. Cardinality estimation accuracy of NDS and our method when $h = 200, l = 120, \alpha = 0.95, \beta = 0.05$

Table 3. Mean relative error for NDS and our method ($h = 200, l = 120, \alpha = 0.95, \beta = 0.05$)

Algorithm	Cardinality					
	All flows	1~100	101~200	201~500	501~1000	1001~
NDS	1.364	1.375	0.121	0.083	0.054	0.032
Ours	1.324	1.334	0.125	0.090	0.075	0.124

Table 4. False positive rate and false negative rate of NDS and our method when $h = 200, l = 120, \alpha = 0.95, \beta = 0.05$

	FPR			FNR		
Cardinality	1~100	101~110	111~120	200~210	211~220	221~
NDS	0	0	0.018	0.029	0.022	0.001
Ours	0	0	0.018	0.039	0	0

actual cardinality; the y-axis represents the estimated cardinality. A point refers to a flow. Therefore, the closer a point is to the line $y = x$, the more accurate the estimation result is. Also, we list the mean relative error of the estimated cardinalities in Table 3. We found out that NDS's estimation accuracy increases when the flow cardinality grows. It achieves the best accuracy when the flow cardinality is much larger than h.

We want to stress that such a high accuracy for the large flows is not desired for high-cardinality flow detection. Different from NDS, our method achieves the best accuracy when flow cardinality is close to threshold T. But when flow cardinality is much larger than T, e.g., 1000, our method will limit the on-chip memory this flow occupies, decreasing the estimation accuracy while ensuring the performance constraints. In Table 4, we compare the proposed method and NDS in terms of FNR and FPR. We found that both methods' FPRs and FNRs satisfy the given constraints, which means our method can show the same performance on high-cardinality identification while using 81.1% on-chip memory that NDS requires.

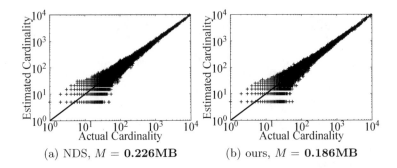

(a) NDS, $M = \mathbf{0.226MB}$ (b) ours, $M = \mathbf{0.186MB}$

Fig. 4. Size estimation accuracy of NDS and the our method when $h = 500, l = 350, \alpha = 0.95, \beta = 0.05$

Table 5. Mean relative error for NDS and our method ($h = 500, l = 350, \alpha = 0.95, \beta = 0.05$)

Algorithm	Cardinality					
	All flows	1~100	101~200	201~500	501~1000	1001~
NDS	1.495	1.506	0.146	0.093	0.063	0.038
Ours	1.466	1.478	0.143	0.094	0.073	0.055

Table 6. False positive rate and false negative rate of NDS and the our method when $h = 500, l = 350, \alpha = 0.95, \beta = 0.05$

	FPR			FNR		
Cardinality	1~330	331~340	341~350	500~510	511~520	521~
NDS	0	0	0	0	0	0
Ours	0	0	0.050	0.111	0	0

In the second sets of experiments, we configure the performance constraints to $h = 500, l = 350, \alpha = 0.95, \beta = 0.05$. At this time, NDS requires 0.226 MB on-chip memory, while our method only needs 0.186 MB, *i.e.*, saving 17.6% on-chip memory. The cardinality estimation results are presented in Fig. 4(a) and Fig. 4(b) and Table 5, where we can obtain similar results as in the first set of experiments. Then we present the FNRs and FPRs of both algorithms in Table 6. It shows that our method can significantly reduce the required on-chip memory while ensuring the same performance constraints, indicating the proposed method's resource efficiency.

6   Related Work

The concept of high-cardinality flow detection is similar to per-flow cardinality estimation but has a different focus. Per-flow cardinality estimation is designed to estimate the cardinalities for all flows. Differently, high-cardinality flow detection only concerns if a flow's cardinality exceeds a predefined threshold.

Sketch-based methods [19–21,25] are often used to implement per-flow cardinality estimation and high-cardinality flow detection. To reduce the memory demand and fit in limited on-chip memory, sketch-based methods use compact data structures, like CM, Bitmap, HLL, to compress the flow traffic and reduce memory usage. Placing all flow traffic in the on-chip memory has the benefit of catching up with the line rate and results in one limitation: they need to scan hundreds or thousands of bits/registers when performing either per-flow cardinality estimation or high-cardinality flow detection. Therefore, these off-chip analysis is only executed periodically.

A different strategy for per-flow cardinality estimation is using sampling. Recently, NDS [16] presents a non-duplicate sampling method based on the on-chip/off-chip design. Unlike sketch-based methods, it only uses on-chip memory to filter the duplicates, then sends and stores the sampled elements in the off-chip memory. The benefit of using off-chip memory is two-fold. First, it maintains a separate counter for each flow in the off-chip memory, reducing the estimation error and estimation time. Second, it can answer online queries since computation for cardinality estimation is low. However, NDS is designed for per-flow cardinality estimation that provides accurate estimation results for all flows. Differently, high-cardinality flow detection only concerns whether a flow's cardinality exceeds a certain threshold. NDS's accurate estimation results for large flows are not desired since it does not affect the identification performance but wastes unnecessary resources.

This motivates us to explore online high-cardinality flow detection. We opt to report the high-cardinality flows when their estimated cardinalities exceed the threshold for the first time. Meanwhile, we want to achieve better resource efficiency than the methods proposed for per-flow cardinality estimation.

7   Conclusion

This paper proposes an online high-cardinality flow detection method over the big network data stream, which reports the high-cardinality flows when their estimated cardinalities exceed the threshold for the first time during the measurement epoch. Based on an on-chip/off-chip design, the proposed method constructs a virtual bitmap sharing module over the physical bitmap, uses on-chip memory to filter the duplicates, and uses off-chip memory to store the flow traffic. With this design, the proposed method can achieve online high-cardinality flow detection with high resource efficiency while meeting the performance constraints. The experimental results based on real Internet traffic traces demonstrate that our solution can achieve higher resource efficiency when bounding

the same constraints for false-positive rates and false-negative rates compared
to the state-of-the-art methods.

Acknowledgements. This research was supported by the National Natural Science
Foundation of China (Grant No. 62072322, 61873177, and U20A20182) and Natural Science Research Project of Jiangsu Higher Education Institution (Grant No. 18KJA520010).

References

1. Yang, T., Zhou, Y., Jin, H., Chen, S., Li, X.: Pyramid sketch: a sketch framework for frequency estimation of data streams. Proc. VLDB Endow. **10**(11), 1442–1453 (2017)
2. Wu, G., et al.: Accelerating real-time tracking applications over big data stream with constrained space. In: Li, G., Yang, J., Gama, J., Natwichai, J., Tong, Y. (eds.) DASFAA 2019. LNCS, vol. 11446, pp. 3–18. Springer, Cham (2019). https://doi.org/10.1007/978-3-030-18576-3_1
3. Huang, H., et al.: An efficient k-persistent spread estimator for traffic measurement in high-speed networks. IEEE ACM Trans. Netw. **28**(4), 1463–1476 (2020)
4. Yang, Z., Zheng, B., Li, G., Zhao, X., Zhou, X., Jensen, C.S.: Adaptive top-k overlap set similarity joins. In: 2020 IEEE 36th International Conference on Data Engineering (ICDE), pp. 1081–1092. IEEE (2020)
5. Zheng, B., et al.: Answering why-not group spatial keyword queries. TKDE **32**(1), 26–39 (2020)
6. Estan, C., Varghese, G.: New directions in traffic measurement and accounting: focusing on the elephants, ignoring the mice. ACM Trans. Comput. Syst. (TOCS) **21**(3), 270–313 (2003)
7. Lieven, P., Scheuermann, B.: High-speed per-flow traffic measurement with probabilistic multiplicity counting. In: Proceedings of the IEEE Conference on Computer Communications (INFOCOM 2010), pp. 1–9 (2010)
8. Yoon, M., Li, T., Chen, S., Kwon Peir, J.: Fit a spread estimator in small memory. In: Proceedings of the IEEE Conference on Computer Communications (INFOCOM 2009), pp. 504–512 (2009)
9. Zhou, Y., Zhou, Y., Chen, M., Xiao, Q., Chen, S.: Highly compact virtual counters for per-flow traffic measurement through register sharing. In: Proceedings of the IEEE GLOBECOM 2016, pp. 1–6 (2016)
10. Ting, D.: Approximate distinct counts for billions of datasets. In: Proceedings of the International Conference on Management of Data (SIGMOD), pp. 69–86. Association for Computing Machinery, New York (2019)
11. Zheng, J., Xu, H., Chen, G., Dai, H.: Minimizing transient congestion during network update in data centers. In: Proceedings of IEEE International Conference on Network Protocols (ICNP 2015), pp. 1–10 (2015)
12. Xu, H., Yu, Z., Qian, C., Li, X., Liu, Z., Huang, L.: Minimizing flow statistics collection cost using wildcard-based requests in SDNs. IEEE ACM Trans. Netw. **25**(6), 3587–3601 (2017)
13. Li, T., Chen, S., Luo, W., Zhang, M.: Scan detection in high-speed networks based on optimal dynamic bit sharing. In: Proceedings of the IEEE Conference on Computer Communications (INFOCOM 2011), pp. 3200–3208 (2011)

14. Hu, C., Liu, B., Wang, S., Tian, J., Cheng, Y., Chen, Y.: ANLS: adaptive non-linear sampling method for accurate flow size measurement. IEEE Trans. Commun. **60**(3), 789–798 (2012)
15. Hao, F., Kodialam, M., Lakshman, T.: ACCEL-RATE: a faster mechanism for memory efficient per-flow traffic estimation. ACM SIGMETRICS Perform. Eval. Revi. **32**, 155–166 (2004)
16. Sun, Y., Huang, H., Ma, C., Chen, S., Du, Y., Xiao, Q.: Online spread estimation with non-duplicate sampling. In: Proceedings of the IEEE Conference on Computer Communications (INFOCOM 2020), pp. 2440–2448 (2020)
17. Heule, S., Nunkesser, M., Hall, A.: HyperLogLog in practice: algorithmic engineering of a state of the art cardinality estimation algorithm. In: Proceedings of the 16th International Conference on Extending Database Technology (EDBT 2013), pp. 683–692 (2013)
18. Yang, T., et al.: A generic technique for sketches to adapt to different counting ranges. In: Proceedings of the IEEE Conference on Computer Communications (INFOCOM 2019), pp. 2017–2025 (2019)
19. Yoon, M., Li, T., Chen, S., Peir, J.K.: Fit a compact spread estimator in small high-speed memory. IEEE ACM Trans. Network. (TON) **19**(5), 1253–1264 (2011)
20. Huang, H., et al.: You can drop but you can't hide: k-persistent spread estimation in high-speed networks. In: Proceedings of the IEEE Conference on Computer Communications (INFOCOM 2018), pp. 1889–1897 (2018)
21. Zhou, Y., Zhou, Y., Chen, S., Zhang, Y.: Highly compact virtual active counters for per-flow traffic measurement. In: Proceedings of the IEEE Conference on Computer Communications (INFOCOM 2018), pp. 1–9 (2018)
22. Zhang, Y.: An adaptive flow counting method for anomaly detection in SDN. In: Proceedings of the Ninth ACM Conference on Emerging Networking Experiments and Technologies, pp. 25–30. Association for Computing Machinery, New York (2013)
23. Cheng, G., Yu, J.: Adaptive sampling for OpenFlow network measurement methods. In: Proceedings of the 12th International Conference on Future Internet Technologies, pp. 1–7. Association for Computing Machinery, New York (2017)
24. CAIDA: The CAIDA UCSD anonymized internet traces (2016). http://www.caida.org/data/passive/passive_2016_dataset.xml. Accessed 28 July 2019
25. Wang, P., Jia, P., Zhang, X., Tao, J., Guan, X., Towsley, D.: Utilizing dynamic properties of sharing bits and registers to estimate user cardinalities over time. In: Proceedings of the IEEE International Conference on Data Engineering (ICDE), pp. 1094–1105 (2019)

SCSG Attention: A Self-centered Star Graph with Attention for Pedestrian Trajectory Prediction

Xu Chen[1], Shuncheng Liu[1], Zhi Xu[1], Yupeng Diao[1], Shaozhi Wu[2], Kai Zheng[1,2], and Han Su[1,2(✉)]

[1] School of Computer Science and Engineering, Chengdu, China
{xuchen,liushuncheng,zhixu023,yupengdiao}@std.uestc.edu.cn,
{zhengkai,hansu}@uestc.edu.cn
[2] Yangtze Delta Region Institute (Quzhou),
University of Electronic Science and Technology of China, Chengdu, China
wszfrank@uestc.edu.cn

Abstract. Pedestrian trajectory prediction enables faster progress in autonomous driving and robot navigation where complex social and environmental interactions involve. Previous models use grid-based pooling or global attention to measure social interactions and use Recurrent Neural Network (RNN) to generate sequences. However, these methods can not extract latent features from temporal and spatial information simultaneously. To address the limitation of previous work, we propose a Self-Centered Star Graph with Attention (SCSG Attention) framework. Firstly, pedestrians' historical trajectories are encoded. Then multi-head attention mechanism plays a role as enhancement of social interaction awareness and simulation of physical attention from human beings. Lastly, the self-centered star graph decoder can aggregate temporal and spatial features and make predictions. Experiments are conducted on public benchmark datasets and measured with uniform standards. Our results show an improvement over the state-of-the-art algorithms by 38% on average displacement error (ADE) and 19% on final displacement error (FDE). Furthermore, it is demonstrated that the star graph has better performance in efficiency of training convergence and ends up with better results in limited time.

Keywords: Pedestrian trajectory prediction · Spatial temporal model · Multi-head attention

1 Introduction

Autonomous driving and robotics that involves human-machine interactions are one of the most promising field of research because there is an unprecedented tendency that let artificial intelligence serve human being and change people

C. S. Jensen et al. (Eds.): DASFAA 2021, LNCS 12681, pp. 422–438, 2021.
https://doi.org/10.1007/978-3-030-73194-6_29

lifestyle. Among all the robotics problems, how to make a machine have a comprehensive understanding of people's motion is a significant issue. Pedestrian trajectory prediction is a branch of the problem from human motion study because it allows robots to plan their own movement to avoid the collision. For example, a self-driving car can make a prediction of pedestrians' movement on the road to make a reasonable adjustment in advance thus avoid the collision. A domestic robot can predict people's trajectory in the room and plan its own movement to minimize the impact on people. Besides, pedestrian trajectory prediction has some important applications in urban city planning and surveillance systems.

Previous work can be classified into four categories: hand-craft rules, grid-based methods, attention-based methods, and graph-based methods. First, the traditional hand-craft rules that simulate human social force [7] have decent results in some circumstances but can not be generalized well on modern datasets because rigid methods are not flexible enough to simulate complex situations in modern datasets. Second, in recent years, Social LSTM [1] is a pioneering work that uses RNN models to make a prediction and also utilizes grid-based pooling to aggregate multiple interactions. After that many different social awareness models are proposed to extract social interactions [3–5]. However, grid-based measurement is not efficient. Sparse grids occupy numerous storage so it needs a large amount of computational power and traversal through grids. Besides, there is a lack of bias on social interaction, which means impacts from different people are considered similarly. Third, to allocate weights to different people and obstacles, global attention models are used in Sophie [14]. However, the drawback of global attention is that it ignores impacts on himself and some potential information from other pedestrians. Forth, recently, SAPTP [6] uses a graph-based method to associate temporal and spatial information and achieve competitive results. Nevertheless, they use a complete graph to extract superfluous features, which will cost redundant computational power.

Fig. 1. Attention to different social interaction (Color figure online)

According to the problems mentioned above, challenges mainly come from two factors: (1) how to extract features that represent social interactions is a difficult task. Let's take Fig. 1 as a running example. The pedestrian k in red changes his path mainly because he wants to hide away from pedestrian 1 in purple while he is less influenced by pedestrian 3 in yellow. It shows that he is influenced by others who are not only close to him but also in conflict direction, fast relative speed, etc. All the potential factors can be influential in social interactions. In addition, multiple external impacts including dynamic and static interactions are supposed to be considered at the same time. The variety of social interactions has not been considered by recent studies. (2) How to aggregate temporal and spatial information simultaneously is also a critical problem. Trajectory prediction can be regarded as a two-dimensional sequence generation issue. Therefore, the chronological order of pedestrians' position is essential. Previous works only consider temporal features at decoder, which is not enough to generate future sequences.

To tackle these challenges, a self-centered star graph based on a multi-head attention model is proposed. We use the multi-head attention model to simulate human being's attention in the real world because it indicates people's reflection on the environment and social impacts. For example, when a person walks on an empty street, he probably will not change his direction and walk in a straight line. On the contrary, when a person walks on a crowded street, he will change his direction and speed to avoid other pedestrians. Therefore, the moving behavior of the pedestrian is mainly influenced by how much attention he should pay to the surroundings. That idea inspires us to build an attention-based model to learn environmental influence. The multi-head attention consists of several layers. Every layer can extract useful latent features respectively thus it is a more comprehensive representation of a human being attention. In addition, the novel self-centered star graph is a data structure designed for pedestrian trajectory prediction. It can combine temporal and spatial information because both of them will flow through the graph. We use the graph at the decoder to capture dynamic and static changes in the environment. Besides, it generates attention simulation from a personal perspective and gets rid of redundant calculations.

Contribution of our paper can be summarized as following:

- We propose the utilization of multi-head attention to simulate pedestrians' attention. Multi-head attention can extract different levels of latent features from social interactions. With a more comprehensive feature representation, our model can find the most possible decision made by the pedestrian.
- A self-centered star graph is proposed to capture temporal and spatial features simultaneously. At the same time, it only takes account of the target pedestrian's interaction with nearby people thus it accelerates training speed.
- Our model is conducted on the benchmark datasets and achieves state-of-the-art accuracy and efficiency of convergence. Extensive experiment results show an improvement over that of previous work by 38% on ADE and 19% on FDE.

The remainder of the paper is structured as follows. Section 2 introduces some important notations and a formal definition of the pedestrian trajectory prediction problem. Our SCSG Attention framework as well as its components are illustrated in Sect. 3. Section 4 presents our evaluation of experimental results and a case study. Lastly, the related work and conclusion are shown in Sect. 5 and Sect. 6.

2 Problem Definitions and Important Notations

Definition 1 (Trajectory sample point). A trajectory sample point p is a location in two-dimensional space, and p_i^t represents the sample point at a specific time stamp t of person i.

Each scene of pedestrians is captured at a fixed frequency in videos, where time can be regarded as frame based on videos. Therefore there is a corresponding time sequence $\{t|t = 1, 2, 3, \ldots, n\}$ where n is final frame. In each frame, every pedestrian will be represented by a two-dimension world coordinate $P = \{(x_i^t, y_i^t)|t = 1, 2, 3, \ldots, n \quad i = 1, 2, 3, \ldots, I\}$ where I is the number of distinct pedestrians of all time. And p_i^t denotes pedestrian i's position at time t.

Definition 2 (Trajectory). Trajectory is a sequence of trajectory sample points, ordered by time stamps t. In this problem, pedestrian i's trajectory is a sequence of two-dimensional trajectory sample points: $T_i = [p_i^1, p_i^2, \ldots, p_i^n]$.

Problem Definition. At any frame t, the problem can be defined as following: from observation of target person k's historical trajectory $T_k^h = [p_k^{t-\lambda+1}, \ldots, p_k^t]$ and his neighbor $T_i^h = [p_i^{t-\lambda+1}, \ldots, p_i^t]$ where $i \neq k$, we want to predict the target pedestrian k's future trajectory $\hat{T}_k^f = [p_k^{t+1}, \ldots, p_k^{t+\delta}]$. λ and δ are historical and future length respectively. Specifically, the target person is denoted by index k in this paper. A static object in the street can be treated as a static pedestrian.

This task can also be viewed as a sequence generation problem, where the input sequence corresponds to the observed positions of a person and we want to generate an output sequence. Our goal is to make predictions $\hat{T}_k^f$ as accurate as possible to the ground truth trajectory T_k^f.

3 Methodology

3.1 SCSG Attention Framework

The overview of model architecture can be shown in Fig. 2. There are three components in our framework, namely (1) spatial and temporal encoder, (2) attention mechanism, (3) self-centered star graph decoder. In the beginning, the historical trajectory of target pedestrian k and his neighboring pedestrians i are encoded by temporal and spatial encoder respectively. Hidden states h_k^t and h_i^t

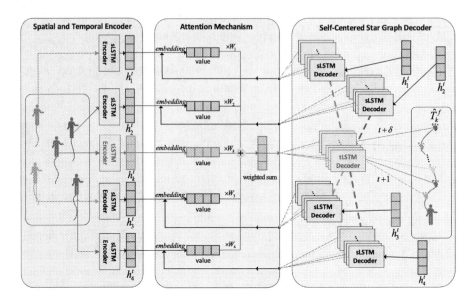

Fig. 2. SCSG attention framework overview

are then entered to our attention mechanism to analog pedestrian k's attention. Finally, the weighted sum of attention vectors is passed through our self-centered star graph decoder to output a predicted location one at a time. At the same time, the neighboring hidden states will be continuously decoded in the graph. Specifically, we calculate attention for every future frame.

3.2 Spatial and Temporal Encoder

Pedestrian location description is based on Cartesian coordinates, thus trajectories in a scene can be shown by Fig. 1. The historical trajectories are represented by a solid line and dash line shows future trajectories. The pedestrian k's historical trajectory contains temporal information and other pedestrian i's historical trajectories are regarded as spatial information. Long Short-Term Memory Networks (LSTM) shows promising functionality in sequence memorization and encoding. For this specific problem, temporal information and spatial information are encoded separately.

For temporal encoding, we defined a dedicated time embedding mapping function to convert historical trajectory T_k^h from locations to a high dimension vector e_k^t as follows:

$$e_k^t = \phi_{temporal}(p_k^t; W) \tag{1}$$

where $\phi_{temporal}(.)$ is a fully connected neural network and p_k^t denotes the location of pedestrian k at frame t, W denotes embedding parameters.

To aggregate historical trajectory features, we define a dedicated temporal LSTM layer to transform temporal embedding e_k^t to a hidden state h_k^t as follows:

$$h_k^t = LSTM(e_k^t, h_k^{t-1}; W) \tag{2}$$

where h_k^{t-1} is the hidden state at last frame, W denotes temporal LSTM parameters. Temporal LSTM layer is executed recursively to obtain the final hidden state h_k^t.

Similarly, a spatial embedding layer is built to transform neighboring pedestrian trajectories T_i^h to high dimension vectors. The embedding layer consists of a fully connected layer. Notably, it does not share parameters with temporal embedding layer but it shares parameters among neighboring pedestrians because neighboring pedestrians together represent context of the target pedestrian. The vector e_i^t is defined as follows:

$$e_i^t = \phi_{spatial}(p_i^t; W) \tag{3}$$

where i is a neighboring pedestrian($i \neq k$) in the scene, p_i^t is location of neighboring pedestrian i at frame t, And W denotes spatial embedding parameters.

We use spatial embedding as input of spatial LSTM in order to incorporate location information of neighboring pedestrians. The spatial LSTM does not share parameters with temporal LSTM. Hidden states h_i^t are defined as follows:

$$h_i^t = LSTM(e_i^t, h_i^{t-1}; W) \tag{4}$$

where i is a neighboring pedestrian($i \neq k$) in the scene, h_i^{t-1} is the pedestrian i's spatial hidden state at previous frame and W denotes spatial LSTM parameters. Spatial LSTM is executed recursively to obtain spatial hidden states of all neighboring pedestrians.

3.3   Attention Mechanism

Human being's attention can allocate bias on different objects thus it makes people focus on useful information. For example, someone who walks on a street will pay more attention to other noticeable pedestrians. Like human beings, attention mechanisms let machines learn useful features thus it makes machines more efficient. In [14], global attention is used to find weights on other pedestrians, but it can not extract features from multiple perspectives. Recently, multi-head attention proposed in [17] vastly boosts the development of the attention mechanism. It is a substitute to simulate physical attention awareness from a personal perspective in our model. To be more specific, multi-head personalized attention is used to mimic physical attention to nearby people in a radius. Therefore, different nearby pedestrians will be measured by unique weights. And multi-head attention can simulate the attention from multiple potential reasoning, which vastly increases the robustness of our model.

The architecture of attention mechanism can be shown in Fig. 3. It shows the calculation process of target pedestrian's attention to himself and his neighboring

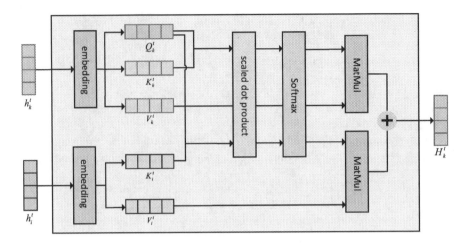

Fig. 3. Attention mechanism

pedestrians in the scene. To calculate attention weights, the output of temporal LSTM h_k^t will be embedded to three vectors namely query vector Q_k^t, key vector K_k^t and value vector V_k^t ($size = d_m$), which are measured as follows:

$$Q_k^t = \phi_t(h_k^t; W_{qt}) \tag{5}$$

$$K_k^t = \phi_k(h_k^t; W_{kt}) \tag{6}$$

$$V_k^t = \phi_t(h_k^t; W_{vt}) \tag{7}$$

where W_{qt}, W_{kt}, W_{vt} in the embedding function are their parameters respectively.

For outputs from spatial LSTM h_i^t, they will be embedded to key vectors K_i^t and value vectors V_i^t. They are defined as follows:

$$K_i^t = \phi_k(h_i^t; W_{ks}) \tag{8}$$

$$V_i^t = \phi_t(h_i^t; W_{vs}) \tag{9}$$

where W_{ks} and W_{vs} in the functions are embedding function parameters.

For target pedestrian k in the scene, the attention values of latent feature j $Score_j^t$ of pedestrian k can be calculated as follows:

$$Score_j^t = \sum_{i=1}^{n} Softmax(\frac{Q_k^t \times K_i^t}{\sqrt{d_m}}) \times V_i^t \tag{10}$$

where index j indicates index of latent features, user can define how many features to calculate.

Finally, this workflow will be repeated for a user-defined fixed number n. Every output is a layer of attention. The benefit of multi-head attention is that every separated attention will extract a feature from a different perspective.

Different from local attention or global attention, it makes a well-rounded consideration to simulate pedestrians' physical attention. The final attention score F_k^t is defined as follows:

$$F_k^t = W_a \times Concat(Score_1^t, Score_2^t, \ldots, Score_n^t) \tag{11}$$

where W_a is matrix representing a linear transformation.

3.4   Self-centered Star Graph Decoder

LSTM can be a sequence generator since it can decode hidden states and predict results one at a time. But it is hard to capture dynamic changes by naively applying LSTM on some real-world problems that are highly dependent on temporal and spatial features. A spatio-temporal graph-based model was proposed in SAPTP [6] to solve pedestrian prediction. While they use a complete graph in their model, a simplified version namely a self-centered star graph is proposed in our model without sacrificing effectiveness and accuracy. The self-centered star graph decreases the number of edges from $O(n^2)$ to $O(n)$, resulting in a faster convergence speed compared to a complete graph.

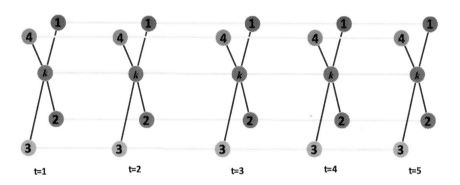

Fig. 4. Self-centered star graph structure

Figure 4 shows the structure of self-centered star graph. It is established in three steps: (1) Add each pedestrian k and i to the vertex set V. So according to the running example in Fig. 1, there are five vertices v_k, v_1, v_2, v_3, v_4 in the beginning. Then add undirected edges from v_k to v_i, denoted by $e(k, i)$. Now a plane self-center star graph is completed (2) Repeat step (1) for $\delta - 1$ times. When $\delta = 5$ (shown in Fig. 4), we get a three-dimensional self-centered star graph. (3) Add undirected grey edges from v_i to v_i and from v_k to v_k respectively from each frame t to $t + 1$, denoted by $e(i, i)$ and $e(k, k)$. That is how the running example in Fig. 1 becomes Fig. 4 topology.

Different edges represent different calculations. From every frame t to $t + 1$, edge $e(i, i)$ represents propagation of spatial information. Specifically, the neighboring pedestrians hidden states h_i^t will be inputted into a spatial decoder as follows:

$$h_i^t = LSTM(h_i^{t-1}; W) \tag{12}$$

where h_i^{t-1} is the hidden state from previous frame and W denotes the spatial LSTM parameter.

Besides, from every frame t to $t + 1$, edge $e(k, k)$ represents propagation of temporal information. Specifically, the target pedestrian hidden states will be decoded by a temporal LSTM as follows:

$$h_k^t = LSTM(F_k^t, h_k^{t-1}; W) \tag{13}$$

where h_k^{t-1} is hidden state of the target pedestrian from previous frame, F_k^t is attention score shown in Eq. 11 and W denotes the temporal LSTM parameters.

At every frame t, edge $e(k, i)$ represents attention from pedestrian k to pedestrian i. The calculation is shown in Eq. 10. After that, the target hidden state will time a matrix W_0 to generate predicted location of the target pedestrian $\hat{p}_k^t$ as follows:

$$\hat{p}_k^t = W_o \times h_k^t \tag{14}$$

Finally, L2 loss is used as our loss function. We have tried both to sample from bivariate Gaussian distribution and to use L2 loss. We find using L2 loss is much beneficial to gradient descent since it boosts the velocity of gradient descent and achieves better results in a shorter time.

4   Experiments and Analysis

In this section, we evaluate our method on several benchmark pedestrian datasets. Besides, our model is compared with selected baselines on two metrics: ADE/FDE.

4.1   Experimental Setup

Datasets: We use two public pedestrian datasets. First, ETH [13] has 750 pedestrians and is divided into two datasets (ETH and hotel) according to two different scenarios. Second, the UCY dataset [10] has 786 pedestrians in total and is divided into three different datasets (zara01, zara02, and univ.) according to different scenarios. Therefore, we used a total of five scenarios to verify our model. These datasets were collected from the real world, including a variety of complex scenes, such as a crowd or two pedestrians walking together. Each pedestrian has a nonlinear trajectory at different speeds.

Evaluation Metrics: According to our baselines, two evaluative metrics are used. The smaller of these two metrics are, the better the model performs.

- Average displacement error(ADE): it calculates the mean square error between all predicted points and ground truth points in a trajectory.

– Final displacement error(FDE): it calculates the distance between the final point of a predicted trajectory and ground truth value of final point.

Evaluation Baselines: We compare our model with the previous competitive models.

– LSTM (vanilla LSTM model): a vanilla LSTM that contain classic encoder-decoder architecture.
– S-LSTM [1]: Social LSTM proposed a grid-based pooling layer, which is designed to model each person via an LSTM with the hidden states being pooled at each time.
– S-GAN [5]: an Generative Adversarial Network(GAN) is used to generate multiple socially-acceptable trajectories. Gupta et al. propose a new grid-based pooling mechanism which encodes the subtle cues for all pedestrians involved in a scene. This model performs well in crowded scenes.
– CF-LSTM [18]: Cascaded Feature-Based Long Short-Term Networks where the feature information of the previous two timestamps is considered as the input of LSTM. Only one pedestrian feature is used in this model.
– SAPTP [6]: a spatio-temporal graph-based model use complete graph to make prediction.
– Global attention model: global attention is used in our model as a comparison to multi-head attention. It only extracts one latent feature and it does not count pedestrian attention to himself.
– Local attention model: local attention is used in our model as a comparison to multi-head attention. Similarly, it only extracts one latent feature. And it generate output attention randomly instead of a weighted sum.

We try our best to reproduce the Cascaded Feature-Based LSTM model (CF-LSTM) following implementation details in the paper [18] and reproduce the vanilla LSTM model to predict the trajectory. Besides, there are three groups of experiments designed to prove our method better than other methods.

Implementation Details: According to all benchmark results, the leave-one-out approach [18] is used to train and validate model parameters. To more specific, every time we train and validate our model on 4 datasets and test on the remaining one. We set $\lambda = 8$ and $\delta = 12$, in other word 3.2 s and 4.8 s respectively. LSTM with 128 units of hidden states is used as encoder and decoder in our model. We use six heads for multi-head attention. Our model is trained with a batch size of 128 for 100 epochs using Adam with a default setting.

4.2 Performance Evaluation

In this subsection, in order to demonstrate the effectiveness of our method, we compare our evaluation metrics with other baselines.

Our Model vs Baselines: Our model is evaluated based on two metrics ADE and FDE against different baselines in Table 1. All baselines use LSTM as the base model since the classic encoder-decoder architecture is powerful in sequence-related problems. The majority of deep learning models in trajectory prediction is based on vanilla LSTM. As expected, all models perform better than it but CF-LSTM is worse than LSTM in FDE. We try our best to reproduce CF-LSTM and fine-tuned parameters on real-world datasets. It is probably because it only extracts limited features from one person while other models make use of multiple people's trajectory. As shown in Table 1, S-GAN performs better than LSTM, CF-LSTM, S-LSTM, and SAPTP, since it revises the pooling layer in S-LSTM and uses generative modeling to produce multiple possible results. Our model outperforms all other models in ADE and FDE since we use a more well-rounded attention mechanism compared to S-GAN. Furthermore, better aggregation graph is used in our model compared to traditional encoder-decoder methods.

Table 1. Quantitative results of baselines and our models on all datasets

Performance (ADE/FDE)					
Type	Base Model	Pooling	Pooling	Individual	Graph	Graph
Datasets	LSTM	S-LSTM	S-GAN	CF-LSTM	SAPTP	Our Model
ETH	1.41/3.13	1.09/2.35	0.87/1.62	1.36/3.40	1.24/2.35	**0.58/1.47**
HOTEL	0.54/1.38	0.79/1.76	0.67/1.37	0.44/1.22	0.48/0.80	**0.20/0.65**
UNIV	1.47/2.83	0.67/**1.40**	0.76/1.52	1.18/2.86	0.69/1.45	**0.53**/1.53
ZARA1	0.41/1.00	0.47/1.0	0.35/0.68	0.38/1.13	0.51/1.15	**0.20/0.62**
ZARA2	0.34/0.93	0.56/1.17	0.42/0.84	0.33/0.96	0.56/1.13	**0.19/0.60**
Average	0.83/1.85	0.72/1.54	0.61/1.21	0.74/1.91	0.70/1.38	**0.34/0.97**

Effectiveness of Multi-head Attention: Attention mechanism plays an important role in our model, and different attention mechanism has different performance. In this set of experiments, we compare the ADE and FDE of different attention mechanisms. As shown in Fig. 5, ADE and FDE from the multi-head attention model are the lowest among all datasets so the multi-head attention model outperforms other methods evidently. An important reason is that the multi-head attention mechanism can pay attention to the subtle cues of pedestrians around. However, global-attention and local-attention both can only pay attention to partial information, leading to negligence of some important information.

Effectiveness of Different Amount of Heads: The amount of the head represents how many times we count attention. It may affect the effectiveness of our attention mechanism since the number of heads partly determines how many features our model can learn. For example, six multi-head means that there are six attention layers to learn six different latent features while hand-craft rules can only extract one rigid feature. These six results will be concatenated and passed to the next layer. In this set of experiments, we compare the ADE and

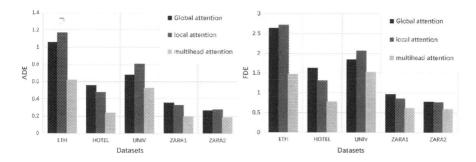

Fig. 5. Comparison of ADE and FDE among attention mechanisms

FDE of different amounts of heads. As shown in Table 2, when the head equals 6, the model outperforms other values of the head. Besides, we only consider the head equal to these values $\{2, 4, 6, 8, 10, 12\}$, the result will be worse with larger number of heads, since it causes overfitting.

Table 2. Multi attention mechanism (Hyparamater)

Datasets	Performance of different head numbers (ADE/FDE)					
	2	4	6	8	10	12
ETH	0.75/2.04	0.64/1.60	**0.58/1.47**	0.62/1.51	0.59/1.54	0.61/1.57
HOTEL	0.22/0.87	0.24/0.91	0.20/0.65	0.21/0.72	**0.17/0.56**	0.20/0.74
UNIV	0.56/1.59	0.55/1.55	**0.53/1.53**	0.54/1.54	0.55/1.53	0.56/1.55
ZARA1	0.21/0.66	0.20/0.64	0.20/**0.62**	0.20/0.63	0.20/0.65	0.20/0.65
ZARA2	**0.19**/0.61	0.20/0.62	**0.19/0.60**	0.20/0.64	0.20/0.62	0.20/0.61
Average	0.39/1.15	0.37/1.06	**0.34/0.97**	0.35/1.01	**0.34**/0.98	0.35/1.02

Effectiveness of the Self-centered Star Graph: The self-centered star graph exhibits two advantages compared with other methods. First, it is designed to capture spatial and temporal features simultaneously, which is shown in the model architecture part. And the average displacement error comparison to the complete graph proves the effectiveness of the star graph. In the experiment (see Table 3), ADE and FDE are compared to show that the effectiveness of star graph is comparable to that of complete graph. Second, less computation is generated in the star graph, which is expected to produce results in less time. According to our experiments, the complete graph model occupies 95% GPU memory while star graph model occupies 83% GPU memory. A stochastic gradient descent loss graph (see Fig. 6) is presented to prove its velocity. Obviously, a faster decreasing tendency can be shown in the graph. Although the loss of a star graph is higher at the beginning, it finally becomes lower in the limited time. Compared with

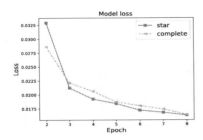

Fig. 6. Loss of star and complete graph

Table 3. ADE and FDE of two graphs

	Performance (ADE/FDE)	
Datasets	Complete graph	Star graph
ETH	0.65/**1.32**	**0.58**/1.47
HOTEL	**0.19/0.57**	0.20/0.65
UNIV	**0.51/1.47**	0.53/1.53
ZARA1	**0.19**/0.62	0.20/**0.62**
ZARA2	**0.13/0.66**	0.19/**0.60**
Average	**0.33/0.93**	0.34/0.97

the complete graph, the self-centered star graph can focus on the interaction from valuable people, and also reduce the amount of calculation.

4.3 Case Study

In this section, according to the results of experiments, some of predicted trajectories are visualized in Fig. 7.

In the first row, the background of these pictures is from ETH, and has two characteristics: simple situations where there are interactions only among people, and there are few noticeable obstacles; most pedestrians move in the same direction. In these four pictures, Figure(a) shows the straight trajectory of a pedestrian with fewer interactions from people around, Figure(b) shows the straight trajectory of a pedestrian with more interactions from people around. Figure(c) shows the crooked trajectory of a pedestrian. Figure(d) shows the trajectory of a pedestrian following multiple pedestrians in the same direction. Obviously, our proposed model has a smaller error and the direction of the predicted trajectory is closer to the ground truth.

In the second row, the background of these pictures is from HOTEL, compared to the first row, these pictures involve complex situations, there are many obstacles in the scene, such as a bench, street lamp, trees. Those static objects will be regarded as static pedestrians so they will be taken into consideration. Besides, these pedestrians move in different directions. In these four pictures, Figure(e) shows that a pedestrian moves in a straight trajectory at a normal speed. Figure(f) shows that a pedestrian walks through the road between a tree and a street lamp. Figure(g) shows that a pedestrian wants to pass the pedestrian ahead. Figure(h) shows the predicted trajectory of two parallel pedestrians. It can be seen that our model has better performance than other models in such complicated situations.

Above all, our model can detect environmental and social interactions. It can also incorporate spatial and temporal features simultaneously with help of attention mechanism and the star graph.

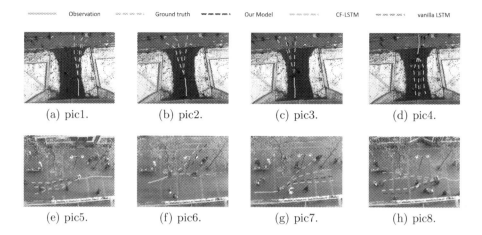

Fig. 7. Visualization of predicted trajectory

5   Related Work

In this section, some important milestones about sequence models, social interaction models, and attention models are introduced. Some of them are used in the real world extensively. State-of-the-art algorithms and techniques are inspired by these previous researches.

5.1   RNN Based Sequence Model

Recurrent Neural Networks(RNNs) are deep learning models used in natural language processing extensively. RNNs are mainly used for sequence processing like machine translation [2,16], image captioning [15] and so on. They are inherently good at sequence memorization and generation since inputs of RNNs are fixed-length sequences and are read step by step. Long Short-Term Memory Networks (LSTM) [8] is a kind of RNNs to avoid gradient exploding and gradient vanishing thus it is capable of encoding more temporal information. With the great success of LSTM in natural language processing [12,19], it is used as temporal information encoded in our model. However, the classic encoder-decoder model is not able to aggregate spatial architecture and temporal information simultaneously. In many real-world applications, problems highly depend on temporal and spatial information. So spatio-temporal LSTM [9,11] are proposed to solve this issue. It uses a spatio-temporal graph to be the abstraction of dynamics information. The edges and vertices are converted to unfolded LSTM layers through shared parameters. The core idea of structured LSTM that can incorporate spatio-temporal encoding is utilized in our model.

5.2   Social Interaction Awareness Model

Pedestrian trajectory prediction has been researched for several decades. Helbing et al. (1995) [7] measured social interaction and inner motivation as social force. The simulations of trajectory prediction are based on some heuristic algorithms. Their algorithms achieved decent results in less complicated circumstances. However, in crowded places, deep learning models tend to perform better [3]. Social LSTM [1] is a pioneering model introducing social interaction in LSTM. Pedestrians' trajectory is described in a grid that can show people relative position and interaction. People in the same grid are then aggregated by pooling layers to obtain synthesized social influence. After Social LSTM, several comparable models are introduced like Convolutional Social Pooling [4], which is a convolutional neural network based on the grid. However, grid-based feature extraction cost a large volume of storage space to cover all interaction around the experiment object. Especially in a sparse environment, which is common in some datasets, the sparse matrix can introduce side effects and redundant storage costs. Since pedestrian trajectory depends on multiple possible factors, Generative Adversarial Networks (GANs) based LSTM was introduced in Social GAN that is able to generate multiple possible results [5]. Their social interaction aggregation is also based on grid pooling. Recently, CF-LSTM [18] predicts pedestrian trajectory without extracting features in social interaction. They make use of the residual network to learn features.

5.3   Attention Model

Attention Mechanism achieves great success in Nature Language Processing (NLP), especially in neural machine translation [16]. It is inherently suitable for sequence generation because it let the generator focus on relevant context instead of considering every information equivalently. In trajectory prediction, pedestrian attention will distribute differently according to different social interactions and spatial situations. In the former models like Social LSTM and Social GAN, weights on every pedestrian are considered equivalently, which is not efficient compared to the attention mechanism. There are three types of attention namely global attention, local attention, and self-attention. Different attention mechanisms will have a substantially different evaluation of social interaction. Haddad et al. [6] use a variant of self-attention to achieve a great result. Their attention is based on historical trajectory, while our model is based on current interaction.

6   Conclusion

In this paper, we focus on predicting the future trajectory of pedestrians in a scene. The self-centered star graph is proposed to make predictions. The pedestrian trajectories will first pass through encoders to become high dimensional vectors. Then these vectors will be extracted latent features by the attention mechanism. Lastly, a self-centered star graph decoder can decode these features and make predictions. We show the efficiency and effectiveness of our model by experiments. Our model proves to work effectively and try to reconstruct complex situations and social norms in real life as much as possible.

Acknowledgment. This work is supported by NSFC (No. 61802054, 61972069, 61836007, 61832017, 61532018), Alibaba Innovation Research (AIR), scientific research projects of Quzhou Science and Technology Bureau, Zhejiang Province (No.2020D010, No.2020D12) and Sichuan Science and Technology Program under Grant 2020JDTD0007. And We thank Qiyang Lyu for his helpful advise.

References

1. Alahi, A., Goel, K., Ramanathan, V., Robicquet, A., Fei-Fei, L., Savarese, S.: Social LSTM: human trajectory prediction in crowded spaces. In: Proceedings of the IEEE Conference on Computer Vision and Pattern Recognition, pp. 961–971 (2016)
2. Bahdanau, D., Cho, K., Bengio, Y.: Neural machine translation by jointly learning to align and translate. arXiv preprint arXiv:1409.0473 (2014)
3. Becker, S., Hug, R., Hübner, W., Arens, M.: An evaluation of trajectory prediction approaches and notes on the TrajNet benchmark. arXiv preprint arXiv:1805.07663 (2018)
4. Deo, N., Trivedi, M.M.: Convolutional social pooling for vehicle trajectory prediction. In: Proceedings of the IEEE Conference on Computer Vision and Pattern Recognition Workshops, pp. 1468–1476 (2018)
5. Gupta, A., Johnson, J., Fei-Fei, L., Savarese, S., Alahi, A.: Social GAN: socially acceptable trajectories with generative adversarial networks. In: Proceedings of the IEEE Conference on Computer Vision and Pattern Recognition, pp. 2255–2264 (2018)
6. Haddad, S., Wu, M., Wei, H., Lam, S.K.: Situation-aware pedestrian trajectory prediction with spatio-temporal attention model. arXiv preprint arXiv:1902.05437 (2019)
7. Helbing, D., Molnar, P.: Social force model for pedestrian dynamics. Phys. Rev. E **51**(5), 4282 (1995)
8. Hochreiter, S., Schmidhuber, J.: Long short-term memory. Neural Comput. **9**(8), 1735–1780 (1997)
9. Jain, A., Zamir, A.R., Savarese, S., Saxena, A.: Structural-RNN: deep learning on spatio-temporal graphs. In: Proceedings of the IEEE Conference on Computer Vision and Pattern Recognition, pp. 5308–5317 (2016)
10. Leal-Taixé, L., Fenzi, M., Kuznetsova, A., Rosenhahn, B., Savarese, S.: Learning an image-based motion context for multiple people tracking. In: Proceedings of the IEEE Conference on Computer Vision and Pattern Recognition, pp. 3542–3549 (2014)
11. Liu, J., Shahroudy, A., Xu, D., Wang, G.: Spatio-temporal LSTM with trust gates for 3D human action recognition. In: Leibe, B., Matas, J., Sebe, N., Welling, M. (eds.) ECCV 2016. LNCS, vol. 9907, pp. 816–833. Springer, Cham (2016). https://doi.org/10.1007/978-3-319-46487-9_50
12. Liu, P., Qiu, X., Huang, X.: Recurrent neural network for text classification with multi-task learning. arXiv preprint arXiv:1605.05101 (2016)

13. Pellegrini, S., Ess, A., Van Gool, L.: Improving data association by joint modeling of pedestrian trajectories and groupings. In: Daniilidis, K., Maragos, P., Paragios, N. (eds.) ECCV 2010. LNCS, vol. 6311, pp. 452–465. Springer, Heidelberg (2010). https://doi.org/10.1007/978-3-642-15549-9_33
14. Sadeghian, A., Kosaraju, V., Sadeghian, A., Hirose, N., Rezatofighi, H., Savarese, S.: SoPhie: an attentive gan for predicting paths compliant to social and physical constraints. In: Proceedings of the IEEE Conference on Computer Vision and Pattern Recognition, pp. 1349–1358 (2019)
15. Soh, M.: Learning CNN-LSTM architectures for image caption generation. Department of Computer Science, Stanford University, Stanford, CA, USA, Technical report (2016)
16. Stahlberg, F.: Neural machine translation: a review. J. Artif. Intell. Res. **69**, 343–418 (2020)
17. Vaswani, A., et al.: Attention is all you need. In: Advances in Neural Information Processing Systems, pp. 5998–6008 (2017)
18. Xu, Y., Yang, J., Du, S.: CF-LSTM: cascaded feature-based long short-term networks for predicting pedestrian trajectory. In: AAAI, pp. 12541–12548 (2020)
19. Young, T., Hazarika, D., Poria, S., Cambria, E.: Recent trends in deep learning based natural language processing. IEEE Comput. Intell. Mag. **13**(3), 55–75 (2018)

Time Period-Based Top-k Semantic Trajectory Pattern Query

Munkh-Erdene Yadamjav[1]($\boxtimes$), Farhana M. Choudhury[2], Zhifeng Bao[1], and Baihua Zheng[3]

[1] RMIT University, Melbourne, Australia
{munkh-erdene.yadamjav,zhifeng.bao}@rmit.edu.au
[2] The University of Melbourne, Melbourne, Australia
farhana.choudhury@unimelb.edu.au
[3] Singapore Management University, Singapore, Singapore
bhzheng@smu.edu.sg

Abstract. The sequences of user check-ins form semantic trajectories that represent the movement of users through time, along with the types of POIs visited. Extracting patterns in semantic trajectories can be widely used in applications such as route planning and trip recommendation. Existing studies focus on the entire time duration of the data, which may miss some temporally significant patterns. In addition, they require thresholds to define the interestingness of the patterns. Motivated by the above, we study a new problem of finding top-k semantic trajectory patterns w.r.t. a given time period and given categories by considering the spatial closeness of POIs. Specifically, we propose a novel algorithm, *EC2M* that converts the problem from POI-based to cluster-based pattern search and progressively consider pattern sequences with efficient pruning strategies at different steps. Two hashmap structures are proposed to validate the spatial closeness of the trajectories that constitute temporally relevant patterns. Experimental results on real-life trajectory data verify both the efficiency and effectiveness of our method.

Keywords: Pattern search · Trajectory queries · Semantic-temporal

1 Introduction

Recommendation systems utilize data analysis techniques to identify items that match the user's preferences and interests. According to Verified Market Research, global recommendation system market is projected to reach \$15.46B by 2026 from \$1.12B in 2018 [1]. In this paper, we focus on finding *top-k semantic trajectory patterns* which is related to a type of recommendation particularly useful for tourism. A typical use case is that Alice, who is going to visit New York City for the first time during Easter holiday, wants to spend quality time at museum, park, and shops. She does not have time to study NYC before her trip. Instead, she relies on the wisdom of the crowd and wants to follow the popular routes people took to visit museums/parks/shops last Easter.

© Springer Nature Switzerland AG 2021
C. S. Jensen et al. (Eds.): DASFAA 2021, LNCS 12681, pp. 439–456, 2021.
https://doi.org/10.1007/978-3-030-73194-6_30

440   M.-E. Yadamjav et al.

Popular location-based services such as Foursquare, Gowalla, and Yelp allow users to upload and update the description of *Points-of-interests (POIs)* and hence lots of POIs are associated with semantic information such as categories. Accordingly, the sequence of check-ins of a specific user over time forms a semantic trajectory, which represents the movement of that user at different timestamps. To support Alice's query, we focus only on the part of the trajectories during last Easter where the visited POIs belong to museums, and/or parks, and/or shops. Ideally, such trajectories of many other users should constitute the most popular routes taken during Easter by users who share similar interests as Alice. Since there can be many spatially close-by POIs of the same category (e.g., shops), the popular routes are expected to include the trajectories that go through close-by POIs of the same category. We study the problem of finding such top-k popular routes, in other words, top-k semantic trajectory patterns.

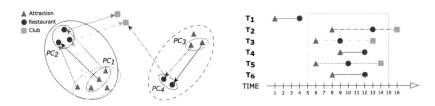

Fig. 1. A running example of six trajectories with their corresponding check-in sequences and timestamps (Color figure online)

Although there are existing studies on semantic trajectory pattern mining [5,18] and top-k frequent pattern mining [9], they suffer from at least one of the following drawbacks. (i) *Finding a threshold that defines the interestingness of a pattern is difficult.* Existing studies [5,18] measure the interestingness of a pattern based on the number of trajectories that exhibit that pattern, namely *support*, and rely on users to specify a minimum support. Hence, the search results highly depend on users' knowledge on the support number of certain patterns, and an improper value may run the risk of missing interesting patterns. (ii) *Not considering time may miss interesting patterns.* Existing studies [5,18] mine the patterns from the entire trajectory time duration. However, it is well-known that the activities people perform and the places people visit vary at different times. Although these mining algorithms can be applied over the subtrajectories w.r.t. the given time period, their efficiency suffers greatly when we consider a set of different category sequences. (more details in Sect. 6). Moreover, the other drawbacks still apply when their work is extended for time dimension. (iii) *Lack of spatial closeness consideration.* The study [9] can only be applied over semantic trajectory sequences by ignoring the spatial closeness between trajectories.

Motivated by the above and to complement existing studies, we propose a new problem of finding top-k semantic trajectory patterns for a given set of categories ψ and time period P. Informally, the query finds k semantic trajectory patterns, where each pattern is represented by a sequence of POI clusters of given categories. A POI cluster in the pattern consists of the POIs of the same category and are spatially close (e.g., shops that are close-by) to overcome the third drawback. The consideration of P addresses the first drawback. The rank of the pattern considers both the number of query categories appearing in the pattern and the number of trajectories that cover the pattern. Thus, we avoid the necessity of specifying any threshold and overcome the second drawback.

To illustrate the patterns of interest, we plot an example in Fig. 1 with six trajectories. Existing work [9] will return the most frequent pattern containing two categories $cs_1 = \langle attraction \rightarrow restaurant \rangle$ with a support of all six trajectories. If the time period is set to $P = \langle 5{:}15 \rangle$, the support of cs_1 changes to T_2, T_3, T_4, T_5, T_6. However, since these trajectories are spatially far-away, such a sequence is not able to suggest any practical route that the user could follow. If we consider the spatial closeness of matching categories, pattern $s_1 = \langle PC_1^{attraction} \rightarrow PC_2^{restaurant} \rangle$ shown with black line ellipse containing trajectories (T_3, T_5, T_6) is a popular route to take w.r.t. P.

To find such top-k semantic trajectory patterns, we have to address two challenges. *First*, how to assign close-by trajectories into the same pattern efficiently and represent the pattern in an informative way? *Second*, how to accelerate the finding of top-k patterns without enumerating all the subsets of given categories, where we can check as few candidates as possible and prune unpromising patterns at an early stage?. In order to address the above challenges, we make the following contributions:

- We address the drawbacks of existing studies that interesting semantic patterns may get missed by proposing the novel *Time Period-based Top-k Semantic Trajectory Pattern* query, which considers a given time period of interest and a set of categories as input and returns top-k patterns w.r.t. the constraints.
- We propose the algorithm *EC2M* that converts the problem from POIs to a cluster-based pattern search problem to limit the search space significantly. We apply a progressive search strategy from shorter to longer patterns to guarantee the return of top-k patterns. We present a hashmap-based data structure, namely *Enclosing Cluster Co-occurrence Map* for efficient pruning at different steps and another structure namely *Neighbors Bitmap* for validating the spatial closeness of trajectories within given time period.
- We conduct extensive experiments to evaluate the efficiency of our query processing algorithm over two real-world datasets and case studies to demonstrate the effectiveness of our top-k semantic trajectory patterns.

2 Problem Formulation

Let $\mathcal{S}$ be a set of semantic categories (e.g., restaurant, park), and $\mathcal{O}$ be a set of POIs, where each $o \in \mathcal{O}$ is a pair $(o.loc, o.cat)$ of a location $o.loc$ and a category $o.cat \in \mathcal{S}$. Let $\mathcal{D}$ be a database of trajectories, where a trajectory $T \in \mathcal{D}$ is represented as a finite sequence of pairs of a POI and a timestamp (o_i, t_i). Here, timestamp t_i corresponds to the time when the POI o_i was visited by T.

A **POI cluster** (PC) is formed by a set of POIs that belong to the same category and meanwhile are located close to each other. A **semantic trajectory pattern (pattern** in short) is a sequence of such POI clusters where each trajectory in the pattern visits at least one POI in every PC of the sequence.

Example 1. Three close-by POIs with the category 'restaurant' (blue dots) in Fig. 1 form a POI cluster PC_2 bounded by a blue ellipse. Note that POIs of the same category inside the blue dotted ellipse are not a part of PC_2 because they are located far away. Three close-by 'attraction' POIs (red triangles) also form a POI cluster PC_1 bounded by a red ellipse. Since trajectories T_3, T_5, T_6 visit at least one POI in PC_1 followed by another POI in PC_2, these trajectories form a pattern $s = \langle PC_1^{attraction} \rightarrow PC_2^{restaurant} \rangle$.

The closeness relationship for clustering depends on the intended application. We use DBSCAN [8] to guarantee the spatial proximity among POIs in a PC, while our problem and approaches are orthogonal to the choice of clustering technique. To find patterns during a specific time period P, we only need to consider the subtrajectories where the corresponding timestamps are within P. Now, we are ready to formally define the trajectory coverage of a pattern.

Definition 1. *Trajectory coverage: Given a time period P and a threshold Δt, a trajectory T covers a pattern $s = \langle PC_1^{cat_1}, \ldots, PC_i^{cat_i} \rangle$ with the category sequence $\langle cat_1, \cdots, cat_i \rangle$ if the following conditions are satisfied: (i) there is a subsequence of POIs in T, denoted as $T' = (o_1, o_2, \ldots, o_i)$, such that $\forall o_i \in T'$, $o_i.cat = cat_i$; (ii) the timestamps of all POIs in T' are within P; and (iii) for any POI $o_j \in T'$ with $j < i$, the time gap between o_j and its subsequent POI o_{j+1} is always bounded by Δt, i.e., $(t_{j+1} - t_j) \leq \Delta t$ always holds.*

Here, the parameter Δt is used to find the patterns where the consecutive POI visits happened within a reasonable time gap (e.g., by setting $\Delta t = 24$ hours). Note that, if a trajectory contains multiple subtrajectories with the same category sequence, that trajectory covers the pattern only once, but all those unique subtrajectories are used to form the clusters of the pattern.

Example 2. A pattern $s = \langle PC_1^{attraction} \rightarrow PC_2^{restaurant} \rangle$ located in the solid black ellipse is covered by trajectories T_3, T_5, and T_6 w.r.t. $P = \langle 5, 15 \rangle$ and $\Delta t = 10$, (Fig. 1). Although T_1 contains POIs belonging to the given categories, its corresponding timestamps are not within P. Hence, it is worth noting that the POIs that belong only to T_1 are not included in the pattern's POI clusters.

When a pattern w.r.t. a time period is covered by many trajectories, it actually implies a "popular" route taken by users when visiting those categories within that time period. Now we introduce a measure of the popularity of a pattern that considers both the number of trajectories covering the pattern, and its semantic importance to a user query.

Definition 2. *Popularity measure:* Given a set of categories ψ, a pattern s where $\forall cat_i \in s$, $cat_i \in \psi$, and the trajectories covering s w.r.t. P and Δt (by Definition 1), the popularity of s, denoted as $Sr(s, \psi)$, is computed by Eq. (1).

$$Sr(s, \psi) = \alpha \frac{|D_P^s|}{|D_P|} + (1 - \alpha) \times \frac{|s|}{|\psi|} \tag{1}$$

Here, D_P^s is the set of trajectories covering s, D_P is the set of trajectories containing at least one POI of any category in ψ, $|s|$ is the number of categories in the pattern s, and $\alpha \in [0, 1]$ is used to set the preference over one component to the other. The second component $\frac{|s|}{|\psi|}$ (denoted as 'category sub-score') quantifies the matching between the categories of s and ψ. Although there are many ways to combine two components in a scoring function, weighted summation is the most common in many spatial-keyword studies [7,10]. Now we introduce our *Time Period-based Top-k Semantic Trajectory Pattern* query.

Definition 3. *Time period-based Top-k Semantic Trajectory Pattern (short):* Given a trajectory database $\mathcal{D}$, a time period P, a time threshold Δt, and categories of interest ψ, the *TkSP* query is to find k highest scoring patterns w.r.t. P and ψ (by Definition 2).

Example 3. A *TkSP* query is given with $k = 1$, $P = \langle 5, 15 \rangle$, $\Delta t = 10$, and $\psi = \langle restaurant, attraction \rangle$. We find six candidate patterns $s_1 = \langle PC_1^{attraction} \rangle$: $\{T_3, T_5, T_6\}$, $s_2 = \langle PC_2^{restaurant} \rangle$: $\{T_3, T_5, T_6\}$, $s_3 = \langle PC_3^{attraction} \rangle$: $\{T_2, T_4\}$, $s_4 = \langle PC_4^{restaurant} \rangle$: $\{T_2, T_4\}$, $s_5 = \langle PC_1^{attraction} \rightarrow PC_2^{restaurant} \rangle$: $\{T_3, T_5, T_6\}$, and $s_6 = \langle PC_3^{attraction} \rightarrow PC_4^{restaurant} \rangle$: $\{T_2, T_4\}$. Their scores for $\alpha = 0.5$ are calculated as, $Sr(s_1, \psi) = Sr(s_2, \psi) = 0.5 \cdot \frac{3}{5} + 0.5 \cdot \frac{1}{2} = 0.55$, $Sr(s_3, \psi) = Sr(s_4, \psi) = 0.5 \cdot \frac{2}{5} + 0.5 \cdot \frac{1}{2} = 0.45$, $Sr(s_5, \psi) = 0.5 \cdot \frac{3}{5} + 0.5 \cdot \frac{2}{2} = 0.8$, $Sr(s_6, \psi) = 0.5 \cdot \frac{2}{5} + 0.5 \cdot \frac{2}{2} = 0.7$. Pattern s_5 that contains both query keywords and 3 trajectories is returned as the result with the highest score.

3   Baseline Method

Since existing approaches do not directly answer *TkSP* query, we tailor the state-of-the-art Top-k sequential pattern mining (*TkS*) [9], denoted as *TkS**, for our baseline. It follows a *retrieval-and-refinement* framework. It first retrieves the POIs of trajectories satisfying ψ and P, with trajectories indexed by their timestamps using a B^+-tree [6]. It then applies *TkS** to find the top-k patterns from these subtrajectories. We use 'trajectory' instead of 'subtrajectory' in the following for simplicity. One may think of enumerating all permutations of query categories and finding the trajectories covering a permutation as a pattern. However, trajectories covering spatially distant patterns with the same category sequence need to be distinguished (Sect. 2). Different from *TkS*, *TkSP* expects multiple patterns corresponding to a category sequence and hence *TkS* is tailored to compute their popularity scores independently.

Generation of Top-k Patterns. *TkS** first generates the POI clusters from the retrieved POIs. Similar to *TkS*, we use a vertical bitmap representation to find trajectories covering a specific POI and create a co-occurrence map to find subsequent POIs visited within Δt timestamps for a specfic POI. If a trajectory contains a POI in the POI cluster, the position that POI of that trajectory in the bitmap is set to 1, otherwise 0. These POI clusters are the 1-length patterns, which are the current candidate patterns. The candidates are maintained in descending order of their popularity scores. In each iteration, if the popularity score of the head pattern hp (i.e., the candidate with the highest score) is larger than that of a current result, the result set is updated accordingly with hp. Since a pattern can be extended by adding a POI cluster of a new category, the popularity score of any longer pattern can increase. Hence, we check what could be the 'potential' score by extending hp. As an extended pattern from hp can have at most the same trajectories as hp and at most $|\psi|$ categories, the potential score is calculated by putting these maximum values in Eq. (1). If this potential score is larger than that of the k-th pattern in the current result set, we extend hp using the POI co-occurrence map w.r.t. Δt constraint and the cluster validity (Sect. 4.1). If the extended patterns of hp have potential popularity scores that are greater than k-th pattern in the current result set, these patterns are added as candidates. Once all candidates are checked, the result set of k patterns is returned.

TkS* is simple but suffers from several major drawbacks: (i) Generating the POI clusters is expensive. (ii) The candidates cannot be pruned if a longer pattern with potential score higher than the current results can be generated from them. As a result, a huge number of candidates need to be checked and the actual trajectories covering those patterns need to be verified in each iteration.

4   Our Approach

To overcome the drawbacks of the baseline, we propose a novel algorithm based on an *Enclosing Cluster Co-occurrence Map*, namely *EC2M*, that takes advantage of the closeness relations of POI clusters and a progressive search technique. We start by presenting some high level key concepts of the algorithm.

1) **POI to cluster conversion.** It is well-known that POIs (e.g., restaurants) do not change their locations often. Hence, we include this pre-processing step where all POIs in the dataset are clustered based on the category and spatial closeness, but without considering the exact trajectories passing through them. We denote these clusters as the '*enclosing clusters*'. In Fig. 1, the green ellipse shows an enclosing cluster corresponding to category 'attraction'.

2) **Pruning search space by pattern length.** *For simplicity, we refer to the number of POI-clusters in a pattern as its 'length'.* The following key concept can significantly prune the search space – **The maximum length of a pattern for any given query is** $|\psi|$. The reason is: the popularity score of pattern s is determined by two sub-scores, where one depends on

the number of covering trajectories and the other depends on its number of categories. Adding a new cluster to s whose category already exists in s does not increase its number of categories. Moreover, for any pattern s' extended from s, the trajectories that cover s' will always cover s. Therefore, extending a pattern from a given pattern s by adding a new cluster whose category already exists in s, the popularity score does not increase. Therefore, we only need to consider the patterns up to length $l = |\psi|$.

3) **Upper bound score.** To enable the filtering of unpromising candidates, we introduce the *enclosing cluster co-occurrence map*. This map has each enclosing cluster as the keys and the list of its covering trajectories w.r.t. Δt and P as values. It is created only when we need to search for patterns greater than 1-length (explained later). The maximum number of the covering trajectories for a pattern is the number of trajectories that are common in all 'enclosing clusters' of that pattern, which can be readily obtained from the map. Such upper bound can be loose if we consider a specific time period. Therefore, we split the time dimension into multiple bins of fixed duration and organize the co-occurrence map accordingly as shown in Fig. 2c. If a trajectory spans two or more consecutive time bins, we store it in all the corresponding bins.

4) **Pattern extension and progressive search.** If a pattern s is extended to s' by adding a category that is not in s, s' may have a higher popularity score than s as its category sub-score increases. Thus, we cannot prune any candidate s without checking all its potential extensions that might outscore it.

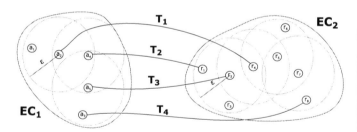

(a) Check-in sequence $\langle attraction \rightarrow restaurant \rangle$ for trajectories T_1, T_2, T_3, T_4 with enclosing clusters ($minPts = 3, \epsilon$)

(b) Neighbours bitmap for EC_2

POI	BITMAP
r_1	01000000
r_2	10110000
r_3	01000000
r_4	01001000
r_5	00010110
r_6	00001000
r_7	00001001
r_8	00000010

EC_1 ┈┈┈┈┈> EC_2

T_2 T_3	T_1 T_4	T_1 T_2 T_3
P_1	P_2	P_3

(c) Covering trajectories of enclosing cluster sequence $EC_1 \rightarrow EC_2$ for different time bins

Fig. 2. Enclosing cluster information

Therefore, we progressively search patterns from length $l = 1$ to length $l = |\psi|$. For each length, we find top-k patterns up to that length. Hence, for pattern s with length i currently under consideration, it will be guaranteed that all the shorter patterns that s might outscore have been already considered. We also compute the maximum possible score that can be obtained by extending current s using the enclosing cluster co-occurrence map. Only when that score is greater than the current results, we consider the possible patterns of length $i + 1$ extended from s (also obtained from the map). This step guarantees that any longer pattern that might outscore a shorter pattern will not be missed.

If the pattern under consideration cannot be pruned based on enclosing cluster based upper bound, we need to find the covering trajectories on the fly to refine its actual POI clusters. All clusters in a pattern need to be checked against the 'cluster validity' using the covering subtrajectories. The cluster validation is presented in more details next.

4.1 Cluster Validity Check

All patterns that are passed to the cluster validity check contain POIs grouped by the enclosing cluster information. However, as the enclosing clusters are formed without considering the exact trajectories actually passing through their POIs, it is possible that some POIs in an enclosing cluster are not part of the actual pattern. The following example illustrates one such scenario.

Example 4. Figure 2a shows a pattern s generated based on the enclosing clusters EC_1 and EC_2. Enclosing clusters are generated based on all POIs w.r.t. spatial closeness. We use DBSCAN clustering w.r.t. the parameters: $minPts$ and ϵ, where at least $minPts$ POIs have to be within ϵ distance from any POI of the same cluster to form a cluster. EC_1 contains five '*attraction*' and EC_2 contains eight '*restaurant*' POIs. Assume the pattern is covered by trajectories T_1, T_2, T_3, and T_4. The attraction POIs of the four trajectories form a cluster, but T_4's restaurant POI cannot form a cluster with the restaurant POIs of other trajectories. This is because none of the restaurant POIs in T_1, T_2, T_3 is within ϵ from the restaurant POI in T_4. Hence, there are actually two patterns (split from EC_1 and EC_2), where one pattern s_1 is covered by T_1, T_2, and T_3, and the other pattern s_2 is covered only by T_4.

Although the number of POIs in an enclosing cluster is much smaller than the total number of POIs, we still need to compute the distances of every pairs of POIs if we re-cluster them. To overcome this limitation, we store a bitmap representation of neighbors for each POI w.r.t. its enclosing cluster. Such structure eliminates the need for spatial distance computation for every pair. Furthermore, the POIs are compared with only the near-by POIs that potentially could form a cluster. Figure 2b shows the neighbor bitmaps for each POI in EC_2. The length of each POI's bitmap equals to the size of the enclosing cluster, e.g., eight in this example. A bit position is assigned to each POI in the cluster. If two POIs are within ϵ distance, then the corresponding bit is 1. For example, the bit sequence 01000000 associated with POI r_1 indicates only r_2 is located within ϵ to it.

Algorithm 1: *EC2M*

Input: B^+-tree *tree*, time period P, time constraint Δt, query categories ψ, k
Output: set of k patterns R

1.1 $R \leftarrow \emptyset$, $PQ \leftarrow$ an empty priority queue
1.2 $Enqueue(PQ$, enclosing clusters satisfying constraints(tree, ψ, P))
1.3 **if** $|\psi| > 1$ **then**
1.4 | $ccMap \leftarrow$ generate co-occuring cluster map(ψ, P, Δt)
1.5 **for** $j \leftarrow 1$ *to* $|\psi|$ **do**
1.6 | $NQ \leftarrow \emptyset$
1.7 | **while** PQ *is not empty* **do**
1.8 | | $cand \leftarrow dequeue(PQ)$
1.9 | | **if** $getScore(cand) > getMinScore(R)$ **then**
1.10 | | | $List \leftarrow getTrajectoryList(cand)$
1.11 | | | $validateClusters(List, j, R)$
1.12 | | **if** $getMaxScore(cand) > getMinScore(R)$ **then**
1.13 | | | $Enqueue(NQ, extendP(cand, ccMap))$
1.14 | $PQ \leftarrow NQ$
1.15 **return** R

4.2 Algorithm

The pseudo code of the *EC2M* algorithm is presented in Algorithm 1. In this approach, the trajectories are indexed by a B^+-tree. A max-priority queue PQ is maintained to keep track of the candidate patterns, where the key is their upper bound popularity score. Result set R keeps k patterns with the highest popularity scores found so far (Line 1.1). At first, the set of enclosing clusters is obtained by retrieving trajectories that pass through at least one query category within P using *tree*. Since each enclosing cluster is a pattern of length 1 (Line 1.2), they are enqueued to PQ. An enclosing cluster co-occurrence map is created w.r.t. P and the query categories ψ by considering time bins that intersect with P. This map is created only when we need to search for longer patterns, i.e., when $|\psi| > 1$ (Lines 1.3–1.4). The map contains the enclosing clusters as keys, and the list of trajectory IDs as values.

The candidate patterns are progressively considered from length $j = 1$ to length $j = |\psi|$ (Line 1.5). For a length j under consideration, another priority queue NQ stores the patterns that may need to be extended to length $(j + 1)$ in next iteration. For a candidate pattern *cand* dequeued from PQ, we perform two actions. First, we compare its upper bound score with the current k-th best score. Note, when R has less than k results, $getMinScore(R)$ returns 0. If *cand* has a higher score, we extract all the trajectories that cover *cand*, and further validate the pattern via the function *validateClusters* (Lines 1.9–1.11). The pseudo-code for cluster validation step is presented in Algorithm 2 and will be explained later. Second, we estimate the maximum possible popularity score of any longer pattern that can be extended from *cand*. If this estimated score is greater than the current k-th best score, we generate the extensions of *cand* of length $j + 1$ using the co-occurrence map, and enqueue them to NQ (Lines 1.12-1.13). At the end of the iteration corresponding to a length j, we replace PQ with NQ for the next length $j + 1$ (Line 1.14). Finally, we return result set R with k most popular patterns (Line 1.15).

Algorithm 2: validateClusters()

Input: list of trajectories LT, length l, current result set R

2.1 $MAP_P \leftarrow createPOIMap(LT)$; $MAP_{ind} \leftarrow createIndexMap(LT)$
2.2 $Q_T \leftarrow getIDs(LT)$
2.3 **while** Q_T *is not empty* **do**
2.4 $\quad$ $LP \leftarrow getPOIs(Q_T.poll(), MAP_{ind}, 1)$; $C \leftarrow clusterPOIs(LP)$
2.5 $\quad$ **foreach** *cluster* $c \in C$ **do**
2.6 $\quad\quad$ $LT' \leftarrow getTrajectory(c, MAP_P)$
2.7 $\quad\quad$ **if** $getScore(l, |LT'|) > getMinScore(R)$ **then**
2.8 $\quad\quad\quad$ $p \leftarrow$ Initialize with cluster c; Update R if the length of p equals l
2.9 $\quad\quad\quad$ **for** $j \leftarrow 2$ *to* l **do**
2.10 $\quad\quad\quad\quad$ $NLP \leftarrow getPOIs(LT', MAP_{ind}, j)$
2.11 $\quad\quad\quad\quad$ $CN \leftarrow clusterPOIs(NLP)$
2.12 $\quad\quad\quad\quad$ **if** $|CN| = 1$ **then**
2.13 $\quad\quad\quad\quad\quad$ $p.add(c' \in CN)$; Update R if the length of p equals l
2.14 $\quad\quad\quad\quad$ **else**
2.15 $\quad\quad\quad\quad\quad$ **foreach** *cluster* $c' \in CN$ **do**
2.16 $\quad\quad\quad\quad\quad\quad$ $NLT' \leftarrow LT' \cap getTrajectory(c', MAP_P)$
2.17 $\quad\quad\quad\quad\quad\quad$ **if** $|LT'| = |NLT'|$ **then**
2.18 $\quad\quad\quad\quad\quad\quad\quad$ $p.add(c')$; Update R if the length of p equals l
2.19 $\quad\quad\quad\quad\quad\quad$ **else**
2.20 $\quad\quad\quad\quad\quad\quad\quad$ $Q_T.add(NLT')$
2.21 $\quad\quad\quad\quad\quad$ **if** $\forall c' \in CN$ *not extends* p **then**
2.22 $\quad\quad\quad\quad\quad\quad$ break

Cluster Validation Algorithm. The function *validateClusters* validates a candidate pattern, with its pseudo-code presented in Algorithm 2. As mentioned in Sect. 3, the candidate patterns are generated from enclosing clusters that do not consider the exact trajectories passing through them, hence the validation procedure is necessary to find the actual patterns with valid POI clusters. Here, two hashmaps MAP_P and MAP_{ind} are created from the input list of trajectories LT (Line 2.1). MAP_P has each POI as a key and the list of trajectories passing through that POI as the value. A trajectory ID in LT and the order of a POI visit is a key (as a tuple) in MAP_{ind} and the list of POIs visited by that trajectory at the corresponding order of visit are the value. A first-come-first-serve queue Q_T is maintained for the set of trajectories that needs to be considered. Q_T is initialized with LT (Line 2.2). For the trajectory set dequeued from Q_T, we find the first POI visited by each trajectory using MAP_{ind}, and store these POIs in LP. We use the neighbors bitmap information (Sect. 4.1) to obtain the actual POI clusters C formed by the POIs in LP (Line 2.4).

For each cluster $c \in C$, we obtain the trajectories covering c using MAP_P (Line 2.6), and store them in a list LT'. Note, $LT' \subseteq LT$. To facilitate pruning of an unpromising candidate at this stage, we compute an upper bound popularity score using c, where the number of covering trajectories is set to $|LT'|$ (as any extended pattern will not have more covering trajectories) and the number of categories is set to l (the maximum categories in a pattern of length l). If this score is lower than the k-th best score, we can safely terminate the examination of c. Otherwise, we initialize a pattern p with c. If $l = 1$, we update R accordingly (Line 2.8). For $l > 1$, we extend p by checking the POIs visited subsequently until its length reaches l. We use parameter j to indicate the visiting order of POIs to be evaluated next (Line 2.9).

We scan each trajectory in LT' to retrieve the j-th visited POIs using MAP_{ind} and store these POIs in NLP (Line 2.10). The neighbors bitmap is used to obtain the actual POI clusters formed by the POIs in NLP, and the resulting clusters are stored in CN (Line 2.11). If CN has only one cluster c' (i.e., the POIs visited next are all contained in one POI cluster), we can extend the current pattern p by appending c'. If the length of p becomes l and it's score is higher than the k-th best score, R is updated (Lines 2.12–2.13). If there are multiple clusters in CN, it indicates that there are multiple options in terms of the next POI cluster to visit from the current p, and we have to explore each $c' \in CN$. For each $c' \in CN$, we obtain the list of trajectories in LT' that also visits a POI in c' in its j-th place, and store them in NLT' (Line 2.16). If $|NLT'| = |LT'|$, c' is added to p and we check if R needs to be updated (Lines 2.17–2.18). Otherwise, we add NLT' to Q_T as a candidate pattern (Lines 2.19–2.20). If no cluster in CN could extend p, we stop the validation of the current c (Lines 2.21–2.22).

5 Experimental Evaluation

In this section, we compare our proposed algorithm with the baseline through an experimental evaluation using real datasets. All algorithms were implemented in Java. Experiments were ran on a 24 core Intel Xeon E5-2630 2.3 GHz using 256 GB RAM, and 1TB 6G SAS 7.2Krpm SFF (2.5-in.) SC Midline disk drives running Red Hat Enterprise Linux Server release 7.5. We test the following methods to answer *TkSP* queries on real-life datasets: (1) *TkS**, a tailored Top-k Sequential Pattern Mining [9] on top of a B^+-tree index as baseline (introduced in Sect. 3); (2) *EC2M*, a *Time Period-based Top-k Semantic Trajectory Pattern* query processing algorithm on top of a B^+-tree index using the enclosing clusters co-occurrence map (introduced in Sect. 4).

Datasets. Foursquare [17] dataset includes check-in data collected from 4 April 2012 to 16 February 2013 in Tokyo, Japan. Yelp dataset includes check-in data between 12 October 2004 and 13 Dec 2019. Each trajectory in the dataset is a sequence of POIs with the corresponding timestamps and semantic categories. Table 1 shows statistics on Foursquare and Yelp datasets.

Table 1. Database statistics

Description	Foursquare	Yelp
# of POIs	61,856	209,393
# of check-ins	573,012	8,016,526
# of users	2,293	1,968,703
# of categories	247	21

Table 2. Experimental parameters

Parameter	Dataset	Values
Time period P	Foursquare	2, 4, **6**, 8, 11 (month)
	Yelp	**3**, 6, 9, 12, 15 (year)
Preference α	Both	0.1, 0.3, **0.5**, 0.7, 0.9
# of categories $|\psi|$	Both	1, 2, **3**, 4

Evaluation and Parameterization. We compared the runtime of all methods by varying the query input parameters as shown in Table 2, where the values in bold represent the default values. For all experiments, a single parameter is varied while other parameters are set to their default values.

Clustering. We evaluated varying parameter combinations for $minPts$ and ϵ to cluster POIs w.r.t. the category using DBSCAN [8] algorithm and chose the following values for our experiment by considering the number of obtained clusters: $minPts = 3$ and $\epsilon = 100$ m.

5.1   Efficiency Study

We conduct experiments to evaluate the efficiency of our proposed algorithm against the baseline. We study the impact of each parameter by running 100 queries and report the average query execution time while varying parameters. The results over Foursquare and Yelp datasets are shown in Figs. 3 and 4 respectively. The performance for multiple runs is shown in boxplots, where the bounding box shows the first and third quartiles; the whiskers show the range, up to 1.5 times of the interquartile range; and the outliers are shown as separate points. The average values are shown as connecting lines.

Effect of the Number of Categories. Figures 3a and 4a show the efficiency studies for varying $|\psi|$ over Foursquare and Yelp datasets, respectively. The execution time gap between the algorithms is small for $|\psi| = 1$ since there is no need to generate longer patterns. *EC2M* outperforms *TkS** in all cases by using the bitmap-based POI neighbors information to cluster POIs of the same category. As we add more categories to the query, the benefit of enclosing cluster co-occurrence map becomes more significant. The performance gap between *TkS** and *EC2M* is larger for Yelp than Foursquare dataset. The reason is, Yelp contains more POIs and less categories than Foursquare, which results in more POIs in the POI clusters of Yelp than Foursquare dataset. Hence, the clustering step in the query processing over Yelp dataset greatly benefits from using the bitmap-based POI neighbors information.

Effect of α. Figures 3b and 4b show the performance for varying α over Foursquare and Yelp, respectively. The performance gap between two algorithms is not big for smaller values of α in Foursquare dataset. The reason is that Foursquare contains only 2,293 trajectories. As we check the query result for those values of α, the result mostly contains patterns of 1-length. The scores of extended patterns are still lower than the k-th best result. The execution time of the baseline declines for values of α that are higher than 0.5. The reason is that the increase of the k-th best score allows the skip of more shorter patterns, leading to less candidate check. In contrast, Yelp has 1,968,703 trajectories and a small value of α can still contribute significantly to the popularity score.

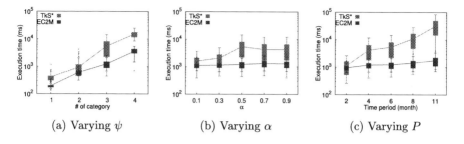

Fig. 3. Efficiency studies on Foursquare dataset

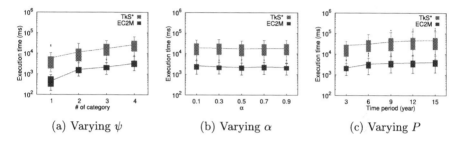

Fig. 4. Efficiency studies on Yelp dataset

Effect of Time Period. As we increase time period P, the runtime increases substantially for the baseline in Foursquare dataset, as shown in Fig. 3c. In contrast, the query execution time gradually increases w.r.t. varying time periods for Yelp dataset. The reason is that time periods for Foursquare dataset are in the unit of months while time periods for Yelp dataset are in the unit of years (Fig. 4c). The number of trajectories covering a certain pattern significantly differs w.r.t. the short time periods. As we expand the time period of interest to a longer time span, eventually we find almost all the trajectories covering a given pattern and the number of new candidate trajectories starts decreasing.

Enclosing Clusters Co-occurrence Map. We obtain 777,018 and 1,513,218 enclosing clusters w.r.t. our clustering settings for Foursquare and Yelp datasets, respectively. Furthermore, we split trajectories covering each co-occurrence into different time bins to see the correlation between time bin size and the co-occurrence map size. The sizes of enclosing cluster co-occurrence maps for Foursquare dataset are 25 MB, 22 MB, 20 MB for time bins of 1-day, 1-month, and 1-year, respectively. The sizes of enclosing cluster co-occurrence maps for Yelp dataset are 54 MB, 79 MB, 48 MB for time bins of 7-day, 3-month, and 1-year, respectively. The size increases for 3-month time bin, likely due to the larger number of covering trajectories that span two consecutive time bins.

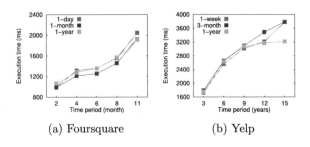

(a) Foursquare   (b) Yelp

Fig. 5. Varying sizes of time bins

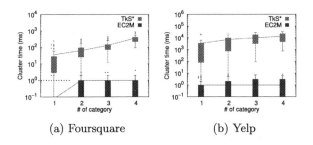

(a) Foursquare   (b) Yelp

Fig. 6. Clustering time

Figures 5a and 5b depict the query execution time of *EC2M* for varying time periods over Foursquare and Yelp, respectively, by using three different time bins of enclosing cluster co-occurrence maps. Overall, queries over Foursquare dataset run slightly faster on time bins of 1-month for all different time periods. Bigger time bins perform better for large Ps, while queries for short time periods perform slightly faster using small time bins. Queries over Yelp dataset consider different time periods of interest, starting from 3 years to 15 years. For larger Ps, the enclosing cluster co-occurrence map on 1-year time bins shows better query performance. However, queries using 1-week time bins perform slightly better for time periods up to 9 years. Here, non-uniform check-in distribution and sparsity in a user's trajectory result in better performance even we use smaller time bins.

Cluster Computation. Our algorithm uses a bitmap representation for clustering close-by POIs. Figures 6a and 6b show the clustering time by running queries for varying number of categories over Foursquare and Yelp datasets, respectively. Since Yelp dataset contains larger clusters, the query performance over Yelp dataset saves significantly more time in the POI clustering than Foursquare dataset. The cluster validity check in the pattern greatly benefits from the bitmap-based clustering algorithm as the length of the pattern increases.

5.2   Case Study

Last, we conduct a case study by presenting the difference between traditional pattern mining over category sequences and semantic trajectory pattern mining

Table 3. Top-5 frequent category sequence over Foursquare

Semantic sequence
Subway → Train Station: 1479
Train Station → Subway: 1452
Train Station → Japanese Restaurant: 1381
Train Station → Ramen/Noodle House: 1366
Japanese Restaurant → Train Station: 1294

Table 4. Top-5 popular semantic trajectory patterns over Foursquare

Semantic trajectory pattern
Train Station → Electronics Store: 465
Electronics Store → Train Station: 377
Train Station: 1517
Train Station → Hobby Shop: 296
Train Station → Electronics Store: 295

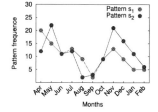
(a) Monthly trajectory coverage of s_1 and s_2

(b) Semantic trajectory pattern s_1

(c) Semantic trajectory pattern s_2

Fig. 7. Case study on the Foursquare dataset

proposed in this paper. Table 3 shows top-5 frequent category sequences of length 2 in Foursquare dataset. The support values depict the number of people (out of 2,293) whose trajectories cover the corresponding sequence. Next, we run our query to find the top-5 semantic trajectory patterns of up to 2-length, which considers the spatial closeness among covering trajectories. The result shown in Table 4 contains 1-length pattern at the third result. The rank is computed by both the number of trajectories and the number of categories.

Typical patterns (e.g. Subway → Train Station) created by people's daily movement can outweigh the potential patterns of interest. Thus, to guide a user to make a better planning based on the interest of places to visit, we accept a set of categories as a user input. Here, we choose the following keywords: Train Station, Hobby Shop, Electronics Store to further explore the region where that pattern is mostly observed. We find two patterns s_1 = { Electronics Store → Hobby Shop → Train Station } and s_2 = { Train Station → Hobby Shop → Electronics Store } shown in Figs. 7b and 7c, respectively. Pattern s_1 is covered by 36 trajectories for the whole database timespan while pattern s_1 is covered by 37 trajectories.

Next, we show the frequency of those two patterns w.r.t. the given time periods. Since Foursquare contains 11-month check-in data, we split data into one-month periods and show the changes in the frequency of each pattern w.r.t. the given month. From the results presented in Fig. 7a, we find the pattern frequencies do change, depending on the time period of interest which can also contribute to the overall ranking of a specific pattern in the result set.

6   Related Work

In this section, we review the studies closely related to our work, including (i) *semantic trajectory pattern mining*, and (ii) *top-k sequential pattern mining*.

Semantic Trajectory Pattern Mining. A semantic pattern defined in [18] is the closest to the pattern we consider in this paper. Specifically, a pattern is defined as a sequence of areas in [18], with each area containing places that are spatially close-by and belong to the same category. A top-down pattern discovery technique called Splitter was proposed. It first generates spatially coarse patterns via a tailored PrefixSpan [12], and then clusters trajectories for each coarse pattern by a variant of the mean shift algorithm. Thus, Splitter works on each category sequence independently, while our work considers different category sequences that can be formed by the given set of categories. Choi et al. [5] find all regional areas where a semantic pattern is expected to be locally frequent in each area. Trajectories that contain each semantic pattern are clustered by a tailored DBSCAN. The subtrajectories covering the pattern form a dense cluster of routes. However, the corresponding categories of the subtrajectories that cover a pattern are not necessarily spatially clustered w.r.t. the category. In contrast, the pattern in our work consists of POI clusters where POIs in a cluster reside spatially close-by and belong to the same category. In addition, the above techniques consider the whole time period in a database and require an input of the minimum support threshold to mine semantic patterns, which is a challenging task for most users. Even if these techniques can be extended by considering subtrajectories w.r.t. a specific time period, the semantic trajectory patterns found for a given category sequence are not necessarily to be same patterns in our problem setting.

Top-k Sequential Pattern Mining. Many studies have been proposed to mine sequential patterns in transactional databases. Majority of them require specifying a threshold for the minimum number of transactions that need to be contained in a frequent pattern. The performance of the mining algorithms can degrade substantially if the support threshold is set to a smaller value, while the patterns of interest can be overlooked by a larger threshold. Top-k sequential pattern mining algorithms [9,13,14] have been proposed to find k most frequent patterns without requiring to specify the threshold. However, these techniques do not consider the spatial property of the trajectories contained in the result pattern. Thus, the trajectories that cover a specific pattern might be scattered over the search space. In contrast, we aim to find semantic patterns where the trajectories that contain each result pattern are spatially close-by. Moreover, none of the existing work supports a query input for the categories of interest.

Other Related Work. Our query is also loosely related to top-k spatial keyword query and collective spatial keyword query. A traditional top-k spatial-keyword query has been studied extensively in literature [7,15,19]. It returns k most similar objects w.r.t. a query location and keywords by considering both spatial and textual similarities. One variant is a *collective spatial keyword query* [2–4,11,16] which aims to fulfil a user request by considering multiple objects collectively instead of a single object. However, none of the methods is applicable in our case as we do not require the result to be close to a query location. Instead, we aim at supporting users who prefer the past travel experience of other users *over* the proximity between the places to be visited and the query location.

7 Conclusion

In this paper, we studied the problem of finding top-k semantic trajectory patterns w.r.t. a set of query categories and a time period of interest. We formally defined the problem and proposed algorithms and data structures that improve the efficiency of the query processing. Experimental study on real-life datasets shows the efficiency and effectiveness of our approach.

Acknowledgement. Zhifeng Bao is supported by ARC DP200102611. Baihua Zheng is supported by the Ministry of Education, Singapore, under its AcRF Tier 2 Funding (Grant No: MOE2019-T2-2-116).

References

1. Global recommendation engine market by type, by application, by geographic scope and forecast to 2026. https://www.verifiedmarketresearch.com/product/recommendation-engine-market/
2. Cao, X., Cong, G., Guo, T., Jensen, C.S., Ooi, B.C.: Efficient processing of spatial group keyword queries. TODS **40**(2), 1–48 (2015)
3. Cao, X., Cong, G., Jensen, C.S., Ooi, B.C.: Collective spatial keyword querying. In: SIGMOD, pp. 373–384 (2011)
4. Chan, H.K.H., Long, C., Wong, R.C.W.: On generalizing collective spatial keyword queries. TKDE **30**(9), 1712–1726 (2018)
5. Choi, D.W., Pei, J., Heinis, T.: Efficient mining of regional movement patterns in semantic trajectories. VLDB **10**(13), 2073–2084 (2017)
6. Comer, D.: Ubiquitous b-tree. ACM Comput. Surv. **11**(2), 121–137 (1979)
7. Cong, G., Jensen, C.S., Wu, D.: Efficient retrieval of the top-k most relevant spatial web objects. Proc. VLDB Endowment **2**(1), 337–348 (2009)
8. Ester, M., Kriegel, H.P., Sander, J., Xu, X.: A density-based algorithm for discovering clusters in large spatial databases with noise. In: SIGKDD, pp. 226–231 (1996)
9. Fournier-Viger, P., Gomariz, A., Gueniche, T., Mwamikazi, E., Thomas, R.: TKS: efficient mining of top-K sequential patterns. In: Motoda, H., Wu, Z., Cao, L., Zaiane, O., Yao, M., Wang, W. (eds.) ADMA 2013. LNCS (LNAI), vol. 8346, pp. 109–120. Springer, Heidelberg (2013). https://doi.org/10.1007/978-3-642-53914-5_10
10. Li, Z., Lee, K.C., Zheng, B., Lee, W.C., Lee, D., Wang, X.: IR-tree: an efficient index for geographic document search. TKDE **23**(4), 585–599 (2010)

11. Long, C., Wong, R.C.W., Wang, K., Fu, A.W.C.: Collective spatial keyword queries: a distance owner-driven approach. In: SIGMOD, pp. 689–700 (2013)
12. Pei, J., et al.: PrefixSpan: mining sequential patterns by prefix-projected growth. In: ICDE, pp. 215–224 (2001)
13. Petitjean, F., Li, T., Tatti, N., Webb, G.I.: Skopus: mining top-k sequential patterns under leverage. Data Min. Knowl. Disc. **30**(5), 1086–1111 (2016). https://doi.org/10.1007/s10618-016-0467-9
14. Tzvetkov, P., Yan, X., Han, J.: TSP: mining top-k closed sequential patterns. KAIS **7**(4), 438–457 (2005). https://doi.org/10.1007/s10115-004-0175-4
15. Wu, D., Cong, G., Jensen, C.S.: A framework for efficient spatial web object retrieval. VLDB J. **21**(6), 797–822 (2012). https://doi.org/10.1007/s00778-012-0271-0
16. Xu, H., Gu, Y., Sun, Y., Qi, J., Yu, G., Zhang, R.: Efficient processing of moving collective spatial keyword queries. VLDB J. **29**(4), 841–865 (2019). https://doi.org/10.1007/s00778-019-00583-8
17. Yang, D., Zhang, D., Zheng, V.W., Yu, Z.: Modeling user activity preference by leveraging user spatial temporal characteristics in LBSNs. Trans. Syst. Man Cybern. Syst. **45**(1), 129–142 (2014)
18. Zhang, C., Han, J., Shou, L., Lu, J., La Porta, T.: Splitter: mining fine-grained sequential patterns in semantic trajectories. VLDB **7**(9), 769–780 (2014)
19. Zhang, C., Zhang, Y., Zhang, W., Lin, X.: Inverted linear quadtree: efficient top k spatial keyword search. TKDE **28**(7), 1706–1721 (2016)

Optimal Sequenced Route Query
with POI Preferences

Wenbin Li[1,2], Huaijie Zhu[1,2(✉)], Wei Liu[1,2], Jian Yin[1,2], and Jianliang Xu[3]

[1] School of Computer Science and Engineering, Sun Yat-Sen University,
Guangzhou, China
`liwb33@mail2.sysu.edu.cn`, `issjyin@mail.sysu.edu.cn`, `xujl@comp.hkbu.edu.hk`
[2] Laboratory of Big Data Analysis and Processing, Guangzhou 510006, China
`zhuhuaijie@mail.sysu.edu.cn`
[3] Hong Kong Baptist University, Kowloon Tong, China
`liuw259@mail.sysu.edu.cn`

Abstract. The optimal sequenced route (OSR) query, as a popular problem in route planning for smart cities, searches for a minimum-distance route passing through several POIs in a specific order from a starting position. In reality, POIs are usually rated, which helps users in making decisions. Existing OSR queries neglect the fact that the POIs in the same category could have different scores, which may affect users' route choices. In this paper, we study a novel variant of OSR query, namely *Rating Constrained Optimal Sequenced Route query (RCOSR)*, in which the rating score of each POI in the optimal sequenced route should exceed the query threshold. To efficiently process RCOSR queries, we first extend the existing TD-OSR algorithm to propose a baseline method, called *MTDOSR*. To tackle the shortcomings of MTDOSR, we try to design a new RCOSR algorithm, namely *Optimal Subroute Expansion (OSE) Algorithm*. To enhance the OSE algorithm, we propose a *Reference Node Inverted Index (RNII)* to accelerate the distance computation of POI pairs in OSE and quickly retrieve the POIs of each category. To make full use of the OSE and RNII, we further propose a new efficient RCOSR algorithm, called *Recurrent Optimal Subroute Expansion (ROSE)*, which recurrently utilizes OSE to compute the current optimal route as the guiding path and update the distance of POI pairs to guide the expansion. The experimental results demonstrate that the proposed algorithm significantly outperforms the existing approaches.

Keywords: Route planning · Optimal sequenced route · Spatial database.

1 Introduction

With the ever-growing popularity of smartphones and other location-based services, various route queries have been studied to cater to users' different needs. Among these route queries, the optimal sequenced route (OSR) query has received

C. S. Jensen et al. (Eds.): DASFAA 2021, LNCS 12681, pp. 457–473, 2021.
https://doi.org/10.1007/978-3-030-73194-6_31

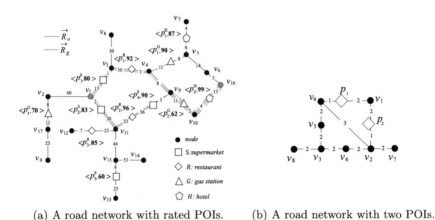

(a) A road network with rated POIs.   (b) A road network with two POIs.

Fig. 1. Running example road networks.

significant research momentum in recent years [12]. It is designed to find the optimal route passing through a sequence of points of interest (POIs) of specific categories (e.g., gas stations, restaurants, and shopping malls) in a particular order. An example of the OSR query in a road network is shown in Fig. 1(a).

The example shows four POI categories, where p_j^S ($j = 1, 2, 3, 4$) denotes POIs of supermarkets (represented by squares), p_j^R ($j = 1, 2, 3$) denotes POIs of restaurants (represented by rhombus), and p_j^G ($j = 1, 2, 3$) denotes POIs of gas stations (represented by triangle), p_j^H ($j = 1, 2$) denotes the hotels (represented by pentagon). There is a number denoting its corresponding rating score, for example, $\langle p_j^S, 70 \rangle$ means the rating score of p_j^S is 70. The black circles denote the nodes in the road network. Given a user u_1 located at v_1, she wants to pass through a sequence of POIs (e.g., restaurant, supermarket) to arrive at the destination node v_{16}, this OSR query returns a route $\{v_1, p_1^S, p_1^R, v_{16}\}^1$ with the cost of 55.

The OSR query is first studied by Sharifzadeh et al. [12,13], followed up by a number of variants [2,8,9,13,17]. However, these prior works assume that the POIs in the same category have the same preference (i.e., they are rated with the same score). In the works [1,4], the weight of POIs is considered as a factor in route cost functions. But in real applications, the rating scores of POIs in the same category could be different, which affects users' route choices. That is, users usually prefer to visit the POIs with a high rating score, while those having a rating score lower than their expectations will not be taken into consideration. For different categories of POIs, the expected rating scores are also different. Take restaurants and gas stations for example. For most customers, they expect a higher rating score for a restaurant than a gas station, but the exact score threshold depends on different individuals.

Motivated by this, in this paper, we propose a new OSR query, namely *Rating Constrained Optimal Sequenced Route* (RCOSR) query, where for each category

---

[1] Between two POIs, the shortest path is used in default, e.g., $\{p_1^S, v_3, p_1^R\}$ is for p_1^S, p_1^R.

of POIs, there is a threshold representing the minimum rating score acceptable to the query user. Given a starting node and a destination node, and a set of sequenced POI categories with the corresponding rating score thresholds, the RCOSR query finds the route with the minimum distance that visits only one POI of each category in order and the rating score of each POI satisfies the user specified threshold. Revisit the example of the RCOSR query illustrated in Fig. 1(a). Given a starting node $v_s = v_1$, and a destination node $v_d = v_{16}$, an RCOSR query $Q(v_1, v_{16}, \langle S, 70 \rangle, \langle R, 90 \rangle, \langle G, 50 \rangle, \langle H, 90 \rangle)$ returns an optimal route in which one of the POI in each category is visited and the rating scores of S, R, G, H are greater than $70, 90, 50, 90$, respectively.

To answer the RCOSR query, one may adopt a greedy search to find the nearest POI satisfying the threshold in each step and generate the route result. Consider the example Fig. 1(a). While this method can quickly find a feasible result $\overrightarrow{R_g} = \{v_1, p_2^S, p_2^R, p_2^G, p_2^H, v_{16}\}$ with a total cost of 114 (i.e., the blue route), but it is not optimal. The optimal solution of this RCOSR query is actually $\overrightarrow{R_o} = \{v_1, p_1^S, p_1^R, p_2^G, p_2^H, v_{16}\}$, which costs 85 only (i.e., the red route). It is obvious that the greedy search may not ensure the optimality for this problem. An idea is to filter the POIs that do not satisfy the rating score threshold to avoid unnecessary expanding exploration. However, the latest optimal sequenced route algorithm TD-OSR [3] cannot be directly applied to this problem, which does not consider the constrained rating score and is not efficient for our problem.

To tackle the RCOSR query problem, we first revise the TD-OSR algorithm as our baseline, named as *MTDOSR*. The TD-OSR algorithm is originally designed for OSR queries in time-dependent road networks, but it can be also extended to address the traditional OSR problem. For our problem, we modify TD-OSR to apply on static road networks and solve RCOSR queries. The main idea of MTDOSR is to use the A* search scheme equipped with an admissible heuristic function. To find the next node to expand the current subroutes during the network expansion, MTDOSR checks the POI to see whether it satisfies the query threshold before inserting it into the path. However, such an expansion scheme fails to exploit all the query categories and generates a large number of candidate subroutes, which consumes a lot of memory to store them in the heap. Further more, the top entry (i.e., subroute) may not be the globally optimal choice to be used to expand. To overcome these two shortcomings, we try to propose a new RCOSR algorithm, *Optimal Subroute Expansion* (OSE) , which iteratively finds the optimal subroute for $Q(v_s, p_j^{c_i}, \langle c_1, \ldots, c_{i-1} \rangle)$ and $p_j^{c_i}$ is one specific POI in category c_i until the optimal route ending at destination node is obtained, which is the query result. However, OSE is very time-consuming to compute the distance of many POI pairs for obtaining the cost of each optimal subroute. To enhance the OSE algorithm, we propose an index, called *Reference Node Inverted Index* (RNII), to accelerate the distance computation and quickly retrieve POIs of each category for POI filtering. To determine the appropriate reference nodes for RNII, we develop a *Greedy Merge* (GM) strategy to determine the reference nodes by maximizing the number of POI pairs they can cover. To effectively utilize RNII and OSE, we propose a new efficient RCOSR algorithm, called *Recurrent Optimal*

Subroute Expansion (ROSE), which iteratively searches the optimal route using OSE. By continuously updating the lower bound distance between POIs computed by RNII with the exact shortest distance, the guiding path that obtained from OSE gets closer to the optimal solution. ROSE terminates when the cost of guiding path is equal to its shortest length.

The contributions of this paper are summarized as follows:

- This paper presents and tackles the rating constrained optimal sequenced route (RCOSR) query problem in which the POIs in the sequenced route should satisfy category rating thresholds.
- We propose the MTDOSR algorithm as our baseline to answer the RCOSR query and explain its inefficiency.
- We propose a new OSE algorithm, which expands the optimal subroutes in dynamic programming scheme.To accelerate the distance computation of POI pairs in OSE, we propose a new index (RNII).
- Based on the OSE and RNII, we further develop a new algorithm (ROSE), which iteratively computes the optimal rating constrained sequenced route as the guiding path to guide the exploration.
- We conduct extensive experiments on synthetic and real road networks to evaluate the proposed algorithms. The experimental results show that ROSE significantly outperforms the baseline by 92.25% in query time and 79.94% in expanded nodes on average.

2   Related Work

This work is relevant to two lines of research, including *optimal route queries*, and *indexes for the road networks*.

2.1   Optimal Route Queries

Li et al. [7] first propose the trip planning query (TPQ) in spatial databases. After that, a variant of the TPQ query, namely the optimal sequenced route query (OSRQ) is studied by Sharifzadeh et al. [12]. In OSRQ, the POI sequence in the optimal route is specified by the user. Three corresponding algorithms (i.e., LORD, R-LORD, and PNE) are developed for both Euclidean and general graphs. Moreover, Sharifzadeh et al. [13] study the OSR query processing algorithm using Voronoi Diagram. Recently, Liu et al. [9] study the top-k optimal sequenced route query, which mainly designs the efficient algorithms for finding the k optimal routes. Yawalkar et al. [16] solve the personalized route preference problem with skyline route queries. Chen et al. [2] study the *multi-rule Partial Sequenced Route* (MRPSR) query, which finds the optimal route via a number of POIs in a partial visiting order defined by the user. Costa et al. [3] study optimal sequenced route queries in time-dependent road networks and propose an effective algorithm called TD-OSR, which is based on the A* scheme with an admissible heuristic function. In this paper, we first modify the TD-OSR to answer our RCOSR query as our baseline.

As for weight constraints on POIs, some research attention has been paid to this area. Dai et al. [4] consider not only road length but also other factors such as POI rating and propose the personalized and sequenced route planning (PSR) query. In PSR, a score is computed for each route for a query, which is determined by both route length and POI rating. Sasaki [11] et al. propose the skyline sequenced route query which searches for all preferred sequenced routes to users by extending the shortest route search with the semantic similarity of POIs in the route. Yao et al. [15] study the multi-approximate-keyword routing query, which returns the shortest route that passes through at least one matching POI for each query keyword. Later in [1], the authors consider weighted POIs in optimal sequenced group trip planning query, where the weight of POIs is computed as utility and the cost of a trip consists of a distance value and utility value.

2.2 Indexes for Road Networks

To accelerate distance computation and nearest neighbor searching, a number of researches on road network indexes have been investigated. R-tree and its variants [5] as the most popular spatial indices have been used in recent years. Zhong et al. [18] propose G-tree for shortest distance computation and nearest neighbor query on large road networks, which splits the road network into multiple sub-networks and then constructs a balanced tree. Thus each node in G-tree corresponds to a sub-network. One of its advantages is that the space complexity is relatively low, thus it can easily scale up to large datasets. Another effective index is Voronoi Diagram [13], which partitions the space into regions named cells, but it uses Euclidean distance of two nodes to split, thus works mostly for Euclidean graph.

Kriegel et al. [6] propose *graph network embedding* to speed up the range and k-nearest neighbor queries. This work is the most related one to our work. Although the idea of graph embedding utilizing reference nodes has been studied to build the filter-refinement architecture, our work is the first attempt to apply the idea of graph embedding in the RCOSR problem. In this paper, we propose a new reference node distribution strategy to guide the determination of the reference nodes.

3 Problem Formalization

In this section, we first introduce some fundamental notations, then formalize RCOSR problem. Generally, a road network is typically represented as a graph.

Definition 1. (Graph). *A graph $G(V, E, P)$ consists of a node-set V and an undirected weighted edge set $E \subseteq V \times V$. P is a set of POIs located on the edges, where each POI belongs to one category, denoted by c_i. The total number of nodes (not including the POIs) is denoted as $|V|$ and the weight on the edge indicates the length of the edge. In addition, $dist(v_i, v_j)$ denotes the shortest distance between nodes v_i and v_j. We assume that the distance satisfies the triangle inequality, and specially $dist(v_i, v_i) = 0$. Besides, each POI is associated with a rating score, which ranges from 0 to 100.*

Definition 2. (Route). *We define a route* $\overrightarrow{R} = \{v_1, v_2, ..., v_n\}$ *where* v_i *(1* $\leq$ *i* $\leq$ *n) is a node/POI in a road network. The cost of a route* $\overrightarrow{R}$, *denoted by* $cost(\overrightarrow{R})$, *is* $\sum_{i=1}^{n} dist(v_{i-1}, v_i)$.

Before defining the RCOSR query, we first define the *feasible rating constrained route* and *optimal rating constrained sequenced route*.

Definition 3. (Feasible rating constrained route). *Given a source-destination pair* (v_s, v_d) *and a category sequence* $C = \{c_1, c_2, \ldots, c_k\}$ *as well as the corresponding rating threshold set* $T = \{t_1, t_2, \ldots, t_k\}$ *(for simplicity, we use* $CT = \{\langle c_1, t_1 \rangle, \langle c_2, t_2 \rangle, \ldots, \langle c_k, t_k \rangle\}$ *and* $|CT| = k$ *to represent the category-threshold pair sequence), a feasible rating constrained route* $\overrightarrow{R}$ *satisfies the following constraints: (i)* $\overrightarrow{R}$ *starts from starting node and ends at the destination node. (ii)* $\overrightarrow{R}$ *passes through at least one POI for each category in* C *and follows the sequence order in* C. *(iii) The rating score of the POIs in* $\overrightarrow{R}$ *should be equal or larger than the corresponding threshold in* T.

According to the definition of feasible rating constrained route, we now define the optimal rating constrained sequenced route below.

Definition 4. (Optimal rating constrained sequenced route). *Given a source-destination pair* (v_s, v_d), *a category-threshold pair sequence* CT, *the feasible route* $\overrightarrow{R_o}$ *is the optimal rating constrained sequenced route with threshold constraint if for any feasible rating constrained route* $\overrightarrow{R'}$, *such that* $cost(\overrightarrow{R_o}) \leq cost(\overrightarrow{R'})$.

Based on the above definitions, we now formally define the RCOSR query.

Definition 5. (RCOSR query). *Given a graph* G, *the RCOSR query is defined as* $Q = (v_s, v_d, CT)$, *where* v_s *and* v_d *are the starting node and destination node, and* CT *is the category-threshold pair sequence. Especially, if the rating score threshold of the category is not specified by the user, it is set as the average rating score of POIs in that category. The query returns the optimal rating constrained sequenced route, as defined in Definition 4.*

When the number of categories $|CT|$ is greater than 1, it can be reduced from TSP [7] that the RCOSR problem is also NP-Hard.

4   Baseline for RCOSR

In this section, we present our baseline algorithm, namely *MTDOSR*, which extends the TD-OSR algorithm [3] to address the RCOSR problem. The main idea of MTDOSR is to utilize A* scheme with an admissible heuristic function to guide the network expansion. To find the most potential node to expand, MTDOSR uses a function $f(v) = g(v) + h(v)$, where $g(v)$ is the distance from starting node to the current node v through the corresponding route, and the heuristic function $h(v)$ is computed as $h(v) = max(dist(v, v_d), dist(v, p_{nn}^c))$, where v_d is the destination node, p_{nn}^c is the nearest POI of node v in category c. To accelerate calculating the heuristic function, before the query comes, an improved TD-NE-A* [3] algorithm is executed to calculate the distance of each node to its nearest POI in each category, i.e., $dist(v, p_{nn}^c)$. When computing $f(v)$, the POIs are checked whether it satisfies the query threshold.

During the expansion, a min-heap H is used to store the intermediate subroutes (i.e., entries). The form of the entry is $(v, g(v), f(v), visitedPOIset)$, where v is the current node, $visitedPOIset$ is the POI sequence visited in this subroute. In the following, we explain the running process for the running example in Fig. 1(a) with the query $Q = (v_1, v_{16}, \langle c_S, 70 \rangle, \langle c_R, 90 \rangle, \langle c_G, 50 \rangle, \langle c_H, 90 \rangle)$. MTDOSR begins to examine the starting node v_1 and computes function $f(v_1)$. Since $g(v_1)$ is 0, $f(v_1)$ is 85 (the shortest distance from v_1 to destination node v_{16}), and the POI sequence in this subroute is empty, then the entry $(v_1, 0, 85, \{\})$ is pushed in H. In the second iteration, v_1 is popped from H and its adjacent nodes v_3, v_{11}, v_2 are found and pushed in H. At the same time, POIs p_1^S (i.e., over the edge $v_3 v_1$) and p_2^S (i.e., over the edge $v_{11} v_1$) are checked and inserted in the corresponding subroute, respectively. The above iterations continue until all the query categories are visited and the destination node v_{16} of the query is found, the algorithm returns the result $R_o = \{p_1^S, p_1^R, p_2^G, p_2^H\}$. During this process, this expansion of subroutes guided by $f(v)$ produces many invalid subroutes, e.g., when expanding on node v_4, node v_5 is de-heap from H and visited, which is not in the optimal route. The main reason is that the heuristic function of v used in MTDOSR only utilizes either the farthest POI or the destination node, which fails to make full use of all the query category sequences in the query, thus MTDOSR is not efficient.

5   Recurrent Optimal Subroute Expansion

As analyzed in the above section, if we can fully make use of the information with respect to all the query categories, the route expansion can be more effective. Inspired by this, we try to design a new RCOSR algorithm, *Optimal Subroute Expansion* (OSE), to search for the optimal solution.

5.1   Optimal Subroute Expansion Algorithm

The main idea of optimal subroute expansion is to iteratively find the optimal rating constrained sequenced route for $Q = (v_s, p_j^{c_i}, \langle c_1, \ldots, c_{i-1} \rangle)$ where $p_j^{c_i}$ is one POI in category c_i, until the *optimal rating constrained route* ending at destination node is obtained, which is the query result. The above idea is naturally implemented using dynamic programming shown as follows.

Definition 6. (Dynamic programming formulation). *Given a query* $Q = (v_s, v_d, CT)$, *we construct a dynamic programming matrix* OS *with* $k + 1$ *rows and* $max(|c_i|)$ *columns, where* $i = (0, 1, \ldots, k)$. *The value of* $OS[i, j]$ *represents the cost of ORCS ending at* $p_j^{c_i}$. *Especially, the* $(k+1)$-*th row represents the ORCS ending at node* v_d. *The dynamic programming formulation is as follows:*

$$OS[i,j] = \begin{cases} 0 & if\ i = 0 \\ \min_{1 \le l \le |c_{i-1}|} \{OS[i-1,l] + dist(p_l^{c_{i-1}}, p_j^{c_i})\} & if\ i > 0 \end{cases} \tag{1}$$

Consider a query $Q = (v_s, v_d, CT)$. The OSE accepts the query Q and a currently feasible route cost $cost_{curr}$, which is initialized with the cost of the greedy route (there is a greedy search after the query comes which generates a greedy route) to be used as a pruning threshold. For each query category c_i, we first retrieve the POIs of c_i and check if the POI satisfies the rating constraint and those unqualified POIs are filtered. Then we construct the optimal subroute of each POI in dynamic programming and check whether the cost of the optimal subroute is greater than the current threshold $cost_{curr}$. If the cost of the current subroute exceeds $cost_{curr}$, it is considered invalid and should be pruned.

When constructing the optimal subroute of each POI in dynamic programming, with the optimal subroutes set S_{ps}, for each subroute $\overrightarrow{R_{os}^{p^{c_{i-1}}}}$ ending at $p^{c_{i-1}}$, we extract the POI $p^{c_{i-1}}$ from the route and compute each new subroute cost $cost_{curr}$ by adding up the distance from p^{c_i} to $p^{c_{i-1}}$ and the cost of $\overrightarrow{R_{os}^{p^{c_{i-1}}}}$, then compares the result with the current minimum cost $minCost$ to find the optimal one.

Notice that in OSE, to calculate the cost of the optimal subroutes, the computation of the exact shortest distance between each two POIs is required, thus it is too expensive to simply use the OSE algorithm to answer our RCOSR query. Therefore, our idea is to adopt the estimated distance (i.e., a lower bound) for each two POIs instead of the exact shortest distance to accelerate the computation. For efficient distance estimation between each two POIs, we next propose a new *Reference Node Inverted Index* (called RNII).

5.2   Reference Node Inverted Index

In this section, we develop our *Reference Node Inverted Index (RNII)*, which is used for distance estimation and POI filtering.

Graph Embedding. As the basis of our RNII, we first recall the idea of graph embedding, which is proposed by Kriegel et al. [6]. Graph embedding is used to perform distance approximation in large datasets. The main idea of graph embedding is to find a set of reference nodes on the graph and computes the shortest distances from each object (i.e., nodes, POIs) to these reference nodes on the graph, then transforming these distances into vectors. When estimating the distance between any two objects, a vector operation is executed to obtain the upper or lower bounds of the road network distance. Compared with regular distance approximation methods such as Euclidean distance estimation, utilizing graph embedding can significantly accelerate the computation, and the accuracy is relatively higher [6].

Given a reference nodes set $\mathcal{RN} = (r_1, r_2, \ldots, r_k)$, for any POI p_i, the distance vectors are $V_{p_i} = (dist(p_i, r_1), \ldots, dist(p_i, r_k))^T$.

As shown in Fig. 1(b), assuming two reference nodes are v_5 and v_6, the distance vector of POI p_1 is computed as $V_{p_1} = (dist(p_1, v_5), dist(p_1, v_6))^T = (3, 6)^T$, and for p_2 it is $V_{p_2} = (dist(p_2, v_5), dist(p_2, v_6))^T = (6, 3)^T$. Once the distance vectors are built, we then perform distance approximation by vector operation. Accordingly, we find the maximum value among each dimension of the vector from the difference (denoted as $LB_{p_1,p_2} = max(|3 - 6|, |6 - 3|) = 3$). Formally, we give the equations of LB_{p_i,p_j}.

$$LB_{p_i,p_j} = \max_{l=1,\ldots,k} (V_{p_i}[l] - V_{p_j}[l]) \tag{2}$$

Where $V_{p_i}[l]$ indicates the l-th element of vector V_{p_i}. According to the triangle inequality, we can infer that $LB_{p_i,p_j} \leq dist(p_i, p_j)$, which represents the lower bound of the distance approximation.

Selecting appropriate reference nodes is important to the accuracy. For example, in the road network in Fig. 1(b), the lower bound distance between p_1 and p_2 is to be estimated. It is better to select v_7 as the reference node than v_1. Such an ideal reference node is desired to obtain the exact or tighter lower bound distance. Furthermore, the accuracy of distance approximation using the reference nodes rely on the number and distribution of the selecting reference nodes. According to these issues above, we discuss the strategy of how to determine the reference nodes below.

Determining the Reference Nodes. The strategy of determining reference nodes significantly influences the accuracy of distance approximation. Notice that in the rating constrained optimal subroute expansion, we just exploit the lower bound to estimate the approximate distance of POIs. Thus, our goal is to derive a tight lower bound of distance approximation instead of the upper bound when designing the strategy of determining reference nodes. A natural and simple idea is to randomly pick up nodes (e.g., $v_1, v_2, \ldots, v_n$)[2] from the road network. Since this strategy is with high uncertainty, thus it cannot guarantee the efficiency of RNII. Accordingly, we propose a new *Greedy Merge* (GM) strategy to guide how to determine the reference nodes.

Consider a reference node r and a pair of POIs $\langle p_i, p_j \rangle$ in the road network. If the LB_{p_i,p_j} computed according to r is *exactly equal to* $dist(p_i, p_j)$, we say r *covers* the pair $\langle p_i, p_j \rangle$, the more pairs it covers, the better lower bound estimation it provides. The main idea of the GM strategy is to find a set of nodes as the reference nodes that maximize the number of POI pairs they can cover. To achieve this goal, we first calculate how many pairs of POI that each node covers, and greedily choose the node which can cover the largest number of POI pairs into the reference node-set iteratively. This process continues until the number of reference nodes reaches the preset value or the total number of POI pairs covered does not increase anymore. Take the road network in Fig. 1(a) as an example. Assume the number of reference nodes is 3, we choose nodes $\{v_7, v_{16}, v_0\}$ as the reference nodes using the GM strategy. This reference node-set can cover 132 pairs of POIs, while the total number of POI pairs in this road network is 144.

---

[2] Theoretically, the reference node can be any point in the road network, but for simplicity we select the nodes as the reference nodes.

RNII Data Structure. Inverted index is the mapping from keywords to documents, is widely used in the search engine, large-scale database index, document retrieval, multimedia retrieval/information retrieval [14]. In this work, we utilize an inverted index to retrieve the POIs by their categories. Each inverted item in RNII consists of two parts: category ID and corresponding POI entities. All the POI entities of category c_i can be retrieved quickly by querying the records of the inverted index with category c_i. A POI entity p is represented as $\langle PID, score(p), V_p \rangle$, where PID is the id of p, $score(p)$ is the rating score of p, and V_p is the corresponding distance vector (i.e. $V_p = dist(p, r_1), dist(p, r_2), \ldots, dist(p, r_k))$, where $dist(p, r_k)$ is the shortest distance from p to the k-th reference node.

5.3 The ROSE Algorithm

Based on the OSE and RNII, we further design a new ROSE algorithm. The ROSE initially adopts the approximate distance (which is computed quickly using RNII) for each two POI and utilizes the OSE algorithm to find the optimal route under the approximate distance. Notice that the route computed by OSE using the approximate distance is not the optimal solution. To find the optimal one, the route is served as the *guiding path* and then we calculate the exact shortest length of this guiding path using the shortest distance algorithm. While calculating the exact shortest length, the exact shortest distances between POIs in the guiding path are obtained and we replace the approximate distance with the exact shortest distance for these POI pairs. After that, we again employ OSE to find a new guiding path with the updated distance information. By continuing the above steps, the cost of the guiding path gets closer to the optimal solution. The algorithm terminates when the cost of the guiding path is equal to the exact cost of the corresponding expanded route, which is the solution to RCOSR.

Algorithm 1. Recurrent Optimal Subroute Expansion (ROSE)

Require: Query $Q = \{v_s, v_d, CT\}$ ($|CT| = k$)
Ensure: $\overrightarrow{R_{os}}$:optimal rating constrained sequenced route for v_d
1: $cost_{guide} \leftarrow 0$, $R_{guide} \leftarrow \varnothing$, $cost_{exact} \leftarrow 0$;
2: obtain the exact cost $cost_{exact}$ by invoking the greedy search;
3: Initialize current optimal cost $cost_{tresh}$ with $cost_{exact}$
4: **while** $cost_{guide} \neq cost_{exact}$ **do**
5:     $\langle R_{guide}, cost_{guide} \rangle \leftarrow OSE(Q, cost_{tresh})$;
6:     $cost_{exact} \leftarrow computeRouteDistance(R_{guide})$;
7:     **if** $cost_{exact} < cost_{tresh}$ **then**
8:         $cost_{tresh} \leftarrow cost_{exact}$;
9:     **end if**
10: **end while**
11: $\overrightarrow{R_{os}} \leftarrow R_{guide}$;
12: return $\overrightarrow{R_{os}}$;

The pseudo-code of ROSE is illustrated in Algorithm 1. Given a query Q, we first find a current feasible route (which is updated later) by invoking the greedy search (Line 2). Next, the recurrent process goes as follows: we find the guiding path and obtain its exact shortest length as the threshold cost of the current feasible route (Lines 5–6). Note that when executing OSE, if the shortest distance of some POI pairs has been explored, we use the shortest distance stored in an unordered map $pairDistMap$, otherwise we use the lower bound distance calculated by RNII. After that, we calculate the exact distance by invoking function $computeRouteDistance()$ (Line 6). $computeRouteDistance()$ can be implemented by any shortest distance algorithms in road networks, such as A* with graph embedding as the heuristic function [6] and H2H-Index [10]. At the same time, the exact distance of the involved POI pairs is stored in a map $pairDistMap$, which is used in further distance estimation. We also update the cost of the current solution (Lines 7–8). This process continues until the cost of the expanded route is the same as the cost of its guiding path.

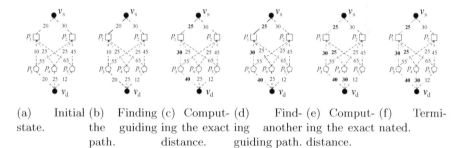

(a) Initial state. (b) Finding the guiding path. (c) Computing the exact distance. (d) Finding another guiding path. (e) Computing the exact distance. (f) Terminated.

Fig. 2. An example of illustrating ROSE.

Example 1. Figure 2 illustrates an example of ROSE. In the initial state, the distance between each two POI is initialized with the lower bound distance, which is represented by the dashed line (Fig. 2(a)). The first iteration computes the optimal route in the current lower bound distance as the guiding path $\{v_s, p_1, p_3, v_d\}$ (i.e., the black bold line in Fig. 2(b)). Then ROSE calculates as well as records the shortest distance from v_s to p_1, the shortest distance from p_1 to p_3, the shortest distance from p_3 to v_d (i.e., the red line in Fig. 2(c)). In the second iteration, another guiding path (i.e., the black bold line in Fig. 2(d)) is found and the exact distance of the corresponding POI pairs is calculated (i.e., the red line in Fig. 2(e)). In the last iteration, the guiding path found by ROSE is $\{v_s, p_1, p_4, v_d\}$, whose cost is equal to the actual length of the corresponding path, thus ROSE terminates and $\{v_s, p_1, p_4, v_d\}$ is returned as the result.

Compared to MTDOSR, ROSE has the following advantages: (1) ROSE utilizes the information of every query category. The guiding path is an optimal route under lower bound distance of the query, which consists of POI in every

query category, thus it can lead to more accurate guidance of the expansion.
(2) ROSE uses an inverted index to manage and pre-filter the unqualified POIs,
rather than filtering the POIs during route expansion. (3) ROSE updates and
uses the cost threshold for pruning in OSE, while MTDOSR has no pruning
strategies.

5.4 Complexity Analysis

In this section, we analyze the time and space complexity of MTDOSR and
ROSE. Let $|V|$ be the size of nodes in the graph, $|E|$ be the number of edges,
$|C|$ be the total categories of POI, r be the number of reference nodes, and
k be the size of query categories. In addition, we assume m is the average
number of qualified POIs in each category. Because a Dijkstra shortest distance algorithm is executed before the main loop, the complexity of MTDOSR
is $O(k|E|logk|V| + |V||E|log|V|)$ [3]. For each iteration in ROSE, we first invoke
the OSE to find the guiding path. In OSE, it constructs the optimal subroute
for each POI using dynamic programming, whose time complexity is $O(mr)$.
Moreover, it requires exploring km POIs. Thus, OSE takes $O(krm^2)$ time for
each iteration. After that, $computeRouteDistance$ takes $O(k|V||E|log|V|)$ to
compute the exact distance of guiding path (using Dijkstra). The total iterations of ROSE depend on the accuracy of lower bound distance computed by
RNII, in the worst case, the iteration goes m^k times for each query. Note that
in reality, due to the lower bound estimation provided by RNII, the iteration
goes far less than m^k times. In theory, the overall time complexity of ROSE is
$O(km^k(m^2r+|V||E|log|V|))$. Besides, the space complexity of RNII is $O(|C|mr)$.

6 Experiments

In this section, we conduct a number of experiments to evaluate the performance
of the proposed algorithms and RNII using both real datasets and synthetic
datasets. All the algorithms are implemented using C++, and the experiments
are conducted on a server with an Intel Core i7-9700 CPU of 3.0 GHz and 16
GB RAM. We use the following real datasets:

- **CA.** The CA dataset is the real road network of California[3]. It contains
  21,048 nodes, 21,693 road edges, and 87,635 POIs belong to 64 categories.
- **OL.** The OL dataset is the real road network of Oldenburg city(see footnote
  3). It contains 6,105 nodes, 7,035 road edges, and 2,404 POIs belong to 26
  categories.

In addition, we generate a synthetic dataset that contains 100,000 nodes,
125,000 road edges, and 10,000 POIs belong to 50 categories and we randomly
set the rating score ranging from 0 to 100 for each POI.

We conduct the performance evaluation in two aspects: (1) evaluating the
efficiency of RCOSR algorithms. We compare the *query time* and *the number*

---

[3] http://www.cs.utah.edu/~lifeifei/SpatialDataset.htm.

of visited nodes (i.e., the number of times the nodes get visited) under various parameters, including *the query size* (i.e., the number of query categories), *the network size* (i.e., $|V|$), *the POI numbers*, shown in Table 1; and (2) evaluating the efficiency of RNII. We compare the query time and cover rate using RNII concerning the reference nodes strategies. The bold number represents the default values. In each experimentation, we vary one parameter at a time and fix the other parameters as the default value, and generate 100 random queries. The reported experimental results are obtained by averaging the query time as well as the number of visited nodes.

6.1   Efficiency of RCOSR Algorithms

In this section, we evaluate the efficiency of MTDOSR and ROSE. While implementing the *computeRouteDistance*() function in ROSE, we use A* algorithm to calculate the shortest distance between POIs and use the lower bound distance as the heuristic function. Moreover, we construct the RNII by adopting the GM strategy with the number of reference nodes as 80 in ROSE.

Table 1. Parameter ranges and default values

Parameters	Query size	Network size (×1000)	POI numbers	Average degree of nodes	Number of reference nodes
Ranges	1, **3**, 5, 7, 9	**10**, 40, 70, 100	**100**, 500, 1000	1.5, **2.5**, 3.5	20, 40, **80**, 160, 320

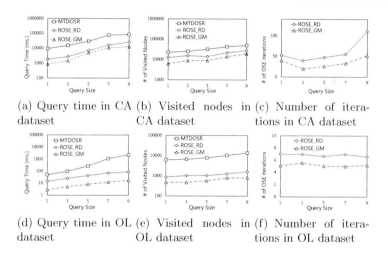

(a) Query time in CA dataset

(b) Visited nodes in CA dataset

(c) Number of iterations in CA dataset

(d) Query time in OL dataset

(e) Visited nodes in OL dataset

(f) Number of iterations in OL dataset

Fig. 3. Performance w.r.t query size in real road network

Algorithm Performance in Real Road Network. The query size refers to the number of POI categories in the query (i.e., the $|CT|$). In this experiment, we compare the efficiency of ROSE using the random strategy (noted as ROSE_RD), ROSE using greedy merge strategy (noted as ROSE_GM), and MTDOSR using the two real datasets. From the results shown in Fig. 3, we can see that the query time and the number of visited nodes for all algorithms increase when the query size increases. ROSE_GM outperforms MTDOSR with regard to the query time (with 85.07% in CA and 98.88% in OL improvement for average query size, respectively) and the number of visited nodes (with 66.8% in CA and 93.08% in OL improvement for average query size, respectively), and ROSE_GM outperforms ROSE_RD with regard to the query time (with 42.96% in CA and 93.8% in OL improvement for average query size, respectively) and the number of visited nodes (with 37.3% in CA and 46.2% in OL improvement for average query size, respectively). For the OL dataset, the superiority of ROSE is more obvious compared to that in the CA dataset. It can be explained that the number of POIs in the OL dataset is smaller than that of CA. That is because ROSE computes the guiding path using all the POI categories information, the ROSE runs faster in a sparser road network. Figures 3(c) and 3(f) shows the number of OSE iterations before the optimal route is found. For the CA dataset, the average numbers of iterations using random and greedy merge strategy are 60.89 and 34.11 respectively, while in the OL dataset the ROSE only needs 6.84 and 5.16 iterations on average to find the optimal route with respect to random and greedy merge strategies.

Effect of the Network Size. Figure 4 shows the query time and the number of visited nodes of ROSE and MTDOSR with respect to different network sizes in the synthetic dataset. As illustrated in Fig. 4, the query time of the two algorithms increases by increasing the network sizes, while ROSE outperforms the MTDOSR significantly. Figure 4 also describes the results of the number of visited nodes by varying the network sizes. When the network size increases, finding the optimal rating constrained sequenced route requires to visit more nodes.

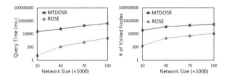

Fig. 4. Time w.r.t. network size

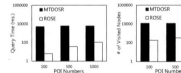

Fig. 5. Time w.r.t. POI numbers

Effect of POI Numbers. As expected, the query time and the number of visited nodes of ROSE increase when increasing the POI percentages in Fig. 5 When the POI number increases, there are more guiding paths that are close to the optimal rating constrained sequenced route, thus ROSE requires to spend more iterations to find the optimal one incurring more computations of the exact distance between the POIs when constructing optimal subroutes. ROSE outperforms MTDOSR with respect to the effect of the POI number, especially when the POI number is small. As the number of POI numbers grows, the query time of both algorithms increase, and ROSE is more sensitive to the number of POI numbers.

Effect of the Average Degree of Nodes. Figure 6 compares the algorithm performance with respect to the average degree of nodes. For the two algorithms, the query time and the number of visited nodes increase slightly when increasing the average degree of nodes, because the network becomes more complex and there is more possibility to search the routes. In addition, we can observe from Fig. 6 that MTDOSR takes more time to visit the same amount of nodes compared to ROSE. This can be explained by the computation cost of the heap that MTDOSR maintains. The heap used in MTDOSR stores the nodes of the entire path, which makes the number of nodes increase as the iteration continues. Moreover, it is very time-consuming to arrange and search the heap, while ROSE uses A* algorithm to compute the distance of two POIs, which only needs to maintain a small number of nodes in the heap.

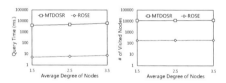

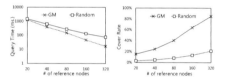

Fig. 6. Time w.r.t. average degree of nodes

Fig. 7. RNII performance w.r.t. number of reference nodes

6.2 Efficiency of RNII Index

As discussed in Sect. 5.2, determining the reference nodes is an essential issue in the construction of RNII. To achieve the desired RNII, we propose the GM strategy. In this section, we evaluate the performance of the GM strategy, in comparison with the Random strategy (i.e., this strategy randomly chooses the reference nodes) under a various number of reference nodes. We use the synthetic network of 10,000 nodes and 1000 POIs, which belong to 10 categories.

Effect of Strategies and Number of Reference Nodes. It is shown in Fig. 7 that GM outperforms Random significantly under a various number of reference nodes. This also indicates that the reference nodes decided by the GM strategy ensure higher accuracy in estimating the lower bound distance than the reference nodes decided by Random. From Fig. 7, we can observe that the more reference nodes we decide, the more POI pairs we can cover, which illustrates that the more reference nodes we use, the more accuracy we gain.

7   Conclusion

In this paper, we formalize and study the *Rating Constrained Optimal Sequenced Route* (RCOSR) query, which constrains the rating score of all POIs in the result and the optimal route should satisfy the user thresholds. To answer the query, we adapt the TD-OSR algorithm as *MTDOSR* to serve as a baseline. Next, we try to propose a new *Optimal Subroute Expansion* (OSE) algorithm to solve the problem. Moreover, we propose a *Reference Node Inverted Index* (RNII) to accelerate the distance computation in OSE. Based on OSE and RNII, we propose a new *Recurrent Optimal Subroute Expansion* (ROSE) algorithm. At last, a comprehensive performance evaluation is conducted to validate the proposed ideas and demonstrate the efficiency and effectiveness of the proposed index and algorithms.

Acknowledgments. This work is supported by the National Natural Science Foundation of China (61902438, 61902439, U1811264, U19112031), Natural Science Foundation of Guangdong Province under Grant (2019A1515011704, 2019A1515011159), Guangdong Basic and Applied Basic Research Foundation (2019B1515130001), National Science Foundation for Post-Doctoral Scientists of China under Grant (2018M643307, 2019M663237), Young Teacher Training Project of Sun Yat-sen University under Grant (19lgpy214,19lgpy223)and Hong Kong RGC Grant 12200817.

References

1. Barua, S., Jahan, R., Ahmed, T.: Weighted optimal sequenced group trip planning queries. In: MDM, pp. 222–227 (2017)
2. Chen, H., Ku, W., Sun, M., Zimmermann, R.: The multi-rule partial sequenced route query. In: SIGSPATIAL, pp. 1–10 (2008)
3. Costa, C.F., Nascimento, M.A., Macêdo, J.A., Theodoridis, Y., Pelekis, N., Machado, J.: Optimal time-dependent sequenced route queries in road networks. In: SIGSPATIAL, pp. 1–4 (2015)
4. Dai, Jian., Liu, Chengfei., Xu, Jiajie, Ding, Zhiming: On personalized and sequenced route planning. World Wide Web **19**(4), 679–705 (2015). https://doi.org/10.1007/s11280-015-0352-2
5. Guttman, A.: R-trees: a dynamic index structure for spatial searching. In: SIGMOD, pp. 47–57 (1984)
6. Kriegel, H., Kröger, P., Kunath, P., Renz, M., Schmidt, T.: Proximity queries in large traffic networks. In: GIS, pp. 1–8 (2007)
7. Li, F., Cheng, D., Hadjieleftheriou, M., Kollios, G., Teng, S.: On trip planning queries in spatial databases. In: SSTD, pp. 273–290 (2005)
8. Li, J., Yang, Y.D., Mamoulis, N.: Optimal route queries with arbitrary order constraints. TKDE **25**(5), 1097–1110 (2013)
9. Liu, H., Jin, C., Yang, B., Zhou, A.: Finding top-k optimal sequenced routes, pp. 569–580 (2018). CoRR abs/1802.08014
10. Ouyang, D., Qin, L., Chang, L., Lin, X., Zhang, Y., Zhu, Q.: When hierarchy meets 2-hop-labeling: efficient shortest distance queries on road networks. In: ICMD, pp. 709–724 (2018)

11. Sasaki, Y., Ishikawa, Y., Fujiwara, Y., Onizuka, M.: Sequenced route query with semantic hierarchy. EDBT **2018**, 37–48 (2018)
12. Sharifzadeh, M., Kolahdouzan, M.R., Shahabi, C.: The optimal sequenced route query. VLDB J **17**(4), 765–787 (2008)
13. Sharifzadeh, M., Shahabi, C.: Processing optimal sequenced route queries using voronoi diagrams. GeoInformatica **12**(4), 411–433 (2008)
14. Singhal, A.: Modern information retrieval: a brief overview. IEEE DEB **24**(4), 35–43 (2001)
15. Yao, B., Tang, M., Li, F.: Multi-approximate-keyword routing in GIS data. In: SIGSPATIAL, pp. 201–210 (2011)
16. Yawalkar, P., Ranu, S.: Route recommendations on road networks for arbitrary user preference functions, pp. 602–613 (2019)
17. Zheng, B., Su, H., Hua, W., Zheng, K., Zhou, X., Li, G.: Efficient clue-based route search on road networks. TKDE **12**(4), 1846–1859 (2017)
18. Zhong, R., Li, G., Tan, K., Zhou, L.: G-tree: an efficient index for KNN search on road networks. In: CIKM, pp. 39–48 (2013)

Privacy-Preserving Polynomial Evaluation over Spatio-Temporal Data on an Untrusted Cloud Server

Wei Song[1]($\boxtimes$), Mengfei Tang[1], Qiben Yan[2], Yuan Shen[1], Yang Cao[3], Qian Wang[4], and Zhiyong Peng[1]

[1] School of Computer Science, Wuhan University, Wuhan, China
{songwei,mengfeitang,yuanshen,peng}@whu.edu.cn
[2] Department of Computer Science and Engineering,
Michigan State University, East Lansing, MI, USA
qyan@msu.edu
[3] Department of Social Informatics, Kyoto University, Kyoto, Japan
yang@i.kyoto-u.ac.jp
[4] School of Cyber Science and Engineering, Wuhan University, Wuhan, China
qianwang@whu.edu.cn

Abstract. Nowadays, with the popularity of location-aware devices, multifarious applications based on the spatio-temporal data come forth in our lives. In these applications, a platform (enterprise) collects the users' spatio-temporal data based on which it recommends the top-k users (passengers) to the registered service providers (drivers). Outsourcing the tremendous scale of spatio-temporal data to cloud provides an economical way for the enterprises to implement their applications. In this paradigm, the third-party cloud server is not completely trustworthy. The collected spatio-temporal data can hold users' privacy, so it's a critical challenge to design a secure and efficient query mechanism for this scenario, such as the ride-hailing or the ride-sharing services. However, the existing solutions for the privacy-preserving kNN queries mainly focus on data privacy protection or computation complexity. There still lacks a practical privacy-preserving polynomial evaluation solution over the spatio-temporal data. In this paper, we propose a virtual road network structure to storage and index the spatio-temporal data in the road network and design a novel homomorphic encryption scheme based on Order-Revealing Encryption to enable an untrusted cloud server to execute the polynomial evaluation over the encrypted spatio-temporal data in the road network. We formally prove the security of the proposed scheme under the random oracle model. Extensive experiments on real world data demonstrate the effectiveness and efficiency of the proposed scheme over alternatives.

1 Introduction

In last decade, we have witnessed the development of multifarious Location Based Services (LBS) based on the spatio-temporal data with the popularity

© Springer Nature Switzerland AG 2021
C. S. Jensen et al. (Eds.): DASFAA 2021, LNCS 12681, pp. 474–490, 2021.
https://doi.org/10.1007/978-3-030-73194-6_32

of location-aware devices. In these applications, the enterprises always have to collect the tremendous scale of spatio-temporal data and deal with a number of users' queries. Outsourcing the spatio-temporal database to cloud is an economical way for them to implement the applications. The collected spatio-temporal data hold the users' privacy, however the third-party cloud server is not completely trustworthy and is possible to infer the users' privacy by analyzing the stored spatio-temporal data and the received query messages. So, it is a problem must be resolved that how to protect user privacy during the services. Encrypting the data before outsourcing is a natural choice to protect user privacy.

An intuitive idea to address this issue is to design a homomorphic encryption algorithm to enable the cloud server to execute query over the encrypted spatio-temporal data or insert noise into the original data. However, both of them are not feasible for the real applications. For the first design, the enormous volume of the spatio-temporal data will inevitably lead to decreased efficiency. For the second design, the users need the accurate query results which are hard to achieve within the noise-based methods. So, the main motivation of this work is to design a privacy-preserving polynomial evaluation scheme over a huge scale of spatio-temporal data. There have been a number of research efforts [1–12] on the privacy-preserving kNN query over road networks. Nonetheless, most of them mainly focus on the data privacy or the computation complexity. There still lacks a practically privacy-preserving polynomial evaluation solution over spatio-temporal data supporting the dynamic evaluation function rather than the fixed one in advance.

Motivated by this, we propose a novel homomorphic encryption scheme based on Order-Revealing Encryption (ORE) to enable the secure polynomial evaluation over the spatio-temporal data in road network. The main contributions of this paper can be summarized as:

- We design a **novel homomorphic encryption scheme**, which contains an extended ORE scheme, to achieve the privacy-preserving polynomial evaluation over the encrypted spatio-temporal data.
- To support the huge scale of the spatio-temporal data, we propose a **virtual road network index structure** to manage the data and enable the untrusted cloud server to efficiently execute the polynomials over them.
- The proposed method is **practical** that can be extended to support any polynomial evaluation function moreover supports the dynamic conditions in the function, which is still an open problem.

2   Related Work

In recent years, many efforts have been spent to address the issue of Privacy-Preserving kNN query (PPkNN), which can be classified into three categories.

PPkNN Query: Wong et al. [1] first proposed an asymmetric scalar-product-preserving encryption (ASPE) scheme as the distance recoverable encryption to implement the secure kNN query. Voronoi diagram is widely used for PPkNN query [2–4]. When meeting the dynamic conditions, the schemes based on Voronoi diagram need to part again and will cause huge maintenance overheads. Lei et al. [5] proposed a PPkNN solution to support two-dimensional data. As a comparison, our scheme can extend to support multi-attributes. All these works can execute only the approximate PPkNN query. To implement the accurate PPkNN query, some methods [6,7] are proposed. But they have low efficiency because they need to evaluate all data. All the above studies did not consider the situation of the road networks with dynamic conditions.

PPkNN Query Over Road Network: If the underlying road network is taken into consideration, the problem of PPkNN query becomes more complex. Palanisamy et al. [8] proposed the mix-zones to preserve the location privacy but it is vulnerable to the background knowledge attack. Yi et al. [9] proposed a generic scheme to support multiple discrete attributes of private location-based queries. Paulet et al. [10] proposed a scheme which combines PIR with a stage of oblivious transfer. These schemes just support the range query. Yang et al. [11] proposed a verifiable privacy-preserving kNN query scheme based on Voronoi diagram. Zeng et al. [12] proposed a privacy-preserving query scheme which only supports boolean query. In general, all the above schemes are not practical enough since they just support the fixed kNN query function.

PPkNN Query over Spatio-temporal Data: Privacy-preserving ride-hailing services were introduced in [13–15]. Pham et al. [13] proposed PrivateRide to provide anonymous ride-hailing services. They improved PrivateRide [14] by adding accountability and enhanced the privacy by increasing the anonymity set. The scheme in [15] recursively divides the area into quad regions with a quadtree but has the huge communication costs because each user needs to send an encryption for each region. There are some studies [16–19] for the ride-sharing services. Sherif et al. [16] proposed a kNN encryption scheme to protect the passenger's and the driver's privacy. Li et al. [17] proposed a scheme using an anonymous authentication scheme to recover the malicious users' real identities. The closest to our work is [20]. In the scheme, the platform can build a directed graph based on the passengers' routes, however it is still not efficient enough and cannot be applied to top-k query directly.

All the above researches do not consider to support the dynamic conditions. In reality, it is high desirable. In a recent work of [21], the scheme supports top-k ranking queries based on the ranking function. However, the scheme is only suitable to the static data because it needs to preprocess all attributes. Another work [22] proposed a privacy-preserving polynomial evaluation scheme. However, 1) when meeting large scale of data, the scheme has low efficiency because it needs to evaluate all data; 2) the scheme doesn't support the evaluation with some dynamic conditions.

So as far now, the existing work can not solve the challenges of the practical privacy-preserving polynomial evaluation over the large scale of spatio-temporal data, not to mention high efficiency.

3   Problem Statement

3.1   System Model

In this paper, we take a ride-hailing application as the example to explain the system model as shown in Fig. 1, which contains five main entities: the platform (enterprise), the computation cloud server ($\mathbf{S}_1$), the en/decryption cloud server ($\mathbf{S}_2$), the service providers (drivers), and the users (passengers).

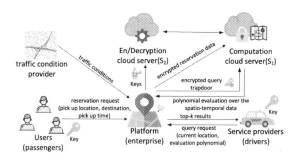

Fig. 1. The system model

In a ride-hailing application, the passenger sends the reservation request including the pick up location, the destination, and the pick up time to the ride-hailing platform. The platform uses a model to calculate a price for each order and encrypts the spatio-temporal data of all the orders then uploads them to the cloud server ($\mathbf{S}_1$). We do not discuss the pricing mechanism in this work. While a driver wants to pick up a user, he firstly generates the query request based on his current location and his preference. To protect the driver's privacy, the platform then encrypts his original query request to generate the query trapdoor and submits it to $\mathbf{S}_1$. Finally, $\mathbf{S}_1$ executes the driver-decided polynomial evaluation over all the encrypted spatio-temporal data and returns top-k results to the driver.

In our model, two cloud servers $\mathbf{S}_1$ and $\mathbf{S}_2$ work on the hybrid cloud mode and are non-colluding. $\mathbf{S}_1$ is a public cloud, which has the strong computing and storage capability but is considered as a semi-trusted entity. $\mathbf{S}_2$ is a private cloud server built by the platform itself to execute the comparing operations over the ciphertext with $\mathbf{S}_1$.

3.2   Threat Model and Design Goals

In our system model, the cloud server $\mathbf{S}_1$ is not completely trusted. It may be vulnerable to the corrupt employees or the external malicious attacks, so we focus on two types of privacy in our paper:

- **Passenger Privacy**: The passenger's sensitive information includes the pick up location, the pick up time and the destination. Our scheme should ensure that S_1 cannot learn any passenger's sensitive information by analyzing the encrypted data and the received messages.
- **Driver Privacy**: The driver's sensitive information mainly is the driver's current location. To protect the driver privacy, the platform encrypts his exact location to generate the query trapdoor. Our scheme ensures that S_1 can not learn the driver's location by analyzing the query trapdoor.

In addition to protecting the privacy, other design goals of our work include:

- **Scalability**: The evaluation function can be an arbitrary polynomial function and decided by the driver, rather than fixed in advance.
- **Practicality**: The evaluation method should comprehensively consider the factors of time and space in the spatio-temporal data, and can efficiently support huge scale of spatio-temporal data. Moreover, the query is based on the road network instead of the Euclidean distance, moreover it can support some dynamic factors in the road network, such as the traffic condition.

4   Privacy-Preserving Polynomial Evaluation over Spatio-temporal Data

4.1   Paillier Homomorphic Encryption

We adopt Paillier to implement the privacy-preserving polynomial evaluation. Suppose $\mathsf{E}_{\mathbf{PK}}(\cdot)$ and $\mathsf{D}_{\mathbf{SK}}(\cdot)$ are the en/decrypting operators of a Paillier cryptosystem in which **PK**, **SK** are the public/secret keys. For convenience, we use $\mathsf{E}(\cdot), \mathsf{D}(\cdot)$ to instead of $\mathsf{E}_{\mathbf{PK}}(\cdot), \mathsf{D}_{\mathbf{SK}}(\cdot)$. Given messages $m_1, m_2 \in \mathbb{Z}_N$, $ct_1 = \mathsf{E}(m_1), ct_2 = \mathsf{E}(m_2)$, it has two important properties which can be exhibited as: 1) $\mathsf{D}(\mathsf{E}(m_1) \times \mathsf{E}(m_2)) = m_1 + m_2 \bmod N^2$, 2)$\mathsf{D}(\mathsf{E}(m_1)^a) = a \times m_1 \bmod N^2, \forall a \in \mathbb{Z}_N$.

4.2   Order-Revealing Encryption

An Order-Revealing Encryption (ORE) [23] scheme is a computable function that compares two ciphertexts with strong security guarantee. It can be defined as a tuple of polynomial algorithms Π=(ORE.Setup, ORE.Encrypt, ORE.Compare):

- ORE.Setup(1^λ) $\rightarrow sk$: On input a security parameter λ, the Setup algorithm outputs a secret key sk.
- ORE.Encrypt(sk, m) $\rightarrow ct$: On input a secret key sk and a message $m \in \mathcal{D}$, the Encrypt algorithm outputs a ciphertext ct for m.
- ORE.Compare(ct_1, ct_2) $\rightarrow b$: On input ciphertexts ct_1, ct_2, Compare algorithm outputs a bit $b \in \{0, 1\}$. If $m_1 < m_2$, output 1. Otherwise, output 0.

Correctness: We say an ORE scheme defined over a well-ordered domain $\mathcal{D}$ is correct if for all message $m_1, m_2 \in \mathcal{D}$, $\Pr[\mathsf{ORE.Compare}(ct_1, ct_2) = 1(m_1 < m_2)] = 1 - negl(\lambda)$ in which $negl(\lambda)$ denotes a negligible function in λ.

4.3 The Proposed Encryption Scheme

We design a novel homomorphic encryption scheme with ORE character to encrypt the spatio-temporal data. We detail it as below:

– Setup$(1^\lambda) \to \{\mathbf{PK}, \mathbf{SK}, \mathsf{MUL}, \mathsf{ADD}, F, \overline{SK}\}$ is run by the platform to initialize the system. It takes a security parameter λ as input and outputs the global security parameters. It first constructs a Paillier cryptosystem with key pair $(\mathbf{PK}, \mathbf{SK})$ and homomorphic additive/multiplicative encryption algorithms ADD and MUL. It stores $\mathbf{SK}$ locally and uploads it to $\mathbf{S}_2$.

$$\mathsf{ADD}(m_1, m_2) = \mathsf{E}(m_1) \times \mathsf{E}(m_2) = \mathsf{E}(m_1 + m_2),$$
$$\mathsf{MUL}(a, m) = \mathsf{E}(m)^a = \mathsf{E}(a \times m).$$

Second, it constructs a secure pseudorandom function [24] $F : \{0,1\}^\lambda \times \{0,1\}^\lambda \to \{0,1\}^\lambda$, and samples n PRF key $k_i \xleftarrow{R} \{0,1\}^\lambda$ for F and n uniform random permutation $\pi_i : [N] \to [N]$ where N is the message space.

Finally, the Setup algorithm outputs the secret key as $\overline{SK} = \{sk_1, \ldots, sk_n\}$, $sk_i = (k_i, \pi_i)$ which are stored by the platform.

– Encrypt$_\mathsf{L}(sk, m) \to \mathsf{ct}_\mathsf{L}(m)$ is the left encryption algorithm run by the platform. Given the secret key $sk = (k, \pi) \in \overline{SK}$, a message $m \in \mathbb{Z}_N$, the Encrypt$_\mathsf{L}$ algorithm outputs the left ciphertext for m as:

$$\mathsf{ct}_\mathsf{L}(m) = (F(k, \pi(m)), \pi(m)).$$

– Encrypt$_\mathsf{R}^{\mathsf{add}}(sk, m) \to \mathsf{ct}_\mathsf{R}^{\mathsf{add}}(m)$ is the right additive encryption algorithm run by the platform. Its inputs are same with those in Encrypt$_\mathsf{L}$. It samples a random element $r \xleftarrow{R} \{0,1\}^\lambda$. For each $i \in [N]$, it computes v_i and outputs the right addition ciphertext $\mathsf{ct}_\mathsf{R}^{\mathsf{add}}(m)$ for m as:

$$\mathsf{ct}_\mathsf{R}^{\mathsf{add}}(m) = (r, v_1, v_2, \ldots, v_N), \quad v_i = \mathsf{ADD}(\pi^{-1}(i), m) + H(F(k, i), r), \quad (1)$$

in which $H : \{0,1\}^\lambda \times \{0,1\}^\lambda \to \mathbb{Z}_p$ is a secure Hash function.

– Encrypt$_\mathsf{R}^{\mathsf{mul}}(sk, m) \to \mathsf{ct}_\mathsf{R}^{\mathsf{mul}}(m)$ is the right multiplicative encryption algorithm run by the platform. For each $i \in [N]$, it computes w_i and outputs the right multiplication ciphertext $\mathsf{ct}_\mathsf{R}^{\mathsf{mul}}(m)$ for m as:

$$\mathsf{ct}_\mathsf{R}^{\mathsf{mul}}(m) = (r, w_1, w_2, \ldots, w_N), \quad w_i = \mathsf{MUL}(\pi^{-1}(i), m) + H(F(k, i), r). \quad (2)$$

– Addition$(\mathsf{ct}_\mathsf{L}(m), \mathsf{ct}_\mathsf{R}^{\mathsf{add}}(n)) \to \mathsf{E}(m+n)$ is the secure addition algorithm run by $\mathbf{S}_1$. It takes $\mathsf{ct}_\mathsf{L}(m)$ and $\mathsf{ct}_\mathsf{R}^{\mathsf{add}}(n)$ as inputs and outputs the ciphertext $\mathsf{E}(m+n)$ as:

It first parses $(s, t) = \mathsf{ct}_\mathsf{L}(m) = (F(k, \pi(m)), \pi(m))$. Then, it picks out:

$$v_t = \mathsf{ADD}(\pi^{-1}(\pi(m)), n) + H(F(k, t), r) = \mathsf{ADD}(m, n) + H(F(k, \pi(m)), r)$$

Finally, the algorithm outputs the result as: $result = v_t - H(s, r) = \mathsf{ADD}(m, n) + H(F(k, \pi(m)), r) - H(F(k, \pi(m)), r) = \mathsf{ADD}(m, n)$.

– Multiplication($\mathsf{ct}_\mathsf{L}(a), \mathsf{ct}_\mathsf{R}^\mathsf{mul}(m)$) → $\mathsf{E}(a \times m)$ is a secure multiplication algorithm executed by $\mathbf{S}_1$. It takes $\mathsf{ct}_\mathsf{L}(a) \leftarrow \mathsf{Encrypt}_\mathsf{L}(sk, a)$, $a \in \mathbb{Z}_N$ and $\mathsf{ct}_\mathsf{R}^\mathsf{mul}(m)$ as inputs and outputs the ciphertext $\mathsf{E}(a \times m)$ as:
It first parses $(s, t) = \mathsf{ct}_\mathsf{L}(a) = (F(k, \pi(a)), \pi(a))$. Then, it picks out:

$$w_t = \mathsf{MUL}(\pi^{-1}(\pi(a)), m) + H(F(k, t), r) = \mathsf{MUL}(a, m) + H(F(k, \pi(a)), r)$$

Finally, the algorithm outputs the result as: $result = w_t - H(s, r) = \mathsf{MUL}(a, m) + H(F(k, \pi(a)), r) - H(F(k, \pi(a)), r) = \mathsf{MUL}(a, m)$.

– Compare($\mathsf{E}(m), \mathsf{E}(n)$) → $\{1, 0, -1\}$ is a secure comparison algorithm run by $\mathbf{S}_1$ and $\mathbf{S}_2$. It takes $\mathsf{E}(m)$ and $\mathsf{E}(n)$ as inputs and outputs 1 while $m > n$, outputs 0 while $m = n$, outputs -1 while $m < n$. Suppose $N = 2^z$, $\mathbf{S}_1$ first executes the bit-decomposition operation $\mathsf{BD}(\cdot)$ [25] to get the encryptions of the individual bits of binary representation of m and n as:

$$\mathsf{BD}(\mathsf{E}(m)) = \langle \mathsf{E}(m_1), \ldots, \mathsf{E}(m_z) \rangle, \quad \mathsf{BD}(\mathsf{E}(n)) = \langle \mathsf{E}(n_1), \ldots, \mathsf{E}(n_z) \rangle,$$

in which m_i, n_i are the ith bits of m and n respectively.

Second, $\mathbf{S}_1$ uses BitMultiply algorithm to do the bit-multiplication over $\mathsf{E}(m_i)$ and $\mathsf{E}(n_i)$ for all $i \in [z]$ with the help of $\mathbf{S}_2$ as below:
(1) $\mathbf{S}_1$ picks two random numbers $r_{m_i}, r_{n_i} \in \mathbb{Z}_N$, computes $h_m = \mathsf{E}(m_i) \times \mathsf{E}(r_{m_i}), h_n = \mathsf{E}(n_i) \times \mathsf{E}(r_{n_i})$ and sends them to $\mathbf{S}_2$.
(2) $\mathbf{S}_2$ decrypts h_m, h_n as $k_m = \mathsf{D}(h_m) = m_i + r_{m_i}, k_n = \mathsf{D}(h_n) = n_i + r_{n_i}$. Then, $\mathbf{S}_2$ encrypts $k_m \times k_n$ as $\mathsf{E}(k_m \times k_n)$ and sends it to $\mathbf{S}_1$.
(3) $\mathbf{S}_1$ computes $\mu = \mathsf{E}(m_i)^{N-r_{n_i}}, \nu = \mathsf{E}(n_i)^{N-r_{m_i}}, \omega = \mathsf{E}(r_{m_i} r_{n_i})^{N-1}$, and gets $\mathsf{E}(m_i \times n_i) = \mathsf{E}(k_m \times k_n) \times \mu \times \nu \times \omega$.
Third, after get $\mathsf{E}(m_i \times n_i)$, $\mathbf{S}_1$ computes $G_i, \hat{G}_i, R_i, \hat{R}_i$ as:

$$G_i = \mathsf{E}(m_i \times (1 - n_i)) = \mathsf{E}(m_i) \times \mathsf{E}(m_i \times n_i)^{N-1}$$
$$\hat{G}_i = \mathsf{E}(n_i \times (1 - m_i)) = \mathsf{E}(n_i) \times \mathsf{E}(n_i \times m_i)^{N-1}.$$

$$R_i = \begin{cases} G_1 & \text{if } i = 1 \\ \mathsf{E}(\mathsf{D}(R_{i-1}) \vee \mathsf{D}(G_i)) & \text{if } i > 1 \end{cases}, \hat{R}_i = \begin{cases} \hat{G}_1 & \text{if } i = 1 \\ \mathsf{E}(\mathsf{D}(\hat{R}_{i-1}) \vee \mathsf{D}(\hat{G}_i)) & \text{if } i > 1. \end{cases}$$

Property 1 (bitwise-or operation). Obviously, $a \vee b = (a + b) - (a \times b)$ always holds. Given two bits a and b which are not known for $\mathbf{S}_1$ and $\mathbf{S}_2$, $\mathbf{S}_1$ holds the ciphertexts of a, b. $\mathbf{S}_1$ can obtain $\mathsf{E}(a \times b)$ by BitMultiply algrithm. So, $\mathbf{S}_1$ can output $\mathsf{E}(a \vee b) = \mathsf{E}(a) \times \mathsf{E}(b) \times \mathsf{E}(a \times b)^{N-1} = \mathsf{E}(a + b - a \times b)$.

Finally, $\mathbf{S}_1$ computes $\Gamma = \prod(R_i) = \mathsf{E}(\sum \mathsf{D}(R_i)), \hat{\Gamma} = \prod(\hat{R}_i) = \mathsf{E}(\sum \mathsf{D}(\hat{R}_i))$ and sends them to $\mathbf{S}_2$. $\mathbf{S}_2$ decrypts $\Gamma, \hat{\Gamma}$ as $\Lambda = \mathsf{D}(\Gamma), \hat{\Lambda} = \mathsf{D}(\hat{\Gamma})$. If $\Lambda > \hat{\Lambda}$, it outputs 1. If $\Lambda < \hat{\Lambda}$, output -1. If $\Lambda = \hat{\Lambda}$, output 0.

4.4   Virtual Road Network

To support huge scale of spatio-temporal data, we design a virtual road network structure in Fig. 2 to store and maintain all the orders in road network. Given a city's road network $RN(R, I)$ where $R = \{R_1, \ldots, R_n\}$ and $I = \{I_1, \ldots, I_l\}$ represent the sets of the roads and the intersections in RN, $O = \{O_1, \ldots, O_m\}$ represents the users' orders. The majority of data including the data of roads and orders are stored at $\mathbf{S}_1$, and the platform only stores the secret key and the basic information for every road.

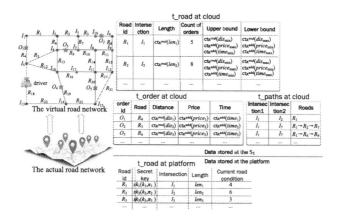

Fig. 2. The virtual road network index stored at $\mathbf{S}_1$ and the platform

We use $R_i(R_i, I_{R_i}, len_i, Count_i, Upper_i, Lower_i)$ to describe a road. R_i is the road's unique identifier. We randomly choose one of R_i's intersections I_{R_i} as its intersection. $len_i, Count_i \in \mathbb{Z}_N$ represent R_i's length and the count of the orders the pick up locations of which are on R_i. The platform encrypts len_i by $\mathsf{Encrypt}_R^{mul}(sk_i, len_i)$. For every road R_i, $Upper_i$ and $Lower_i$ records the boundary values of distance, price, and pick up time of all the orders on R_i. They decide the possible max/min evaluation values for all the orders on R_i. The price is calculated by the platform based on the pick up location and destination. The platform randomly picks a key $sk_i = (k_i, \pi_i) \in \overline{SK}$ as R_i's secret key and stores $R_i(R_i, sk_i, I_{R_i}, len_i, cond_i)$ at local. According to Baidu Map, $cond_i \in [1, 6]$ represents R_i's current traffic condition. The larger $cond_i$ value represents the worse traffic condition.

$\mathbf{S}_1$ stores all the orders as $O_i(O_i, R_j, O_i.dis, O_i.price, O_i.time)$ in which O_i is the order's identifier and R_j is the road covers O_i's pick up location. $O_i.dis$, $O_i.price$, $O_i.time \in \mathbb{Z}_N$ represent the distance from the location to I_{R_j}, the price value, and the pick up time. To protect the passenger's privacy, $O_i.dis, O_i.price$, and $O_i.time$ are encrypted.

There are a number of available paths between two intersections. In this paper, we do not discuss how to select the optimal path. $\mathbf{S}_1$ stores several path candidates between two intersections. For example, two candidate paths from I_1

to I_7 are stored in $\mathbf{S}_1$ in Fig. 2. During query, the platform will select an optimal path from these candidates based on the current traffic conditions.

4.5 Encrypting Spatio-temporal Data over Virtual Road Network

By the proposed encryption scheme, the platform encrypts the data (i.e., the road data, the order data) before uploading to protect the passenger's privacy.

Encrypting Road Data: For a road R_i, the platform calls $\mathsf{Encrypt}_R^{\mathsf{mul}}(sk_i, len_i)$ to encrypt its length ($R_i.len$). Then, it sets the count of orders, the upper bound, the lower bound as NULL to finish the road initialization. Note that the road network encryption is only executed once, so it will not overburden the platform.

Encrypting Order Data: For a passenger's order $O_i(O_i, R_j, O_i.dis, O_i.price, O_i.time)$, the platform encrypts $O_i.dis$, $O_i.price$, and $O_i.time$ by R_j's secret key sk_j and uploads $(O_i, R_j, \mathsf{ct}_R^{\mathsf{mul}}(O_i.dis), \mathsf{ct}_R^{\mathsf{add}}(O_i.price), \mathsf{ct}_R^{\mathsf{add}}(O_i.time))$ to $\mathbf{S}_1$.

In our system, when a passenger submits a new order, the platform will encrypt this order and upload it to $\mathbf{S}_1$. After that, the platform updates the count of orders, the upper bound, and the lower bound for this road as well.

4.6 Executing Polynomials over Encrypted Spatio-temporal Data

Given a driver d at the point $Q(R_k, Q.dis)$ on the road R_k in which $Q.dis$ is the distance from Q to R_k's intersection I_{R_k}, the order $O_i(O_i, R_j, \mathsf{ct}_R^{\mathsf{mul}}(O_i.dis), \mathsf{ct}_R^{\mathsf{add}}(O_i.price), \mathsf{ct}_R^{\mathsf{add}}(O_i.time))$ is stored at $\mathbf{S}_1$, d decides the polynomial evaluation function as in Eq. (3):

$$Eval(Q, O_i) = \alpha \times O_i.price - \beta \times \sum_{R_i \in Path(Q, O_i)} cond_i \times R_i.len - \gamma \times O_i.time, \quad (3)$$

in which $\alpha, \beta, \gamma \in \mathbb{Z}_N$ is selected by the driver according to his preference, $Path(Q, O_i)$ is the path from Q to O_i, and $cond_i$ is R_i's traffic condition. Note that we use the above polynomial to explain our design but our method can support an arbitrary polynomial function. Based on the proposed virtual road network, we assume that the driver Q always goes to I_{R_k} first, and then goes to I_{R_j}, and picks up the passenger at O_i at last. As in Fig. 2, the path from the driver on $R_{18}(I_5)$ to O_1 on $R_4(I_1)$ is $driver \to I_5 \to I_1 \to O_1$.

Obviously, $\beta \times cond_{R_k} \times Q.dis$ are same for the evaluations over any order. So, the platform removes this part as Eq. (4) to simplify the computation and hide the driver's exact location. It will not affect the correctness of evaluation.

$$Eval(Q, O_i) = \alpha \times O_i.price - \beta \sum_{R_i \in Path(I_{R_k}, O_i)} cond_i \times R_i.len - \gamma \times O_i.time. \quad (4)$$

The driver first submits an original request $\mathcal{Q}(Q, \alpha, \beta, \gamma, k)$ to the platform. Then, the platform randomly picks two random numbers $r_1, r_2 \in \mathbb{Z}_N$ and generates a query trapdoor $\mathcal{Q}$ as Eq. (5) for all the orders on the road R_j with the secret key $sk_j(k_j, \pi_j)$. The platform periodically updates the traffic condition for every road from a traffic condition provider.

$$\mathcal{Q} = (\mathsf{ct_L}(r_1 \times \beta \times cond_i^j), \mathsf{ct_L}(r_2), r_1 \times \alpha, r_1 \times \gamma). \tag{5}$$

Before detailing the top-k polynomial evaluation method, we explain how $\mathbf{S}_1$ decides the optimal path from I_{R_k} to O_i based on the current traffic condition. Given an order O_i on R_j, $C = \{C_1, \ldots, C_w\}$ in which $C_i = \{R_i^1, R_i^2, \ldots\}$ is the ith candidate path from I_{R_k} to O_i and R_i^j is the jth road in C_i, $\mathbf{S}_1$ decides the optimal path with Algorithm 1, in which $sk_i^j, cond_i^j$ represent R_i^j's secret key and the current traffic condition value.

Algorithm 1. EvalDistance$(r_1, I_{R_k}, O_i) \to \mathsf{E}(r_1 \times \beta \times \sum cond_i \times R_i.len))$

Input: I_{R_k}, $O_i(O_i, R_j, \mathsf{ct_R^{mul}}(O_i.dis), \mathsf{ct_R^{add}}(O_i.price), \mathsf{ct_R^{add}}(O_i.time)), r_1 \in \mathbb{Z}_N$
Output: $\mathsf{E}(r_1 \times \beta \times \sum cond_i \times R_i.len)$

1: Platform:
2: **for** $i = 1; i \le w; i{+}{+}$ **do**
3:     **for** $j = 1; j \le |C_i|; j{+}{+}$ **do**
4:         $\mathsf{Encrypt_L}(sk_i^j, r_1 \times \beta \times cond_i^j) \to \mathsf{ct_L}(r_1 \times \beta \times cond_i^j)$;
5:     **end for**
6: **end for**
7: Send $\mathsf{ct_L}(r_1\beta cond_j)$, $\mathsf{ct_L}(r_1\beta cond_i^j), r_1\alpha, \mathsf{ct_L}(r_2)$ to $\mathbf{S}_1$.
8: $\mathbf{S}_1$:
9: **for** $i = 1; i \le w; i{+}{+}$ **do**
10:     **for** $j = 1; j \le |C_i|; j{+}{+}$ **do**
11:         $\mathsf{Multiplication}(\mathsf{ct_L}(r_1\beta cond_i^j), \mathsf{ct_R^{mul}}(R_i^j.len)) \to \mathsf{E}(r_1 \times \beta \times cond_i^j \times R_i^j.len)$
12:     **end for**
13:     $\mathsf{E}(C_i) = \prod \mathsf{E}(r_1 \times \beta \times cond_i^j \times R_i^j.len) = \mathsf{E}(r_1 \times \beta \times \sum cond_i^j \times R_i^j.len)$
14: **end for**
15: $\mathsf{E}(C_i) = \mathsf{E}(C_i) \times \mathsf{Multiplication}(\mathsf{ct_L}(r_1\beta cond_j), \mathsf{ct_R^{mul}}(O_j.dis))$
16: **return min**$(\mathsf{E}(r_1 \times \beta \times \sum cond_i \times R_i.len))$ by $\mathsf{Compare}(\mathsf{E}(m), \mathsf{E}(n))$.

By Algorithm 1, the cloud server is able to evaluate the spatial factor in the polynomial function i.e., the second part in Eq. (4), then $\mathbf{S}_1$ evaluates the temporal and other factors, i.e., the first and third parts in Eq. (4), to finish the evaluation by Algorithm 2.

Algorithm 2. Eval$(\mathcal{T}_{\mathcal{Q}}, O_i) \to Eval(Q, O_i)$

Input: $\mathcal{Q} = (\mathsf{ct_L}(r_1 \times \beta \times cond_i^j), \mathsf{ct_L}(r_2), r_1 \times \alpha, r_1 \times \gamma), O_i$
Output: $Eval(Q, O_i)$, which is the result of Eq. (4)

1: $\mathsf{Addition}(\mathsf{ct_L}(r_2), \mathsf{ct_R^{add}}(O_i.price)) \to \mathsf{E}(r_2 + O_i.price)$
2: $\mathsf{Addition}(\mathsf{ct_L}(r_2), \mathsf{ct_R^{add}}(O_i.time)) \to \mathsf{E}(r_2 + O_i.time)$
3: $result = \mathsf{E}(r_2 + O_i.price)^{r_1\alpha} \times \underbrace{\mathsf{E}(r_1\beta \sum cond_i \times R_i.len)}_{\text{output of Algorithm 1}}^{N-1} \times \mathsf{E}(r_2 + $

    $O_i.time)^{N-r_1\gamma}$
4: **return** result

Proof (correctness of evaluation). For a query Q and any two passengers' orders O and O', if we have $\mathsf{D}(Eval(\mathcal{T}_Q), O) > \mathsf{D}(Eval(\mathcal{T}_Q), O')$ which are the outputs of Algorithm 2, we must have $Eval(Q, O) > Eval(Q, O')$ by Eq. (4). Based on the properties of Paillier cryptosystem, we have:

$$result = \mathsf{E}(r_1\alpha(r_2 + O.price) - r_1\beta \sum cond_i \times R_i.len - r_1\gamma \times (r_2 + O.time))$$
$$= \mathsf{E}(r_1 \times Eval(Q, O) + r_1 \times r_2 \times (\alpha - \gamma)).$$

$r_1 r_2 (\alpha - \gamma)$ is same for all orders, so the correctness of evaluation always holds.

By Algorithm 2, $\mathbf{S}_1$ can execute the polynomials over the encrypted spatio-temporal data. However, it has to execute the evaluations over all the orders. Obviously, it is inefficient. We improve the proposed method based on the designed virtual road network. Given a driver at $Q(R_k, Q.dis)$ asks the platform to execute top-k polynomial evaluation for him, the platform first selects several roads near to I_{R_k} as the road candidates for this query. The sum of the orders on the candidate roads is larger than k (we let the sum is larger than $3k$). In Fig. 3, the platform selects 7 roads in red as the candidates for a top-10 query.

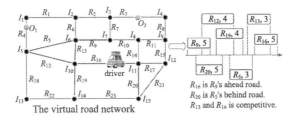

Fig. 3. Top-k ($\mathrm{k} = 10$) polynomial evaluation over the virtual road network

After the platform generates the trapdoor $\mathcal{T}_Q$ and sends it to $\mathbf{S}_1$. $\mathbf{S}_1$ calls $\mathsf{Eval}(\mathcal{T}_Q, O_i)$ to evaluate all the upper bounds and lower bounds of the road candidates and ranks them by the $\mathsf{Compare}(\mathsf{E}(m), \mathsf{E}(n))$ algorithm. Intuitively, the evaluation values of the orders on a road are between its upper bound and lower bound. We divide the relationship of two roads R_1 and R_2 into three categories: if R_1's upper bound is less than R_2's lower bound, we say R_1 is R_2's 'behind road' and R_2 is R_1's 'ahead road'; if R_1's range overlaps R_2's range, we say they are competitive. Then, $\mathbf{S}_1$ executes the 1st round top-k query as:

1. $\mathbf{S}_1$ removes all the roads from the candidates, if the sum of orders on their ahead roads is larger than k. The roads R_9, R_{20} are removed, because the sum of orders on R_5, R_{13}, R_{16} is larger than k.

2. If the sum of the orders on R_i, its competitive roads, and its ahead roads is less than k, put all the orders on R_i into **result** without evaluating. For example, 5 reservation orders on R_{16} is put into **result** without evaluation.
3. Assume there are a orders in **result** after Step 2. $\mathbf{S}_1$ evaluates other orders in the candidates and ranks them by Compare algorithm. $\mathbf{S}_1$ selects top-$(k-a)$ orders into **result** and chooses the minimum one as the 'symbol order'.

$\mathbf{S}_1$ sends the 'symbol order' O_s to the platform which decrypts and calculates O_s's evaluation value. Then, $\mathbf{S}_1$ executes the 2nd round top-k query as:

1. $\mathbf{S}_1$ evaluates the upper bound of the remaining roads and removes those the upper bound evaluation values by Algorithm 2 of which are less than O_s's.
2. $\mathbf{S}_1$ combines the remaining roads and the 1st round query remaining road candidates as the final road candidates.

Finally, $\mathbf{S}_1$ re-executes the 1st round query over the final candidates and returns the final **result** to the platform which decrypts and returns it to the driver to finish the top-k polynomial evaluation.

5   Security and Performance Analysis

5.1   Security Analysis

Theorem 1 (security of our scheme). *Our scheme Π is secure with the best possible leakage function $\mathcal{L}(\cdot)$ under the random oracle model.*

Proof. Let Π be the encryption scheme defined in Sect. 4.3, $\mathcal{A} = (\mathcal{A}_1, \ldots, \mathcal{A}_q)$ be an adversary for $q = polynomial(\lambda)$, $\mathcal{S} = (\mathcal{S}_0, \ldots, \mathcal{S}_q)$ be a simulator, $\mathcal{L}(\cdot)$ be a leakage function. We say that Π is secure with $\mathcal{L}(\cdot)$ if for all $\mathcal{A} = (\mathcal{A}_1, \ldots, \mathcal{A}_q)$, there exists a simulator $\mathcal{S} = (\mathcal{S}_0, \ldots, \mathcal{S}_q)$ such that the outputs of the experiments $\mathsf{REAL}_{\mathcal{A}}(\lambda)$ and $\mathsf{SIM}_{\mathcal{A}, \mathcal{S}, \mathcal{L}}(\lambda)$ are computationally indistinguishable.

Let $m \in [N]$ be a message. We prove that the adversary's view in the experiment is independent of $f(\pi(m))$ with a PRF f. Consider $\mathsf{ct}(m') = \mathsf{ct}_R^{add}(m')$ or $\mathsf{ct}_R^{mul}(m')$ $\mathcal{A}$ obtains when it requests some encrypted messages $m' \neq m$. For $\mathsf{ct}_R^{add}(m') = (r', v'_1, \ldots, v'_N)$, r' is distributed independently of $f(\pi(m))$, so $\forall i \in [N]$, $v'_i = \mathsf{ADD}(\pi^{-1}(i), m') + H(f(i), r')$ is also independent of $H(f(i), r')$ with the collision resistant hash function H. So, the ciphertext of the order $O_i(\mathsf{ct}_R^{mul}(O_i.dis), \mathsf{ct}_R^{add}(O_i.price), \mathsf{ct}_R^{add}(O_i.time))$ also is distributed independently of $f(\pi(m))$. We let $q_1, \ldots, q_z, z = polynomial(\lambda)$ be the adversary's queries on the random oracle model before $\mathcal{A}$ requests for an encryption of m. The probability that $q_i = \mathsf{ADD}(f(\pi(m)), m)$ is $z/2^\lambda = negl(\lambda)$ which is negligible. Any adversary can not decide which experiments ($\mathsf{REAL}_{\mathcal{A}}(\lambda)$ and $\mathsf{SIM}_{\mathcal{A}, \mathcal{S}, \mathcal{L}}(\lambda)$) he is in with a non-negligible probability. We have proved Theorem 1, so we conclude that the proposed encryption scheme Π is secure under the random oracle model.

5.2 Performance Analysis

Computing Cost. The computing cost of generating a left ciphertext is $T_F + T_\pi$ in which T_F, T_π represent the operations of one pseudorandom function and one uniform random permutation. For the platform, the main computing cost is to generate the trapdoor T_Q. So, the computing cost at the platform is $(L+3)T_{mul} + (L+1)(T_F + T_\pi)$ in which L is the count of the roads involved in the query, T_{mul} represents one multiplication operation over $\mathbb{Z}_N$.

The main computing costs at $\mathbf{S}_1$ include the costs of evaluating the boundary values of road candidates and evaluating all the necessary orders. The computing cost of evaluating one order by Algorithm 1 and 2 is $(L + 2)(T_{Hash} + T_{add2}) + (L + 1)T_{mul2}$ in which T_{Hash}, T_{add2}, and T_{mul2} represent one Hash operation, one addition operation and one multiplication operation over $\mathbb{Z}_{N^2}$.

Overheads of Storage and Communication. The size of the query trapdoor T_Q is $4 \times S_{int}$ in which S_{int} represents the size of the element in $\mathbb{Z}_N$. So, the total communication cost for a query is $4L \times S_{int}$.

By the proposed virtual road network, $\mathbf{S}_1$ stores the roads, the orders and the paths. The total storage overheads at $\mathbf{S}_1$ are $C_{road}(2S_{id} + 8S_{int} + 7NS_{int2}) + C_{order}(2S_{id} + 3S_{int} + 3NS_{int2}) + C_{path} \times (2 \times l_{road} \times S_{id})$ in which C_{road}, C_{order}, C_{path} represent the counts of the roads, the orders and the paths, l_{road} represents the average count of roads in a path, $S_{id}, S_{int}, S_{int2}$ represent the size of the identifier, an element in $\mathbb{Z}_N$, and an element in $\mathbb{Z}_{N^2}$, N represents the size of the message space. And the total storage overheads at the platform are $C_{road}(2S_{id} + 3S_{int} + S_\pi)$ in which S_π is the size of the random permutation.

6 Experimental Study

6.1 Experimental Setup

We utilize jPaillier to implement the proposed encryption scheme and use the real-world data set (https://uofi.app.box.com/NYCtaxidata) of NYC taxi rides in New York with records which contains pick up location, destination, and time as the experimental data. The default size of secret key $\mathbf{SK}$ is 1024 bits. We run the experiments on a machine with a 3.6 GHz processor and 16 GB memory. There is not a solution of privacy-preserving polynomial evaluation over spatio-temporal data in the existing literatures, so we choose the related studies [9,10,20–22] as the comparison basis. We randomly pick three numbers from 1 to 10 as the parameters α, β, and γ in Eq. (4) as the evaluation function. We also pick a random number in [1,6] as the traffic condition for every road. For every experiment, we randomly pick 100 points in the road network as the driver's location to launch the queries.

6.2 Encrypting Costs

We measure the time of encrypting data at the platform with the different scales of passenger orders from 10,000 to 100,000. The platform can build the index structure before encrypting so we only measure the time costs of encrypting processes. The experimental results are shown in Fig. 4(a). The time costs of encrypting data in our scheme range from 2.3 to 10.5 s. Overall, the encrypting costs of our scheme are comparable to others. That clearly shows that the platform in our scheme is able to efficiently encrypt the new order data.

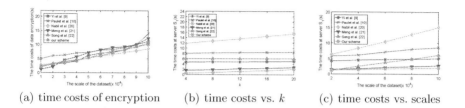

(a) time costs of encryption   (b) time costs vs. k (c) time costs vs. scales

Fig. 4. The time costs of encrypting data, executing the query on server side

6.3 Time Costs of Executing Polynomial Evaluation

To evaluate the efficiency of our scheme, we carry out the experiments to respectively measure the time costs at the server side and the platform side.

Server Side, since most of the computing operations are executed at S_1, we only measure the time costs of executing polynomial evaluation at S_1. The results are illustrated in Fig. 4(b) and Fig. 4(c) respectively. In Fig. 4(b), order scale is 100,000 and our scheme has the fewer time costs than others. In Fig. 4(c), the k is 20 and the time cost of our scheme keeps a minor increase with the scale of orders increasing because our scheme doesn't need the computations on all data comparing to [10, 20, 22]. In general, our solution has the smallest time overhead and maintains a small increase in the face of large data volumes. Moreover, our scheme supports the privacy-preserving polynomial evaluation with dynamic conditions over the spatio-temporal data.

We also evaluated the impact of different key size on query efficiency. As shown in Fig. 5(a), with the key size **SK** increasing, the query time of the method increases exponentially, and our solution is still optimal.

Platform Side, We carry out the similar experiments to measure the time costs at the platform. As the experimental results shown in Fig. 5(b) and 5(c), the time costs at the platform side are less than others. Based on the proposed virtual road network index, most of computation of the evaluation are delegated to cloud (S_1), the computing overheads at the platform is very little. So, our scheme can make full use of the advantage of the cloud outsourced service model.

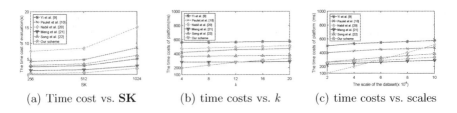

(a) Time cost vs. **SK**   (b) time costs vs. k (c) time costs vs. scales

Fig. 5. The time costs of evaluation with varying **SK** and at the platform

6.4 Overheads of Communication

To evaluate the communication costs of a query between cloud and platform, we measure the communication costs with the varying k values(data scale is 100,000) and data scales(k is 20). The results shown in Fig. 6(a) and Fig. 6(b) respectively. Because just our scheme and [21] utilize the model of two clouds, we also evaluate the communication costs between the clouds. As the experimental results shown, the communication costs of our scheme are heavily influenced by k, but values are small and still will not affect the efficiency of our scheme.

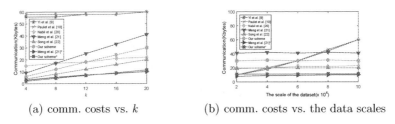

(a) comm. costs vs. k (b) comm. costs vs. the data scales

Fig. 6. The communication costs. Note that the first six figures in the legend are the communication costs between server and platform and the last two ones with * illustrate the communication costs between two cloud servers, their units are Mbytes.

7   Conclusion

In this paper, we examined an important problem of privacy-preserving polynomial evaluation over the spatio-temporal data in road networks with dynamic conditions. To address this challenging problem, we proposed an encryption scheme which combines the properties of the homomorphic encryption and the order-revealing encryption. The homomorphic property enables the cloud server to execute polynomials over the encrypted spatio-temporal data. And the property of order-revealing encryption enables the cloud server to securely compare the evaluation values. As a future work, we will extend our research to other complex operations over the encrypted spatio-temporal data in the outsourced database paradigm.

Acknowledgements. This work is partially supported by National Key Research and Development Project of China Nos. 2020YFC1522602, 2020AAA0107700, National Natural Science Foundation of China Nos. 62072349, U1811263, 61572378, 61822207, U20B2049, Technological Innovation Major Program of Hubei Province No. 2019AAA072, JSPS KAKENHI No.19K20269, and CCF-Tencent Open Fund WeBank Special Fund.

References

1. Wong, W.K., Cheung, D.W., Kao, B., Mamoulis, N.: Secure kNN computation on encrypted databases. In: SIGMOD, pp. 139–152 (2009)
2. Yao, B., Li, F., Xiao, X.: Secure nearest neighbor revisited. In: ICDE, pp. 733–744 (2013)
3. Choi, S., Ghinita, G., Lim, H.S., Bertino, E.: Secure kNN query processing in untrusted cloud environments. TKDE **26**(11), 2818–2831 (2014)
4. Cui, N., Yang, X., et al.: SVkNN: efficient secure and verifiable k-nearest neighbor query on the cloud platform. In: ICDE, pp. 253–264 (2020)
5. Lei, X., Liu, A.X., Li, R., Tu, G.-H.: SecEQP: a secure and efficient scheme for SkNN query problem over encrypted geodata on cloud. In: ICDE (2019)
6. Rodrigo, A., Dayarathna, M., Jayasena, S.: Latency-aware secure elastic stream processing with homomorphic encryption. Data Sci. Eng. **4**(3), 223–239 (2019). https://doi.org/10.1007/s41019-019-00100-5
7. Elmehdwi, Y., Samanthula, B.K., Jiang, W.: Secure k-nearest neighbor query over encrypted data in outsourced environments. In: ICDE, pp. 664–675 (2014)
8. Palanisamy, B., Liu, L.: MobiMix: protecting location privacy with mix-zones over road networks. In: ICDE, pp. 494–505 (2011)
9. Yi, X., Paulet, R., Bertino, E., Varadharajan, V.: Practical approximate k nearest neighbor queries with location and query privacy. TKDE **28**(6), 1546–1559 (2016)
10. Paulet, R., Kaosar, M.G., Yi, X., Bertino, E.: Practical approximate k nearest neighbor queries with location and query privacy. TKDE **26**(5), 1200–1210 (2014)
11. Yang, S., Tang, S., Zhang, X.: Privacy-preserving k nearest neighbor query with authentication on road networks. JPDC **134**, 25–36 (2019)
12. Zeng, M., Zhang, K., Chen, J., Qian, H.: P3GQ: a practical privacy-preserving generic location-based services query scheme. PMC **51**, 56–72 (2018)
13. Pham, A., Dacosta, I., et al.: PrivateRide: a privacy-enhanced ride-hailing service. Priv. Enhancing Technol. **2017**(2), 38–56 (2017)
14. Pham, A., Dacosta, I., et al. ORide: a privacy-preserving yet accountable ride-hailing service. In: USENIX Security, pp. 1235–1252 (2017)
15. Wang, F., Zhu, H., et al.: Efficient and privacy-preserving dynamic spatial query scheme for ride-hailing services. IEEE Trans. Veh. Technol. **67**(11), 11084–11097 (2018)
16. Sherif, A., Rabieh, K., et al.: Privacy-preserving ride sharing scheme for autonomous vehicles in big data era. IEEE Internet Things J. **4**(2), 611–618 (2016)
17. Li, M., Zhu, L., Lin, X.: Efficient and privacy-preserving carpooling using blockchain-assisted vehicular fog computing. IEEE Internet Things J. **6**(3), 4573–4584 (2018)
18. Song, W., Wang, B., Wang, Q., Shi, C., Lou, W., Peng, Z.: Publicly verifiable computation of polynomials over outsourced data with multiple sources. TIFS **12**(10), 2334–2347 (2017)

19. Xu, Y., Tong, Y., Shi, Y., Tao, Q., Xu, K., Li, W.: An efficient insertion operator in dynamic ridesharing services. In: TKDE (2020)
20. Nabil, M., Sherif, A., et al.: Efficient and privacy-preserving ridesharing organization for transferable and non-transferable services. TDSC **PP**, 1 (2019)
21. Meng, X., Zhu, H., Kollios, G.: Top-k query processing on encrypted databases with strong security guarantees. In: ICDE, pp. 353–364 (2018)
22. Song, W., Shi, C., Shen, Y., Peng, Z.: Select the best for me: privacy-preserving polynomial evaluation algorithm over road network. In: Li, G., Yang, J., Gama, J., Natwichai, J., Tong, Y. (eds.) DASFAA 2019. LNCS, vol. 11447, pp. 281–297. Springer, Cham (2019). https://doi.org/10.1007/978-3-030-18579-4_17
23. Lewi, K., Wu. D.J.: Order-revealing encryption: new constructions, applications, and bounds. In: CCS, pp. 1167–1178 (2016)
24. Goldreich, O., Goldwasser, S., Micali, S.: How to construct random functions. J. ACM **33**(4), 792–807 (1986)
25. Samanthala, B.K., Chun, H., Jiang, W.: An efficient and probabilistic secure bit-decomposition. In: AsiaCCS, pp. 541–546 (2013)

Exploiting Multi-source Data for Adversarial Driving Style Representation Learning

Zhidan Liu[1(✉)], Junhong Zheng[1], Zengyang Gong[2], Haodi Zhang[1], and Kaishun Wu[1]

[1] Shenzhen University, Shenzhen, China
{liuzhidan,hdzhang,wu}@szu.edu.cn,
zhengjun4hong2019@email.szu.edu.cn
[2] Hong Kong University of Science and Technology,
Clear Water Bay, Hong Kong
zgongae@cse.ust.hk

Abstract. Characterizing human driver's driving behaviors from GPS trajectories is an important yet challenging trajectory mining task. Previous works heavily rely on high-quality GPS data to learn such driving style representations through deep neural networks. However, they have overlooked the driving contexts that greatly govern drivers' driving activities and the data sparsity issue of practical GPS trajectories collected at a low-sampling rate. To address the limitations of existing works, we present an adversarial driving style representation learning approach, named `Radar`. In addition to summarizing statistic features from raw GPS data, `Radar` also extracts contextual features from three aspects of road condition, geographic semantic, and traffic condition. We further exploit the advanced semi-supervised generative adversarial networks to construct our learning model. By jointly considering statistic features and contextual features, the trained model is able to efficiently learn driving style representations even from sparse trajectories. Experiments on two benchmark applications, *i.e.*, driver number estimation and driver identification, with a large real-world GPS trajectory dataset demonstrate that `Radar` can outperform the state-of-the-art approaches by learning more effective and accurate driving style representations.

Keywords: GPS trajectory · Multi-source data · Driving style representation · Generative adversarial networks

1 Introduction

The advances of GPS and wireless communication techniques have enhanced the ability of various systems in collecting the spatio-temporal vehicular trajectories. The massive GPS trajectories stimulate a number of trajectory mining tasks for better understanding human mobility patterns and behaviors [29], among which

C. S. Jensen et al. (Eds.): DASFAA 2021, LNCS 12681, pp. 491–508, 2021.
https://doi.org/10.1007/978-3-030-73194-6_33

characterizing human driver's driving behaviors is an important yet challenging task. Similar as the bio-metrics, it is believed that each driver also has a distinguishable pattern of driving, which is referred as *driving style* [13]. Specifically, driving style reflects a driver's fine-grained behavioral habits of steering and speed control and their temporal combinations [2]. Learning drivers' driving style representations from their trajectories can benefit many intelligent applications, *e.g.*, driving assessment and assistance [25], driver-vehicle interaction [13], autonomous driving [12], and *etc.* In addition, auto insurance companies have been interested in utilizing the driving style information for risk assessments and personalized insurance pricing [9].

In the literature, some valuable efforts have been made to derive the driving style representations. Traditional approaches heavily rely on the data collected from automobile sensors (*e.g.*, controller area network buses) [6] or dedicated sensors (*e.g.*, high-definition cameras) [8] for driving style learning. However, it is difficult to retrieve data from automobile sensors while dedicated devices will incur installation costs. Recent studies [2,10,30] turn to leverage deep learning models to process GPS trajectories for learning the driving style representations. Compared to automobile and dedicated sensors, GPS sensor data are often easier to access and thus are more popular in large-scale study [2,29]. These works, however, require high-frequency rate of GPS data collections, which may be prohibited due to privacy and energy consumption [15]. Furthermore, these works merely focus on feature extractions from GPS data, but have overlooked the instant driving context information, such as road conditions and traffic conditions. As a result, they are inadequate to acquire accurate driving style representations.

Despite these research efforts, it is still non-trivial to efficiently learn driving style representation from GPS trajectories, mainly due to following challenges. First, practical GPS data are usually collected at a low-sampling rate, *e.g.*, 1 sample per 30 s [15], and are probably sparse, *i.e.*, there may be insufficient qualified data to train a deep learning model [10]. Second, driving is a complex activity and the resultant driving style is influenced by many factors. The GPS trajectory data themselves cannot capture the complete view of a driver's driving style, and hence the external context information should be taken into account. However, how to properly integrate the features from GPS data and context information into one model needs to be well designed and thus is challenging.

In this paper, we present an adve*r*sarial *d*riving style represent*a*tion lea*r*ning approach, named Radar, which extracts comprehensive features from multi-source data and builds a semi-supervised generative adversarial networks (SGAN) based model to learn driving style representations from these extracted features. To better describe a driver's driving behaviors, Radar not only transforms raw GPS trajectory data to fine-grained statistic features about driver's habits of steering and speed control, but also additionally considers each GPS trajectory's contextual features, which are captured by three aspects of road condition, geographic semantic, and traffic condition. In particular, different from the specific GPS locations, geographic semantic encodes high-level geographic features of a trajectory by mapping it to the whole city area. These driving contexts greatly govern a driver's driving activity, and thus are important complements for learning driving styles. To tackle data sparsity issue, `Radar` makes use of SGAN to construct the learning model, which equally treats statistic features and contextual features as the input to learn the driving style representations. Our learning model consists of three different components: generator, discriminator, and classifier, which work together to not only classify drivers from inputted trajectories but also generate fake samples close to the training data. As a result, `Radar`' learning model can achieve better generalization ability through both data augmentation and the competition between generator and discriminator.

In summary, the contributions of our work are as follows:

- To the best of our knowledge, we are the first to consider the problem of context-aware driving style representation learning from sparse trajectories, which improves existing works by considering the driving contexts.
- We propose an adversarial driving style representation learning approach – `Radar`, which exploits multi-source data and a SGAN based learning model to efficiently learn driving style representations from practical trajectories.
- We conduct extensive experiments with two benchmark applications, namely *driver number estimation* and *driver identification*, using a large real-world trajectory dataset. Experimental results demonstrate `Radar` outperforms state-of-the-art approaches, *e.g.*, on average improving the accuracy of driver number estimation and driver identification by 9.6% and 5.6%, respectively.

The remainder of the paper is organized as follows. We review related works in Sect. 2. The problem statement is presented in Sect. 3. We elaborate and evaluate our proposed approach in Sect. 4 and Sect. 5, respectively. Finally, Sect. 6 concludes this paper.

2 Related Work

The related works can be grouped into two categories: driving behavior analysis and trajectory mining. We review and discuss these works as follows.

Driving Behavior Analysis. Extensive studies have been conducted on the driving behavior analysis. Previous works primarily rely on the data collected from automobile sensors, *e.g.*, on-board diagnostic systems [9], controller area network buses [6], and digital cameras [8], to analyze drivers' driving behaviors. For example, Ezzini *et al.* utilize the measurements taken from various in-vehicle sensors to realize driver identification and fingerprinting [3]. However, it is relatively difficult to collect data from these automobile sensors, while dedicated devices like cameras bring installation costs. These constraints greatly limit their usability. Some recent works [1] resort to collect driving data using the internal sensors in smartphones and analyze the sensing data for monitoring drivers' behaviors. These works concern about driving safety, rather than driving styles.

Compared to automobile sensors, GPS sensor data are much easier to collect and GPS trajectory based driving behavior analysis has attracted many research efforts [2,10,25,27] in recent years. For example, Yang *et al.* analyze GPS traces of peer vehicles to proactively alter drivers of the vehicles with dangerous behaviors nearby [27]. By jointly modeling the peer and temporal dependencies of driving trajectories, Wang *et al.* enable the applications of driving score prediction and risk area detection [25]. The two works, however, mainly concern the identification of dangerous driving behaviors, rather than capturing a driver's latent driving styles. Instead, Dong *et al.* propose an autoencoder regularized deep neural network and a trip encoding framework to learn drivers' driving styles directly from GPS trajectories [2]. Tung *et al.* propose a trajectory-to-image representation framework that encodes both geographic features and driving behaviors of trajectories into multi-channel images [10]. Although the two works can achieve remarkable performances, they are still not sufficiently efficient and practical. First, they require high-quality GPS trajectories that are collected at a high-sampling rate such 1 Hz, while most practical GPS trajectories are collected at a low frequency, *e.g.*, 1 sample per 30 s, due to concerns of energy consumption and privacy [15]. Second, they merely extract features from GPS data while overlooking the driving contexts, within which a trajectory has been generated. Our approach overcomes these limitations by utilizing multi-source data to fully describe driving behaviors and the advanced SGAN modeling, and thus can learn more effective and accurate driving style representations.

Trajectory Mining. The wide availability of GPS trajectories has inspired a wide range of applications [29], *e.g.*, urban traffic estimation [17] and prediction [16], personalized recommender systems [28], and ridesharing [14]. To enable such applications, various trajectory mining tasks have been widely studied, *e.g.*, trajectory pattern mining [11], trajectory-user linking [30], and *etc.*. In particular, trajectory-user linking, which links trajectories to users who produce them, is quite relevant to our work. Existing works on this problem mainly analyze mobility trajectories by exploiting various deep learning models to learn the semantic trajectory representations [4,19,22,30]. For example, Feng *et al.* present a deep learning framework to link heterogeneous mobility data, which are collected from different online services, to the users [4]. Ren *et al.* build a spatio-temporal Siamese network model to predict whether an income set of trajectories belong to a certain agent based on historical trajectory data [22]. In addition, Miao *et al.* utilize recurrent networks with attention mechanism to solve the trajectory-user linking problem [19]. Different from these works, we aim to learn drivers' driving style representations from practical GPS trajectories, which involves more complex human behaviors and thus is more challenging.

3   Problem Statement

3.1   Definitions and Notations

The GPS trajectory data are collected when a set of drivers $\mathbb{U} = \{u_1, \cdots, u_{|\mathbb{U}|}\}$ drive their vehicles, which have been equipped with GPS sensors, on a road network. The GPS trajectory set $\mathbb{T}_{u_i}$ generated by driver u_i implicitly encodes u_i's driving style. Accurately learning the driving style representation can benefit many potential applications, such as driving assessment and assistance [25], driver-vehicle interaction [13], autonomous driving [12], and so on.

Definition 1 (GPS trajectory). *Let $\mathcal{T}_j^i \in \mathbb{T}_{u_i}$ denotes the j-th trajectory generated by driver u_i. Specifically, $\mathcal{T}_j^i = \{g_1, \cdots, g_{|\mathcal{T}_j^i|}\}$ is a time-ordered sequence of GPS records, where each record is denoted as a tuple $< ts, lat, lng, v, dir >$, indicating that u_i's vehicle located at latitude lat and longitude lng at time ts, with instant travel speed v in direction dir.*

Due to GPS localization errors, we have to map raw GPS locations to their actual locations on the roads through map matching techniques [20]. Therefore, a trajectory $\mathcal{T}_j{}^1$ could be mapped to a travel route $\mathcal{R}_j$ on the road network $\mathbb{G}$.

Definition 2 (Road network). *A road network is modelled as graph $\mathbb{G} = \{\mathbb{V}, \mathbb{E}\}$, where $\mathbb{V}$ represents the set of road intersections and $\mathbb{E}$ represents the set of road segments in a city. In addition, each road segment has following attributes: ID of road segment, road type, number of lanes, and one-way indicator.*

Definition 3 (Travel route). *The travel route $\mathcal{R}_j$ for a GPS trajectory $\mathcal{T}_j$ is denoted by a sequence of road segments, i.e., $\mathcal{R}_j = \{e_1, e_2, \cdots, e_{|\mathcal{R}_j|}\}$, on road network $\mathbb{G}$, where $e_i \in \mathbb{E}$ is a road segment in route $\mathcal{R}_j$ and $|\mathcal{R}_j|$ is the number of all traveled road segments. Note that end point of e_i is the start point of e_{i+1}.*

3.2   Problem Statement

Definition 4 (Context-aware driving style representation learning). *Given a set of GPS trajectories generated by drivers in $\mathbb{U}$, we aim to learn driving style representations for drivers in $\mathbb{U}$ by exploiting necessary context information, so as to support applications like driver number estimation and driver identification.*

Different from previous works [2,10] that heavily rely on high-quality GPS trajectories, we should devise an approach that works well for practical trajectories and incorporates contextual information for much better driving style representation learning. To that end, we have to address the following challenges.

---

[1] We omit the upper-script if the context is clear.

(1) *Data sparsity.* This challenge is arised from two aspects. On one hand, in practice GPS data are usually collected at a low-sampling rate, *e.g.*, 0.1Hz. On the other hand, trajectories are of different lengths and may contain deficient driving behavior information, resulting in insufficient qualified trajectories. These factors will lead to low-quality data for training the deep learning models and thus impair their performances.

(2) *Balanced integration of features from GPS data and contextual information.* Although driving contexts would benefit driving style representation learning, how to efficiently encode these contextual information and further gracefully integrate features extracted from raw GPS data and driving contexts should be wisely designed. The driving contexts involve various information, and the resultant feature vectors may be of different dimensions.

4   The Design

4.1   Overview

Figure 1 illustrates the architecture of our approach, which consists of three major modules: *GPS data transformation, driving context representation*, and *learning model*. At high-level, `Radar` takes raw GPS trajectories and road map as the input, and exploits the modules of *GPS data transformation* and *driving context representation* to extract features from a GPS trajectory and the corresponding contexts. The integrated feature tensors are fed into *learning model* to compute driving style representations, which can support many trajectory mining applications, *e.g.*, driver number estimation and driver identification.

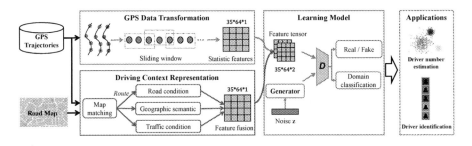

Fig. 1. The architecture of `Radar`.

Specifically, *GPS data transformation* utilizes a sliding window to calculate various statistics of the GPS data, which finally form the statistic feature matrix. For the *driving context representation* modules, it firstly applies map matching technique to transform each GPS trajectory to an actual travel route. With this route, `Radar` derives context information about road conditions, geographic semantic (*i.e.*, geographical distribution of the route over the whole city area), and traffic conditions. These context information are fused to form a contextual feature matrix. Lastly, both statistic feature matrix and contextual feature matrix are integrated as the input for the *learning model*. In particular, we adopt the emerging semi-supervised generative adversarial network architecture [21] for building our model to learn effective and accurate driving style representations.

4.2   GPS Data Transformation

Instead of inputting raw GPS data to deep learning models, we will transform each GPS trajectory into more stable statistic features. Similar as previous work [2], we divide a GPS trajectory into segments of a fixed length L_s, with a shift of $\frac{L_s}{2}$ to avoid much information loss between any two adjacent segments. We employ five basic features to capture the instantaneous vehicular movement features, namely *speed norm, difference of speed norm, acceleration norm, difference of acceleration norm*, and *angular speed*. To reduce the possible impact of outliers, we further divide a segment into frames of a fixed size L_f, with a shift $\frac{L_f}{2}$. For each frame, we calculate seven statistics for each basic feature, including mean, minimum, maximum, 25%, 50% and 75% quartiles, and standard deviation. For each trajectory $\mathcal{T}_j$ consisting of a sequence of time-ordered GPS records in the form of $g_i = < ts, lat, lng, v, dir >$, we can easily calculate speed statistics using travel speed v, acceleration statistics with location (lat, lng), and angular statistics with travel direction dir, respectively. As a result, we can derive a set of statistic feature matrices, each of which consists of $5 \times 7 = 35$ rows and $2 \times \lfloor \frac{L_s}{L_f} \rfloor$ columns. A statistic feature matrix encodes the driving behavior information of a trajectory segment, and serves as partial input to the learning model with its class label (*i.e.*, the driver identifier) as the original GPS trajectory $\mathcal{T}_j$.

In our implementation, we set $L_s = 195$ and $L_f = 6$ for the best performance. Therefore, we obtain a set of statistic feature matrices of size 35×64 for each trajectory. In particular, if a trajectory segment is shorter than L_s, we will pad zeros into the matrix, so as to unify the size of all statistic feature matrices. In principle, long trajectories contain more information about the driving behaviors, and thus are more preferable for the model training.

4.3   Driving Context Representation

Since driving activities will be implicitly governed by the surrounding driving environment, thus Radar also takes driving context information into consideration to let machines deeply "understand" drivers' behaviors especially under certain circumstances. In the design of Radar, we particularly consider the three contexts of road conditions, geographic semantic, and traffic conditions.

Figure 2 illustrates how Radar processes each raw GPS trajectory to generate the contextual features. For each GPS trajectory $\mathcal{T}_j$, we firstly recover the travel route $\mathcal{R}_j$ through map matching techniques [20]. Since GPS data transformation module outputs one statistic feature matrix for each trajectory segment, thus the driving context representation module operates on trajectory segment and

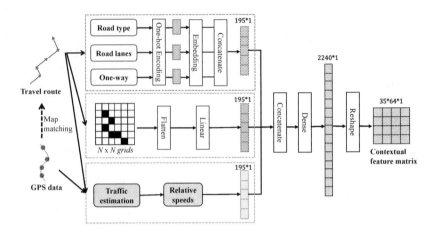

Fig. 2. The framework of driving context representation module.

its associated travel route segment as well, and accordingly produces one contextual feature matrix. Based on road network $\mathbb{G}$, the s-th trajectory segment $\mathcal{T}_{js}$ and its travel route segment $\mathcal{R}_{js}$, we derive each context representation as follows.

Road Condition. We utilize static road attributes of road type, number of lanes, and one-way indicator to characterize road conditions. Let n_t, n_ℓ, and n_o to represent the numbers of possible values in the three types of categorical attributes, we thus employ three attribute vectors of length n_t, n_ℓ, and n_o, respectively, to encode the attributes of each road segment, respectively. Specifically, one-hot encoding is adopted to generate the attribute vectors. Given a travel route $\mathcal{R}_{js}$, we derive road type vectors of road segments covered by $\mathcal{R}_{js}$, and sequentially connect them into one vector, which describes the road types a vehicle had traveled when generating trajectory $\mathcal{T}_{js}$. In addition, we adopt an embedding layer to reduce the dimensionality of the sparse attribute vector. Similarly, we apply the same operations to the attributes of road lanes and one-way, and derive their attribute vectors for route $\mathcal{R}_{js}$, respectively. Finally, we concatenate the three embedding vectors into one vector of size 195×1.

Geographic Semantic. The GPS data only reflect the instantaneous driving statuses, but not capture the high-level geographic semantic of a trajectory, e.g., origin, destination, and traveled regions. Thus, Radar maps each GPS trajectory segment to the whole city area to derive its geographic semantic representation, which is formally defined as follows.

Definition 5 (Geographic semantic representation). *We partition the city area into $N \times N$ grids. For each trajectory $\mathcal{T}_j$, we compute a geographic semantic* *representation matrix* $\mathbf{M}_j$, *where we set* $\mathbf{M}_j[a,b] = 1$ *if the travel route* $\mathcal{R}_j$ *of* $\mathcal{T}_j$ *intersects with the grid* $[a,b]$; *otherwise* $\mathbf{M}_j[a,b] = 0$.

As shown in Fig. 2, we further flatten the matrix $\mathbf{M}_j$ as a vector, which is fed into a linear layer for reducing the dimensionality. In our design, we set the final geographic semantic vector of size 195×1. It is worthy noting that we generate such a vector for each trajectory segment $\mathcal{T}_{js}$ as well.

Traffic Condition. In addition to road conditions, another factor that has great impact on driving activities is real-time traffic conditions. Considering both vehicle's instantaneous movements and surrounding traffic conditions could better define a driver's driving behaviors. Therefore, we use the *relative speed*, which is calculated as the ratio between vehicle's travel speed and average travel speed of the vehicle's locating road segment, to represent traffic condition context.

To that end, we make use of all available GPS data to estimate the real-time traffic conditions. For each road segment, its traffic condition can be approximated as the average travel speed of all vehicles passing by within a time slot Δt. Therefore, we classify all GPS records to road segments according to their map matching results. For a given road segment, we calculate its average travel speed of a specific time slot using the GPS records falling into that time slot. Due to data sparsity, we may not derive a complete traffic conditions of the whole road network $\mathbb{G}$ over all time slots. For simplicity, we directly apply temporal-spatial interpolations to infer the traffic conditions of uncovered road segments by leveraging the inherent traffic correlations among roads. In fact, some advanced traffic estimation methods [15] can be adopted to compute the complete real-time traffic conditions. Once we obtain the traffic conditions of all road segments, we calculate the relative speeds for the road segments covered by travel route $\mathcal{R}_{js}$ of a trajectory segment $\mathcal{T}_{js}$. These relative speeds then form $\mathcal{T}_{js}$'s traffic condition representation.

As shown in Fig. 2, when the three representations of driving contexts are ready, Radar concatenates them into one vector, which is then fed into a dense layer to derive a vector of size 2240×1. To be compatible with the statistic feature matrix, we reshape it into a contextual feature matrix of size $35 \times 64 \times 1$.

4.4 Learning Model

To tackle the poor data quality issue, we employ generative adversarial networks (GAN) [7] to construct the learning model. Essentially, GAN operates by training two neural networks that play a *min-max* game: a *discriminator* is trained to discriminate real samples from fake ones, while a *generator* tries to generate fake training data to fool the discriminator. Therefore, GAN is able to generate samples very similar to real trajectories for training data augmentation and as a result improves the generalization ability of the derived model.

In particular, we adopt the emerging semi-supervised GAN (*SGAN*) architecture [21] to build our learning model, which mainly consists of a generator $\mathcal{G}$ and a discriminator $\mathcal{D}$, as shown in Fig. 3. In SGAN, discriminator $\mathcal{D}$ can also

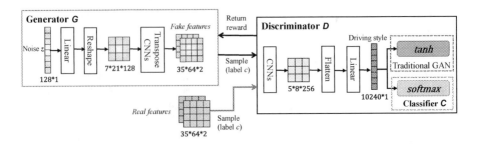

Fig. 3. The framework of learning model.

act as a classifier $\mathcal{C}$ to classify each input sample into one of the predefined $(k+1)$ classes, where k is the number of classes and the additional class label is added for a new "*fake*" class. The competition and interaction (via *reward*) between generator and discriminator will improve the quality of resultant driving style representations. Therefore, our model can not only classify drivers according to the learned driving styles, but also for a given class c generates corresponding fake driving style features, which are similar to training samples belonging to class c. To achieve this goal, the model training will involve both traditional unsupervised GAN task and supervised classification task simultaneously. Training in unsupervised mode allows our model to learn useful feature extraction capabilities from unlabeled samples, whereas training in supervised mode allows the model to use the extracted features and apply classifications.

Discriminator $\mathcal{D}$ (and classifier $\mathcal{C}$). As shown in Fig. 3, discriminator takes either real samples, generated from GPS data and context information, or fake samples, produced by generator $\mathcal{G}$, as the input, which are further processed by a neural network to derive driving style representations. Discriminator $\mathcal{D}$ is trained in both unsupervised mode and supervised mode.

- *Unsupervised mode.* In this mode, discriminator $\mathcal{D}$, with parameter θ_d, predicts whether a sample is true (sampled from real trajectory data) or fake (generated by the generator $\mathcal{G}$) by calculating the probability score $\mathcal{D}(x|\theta_d)$ that the sample x is true. We train our learning model like traditional GANs by maximizing the score for real samples and minimizing it for fake ones. We achieve this objective by minimizing $\mathcal{L}^{(\mathcal{D})}$, which is defined as follows.

$$\mathcal{L}^{(\mathcal{D})} = -[\mathbb{E}_{x \sim p_r(x)} \log \mathcal{D}(x|\theta_d) + \mathbb{E}_{x \sim \mathcal{G}} \log(1 - \mathcal{D}(x|\theta_d))], \tag{1}$$

where $p_r(x)$ represents the distribution of real samples from trajectory data.
- *Supervised mode.* In this mode, discriminator $\mathcal{D}$ acts as classifier $\mathcal{C}$ to complete a multi-class classification problem. For each sample, classifier $\mathcal{C}$, with parameter θ_c, predicts if the sample belongs to one of the predefined $(k+1)$ classes. Because the label of driving style features generated by the generator $\mathcal{G}$ is known, classifier $\mathcal{C}$ can also utilize the labels of fake samples for training. Thus the generalization ability of the model could be improved. In addition,

classifier $\mathcal{C}$'s classification on both real and fake samples can be used as feedback (via reward) to improve generator $\mathcal{G}$, i.e., higher classification accuracy will bring more returns. To train the classifier $\mathcal{C}$, we aim to minimize the classifier loss $\mathcal{L}^{(\mathcal{C})}$, i.e., the cross entropy loss on true labeled samples that is computed using the overall classifier score.

$$\mathcal{L}^{(\mathcal{C})} = -\mathbb{E}_{p(x_c, c)}[\log \mathcal{C}(c|x_c, \theta_c)], \tag{2}$$

where x is a sample of class c, and $\mathcal{C}$ should correctly classify it as class c.

We implement above two modes in one unified framework, as shown in Fig. 3. Discriminator $\mathcal{D}$ and classifier $\mathcal{C}$ share the same feature extraction layers, but have different output layers. Specifically, we use a stack of convolution layers with LeakyReLu to process each input sample. After a series of convolutions, we get a feature tensor that is flatten and inputted to a dense layer to derive the driving style representation vector. For traditional GAN task, the vector is fed into tanh to discriminate real samples and fake ones. For classifier $\mathcal{C}$, the vector is fed into softmax to obtain classification probabilities of the $(k + 1)$ classes.

Generator $\mathcal{G}$. Given the distribution $p_r(x)$ of real samples and k class labels from real training data, generator $\mathcal{G}$ aims to find the parameterized conditional distribution $\mathcal{G}(\mathbf{z}, c, \theta_g)$ that is close to the real distribution $p_r(x)$. The generated fake samples are conditioned on the network parameters θ_g, noise vector $\mathbf{z}$, and class label c, which are sampled from prior distribution p_z and p_c, respectively. Label c of a fake sample y can be known when the generator $\mathcal{G}$ generates y, so that the actual classification label of each generated sample is retained for training classifier $\mathcal{C}$. Following the feature matching technique proposed to addresses the instability of GANs [23], we train $\mathcal{G}$ by minimizing loss $\mathcal{L}^{(\mathcal{G})}$ expressed as:

$$\mathcal{L}^{(\mathcal{G})} = ||\mathbb{E}_{x \sim p_r(x)} \mathbf{f}(x, \theta_f) - \mathbb{E}_{\mathbf{z} \sim p_z} \mathbf{f}(\mathcal{G}(\mathbf{z}, c, \theta_g), \theta_f)||_2^2, \tag{3}$$

where $\mathbf{f}(\cdot)$ denotes activation on an intermediate layer (e.g., the stack of convolution layers) of discriminator $\mathcal{D}$, θ_f is the parameter subset of θ_d corresponding to the intermediate layer of discriminator $\mathcal{D}$, and c is the class label of real sample x. The objective of generator training is thus to minimize the discrepancy between the real and generated data distributions in feature space.

As shown in Fig. 3, generator $\mathcal{G}$ is implemented with four deconvolution layers, which transform noise vector $\mathbf{z}$ into fake driving style features. In particular, each deconvolution layer is followed by a nonlinear activation based on batch normalization and rectified linear unit (ReLU). $\mathbf{z}$ is a 128 dimensional vector sampled from a uniform distribution p_z, and it is processed by dense and reshape layers before inputting to the deconvolution layers. Finally, generator $\mathcal{G}$ outputs a $35 \times 64 \times 2$ feature tensor as the same size of real feature tensors. The values of tensor items are shape squashed within $[-1, 1]$ through tanh function.

5   Performance Evaluation

5.1   Experimental Setup

Dataset. We use a large real-wold anonymized GPS trajectory dataset for the experiments. This dataset contains 1.3 billions GPS records that are collected from 10000 drivers in Shanghai city, China, during a six-month period in 2015. The GPS records are collected at a low-sampling rate as 0.1 Hz (*i.e.*, one sample per ten seconds). Each GPS record includes the driver identifier, a timestamp, location with longitude and latitude, travel speed, and travel direction. Furthermore, we download the road network of the city area covered by GPS records from OpenStreetMap (OSM)[2], and model the road network as a graph $\mathbb{G}(\mathbb{V}, \mathbb{E})$, which has 159386 vertices and 30336 edges (*i.e.*, road segments) in total. In addition, we obtain the attributes of each road segment from OSM as well. After map matching, we have 430 trajectories for each driver on average.

Baseline Approaches. We compare Radar with following baseline approaches, which can also learn driving style representations from GPS trajectories.

- *ARNet* is one of the state-of-the-art approaches. *ARNet* proposes an autoencoder regularized neural network for driving style representation learning, merely from raw GPS data [2].
- *T2INET* is one of the state-of-the-art approaches as well. *T2INET* represents a GPS trajectory as the multi-channel images that capture both geographic and driving behavior features using a sequence of convolution layers [10].
- Radar-C serves as one variant of our approach Radar by disabling the driving context representation module. As a result, Radar-C only takes the statistic features extracted from GPS records as input for the learning model to compute driving style representations.

Implementation. We implement Radar and all baseline approaches in Python 3.7.3 with Keras[3] 2.3.1 and TensorFlow[4] 2.2.0 for building various machine/deep learning models. We set Radar's parameters as follows. We set $L_s = 195$ and $L_f = 6$ for GPS data transformation. The city area is partitioned into 80×80 grids for geographic semantic representation. In graph $\mathbb{G}$, we have $n_t = 5$ road types, maximum number of lanes $n_\ell = 6$, and $n_o = 2$ for indicating one-way or not. We estimate traffic conditions with time slot $\Delta t = 30$ min. For the learning model, we use Adadelta as the optimizer, and set learning rates for generator $\mathcal{G}$ and discriminator $\mathcal{D}$ as 0.0001 and 0.0004, respectively. We set batch size as 128 and the epochs as 5000. Besides, we directly adopt the implementations of *ARNet* [2] and *T2INET* [10], which are provided by the authors respectively, and tune their parameters with our data to achieve their best performances.

We evaluate these approaches with two benchmark applications, *i.e.*, *driver number estimation* and *driver identification*, on a server, which is equipped with

---

[2] OpenStreetMap: https://www.openstreetmap.org/.
[3] Keras: https://keras.io/.
[4] TensorFlow: https://www.tensorflow.org/.

Intel Core i9-9900K CPU@3.60 GHz, NAVIDA GeForce RTX 2080 Ti GPU, and 32 GB memory. We repeat each experiment setting 10 times, and only the average results are reported in this section.

5.2   Driver Number Estimation

This application aims to estimate the number of drivers from a set of anonymous trajectories. To solve this problem, we train the driving style representation learning models with a set of labeled trajectories (*i.e.*, with known driver identifiers), and exploit trained models to represent each testing trajectory as a driving style representation vector. Then, we employ the affinity propagation [5] clustering algorithm to classify all representation vectors into clusters. In theory, a desired model should effectively learn drivers' driving styles, and would classify the testing trajectories generated by a specific driver into the same cluster. Finally, the number of clusters is regarded as the number of drivers.

Training and Testing. We randomly select 10 drivers from the driver set $\mathbb{U}$ and take their labeled trajectories as the training data. In addition, we randomly select κ drivers from the remaining drivers, who are absent in the training data, to form a group, denoted by *Group κ*. We vary κ from 1 to 10. For each group, we randomly sample 50 trajectories from the κ drivers, and use these trajectories as the testing data. We repeat 10 runs for each κ value and report average results.

Performance Metrics. We compare different approaches on the following two performance metrics: (1) the mean absolute error (*MAE*), which is the difference between the ground truth of driver number and the estimation; (2) the adjusted mutual information score (*AMI*) [24] that measures the clustering quality. The AMI values fall in the range of $[0, 1]$, and larger AMI values are preferable.

Table 1. Performance comparisons on MAE for driver number estimation.

Group κ	ARNet	T2INET	Radar-C	Radar
1	**0.64** $\pm$ 0.60	0.70 $\pm$ 0.68	0.80 $\pm$ 0.64	0.78 $\pm$ 0.65
2	**0.82** $\pm$ 0.80	0.88 $\pm$ 0.74	0.92 $\pm$ 1.20	0.84 $\pm$ 0.97
3	1.08 $\pm$ 1.26	1.22 $\pm$ 1.48	1.02 $\pm$ 1.24	**0.98** $\pm$ 1.24
4	1.18 $\pm$ 1.40	1.04 $\pm$ 1.46	**0.92** $\pm$ 1.02	1.02 $\pm$ 1.07
5	0.98 $\pm$ 1.24	**0.88** $\pm$ 1.56	1.20 $\pm$ 0.90	1.02 $\pm$ 0.88
6	1.24 $\pm$ 0.98	1.24 $\pm$ 0.96	1.04 $\pm$ 1.24	**1.04** $\pm$ 1.06
7	1.60 $\pm$ 1.24	1.42 $\pm$ 1.64	1.42 $\pm$ 1.44	**1.24** $\pm$ 1.12
8	1.48 $\pm$ 1.46	1.46 $\pm$ 1.45	1.46 $\pm$ 1.50	**1.38** $\pm$ 1.24
9	1.74 $\pm$ 1.48	1.82 $\pm$ 1.46	1.62 $\pm$ 1.42	**1.56** $\pm$ 1.48
10	2.32 $\pm$ 1.50	2.10 $\pm$ 1.68	1.94 $\pm$ 1.54	**1.82** $\pm$ 1.46
Average	1.308	1.276	1.234	**1.168**

Experimental Results. Table 1 shows the MAE results and deviations, where the best result of each group is marked in bold. When κ increases, the driver number estimation problem becomes harder, and thus the MAE is larger. Among all the experiments, we see our approach (`Radar` and `Radar-C`) wins 7 best results (*i.e.*, the smallest MAE) out of ten tests. For the three lost cases, our approach falls behind with marginal differences, *e.g.*, 0.14 at most. As shown by the average experiment results in the last row of Table 1, `Radar-C` achieves slightly better performance than the state-of-the-art approaches, *i.e.*, *ARNet* and *T2INET*. It implies that our learning model is more effective on capturing driving style features from raw GPS data. By incorporating the driving context information, `Radar` further improves `Radar-C` by reducing average MAE from 1.234 to 1.168. Overall, our approach `Radar` can improve *ARNet* and *T2INET* on the performance metric of MAE by 10.7% and 8.5%, respectively.

Table 2 presents the AMI results and deviations, where we also mark the best AMI of each group in bold. Similarly, we find that `Radar` outperforms other two baselines in most cases, with six wins out of ten tests. The results in Table 2 are in accordance with the results in Table 1. In general, a better clustering quality (*i.e.*, a larger AMI) potentially leads to a better estimation of driver number (*i.e.*, a smaller MAE). The average AMI values of the four approaches are 0.212, 0.234, 0.225, and 0.239, respectively. The results in both Table 1 and Table 2 demonstrate that `Radar` is capable of learning more effective and accurate driving style representations, which thus well support the application of driver number estimation, with smaller MAE and larger AMI.

Table 2. Performance comparisons on AMI for driver number estimation.

Group κ	ARNet	T2INET	Radar-C	Radar
1	**0.34** ± 0.06	0.32 ± 0.12	0.27 ± 0.06	0.25 ± 0.09
2	**0.37** ± 0.08	0.36 ± 0.07	0.25 ± 0.08	0.28 ± 0.03
3	0.21 ± 0.04	0.21 ± 0.08	0.26 ± 0.08	**0.27** ± 0.03
4	0.16 ± 0.08	0.24 ± 0.05	0.22 ± 0.05	**0.25** ± 0.04
5	0.19 ± 0.06	**0.23** ± 0.08	0.19 ± 0.08	0.18 ± 0.06
6	0.18 ± 0.05	0.22 ± 0.04	**0.26** ± 0.06	**0.26** ± 0.07
7	0.17 ± 0.07	0.17 ± 0.05	0.20 ± 0.08	**0.23** ± 0.02
8	**0.19** ± 0.06	**0.19** ± 0.06	0.14 ± 0.05	0.18 ± 0.04
9	0.15 ± 0.08	0.14 ± 0.08	0.20 ± 0.04	**0.22** ± 0.08
10	0.16 ± 0.07	0.26 ± 0.04	0.26 ± 0.05	**0.27** ± 0.04
Average	0.212	0.234	0.225	**0.239**

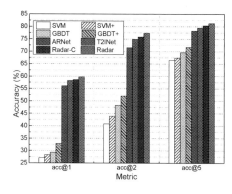
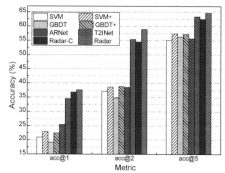

Fig. 4. Performance comparisons on Top-n accuracy with the long trajectories.

Fig. 5. Performance comparisons on Top-n accuracy with the short trajectories.

5.3   Driver Identification

The driver identification problem aims to identify the driver of a given unlabeled trajectory, which belongs to the supervised multi-class classification problem.

Training and Testing. In each experiment, we randomly select 10 drivers and use their GPS trajectories for model training and testing. Specifically, 70% of the trajectory data are used for training, 10% for validation, and the remaining 20% for testing. The models of all approaches are trained with labeled trajectories, and for a testing trajectory the models should predict its driver identifier.

Performance Metric. We employ the top-n accuracy (denoted by $acc@n$) to evaluate the prediction performances of all approaches. In particular, $acc@n$ is calculated as the percentage of testing trajectories for which the ground truth drivers are in the top n predictions. For a testing trajectory, we rank the predicted driver identifiers in the descending order of probability values.

Experimental Results. In addition to the aforementioned three baselines, we further include two typical supervised learning models, *i.e.*, support vector machines (SVM) [26] and gradient boosting decision trees (GBDT) [18], for performance comparisons. More specifically, SVM and GBDT take the statistic features produced by Radar as input for the predictions, while SVM+ and GBDT+ make use of both statistic and contextual features generated by Radar for the predictions. Furthermore, we partition drivers' trajectories into two sets: long trajectories (with duration more than 1950 s) and short trajectories (with duration less than 1950 s). We conduct experiments on each set of trajectories separately, and present the results in Fig. 4 and Fig. 5, respectively.

As shown in Fig. 4, when n increases, the top-n accuracy of each approach becomes higher. Our approach achieves the highest $acc@5$ accuracy as 81.3%. These deep learning models, *i.e.*, ARNet, T2INET, Radar-C and Radar, always have better predictions than traditional supervised learning models, *i.e.*, SVM and GBDT and their variants, with the largest performance gap as 36.6% on *acc*@2. It implies that deep learning models are indeed powerful at representation learning, and thus can support various applications better. On the other hand, by comparing the performances of traditional models, we find that SVM+/GBDT+ outperform SVM/GBDT, *e.g.*, with *acc*@1 accuracy improvement by 1.2% and 3.6%, respectively. Hence, it is necessary to include contextual features for better modeling. Compared to state-of-the-art *ARNet* and *T2INET*, our approach Radar has more accurate predictions, *e.g.*, on average improving them by 2.6%, 4.2%, and 2.4% for *acc*@1, *acc*@2, and *acc*@5, respectively.

The prediction results on short trajectory set are plotted in Fig. 5. Since short trajectories contain less information, and thus the prediction performances of all approaches have been seriously deteriorated. However, we find that the performance gap between *ARNet*/*T2INET* and our approach becomes even larger, *i.e.*, on average Radar improves the two advanced approaches by 7.1%, 12.0%, and 5.2% for *acc*@1, *acc*@2, and *acc*@5, respectively. These comparisons reflect that Radar is able to extract more useful and accurate features from low-quality trajectory data, and thus can still achieve reasonably high prediction accuracy.

6   Conclusion

In this paper, we present an adversarial driving style representation learning approach – Radar. Different from previous works, Radar not only extracts statistic features from raw GPS data, but also builds contextual features by jointly considering road conditions, geographic semantics, and traffic conditions. We further exploit an advanced semi-supervised GAN architecture to construct the learning model to compute more effective and accurate driving style representations. Experiment results from a large GPS trajectory dataset demonstrate that Radar outperforms state-of-the-art approaches on two benchmark applications.

Acknowledgments. This work was supported in part by the National Science Foundation of China (NSFC) under Grant Nos. 61802261, 61806132, 61872248, and U2001207, the grant of Guangdong Basic and Applied Basic Research Foundation (No. 2020A1515011502), Tencent Rhino-Bird Young Faculty Open Fund, Guangdong NSF No. 2017A030312008, Shenzhen Science and Technology Foundation (No. ZDSYS20190902092853047), the Project of DEGP (No. 2019KCXTD005),the Guangdong "Pearl River Talent Recruitment Program" under Grant No. 2019ZT08X603, Guangdong Science and Technology Foundation (Nos. 2019B111103001, 2019B020209001).

References

1. Castignani, G., Derrmann, T., Frank, R., Engel, T.: Driver behavior profiling using smartphones: a low-cost platform for driver monitoring. IEEE Intell. Transp. Syst. Mag. **7**(1), 91–102 (2015)
2. Dong, W., Yuan, T., Yang, K., Li, C., Zhang, S.: Autoencoder regularized network for driving style representation learning. In: IJCAI (2017)

3. Ezzini, S., Berrada, I., Ghogho, M.: Who is behind the wheel? Driver identification and fingerprinting. J. Big Data **5**(1), 9 (2018)
4. Feng, J., et al.: User identity linkage via co-attentional neural network from heterogeneous mobility data. IEEE Trans. Knowl. Data Eng. **1**, 1–15 (2020)
5. Frey, B.J., Dueck, D.: Clustering by passing messages between data points. Science **315**(5814), 972–976 (2007)
6. Fugiglando, U., et al.: Driving behavior analysis through CAN bus data in an uncontrolled environment. IEEE Trans. Intell. Transp. Syst. **20**(2), 737–748 (2018)
7. Goodfellow, I., et al.: Generative adversarial nets. In: NeurIPS (2014)
8. Guangyu Li, M., et al.: DBUS: human driving behavior understanding system. In: IEEE ICCV Workshops (2019)
9. He, B., Zhang, D., Liu, S., Liu, H., Han, D., Ni, L.M.: Profiling driver behavior for personalized insurance pricing and maximal profit. In: IEEE Big Data (2018)
10. Kieu, T., Yang, B., Guo, C., Jensen, C.S.: Distinguishing trajectories from different drivers using incompletely labeled trajectories. In: ACM CIKM (2018)
11. Kim, Y., Han, J., Yuan, C.: TOPTRAC: topical trajectory pattern mining. In: ACM SIGKDD (2015)
12. Kuderer, M., Gulati, S., Burgard, W.: Learning driving styles for autonomous vehicles from demonstration. In: IEEE ICRA (2015)
13. Lin, N., Zong, C., Tomizuka, M., Song, P., Zhang, Z., Li, G.: An overview on study of identification of driver behavior characteristics for automotive control. Math. Probl. Eng. **2014**, 1–15 (2014)
14. Liu, Z., Gong, Z., Li, J., Wu, K.: Mobility-aware dynamic taxi ridesharing. In: IEEE ICDE (2020)
15. Liu, Z., Li, Z., Li, M., Xing, W., Lu, D.: Mining road network correlation for traffic estimation via compressive sensing. IEEE Trans. Intell. Transp. Syst. **17**(7), 1880–1893 (2016)
16. Liu, Z., Li, Z., Wu, K., Li, M.: Urban traffic prediction from mobility data using deep learning. IEEE Network **32**(4), 40–46 (2018)
17. Liu, Z., Zhou, P., Li, Z., Li, M.: Think like a graph: real-time traffic estimation at city-scale. IEEE Trans. Mob. Comput. **18**(10), 2446–2459 (2018)
18. Mason, L., Baxter, J., Bartlett, P.L., Frean, M.R.: Boosting algorithms as gradient descent. In: NeurIPS (2000)
19. Miao, C., Wang, J., Yu, H., Zhang, W., Qi, Y.: Trajectory-user linking with attentive recurrent network. In: ACM AAMAS (2020)
20. Newson, P., Krumm, J.: Hidden Markov map matching through noise and sparseness. In: ACM SIGSPATIAL (2009)
21. Odena, A.: Semi-supervised learning with generative adversarial networks. arXiv preprint arXiv:1606.01583 (2016)
22. Ren, H., Pan, M., Li, Y., Zhou, X., Luo, J.: ST-SiameseNet: spatio-temporal siamese networks for human mobility signature identification. In: ACM SIGKDD (2020)
23. Salimans, T., Goodfellow, I., Zaremba, W., Cheung, V., Radford, A., Chen, X.: Improved techniques for training GANs. In: NeurIPS (2016)
24. Vinh, N.X., Epps, J., Bailey, J.: Information theoretic measures for clusterings comparison: variants, properties, normalization and correction for chance. J. Mach. Learn. Res. **11**, 2837–2854 (2010)
25. Wang, P., Fu, Y., Zhang, J., Wang, P., Zheng, Y., Aggarwal, C.: You are how you drive: peer and temporal-aware representation learning for driving behavior analysis. In: ACM SIGKDD (2018)

26. Wu, T.F., Lin, C.J., Weng, R.C.: Probability estimates for multi-class classification by pairwise coupling. J. Mach. Learn. Res. **5**(Aug), 975–1005 (2004)
27. Yang, S., Wang, C., Zhu, H., Jiang, C.: APP: augmented proactive perception for driving hazards with sparse GPS trace. In: ACM MobiHoc (2019)
28. Zhao, K., et al.: Discovering subsequence patterns for next POI recommendation. In: AAAI (2020)
29. Zheng, Y.: Trajectory data mining: an overview. ACM Trans. Intell. Syst. Technol. **6**(3), 1–41 (2015)
30. Zhou, F., Gao, Q., Trajcevski, G., Zhang, K., Zhong, T., Zhang, F.: Trajectory-user linking via variational AutoEncoder. In: IJCAI (2018)

MM-CPred: A Multi-task Predictive Model for Continuous-Time Event Sequences with Mixture Learning Losses

Li Lin, Zan Zong, Lijie Wen$^{(\boxtimes)}$, Chen Qian, Shuang Li, and Jianmin Wang

School of Software, Tsinghua University, Beijing, China
{lin-l16,zongz17,lsa18}@mails.tsinghua.edu.cn,
{wenlj,jimwang}@tsinghua.edu.cn

Abstract. Sequence prediction is a well-defined problem with a proliferation of applications, such as recommendation systems, social media monitor, economic analysis, etc. Recently, RNN-based methodologies have shown their superiority in time-series data analysis and sequence modeling. The question of which event would happen next is not difficult to answer anymore, but the prediction of when it would happen is still a mountain to climb. In this paper, we propose a **Multi-task** model to predict both event and their **continuous** timestamps at the same time. Specifically, (1) we design a two-layer RNN encoder for event sequences and a CNN encoder for time sequences, both equipped with multi-head self-attention to align history features; (2) we form multiple generative adversarial models for predicting future time sequences to solve the problem of multi-modal time distribution; (3) **Mixture** learning losses are adopted to conduct a 3-step learning strategy for training our model, the cross-entropy loss for events, Huber loss and adversarial classification loss which induces the Wasserstein distance for times. Due to these characteristics, we name it **MM-CPred**. The experiments on 4 real-life datasets confirmed its improvements compared with the baselines.

Keywords: Multi-task prediction · Neural networks · Continuous-time sequence · Mixture learning

1 Introduction

Sequence prediction is a well-defined problem with wide applications in recommendation systems [15] and information management systems [14].

In the field of recommendation systems, some item-to-item methods [20] and matrix factorization machines [17] have shown successful applications. Such rule-based recommendation algorithms can efficiently capture the relationships between items but the sequential nature is not considered. Some Markov Chain-based methods [22] were proposed to capture the chronological dependence in event sequences. But Markov Decision Process is limited to short time order information, and cannot deal with either long or non-stationary sequences.

© Springer Nature Switzerland AG 2021
C. S. Jensen et al. (Eds.): DASFAA 2021, LNCS 12681, pp. 509–525, 2021.
https://doi.org/10.1007/978-3-030-73194-6_34

Recently, Deep Learning (DL) has shown strong competitiveness in data analysis. Taking advantages of Recurrent Neural Networks (RNNs), sequence modeling has profound developments in many fields. RNNs are also widely referred to as a basic model of end-to-end predictor for event sequences, because its recursive computation can naturally capture the history information and the cascade of nonlinear computation layers helps understand the complex relations between given histories and the target.

Either traditional event sequence models or neural networks for modeling event sequences are trying to capture features from event data and its discrete attributes [14]. They seldom considered the effects of timestamps. These years, the importance of *time* for events arouses researchers' interests and widespread concern [27,28]. In this paper, we concentrate on predicting event sequence and the timestamp for each event simultaneously through a multi-task predictive model.

The simultaneous prediction of events and their corresponding timestamps is a challenge because (1) discrete event sequence and real-value time have their own characteristics; thus it's not appropriate to encode the event sequence and timestamps using a same or shared encoder as most multi-task learning models do and (2) the probability density distribution of time tends to contain multiple modes, which is difficult to be described by a single decoder. Meanwhile existing sequence predictors take history events as input but just output one next event in the future, which lacks foresight. Even some of them proposed to predict a set of candidates for users. However the order information and causality between this candidates are still not considered. To bridge these gaps, we propose MM-CPred (A **M**ulti-task **Pred**ictive model for **C**ontinuous-time event sequences via **M**ixture learning losses), which is able to predict future event sequence and the timestamps of events at the same time rather than output only the next one event. The main contributions are summarized as follows.

- We design a two-layer RNN encoder for event sequences to capture the sequential information and a CNN-based encoder for time sequences to capture the local numerical distributions.
- Self-attention mechanism is applied to provide weights calculation for history events and time features. Then the unified features will be pushed into the decoders to generate event sequence and the timestamps, respectively.
- Inspired by that the Gaussian Mixture Model can approximate almost any arbitrary density functions with multiple Gaussian Distributions. Similarly, we construct multiple generators for time sequences to deal with the multi-modal problem of time distributions.
- We take mixture learning losses, which consist of cross-entropy, Huber loss, and adversarial learning loss, to conduct a 3-step training strategy for MM-CPred.

In the experiments, we evaluate the performance of MM-CPred on four different datasets. Empirical results show that the proposed model achieved the state-of-the-art on sequence prediction tasks.

2   Related Work

Event sequence learning tries to understand and model the sequential user events, like the interactions between users and items, and the evolution of users' preferences and item popularity over time. Many researchers have made efforts in this regard. [19] presented a music recommendation system exploiting a dataset containing the listening histories of users who posted what they were listening to at the moment on the microblogging platform Twitter. In [18], the problem of future event prediction in video: if and when a future event will occur was considered. [5] provided a predictive model to be used in narrative generation systems. While in [12], an event graph was constructed to better utilize the event network information for script event prediction.

However when faced with a sequence of timestamps in continuous space, many researchers preferred to discretize the real-value timestamps to a given range to adapt the characteristic of RNNs. JUMP was proposed as a joint predictor for user click and interval to enhance the performance of recommendation [27]. In [28], several LSTM variants were proposed, i.e., Time-LSTMs, to model users' sequential actions, equipping LSTM with specifically designed time gates to model time intervals. The discretization of timestamps relies on prior knowledge of the data. For example, in [27] and [28], authors truncated sequences to alleviate the incremental size of the look-up table for long-range timestamps. Besides, when the interval between two adjacent events is out of range predefined, they cut the sequence into two pieces.

Discretizing the real-value times couldn't capture the true features of continuous-time sequences. Because the prediction of future timestamps was established as a classifier but not a generator, which loses the continuity of time. To construct a general method for continuous-time sequence prediction, we should take advantages of generative models and CNNs [10]. Because CNNs sre designed for capturing local dependencies among item values on adjacent locations, without considering these values are continuous or discrete.

Recently, researchers begin to find out various applications of GANs [4] on sequence modeling. A set of experiments was compared on different GAN models to find out the potential of GANs in language modeling [21]. SeqGAN [26] was proposed to bypass the generator differentiation problem by directly performing gradient policy update by modeling the data generator as a stochastic policy in reinforcement learning (RL). Most of the existing generative adversarial networks for sequence generation suffer from the instability of policy gradient. Since loss functions of supervised learning like MLE (*Maximum Likelihood Estimation*) can dominate the samples tending to satisfy the real data, which can balance out the unstable performance from the generative model when combined with GANs. LeakGAN [6], MaskGAN [3], RankGAN [13], DPGAN [24] all guided the generative adversarial networks with the rewards information from discriminative models to stably generate the discrete sequences. Such applications show that the Mixture of MLE loss and adversarial learning loss has gained great success in text generation and language learning.

Thus, the overall learning strategy for our model is a mixture of supervised learning and unsupervised learning. Supervised learning methods for event sequence and time sequence take the real future items followed up as labels, to make the predicted sequences and the ground truth approximate. Meanwhile, the unsupervised learning algorithm assists in holding the consistency of time distribution density between the predicted time sequence and the sampled ones.

3   Continuous-Time Event Sequence

Discrete-time event sequence prediction tasks consider event sequence as a series of *tokens* (events) with order information, evolving synchronously in unit time step, e.g., sentences of words. Continuous-time event sequence prediction attends to the sequence where the events occur asynchronously, i.e., the *intervals* between consecutive events vary a lot.

Time prediction is an important task along with event prediction. Some studies considered that the timestamps of history events have effects on the probabilities of the next event to be elevated or decreased. As most of them do [11,16,25], we subtract the timestamps of two adjacent events to get *intervals* between events to construct the time sequences.

Given the event set $\mathbb{E}$, we define the history event sequence $S_e = \langle e_1, e_2, \ldots, e_n \rangle, e_i \in \mathbb{E}$ and the corresponding time sequence $S_t = \langle t_1, t_2, t_3, \ldots, t_n \rangle, t_i \in \mathbb{R}$, where $t_1 = 0$ because there is no prefix for e_1. A continuous-time event sequence predictor F_θ will be learnt to predict the most possible event sequence and its time sequence $P_e = \langle e_{n+1}, e_{n+2}, \ldots, e_{n+m} \rangle$ and $P_t = \langle t_{n+1}, t_{n+2}, \ldots, t_{n+m} \rangle$, where m is a hyper-parameter which decides the length of the output sequence.

$$(P_e, P_t) = F_\theta(S_e, S_t; \theta), P_e \subseteq \mathbb{E}^*, P_t \subseteq (R^+)^* \tag{1}$$

4   Methodology

To solve the simultaneous prediction issue for events and times, we propose MM-CPred, a multi-task predictive model with mixture learning losses. In this section, we will explain the details of our model.

The architecture of MM-CPred is shown in Fig. 1. We use a lookup layer to transform an event into an embedding $\mathbf{x}_e^i \in \mathbb{R}^{l_e}$. The input of time encoder is raw real-value time sequences $\mathbf{x}_t^i = t_i$. Event encoder and time encoder compress the event sequences to compact representations and expand 1-dimensional time sequences to high-dimension vectors respectively. We apply the multi-head attention mechanism to align the history features into combined vectors for follow-up generators. The generators are RNN-based which can output sequences benefiting from their recurrent nature. Based on the characteristic of discrete event sequences, cross-entropy loss is applied to train the event generator, while for time generators, Huber loss is designed to force the numerical approximation of times. Meanwhile, the adversarial learning loss (i.e. Wasserstein distance) is used to constrain the consistency of time distributions.

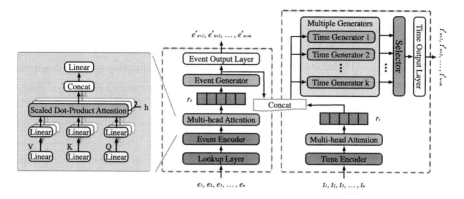

Fig. 1. The architecture of MM-CPred. Three main components are respective Encoders for events and times, Multi-Head Attentions to align history features and Multiple time generators with a selector proposed aiming to capture multimodal time distributions.

4.1 RNN Encoder and CNN Encoder

Given a history event sequence $\langle \mathbf{x}_e^1, \mathbf{x}_e^2, \ldots, \mathbf{x}_e^n \rangle$ and its time sequence $\langle \mathbf{x}_t^1, \mathbf{x}_t^2, \ldots, \mathbf{x}_t^n \rangle$. We design two encoders for capturing the order information of them respectively. In this paper, we use GRU (Gated Recurrent Unit) as the basic cell of RNN Encoder for event sequences. Due to space limitations, we won't talk about the details about GRU. The computation of GRU cell we used in this paper is the same as that defined in [2]. In this paper, the event encoder was established with 2-layer GRU networks.

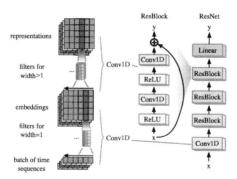

Fig. 2. The definition of Conv1D operation and ResBlock

But for time sequences, they are series of 1-dimensional real values, and adjacent times usually present local consistency. Therefore we design a Conv1D operator for time sequences, which conducts convolutions along the orders of items only. Then a CNN encoder for time sequences was established based on ResNet [7]. The reasons why we choose ResNet are: (1) the layers are reformulated as learning residual functions with reference to the layer inputs instead of learning unreferenced functions, which takes raw inputs as a part of outputs preventing forgetting history information; (2) the architecture is simple and practicable, where deep stacking of layers can increase the effective history length. Conv1D unit and the basic model of ResBlock are presented in Fig. 2. Same as CNN, the core of Conv1D is a filter weight $\mathbf{F} \in \mathbf{R}^{d_t \times d_F}$. Sliding the filter across a sequence $\langle \mathbf{x}_t^1, \mathbf{x}_t^2, \ldots, \mathbf{x}_t^n \rangle$ via a same padding trick can divide the entire sequence into n subsequences (windows) as following shows: $\mathbb{X} = [X_1, X_2, \ldots, X_n]$, where $X_i = [\mathbf{x}_t^i, \mathbf{x}_t^{i+1}, \ldots, \mathbf{x}_t^{i+k-1}]$ is the i-th window. As Fig. 2 shows, the size of filters $k_F = 3$. We omit the pooling operation in Conv1D, thus the output of convolutional layer is a vector feature $\mathbf{u}_i = \mathbf{F}X_i + \mathbf{b}_F$. The $\mathbf{u}_i \in \mathbb{R}^{l_F}$ represents the higher-level representation that aggregates local information inside i-th window. Its goal is to capture the local context dependencies among time sequences. At the end, Conv1D transforms the input sequence to a feature sequence with equal length.

ResBlock we used in this paper was constructed with cascaded units consisting of a ReLU layer and a Conv1D layer. Both event encoder and time encoder bring out feature sequences of the same length as the input sequences. As most encoder-decoder models do, we also add an attention block between the encoder and generator: the Multi-Head Self-attention [23]. The attention function can be described as a mapping from a query and a set of key-value pairs to an output, where the query($\mathbf{Q}$), keys($\mathbf{K}$), values($\mathbf{V}$), and outputs($\mathbf{O}$) are all vectors. We use the bold uppercase to represent vectors in Multi-Head Attention, especially to keep pace with the original definition in [23]. The output is computed as a weighted sum of the values, where the weight assigned to each value is computed by a compatibility function of $\mathbf{Q}$ with the corresponding $\mathbf{K}$.

$$\begin{bmatrix} \mathbf{Q} \\ \mathbf{K} \\ \mathbf{V} \end{bmatrix} = \begin{bmatrix} \mathbf{W}_q \\ \mathbf{W}_k \\ \mathbf{W}_v \end{bmatrix} x_i + \begin{bmatrix} \mathbf{b}_q \\ \mathbf{b}_k \\ \mathbf{b}_v \end{bmatrix} \tag{2}$$

where $\mathbf{W}_q \in \mathbb{R}^{d_{model} \times d_q}$, $\mathbf{W}_k \in \mathbb{R}^{d_{model} \times d_k}$, $\mathbf{W}_v \in \mathbb{R}^{d_{model} \times d_v}$ and $\mathbf{b}_q, \mathbf{b}_k, \mathbf{b}_v$ are parameter matrices to be learned.

Then the computation of attention is:

$$\text{Attention}(\mathbf{Q}, \mathbf{K}, \mathbf{V}) = \text{softmax}(\frac{\mathbf{Q}\mathbf{K}^T}{\sqrt{d_k}})\mathbf{V} \tag{3}$$

Linear projection in Eq. 2 is conducted on the queries, keys and values h times respectively. On each of these projected versions of queries, keys, and values, the attention function was performed in parallel, yielding d_v-dimensional output values. The output of the multi-head attention can be calculated in the following formula:

$$\mathbf{O} = \text{MultiHead}(\mathbf{Q}, \mathbf{K}, \mathbf{V}) = \text{concat}(\text{head}^1, \text{head}^2, \ldots, \text{head}^h)\mathbf{W}_o \tag{4}$$

where $\text{head}^i = \text{Attention}(\mathbf{Q}\mathbf{W}_q^i, \mathbf{K}\mathbf{W}_k^i, \mathbf{V}\mathbf{W}_v^i)$. In this paper, we take the same size for queries, keys, and values as d_{model}/h.

4.2   Generators and Discriminator

Different from traditional predictive models that output the event keeping in *only one* step given history event sequence, MM-CPred takes hidden representations into generators to output future sequences. Additionally, we design multiple generators for time sequences to solve the multi-modal issue in distributions of times.

Event Generator. We apply two-layer GRU networks to decode the hidden representation $\mathbf{r}_e$ to generate future event sequences. The probabilities $\mathbb{P}_e$ of future events over $\mathbb{E}$ can be caught from the output layer.

Multiple Generators. In order to ensure the distribution consistency between the predicted times and the targets, we intend to combine regression loss and generative adversarial loss when training time sequence generators. We observe the training of generative models is still challenging because it can be easily trapped into the mode collapsing problem where the generator only concentrates on producing samples lying on a few modes instead of the whole data space while the multi-modal distributions of times are ubiquitous in contrast.

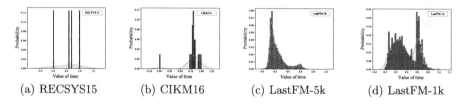

(a) RECSYS15   (b) CIKM16   (c) LastFM-5k   (d) LastFM-1k

Fig. 3. The distribution of time values in a sampled sequence from four different datasets respectively. X-axis: the min-max normalized values of $log(t)$, Y-axis: the probability density.

Figure 3 shows the time distributions from four different datasets. We visualize the probability density distribution of min-max normalized logarithmic values of times. It can be clearly seen that there are multiple distinct peaks in the distributions. To better model the multi-modal distributions, we propose multiple generators $G_t^1, G_t^2, G_t^3, \ldots, G_t^k$ to produce times $\{t_i^1, t_i^2, t_i^3, \ldots, t_i^k\}, i \in \{1, 2, 3, \ldots, m\}$ at each step i but only one of them can be selected by the selector and injected to the discriminator.

$$t_i^j = G_t^j(P_e, \mathbf{r}_t; \theta_{G_t^j}), i \in \{1, 2, \ldots, m\} \& j \in \{1, 2, \ldots, k\} \tag{5}$$

For the selector at each time step i, the following formulations tell how to choose the output.

$$\mathbf{a} = \text{softmax}(\mathbf{W}_a \cdot \mathbf{r}_t + \mathbf{b}_a)$$

$$v = \text{index}(max(\mathbf{a})), v \in \{1, 2, \ldots k\}$$

where v is the index of time generators selected. It means the predicted time at i-step is $t_i = G_t^v(P_e, \mathbf{r}_t; \theta_{G_t^v})$. But selecting the max from softmax distribution makes the model non-differentiable. To address this issue, the Straight-Through Gumbel-Softmax [9] function is used for training the selector:

$$q_i = \frac{\exp(\frac{a_i + g_i}{\tau})}{\sum_{j=1}^{k} \exp(\frac{a_i + g_i}{\tau})} \tag{6}$$

where g_i is the Gumbel noise which can be sampled from Gumbel distribution and τ is the temperature parameter which controls the smoothness of the vector $\mathbf{q}$. If τ is close to zero, $\mathbf{q}$ will reach the one-hot vector corresponding to **One_Hot**$(v) \in \mathbb{R}^k$. We set $\tau = 0.0001$ in our experiments. In the backward pass, the function approximates the gradients of Selector v by using the gradients of $\mathbf{q}$ to ensure end-to-end model training. While the conventional *softmax* function is used to produce the index of output event. After m steps of generation, we finally have the predicted time sequence $P_t = \langle t_1, t_2, t_3, \ldots, t_m \rangle$.

To train the time predictor from supervised learning perspective, we perform the regression error *Huber loss* [8] which is defined as follows:

$$\text{loss}_{huber}(x) = \begin{cases} 0.5x^2 & |x| \le d \\ 0.5x^2 + d(|x| - d) & |x| > d \end{cases}$$

where $d \in \mathbb{R}^+$ is a hyperparameter. It is tempting to look at this loss as the log-likelihood function of the underlying heavy-tailed error distribution.

Besides Huber loss, the discriminator will give out the Wasserstein distance between generated time sequence and samples from real-time sequences to constrain the consistency of predicted and real-time distributions from the perspective of unsupervised learning. When the discriminator cannot distinguish generated sequences and real samples, the adversarial training tends to get an equilibrium.

Discriminator. We concatenate the raw inputs of times S_t and predicted ones P_t as a complete-time sequence marked as T_g. Time samples T_r with the same length as T_g are randomly sampled from the dataset. The discriminator is in charge of distinguishing real samples T_r from generated time sequences T_g. Same as the time sequence encoder, we take the basic ResNet illustrated in Fig. 2 as the architecture of the discriminator, which concentrates on capturing the local dependencies among the sliding window when scanning time sequences.

4.3 Training Strategy

In this section, we provide details about training methods. We will define the loss function of each task and training methods in this subsection. As described above, we have

$$\mathbf{r}_e = \text{Encoder}_e(\langle \mathbf{x}_e^1, \mathbf{x}_e^2, \ldots, \mathbf{x}_e^n \rangle; \theta_{E_e}) \tag{7}$$

$$\mathbf{r}_t = \text{Encoder}_t(\langle \mathbf{x}_t^1, \mathbf{x}_t^2, \ldots, \mathbf{x}_t^n \rangle; \theta_{E_t}) \tag{8}$$

Cross-entropy Loss. For event generation, we use cross-entropy to measure the similarity between predicted event sequences and target sequences. The computation is in the following:

$$\text{loss}_e = -\frac{1}{m} \sum_{i=1}^{m} log\mathbb{P}(e_{n+i})\text{Generator}_e(\mathbf{r}_e; \theta_{E_e}, \theta_{G_e})_i \tag{9}$$

Algorithm 1. A Multi-task Predictive Model for Continuous-time Event Sequence with Mixture Learning Losses

---

Require: dataset $\mathcal{X} = \{\mathbf{x}_e^i, \mathbf{x}_t^i\}_{i=1}^N$, initialization of $\gamma, \alpha = 1.0$, learning rates $lr_i = 1e-4, i \in \{1,2,3\}$, number of time generators $n_g = 3$

Ensure: future events and respective time sequence P_e, P_t

1: **repeat**
2:   **for** every batch of data **do**
3:     Train event predictor: minimize cross-entropy loss in (9), update $\theta_{E_e}, \theta_{G_e}$.
4:     Forward the events to time generators.
5:     Train time predictor: minimize loss function in (10), update $\theta_{E_t}, \theta_{G_t}$.
6:     **if** this is discriminative epoch **then**
7:       Do sampling and train discriminator to minimize loss function in (12), update w.
8:     **else**
9:       Jointly train the whole model to maximize loss function in (11) , update $\theta_{E_e}, \theta_{E_t}, \theta_{G_e}, \theta_{G_t}$.
10:    **end if**
11:    Compute losses on validate data, and update γ, α using the *loss ratios* to balance the different losses.
12: **end for**
13: **Until** reach the max epochs

Huber Loss. For supervised training of time generation, we use Huber loss to measure the regression error between predicted and target times.

$$\text{loss}_t = \sum_{i=1}^{m} \text{loss}_{huber}(t_{n+i} - \text{Generator}_t(O_e, \mathbf{r}_t; \theta_{E_t}, \theta_{G_t})_i) \tag{10}$$

where $\theta_{G_t} = \{\theta_{G_t^1}, \theta_{G_t^2}, \ldots, \theta_{G_t^k}\}$.

Adversarial Training Loss. Except the regression loss, we also use adversarial training loss to constrain the consistency of time distributions.

$$\min_{\theta} \max_{w} V(D_w, G_\theta) = \mathbb{E}_{t \sim \mathbb{P}_t}[f_w(T_r)] - \mathbb{E}_{r_t}[f_w(T_g))]$$

where θ is the parameter set for generators consisting of θ_{E_e}, θ_{G_e}, θ_{E_t}, θ_{G_t}, w represents the parameters of the discriminator, T_r is randomly sampled from datasets and T_g is the concatenation of input time sequence and predicted sequence from time generators. In order to have parameters w lie in a compact space, we clamp the weights to a fixed box (say $w = [-0.01, 0.01]$) after each gradient update [1].

Finally, we conduct a 3-step learning strategy to train MM-CPred as Algorithm 1 shows: first, we train the event predictor using loss in Eq. 9; second, we train the time predictor to minimize Huber loss in Eq. 10; third, we jointly train the whole model according to the combined loss in Eqs. 11 and 12. In this paper, we train the discriminator once after every 5-epochs of generative training. The reasons we adopt the 3-step training strategy are that (1) event sequence contains the causal information of users' behavior, which decides the future events of interests; (2) when would an event happen relies on the type of event; thus the prediction of times takes advantage of predicted events and history time sequences' features at the same time, and (3) taking the linear combination of above losses as the last step helps to regularize different losses and jointly update the parameters to keep the global optimization.

$$\theta = \text{argmin } V(D_w, G_\theta) + \gamma l\mathbf{loss}_t + \alpha\mathbf{loss}_e \tag{11}$$

$$w = \text{argmin } - V(D_w, G_\theta) \tag{12}$$

γ, α are hyperparameters that regularize the combined loss function, which can be automatically tuned according to the loss values on validation datasets.

5   Experiments

5.1   Datasets

We conduct experiments on three public datasets including RECSYS15[1], CIKM16[2], and LastFM[3]. RECSYS15 is used for RecSys Challenge 2015. It contains click streams with timestamps collected from a commerce site. CIKM16 is published by CIKMCup 2016. It contains sequences of anonymous transactions provided by DIGINETICA. The LastFM dataset consists of the keywords artist, title, timestamp, similars, and tags. We extract the events and timestamps from the above datasets and re-organize the events and times (intervals between adjacent events) along with time order. Since LastFM has a huge amount of low-frequency events, as well as we expect to verify the effectiveness of the model on datasets of different sizes, we extract most k-frequent events from LastFM to construct two Small scale datasets: LastFM-5k and LastFM-1k. The statistics of these four datasets are illustrated in Table 1. The original time unit in RECSYS15 (resp. CIKM16, LastFM-5k, and LastFM-1k) is *second* (resp. *milesecond*). To keep it consistent, we show the intervals using *second (s)* as the time unit.

Table 1. Basic statistics of the datasets

Dataset	RECSYS15	CIKM16	LastFM-5k	LastFM-1k
Number of items	52,739	122,911	5,000	1,000
Number of seqs	9.25M	310,062	983	977
Average length	3.57	3.98	828.40	685.29
Max interval (s)	3,600	1.19M	121.21M	160.72M
Average interval (s)	249.48	0.17M	12.75M	15.68M

[1] http://recsys.yoochoose.net.
[2] http://cikm2016.cs.iupui.edu/cikm-cup.
[3] http://www.last.fm/api.

For each dataset, 70% sequences are randomly selected as training data. The remaining sequences are split into a 2:1 ratio, where two-thirds are for validation, and the left is test data. We didn't do any other pre-processing on RECSYS15 and CIKM16, including large word removing or session clipping. For extracting frequent events from LastFM, we conduct the same pre-processing as [28] did.

5.2   Compared Methods

We compare MM-CPred with the following methods.

NHP (Neural Hawkes Process) [16]. It models streams of discrete events in continuous time, by constructing a neurally self-modulating multivariate point process in which the intensities of multiple event types evolve according to a novel continuous-time LSTM.

T-LSTMs [28]. These are several new LSTM variants, i.e., Time-LSTMs, to model users' sequential actions, equipping LSTM with specifically designed time gates to model time intervals. But they only concentrate on the prediction of event sequences rather than time intervals. We take the best variant T-LSTM3 as a baseline.

GRU4e, GRU4t. These two are basic RNN models for events and times separately. The encoder and decoder were constructed with 2-layer GRU cells.

GAN4t. Similar to GRU4t, this is an encoder-decoder model constructed with 2-layer GRUs. Discriminator was applied to jointly learning the numerical approximation and distributional consistency.

MM-CPred-iG. The variants of proposed MM-CPred equipped with different numbers of time generators.

5.3   Metrics

The metrics we used to evaluate the event prediction performance are MRR@10 and Recall@10.

Mean Reciprocal Rank (MRR) is a statistic measure for evaluating the process that produces a list of possible items. It takes into account the rank of the items. The specific formulation is:

$$\text{MRR@}k = \frac{1}{|Q|} \sum_{i=1}^{|Q|} \frac{1}{rank_i} \tag{13}$$

where Q denotes the desired events and $rank_i$ is event's rank in the corresponding result list. And the reciprocal rank is set to zero if the rank is over k.

Recall@k is the proportion of relevant items found in the top-k recommendations which can be written as:

$$Recall@k = \frac{\#relevant\ recommended\ events@k}{\#total\ relevant\ items} \tag{14}$$

5.4 Experiments on Event Prediction

In this subsection, we compare the *next one* event prediction performance of our model with the baselines. We evaluate MRR@10, Recall@10 of these methods on RECSYS15, CIKM16, LastFM-5k, and LastFM-1k. For NHP and T-LSTM3, we use the same settings as they were proposed. For GRU4e and MM-CPred, the dimension of the event embedding vectors and all hidden states of GRU is set to 200. At the beginning of training, we initialize the linear regularization factor for joint loss function with $\gamma = 1.0, \alpha = 1.0$. The learning rates for all three steps of training are $1e-4$.

Table 2. The accuracy(%) of predicting the next one event on four datasets. (M@10:MRR@10, R@10:Recall@10)

Models	RECSYS15		CIKM16		LastFM-5k		LastFM-1k	
	M@10	R@10	M@10	R@10	M@10	R@10	M@10	R@10
NHP	-	-	-	-	6.3	10.1	11.2	28.8
T-LSTM3	1.2	5.4	0.5	4.7	15.2	30.2	19.2	36.8
GRU4e	**10.5**	26.4	3.6	8.6	7.3	15.6	21.0	35.4
MM-CPred	10.2	**30.8**	**4.1**	**10.2**	**19.4**	**35.8**	**24.6**	**40.5**

The results are shown in Table 2. We can see MM-CPred outperforms almost all the baselines on four datasets, except MRR@10 on RECSYS15, which is 0.3% lower than GRU4e. The reason for the decrement is probably that joint training of MM-CPred in the third step sacrifices the accuracy in event prediction to facilitate the improvement on time prediction, which is quite normal in multi-task learning models. This phenomenon also reflects that time features are not always helpful for predicting what events would happen in the future. Besides, NHP failed in predicting the next event because the sizes of events in REC-SYS15 and CIKM16 are beyond the capability of modeling. Thus we cannot get MRR@10 and Recall@10 on these two datasets.

5.5 Experiments on Time Prediction

We validate the accuracy of time prediction of MM-CPred comparing with NHP, T-LSTM3, GRU4t and GAN4t. NHP and MM-CPred are able to predict events and times simultaneously, so the predicted times are brought out from above trained models. For GRU4t and GAN4t, we take the same setting as introduced in previous subsection. Since T-LSTM3 was not designed for time prediction at all, we re-implement the T-LSTM3 cell and apply it into GAN4t to get the predicted times. The accuracy of time prediction is measured by Mean Absolute Error (MAE). From the results reported in Table 3, we can see MM-CPred outperforms all baselines on time prediction. At the bottom of Table 3, we show the maximum relative differences between our model and baselines. On LastFM-5k, the model achieved a 39.7% error reduction compared with NHP.

Table 3. MAE between predicted and real times(s). The bottom row shows the maximum relative differences compared with baselines.

Model	RECSYS15	CIKM16	LastFM-5k	LastFM-1k
NHP	98.10	63.22	34.87	61.03
T-LSTM3	84.11	47.32	21.19	50.88
GRU4t	89.29	49.72	26.99	52.30
GAN4t	85.22	48.58	21.58	51.72
MM-CPred	**83.44**	**45.23**	**21.02**	**50.88**
	▽14.9%	▽28.5%	▽39.7%	▽16.6%

Table 4. Time MAE comparison of MM-CPred-iGs

Model	RECSYS15	CIKM16	LastFM-All
MM-CPred-1G	88.69 (s)	46.17 (s)	114.02 (ms)
MM-CPred-3G	**83.44** (s)	**45.23** (s)	**107.77** (ms)
MM-CPred-6G	88.28 (s)	45.89 (s)	109.84 (ms)

5.6   Ablation Study

To clarify the contributions of mixture learning losses, we design GRU4e and GRU4t, which only captures the order information of events and times respectively. Table 2 shows without the third step, joint learning for events and times, the accuracy of events prediction decreased on most datasets. GAN4t was established by equipping GRU4t with the adversarial training. Compare with MM-CPred, GAN4t skipped the first step of training, which means that there is no event information used in it. MAE of times in Table 3 shows that the improvements of MM-CPred on time prediction benefit from adversarial training and multi-task learning.

The number of time generators (n_g) controls the multi-modal of time distributions. If we use a large n_g, time distribution density will be flattened, which leads to that the predicted mean value would be greater than the real mean. But if n_g is small, time distribution density will be sharp, which leads to mode collapse. To depict this, we vary the number of time generators and report the prediction accuracies of the next event's time in Table 4. We see that when applying 3 time generators, MM-CPred-3G gets the best performance on time prediction. To verify the contribution of mixture generators for time, we visualize the distribution densities of two samples from RECSYS15 in Fig. 4. From the comparison of time distribution densities between Fig. 5(a) and 5(b), we can see when 3 time generators were applied in MM-CPred, the distribution of predicted times appears 3 obvious peaks, which increased the overlap area with the real-time distribution.

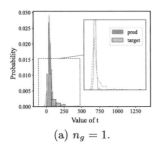 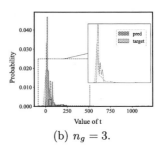

(a) $n_g = 1$.   (b) $n_g = 3$.

Fig. 4. Time distributions from varied number of generators

5.7 Discussion of Sequence Length

MM-CPred is proposed to predict future events and their timestamps, in other words, it outputs event sequences and the relative time sequences at the same time rather than just one next event. In Fig. 5, we show the performance of our model using varied output length. In this experiment, the length of sequences injected into MM-CPred is 20. We set the output length with 1, 5, 10, 15, and 20 respectively. The MAE between predicted and real times decreases as the length of output sequences increases, which is the opposite of common RNN-based sequence prediction models. The reason is that the adversarial learning algorithm aims to hold the consistency of distribution densities between the predicted times and real samples. The longer sequence we have, the more precise distance we can obtain. Thus the mean errors decreased. MRR@10 for events maintains stable when the length increases, which shows the 3-step training strategy is effective even for long sequence prediction.

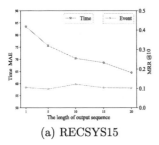 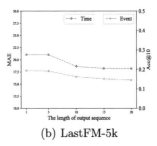

(a) RECSYS15   (b) LastFM-5k

Fig. 5. Experimental results varying the length of output sequences on the dataset RECSYS15 and LastFM-5k in terms of Time MAE and MRR@10

5.8 Experimental Details

We implement MM-CPred using Tensorflow v1.14.0. The models are trained on single GeForce GTX 1080ti GPU. The lookup table for events is constructed by a uniform initializer. We set the hidden size of vectors in our model as the same with the dimension of event embeddings d_e, which means $d_{model} = d_e = 200$. The batch size is 512, and the input length is set to 5 when predicting next one event. We employ $h = 4$ parallel attention layers (heads) in multi-head attention. For each of these we use $d_k = d_v = d_{model}/h = 50$. The initialization of regularizer parameters for mixture loss $\gamma, \alpha = 1.0$, but they can be updated by the ratio of losses on the validation set. We set the learning rate for each step of training as 0.0001. RMSPropOptimizer is used to backward the gradients in our model. Since we train the model linearly following the 3-step training strategy, though the architecture itself is not very deep, the time cost for one batch training must be considered. In the experiments, we scale the batch size to 512 and decrease the number of batches. When training MM-CPred, the time cost for training one batch of data is 8.9 s.

6 Conclusion and Future Work

We explore using multiple generative models to keep the consistency of predicted time distribution with real-time samples. Mixture losses are used to conduct a 3-step training strategy on MM-CPred, taking into account the different characteristics between discrete event sequences and continuous timestamps. The joint loss function in our model is an aggregate of MLE, regression, and adversarial loss. The RNN encoder for event sequences and the CNN encoder for time sequences can capture the information of them separately. The introduction of the multi-head attention mechanism helps align history features effectively. Empirical results validate the performance of MM-CPred not only on event sequence prediction but also on time sequence prediction.

We are looking forward to the future of MM-CPred and plan to apply them to sequence generation. We plan to combine the adversarial learning with MLE estimation on event prediction to generate complete event sequences. Though the multiple generators with a selector can capture the multi-modal characteristics of probabilities, the stability is still a weak point because they sometimes sharpen the indistinct distribution in real data, which may also cause the model not to converge.

Acknowledgment. The work was supported by the National Key Research and Development Program of China (No. 2019YFB1704003), the National Nature Science Foundation of China (No. 71690231), Tsinghua BNRist.

References

1. Arjovsky, M., Chintala, S., Bottou, L.: Wasserstein GAN (2017). arXiv:abs/1701.07875
2. Cho, K., et al.: Learning phrase representations using RNN encoder-decoder for statistical machine translation (2014). arXiv preprint: arXiv:1406.1078
3. Fedus, W., Goodfellow, I., Dai, A.M.: MaskGAN: better text generation via filling in the blank. In: International Conference on Learning Representations (2018)

4. Goodfellow, I.J., et al.: Generative adversarial nets. In: NIPS (2014)
5. Granroth-Wilding, M., Clark, S.: What happens next? Event prediction using a compositional neural network model. In: Thirtieth AAAI Conference on Artificial Intelligence (2016)
6. Guo, J., Lu, S., Cai, H., Zhang, W., Yu, Y., Wang, J.: Long text generation via adversarial training with leaked information (2017). arXiv preprint: arXiv:abs/1709.08624
7. He, K., Zhang, X., Ren, S., Sun, J.: Deep residual learning for image recognition. In: Proceedings of the IEEE Conference on Computer Vision and Pattern Recognition, pp. 770–778 (2016)
8. Huber, P.J.: Robust estimation of a location parameter. In: Kotz, S., Johnson, N.L. (eds.) Breakthroughs in Statistics. Springer Series in Statistics (Perspectives in Statistics)Springer Series in Statistics (Perspectives in Statistics), pp. 492–518. Springer, New York (1992). https://doi.org/10.1007/978-1-4612-4380-9_35
9. Jang, E., Gu, S., Poole, B.: Categorical reparameterization with gumbel-softmax. In: ICLR (2017)
10. LeCun, Y., Bengio, Y., et al.: Convolutional networks for images, speech, and time series. The Handbook of Brain Theory and Neural Networks **3361**(10), 1995 (1995)
11. Li, Y., Du, N., Bengio, S.: Time-dependent representation for neural event sequence prediction (2017). arXiv preprint arXiv:1708.00065
12. Li, Z., Ding, X., Liu, T.: Constructing narrative event evolutionary graph for script event prediction. In: IJCAI (2018)
13. Lin, K., Li, D., He, X., Zhang, Z., Sun, M.T.: Adversarial ranking for language generation. In: NIPS (2017)
14. Lin, L., Wen, L., Wang, J.: MM-Pred: a deep predictive model for multi-attribute event sequence. In: Proceedings of the 2019 SIAM International Conference on Data Mining, pp. 118–126. SIAM (2019)
15. Linden, G., Smith, B., Bullet, J.L.: Recommendations item-to-item collaborative filtering (2001)
16. Mei, H., Eisner, J.M.: The neural Hawkes process: a neurally self-modulating multivariate point process. In: Advances in Neural Information Processing Systems, pp. 6754–6764 (2017)
17. Musto, C., Semeraro, G., Degemmis, M., Lops, P.: Word embedding techniques for content-based recommender systems: an empirical evaluation. In: RecSys Posters (2015)
18. Neumann, L., Zisserman, A., Vedaldi, A.: Future event prediction: if and when. In: Proceedings of the IEEE Conference on Computer Vision and Pattern Recognition Workshops (2019)
19. Pichl, M., Zangerle, E., Specht, G.: Now playing on spotify: leveraging spotify information on twitter for artist recommendations. In: ICWE Workshops (2015)
20. Sarwar, B.M., Karypis, G., Konstan, J.A., Riedl, J.: Item-based collaborative filtering recommendation algorithms. In: WWW (2001)
21. Semeniuta, S., Severyn, A., Gelly, S.: On accurate evaluation of GANs for language generation (2018). ArXiv: arXiv:abs/1806.04936
22. Tavakol, M., Brefeld, U.: Factored MDPs for detecting topics of user sessions. In: RecSys (2014)
23. Vaswani, A., et al.: Attention is all you need. In: Advances in neural information processing systems, pp. 5998–6008 (2017)
24. Xu, J., Ren, X., Lin, J., Sun, X.: Diversity-promoting GAN: a cross-entropy based generative adversarial network for diversified text generation. In: EMNLP (2018)

25. Yi, X., Hong, L., Zhong, E., Liu, N.N., Rajan, S.: Beyond clicks: dwell time for personalization. In: Proceedings of the 8th ACM Conference on Recommender Systems, pp. 113–120. ACM (2014)
26. Yu, L., Zhang, W., Wang, J., Yu, Y.: SeqGAN: sequence generative adversarial nets with policy gradient (2016). ArXiv:abs/1609.05473
27. Zhou, T., Qian, H., Shen, Z., Zhang, C., Wang, C., Liu, S., Ou, W.: Jump: a jointly predictor for user click and dwell time. In: IJCAI (2018)
28. Zhu, Y., et al.: What to do next: modeling user behaviors by time-LSTM. In: IJCAI, pp. 3602–3608 (2017)

Modeling Dynamic Social Behaviors with Time-Evolving Graphs for User Behavior Predictions

Tianzi Zang[1], Yanmin Zhu[1(✉)], Chen Gong[1], Haobing Liu[1], and Bo Li[2]

[1] Department of Computer Science and Engineering, Shanghai Jiao Tong University,
Shanghai, China
{zangtianzi,yzhu,gongchen,liuhaobing}@sjtu.edu.cn
[2] Hong Kong University of Science and Technology, Kowloon, Hong Kong SAR
bli@cse.ust.hk

Abstract. The full coverage of Wi-Fi signals and the popularization of intelligent card systems provide a large volume of data that contain human mobility patterns. Effectively utilizing such data to make user behavior predictions finds useful applications such as predictive behavior analysis, personalized recommendation, and location-aware services. Existing methods for user behavior predictions mercly capture temporal dependencies within individual historical records. We argue that user behaviors are largely affected by friends in their social circles and such influences are dynamic due to users' dynamic social behaviors. In this paper, we propose a model named SDSIM which consists of three independent and complementary modules to jointly model the influences of user dynamic social behaviors, user demographics similarities, and individual-level behavior patterns. We construct time-evolving graphs to indicate user dynamic social behaviors and design a novel component named DSBcell which captures not only the social influences but also the regularity and periodicity of user social behaviors. We also construct a graph based on user similarities in demographics and generate a representation for each user. Experiments on two real-world datasets for multiple user behavior-related prediction tasks prove the effectiveness of our proposed model compared with state-of-the-art methods.

Keywords: User behavior prediction · Time-evolving graphs · Dynamic graph convolution network

1 Introduction

With the rapid development of informatization and digitalization, a large number of user behavior logs (e.g., Wi-Fi records and smart card records) continuously generate at all times [5,29]. The availability of such data offers a good opportunity to depict user behavior patterns which has a variety of applications including marketing strategy formulation, rational allocation of resources, personalized recommendation, and thus has attracted a lot of attentions [20,29].

© Springer Nature Switzerland AG 2021
C. S. Jensen et al. (Eds.): DASFAA 2021, LNCS 12681, pp. 526–541, 2021.
https://doi.org/10.1007/978-3-030-73194-6_35

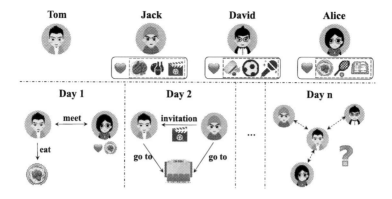

Fig. 1. An example of the social influences and users' dynamic social behaviors.

In this paper, we focus on multiple user behavior-related prediction tasks. Specifically, we study the behaviors of individuals with respect to three important aspects, namely, geographical location (**where** will a user visit), time (**when** will a user visit locations of a given category), and payment amount (**how much** will a user spend). These tasks investigate user mobility predictions in the spatial-temporal dimensions and have a deep sight in the consumption-ability of users, which are of great significance for providing personalized recommendations at the right place, time, and price.

Through an empirical study on real-world datasets, we find that user behaviors are influenced not only by historical behaviors but also by behaviors of others in their social circles. Furthermore, users' social behaviors are dynamic, which results in users being influenced by different friends each day. Figure 1 shows an example. As we can see, Tom has three friends (i.e., Jack, David, and Alice) with different interests and hobbies. On the first day, Tom and Alice meet at a conference, so Tom may eat pasta with Alice at her usual restaurant as Alice likes pasta. At that time, Tom's behaviors are mainly influenced by Alice and have nothing to do with Jack and David. On the second day, Tom receives Jack's invitation to see a movie, and Tom's behaviors will be more influenced by Jack than Alice and David. Besides, we propose that demographics also affect user behaviors, and users with similar demographics tend to have similar behaviors. A common scenario is that women generally shop more frequently than men while men eat more than women leading to a higher dining payment amount.

Existing research on user behavior modeling can be divided into two categories (i.e., probability graphical-based methods and recurrent neural network-based (RNN) methods). The probability graphical-based methods [6,20,25,26] regard multiple predictions such as spatial, temporal, payment, and category as different variables and build probability graphs to describe relationships among the variables. These methods are based on known casual relationships and conditional independence among variables, thus cannot model complex latent dependencies among variables. The RNN-based methods [1,10,17,28] utilize deep learning methods and treat user behavior predictions as sequence prediction tasks. There are multiple variants (e.g., bidirectional RNN [11] and Time-LSTM [29]) being applied to capture temporal dependencies within a sequence. However, these methods assume users behave independently and ignore the influences of user dynamic social behaviors as well as the demographics described above.

There are three main challenges to capturing the influences of user dynamic social behaviors and demographic similarities. First, in many cases, there are no explicit user-user relationships or observed social circles. The challenge is how to infer social relationships among users based on other signals in the data. Second, as shown in Fig. 1, users dynamically interact with different people within their social circles, thus different friends affect users' behaviors each day. However, there is no ready-made way to simultaneously capture such social influences as well as the regularity of users' dynamic social behaviors. Third, the demographics contain much information fields with redundancies and noises, the challenge is how to effectively utilize such information for user behavior predictions.

In response to the challenges mentioned above, we propose a model named **SDSIM** which consists of three independent and complementary modules to simultaneously capture the influences of user dynamic social behaviors, user similarities in demographics, and individual-level behavior patterns. First, to capture the influences of user dynamic social behaviors, we treat user co-occurrence behaviors as indications of social relationships and construct a graph to represent social circles among users. Based on this graph, we further construct time-evolving graphs in days to represent the dynamic social behaviors of users on each day. Then we design a novel component named DSBcell to capture the social influences from friends as well as the regularity and periodicity of users' social behaviors. Second, to capture the influences of user demographics similarities, instead of taking the information directly as input, we construct a graph based on user similarities in demographics and apply a graph embedding method mapping it into a dense latent space to reduce noises and redundancies. Third, we capture the influences of individual-level behavior patterns at different temporal granularities by an improved 1D dilated convolutional network. After that, we obtain predicted results based on the outputs of these three modules.

Our main contributions can be summarized as follows:

- We focus on multiple user behavior-related predictions and propose a model named SDSIM which jointly captures influences of user dynamic social behaviors, similarities in demographics, and individual-level behavior patterns.
- We construct time-evolving graphs in days based on co-occurrence behaviors among users to indicate user dynamic social behaviors. We design a component named DSBcell to capture the social influences from friends as well as the regularity and periodicity of user dynamic social behaviors.
- We demonstrate the effectiveness of our proposed SDSIM model in various applications. The experiments are conducted on two real-world datasets and results show that SDSIM consistently outperforms competitive baselines.

The remainder of this paper is organized as follows. After surveying the related work in Sect. 2, we formally define the user behavior prediction problem in Sect. 3. Section 4 introduces our proposed SDSIM model in detail. Section 5 presents the evaluation datasets, the baselines, and the experimental results. Finally, we conclude this paper in Sect. 6.

2   Related Work

2.1   User Behavior Modeling

Probabilistic Graphical-Based Models. Probabilistic graphical-based models [6,20,25,26] predict the mobility of an individual from multiple aspects such as spatial, temporal, payment, and category. They regard multiple predictions as different variables and construct graphs in which nodes denote variables and edges describe the conditional independence between variables. Wen et al. [20] modeled customer behaviors using a payment dataset from banks. Yuan et al. [25] studied individuals' mobility behaviors from aspects of the user, geographic information, time, and text contents from Twitter data. Cho et al. [6] took effects of social ties on human mobility into consideration. However, these methods are based on known casual relationships and conditional independence among variables. They cannot model complex latent dependencies among variables.

Recurrent Neural Network-Based Models. RNN-based models utilize deep learning methods and treat user behavior predictions as sequential modeling tasks. RNN and its variants (e.g., GRU and LSTM) have been widely proven to be powerful in modeling temporal dependencies within a sequence. In recent years, some work tries to improve the structure of the cells or adds attention mechanisms to enhance performance. Specifically, Ma et al. [11] utilized bidirectional RNN and attention mechanisms to predict patients' future health information using electronic health records. Time-LSTM [29] models time intervals between users' actions to improve recommendation performance. Liu et al. [10] proposed a profile-aware LSTM for multiple predictions about students. Yang et al. [21,22] predicted the time a user will visit a specific location using data collected from location-based social networks. However, these methods assume users behave independently and ignore the fact that users are often influenced by the activities of their friends and peers.

2.2   Graph Convolution Network-Based Prediction Models

Graph Convolutional Networks (GCNs) focus on integrating or aggregating signals from neighbor nodes by defining weighted average functions [12,13] or multiplication of signals in the Fourier domain [4,7]. They can model interactions between connected nodes and are suitable structures to capture social influences among users. To capture the dynamics of graphs in real-world scenarios, GCRN [16] combined the LSTM network with ChebNet. DCRNN [9] incorporated a proposed diffusion graph convolutional layer into a GRU network. STGCN [23] and ASTGCN [8] interleaved 1D-CNN layers with graph convolutional layers to jointly learn spatiotemporal dependencies. These methods only focus on changes in the inputs while cannot capture changes in the graph structures. A recently proposed approach named EvolveGCN [14] captured the dynamism of the graph sequence by using an RNN to evolve the GCN parameters. Our proposed DSBcell is different from theirs as it captures both the social influences from friends and the regularity and periodicity of users' dynamic social behaviors in the time-evolving graphs.

3   Problem Statement

In this section, we introduce relevant definitions and formulate the problem of user behavior predictions.

User Mobility Record. One user mobility record is denoted as $R_u = \{user\_id, time, loc\_id, context\}$ in which $user\_id$ is the identification of user u. $time$ and loc_id indicate when and where the record is generated. $context$ contains other related information such as the payment amount. Each loc_id represents a unique location which can be classified into different categories, such as canteen, theater, and workspace, based on functional descriptions.

Demographics. Demographics involve some basic information of a user which is denoted as $D_u - \{d_{u,1}, d_{u,2}, ..., d_{u,z}\}$. z denotes the total number of information fields that differ in different scenarios and datasets. Typically, the information fields include *gender*, *age*, *nationality*, and *occupation*.

Problem Formulation. The problem of user behavior predictions is defined as follows: given a user' past mobility records $\mathcal{R}_u^T = \{R_u^{T-M+1}, ..., R_u^{T-1}, R_u^T\}$ of previous M days and her demographics D_u, we aim to predict multiple behavior-related values (e.g., the next visit time $T_{u,c}^{next}$ to locations of a specific category c, that is, the $time$ in one record R_u; the next dining position P_u^{next} that is the loc_id in R_u; the payment amount A_u^{next} of a meal which is contained in the $context$ of R_u).

4   Proposed Model

In this section, we first show the overall structure of the proposed SDSIM model and then introduce the implementations of each module in detail.

4.1   Overview

Figure 2 shows the overall structure of our proposed SDSIM model which consists of three independent and complementary modules. The first module aims to capture the influences of user dynamic social behaviors. We construct time-evolving graphs in days based on user co-occurrence behaviors and design a novel component named DSBcell to capture such dynamic influences. The second module is designed to capture the influence of user similarities in demographics. The third module hierarchically captures the influences of individual-level behavior patterns. We concatenate the output of each module to generate predict values.

In the following, we introduce the functionality of each module as well as the implementations in detail.

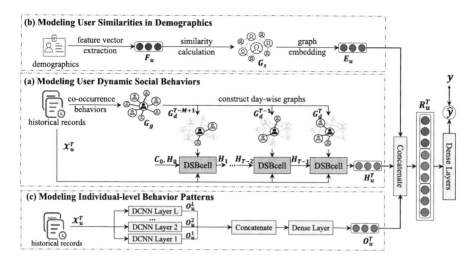

Fig. 2. The structure of SDSIM. It consists of three modules to jointly modeling the influences of user dynamic social behaviors, user similarities in demographics, and individual-level behavior patterns.

4.2 Modeling User Dynamic Social Behaviors

In many cases, social relationships among users are hidden without explicit description. Moreover, due to users' dynamic social behaviors, a user may be influenced by different friends each day. Therefore, as shown in Fig. 2 (a), we propose to construct time-evolving graphs to indicate user dynamic social behaviors and design a component named DSBcell to capture the influences from friends as well as the regularity and periodicity of user dynamic social behaviors.

Constructing Time-Evolving Graphs. When explicit social relationships are unavailable, we propose to represent user social circles based on co-occurrence behaviors among users as companionship can be seen as an indicator of social relationships. A co-occurrence behavior between user u_i and u_j is defined as they, respectively, have a mobility record R_{u_i} and R_{u_j} which satisfy that $loc\_id_{u_i} = loc\_id_{u_j}$ and $|time_{u_i} - time_{u_j}|$ is smaller than a fixed interval I.

We count the number of co-occurrence behaviors $N(u_i, u_j)$ in the training set and get a graph $G_g(V, E)$ where each node $u \in V$ denotes a user and the adjacent matrix A_g represents the social relationships among users. The values of entries in A_g are determined according to Eq. (1).

$$A_g(u_i, u_j) = \begin{cases} 0, & \frac{N(u_i, u_j)}{N_{tr}(u_i, u_j)} < \alpha \\ 1, & \frac{N(u_i, u_j)}{N_{tr}(u_i, u_j)} \geqslant \alpha, \end{cases} \tag{1}$$

$$N_{tr}(u_i, u_j) = |\mathcal{R}_{u_i}| + |\mathcal{R}_{u_j}|.$$

$|\mathcal{R}_u|$ is the number of mobility records in the training set of user u. Though one co-occurrence behavior may happen just due to chance, co-occurrence behaviors more than a proportion among users can be seen as an indication of social relationships. Therefore, a threshold α is set to mitigate the effects of chance events. $A_g(u_i, u_j) = 1$ means that two users are in a social circle while $A_g(u_i, u_j) = 0$ means not.

To represent the dynamic social behaviors of users, we further construct day-wise graphs $G_d^T(V^T, E^T)$. An edge $e_{ij}^T \in E^T$ is the co-occurrence times $N^T(u_i, u_j)$ on that day between two users who have an edge in G_g. Formally,

$$e_{ij}^T = \begin{cases} N^T(u_i, u_j), & A_g(u_i, u_j) = 1 \\ 0, & A_g(u_i, u_j) = 0. \end{cases} \tag{2}$$

As only co-occurrence times of users with an edge in G_g are counted, it is guaranteed that the social relationships contained in the day-wise graphs are subsets of users' social circles. Therefore, each day-wise graph reflects the social behaviors of users on a day and a sequence of such time-evolving graphs contains the regularity and periodicity of user social behaviors.

Designing DSBcell. To capture the influences of dynamic social behaviors in the time-evolving graphs, we design a component named DSBcell which incorporates graph convolution operations into the LSTM structure. Figure 3 shows an illustration of DSBcell.

The adjacent matrix $\mathbf{A}_t$ of time-evolving graphs and the record matrix $\mathbf{X}^t$ on that day take turns as input to DSBcell. Each row in $\mathbf{X}^t$ is a record vector X_u^t generated from mobility records R_u^t of a user. Taking predicting when a user will visit locations loc_c of a specific category as an example. We divide a day into S time slots and generate $X_u^t \in \mathbb{R}^S$ where $X_u^t[s]$ represents the number of times that u had visited loc_c between the s-th and $(s+1)$-th time slot on that day. The record vectors of all users are stacked together to form $\mathbf{X}^t$.

When update states, instead of just considering previous states, DSBcell first performs graph convolution operations on $\mathbf{X}^t$ to capture the influences from connected friends. Besides, the RNN-based structure makes DSBcell able to capture the periodicity and regularity of users' social behaviors contained in the sequence of time-evolving graphs. The updating formulas are as follows.

$$i_t = \sigma(W_i A_t + g_{\theta_{hki}} *_G [H_{t-1} \oplus X_u^t] + b_i), \tag{3}$$

$$f_t = \sigma(W_f A_t + g_{\theta_{hkf}} *_G [H_{t-1} \oplus X_u^t] + b_f), \tag{4}$$

$$o_t = \sigma(W_o A_t + g_{\theta_{hko}} *_G [H_{t-1} \oplus X_u^t] + b_o), \tag{5}$$

$$C_t = f_t \circ C_{t-1} + i_t \circ tanh(W_c A_t + g_{\theta_{ck}} *_G [H_{t-1} \oplus X_u^t] + b_c), \tag{6}$$

$$H_t = o_t \circ tanh(C_t), \tag{7}$$

where i_t, f_t and o_t denote the input gate, the forget gate, and the output gate and C_t, H_t are the memory and hidden states. g_θ are convolution filters and b are biases. $*_G$ denotes the spectral graph convolution operation. Formally,

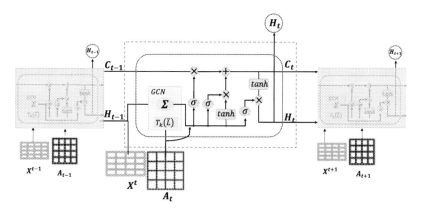

Fig. 3. An illustration of DSBcell which integrates graph convolution operation into the structure of LSTM. It captures the social influences from users as well as the regularity and periodicity of users' dynamic social behaviors.

$$g_0 *_G x - g_\theta(L)x = \sum_{k=0}^{K} \theta_k T_k(\tilde{L})x, \tag{8}$$

where x is the input and $\theta_k \in \mathbb{R}^K$ is a vector of polynomial coefficients that act as convolutional filters. $\tilde{L} = \frac{2}{\lambda_{max}}L - I_N$, where L is the normalized Laplacian matrix and I_N is a unit matrix. The Chebyshev polynomial is a recursive definition where $T_k(a) = 2aT_{k-1}(a) - T_{k-2}(a)$ and $T_0(a) = 1$, $T_1(a) = a$. The hyper-parameter K determines that the 0 to $(K-1)$-th order neighbors are considered when generate the predict values.

The hidden state H_T in the last step is treated as the generated representation of all users.

4.3   Modeling User Similarities in Demographics

To model user similarities in demographics, we first extract user feature vectors from demographics and construct a graph based on the feature vector similarities of every pair of users. We then apply a graph embedding method to generate a dense representation for each user where similar users have closer representations. We illustrate this process in Fig. 2(b).

Constructing Similarity Graph. Since demographics contain much information fields with redundancies and noises, to integrate them and facilitate predictions, we first extract the feature vector $F_u \in \mathbb{R}^{L_d}$ of each user by encoding discrete values with one-hot coding and concatenate them with continuous values. We then calculate the similarities between user feature vectors and get a static graph $G_s(V^s, E^s)$ in which each node $u \in V^s$ is a user while an edge $e_{ij}^s \in E^s$ denotes the cosine similarity between u_i and u_j. Formally,

$$e_{ij}^s = \frac{F_{u_i} \cdot F_{u_j}}{\|F_{u_i}\|\|F_{u_j}\|} = \frac{\sum_{l=1}^{L_d}(F_{u_i}^l \times F_{u_j}^l)}{\sqrt{\sum_{l=1}^{L_d}(F_{u_i}^l)^2} \times \sqrt{\sum_{l=1}^{L_d}(F_{u_j}^l)^2}}, \qquad (9)$$

where L_d denotes the total length of the feature vector. $F_{u_i}^l$ is the l-th value in the feature vector of u_i.

Generating User Representations. We apply a graph embedding method called LINE [18] to encode each user into a low-dimensional dense representation. It has a carefully designed objective function that preserves both the first-order and second-order proximities. Given the graph $G_s(V^s, E^s)$, for each node (i.e., a user), it outputs a dense latent representation $E_u \in \mathbb{R}^{L_s}$ where $L_s \ll L_d$ that preserves the structure of the graph. Therefore, the generated representations reflect similarities among users by forcing similar users to have closer representations. Note that LINE can be replaced with any other graph embedding methods, and we leave this for future work.

4.4   Modeling Individual-Level Behavior Patterns

We argue that there exist individual-level behavior patterns on different temporal scales contained in user mobility records. In general, user behaviors have a periodicity in days or weeks. Existing RNN-based methods tend to capture only temporal dependencies without explicitly modeling such periodicity. Therefore, we perform L dilated convolution layers (DCNN) [19,24] with different dilated rates on the input hierarchically to capture individual-level patterns at different temporal scales. For example, when the dilated rate is set to 7, this layer extracts dependencies between the same days of adjacent weeks.

Given $\mathcal{X}_u^T = \{X_u^{T-M+1}, ..., X_u^{T-1}, X_u^T\} \in \mathbb{R}^{M \times S}$ as input, the formula of DCNN is shown in the following.

$$O_u^l = \theta_d^l *_D \mathcal{X}_u^T, \qquad l = 1, 2, ..., L, \qquad (10)$$

where $*_D$ denotes the dilated convolutional operation. $\theta_d^l \in \mathbb{R}^{M_l \times S}$ and $O_u^l \in \mathbb{R}^{M \times S}$ are convolution filters and outputs of the l-th layer where M_l is the dilated rate. We applied padding operation to keep the shape of output consistent. A more detailed explanation of the DCNN network can be found in [19,24].

4.5   Generating Prediction Results

At each timestamp T and for each user u, we obtain three representations: H_u^T which contains the influences of user dynamic social behaviors, O_u^T which captures individual-level behavior patterns at different temporal scales, and E_u which preserves user similarities in demographics. We concatenate them as $R_u^T = E_u \oplus H_u^T \oplus O_u^T$ and add two fully-connected layers converting R_u^T into predicted values. Formally,

$$\hat{y}_u^{T+1} = \sigma(W_2(relu(W_1 R_u^T + b_1)) + b_2), \tag{11}$$

where W_1, W_2, b_1 and b_2 are learnable parameters. $\hat{y}_u^{T+1}$ is the predicted value. $relu$ and σ are activation functions.

4.6 Model Learning

Depending on whether the target values are continuous or discrete, user behavior predictions can be either regression tasks (e.g., predicting next visit time $T_{u,c}^{next}$ and predicting payment amount A_u^{next}) or classification tasks (e.g., predicting next dining position P_u^{next}). We learn parameters for regression tasks and classification tasks by, respectively, minimizing the root of the mean squared loss (denoted as $\mathcal{L}_1$) and the cross-entropy loss (denoted as $\mathcal{L}_2$) between predicted values $\hat{y}_u$ and true values y_u. The loss functions are as follows:

$$\mathcal{L}_1(\theta) = \sqrt{\frac{1}{U} \sum_{u=1}^{U}(y_u - \hat{y}_u)^2}, \qquad \mathcal{L}_2(\theta) = -\sum_{u=1}^{U}(y_u * log(\hat{y}_u)), \tag{12}$$

where θ are all learnable parameters in the model. U is the number of samples in a batch. Note that, for the sake of generality, we use $\hat{y}_u$ and y_u to represent predicted values and true values for both classification and regression tasks. In a specific task, y can be $T_{u,c}^{next}$, A_u^{next}, P_u^{next} or other user behavior-related values. We adopt the mini-batch gradient descent optimization method and the Adam optimizer to learn the parameters.

5 Experiments

In this section, based on two real-world datasets, we compare our SDSIM model with several models concerning both regression and classification tasks. We first describe the datasets followed by baselines, evaluation metrics, and settings of hyperparameters. Finally, we present experimental results in detail.

5.1 Dataset

We evaluate our proposed SDSIM model on two real-world datasets. The first dataset, **WIFI**[1], contains Internet access records of more than 20000 users in the community of a big company from September 1st, 2014 to January 31th, 2015. Each record includes user identification (anonymized), location, session start time, and session duration. There are more than 12,000,000 records. According to functional descriptions, positions are divided into different categories (e.g., canteen, gym, workspace, public activity building). We use this dataset to predict

---

[1] https://www.kesci.com/home/competition/55d1ca96fc5e031af03ddc65/content/1.

536   T. Zang et al.

users' next visit time to canteens and workspaces. The records of the first 4 months are used as the training set while the records of the last month are treated as the test set.

The second dataset, **Smart Card**, is collected from the campus smart card system of a university. Smart cards are used by students for making food payments. This dataset covers records of more than 1000 students from July 1st, 2016 to June 30th, 2017, and there are more than 600,000 records. Each record contains student id (anonymized), dining time, payment amount, and dining position. This dataset is used to make predictions about the dining position and payment amount. We use records of the first 9 months as the training set and the last 3 months as the test set.

5.2   Baselines and Evaluation Metrics

The baselines used in our experiments are given below:

- **HA**: Historical Average (HA) gives predictions based on the average values of historical observations.
- **BRR/LR** [27]: Bayesian Ridge Regression (BRR) and Logistic Regression (LR) are two generalized linear models.
- **SVR/SVM** [27] : Support Vector Regression (SVR) is a variant of Support Vector machine (SVM) for supporting regression tasks.
- **GBDT**: Gradient Boosting Decision Tree (GBDT) is an ensemble method based on a "boosting" idea.
- **FNN** [15]: A neural network stacked by several fully connected layers.
- **Time-LSTM** [29]: An improved structure of long-short term memory network (LSTM) which takes time interval into consideration.
- **ASTGCN** [8]: A spatial-temporal graph convolution model with an attention mechanism to capture both spatiotemporal dependencies.
- **DCRNN** [9]: The diffusion process is used to simulate the interactions between nodes in a graph.
- **EvolveGCN** [14]: A model that tries to capture dynamic interactions by combining GCN with GRU.

For regression tasks, we use the Mean Absolute Error (MAE) and the Root Mean Square Error (RMSE) as evaluation metrics. For classification tasks, we adopt three widely used evaluation metrics [2,3]: precision, recall, and macro-$F1$ score. Due to space limitations, we omit specific calculation formulas.

5.3   Experimental Setting

When predicting user next visit time to canteens, dining payment amount, and dining position, a day is divided into 18 time slots and each slot is half an hour as canteens only open between $6-9$, $11-14$ and $17-20$ each day. When predicting a user's next visit time to the workspace, we divide a day into 24 time slots and each time slot is an hour as workspaces are accessible at any time of day. We

Table 1. Performance comparison between SDSIM and baselines.

Dataset	WIF				Smart Card				
Model	Canteen		Workspace		Payment		Position		
	MAE	RMSE	MAE	RMSE	MAE	RMSE	Precision	Recall	Macro-F1
HA	0.424	0.538	0.313	0.386	1.658	2.727	–	–	–
BRR/LR	0.350	0.474	0.241	0.275	1.589	2.603	0.345	**0.302**	0.310
GBDT	0.342	0.382	0.222	0.251	1.552	2.560	0.312	0.289	0.299
SVR/SVC	0.359	0.427	0.261	0.282	1.669	2.581	0.324	0.298	0.308
FNN	0.358	0.439	0.238	0.264	1.218	2.084	0.319	0.287	0.294
Time-LSTM	0.334	0.382	0.215	0.240	1.150	2.078	0.344	0.274	0.309
ASTGCN	0.329	0.377	0.212	0.237	1.136	2.058	0.384	0.276	0.299
DCRNN	0.327	0.379	0.209	0.235	1.134	2.055	0.388	0.280	0.301
EvolveGCN	0.326	0.363	0.199	0.231	1.121	2.034	0.399	0.283	0.303
SDSIM	**0.310**	**0.363**	**0.199**	**0.231**	**1.073**	**2.007**	**0.451**	0.286	**0.319**

append two-dimensional metadata information, indicating the day of a week and weekday/weekend. For the dining payment amount and position predictions, we further append the payment amount and position of previous records as input.

In SDSIM, the state size of DSBcell is set to 64 and the length of user representation generated by LINE is set to 32. For each sample, we utilize records of the previous 14 days to predict the target values. We stack two-layer dilated $1D$ convolution networks with dilated rates of 1 and 7 to capture periodicities of days and weeks. When counting co-occurrence behaviors among users, the time interval is set to 10 minutes. When training the model, the learning rate is set to 0.001. We tune the parameters of baselines to get the best performance.

5.4 Experimental Results

In the following, we show the experimental results of SDSIM and baselines to evaluate the performance.

Performance Comparison. Table 1 shows the performance comparison between SDSIM and baselines. We can see that, except for the recall score in the dining position prediction task, SDSIM consistently achieves the best performance. The suboptimal performance is because higher precision tends to result in lower recall in theory. HA gets the worst performance as it does not capture the dependencies within the input. The performance of traditional machine learning methods (i.e., BPR, LR, GBDT, SVR, and SVC) is relatively poor resulted from their deficiencies in capturing complex dependencies. FNN and Time-LSTM get better performance as they capture non-linear dependencies among inputs. As both the graph convolution operation in ASTGCN and the diffusion process in DCRNN can capture social influences among users, they get satisfied performance on all the tasks. EvolveGCN gets the best performance among the

Table 2. Ablation analysis of SDSIM

Variants	MAE	RMSE	Precision	Recall	Macro-F1
SDSIM-I	1.245	2.272	0.306	0.277	0.294
SDSIM-S	1.134	2.054	0.391	0.273	0.296
SDSIM-D	1.127	2.031	0.420	0.281	0.301
SDSIM-DS	1.116	2.020	0.427	0.284	0.307
SDSIM	**1.073**	**2.007**	**0.451**	**0.286**	**0.319**

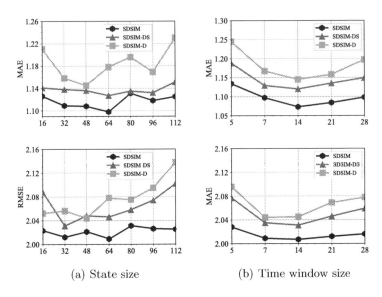

(a) State size   (b) Time window size

Fig. 4. Parameter sensitivity of SDSIM and its variants.

baselines as it captures user dynamic social behaviors. The best performance achieved by SDSIM on both regression and classification tasks demonstrates the reasonability of our design and the effectiveness to capture the influences of both the user dynamic social behaviors and demographics similarities.

Ablation Analysis. We study the effects of different modules in SDSIM. We only show the results on the **Smart Card** dataset while the results on the **WIFI** dataset are similar. To evaluate the impact of each module, we remove one factor at a time and get four variants. Among them, SDSIM-S is a variant that does not consider the influences of social behaviors. SDSIM-D is a variant that does not utilize demographics. SDSIM-I is a variant without modeling individual-level behavior patterns. SDSIM-DS only captures static social influences which means we dos not construct day-wise graphs. The performance is reported in Table 2. As we can see, SDSIM-I gets the worst performance, demonstrating individual-level behavior patterns are of the most crucial and central significance for user behavior predictions. The better performance of SDSIM-DS than SDSIM-S shows the necessity of capturing the influences of user social behaviors. Meanwhile, compared with SDSIM, the suboptimal performance of SDSIM-DS demonstrates the effectiveness of considering the dynamics of user social behaviors when capturing social influences. None of these variants perform as well as SDSIM showing the validity of our proposed model and the importance of simultaneously capturing the three types of influences.

Parameter Sensitivity. Figures 4(a) and 4(b) show the effects of two hyperparameters in our model, namely, the state size in DSBcell and time window size which determines how many days of records we utilize when generating a training sample. We show the performance changes of SDSIM and the two variants (i.e., SDSIM-DS and SDSIM-D). We vary the state size from 16 to 112. From Fig. 4(a), we can see that the state size does not affect the performance of the model significantly and we get both the smallest MAE and RMSE when we set the state size to 64. We then vary the time window size and respectively use records of the previous 5, 7, 14, 21, and 28 days to predict the target values. From Fig. 4(b), we can see when we set the time window size to 5, we get the worst performance. This may because that periodicity of weeks cannot be captured just from the records of the previous 5 days. The performance begins to improve as we use more records and get the best performance when we set the time window size to 14. However, when the time window size is too long and the dataset is relatively small, the number of training samples will be reduced and the performance of the model begins to decline. Therefore, in our experiment, we set the time window size to 14. Comparing the curve of SDSIM and the variants, we can find that SDSIM achieves optimal performance under all settings with the highest robustness and the lowest parameter sensitivity.

6   Conclusion and Future Work

In this paper, we focus on user behavior predictions. We propose a model named SDSIM which jointly captures the influences of user dynamic social behaviors, demographics similarities, and individual-level behavior patterns. We propose to construct day-wise graphs and design a novel component called DSBcell to capture the influences from friends as well as the regularity and periodicity of user social behaviors. We capture user similarities in demographics by constructing a user similarity graph and generating a static representation of each user. Experiments on two real-world datasets demonstrate the effectiveness of our model.

The proposed SDSIM model can be easily applied to other datasets such as location-based social network datasets with available social relationships and do not need to refer them. Besides, our SDSIM model can be easily extended to other behavior prediction tasks. When predicting some low-frequency behaviors, such as going to the gym or going to the theater, we can construct week-wise or month-wise time-evolving graphs to adapt to the new scenarios. In the future, we will try to improve the way of graph construction to get better performance.

Acknowledgments. This research is supported in part by the 2030 National Key AI Program of China 2018AAA0100503 (2018AAA0100500), National Science Foundation of China (No. 62072304, No. 61772341, No. 61472254), Shanghai Municipal Science and Technology Commission (No. 18511103002, No. 19510760500, and No. 19511101500), the Program for Changjiang Young Scholars in University of China, the Program for China Top Young Talents, the Program for Shanghai Top Young Talents, SJTU Global Strategic Partnership Fund (2019 SJTU-HKUST), the Oceanic Interdisciplinary Program of Shanghai Jiao Tong University (No. SL2020MS032) and Scientific Research Fund of Second Institute of Oceanography (No. SL2020MS032).

References

1. Bai, T., Zhang, S., Egleston, B.L., Vucetic, S.: Interpretable representation learning for healthcare via capturing disease progression through time. In: SIGKDD, pp. 43–51 (2018)
2. Bi, W., Kwok, J.T.: Multi-label classification on tree-and dag-structured hierarchies. In: ICML, pp. 17–24 (2011)
3. Braytee, A., Liu, W., Catchpoole, D.R., Kennedy, P.J.: Multi-label feature selection using correlation information. In: CIKM, pp. 1649–1656 (2017)
4. Bruna, J., Zaremba, W., Szlam, A., LeCun, Y.: Spectral networks and locally connected networks on graphs. In: ICLR (2014)
5. Chen, C., et al.: Predictive analysis by leveraging temporal user behavior and user embeddings. In: CIKM, pp. 2175–2182 (2018)
6. Cho, E., Myers, S.A., Leskovec, J.: Friendship and mobility: user movement in location-based social networks. In: SIGKDD, pp. 1082–1090 (2011)
7. Defferrard, M., Bresson, X., Vandergheynst, P.: Convolutional neural networks on graphs with fast localized spectral filtering. In: NIPS, pp. 3844–3852 (2016)
8. Guo, S., Lin, Y., Feng, N., Song, C., Wan, H.: Attention based spatial-temporal graph convolutional networks for traffic flow forecasting. In: AAAI, pp. 922–929 (2019)
9. Li, Y., Yu, R., Shahabi, C., Liu, Y.: Diffusion convolutional recurrent neural network: data-driven traffic forecasting. In: ICLR (2018)
10. Liu, H., Zhu, Y., Xu, Y.: Learning from heterogeneous student behaviors for multiple prediction tasks. In: Nah, Y., Cui, B., Lee, S.-W., Yu, J.X., Moon, Y.-S., Whang, S.E. (eds.) DASFAA 2020. LNCS, vol. 12113, pp. 297–313. Springer, Cham (2020). https://doi.org/10.1007/978-3-030-59416-9_18
11. Ma, F., Chitta, R., Zhou, J., You, Q., Sun, T., Gao, J.: Dipole: diagnosis prediction in healthcare via attention-based bidirectional recurrent neural networks. In: SIGKDD, pp. 1903–1911 (2017)
12. Monti, F., Boscaini, D., Masci, J., Rodola, E., Svoboda, J., Bronstein, M.M.: Geometric deep learning on graphs and manifolds using mixture model CNNs. In: CVPR, pp. 5115–5124 (2017)
13. Niepert, M., Ahmed, M., Kutzkov, K.: Learning convolutional neural networks for graphs. In: ICML, pp. 2014–2023 (2016)
14. Pareja, A., et al.: EvolveGCN: Evolving graph convolutional networks for dynamic graphs. In: AAAI (2020)
15. Pi, Q., Bian, W., Zhou, G., Zhu, X., Gai, K.: Practice on long sequential user behavior modeling for click-through rate prediction. In: SIGKDD, pp. 2671–2679 (2019)

16. Seo, Y., Defferrard, M., Vandergheynst, P., Bresson, X.: Structured sequence modeling with graph convolutional recurrent networks. In: Cheng, L., Leung, A.C.S., Ozawa, S. (eds.) ICONIP 2018. LNCS, vol. 11301, pp. 362–373. Springer, Cham (2018). https://doi.org/10.1007/978-3-030-04167-0_33
17. Su, Y., et al.: Exercise-enhanced sequential modeling for student performance prediction. In: AAAI, pp. 2435–2443 (2018)
18. Tang, J., Qu, M., Wang, M., Zhang, M., Yan, J., Mei, Q.: Line: Large-scale information network embedding. In: WWW, pp. 1067–1077 (2015)
19. Wang, P., et al.: Understanding convolution for semantic segmentation. In: WACV, pp. 1451–1460 (2018)
20. Wen, Y.T., Yeh, P.W., Tsai, T.H., Peng, W.C., Shuai, H.H.: Customer purchase behavior prediction from payment datasets. In: WSDM, pp. 628–636 (2018)
21. Yang, G., Cai, Y., Reddy, C.K.: Recurrent spatio-temporal point process for check-in time prediction. In: CIKM, pp. 2203–2211 (2018)
22. Yang, G., Cai, Y., Reddy, C.K.: Spatio-temporal check-in time prediction with recurrent neural network based survival analysis. In: IJCAI, pp. 2976–2983 (2018)
23. Yu, B., Yin, H., Zhu, Z.: Spatio-temporal graph convolutional networks: a deep learning framework for traffic forecasting. In: IJCAI, pp. 3634–3640 (2018)
24. Yu, F., Koltun, V.: Multi-scale context aggregation by dilated convolutions. In: ICLR (2016)
25. Yuan, Q., Cong, G., Ma, Z., Sun, A., Thalmann, N.M.: Who, where, when and what: discover spatio-temporal topics for twitter users. In: SIGKDD, pp. 605–613 (2013)
26. Yuan, Q., Zhang, W., Zhang, C., Geng, X., Cong, G., Han, J.: PRED: Periodic region detection for mobility modeling of social media users. In: WSDM, pp. 263–272 (2017)
27. Zhang, Y., Pennacchiotti, M.: Predicting purchase behaviors from social media. In: WWW, pp. 1521–1532 (2013)
28. Zhang, Y., et al.: Sequential click prediction for sponsored search with recurrent neural networks. In: AAAI, pp. 1369–1375 (2014)
29. Zhu, Y., et al.: What to do next: modeling user behaviors by time-LSTM. In: IJCAI, pp. 3602–3608 (2017)

Memory-Efficient Storing of Timestamps for Spatio-Temporal Data Management in Columnar In-Memory Databases

Keven Richly[✉]

Hasso Plattner Institute, University of Potsdam, Potsdam, Germany
keven.richly@hpi.uni-potsdam.de

Abstract. Vast amounts of spatio-temporal data are continuously accumulated through the wide distribution of location-acquisition technologies. Concerning the increased performance requirements of spatio-temporal data mining applications, in-memory database systems are used to store and process the data. As DRAM capacities are limited and expensive, the efficient utilization of the available resources is necessary. In contrast to storing the positions of moving objects, there is less focus on optimized storage concepts for the temporal component. However, it has a significant impact on the memory footprint and the overall system performance. Especially for columnar databases, the memory-efficient storing of timestamps is challenging as numerous compression approaches are optimized for contradicting data characteristics (e.g., low number of distinct values, sequences of equal values). In this paper, we present and compare different data layouts for columnar in-memory databases to store timestamps. Additionally, we propose an optimized approach for range queries with standard access ranges that uses multiple columns. We evaluate the memory consumption and performance of different compression techniques for specific access patterns. Based on the results, we introduce a workload-aware heuristic approach for the selection of performance and cost balancing data layouts. Further, we demonstrate that workload-driven optimizations for timestamps can significantly reduce the data footprint and increase the performance of spatio-temporal data management.

Keywords: In-memory databases · Data layout optimization · Spatio-temporal data

1 Introduction

In recent years, the wide distribution of location-acquisition technologies has led to large amounts of spatio-temporal data. Positioning systems such as GPS enable the tracking of different moving objects (e.g., persons or vehicles) [25]. Various applications use the generated trajectory data to analyze and optimize

© Springer Nature Switzerland AG 2021
C. S. Jensen et al. (Eds.): DASFAA 2021, LNCS 12681, pp. 542–557, 2021.
https://doi.org/10.1007/978-3-030-73194-6_36

movement patterns (e.g., ride-sharing) [13]. The trajectory of a moving object is represented by a chronologically ordered sequence of timestamped coordinates in a geographical reference system. Due to the massive volumes of continuously captured data, the storing and processing of trajectory data is challenging [23]. To address the performance requirements of modern spatio-temporal data mining applications, in-memory architectures are used for data management [17,24]. Especially for main-memory optimized databases that keep the most data in relatively limited and expensive DRAM, operating costs can be reduced by utilizing the available resources more efficiently [2]. The most common data format to store spatio-temporal data is the sample point format, in which each observed location is stored as a tuple with a set of attributes. This format is well suited for the relational schema of modern database systems. In addition to standalone storage systems specialized for trajectory data, database systems enable a simplified integration of further data sources (e.g., business data) [14]. By integrating spatio-temporal data management into relational database systems, the data querying benefits from the highly optimized data processing capabilities and the advanced compression techniques of such systems.

In the context of spatio-temporal data, the temporal component is mostly represented by timestamps to reflect different and varying sample rates. Similar to the location of a moving object, the memory-efficient storing of timestamps is challenging as numerous compression approaches in columnar databases are optimized for contradicting data characteristics (e.g., low number of distinct values or sequences of equal values). There are several research approaches, which focus on storing spatial coordinates in columnar databases [6,8]. In contrast, there is less focus on optimized data structures for the temporal component. However, it has a significant impact on the memory footprint and the overall system performance. By still applying storage concepts designed for row-oriented databases, we do not leverage the columnar data layout's full potential. Here, optimized data layouts can improve the runtime performance and reduce the memory footprint [2,3]. Another aspect is that in various spatio-temporal data mining applications, the queries request specific time ranges (e.g., all values of a day or an hour) predominantly, which only address parts of the temporal information. By considering these access patterns, we can apply optimized storage formats for timestamps, which are able to reduce the memory traffic and only transfer the relevant parts of the timestamp.

In this paper, we present and compare different data layouts for columnar in-memory databases to store timestamps. We evaluate the memory consumption and runtime performance of four data layouts in combination with different compression techniques. As the different layouts have advantages and disadvantages for specific requirements (e.g., memory limitations, performance constraints) and workload characteristics, we present a heuristic approach for the workload-driven joint selection of a data layout and compression scheme. Based on the weighted ratio between memory consumption and runtime performance, the selection can reflect different requirements. Additionally, we describe a multiple column data layout for timestamps, which can reduce memory traffic for access patterns that focus on parts of the timestamp. Furthermore, we introduce a linear programming approach to select a compression scheme that optimizes the compression rate and runtime performance of the multiple column data layout. Our contributions are the following:

- We evaluate the memory consumption and runtime performance of different data layouts and for timestamps in columnar in-memory databases.
- We present an optimized data layout for temporal range queries with standard access ranges that use multiple columns to store the different components.
- We introduce a heuristic approach for the workload-driven optimization of data structures for timestamps.

This paper is organized as follows. We describe related work in Sect. 2. In Sect. 3, we present four data layouts to represent timestamps in database systems. In Sect. 4, we evaluate the memory consumption and runtime performance of the data layouts for different compression techniques based on a real-world dataset of a transportation network company. Afterward, in Sect. 5, we present a workload-aware selection heuristic and introduce further optimizations concepts. We conclude the paper in Sect. 6.

2 Related Work

There are different approaches to store the timestamps of spatio-temporal and time-series data more efficient. One approach is to calculate the timestamp based on the position of an observed location in the sequence of chronologically ordered locations [13]. Here, no main memory is used to store the temporal component, which reduces the memory footprint of spatio-temporal data significantly. However, this approach does not apply to applications where the moving objects have different sample-rates, the locations are tracked without a fixed sample rate, or the transmission of each observed location can not be guaranteed. Pelkonen et al. [11] introduced a concept to store timestamps for time-series data as the delta of deltas. Here, the offset to the standard delta between two data points is stored with a variable amount of bits depending on the offset. Timestamps with the same delta can be stored with one bit, which makes the approach memory efficient. However, this storage concept is optimized for the analysis and visualization of time-series data. For spatio-temporal workloads with trajectory-based and spatio-temporal range queries, additional overhead is introduced due to the computation of an observed location's timestamp based on the deltas, which has to be performed for various moving objects. Wang et al. [20,21] presented an I/P frame-based structure to store trajectory data in in-memory column stores. The authors divide the data into frames of a fixed period (e.g., one minute) and create for each of these frames two table columns, which store the two-dimensional coordinates of one representative location for each moving object in this timeframe. In this approach, the determination of the timespan of one frame is challenging for use cases with varying sample rates. However, the time-frames introduce additional uncertainty. A disadvantage of this data layout is that many columns have to be scanned for spatial range queries.

Also, different research papers focus on the selection of optimized compression schemes. Abadi et al. [1] presented a decision tree-based approach to determine compression scheme in C-store. The selection is based on workload and data properties but does not adapt to changing environments nor considers memory budgets. Boissier et al. [2] introduced a workload-driven selection of compression configurations with memory constraints for columnar databases. The approach uses a greedy heuristic to determine configurations based on data characteristics and estimated runtime performances based on regression models. Lang et al. [7] presented data blocks for HyPer. It selects compression schemes statically based on data characteristics for the data blocks. These approaches focus on the optimized selection of compression scheme based on data and workload characteristics but do not consider that different data layouts can be applied to store specific data types. Based on the applied data layout, columns' data characteristics can change significantly and consequently create further optimization potentials.

3   Data Layouts for Timestamps in Columnar Databases

In various spatio-temporal applications, we observed that the provided temporal data types of columnar databases are not used. Although these data types offer many features to query and process temporal information, the applications use customized data layouts based on standard data types due to performance or memory budget constraints. In this section, we analyze the common data layouts (i) string format, (ii) Unix timestamps, and (iii) separated date and time columns concerning their applicability in columnar databases. Additionally, we introduce a multiple column approach to store timestamps.

3.1   Common Data Layouts for Timestamps

The first format based on standard data types that we observed in spatio-temporal data mining applications is the string format. It stores a timestamp in a single column of the data type string. Based on the ISO 8601 guidelines, the different time units (e.g., year, month, or day) are stored in descending order and divided by specific delimiters. The format applied by most applications is *YYYY-MM-DD HH:MM:SS*. Due to the specified delimiter symbols, it is possible to query parts of the timestamp (e.g., all observed locations in a specific hour over multiple days) via SQL LIKE statements. Due to the defined order of the time units, the chronological order of the timestamps is preserved. SQL statements that require such an order (e.g., BETWEEN) can be realized via string comparisons. A disadvantage of this approach is that a relatively high amount of memory is necessary to store a single element. Consequently, we have high memory traffic to process the data even if only parts of the timestamp are queried.

Another approach is the usage of Unix timestamps, which is widely used by operating systems and file formats. Here, each timestamp is stored as an

String (string)	Unix Timestamp (integer)	Date/Time (string)		Multiple Columns (integer)					

TIMESTAMP	TIMESTAMP	DATE	TIME	YEAR	MONTH	DAY	HOUR	MINUTE	SECOND
2018-11-01 00:00:11	1541030411	2018-11-01	00:00:11	2018	11	1	0	0	11
2018-11-01 00:00:26	1541030426	2018-11-01	00:00:26	2018	11	1	0	0	26
2018-11-01 00:01:42	1541030502	2018-11-01	00:01:42	2018	11	1	0	1	42
2018-11-01 17:00:16	1541091616	2018-11-01	17:00:16	2018	11	1	17	0	16
2018-11-03 06:14:40	1541225680	2018-11-03	06:14:40	2018	11	3	6	14	40

Fig. 1. Data layout of the different approaches to represent timestamps in databases.

integer value in a single column. The integer value represents the number of seconds that have elapsed since the Unix epoch, which is defined as the 00:00:00 UTC on the first of January 1970. Compared to the string format, we need less memory to store a single timestamp. Additionally, modern CPUs are optimized to process integers values efficiently [22]. This approach is well-suited for queries that request a continuous period in the data. Due to leap seconds and years, the calculation of specific recurring periods in a larger timeframe is not trivial. For that reason, several database systems convert the Unix timestamps into the string format to process these types of queries (e.g., all observed locations in an hour on various days). This entire process is time-consuming, especially for high spatio-temporal data volumes.

Several efficient compression techniques for columnar databases benefit from a relatively low number of distinct values (e.g., dictionary-encoding) or a high number of equal consecutive values (e.g., run-length encoding) [12]. Both presented data layouts led to a high number of distinct values. To address this issue, the timestamp is split into a date and time part and stored in separate columns in some applications. By storing date and time in different columns, we are able to reduce the number of distinct values. Especially for fine-grained tracking applications, which store the position of a moving object several times per minute, we have large sequences of equal values in the date column. Additionally, we can reduce the memory traffic for queries that only address the date or time unit of a timestamp. Correspondingly, we have to perform two scan operations for queries that access both components.

3.2   A Multiple Column Approach to Store Timestamps

Based on the concept of the separation of date and time, we propose a data layout that stores each time unit in a single integer column. This data layout has obvious disadvantages for workloads dominated by selects of random timestamps or timeframes between two random timestamps as multiple columns have to be scanned. In contrast, the number of columns that have to be considered is less for database queries that analyze only parts of the timestamp (e.g., all observed locations of a month). By applying this data layout, the data characteristic of the different columns changes (see Fig. 1). All columns have a limited low number of distinct values. Additionally, for most of the columns, we have larger sequences of equal values (e.g., the year column), which is beneficial for different data encodings. For dictionary-encoding, we also have the advantage of stable dictionaries. Thus, the number of bits necessary to store the dictionary position will not increase anymore (except the year column). For that reason, this approach is suited for massive amounts of spatio-temporal data. Due to simplification reasons, we ignore in this overview fractional seconds, which can be represented by additional columns.

4 Evaluation

Modern main-memory optimized databases employ a variety of compression techniques [2]. In this section, we demonstrate the impact on the memory consumption and runtime performance of the presented data layouts (cf. Sect. 3) in combination with different compression approaches. To evaluate the data layouts, we use a real-world dataset of a transportation company introduced in Sect. 4.1 and the in-memory research database *Hyrise*, which is optimized for columnar data layouts [4]. All measurements have been executed on a four-socket server equipped with Intel Xeon E7-4880v2 CPUs (2.50 GHz, 30 logical cores). We use the numactl command to bind the thread and memory to one node to avoid NUMA effects.

4.1 Dataset

For the evaluation, we use a real-world dataset of a transportation network company (e.g., Uber or Lyft). The dataset includes the timestamps of ten million dispatch process-related observed locations of drivers for three consecutive days in the City of Dubai. In comparison to other passenger transportation datasets (e.g., NYC Taxi Rides [18]), the dataset has a significantly higher granularity as the position of a driver is tracked every five seconds. Besides the timestamp, latitude, longitude, and the driver's identifier, a status attribute is tracked for each sample point. This status indicates the driver's occupation status [15]. Based on the insertion order, a certain temporal ordering of the sample points exists, but we cannot guarantee that the timestamp column is sorted due to transmission problems and delayed transmissions.

4.2 Impact of Different Compression Techniques on the Memory Consumption of the Data Layouts

To analyze the impact on the data footprint of different compression techniques in combination with the data layouts (cf. Sect. 3), we applied the three compression techniques (i) dictionary-encoding, (ii) LZ4, and (iii) run-length encoding on the ten million timestamps of the dataset. As displayed in Fig. 2, the applied compression approach has a significant impact on the memory size that is necessary to store the timestamps. For example, the Unix timestamp column consumes with applied LZ4 compression 1.3 MB, which is only 3.2% of the main memory to store the Unix timestamp column with applied dictionary encoding. Additionally, we observe that data layouts storing the data as integer values have a better compression rate for LZ4 and run-length encoding. Although the timestamps are split in the multiple column approach into six different columns, we only need 1.7 MB to store the data. Also, the measurements show that different compression techniques are better suited for specific data layouts. For example, the separated date and time approach consumes around two-thirds of the string approach for applied run-length encoding. In contrast, for LZ4 encoding, the string data layout needs 37.5 MB, which is significantly less memory than the separated date and time approach (66.88 MB). This is because, through the separation, the date column contains long sequences of equal values, which can be efficiently compressed with run-length encoding.

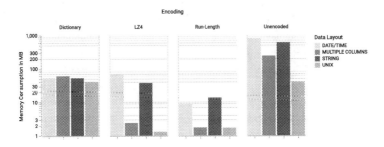

Fig. 2. Comparison of the memory consumption of the different data layouts in combination with various compression techniques.

A remarkable observation is that the multiple column approach has the highest memory consumption of the four data layouts for dictionary-encoding, even though the data characteristics should be beneficial for this compression technique. The reason for this is that the used columnar database does not support bit-packing mechanisms [22]. Consequently, one byte is used to store the dictionary position for each value instead of the minimal number of bits that are necessary to specify the dictionary entry. By applying such bit-packing mechanisms, we can reduce the data footprint significantly. Figure 3 shows the memory consumption for the six columns of the multiple columns data layout with and without bit-packing. Note, the memory consumption is calculated for the bit-packing columns. The bit-optimized storing of the values is particularly efficient for the YEAR and MONTH column in our dataset. As both columns only contain one distinct value, we need a single bit to store each value's dictionary position. Consequently, we can reduce the necessary data footprint of these columns to about 1.25 MB. By applying a bit-packing mechanism, the overall memory consumption of the multiple columns data layout could be reduced by nearly 50% to about 30.5 MB. With a memory consumption of about 30.5 MB the multiple column data layout would be the data layout with the smallest memory size compared to the other data layouts with applied dictionary encoding. Additionally, the space efficiency of the storage layout increases significantly [5].

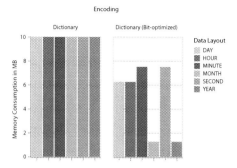

Fig. 3. Comparison of the memory consumption for the different columns of the multiple column data layout with (right) and without (left) applied bit-packing mechanism.

4.3 Impact of Different Compression Techniques on the Runtime Performance of the Data Layouts

Data compression is usually a tradeoff between compression rate and runtime performance [14]. We defined a set of five benchmark queries to analyze the four data layouts' runtime performance in combination with different compression techniques. To consider a broad spectrum of access patterns, the benchmark contains different commonly establish query types, including single value access and range scans with varying selectivity values as well as query ranges. The first query (Q0) is an equal scan that selects one specific timestamp. The second query (Q1) and the third query (Q2) select all observed locations between two timestamps. Q1 queries all observed locations between the 1st of November 2018, 21:01:43, and the 3rd of November 2018, 03:15:06, which include five million data points. Q2 queries all observed locations in a timeframe of 180 s. The queries Q3 and Q4 access only parts of the timestamps. Q3 queries only the date part and returns all data entries of the 2nd of November 2018 and Q4 returns for all three days all data entries between 6 am and 6:59 am.

As displayed in Fig. 4, depending on the applied compression technique, there are significant differences in the query runtime. Overall, dictionary-encoding has the lowest query runtimes. Also, the string and the approach that separates the date and time column have a comparable performance for dictionary encoding compared to the multiple columns and Unix timestamp approach. For all other compression techniques, these two approaches have significantly slower query runtimes. We can observe that the separated date and time approach has a better query runtime than the string data layout except for query Q2. Also, the multiple columns approach has an equal or better performance compared to the Unix timestamp except for query Q2. This is because multiple scan operations that depend on the previous column scans' results have to be performed for the multiple columns approach. In contrast, the multiple columns data layout has a better scan performance for Q4 compared to the Unix timestamp. Finally, we can observe that the selection of the data layout and compression technique strongly depends on the given workload characteristics.

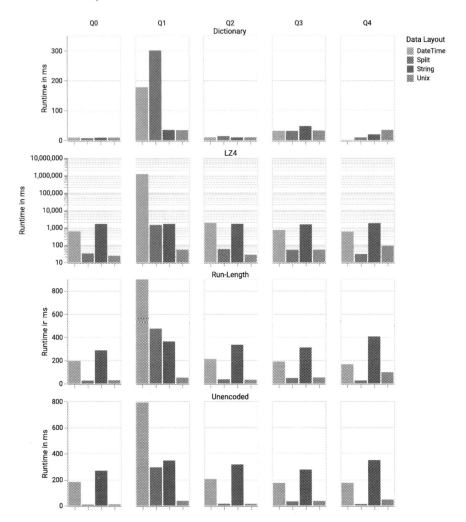

Fig. 4. Comparison of the runtime performance of the different data layouts in combination with different compression techniques.

5 Workload-Aware Optimizations of Data Layouts for Timestamps in Columnar In-Memory Databases

As described in the previous section, the different configurations to store timestamps have significant differences in memory consumption and runtime performance for various workload characteristics. The selection of cost and performance balancing configurations is challenging due to the number of possible configurations and the fact that usually spatio-temporal applications with query characteristics operate on the same database table. To optimize the storage layout of timestamps, we present two different workload-driven approaches. In

Sect. 5.1, we describe a heuristic approach for the workload-aware selection of a timestamp storage configuration consisting of a data layout and compression technique. In Sect. 5.2, we introduce a linear programming model to select a optimized compression scheme for the multiple column data layout.

5.1   Workload-Driven Combined Data Layout and Compression Scheme Optimization for Timestamps

To select a configuration for storing timestamps, we can apply different heuristic approaches. As the selection decisions of a data layout and compression technique are mutually dependent (cf. Sect. 4), we propose a joint optimization approach. For each data layout $n \in N$, where N describes the set of available data layouts, and compression technique $e \in E$, where E defines the set of available data encodings, we specify $b_{n,e}$ as the memory consumption to store the timestamps of the dataset with the data layout n and the encoding e. For applications with strict memory requirements, we can select the configuration with minimal memory consumption, $min_{e,n}(k_{n,e})$ for $e \in E, n \in N$. Equally, we can select the configuration based on the maximum performance for a given workload without any data footprint considerations. We specify a workload Q as a set of queries $q \in Q$. The performance of a workload Q is determined as the sum of the costs of each query $q \in Q$, which are calculated based on the costs of the query execution $d_{n,e,q}$ multiplied with the frequency of the query f_q,

$$\sum_{q \in Q} d_{q,n,e} \cdot f_q. \tag{1}$$

We calculate the benefit $r_{n,e}$ based on a weighted ratio between memory consumption and runtime performance to optimize the timestamps' storage configuration. A similar approach is used by Valentin et al. [19] in the context of the index selection problem. For each combination of a data layout $n \in N$ and data encoding $e \in E$, we define the benefit r as ($\alpha \geq 0$):

$$r_{n,e} = 1/\left(k_{n,e} \cdot \left(\sum_{q \in Q} d_{q,n,e} \cdot f_q\right)^{\alpha}\right). \tag{2}$$

The α value defines the proportional balancing of memory consumption and runtime performance for the optimization objective. This factor enables the database administrator to adapt the optimization process to different application requirements. To illustrate the impact of various α values, we analyze the determined configurations for three chosen α values. In Fig. 5, we visualize the data layout and compression techniques for a given α value and workload distribution. The workload is defined based on the queries introduced in Sect. 4.3. For a value $\alpha \leq 1$, the memory consumption is higher weighted than the runtime performance. Consequently, for all three workload distributions, the Unix timestamp data layout in combination with LZ4 encoding is selected for $\alpha = 0.1$, as this combination has the lowest memory footprint (cf. Fig. 2). The first workload is dominated by the runtime of Q2 (cf. Fig. 4). For that reason, for all three α values, the Unix timestamp data layout is selected. As for $\alpha = 5$ the performance is significantly higher weighted as the memory consumption dictionary encoding is chosen compared to LZ4 encoding for $\alpha = 2$. For the two other workload distributions, the separated date and time format is selected for $\alpha = 5$ based on the superior runtime performance for query Q0, Q2, and Q4 with applied dictionary encoding. As this data layout has a relatively high memory consumption, the multiple columns approach is selected for $\alpha = 2$.

Workload Distribution in Percent					Selected Data Layout and Encoding		
Q0	Q1	Q2	Q3	Q4	$\alpha = 0.1$	$\alpha = 2$	$\alpha = 5$
20	20	20	20	20	Unix LZ4	Unix LZ4	Unix Dictionary
15	1	34	40	10	Unix LZ4	Multiple Columns Run-Length	DateTime Dictionary
50	1	9	20	20	Unix LZ4	Multiple Columns Run-Length	DateTime Dictionary

Fig. 5. Visualization of the selected data layout and compression technique by the heuristic, cf. (2), for different workloads based on the queries introduced in Sect. 4.3.

Modern databases divide database tables into partitions of fixed size to benefit from pruning during query execution, more efficient workload distribution, and simplified data tiering [4,7,9,10]. An optimized adaptation of the storage layout to spatio-temporal access patterns can be archived by applying optimized configurations individually for each partition [16]. Here, we have to calculate the benefit for each partition $r_{n,e,p}$ for $n \in N$, $e \in E$, and $p \in P$, where P describes a set of partitions. Consequently, we have to adapt the equation:

$$r_{n,e,p} = 1/\left(k_{n,e,p} \cdot \left(\sum_{q \in Q} d_{q,n,e} \cdot f_q \cdot a_{p,q}\right)^\alpha\right). \tag{3}$$

In Eq. 3, we determine the memory consumption $b_{n,e,p}$ separately for each partition. Additionally, we have to introduce a parameter $a_{p,q}$, which determines the proportional share of the partition p of the costs $c_{q,n,e}$ for a given query $q \in Q$.

5.2 Workload-Driven Compression Scheme Selection for Storing Timestamps in a Multiple Column Data Layout

By storing the time units of a timestamp in multiple columns, we have further potentials for optimizations. As displayed in Fig. 6, the columns in the multiple columns data layout have different data characteristics. For example, in our dataset of a transportation network company (cf. Sect. 4.1), the YEAR column has only one distinct value. Consequently, this column is well-suited for run-length encoding or could also be stored as a single default value. In contrast, the SECOND column has significantly shorter sequences of equal values, which leads to a lower compression rate for run-length encoding. To address this issue, we propose a linear programming approach to determine optimized compression schemes for given workloads and memory budgets. By introducing fixed memory budgets, we are able to specify the amount of memory that should be used for the timestamp.

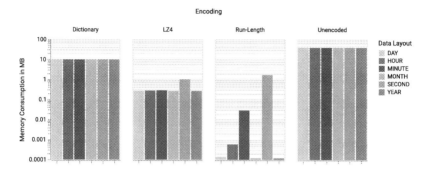

Fig. 6. Comparison of the memory consumption of the different columns in the multiple columns data layout for timestamps.

To select the best compression technique for each column based on a given workload Q, we have to determine the proportional costs of each database operation (e.g., scan operation) for each database column. For that reason, we define the function $D(q)$, which returns all operations of a query q with $q \in Q$. As spatio-temporal workloads are often dominated by range queries and trajectory-based queries [13], we are focusing on scan operations. Correspondingly, S describes the set of all scan operations of a given workload:

$$S = \bigcup_{q \in Q} D(q). \tag{4}$$

The costs of a scan operation s on a column is denoted as $c_{s,e}$ and determined by the column's encoding $e \in E$. For $s \in S, e \in E$, we define:

$$c_{s,e} := p_{s,e} \cdot \omega_s \cdot f_s \cdot u_{s,e}. \tag{5}$$

The parameter $p_{s,e}$ defines the measured scan performance of an isolated scan operation with a similar scan selectivity value on the scan column of n_s with applied encoding e. Further, the parameter ω_s denotes the accumulated selectivity of the previous operations of the same query and f_s the frequency of the scan operation. We use the successive scan penalty $u_{s,e}$ as we observed that consecutive scans are slower than single scan operations, depending on the applied compression technique e.

The objective of the model is to minimize the costs (in this case, the runtime) for a given set of scan operations S,

$$min \sum_{s \in S, e \in E} x_{n_s,e} \cdot c_{s,e} \qquad (6)$$

$$\sum_{n \in N, e \in E} x_{n,e} \cdot b_{n,e} \leq B, \qquad (7)$$

where the binary variable $x_{n_s,e}$ describe whether a certain encoding e is applied ('1') or not ('0') on column n_s. Here, $n_s \in N$ is the column in the set of all columns N that is scanned by the given scan operation s. For the model, we define two constraints. The first constraint guarantees that the accumulated memory consumption of all columns $n \in N$ with their selected encoding does not exceed the given memory budget B, i.e.,

To guarantee that for each column n exact one compression approach $e \in E$ is selected, we specify the second constraint:

$$\sum_{e \in E} x_{n,e} = 1 \quad \forall n \in N. \qquad (8)$$

To evaluate our linear programming model, we defined a set of overall 100 benchmark queries based on the query templates introduced in Sect. 4.3. In the defined benchmark, we have the following query distribution: all observed timestamps of a specified day are selected by 40% of the queries, a specific timestamp is selected by five percent of the queries, and a specific timeframe of 20 s is also selected by five percent of the queries. Additionally, the second half of the benchmark consists of a set of queries that returns all observed timestamps at all days in a timeframe of 30 min (20%), before noon (15%), and in a timeframe of two hours (15%).

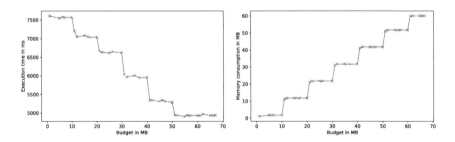

Fig. 7. Comparison of the measured runtime performance (left) and memory consumption (right) of the applied compression scheme determined based on the given workload for different memory budgets B.

We used the model to determine the compression configuration for six different columns of the timestamp for various memory budgets B and the described workload. As displayed in Fig. 7, we applied the configurations on the dataset of ten million timestamps and measured the runtime performance and memory consumption. As expected, the runtime of the benchmark queries decreases for increased memory budgets. The step-wise increases in the runtime performance are based on replacing the applied encoding of a column from run-length or LZ4 encoding to dictionary encoding, which has a faster scan performance. We could observe that the increases of the consumed memory size have no significant impact after a memory budget of 50 MB. Here, we use 10 MB to compress the SECOND column with dictionary encoding, which has no considerable impact on the overall performance.

		Compression Schema					Measurements	
	YEAR	MONTH	DAY	HOUR	MINUTE	SECOND	Memory Consumption	Runtime
1.1 MB	Run-Length	Run-Length	Run-Length	Run-Length	Run-Length	LZ4	1.05 MB	7609.68 ms
6.1 MB	Run-Length	Run-Length	Run-Length	Run-Length	Run-Length	Run-Length	1.73 MB	7580.04 ms
11.1 MB	Run-Length	Run-Length	Dictionary	Run-Length	Run-Length	LZ4	11.05 MB	7205.69 ms
16.1 MB	Run-Length	Run-Length	Dictionary	Run-Length	Run-Length	Run-Length	11.73 MB	7054.54 ms
51.1 MB	Dictionary	Dictionary	Dictionary	Dictionary	Dictionary	LZ4	51.02 MB	4943.79 ms
61.1 MB	Dictionary	Dictionary	Dictionary	Dictionary	Dictionary	Dictionary	60.00 MB	4927.80 ms
	Dictionary	Dictionary	Dictionary	Dictionary	Dictionary	Dictionary	60.00 MB	4927.80 ms
	LZ4	LZ4	LZ4	LZ4	LZ4	LZ4	2.39 MB	14429.28 ms
	Run-Length	Run-Length	Run-Length	Run-Length	Run-Length	Run-Length	1.73 MB	7534.34 ms
	Unencoded	Unencoded	Unencoded	Unencoded	Unencoded	Unencoded	240.00 MB	4967.31 ms

Fig. 8. Evaluation of the measured runtime performance and memory consumption for different compression scheme determined by the linear programming models for the memory budgets B (top) compared to the standard compression scheme, where all columns are encoded with the same compression technique (bottom).

Figure 8 provides more details about the selected compression scheme for different memory budgets. The evaluation shows that for low memory budgets, the linear programming model's compression scheme reduces the data footprint to store the ten million timestamps to 1.05 MB. This represents a reduction of over one-third compared to the scheme that uses run-length encoding for all columns, which has the lowest memory consumption of the four single encoding compression scheme with 1.74 MB. With increasing memory budgets, we can observe that the linear programming model replaces run-length encoded columns with dictionary-encoded columns, which have a better runtime performance. Here, we selected the memory budgets in such a way that the model can not only replace run-length encoded columns with dictionary-encoded columns in the compression configurations. As displayed in the configuration for a memory budget of $B = 11.1$ MB, to apply dictionary encoding on the DAY column and do not violate the memory budget constraint, cf. (7), the consumed memory of the SECOND column is reduced by applying LZ4 encoding compared to the configuration for $B = 6.1$ MB. This is done due to the fact that the overall benchmark performance benefits more from faster scan operations on the DAY column compared to the SECOND column. The DAY column's scan performance has the proportional highest impact for the given workload, as 80% of the queries access this column. Consequently, this is the first column the model determines to apply dictionary encoding.

6   Conclusions and Future Work

This paper demonstrates the impact of different data layouts and compression approaches for timestamps on data footprint and runtime performance. We pointed out that the presented commonly established approaches have advantages and disadvantages for different workload characteristics. Furthermore, we introduce a data layout that uses multiple columns to store a single timestamp. This optimized data layout for columnar databases is beneficial for various workloads and data compression techniques. Additionally, it enables the enhanced selection of a compression scheme that incorporates the specific workload and data characteristics (e.g., limited number of distinct values per column). To determine an optimized compression scheme for a given workload and memory budget, we introduce a linear programming model. This model enables database administrators to restrict the used memory for timestamps and evaluate the anticipated performance decreases.

For the joint workload-aware selection of a superior compression scheme and data layout, we present and evaluate a heuristic approach, which enables the balancing of memory consumption and performance requirements. We describe an extension of the heuristic for partitioned data tables to reflect time-specific access patterns by applying different configurations for various data partitions. In future work, the approaches can be extended to include the effects of auxiliary data structures such as indexes. Additionally, the database should be able to optimize the storage configuration of timestamps for specific workloads and requirements autonomously. It should provide a unified interface (e.g., the timestamp data type provided by various database systems) that the applied storage optimizations are transparent for applications. Overall, we demonstrate that workload-aware data layout and compression scheme optimizations can significantly reduce memory consumption and improve performance.

References

1. Abadi, D.J., et al.: Integrating compression and execution in column-oriented database systems. In: Proceedings of SIGMOD, pp. 671–682 (2006)
2. Boissier, M., Jendruk, M.: Workload-driven and robust selection of compression schemes for column stores. In: Proceedings of EDBT, pp. 674–677 (2019)
3. Boncz, P.A., et al.: Database architecture optimized for the new bottleneck: memory access. In: VLDB, pp. 54–65 (1999)
4. Dreseler, M., et al.: Hyrise re-engineered: an extensible database system for research in relational in-memory data management. In: Proceedings of EDBT, pp. 313–324 (2019)
5. Dyreson, C.E., Snodgrass, R.T.: Timestamp semantics and representation. Inf. Syst. **18**, 143–166 (1993)

6. Kazmaier, G.S., et al.: Managing and querying spatial point data in column stores, Patent app. 13/962,725
7. Lang, H., et al.: Data blocks: hybrid OLTP and OLAP on compressed storage using both vectorization and compilation. In: Proceedings of SIGMOD, pp. 311–326 (2016)
8. Pandey, V., et al.: High-performance geospatial analytics in hyperspace. In: Proceedings of SIGMOD, pp. 2145–2148 (2016)
9. Patel, J.M., et al.: Quickstep: a data platform based on the scaling-up approach. Proc. VLDB **11**, 663–676 (2018)
10. Pavlo, A., et al.: Self-driving database management systems. In: CIDR (2017)
11. Pelkonen, T., et al.: Gorilla: a fast, scalable, in-memory time series database. Proc. VLDB Endow. **8**, 1816–1827 (2015)
12. Plattner, H.: The impact of columnar in-memory databases on enterprise systems: implications of eliminating transaction-maintained aggregates. Proc. VLDB Endow. **7**, 1722–1729 (2014)
13. Richly, K.: A survey on trajectory data management for hybrid transactional and analytical workloads. In: IEEE Big Data, pp. 562–569 (2018)
14. Richly, K.: Optimized spatio-temporal data structures for hybrid transactional and analytical workloads on columnar in-memory databases. In: Proc. VLDB, Ph.D. Workshop (2019)
15. Richly, K., et al.: Predicting location probabilities of drivers to improve dispatch decisions of transportation network companies based on trajectory data. In: ICORES, pp. 47–58 (2020)
16. Richly, K., et al.: Joint index, sorting, and compression optimization for memory-efficient spatio-temporal data management. In: Proceedings of ICDE (2021)
17. Shang, Z., et al.: DITA: Distributed in-memory trajectory analytics. In: Proceedings of SIGMOD, pp. 725–740 (2018)
18. Taxi, N., (TLC), L.C.: Trip record data (2020). https://www1.nyc.gov/site/tlc/about/tlc-trip-record-data.page
19. Valentin, G., et al.: DB2 advisor: an optimizer smart enough to recommend its own indexes. In: Proceedings of ICDE, pp. 101–110 (2000)
20. Wang, H., et al.: SharkDB: an in-memory column-oriented trajectory storage. In: Proceedings of CIKM, pp. 1409–1418 (2014)
21. Wang, H., et al.: Storing and processing massive trajectory data on SAP HANA. In: Sharaf, M.A., Cheema, M.A., Qi, J. (eds.) ADC 2015. LNCS, vol. 9093, pp. 66–77. Springer, Cham (2015). https://doi.org/10.1007/978-3-319-19548-3_6
22. Willhalm, T., et al.: SIMD-scan: ultra fast in-memory table scan using on-chip vector processing units. Proc. VLDB Endow. **2**, 385–394 (2009)
23. Xie, X., Mei, B., Chen, J., Du, X., Jensen, C.S.: Elite: an elastic infrastructure for big spatiotemporal trajectories. VLDB J. **25**(4), 473–493 (2016). https://doi.org/10.1007/s00778-016-0425-6
24. Zhang, Z., Jin, C., Mao, J., Yang, X., Zhou, A.: TrajSpark: a scalable and efficient in-memory management system for big trajectory data. In: Chen, L., Jensen, C.S., Shahabi, C., Yang, X., Lian, X. (eds.) APWeb-WAIM 2017. LNCS, vol. 10366, pp. 11–26. Springer, Cham (2017). https://doi.org/10.1007/978-3-319-63579-8_2
25. Zheng, Y.: Trajectory data mining: an overview. ACM TIST **6**, 1–41 (2015)

Personalized POI Recommendation: Spatio-Temporal Representation Learning with Social Tie

Shaojie Dai, Yanwei Yu$^{(\boxtimes)}$, Hao Fan, and Junyu Dong

Department of Computer Science and Technology,
Ocean University of China, Qingdao, China
daishaojie@stu.ouc.edu.cn, {yuyanwei,fanhao,dongjunyu}@ouc.edu.cn

Abstract. Recommending a limited number of Point-of-Interests (POIs) a user will visit next has become increasingly important to both users and POI holders for Location-Based Social Networks (LBSNs). However, POI recommendation is a challenging task since complex sequential patterns and rich contexts are contained in extremely sparse user check-in data. Recent studies show that embedding techniques effectively incorporate POI contextual information to alleviate the data sparsity issue, and Recurrent Neural Network (RNN) has been successfully employed for sequential prediction. Nevertheless, existing POI recommendation approaches are still limited in capturing user personalized preference due to separate embedding learning or network modeling. To this end, we propose a novel unified spatio-temporal neural network framework, named PPR, which leverages users' check-in records and social ties to recommend personalized POIs for querying users by joint embedding and sequential modeling. Specifically, PPR first learns user and POI representations by joint modeling User-POI relation, sequential patterns, geographical influence, and social ties in a heterogeneous graph, and then models user personalized sequential patterns using the designed spatio-temporal neural network based on LSTM model for the personalized POI recommendation. Extensive experiments on three real-world datasets demonstrate that our model significantly outperforms state-of-the-art baselines for successive POI recommendation in terms of Accuracy, Precision, Recall and NDCG. The source code is available at: https://github.com/dsj96/PPR-master.

Keywords: POI recommendation · Location-based social network · Spatio-temporal neural network · Heterogeneous graph

1 Introduction

Newly emerging LBSNs has become an important mean for people to share their experience, write comments, or even interact with friends. With the prosperity of LBSNs, many users check in at various POIs via mobile devices in real time. Therefore, a large amount of check-in data is being generated, which is crucial to understand the users' preferences and behaviors. POI recommendation not only helps

© Springer Nature Switzerland AG 2021
C. S. Jensen et al. (Eds.): DASFAA 2021, LNCS 12681, pp. 558–574, 2021.
https://doi.org/10.1007/978-3-030-73194-6_37

users explore attractive and interesting places, but also gives guidance to location-based service providers, where to launch advertisements to target customers for marketing. Due to the great significance to both of users and businesses, how to use spatio-temporal information effectively, and recommend a limited number of POIs users more likely visit next have been attracting increasing attention in both industry and academia.

In particular, several studies [2,10,14,20,24] have been conducted to recommend successive POIs for users based on users' spatio-temporal check-in sequence in LBSNs. Based on Markov chain model, LORE [24] and NLPMM [2] explore users' successive check-in patterns by considering temporal and spatial information. ST-RNN [10] employs RNN to capture the users' sequential check-in behaviors. In a follow-up work, STGN [28] carefully designs the time gates and distance gates in LSTM to model users' sequential visiting behaviors by enhancing long short term memory. Additionally, some models [1,4] based on Word2Vec [15] framework to capture the preference and mobility pattern of users and the relationship among POIs also achieved decent performance. GE [20] uses graph embedding to combine the sequential effect, geographical influence, temporal effect and semantic effect in a unified way for location-based recommendation. Recently, SAE-NAD [14] utilizes a self-attentive encoder to differentiate the user preference and a neighbor-aware decoder to incorporate the geographical context information for POI recommendation.

However, location-based POI recommendation still faces three major challenges. First, *data sparsity*, unlike the general e-commerce, music and movie recommendation, which can be collected and verified just online, location-based POI recommendation systems usually associate with the POI-entities. Only when a user visits a POI-entity, a check-in record is generated. Therefore, the check-in records in the POI recommendation task is much sparser. This issue has plagued many POI recommendation models based on the collaborative filtering. Furthermore, data sparsity problem in check-in data makes it difficult to capture user's sequential pattern, because the check-in sequence is very short or is not continuous in time. Second, *contextual factors*, POI recommendation may be affected by various contextual factors, including social tie influence, geographical influence, temporal context, and so on. In fact, social ties are often available in LBSNs, and recently studies show that social networks associated with users are important in POI recommendation task since users are more likely to be influenced by their close friends (Who keeps company with the wolf will learn to howl). In this work, we incorporate social ties, check-in time interval, sequential and geographical effect into user-POI interaction graph to joint learn user and POI representations. Lastly, *dynamic and personalized preferences*, users' preferences are changing dynamically over time. At different time and circumstances, users may prefer different POIs. For example, some users prefer to visit gourmet restaurants in the local area, but when they go to a new city, some prefer to visit the cultural landscapes, while some prefer the natural landscapes. Dynamically and accurately capturing this trend has been proved to be essential for personalized POI recommendation task. However, effectively modeling the personalized sequential transitions from the sparse check-in data is challenging.

To address the aforementioned challenges, in this work, we stand on advances in embedding technique and RNN network, and propose our model, named **PPR**, which is a *spatial-temporal representation learning* framework for personalized and successive POI recommendation. First, we jointly model the user-POI relation, sequential effect, geographical influence and social ties by constructing a heterogeneous graph, and then develop a densifying trick by adding second-order neighbors to nodes with low in/out-degrees to alleviate the data sparsity issue. Then, we learn user and POI representations by embedding the densified heterogeneous graph into a shared low-dimensional space. Furthermore, to better capture the user dynamic and personalized preference, we also design a spatio-temporal neural network by concatenating user embedding, POI embedding and POI category as personalized sequence input to feed the network.

The main contributions of this paper are summarized as follows:

- We propose a novel PPR model for personalized POI recommendation, which incorporates users' check-in records and social ties. We construct a heterogeneous graph by jointly taking user-POI relation, sequential pattern, geographical effect and social ties into consideration to learn the representations of users and POIs.
- We propose a spatio-temporal neural network to model users' dynamic and personalized preference by concatenating user, POI embedding and POI category to generate personalized behavior sequence.
- We conduct extensive experiments to compare our method with state-of-the-art baselines, and our method significantly outperforms state-of-the-art baselines for successive POI recommendation task.

2   Related Work

General POI Recommendation. The most well-known approaches of personalized recommendation are collaborative filtering (CF) and Matrix Factorization (MF). The conventional CF techniques have been widely studied for POI recommendation. LARS [6] employs item-based CF to make POI recommendation with the consideration of travel penalty. FCF [22] is a friend-based CF model based on the common visited POIs among friends, which considers the social influence. UTE [23] is a collaborative recommendation model that incorporates with temporal and geographical information. However, such methods suffer the data sparsity problem, leading them difficult to identify similar users.

Recommendation models based on MF and embedding learning [7,11,12] have been intensively studied. Rank-GeoFM [9] fits the users' preference rankings for POIs to learn the latent embeddings. By incorporating the geographical context, it utilizes a geographical factorization method for calculating the recommendation score. TSG-MF [26] models the multi-tag influences via extracting a user-tag matrix and the social influences via social regularization, and uses a normalized function to model geographical influences.

Next POI Recommendation. In the literature, next POI recommendation issues have been studied in [19,28], in which the main objective is to exploit the user's check-in sequence between different POIs and dynamic preference.

Markov Chains (MC) Based Methods. MC based models aim to predict the next behavior according the historical sequential behaviors. FPMC-LR [3] considers first-order Markov chain for POI transitions and distance constraints. HMM [21] exploits check-in category information to capture the latent user movement pattern by using a mixed hidden Markov chain. LORE [24] incrementally mines sequential patterns and represents it as a dynamic location-location transition graph. By utilize an additive Markov chain, LORE fuses the sequential, geographical and social influence in a unified way.

Graph-Based Methods. Graph-based approaches are exploring in the literature of next POI recommendation. GE [20] jointly captures the latent relations among the POI, region, time slot and words related to the POIs by constructing four bipartite graphs. HME [5] projects the entities into a hyperbolic space after study multiple contextual subgraphs. Although the above approaches achieve promising performance, they can not model the sequential patterns effectively.

RNN-Based Methods. Recently, RNNs such as LSTM or GRU have demonstrated groundbreaking performance on predicting sequential problem. ST-RNN [10] utilizes RNN structure to model the temporal contexts by carefully designing the time-specific and distance-specific transition matrices. NEXT [25] encodes the sequential relations within the pre-trained POI embeddings by adopting DeepWalk [16] technique. Time-LSTM [30] employs LSTM with time gates to capture time interval among users' behaviors. CAPE [1] first uses a check-in context layer to capture the geographical influence of POIs and a text content layer to model the characteristics of POIs from text content. Then, CAPE employs RNN as recommendation component to predict successive POIs. PEU-RNN [13] proposes a LSTM based model that combines the user and POI embeddings, which are learned from Word2Vec. ASPPA [27] proposes to identify the semantic subsequences of POIs and discover their sequential patterns. Recently, STGN [28] extends the LSTM gating mechanism with the spatial and temporal gates to capture the user's space and time preference. However, these approaches fail to capture users' personalized preferences.

3   Problem Definition

In this section, we first give the key concepts used in this paper. Then, the problem definition for personalized POI recommendation is formulated.

Definition 1 (POI). *A POI is a uniquely identified venue in the form of $\langle p, \ell, cat \rangle$, where p is the POI identifier, cat denotes the category of the POI, and ℓ represents the geographical coordinates of the POI (i.e., longitude and latitude).*

Definition 2 (Check-in record). *A check-in record is a triple $c = \langle u, v, t \rangle$ that represents user u visiting POI v at timestamp t.*

The collection of all users is denoted as U, and the collection of all POIs is denoted as V.

Definition 3 (Trajectory). *The trajectory of a user u is a sequence of all check-in records $(\langle u, v_1, t_1 \rangle, \langle u, v_2, t_2 \rangle, \ldots, \langle u, v_n, t_n \rangle)$ made by user u in chronological order. We denote it as T_u.*

Definition 4 (Social Ties). *Social ties among users is defined as a graph $\mathcal{G}_u = (U, \mathcal{E}_u)$, where U is the set of users, and $\mathcal{E}_u$ is the set of edges between the users. Each edge $e_{ij} \in \mathcal{E}_u$ represents users u_i and u_j being friends in LBSNs and is associated with a weight $w_{ij} > 0$, which indicates their tie strength.*

Problem 1 (Successive POI Recommendation). *Given users' check-in records and their social ties, and a querying user u with his/her current check-in $\langle u, v, t \rangle$, our goal is to recommend top-k POIs that user u would be interested in in the next τ time period.*

4   Methodology

In this section, we first present the details of the proposed framework PPR. Then we introduce our model how to utilize PPR model to make personalized POI recommendation.

4.1   Heterogeneous Graph Construction

We first introduce the heterogeneous User-POI graph to model users' sequential check-ins and social relationships. Specifically, we employ a heterogeneous graph $\mathcal{G} = (V, U, \mathcal{E}, W)$ to jointly model the multiple relations between users and POIs. U and V are the user collection and POI collection respectively, and $\mathcal{E}$ is the set of all edges between nodes in $\mathcal{G}$, which are categorized into three edge types, *i.e.*, $\mathcal{E}_u$, $\mathcal{E}_v$, and $\mathcal{E}_{u,v}$. As mentioned in Definition 4, each edge $e_{i,j} \in \mathcal{E}_u$ represents that user u_i and u_j are friends. Each edge $e_{i,j} \in \mathcal{E}_v$ denotes that there exists at least one user visits POI v_j after visiting POI v_i, and each edge $e_{i,j} \in \mathcal{E}_{u,v}$ indicates that user u_i visits POI v_j at least one time. Notice that each edge $e \in \mathcal{E}_u \cup \mathcal{E}_{u,v}$ is a bi-directed edge and each edge $e \in \mathcal{E}_v$ is a directed edge, and each edge is associated with a weight $w \in W(w > 0)$, which indicates the strength of the relation.

Modeling User-POI Relation. Intuitively, we consider that if user u_i visit POI v_j more frequent, u_i and v_j have a stronger relation than with other POIs. Therefore, we formulate the weight between user u_i and POI v_j as:

$$w_{i,j} = freq(u_i, v_j), \tag{1}$$

where $freq(,)$ denotes check-in frequency of user u_i visiting POI v_j. Since we aim to build a directed graph to accommodate the following work, we define $w_{i,j} = w_{j,i}$ for the bi-directed edge $e_{i,j} \in \mathcal{E}_{u,v}$ between user u_i and POI v_j.

Modeling Sequential and Geographical Effect. Compared with general POI recommendation, successive POI recommendation pays more attention to sequential pattern. The impact of user's recent check-in behaviors are greater than those of a long time ago when making POI recommendations [20]. To further model the sequential effect, we carefully design a weighting strategy for the edges in $\mathcal{E}_v$.

Let $\Delta t^u_{k,k+1}$ be the time interval between two consecutive check-in records in the trajectory T_u of user u. $l^u_{k,k+1}$ is the flag that indicates the status of a pair of consecutive check-in records in the trajectory T_u, which is defined as:

$$l^u_{k,k+1} = \begin{cases} 1 & if \ \Delta t^u_{k,k+1} < \theta \\ 0 & else \end{cases}, \tag{2}$$

where θ is a predefined time threshold.

Given an edge $e_{i,j} \in \mathcal{E}_v$ from POI v_i to POI v_j, the sequential weight $w^{(seq)}_{i,j}$ for the edge $e_{i,j}$ is defined as:

$$w^{(seq)}_{i,j} = \sum_{u \in U} \sum_{k=1}^{|T_u|-1} l^u_{k,k+1}, \ if \ v_k = v_i \ and \ v_{k+1} = v_j. \tag{3}$$

Namely, the weight $w^{(seq)}_{i,j}$ for the edge from POI v_i to POI v_j is the total number of times that all users visit v_i first and then v_j in their trajectories.

Furthermore, geographical influence indicates the impact of geographical distance to the users' spatial behaviors. According to [5,8], the distribution of the geographical distance between two successive POIs follows the power-law distribution, which means users are more willing to visit POIs close to the current location. Therefore, we incorporate the geographical distance into our model as follows:

$$w^{(geo)}_{i,j} = \frac{d^\kappa_{i,j}}{\sum\limits_{v_k \in N(v_i)} d^\kappa_{i,k}}, \tag{4}$$

where $N(v_i)$ represents the set of out-neighbor POIs of POI v_i in $\mathcal{E}_v$, $d_{i,j}$ denotes the Euclidean distance between POIs v_i and v_j, and κ is the negative exponent (*i.e.*, $\kappa < 0$). Finally, we combine the sequential and geographical influence as follows:

$$w_{i,j} = w^{(seq)}_{i,j} \cdot w^{(geo)}_{i,j}. \tag{5}$$

In such way, the sequential, time interval and geographical information are all reflected in graph $\mathcal{G}$.

Modeling Social Tie Strength. Users in an LBSN have multiple types of relations with other users, such as friends, family and colleagues. The preference of a user in social network are easily affected by his/her close friends or other users which has some kind of relations with them. Recently, these social ties are incorporated into the POI recommendation system [25] to improve the recommendation performance. In this work, we propose to assign the weight between the users based on their historical check-in interactions. Specifically, for two socially connected users u_i and u_j, we assign the edge weight $w_{i,j}$ as:

$$w_{i,j} = \frac{\varepsilon + \sum\limits_{v \in V} \min(f_{u_i,v}, f_{u_j,v})}{|T_{u_i} \cap T_{u_j}| + 1}, \tag{6}$$

where ε is a very small float number to avoid two users have connection but no common visited POIs, $f_{u_i,v}$ denotes the frequency of user u_i visiting at POI v, and $|T_{u_i} \cap T_{u_j}|$ represents the number of the common visited POIs for user u_i and u_j. Therefore, the common preferences between socially connected users are also taken into account in the User-POI graph $\mathcal{G}$.

Densifying Graph. Most recommendation models need to take the data sparsity into consideration, but the check-in data in POI recommendation area is much sparser. To address the data sparsity issue, we propose to construct a dense graph based on the graph $\mathcal{G}$. Specifically, we regard each user and POI as a node, and expand the neighbors of those nodes with low in/out degrees by adding higher order neighbors. In this work, we only consider expanding second-order neighbors to every node. If the out-degree of a node in $\mathcal{G}$ is less than a predefined threshold ρ, we create an edge from node v_i to its second-order out-neighbor node v_j and assign the weight as follows:

$$w_{i,j} = \sum_{v_k \in N(v_i)} w_{i,k} \frac{w_{k,j}}{d_k^{(o)}}, \tag{7}$$

where $N(v_i)$ is the set of out-neighbors of node v_i, and $d_k^{(o)}$ is the out-degree of the node v_k. The densifying method for nodes with a low in-degree less than ρ is same. After densifying the User-POI graph, we can get the a more dense network, denoted by $\mathcal{G}_{dense}$. Then we use $\mathcal{G}_{dense}$ instead of $\mathcal{G}$ and exploit embedding technique to learn the nodes' representation vectors.

4.2   Learning Latent Representation

Inspired by LINE [17], which learns the first- and second-order relations representations of homogeneous networks. We develop it to learn heterogeneous node representations on our constructed heterogeneous graph $\mathcal{G}_{dense}$.

Specifically, we regard each user or POI as a node v and ignore their node type. In graph $\mathcal{G}_{dense}$, each node plays two roles: the node itself and a specific "context" of other nodes. We use $\overrightarrow{v_i}$ to denote the embedding vector of node v_i when it is treated as a node, and $\overrightarrow{v_i}'$ to denote the embedding vector of v_i when it is treated as a specific "context". In particular, we use a binary cross-entropy loss to encourage nodes and their "context" connected with an edge, to have similar embeddings. Therefore, we minimize the following objective function:

$$\mathcal{O} = -\sum_{e_{i,j} \in \mathcal{E}} \left(\log \left(\sigma(\overrightarrow{v_j}'^{\mathrm{T}} \cdot \overrightarrow{v_i}) \right) + w_n \sum_{v_n \in Neg(v_i)} \log \left(1 - \sigma(\overrightarrow{v_n}'^{\mathrm{T}} \cdot \overrightarrow{v_i}) \right) \right), \tag{8}$$

where $\sigma()$ is the sigmoid function, $\vec{v_j}'^{\mathrm{T}}$ denotes vector transpose, $Neg(v_i)$ is a negative edge sampling $w.r.t.$ node v_i in $\mathcal{G}_{dense}$, and w_n denotes the negative sampling ratio, which is a tunable hyper-parameter to balance the positive and negative samples.

By minimizing the objective function $\mathcal{O}$ with ASGD (asynchronous stochastic gradient) optimization and edge sampling technique, we can learn a d-dimensional embedding vector for each user and POI in $\mathcal{G}_{dense}$. Additionally, the representation learning is highly efficient and is able to scale to very large graphs because of the use of edge sampling technique.

4.3   Modeling User Dynamic and Personalized Preference

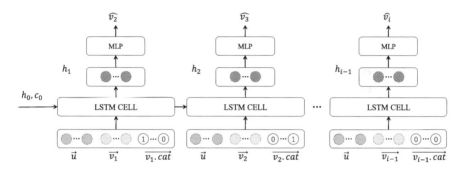

Fig. 1. Architecture of the proposed model

After representation learning, all users and POIs are mapped into a low dimensional space. However, the latent representations only capture the users' preferences or POIs' characteristics in a general way. Although it can model sequence transition patterns and geographical influence, some personalized preference may not be preserved in the node representations.

Furthermore, the categories of POIs are very useful to make a better representation of venues and improve the recommendation performance. In order to model user dynamic and personalized preference, we propose to concatenate user embedding, POI embedding and POI category to generate a new and more personalized embedding to represent a check-in record. More concretely, we use one-hot encoding to represent the POI category information.

Additionally, to better model user dynamic preference and sequential behavior patterns, we utilize LSTM model to construct a spatio-temporal neural network. As illustrated in Fig. 1, h_t and c_t denote the hidden state and cell state of LSTM at time t respectively. Given a user u and his/her trajectory sequence T_u, first, we concatenate the user embedding, POI embeddings with POI categories that he/she visited, and we can get a new embedding sequence. Second, we feed LSTM network with these new embedding sequences of all users. Specifically, we utilize the first $i - 1$ POIs as input to train the network, and predict the

$(i + 1)$-th POI as the recommended POI based on the current i-th POI. At the output layer, we also connect a multi-layer perceptron (MLP). Therefore, we use the following objective function to train the model:

$$\mathcal{O}_{lstm} = \sum_{t=1}^{i-1} MSE(MLP(h_t), \overrightarrow{v_{t+1}}), \tag{9}$$

where h_t is hidden representation at time step t, $MSE(\cdot, \cdot)$ is a criterion that measures the mean squared error (*e.g.*, squared L2 norm) between each element.

4.4   Personalized POI Recommendation

As described in Sect. 4.3, the user embedding and the first i POI embedding sequence are used to train the spatio-temporal neural network. For the querying user u, the embedding vector of the $(i + 1)$-th POI can be predicted by the current POI v_i as:

$$\widehat{v_{i+1}} = MLP(\mathrm{h}_i). \tag{10}$$

Therefore, for each POI v, we calculate its recommendation score as follows:

$$Score(v|\widehat{v_{i+1}}, u, T_u) = 1 - MSE(\widehat{v_{i+1}}, \overrightarrow{v}). \tag{11}$$

Finally, we rank all POIs by their recommendation scores and select top-k POIs as the candidate that user u is more likely to visit in the next τ time period.

5   Experiments

5.1   Datasets

We conduct extensive experiments on three public real-world large-scale datasets: Foursquare[1], Gowalla[2] and Brightkite[3]. The basic statistics of these three datasets are summarized in Table 1. Notice that we preprocess these datasets utilizing the same method of [29] by filtering the POIs visited by less than five users and the users with less than ten check-in records.

- **Foursuqare:** This dataset contains 483,813 check-in records generated by 4,163 users who live in California from December 2009 to July 2013.
- **Gowalla:** Gowalla is a location-based social networking website where users share their locations by checking-in. We choose data from Asian area for our experiments. It includes 251,378 check-in records generated by 6,846 users over the period of February 2009 to October 2010.
- **Brightkite:** Brightkite is also a location-based social networking service provider. We use the same selection strategy to obtain the check-in records generated by Asian users, which contains 572,739 records of 5,677 users.

Notice that there are 35 POI categories in Foursquare, and no category information is attached to Gowalla and Brightkite datasets.

---

[1] https://sites.google.com/site/dbhongzhi/.
[2] http://snap.stanford.edu/data/loc-Gowalla.html.
[3] http://snap.stanford.edu/data/loc-Brightkite.html.

Table 1. Basic statistics of three datasets

Dataset	Foursquare	Gowalla	Brightkite
# of users	4,163	6,846	5,677
# of POIs	121,142	74,856	128,799
# of check-ins	483,813	251,378	572,739
# of categories	35	/	/
Time span	Dec. 2009–Jul. 2013	Feb. 2009-Oct. 2010	Apr. 2008 - Oct. 2010

5.2   Evaluation Metrics

To evaluate the recommendation model performance, we use four widely-used metrics, *i.e.*, Accuracy ($Acc@k$), Precision ($Pre@k$), Recall ($Rec@k$) and Normalized Discounted Cumulative Gain ($NDCG@k$), which are also used to evaluate top-k POI recommendation in [1,18,27,28].

Let $\#hit@k$ denote the number of hits in the test set, and $|D_{Test}|$ is the number of all test records. $Acc@k$ is defined as:

$$Acc@k = \frac{\#hit@k}{|D_{Test}|}. \tag{12}$$

Let R_k denote the top-k POIs with the highest recommendation score, and T_k be the ground truth of the corresponding record, respectively. $Pre@k$ and $Rec@k$ are defined as:

$$Pre@k = \frac{1}{|D_{Test}|} \sum \frac{|R_k \cap T_k|}{|R_k|}, \tag{13}$$

$$Rec@k = \frac{1}{|D_{Test}|} \sum \frac{|R_k \cap T_k|}{|T_k|}. \tag{14}$$

To better measure the ranking quality, we further utilize $NDCG@k$, which assigns higher scores to POIs at top position ranks, to evaluate the model. The $NDCG@k$ for each test case is defined as:

$$NDCG@k = \frac{DCG@k}{IDCG@k}, \tag{15}$$

where $DCG@k = \sum_{i=1}^{k} \frac{2^{rel_i}-1}{\log_2(i+1)}$, $IDCG@k = \sum_{i=1}^{k} \frac{1}{\log_2(i+1)}$ and $rel_i = 1$ refers to the graded relevance of result ranked at position i. We use the binary relevance in our experiments, *i.e.*, $rel_i = 1$ if the recommended POI is in the ground truth, otherwise, $rel_i = 0$.

5.3   Baselines

We compare our model against the following baselines for successive POI recommendation:

- **Rank-GeoFM** [9]: It is a ranking based geographical factorization model, which earns the embeddings of users and POIs by combining geographical and temporal influence in a weighting scheme.
- **ST-RNN** [10]: ST-RNN is a RNN-based model with spatial and temporal contexts for next POI recommendation.
- **GE** [20]: GE jointly learns the embedding of POIs, regions, time slots and word into a shared low dimensional space by constructing four bipartite graphs.
- **PEU-RNN** [13]: It is a LSTM based model that combines the user and POI embeddings, which are learned from Word2Vec, for modeling the dynamic user preference and successive transition influence.
- **SAE-NAD** [14]: SAE-NAD exploits the self-attentive encoder to differentiate the user preference and the neighbor-aware decoder to incorporate the geographical context information for POI recommendation.

Notice that STGN [28] and ASPPA [27] are not compared in our experiment due to no publicly available source code. However, our PPR consistently outperforms ASPPA and STGN in terms of $Acc@k$ on both Foursquare and Gowalla datasets according to the experimental results reported in [27] (*e.g.*, PPR vs. STGN vs. ASPPA: 0.3008: 0.2: 0.2796 in $Acc@5$, 0.3935: 0.2592: 0.3371 in $Acc@10$ on Foursquare; PPR vs. STGN vs. ASPPA: 0.3835: 0.1947: 0.2363 in $Acc@5$, 0.4905: 0.2367: 0.2947 in $Acc@10$ on Gowalla).

To further validate the effectiveness of each component in our model, we design four variations of PPR:

- **PPR-RL**: This is a simplified version of PPR, which do not use LSTM network for personalized preference modeling. After representation learning on $\mathcal{G}_{dense}$, we use $Score(v|v_c, u) = \vec{u} \cdot \vec{v} + \vec{v_c} \cdot \vec{v}$ to calculate the recommendation score, where v_c is the current location of the querying user u.
- **PPR-Seq**: This variation do not model the sequential and geographical effect (*i.e.*, ignore POI-POI edges) in graph $\mathcal{G}_{dense}$, and the other components remain the same.
- **PPR-Den**: This variation directly learns representations for users and POIs on graph $\mathcal{G}$, which do not densify the graph. And the other components remain the same.
- **PPR-GRU**: In this variation, we use GRU to replace LSTM in user personalized preference modeling, and the other components remain the same.

5.4   Parameter Setting

Following [24,28,29], we utilize the first 80% chronological check-ins of each user as the training set, the remaining 20% as the test data.

We use the source code released by their authors for baselines. We set learning rate to 0.0025 in graph embedding, embedding dimension d to 128, the number of negative samples to 5, threshold θ to 24 h, κ to -2, ε to 0.5 and in/out-degree threshold ρ to 400. Following [5], we uniformly set the next time period as

$\tau=$ 6 h for all methods unless stated otherwise, and other parameters of all baselines are tuned to be optimal. In the experiment, we use a two-layer stacked LSTM, the hidden state size is 128. The learning rate of LSTM is set to 0.001 with epoch decay, which makes the learning rate becomes 1/10 of the original value when the number of training rounds reaches 75%.

Table 2. Performance comparison on Foursquare dataset

Methods	Acc@5	Acc@10	Pre@5	Pre@10	Rec@5	Rec@10	NDCG@5	NDCG@10
Rank-GeoFM	0.2456	0.2983	0.0618	0.0413	0.0509	0.0669	0.0683	0.0468
ST-RNN	0.1642	0.2150	0.0167	0.0118	0.1207	0.1790	0.0175	0.0152
GE	0.1357	0.3100	0.0378	0.0342	0.1579	0.1919	0.0431	0.0362
PEU-RNN	0.2021	0.2775	0.0495	0.0276	0.1888	0.2848	0.0494	0.0375
SAD-NAE	0.2429	0.3221	0.0588	0.0442	0.0333	0.0505	0.0672	0.0542
PPR	**0.3008**	**0.3935**	**0.0698**	**0.0501**	**0.2471**	**0.3387**	**0.0802**	**0.0628**

Table 3. Performance comparison on Gowalla dataset

Methods	Acc@5	Acc@10	Pre@5	Pre@10	Rec@5	Rec@10	NDCG@5	NDCG@10
Rank-GeoFM	0.2162	0.2643	0.0647	0.0453	0.0887	0.1180	0.0696	0.0499
ST-RNN	0.1865	0.2246	0.0278	0.0217	0.0817	0.1075	0.0606	0.0574
GE	0.1763	0.4060	0.0391	0.0203	0.1363	0.3135	0.0813	0.0157
PEU-RNN	0.3329	0.3766	0.0663	0.0390	0.2504	**0.3613**	0.0919	0.0627
SAD-NAE	0.3273	0.4300	0.0849	0.0645	0.1102	0.1600	0.0956	0.0777
PPR	**0.3835**	**0.4905**	**0.0936**	**0.0687**	**0.2573**	0.3430	**0.1055**	**0.0840**

5.5 Performance Comparison

Table 4. Performance comparison on Brightkite dataset

Methods	Acc@5	Acc@10	Pre@5	Pre@10	Rec@5	Rec@10	NDCG@5	NDCG@10
Rank-GeoFM	0.3681	0.4270	0.0968	0.0618	0.2497	0.2983	0.1058	0.0700
ST-RNN	0.2396	0.3540	0.0389	0.0394	0.2279	0.3400	0.1166	0.1074
GE	0.1903	0.4259	0.0869	0.0483	0.1303	0.4119	0.1313	0.1217
PEU-RNN	0.7187	0.7383	0.1437	0.0720	0.6944	0.7204	0.2348	0.1538
SAD-NAE	0.2578	0.3383	0.0645	0.0499	0.0703	0.1047	0.0708	0.0584
PPR	**0.8717**	**0.8966**	**0.1788**	**0.0927**	**0.8485**	**0.8741**	**0.2875**	**0.1889**

First, we evaluate the overall performance of our model PPR compared with five baselines on three real-world datasets. We repeat 10 runs for all methods on each dataset and report average $Acc@k$, $Pre@k$, $Rec@k$ and $NDCG@k$ in Table 2, Table 3 and Table 4, respectively.

From Table 2, we observe that PPR is significantly better than all baselines in terms of four evaluation metrics on Foursquare dataset. Specifically, PPR achieves 0.3008 in $Acc@5$ and 0.3935 in $Acc@10$, improving 22.5% and 22.2% over second-best baseline Rank-GeoFM and SAD-NAE, respectively. Additionally, our PPR slightly outperforms the strong baselines (*e.g.*, SAD-NAE) in $Pre@k$, but it is significantly better than the strong baselines in $Rec@k$.

As depicted in Table 3, our PPR also significantly outperforms all baselines in terms of $Acc@k$, $Pre@k$, $Rec@k$ and $NDCG@k$ on Gowalla dataset. In particular, PPR performs better than the second-best baseline by 14.6% in $Acc@k$ and 9.2% in $NDCG@k$ on average. PPR shows slightly poor performance compared to PEU-RNN in terms of $Rec@10$. This phenomenon can be explained that PEU-RNN uses a distance constraint, which may significantly reduce the potential POIs as k increases.

As we can see in Table 4, PPR consistently significantly outperforms all baselines in terms of all evaluation metrics on Brightkite dataset. PPR achieves the state-of-the-art performance, *e.g.*, 0.8717 in $Acc@5$ and 0.8485 in $Rec@5$. More specifically, our PPR achieves about 21.3%, 24.4%, 22.2% and 22.4% improvement compared to state-of-the-art RNN-based method PEU-RNN in terms of $Acc@5$, $Pre@5$, $Rec@5$ and $NDCG@5$, respectively. Furthermore, all methods achieve better performance on Brightkite than the other datasets. This is because users in Brightkite have more check-in records than users in Foursquare and Gowalla on average, which may enable all methods to model users' behavior and preference more accurately.

5.6 Ablation Study

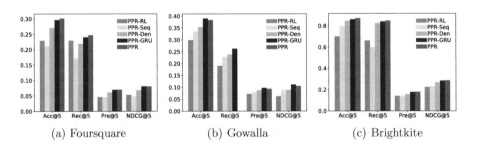

(a) Foursquare   (b) Gowalla   (c) Brightkite

Fig. 2. Performance comparison of variations

To explore the benefits of incorporating the sequential and geographical effect, densifying technique and modeling personalized preference into PPR respectively, we compare our model with four carefully designed variations, *i.e.*, PPR-RL, PPR-Seq, PPR-Den and PPR-GRU. We show the results in terms of $Acc@5$, $Pre@5$, $Rec@5$, and $NDCG@5$ on three datasets in Fig. 2.

Based on the results, we have the following observations: First, PPR achieves the best performance in most cases on three datasets, indicating that PPR benefits from simultaneously considering the various contextual factors and personalized preference in a joint way. Second, the contributions of different components to recommendation performance are different. Sequential and geographical effect and modeling personalized preference have comparable importance, specifically, the later contributes more on Gowalla, and the former contributes more on Foursquare. And both of them are necessary for improving performance. Furthermore, through the comparison of PPR and PPR-Den, it is obvious that the densifying trick works for alleviating the data sparse issue. Third, our PPR and PPR-GRU exhibit a decent performance compared to other variations, which indicates that sequential pattern and users' dynamic and personalized preference play an important role in location-based recommendation.

5.7   Sensitivity of Hyper-parameters

We now investigate the sensitivity of our model compared against three strong baselines (*i.e.*, Rank-GeoFM, PEU-RNN, and SAE-NAD) with respect to the important parameters, including embedding dimension d, the number of recommended POIs k, and next time period τ. To clearly show the influence of these parameters, we report $Acc@5$ with different parameter settings on Foursquare and Gowalla datasets. Figure 3 and Fig. 4 show the experimental results.

As shown in Figs. 3(a) and 4(a), PPR achieves best performance compared to the three strong baselines with the increasing number of dimension d. Meanwhile, PPR achieves the best result when $d = 128$, and then begins to decline as d further increases. From the results in Figs. 3(b) and 4(b), we can see that the recommendation accuracy of all methods increases as k increases. This is expected, because the more results are recommended, the easier they are to fall into the ground truth. However, we also observe that our PPR exhibits an increasing performance improvement compared to all baselines, as k increases. In Figs. 3(c) and 4(c), as τ increases, our PPR is also consistently better than the strong baselines. More specifically, PPR improves the recommendation accuracy more significantly for near future prediction (*e.g.*, $\tau = 2$ vs. $\tau = 12$), indicating that our PPR can effectively capture users' personalized preferences, especially short-term preferences.

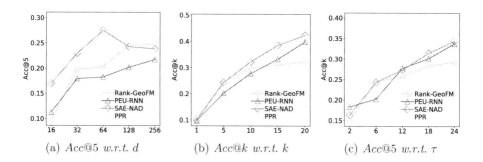

(a) $Acc@5$ *w.r.t.* d (b) $Acc@k$ *w.r.t.* k (c) $Acc@5$ *w.r.t.* τ

Fig. 3. Parameter sensitivity *w.r.t.* parameter d, k and τ on Foursquare

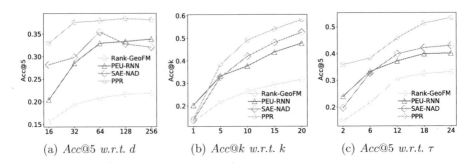

(a) *Acc@5 w.r.t. d*   (b) *Acc@k w.r.t. k*   (c) *Acc@5 w.r.t. τ*

Fig. 4. Parameter sensitivity *w.r.t.* parameter *d*, *k* and *τ* on Gowalla

6   Conclusion

In this work, we propose a novel spatio-temporal representation learning model for personalized POI recommendation. By incorporating the user-POI relation, sequential effect, geographical effect and social ties, we construct a heterogeneous network. Afterwards, we exploit the embedding technique to learn the latent representation of users and POIs. In light of recent success of RNN on sequential prediction problem, we feed the spatio-temporal network with concatenated user and POI embedding sequences for capturing the users' dynamic and personalized preference. The results on three real-world datasets demonstrate the superiority of our proposal over state-of-the-art baselines. Furthermore, we explore the importance of each factor in improving recommendation performance. We observe that sequential effect, geographical effect, and users' dynamic and personalized preference play a vital role in POI recommendation task.

Acknowledgments. This work is partially supported by the National Natural Science Foundation of China under grant Nos. 61773331, U1706218 and 41927805, the National Key Research and Development Program of China under grant No. 2018AAA0100602, and the Natural Science Foundation of Shandong Province under grant No. ZR2020QF030.

References

1. Chang, B., Park, Y., Park, D., Kim, S., Kang, J.: Content-aware hierarchical point-of-interest embedding model for successive poi recommendation. In: IJCAI, pp. 3301–3307 (2018)
2. Chen, M., Liu, Y., Yu, X.: NLPMM: a next location predictor with Markov modeling. In: Tseng, V.S., Ho, T.B., Zhou, Z.H., Chen, A.L.P., Kao, H.Y. (eds.) PAKDD'2014. LNCS, vol. 8444, pp. 186–197. Springer, Cham (2014). https://doi.org/10.1007/978-3-319-06605-9_16
3. Cheng, C., Yang, H., Lyu, M.R., King, I.: Where you like to go next: successive point-of-interest recommendation. In: IJCAI (2013)

4. Feng, S., Cong, G., An, B., Chee, Y.M.: POI2VEC: geographical latent representation for predicting future visitors. In: AAAI, pp. 102–108 (2017)
5. Feng, S., Tran, L.V., Cong, G., Chen, L., Li, J., Li, F.: HME: a hyperbolic metric embedding approach for next-poi recommendation. In: SIGIR, pp. 1429–1438 (2020)
6. Levandoski, J.J., Sarwat, M., Eldawy, A., Mokbel, M.F.: LARS: a location-aware recommender system. In: ICDE, pp. 450–461. IEEE (2012)
7. Li, K., Lu, G., Luo, G., Cai, Z.: Seed-free graph de-anonymiztiation with adversarial learning. In: CIKM, pp. 745–754 (2020)
8. Li, X., Han, D., He, J., Liao, L., Wang, M.: Next and next new POI recommendation via latent behavior pattern inference. ACM Trans. Inf. Syst. (TOIS) **37**(4), 1–28 (2019)
9. Li, X., Cong, G., Li, X.L., Pham, T.A.N., Krishnaswamy, S.: Rank-GeoFM: a ranking based geographical factorization method for point of interest recommendation. In: SIGIR, pp. 433–442 (2015)
10. Liu, Q., Wu, S., Wang, L., Tan, T.: Predicting the next location: a recurrent model with spatial and temporal contexts. In: AAAI (2016)
11. Liu, Z., Huang, C., Yu, Y., Fan, B., Dong, J.: Fast attributed multiplex heterogeneous network embedding. In: CIKM, pp. 995–1004 (2020)
12. Liu, Z., Huang, C., Yu, Y., Song, P., Fan, B., Dong, J.: Dynamic representation learning for large-scale attributed networks. In: CIKM, pp. 1005–1014 (2020)
13. Lu, Y.-S., Shih, W.-Y., Gau, H.-Y., Chung, K.-C., Huang, J.-L.: On successive point-of-interest recommendation. World Wide Web **22**(3), 1151–1173 (2018). https://doi.org/10.1007/s11280-018-0599-5
14. Ma, C., Zhang, Y., Wang, Q., Liu, X.: Point-of-interest recommendation: exploiting self-attentive autoencoders with neighbor-aware influence. In: CIKM, pp. 697–706 (2018)
15. Mikolov, T., Chen, K., Corrado, G., Dean, J.: Efficient estimation of word representations in vector space. arXiv preprint arXiv:1301.3781 (2013)
16. Perozzi, B., Al-Rfou, R., Skiena, S.: DeepWalk: Online learning of social representations. In: KDD, pp. 701–710 (2014)
17. Tang, J., Qu, M., Wang, M., Zhang, M., Yan, J., Mei, Q.: LINE: large-scale information network embedding. In: WWW, pp. 1067–1077 (2015)
18. Wang, Q., Yin, H., Chen, T., Huang, Z., Wang, H., Zhao, Y., Viet Hung, N.Q.: Next point-of-interest recommendation on resource-constrained mobile devices. In: WWW, pp. 906–916 (2020)
19. Wu, Y., Li, K., Zhao, G., Xueming, Q.: Personalized long-and short-term preference learning for next POI recommendation. IEEE Trans. Knowl. Data Eng. (2020)
20. Xie, M., Yin, H., Wang, H., Xu, F., Chen, W., Wang, S.: Learning graph-based poi embedding for location-based recommendation. In: CIKM, pp. 15–24 (2016)
21. Ye, J., Zhu, Z., Cheng, H.: What's your next move: user activity prediction in location-based social networks. In: SDM, pp. 171–179. SIAM (2013)
22. Ye, M., Yin, P., Lee, W.C.: Location recommendation for location-based social networks. In: SIGSPATIAL, pp. 458–461 (2010)
23. Yuan, Q., Cong, G., Ma, Z., Sun, A., Thalmann, N.M.: Time-aware point-of-interest recommendation. In: SIGIR, pp. 363–372 (2013)
24. Zhang, J.D., Chow, C.Y., Li, Y.: LORE: exploiting sequential influence for location recommendations. In: SIGSPATIAL, pp. 103–112 (2014)
25. Zhang, Z., Li, C., Wu, Z., Sun, A., Ye, D., Luo, X.: Next: a neural network framework for next POI recommendation. Front. Comput. Sci. **14**(2), 314–333 (2020)

26. Zhang, Z., Liu, Y., Zhang, Z., Shen, B.: Fused matrix factorization with multi-tag, social and geographical influences for POI recommendation. World Wide Web **22**(3), 1135–1150 (2019)
27. Zhao, K., Zhang, Y., Yin, H., Wang, J., Zheng, K., Zhou, X., Xing, C.: Discovering subsequence patterns for next POI recommendation. In: IJCAI, pp. 3216–3222 (2020)
28. Zhao, P., et al.: Where to go next: a spatio-temporal gated network for next POI recommendation. IEEE Trans. Knowl. Data Eng. (2020)
29. Zhao, S., Zhao, T., King, I., Lyu, M.R.: Geo-Teaser: geo-temporal sequential embedding rank for point-of-interest recommendation. In: WWW, pp. 153–162 (2017)
30. Zhu, Y., et al.: What to do next: modeling user behaviors by time-LSTM. In: IJCAI, pp. 3602–3608 (2017)

Missing POI Check-in Identification Using Generative Adversarial Networks

Meihui Shi, Derong Shen$^{(\boxtimes)}$, Yue Kou, Tiezheng Nie, and Ge Yu

College of Computer Science and Engineering, Northeastern University,
Shenyang 110169, China
shimeihui@stumail.neu.edu.cn,
{shenderong,kouyue,nietiezheng,yuge}@cse.neu.edu.cn

Abstract. The missing point-of-interest (POI) check-ins in real-life mobility data prevent advanced analysis of users' preferences and mobile patterns. Existing approaches for missing POI check-in identification mainly focus on modelling spatio-temporal dependencies and memorising transition patterns through users' check-in sequences. However, these methods cannot ensure that the generated missing records obey the same distribution as the observed check-ins. To this end, we propose a novel Bi-G^{2}AN model, which fuses the merits of generative adversarial network (GAN) and bi-directional gated recurrent unit (GRU), to identify the missing POI check-ins. Specifically, we develop a GAN-based method to mimic the overall distribution of a given check-in dataset, and it is further utilized to generate more reasonable missing POI check-ins. In order to capture bi-directional dependencies and historical impact, a modified bi-directional GRU is utilized in GAN. Moreover, both spatio-temporal influence and local motion information are employed to learn users' dynamic preferences. Finally, experiments conducted on three real datasets demonstrate the competitiveness of the Bi-G^{2}AN model, outperforming state-of-the-art approaches.

Keywords: Missing POI check-ins · Generative adversarial network · Gated recurrent unit · Time decay

1 Introduction

With the rapid development of mobile technologies, numerous location-based social network (LBSN) services, such as Foursquare, Instagram and Facebook place, have become pervasive in our daily lives. These services enable millions of users to check in at real-world locations and share life experiences with friends, resulting in a huge amount of mobility data. Data collected from LBSNs provides great opportunities to understand users' preferences and mobile patterns. The analysis of mobility data can lead to improvements in user experiences and quality of services. Besides, it facilitates targeted advertising to help merchants attract more potential customers.

© Springer Nature Switzerland AG 2021
C. S. Jensen et al. (Eds.): DASFAA 2021, LNCS 12681, pp. 575–590, 2021.
https://doi.org/10.1007/978-3-030-73194-6_38

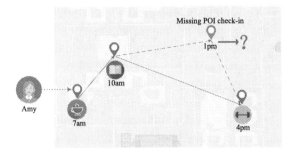

Fig. 1. An illustration of missing POI check-in identification task in this paper, where blue locations are actual check-ins. Assume that there is a long time gap between two successive observed check-ins. We aim to identify possible missing POI check-ins of the target user.

However, missing point-of-interest (POI) check-ins in mobility data pose a challenge for engineers and researchers. In reality, users' mobility data is typically incomplete, due to the spatial information missing and personal privacy. Spatial information missing frequently occurs in actual mobility data. For instance, users are less likely to check in every time they visit a location, and as a result, some movement records may be missing. Furthermore, due to personal privacy, users are unwilling to disclose certain check-ins, resulting in unreal data being recorded and real POI check-in missing. Such missing POI check-ins in mobility data hide useful information, which may have a negative impact on further analysis of users' preferences and mobile patterns. Therefore, missing POI check-in identification is quite a vital task for user understanding.

As illustrated in Fig. 1, user Amy visited the cafe at 7am and the library at 10am successively. She then checked in at the gym at 4pm, and there was a long time gap between the last two observed check-ins. Based on such check-in sequence, the existing studies mainly focus on recommending POIs that Amy is more willing to explore in the future. By contrast, we aim to identify possible POIs that Amy has visited during that long time gap without check-in records.

In the literature, many efforts have been made for missing data imputation that are most relevant to missing POI check-in identification. Specifically, the missing data imputation methods have evolved from statistical imputation [17] to machine learning [11]. Inspired by the great success of deep learning techniques in various research areas, recurrent neural network (RNN) and generative adversarial network (GAN) have also been commonly utilized in the field of missing data imputation [2,26]. Nevertheless, the above methods cannot be directly extended to missing POI check-in identification, since they are not designed for the spatial-aware issues. Besides, the methods of POI recommendation [5,14] can also be exploited to identify missing POI check-ins. However, the POI recommendation methods have inherent disadvantages, i.e., they only utilize check-ins before the specific time. Check-in records after the missing POI should also be taken into account.

Though missing POI check-ins are important for user understanding, only a few existing studies have investigated missing POI check-in identification.

A successful example is the Bi-STDDP [22] which identified missing check-in records by integrating spatio-temporal influence and users' dynamic preferences. PA-Seq2Seq [13] developed an attention-based sequence-to-sequence model to capture the actual users' trajectories. These existing solutions have made inspiring progress, but they have limitations: they cannot ensure that the distribution of the generated missing POI check-ins obeys that of the actual records. Moreover, the influence of historical observations will disappear over time. Previous works fail to capture historical impact. In addition, the current studies ignore local motion information, such as speed and direction, that may have an impact on revealing user characteristics.

To this end, we propose a novel Bi-G^{2}AN model for missing POI check-in identification, which combines GAN and bi-directional gated recurrent unit (GRU) to generate possible missing check-in records. We summarize the main contributions and innovations of this paper as follows:

- Towards generating robust missing POI check-ins, we develop a GAN-based framework. It can reconstruct POI check-in records following the distribution of actual check-ins.
- We design a bi-directional GRU cell with time decay, namely Bi-GTD, to capture bi-directional dependencies and historical impact.
- We integrate spatio-temporal transitions and local motion information to learn users' preferences and mobile patterns, which can achieve better missing POI check-in identification performance.
- We conduct extensive experiments on three real-world datasets. Experimental results demonstrate that the proposed Bi-G^{2}AN model outperforms state-of-the-art approaches.

2 Related Work

In this section, we investigate recent advance of spatial missing data imputation and POI recommendation, which are relevant to missing POI check-in identification.

2.1 Spatial Missing Data Imputation

The missing data negatively affect data analyses results. To ease the issue, researchers have proposed a set of classical missing data imputation approaches. These approaches can be classified into statistical imputation and machine learning based imputation. Statistical imputation methods mainly include mean or mode [1,17], expectation management [6], linear regression [19] and least squares [16]. Although methods based on statistical techniques are easy to implement, they have limited imputation performance. Many machine learning based methods [11,15] have been applied for missing data imputation. Moreover, there are some RNN-based [2] and GAN-based imputation methods [26]. Though the above approaches can be utilized for spatial missing data imputation, they have disadvantages, that is, they are not developed for the spatial-aware issues. To deal with spatial missing data, collaborative filtering and neighbourhood based approaches are the two main data imputation approaches. A spatio-temporal multi-view based learning model (ST-MVL) [24] was proposed to fill missing data in a set of geo-aware time series data. Furthermore, some spatio-temporal models have been designed for time series imputation. Besides, recent studies designed models to identify missing location information. An attention-based sequence-to-sequence model (PA-Seq2Seq) [13] exploited encoder-decoder framework to capture the actual users' trajectories. Bi-directional spatial and temporal dependencies and users' dynamic preferences model (Bi-STDDP) [22] integrated local temporal factor and global spatial influence with users' preferences to identify missing POI check-ins. However, the above methods cannot ensure the distribution of the generated missing data obeys that of the actual data.

2.2 POI Recommendation

POI recommendation, which is proposed to predict a location that a certain user prefers to visit in the future, has attracted wide attention from both academia and industry [8,20,21]. Both spatio-temporal context and sequential influence are crucial for POI recommendation. Factorizing personalized Markov chain (FPMC) [18] was proposed for sequential prediction, and it has been extended by embedding personalized transitions and localized regions for successive POI recommendation task [3]. Personalized ranking metric embedding model (PRME) [5] employed a pair-wise metric embedding method to jointly consider sequential information and geographical influence. Moreover, Graph-based embedding learning model (GE) [23] adopted bipartite graph to facilitate the recommendation performance. In addition, other auxiliary factors include spatial proximity [25], temporal influence [30], category information [9] and social relationships [29] have been studied accordingly. In recent years, due to the success of neural network based methods in various fields, many researchers have extended neural networks to deal with POI recommendation. Spatial temporal recurrent neural networks (ST-RNN) [14] employed time-specific and distance-specific transition matrices to capture the spatio-temporal influence. Hierarchical spatial-temporal long-short term memory (HST-LSTM) [12] incorporated spatio-temporal effects in LSTM for modelling contextual visiting history. A category-aware deep model (CatDM) [27] integrated POI categories and spatial information. Attentional recurrent neural network (ARNN) [7] jointly modelled transition regularities and sequential regularities of neighbours. Besides, users' check-ins are typically fuzzy. Therefore, some studies focus on alleviating the transition pattern vanishing issue for more accurate POI recommendation. Interactive multi-task learning model (iMTL) [28] introduced the interplay between activity and location preferences to improve the performance of POI recommendation with uncertain records. The above POI recommendation methods learn sequential patterns based on past check-in information, thus failing to exploit check-ins after the specific time when dealing with missing POI check-in identification task.

3   Problem Formulation

Let $\mathcal{U} = \{u_1, u_2, \ldots, u_{|\mathcal{U}|}\}$ be a set of users and $\mathcal{V} = \{v_1, v_2, \ldots, v_{|\mathcal{V}|}\}$ be a set of POIs, where $|\cdot|$ denotes the size of an arbitrary set and each POI v is associated with its coordinate (lat_v, lng_v).

Definition 1. *Check-in sequence. A check-in sequence is a temporally ordered sequential check-in records, i.e., $cs = \{c_{t_1}, c_{t_2}, \ldots, c_{t_T}\}$. Each check-in c_{t_i} is a user-POI-time tuple (u, v, t_i).*

Definition 2. *Local motion information. Local motion information contains user's local speed and direction at the target time.*

Definition 3. *Time gap. Time gap δ_i is the length of time between the current check-in c_{t_i} and the last observed check-in.*

We further define forward and backward time gap vectors to record the time gap. Specifically, we first define a masking vector $\boldsymbol{m} = [m_1 \ldots m_T]^{\mathrm{T}}$ to indicate which records in a check-in sequence are missing.

$$m_i = \begin{cases} 1, & if\ c_{t_i}\ exists \\ 0, & \text{otherwise} \end{cases} \tag{1}$$

Then, the time gap vector in forward direction $\boldsymbol{\delta}^f = [\delta_1^f \ldots \delta_T^f]$ is calculated as:

$$\delta_i^f = \begin{cases} 0, & i = 1 \\ \triangle t_{i-1,i}, & i > 1, m_{i-1} = 1 \\ \triangle t_{i-1,i} + \delta_{i-1}^f, & i > 1, m_{i-1} = 0 \end{cases} \tag{2}$$

where $\triangle t_{i-1,i} = t_i - t_{i-1}$ denotes the time length between c_{t_i-1} and c_{t_i}. For calculating the backward time gap vector $\boldsymbol{\delta}^b$, we read the sequence in reverse order, i.e., $\{c_{t_T}, c_{t_{T-1}}, \ldots, c_{t_1}\}$. The backward time gap vector $\boldsymbol{\delta}^b$ is calculated in a similar way to that of the forward time gap vector. For instance, the masking vector of the check-in sequence illustrated in Fig. 1 is $[1\ 1\ 0\ 1]^{\mathrm{T}}$, the forward time gap vector is $[0\ 3\ 3\ 6]^{\mathrm{T}}$, and the backward time gap vector is $[3\ 6\ 3\ 0]^{\mathrm{T}}$.

Missing POI check-in identification task is formulated as: for a user $u \in \mathcal{U}$, assuming that the t-th check-in record c_{t_t} is missing, the target of missing POI check-in identification is to identify possible POIs that user u visited at time t_t, with the help of check-in records before and after t_t.

4   Methodology

In this section, we give an overview of our proposed model Bi-G^{2}AN, which is designed for missing POI check-in identification. At a high level, Bi-G^{2}AN fuses the merits of GAN and bi-directional GRU. Our model is composed by a generator module and a discriminator module. Nevertheless, it is different from the standard GAN. We present the overall framework of Bi-G^{2}AN in Fig. 2.

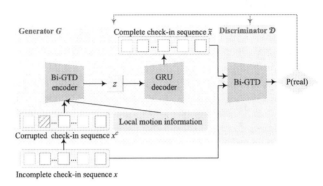

Fig. 2. Framework of Bi-G^{2}AN. The model contains a generator G and a discriminator D. G is a denoising auto-encoder. D is a modified bi-directional GRU to obtain the degree of authenticity. The blank box indicates a missing POI check-in.

In brief, the goal of the GAN is to generate synthetic check-ins following the distribution of actual check-ins, through the adoption of a minimax game between the generator and the discriminator. With the help of the modified bi-directional GRU (Bi-GTD) and denoising auto-encoder, the generator reconstructs a complete check-in sequence $\tilde{x}$ to fool the discriminator. Meanwhile, we introduce local motion information for behaviour analysis. The discriminator tries to distinguish the actual sequence and the generated one by adopting Bi-GTD cells. We take the corresponding check-ins of $\tilde{x}$ as the missing POI check-ins. The formula is shown as follows:

$$x_{identified} = \tilde{x}(1 - m) \tag{3}$$

4.1 Bi-directional GRU Cell with Time Decay

We adopt GRU as the recurrent unit to capture long-term dependencies, due to its powerful capability of memorizing sequential information. Note that LSTM can also be exploited to learn high-order sequential regularities, we choose GRU since it is simpler. In order to utilize more context information, our model encodes high-dimensional incomplete check-in sequences into a low-dimensional space with the help of bi-directional GRU. Towards dealing with missing POI check-in issue, we notice that if the current check-in record was missed a long time ago, the influence of historical observations will gradually disappear over time. Thus, we propose a variant of bi-directional GRU cell, namely Bi-GTD, which introduces a temporal decay mechanism to capture historical impact.

Specifically, we first extend the standard GRU by utilizing temporal decay factor β_i. β_i is calculated based on the irregular time gap between two check-ins. We exploit the logistic decay function to simulate the evolution of historical influence. The decay function is defined as follows:

$$\beta_i = 1/(1 + exp(\alpha\delta_i)) \tag{4}$$

where α is a decay parameter. δ_i is the time gap which has already been described in Sect. 3. By introducing temporal decay factor β_i, we obtain the confidence of historical information. Taking $\{x_i\}_{i=1}^T$ as input, the update functions of GRU with time decay can be summarized as follows:

$$
\begin{aligned}
h'_{i-1} &= \beta_i \odot h_{i-1} \\
z_i &= \sigma(W_z x_i + U_z h'_{i-1}) \\
r_i &= \sigma(W_r x_i + U_r h'_{i-1}) \\
\tilde{h}_i &= tanh(W_h x_i + U_h(r_i \odot h'_{i-1})) \\
h_i &= (1 - z_i) \odot h'_{i-1} + z_i \odot \tilde{h}_i
\end{aligned}
\tag{5}
$$

where $W_z, U_z, W_r, U_r, W_h, U_h$ are training parameters, σ is the logistic function. z_i and r_i are respectively update gate and reset gate. The term $tanh$ represents the hyperbolic tangent function, and $\odot$ shows the dot product operation.

Sequentially, we design the bi-directional GRU cells. Bi-directional GRU consists of forward and backward GRUs. With the help of forward and backward time gap vectors, i.e., δ^f and δ^b, we can calculate forward temporal decay factors $\{\beta_i^f\}_{i=1}^T$ and backward temporal decay factors $\{\beta_i^b\}_{i=1}^T$ with Eq. 4. The forward GRU reads the sequence in its original order, and generates forward hidden states $\{h_1^f, ..., h_T^f\}$ based on $\{\beta_i^f\}_{i=1}^T$. The backward GRU reads the sequence in reverse order, and calculates backward hidden states $\{h_1^b, ..., h_T^b\}$ by using $\{\beta_i^b\}_{i=1}^T$. Then we concatenate forward hidden state h_i^f and backward hidden state h_i^b to obtain an representation for each x_i, i.e., $\bar{h}_i = [h_i^f, h_i^b]^\mathrm{T}$.

4.2   The Generator Module

Inspired by the reconstruction capability of auto-encoder (AE), we exploit it to generate complete check-in sequences. Compared with classical AE, denoising auto-encoder (DAE) reconstructs each sequence from its corrupted version, which allows the hidden layer to capture more robust features. Mask-out/dropout noise is a common method to destroy the original sequence. Therefore, for missing POI check-in identification task, we get corrupted sequence x^c by using mask-out corruption. The main idea is to randomly mask some original records. Then we train a denoising auto-encoder.

$$
G(x^c) = \tilde{x}
\tag{6}
$$

To ensure that the reconstructed complete check-in sequence $\tilde{x}$ produced by the generator is similar to the actual sequence x, we introduce a reconstruction loss L_r, which is a squared error between the actual and the generated sequences.

$$
L_r = ||x \cdot m - \tilde{x} \cdot m||^2
\tag{7}
$$

The generator attempts to provide high-quality synthetic check-ins that approximate the true distribution. Specifically, the classification loss L_c can be formulated as follows.

$$
L_c = log(1 - D(\tilde{x}))
\tag{8}
$$

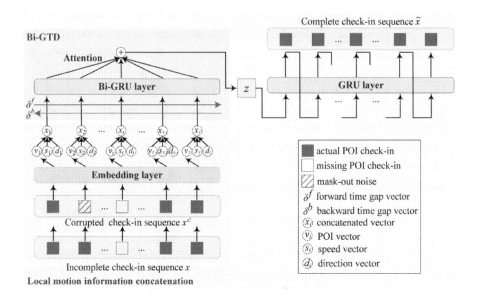

Fig. 3. Architecture of the generator. The input is a corrupted incomplete check-in sequence and the output is a reconstructed complete check-in sequence. The concatenation operation fuses POI vector v_i, speed vector s_i and direction vector d_i into x_i. By utilizing the Bi-GTD cells, the encoder compresses the concatenated input into a low-dimensional vector z and then feeds z to GRU decoder.

The loss function for the generator is formulated as:

$$L_G = \lambda L_r + L_c \tag{9}$$

where λ is a parameter that helps to control the influence of the reconstruction loss and the classification loss.

The generator module of Bi-G^{2}AN is presented in Fig. 3. We first encode the location information of a corrupted check-in sequence at each time step. Since local motion information has an impact on revealing user characteristics and is helpful for analysing behaviours, we integrate each check-in record with local speed and direction. Here, we utilize a coordinate transformation to represent these local motion information. Specifically, the speed vector s_i is related to $\triangle t_{i-1,i}$ and the corresponding distance $\triangle d_{i-1,i}$. The direction vector d_i depends on $lat_{v_i} - lat_{v_{i-1}}$ and $lng_{v_i} - lng_{v_{i-1}}$. Then, the concatenated vector $x_i = [v_i, s_i, d_i]^{\mathrm{T}}$ is fed into a Bi-GTD cell. Sequentially, we use the Bi-GTD cells to obtain hidden states $\{\bar{h}_i\}_{i=1}^{T}$. An attention layer is further utilized to calculate the weighted sum of these hidden states. After taking the sum as the input of a fully connected layer, we obtain a low-dimensional vector z. Next, z is transmitted to another fully connected layer. This output is taken as the initial input of a standard GRU layer, and the predicted check-in will be fed into the next iteration. Finally, the generated sequence is the combination of all outputs.

Algorithm 1. Bi-G^{2}AN

Require: generator G; discriminator D; LBSN dataset L
1: Initialize G and D with random weights
2: Pre-train G utilizing mean squared error loss
3: Reconstruct check-in sequences utilizing the pre-trained G
4: Pre-train D with Equation 10 utilizing actual sequences as positive samples and reconstructed sequences as negative samples
5: **while** not convergent **do**
6:     **for** g in g-steps **do**
7:         Update G with Equation 9
8:     **end for**
9:     **for** d in d-steps **do**
10:         Reconstruct check-in sequences utilizing G
11:         Train D with Equation 10
12:     **end for**
13: **end while**

4.3 The Discriminator Module

The goal of the discriminator is to distinguish the reconstructed complete check-in sequence $\tilde{x}$ from the ground truth x as accurately as possible. Therefore, it tries to rank the true sequence before the fake one. The discriminator also consists of a Bi-GTD based RNN layer and a fully connected layer. For the discriminator, the output is the probability of $\tilde{x}$ being sampled from the underlying distribution of real sequences. We train the discriminator to obtain a set of parameters that maximize the probability of correctly classifying sequences. Therefore, the loss function for the discriminator is:

$$L_D = -log(D(\boldsymbol{x})) - log(1 - D(\tilde{\boldsymbol{x}})) \tag{10}$$

We feed the Bi-GTD cells with reconstructed sequence $\tilde{x}$ or actual sequence x, and the corresponding time gap vectors ($\boldsymbol{\delta}^f$ and $\boldsymbol{\delta}^b$). After processing the input utilizing a bi-directional GRU layer, the weighted sum of hidden states further flows to a fully connected layer. The final output of discriminator is truth probability.

The overall procedure of Bi-G^{2}AN is summarized in Algorithm 1. Initially, the generator and the discriminator are initialized by pre-trained models (lines 1–4). Then the generator and the discriminator are trained alternatively during the adversarial training phase (lines 5–13).

5 Experiments

In this section, we evaluate the Bi-G^{2}AN with the goal of answering the following questions. **RQ1**: Does our model outperform state-of-the-art methods? **RQ2**: How do different components affect Bi-G^{2}AN? **RQ3**: How do hyper-parameters affect our model performance?

5.1   Experimental Setup

Datasets. We train and evaluate our model on three real-world LBSN datasets, NYC[1], TKY[1] and Gowalla[2], with density 0.548%, 0.404% and 0.185%. NYC dataset is collected from Foursquare, which contains 227,428 check-ins made within New York. They are made by 1,083 users on 38,333 POIs from April 2012 to February 2013. As for the TKY dataset, it is a dataset similar to NYC, but it is collected from Tokyo. TKY contains 2,293 users and 61,858 POIs, and the total number of check-ins is 573,703. Gowalla dataset contains 456,988 check-ins of 10,162 users and 24,250 POIs collected from February 2009 till October 2010. The statistics of datasets are summarized in Table 1. Following the previous work [22], users with fewer than 10 check-in records and POIs visited by fewer than 10 users are removed.

Table 1. Statistics of the three datasets.

Dataset	#User	#POI	#Check-in	Density
NYC	1,083	38,333	227,428	0.548%
TKY	2,293	61,858	573,703	0.404%
Gowalla	10,162	24,250	456,988	0.185%

Baselines. We compare our Bi-G^{2}AN with the following baseline methods for missing POI check-in identification.

- FTP: This is a counting-based method which directly takes the forward transition probability among POIs as prediction.
- BTP: This is a method similar to FTP except it utilizes the probability of backward transition.
- NN: This is a linear interpolation method that chooses the nearest neighbours around the midpoint of two successive observed check-ins.
- POP: This is another linear interpolation method that selects the most popular POIs according to the midpoint of two successive observed check-ins.
- PRME [5]: A metric embedding based model which is utilized to capture user transition patterns.
- PRME-G [5]: It integrates spatial influence and users' preferences on the basis of PRME.
- LSTM [10]: A variant of RNN, which contains a memory cell, an input gate, an output gate and a forget gate. It helps to learn long-term dependencies.
- GRU [4]: This is another variant of RNN, which has two gating mechanisms and is simpler than LSTM.
- ST-RNN [14]: A RNN-based model that captures spatial and temporal contexts with time-specific and distance-specific transition matrices.

---

[1] https://sites.google.com/site/yangdingqi/home/foursquaredataset.
[2] http://snap.stanford.edu/data/loc-gowalla.html.

Table 2. Performance of various methods on NYC.

	Rec@1	Rec@5	Rec@10	F1@1	F1@5	F1@10	MAP
FTP	0.1001	0.2411	0.2765	0.1001	0.0804	0.0503	0.1626
BTP	0.1013	0.2387	0.2786	0.1013	0.0796	0.0507	0.1664
NN	0.0890	0.2406	0.3095	0.0890	0.0802	0.0563	0.1773
POP	0.1069	0.2887	0.3717	0.1069	0.0962	0.0676	0.1907
PRME	0.0965	0.2687	0.3479	0.0965	0.0896	0.0633	0.1750
PRME-G	0.1049	0.2768	0.3677	0.1049	0.0923	0.0669	0.1896
LSTM	0.1216	0.3123	0.3905	0.1216	0.1041	0.0708	0.2097
GRU	0.1250	0.3159	0.3965	0.1250	0.1053	0.0721	0.2105
ST-RNN	0.1275	0.3202	0.4013	0.1275	0.1067	0.0730	0.2213
PA-Seq2Seq	0.1749	0.3475	0.4196	0.1749	0.1158	0.0763	0.2502
Bi-STDDP	0.1703	0.3421	0.4125	0.1703	0.1140	0.0749	0.2455
Bi-G^{2}AN	**0.1830**	**0.3686**	**0.4358**	**0.1830**	**0.1229**	**0.0792**	**0.2674**

- PA-Seq2Seq [13]: An attention-based sequence-to-sequence method that utilizes a stacked-LSTM structure and the encoder-decoder framework.
- Bi-STDDP [22]: This a novel model that captures bi-directional spatio-temporal dependencies and users' dynamic preferences.

Evaluation Metrics. Following the existing works [22, 28], we employ three well-known metrics to evaluate the performance of all methods, i.e., Recall (Rec@K), F1-score (F1@K) and Mean Average Precision (MAP). Rec@K measures the presence of the ground truth in the top-K ranked list. F1@K is a comprehensive index reflecting both precision and recall. Note that instead of taking only the POI with the highest probability as the result, we return the K highest-ranked POIs. Empirically, we set K to 1, 5 and 10. MAP is a global precision metric for evaluating ranking performance.

Implementation Details. The embedding size is set to 120/160/160 for NYC, TKY and Gowalla, respectively. Initially, the generator is trained via a mean squared error loss. The weights of the pre-trained model will be utilized for GAN training. The mask-out rate increases iteratively from 0.1 to 0.5. We sort the check-ins of each user by time. For all the experiments, we utilize the earliest 80% as the training set, and the latest 10% as the test set and the remaining 10% as the validation set.

5.2 Performance Comparison

The performance evaluated by Rec@K, F1@K and MAP on the three datasets are summarized in Tables 2, 3 and 4, respectively.

Table 3. Performance of various methods on TKY.

	Rec@1	Rec@5	Rec@10	F1@1	F1@5	F1@10	MAP
FTP	0.1096	0.2580	0.3158	0.1096	0.0860	0.0574	0.1812
BTP	0.1270	0.2739	0.3269	0.1270	0.0913	0.0596	0.1957
NN	0.1157	0.2665	0.3214	0.1157	0.0889	0.0584	0.2123
POP	0.1308	0.3217	0.3987	0.1308	0.1072	0.0725	0.2218
PRME	0.0438	0.1142	0.1475	0.0438	0.0381	0.0268	0.0839
PRME-G	0.0861	0.1981	0.2539	0.0861	0.0660	0.0462	0.1415
LSTM	0.1430	0.3555	0.4377	0.1430	0.1185	0.0796	0.2410
GRU	0.1504	0.3502	0.4429	0.1504	0.1167	0.0805	0.2424
ST-RNN	0.1591	0.3625	0.4469	0.1591	0.1208	0.0813	0.2529
PA-Seq2Seq	0.2032	0.4098	0.4825	0.2032	0.1366	0.0877	0.2931
Bi-STDDP	0.1996	0.4057	0.4776	0.1996	0.1352	0.0814	0.2855
Bi-G^{2}AN	**0.2145**	**0.4367**	**0.4908**	**0.2145**	**0.1456**	**0.0892**	**0.3120**

Table 4. Performance of various methods on Gowalla.

	Rec@1	Rec@5	Rec@10	F1@1	F1@5	F1@10	MAP
FTP	0.0844	0.1794	0.2256	0.0844	0.0598	0.0410	0.1273
BTP	0.0931	0.1889	0.2309	0.0931	0.0629	0.0419	0.1324
NN	0.0568	0.1335	0.1802	0.0568	0.0445	0.0328	0.1005
POP	0.0583	0.1341	0.1795	0.0583	0.0447	0.0321	0.0989
PRME	0.0306	0.0691	0.0914	0.0306	0.0231	0.0166	0.0512
PRME-G	0.0401	0.0885	0.1128	0.0401	0.0295	0.0205	0.0654
LSTM	0.0663	0.1557	0.2076	0.0663	0.0519	0.0377	0.1137
GRU	0.0665	0.1633	0.2173	0.0665	0.0545	0.0395	0.1180
ST-RNN	0.0674	0.1681	0.2213	0.0674	0.0560	0.0402	0.1289
PA-Seq2Seq	0.0987	0.2295	0.2906	0.0987	0.0765	0.0528	0.1593
Bi-STDDP	0.1069	0.2357	0.2925	0.1069	0.0786	0.0532	0.1636
Bi-G^{2}AN	**0.1145**	**0.2508**	**0.3124**	**0.1145**	**0.0836**	**0.0568**	**0.1764**

It can be seen that, Bi-G^{2}AN outperforms the baseline methods for all these cases on the three datasets. Compared with the strongest competitor, Bi-G^{2}AN is consistently better, obtaining average relative improvements of 5.76% for Rec@1, 6.35% for Rec@5, 4.13% for Rec@10 and 7.05% for MAP, respectively. The significant improvement margins indicate that Bi-G^{2}AN has a strong ability to identify missing POI check-ins.

Moreover, we find that RNN based methods generally performs better than non-RNN based baselines, demonstrating the sequence modelling capability of RNN. As for non-RNN based baselines, counting-based methods (FTP and BTP) have good performance on the three datasets. This is because users' transition

Table 5. Comparison of the variant models of Bi-G^{2}AN.

Variants	Bi-GTD	Local motion information		GAN
		Speed	Direction	
G^{2}AN	✗	✓	✓	✓
Bi-G^{2}AN$_{NS}$	✓	✗	✓	✓
Bi-G^{2}AN$_{ND}$	✓	✓	✗	✓
Bi-GAE	✓	✓	✓	✗

regularities have commonalities. Similarly, linear interpolation methods (NN and POP) also have acceptable performance, which validates the effectiveness of spatial influence and social information in modelling users' preferences. Both counting-based methods and linear interpolation methods achieve better performance than metric embedding based methods (PRME and PRME-G). PRME is outperformed by PRME-G slightly. This further indicates that modelling spatial impact is essential for missing POI check-in identification. RNN-based methods (LSTM, GRU) deliver decent results due to the capability of learning long-term dependencies. ST-RNN can achieve better performance in predicting missing records by utilizing spatial and temporal contexts. In general, PA-Seq2Seq and Bi-STDDP perform better than other baselines by taking into account forward and backward information, while others only consider check-ins before target time. In addition, Bi-G^{2}AN performs best among all methods, which verifies the generative adversarial network helps identify missing POI check-ins. Bi-G^{2}AN exploits Bi-GTD and local motion information to further improve the performance.

5.3 Ablation Analysis

To study the effectiveness of different components, we conduct ablation tests with four variants: (1) G^{2}AN, which replaces the Bi-GTD cells with GRU cells; (2) Bi-G^{2}AN$_{NS}$, which ignores the impact of local speed information on depicting user's preferences; (3) Bi-G^{2}AN$_{ND}$, which doesn't consider the direction context when learning users' mobile patterns; (4) Bi-GAE, which removes the discriminator module, is equivalent to a denoising auto-encoder based on a Bi-GTD encoder and a GRU decoder. We list the characteristics of the variant models in Table 5.

The compared results w.r.t. Rec@1 and MAP on the three datasets are shown in Fig. 4. We notice that our proposed Bi-G^{2}AN performs significantly better than all the variants, implying that Bi-GTD, local motion information and the GAN-based framework indeed improve the model performance. These components do not conflict with each other and can be exploited to collaboratively identify missing POI check-ins. Besides, the performance decrease of Bi-GAE far exceeds that of other variants, which demonstrates that the GAN-based framework is significant for generating robust missing POI check-ins. Generally, G^{2}AN is outperformed by both Bi-G^{2}AN$_{NS}$ and Bi-G^{2}AN$_{ND}$, indicating the

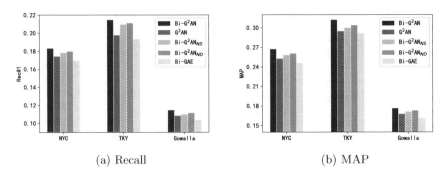

Fig. 4. Performance comparison with variants of Bi-G^{2}AN on the three datasets.

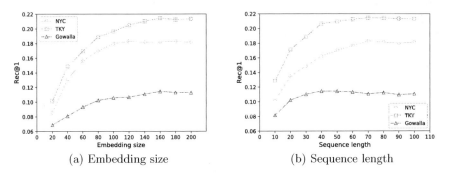

(a) Embedding size   (b) Sequence length

Fig. 5. Parameter sensitivity analysis.

advantages of Bi-GTD. Moreover, Bi-G^{2}AN$_{NS}$ inter underperform Bi-G^{2}AN$_{ND}$, implying that the speed factor is more important the direction factor. In summary, Bi-G^{2}AN benefits from these components.

5.4   Parameter Analysis

Figure 5 describes the results w.r.t. Rec@1 of parameter sensitivity analysis. Note that we can observe similar trends for other metrics. We first illustrate the performance under various settings of embedding size while keeping other optimal hyper-parameters fixed as shown in Fig. 5(a). A grid search is applied to find the optimal settings for embedding size. We can see that the performance of Bi-G^{2}AN improves as the embedding size increases, and gradually becomes stable when the embedding size increases to a certain level. Therefore, we select 120/160/160 for NYC, TKY and Gowalla as the embedding size, respectively. Since NYC dataset has fewer locations, the optimal dimension size of NYC is lower. Furthermore, to investigate the effects of the sequence length, we depict the performance change with diverse sequence lengths. As shown in Fig. 5(b), we observe that the optimal performance is achieved with different sequence lengths on the three datasets. It shows that Bi-G^{2}AN can capture short-term and long-term dependencies well. To summarize, our proposed model has powerful capabilities to process dimensionality and capture mobile patterns.

6 Conclusion

In this paper, we proposed a Bi-G^{2}AN model to generate more reasonable check-ins for missing POI check-in identification. In particular, we designed a GAN-based method that can generate missing POI check-ins obey the same distribution as actual check-ins. Moreover, we devised the Bi-GTD cells to capture bi-directional dependencies and historical impact. In addition, local motion information was incorporated into sequential patterns to learn users' preferences. Experimental results on three real-world datasets demonstrated substantial performance improvements of Bi-G^{2}AN over multiple state-of-the-art approaches. Enlightened by the successful application of reinforcement learning in various research areas, we would like to adopt reinforcement learning to further improve our method in the future.

Acknowlegements. This work was supported by the National Key R&D Program of China [2018YFB1003404]; and the National Natural Science Foundation of China [61672142, 62072086, 62072084, U1811261].

References

1. Amiri, M., Jensen, R.: Missing data imputation using fuzzy-rough methods. Neurocomputing **205**, 152–164 (2016)
2. Cao, W., Wang, D., Li, J., Zhou, H., Li, L., Li, Y.: BRITS: bidirectional recurrent imputation for time series. In: NeurIPS, pp. 6776–6786 (2018)
3. Cheng, C., Yang, H., Lyu, M.R., King, I.: Where you like to go next: successive point-of-interest recommendation. In: IJCAI, pp. 2605–2611 (2013)
4. Chung, J., Gülçehre, Ç., Cho, K., Bengio, Y.: Empirical evaluation of gated recurrent neural networks on sequence modeling. arXiv:1412.3555 (2014)
5. Feng, S., Li, X., Zeng, Y., Cong, G., Chee, Y.M., Yuan, Q.: Personalized ranking metric embedding for next new POI recommendation. In: IJCAI, pp. 2069–2075 (2015)
6. Ghorbani, S., Desmarais, M.C.: Performance comparison of recent imputation methods for classification tasks over binary data. Appl. Artif. Intell. **31**(1), 1–22 (2017)
7. Guo, Q., Sun, Z., Zhang, J., Theng, Y.: An attentional recurrent neural network for personalized next location recommendation. In: AAAI, pp. 83–90 (2020)
8. Han, P., Li, Z., Liu, Y., Zhao, P., Li, J., Wang, H., Shang, S.: Contextualized point-of-interest recommendation. In: IJCAI, pp. 2484–2490 (2020)
9. He, J., Li, X., Liao, L.: Category-aware next point-of-interest recommendation via listwise bayesian personalized ranking. In: IJCAI, pp. 1837–1843 (2017)
10. Hochreiter, S., Schmidhuber, J.: Long short-term memory. Neural Comput. **9**(8), 1735–1780 (1997)
11. Huang, J., et al.: Cross-validation based K nearest neighbor imputation for software quality datasets: an empirical study. J. Syst. Softw. **132**, 226–252 (2017)

12. Kong, D., Wu, F.: HST-LSTM: a hierarchical spatial-temporal long-short term memory network for location prediction. In: IJCAI, pp. 2341–2347 (2018)
13. Li, Y., Luo, Y., Zhang, Z., Sadiq, S.W., Cui, P.: Context-aware attention-based data augmentation for POI recommendation. In: ICDE, pp. 177–184 (2019)
14. Liu, Q., Wu, S., Wang, L., Tan, T.: Predicting the next location: a recurrent model with spatial and temporal contexts. In: AAAI, pp. 194–200 (2016)
15. Nishanth, K.J., Ravi, V.: Probabilistic neural network based categorical data imputation. Neurocomputing **218**, 17–25 (2016)
16. Pati, S.K., Das, A.K.: Missing value estimation for microarray data through cluster analysis. Knowl. Inf. Syst. **52**(3), 709–750 (2017). https://doi.org/10.1007/s10115-017-1025-5
17. Purwar, A., Singh, S.K.: Hybrid prediction model with missing value imputation for medical data. Expert Syst. Appl. **42**(13), 5621–5631 (2015)
18. Rendle, S., Freudenthaler, C., Schmidt-Thieme, L.: Factorizing personalized markov chains for next-basket recommendation. In: WWW, pp. 811–820 (2010)
19. de Souto, M.C.P., Jaskowiak, P.A., Costa, I.G.: Impact of missing data imputation methods on gene expression clustering and classification. Bioinformatics **16**, 64:1–64:9 (2015)
20. Sun, K., Qian, T., Chen, T., Liang, Y., Nguyen, Q.V.H., Yin, H.: Where to go next: modeling long- and short-term user preferences for point-of-interest recommendation. In: AAAI, pp. 214–221 (2020)
21. Wu, S., Zhang, Y., Gao, C., Bian, K., Cui, B.: GARG: anonymous recommendation of point-of-interest in mobile networks by graph convolution network. Data Sci. Eng. **5**(4), 433–447 (2020)
22. Xi, D., Zhuang, F., Liu, Y., Gu, J., Xiong, H., He, Q.: Modelling of bi-directional spatio-temporal dependence and users' dynamic preferences for missing POI check-in identification. In: AAAI, pp. 5458–5465 (2019)
23. Xie, M., Yin, H., Wang, H., Xu, F., Chen, W., Wang, S.: Learning graph-based POI embedding for location-based recommendation. In: CIKM, pp. 15–24 (2016)
24. Yi, X., Zheng, Y., Zhang, J., Li, T.: ST-MVL: filling missing values in geo-sensory time series data. In: IJCAI, pp. 2704–2710 (2016)
25. Yin, H., Wang, W., Wang, H., Chen, L., Zhou, X.: Spatial-aware hierarchical collaborative deep learning for POI recommendation. IEEE Trans. Knowl. Data Eng. **29**(11), 2537–2551 (2017)
26. Yoon, J., Jordon, J., van der Schaar, M.: GAIN: missing data imputation using generative adversarial nets. In: ICML, pp. 5675–5684 (2018)
27. Yu, F., Cui, L., Guo, W., Lu, X., Li, Q., Lu, H.: A category-aware deep model for successive POI recommendation on sparse check-in data. In: WWW, pp. 1264–1274 (2020)
28. Zhang, L., et al.: An interactive multi-task learning framework for next POI recommendation with uncertain check-ins. In: IJCAI, pp. 3551–3557 (2020)
29. Zhang, Z., Liu, Y., Zhang, Z., Shen, B.: Fused matrix factorization with multi-tag, social and geographical influences for POI recommendation. World Wide Web **22**(3), 1135–1150 (2018). https://doi.org/10.1007/s11280-018-0579-9
30. Zhao, S., Zhao, T., Yang, H., Lyu, M.R., King, I.: STELLAR: spatial-temporal latent ranking for successive point-of-interest recommendation. In: AAAI, pp. 315–322 (2016)

Efficiently Discovering Regions of Interest with User-Defined Score Function

Qiyu Liu[1], Libin Zheng[1(✉)], Xiang Lian[2], and Lei Chen[1]

[1] The Hong Kong University of Science and Technology, Kowloon, Hong Kong
{qliuau,lzhengab,leichen}@cse.ust.hk
[2] Kent State University, Kent, OH, USA
xlian@kent.edu

Abstract. Region of Interest (ROI) queries are of great importance in many location based services. However, the previous studies on ROI queries usually adopt either a simple spatial data model or a non-flexible enough query geometry, e.g., fixed-size rectangle. In this paper, to fix these drawbacks, we propose a new ROI search operator called _Radius Bounded ROI_ (RBR) query. An RBR query retrieves a subset of spatial objects satisfying co-location constraints and maximizing a user-configurable score function at the same time. We formally prove that answering an RBR query is 3SUM-hard, which implies that it is unlikely to find a sub-quadratic solution. To answer the RBR queries efficiently, we propose three algorithms, PairEnum, BaseRotation and OptRotation based on novel geometric findings. In addition, the query processing technique we proposed can be easily extended to other related problems like top-k ROI search. To demonstrate both efficiency and effectiveness of our proposed algorithms, we conduct extensive experimental studies on both real-world datasets and synthetic benchmarks, and the results show that OptRotation, our most efficient algorithm, achieves more than $10^3 \times$ efficiency improvement on both real and synthetic datasets compared with the baseline algorithm.

Keywords: ROI queries · Spatial database · Computational geometry

1 Introduction

In recent years, a number of location-based service (LBS) providers have emerged to retrieve location-related heterogeneous information for mobile users. For example, Google Map [1] and Yelp [3] enable users to query collections of objects which simultaneously contain geographical locations, textual attributes and numeric attributes. However, due to the increasing prevalence of mobile devices, they also face the challenge of managing a booming number of multi-attribute spatial objects. Effective and efficient query processing techniques are needed for such platforms to communicate with their users in real-time.

As a typical spatial operator in LBS, Region of Interest (ROI) queries retrieve one (or several) region(s) such that the enclosed spatial objects optimize a certain objective function (e.g., the total number of objects). As it is useful in

© Springer Nature Switzerland AG 2021
C. S. Jensen et al. (Eds.): DASFAA 2021, LNCS 12681, pp. 591–608, 2021.
https://doi.org/10.1007/978-3-030-73194-6_39

Fig. 1. An application scenario of *Radius Bounded ROI* (RBR) queries.

many applications like location-based advertising and geo-tagged event detection, continuous research efforts have been paid to answering the ROI query [5,6,10,11,14,15,23]. Existing studies usually adopt a simple spatial object model (e.g., an object is solely associated with either a weight [10] or a document [7]). However, spatial data objects in reality are usually more complex (i.e., with heterogeneous attributes), which makes it infeasible to directly apply existing methods. Thus, it is important to devise new ROI operators and solutions to deal with such enriched spatial objects.

In this paper, we study a novel spatial operator called <u>R</u>adius <u>B</u>ounded <u>R</u>OI *Query* (RBR) over an enriched spatial database where each object has both textual and numeric attributes. The RBR query retrieves a circular region which owns the maximum value of a user-configured score function. More specifically, an RBR query takes a collection of spatial objects $\mathcal{O}$ as input where each object $o \in \mathcal{O}$ has three attributes: geographic location $o.\ell$, numeric attribute $o.w$ and textual attribute $o.d$. It outputs a subset of objects $\mathcal{O}' \subseteq \mathcal{O}$ such that a user-configurable score function defined on $\mathcal{O}'$ is maximized while the radius of the minimum bounding circle (MBC) of point set $\{o.\ell | o \in \mathcal{O}'\}$ is no larger than a threshold r_{max}. Two motivation examples are shown as follows to demonstrate the function of RBR queries.

Example 1 (Yelp's ROI Recommendation). Figure 1 illustrates a toy example of an RBR query over four spatial objects $O_1 \sim O_4$. These objects represent 4 restaurants and each of them is associated with a textual attribute "Reviews" and a numeric attribute "Stars". Regions R_1 and R_2 are the minimum bounding circles of $\{O_1, O_2, O_3\}$ and $\{O_3, O_4\}$, respectively. Consider a query Q_1: find a region with as many good sushi places as possible. The RBR query is expected to return region R_1 by considering the text relevance between the query keyword (sushi) and the set of covered restaurants. Similarly, queries Q_2 and Q_3, shown in Fig. 1, should return R_2 and R_1, respectively.

Example 2 (Epidemic Outbreak Detection from Social Media). People would like to share their life on social media like Twitter, including some upset sick experience, which provides us chance to detect the region with densely distributed topics about some specific epidemic, e.g., flu, from a large amount of geo-tagged tweets. For example, user CamiloidG posted on Twitter "Damn flu... I would

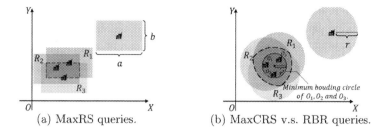

(a) MaxRS queries.   (b) MaxCRS v.s. RBR queries.

Fig. 2. Comparison among MaxRS, MaxCRS, and RBR queries.

love to have someone that can make me a chicken soup..". With text relevance between user tweets and "flu" as score function, the RBR query can be periodically executed to detect regions with frequent occurrence of topic "flu".

As shown in Example 1 and Example 2, an RBR query supports customized scoring function over the enclosed spatial objects, e.g., textual relevance, numeric aggregates like SUM and COUNT, or their arbitrary linear combination. In this work, our RBR queries support *any* monotonic and self-decomposible set function as a legal region goodness measurement, which differs from existing region queries like *Maximum Range-Sum Queries* (MaxRS) [10,11,20] where only SUM and COUNT are supported.

In addition, instead of a fixed query geometry like a given-size rectangle, which is adopted by most region search problems like [4,5,9,14,15], our RBR query retrieves a subset of spatial objects satisfying the minimum bounding circle constraint. This constraint requires that the radius of the minimum bounding circle of the covered objects is not larger than r_{max}. Compared with using fixed-size query geometries, either rectangle or circle, it has three major merits: 1) it requires fewer user efforts to specify the parameters; 2) it makes the query more robust by avoiding covering unnecessary marginal areas (i.e., a big region contains only few points in the middle); 3) the found optimal region is tight and unique, whereas there might be *infinite* optimal solutions for fixed-size geometry queries (e.g., it does not change the covered objects when we slightly move a loose rectangle). Example 3 demonstrates the aforementioned merits.

Example 3 (Minimum Bounding Circle). Figure 2a is an instance of a MaxRS query, which finds the optimal location to place a rectangle of size $a \times b$ such that the weight (we use count here) of covered spatial objects is maximized. Similarly, Fig. 2b demonstrates the MaxCRS query, the circular version of the MaxRS query. It is not hard to prove that for fixed-size geometries, either rectangle of size $a \times b$ or circle of radius r, the solution set is actually infinite as a rectangle (or a circle) centering in the region bounded by dashed lines in Fig. 2a (or Fig. 2b) always covers most spatial objects. As a comparison, the optimal minimum bounding circle for this instance is plotted with a solid line in Fig. 2b, which is more informative to users than $R_1 \cdots R_3$ since it covers less unnecessary area and remains stable even if r_{max} is set improperly large.

However, answering the RBR query is non-trivial. Existing region search techniques cannot be directly applied or extended to solve the RBR query due to their simple data models or the fixed-size geometries. The challenge lies in that we need to determine the circle size as well as its location during the region searching process. Note that all circles with radius less than r_{max} are qualified. Then, dynamically determining the circle size regarding the objects in the current search area brings many difficulties, which disables applying existing methods.

In more detail, the major technical contributions we made in this paper are summarized below.

- We formulate the *Radius Bounded ROI* (RBR) queries over a multi-attributed spatial database supporting user-defined score functions, and prove that answering RBR queries is 3SUM-hard. (▷ See Sect. 2)
- We propose three exact query processing algorithms based on geometric properties together with non-trivial pruning strategies, which is near-optimal in time complexity considering the 3SUM-hardness. (▷ See Sect. 3)
- We conduct extensive experimental studies on both real-world data and synthetic benchmark. Compared to baselines, our framework achieves more than $10^3 \times$ improvement in terms of the time cost. (▷ See Sect. 4)

2   Problem Statement

2.1   Preliminaries

Definition 1 (Spatial Object). *A spatial object is represented by a triple $o = (\ell, d, w)$, where $\ell \in \mathbb{R}^2$ denotes the geographic location of o, d is the document associated with o, and w is a non-negative weight value.*

Our definition of spatial object extends the spatial-textual object, commonly adopted in spatial-keyword queries [6,7,17,18], by associating a weight value, which enables us to model more real-world applications (e.g., the Yelp's service illustrated in Example 1). We then define the query geometry and region score.

Definition 2 (Minimum Circle Bounded Region). *Given a subset of spatial objects $\mathcal{O}' \subseteq \mathcal{O}$, the Minimum Circle Bounded Region (MCBR) of $\mathcal{O}'$, denoted by $R_{\mathcal{O}'}$, is a circle covering all objects $o \in \mathcal{O}'$ with the smallest radius. Let $r(R_{\mathcal{O}'})$ denote the radius of $R_{\mathcal{O}'}$.*

Definition 3 (Region Score). *Given a subset of spatial objects $\mathcal{O}' \subseteq \mathcal{O}$, the Region Score of the corresponding MCBR $R_{\mathcal{O}'}$, denoted by Score($\mathcal{O}'$), is defined as a set function $f : 2^{\mathcal{O}} \to \mathbb{R}^+$ satisfying:*

- *Monotonicity: $f(\mathcal{O}') \geq f(\mathcal{O}'')$ holds for any $\mathcal{O}'' \subseteq \mathcal{O}'$ where $\mathcal{O}'$ and $\mathcal{O}''$ are two subsets of $\mathcal{O}$;*
- *Self-decomposability: $f(\mathcal{O}' \cup \mathcal{O}'') = f(\mathcal{O}') \diamond f(\mathcal{O}'')$ holds for any disjoint sets $\mathcal{O}'$ and $\mathcal{O}''$ where $\diamond$ refers to a merge operator. For example, aggregates COUNT and SUM are self-decomposible, whereas MEDIAN is not.*

The monotonicity constraint guarantees that a region covering $\mathcal{O}'$ is not worse than another region covering a subset of $\mathcal{O}'$. The self-decomposibility constraint enables the region score to be computed in a decomposing manner.

2.2 Radius Bounded ROI (RBR) Queries

Definition 4 (RBR Query). *Given a collection of spatial objects $\mathcal{O}$, a region score function* Score($\cdot$), *and the maximum radius of query region r_{max}, an RBR query $\mathcal{Q}$, retrieves a subset $\mathcal{O}^* \subseteq \mathcal{O}$ with the highest region score satisfying that the radius of MCBR of $\mathcal{O}^*$ is no larger than r_{max}, i.e.,*

$$\mathcal{O}^* = \arg\max_{\mathcal{O}' \subseteq \mathcal{O}} \mathsf{Score}(\mathcal{O}') \ s.t. \ r(R_{\mathcal{O}'}) \leq r_{max}. \tag{1}$$

Answering RBR queries is polynomial-time solvable by providing a straightforward exact algorithm. An observation about MCBR is that given a collection of 2D points P, the minimum covering circle of P can be determined by either 2 or 3 points on the boundary of a region. If it is determined by 2 points, then these two points must form the diameter of this MCBR [22]. Thus, we can simply find the RBR query result through enumerating all circles determined by any three points and any two points satisfying the maximum radius constraint. By assuming the computation of the score function is in $O(1)$, such a naive enumeration takes time $O(n^3 + n^2) = O(n^3)$, which is polynomial. The intrinsic complexity of the RBR problem is shown in Theorem 1.

Theorem 1 (Hardness). *The RBR problem is 3SUM-hard.*

The complexity class 3SUM [16] implies that there exists no worst-case subquadratic algorithm, i.e., $O(n^{2-\epsilon})$ for $\forall \epsilon > 0$, to answer RBR queries exactly. To expedite the query processing, we propose an efficient framework based on novel geometric findings, which turns out to be near-optimal considering the quadratic lower bound on time complexity.

2.3 Discussion of the Region Score

Our RBR queries enable user-configured functions for scoring a region. In this paper, we adopt a score function, as shown in Eq. (2), considering both weights and documents associated on spatial objects.

$$\mathsf{Score}(\mathcal{O}') = \alpha \cdot \mathsf{Score\_w}(\mathcal{O}') + (1 - \alpha) \cdot \mathsf{Score\_t}(\mathcal{O}', q) \tag{2}$$

where Score_w($\cdot$) is a user-specified aggregation function (e.g., SUM, COUNT) defined on the weight set $\{o.w | o \in \mathcal{O}'\}$, Score_t($\cdot$) is a function to measure the relevance of the document set $\{o.d | d \in \mathcal{O}'\}$ to user-specified query keywords q, and α is a hyper-parameter to tune the relative importance between Score_w($\cdot$) and Score_t($\cdot$). For Score_t($\cdot$), we follow the convention of spatial-keyword queries [8] and adopt the Vector Space Model [24] to evaluate the textual similarity between query keywords and spatial objects. Note that, other information retrieval models, like *Language Model* and *Bag-of-words Model*, can also be adopted, which does not require any changes to our proposed techniques. We can verify that Eq. (2) satisfies the two properties of a qualified region score function when Score_w($\cdot$) is set to SUM (or COUNT, AVG, etc.) and Score_t($\cdot$) is set according to [24]. Example 4 illustrates three different settings of Score($\cdot$).

Algorithm 1: PairEnum

Input: An RBR query $\mathcal{Q} = \{\mathcal{O}, q, \mathsf{Score}, r_{max}\}$
Output: The result region of query $\mathcal{Q}$

1 $s^* \leftarrow 0, \mathcal{C}^* \leftarrow \{\}$;
2 **for** $i = 1, \cdots, |\mathcal{O}|$ **do**
3   **for** $j = i + 1, \cdots, |\mathcal{O}|$ **do**
4   **if** $\|o_i.\ell - o_j.\ell\|_2 < 2r_{max}$ **then**
5              $R \leftarrow$ circle determined by $o_i.\ell, o_j.\ell, r_{max}$;
6              $\mathcal{O}_R \leftarrow \{o | o.\ell \in R\}$;
7   **if** $\mathsf{Score}(\mathcal{O}_R) > s^*$ **then**
8                  $s^* \leftarrow \mathsf{Score}(\mathcal{O}_R), \mathcal{C}^* \leftarrow \mathcal{O}_R$;
9 **return** $\mathsf{MCBR}(\mathcal{C}^*)$;

Example 4. We illustrate three different configurations of our RBR queries over the spatial database shown in Example 1.

1. $\alpha = 0$: Only the textual relevance to query keywords is considered, which corresponds to queries Q_1 and Q_2 in Fig. 1. Specifically, for query Q_1, the keyword is "sushi" and apparently region R_1 is more textually relevant to this keyword than region R_2.
2. $\mathsf{Score\_w}(\cdot) = \mathsf{SUM}, \alpha = 1$: Only the object weights are considered. In Fig. 1, Q_3 adopts this setting which implies the user only wants to find a region with highly-rated restaurants without food preference. In this case, our RBR query performs like MaxRS queries with the MCBR as the query geometry.
3. $\mathsf{Score\_w}(\cdot) = \mathsf{SUM}, \alpha \in (0,1)$: Both keyword relevance and object weight values are considered, which is commonly used for queries like *"find a region covering many sushi places with high ratings (stars)"*.

3   Query Processing Algorithms

3.1   Baseline Algorithm: PairEnum

Recall that given a set of spatial objects $\mathcal{O}$, we can generate all possible minimum bounding circles by enumerating circles determined by any pair/triple of the objects. However, since the maximum radius r_{max} is given, we only need to check all pairs, instead of pairs and triples, of objects to enumerate the circles. Based on this intuition, we propose the baseline algorithm PairEnum as shown in Algorithm 1. Line 1 initializes s^* and $\mathcal{C}^*$ for tracking the score and the covered objects of the current optimal solution. Lines 2–8 find the optimal region among the circles with radius r_{max} and passing any pair of objects. After the enumeration, Line 9 invokes a subroutine $\mathsf{MCBR}(\mathcal{C}^*)$ to get the minimum bounding circle of $\mathcal{C}^*$. The correctness of Algorithm 1 is demonstrated in Theorem 2.

Theorem 2. *Let ALG and OPT be the scores of the region found by PairEnum and the optimal region, respectively. Then, ALG = OPT.*

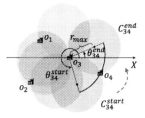

Fig. 3. Illustration of circle rotation.

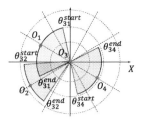

Fig. 4. Polar system transformation.

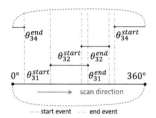

Fig. 5. Angular scan example.

Angle	Cover Set	Score
θ_{31}^{start}	$\{o_1, o_3\}$	2
θ_{32}^{start}	$\{o_1, o_2, o_3\}$	3
θ_{31}^{end}	$\{o_2, o_3\}$	2
θ_{32}^{end}	$\{o_3\}$	1
θ_{34}^{start}	$\{o_3, o_4\}$	2
θ_{34}^{end}	$\{o_3\}$	1

Fig. 6. Table of angle scan result.

We then analyze the time complexity of Algorithm 1. The worst case occurs when the spatial objects are extremely densely distributed and all the n^2 ($n = |\mathcal{O}|$) pairs of objects need to be checked. Each object pair requires one range query and one computation of region score. The subroutine $\mathsf{MCBR}(\mathcal{C}^*)$ is implemented by [19], which takes linear amount of time. Suppose that spatial index like R-tree is used to index all spatial objects, then each range query takes time $O(\log n + c)$ where c is the result size. Let $O(q)$ be the complexity of region score computation, then the total time complexity is $O(n^2(\log n + \max c + q)) = O(n^3 + n^2 q)$, since we have $\max c = O(n)$ in the worst case.

3.2   Circle Rotation and Angle Scan

PairNum finds the optimal region but has a cubic time complexity. In this section, we aim to further reduce its time complexity. Existing rectangle-based MaxRS query studies [10,11] can efficiently generate the candidate regions by scanning with a sweep line. However, this sweep-line based region generation method cannot be applied to our RBR queries which target circular regions. To address this challenge, we propose novel circular region generation method which works with *circle rotation* and *angle scanning* as introduced below. For the ease of presentation, we first define two notations:

– $\angle_{A,B,C}$: the angle of counterclockwise rotation surrounding point A from direction $\overrightarrow{AB}$ to $\overrightarrow{AC}$;
– $O(C)$: the center point of a circle C.

Definition 5 (o_i-Bounded Circle). *Given a radius upper bound r_{max} and a spatial object o_i, an o_i-Bounded Circle, denoted by C_i, refers to **any** circle with radius r_{max} and $o_i.\ell$ on its boundary. Notably, the union of all C_i's is the circular region centering at $o_i.\ell$ with radius $2r_{max}$.*

The circle rotation process is to fix an object o_i as a *reference point* and scan the o_i-bounded circles surrounding $o_i.\ell$ in counterclockwise order. To introduce this procedure, we define the polar system (Definition 6) to locate a o_i-bounded circle and the start/end events (Definition 7) to describe the change of object coverage.

Definition 6 (o_i-Polar System). *For a reference point o_i, the o_i-polar system, denoted by $\mathcal{X}(o_i)$, is the polar coordinate system with pole $o_i.\ell$ and polar axis $\overrightarrow{o_i.\ell, X}$ where $\overrightarrow{o_i.\ell, X}$ is a ray from $o_i.\ell$ horizontally to the right.*

With the concept of the o_i-polar system, we can use the center coordinate under $\mathcal{X}(o_i)$ to uniquely represent an o_i-bounded circle. Let $C_i(\theta)$ be the o_i-bounded circle whose center coordinate is (r_{max}, θ) under $\mathcal{X}(o_i)$. Then, the circle rotation over o_i can be described as scanning $C_i(\theta)$ by varying $\theta \in [0, 2\pi]$. Instead of continuously rotating a circle, we are more interested in some key events which incur change of the objects covered by $C_i(\theta)$. We then define the *start event* and *end event* as below.

Definition 7 (Start Event and End Event). *Given the reference point o_i, let o_j be another spatial object satisfying $||o_i.\ell - o_j.\ell||_2 \leq 2r_{max}$, the **start event** and **end event** for object pair (o_i, o_j) are defined as the moments that $C_i(\theta)$ first **meets** and **leaves** $o_j.\ell$ when rotating it counterclockwise. We denote θ_{ij}^{start} and θ_{ij}^{end} as the angles of $C_i(\theta)$ for the corresponding start and end events, respectively, and denote Φ_i as the set of the angles of all events when rotating circle $C_i(\theta)$, i.e., $\Phi_i = \{\theta_{ij}^{start}\} \cup \{\theta_{ij}^{end}\}$.*

With some simple deductions, θ_{ij}^{start} and θ_{ij}^{end} can be calculated as

$$\theta_{ij}^{start}, \theta_{ij}^{end} = \theta_{ij} \pm \cos^{-1}(\frac{||o_i.\ell - o_j.\ell||_2}{2r_{max}}) \tag{3}$$

where $\theta_{ij}{}^{1}$ is the angular coordinate of $o_j.\ell$ in system $\mathcal{X}(o_i)$, i.e., $\angle_{o_i.\ell, X, o_j.\ell}$. We define the angular pair $\Theta_{ij} = (\theta_{ij}^{start}, \theta_{ij}^{end})$ as the *covering interval* for object o_j when rotating circle C_i. There are two observations about Θ_{ij}:

1. $C_i(\theta)$ always covers $o_j.\ell$ when $\theta \in \Theta_{ij}$;
2. the length of Θ_{ij}, i.e., $|\theta_{ij}^{end} - \theta_{ij}^{start}| = 2\Delta_{ij}$, is no larger than π, and $2\Delta_{ij} = \pi$ i.f.f. $||o_i.\ell - o_j.\ell||_2 = 2r_{max}$.

---

[1] All angles are in the range $[0, 2\pi]$ to make the representation unique.

Note that we may have $\theta_{ij}^{start} > \theta_{ij}^{end}$ (e.g., Fig. 3), as we limit the angles into the range $[0, 2\pi]$. Example 5 illustrates the concepts introduced above.

Example 5. In the example shown in Fig. 3, the object o_3 is selected as the reference point, and $o_1, o_2, \& o_4$ are three objects with distances less than $2r_{max}$ from o_3. All six circles in Fig. 3 are o_3-bounded circles. By selecting o_3 as the pole and $\overrightarrow{o_3, X}$ as the polar axis, we can construct the corresponding o_3-polar system $\mathcal{X}(o_3)$, which is shown in Fig. 4. The two blue circles, i.e., $C_3(\theta_{34}^{start})$ and $C_3(\theta_{34}^{end})$, correspond to the start and end events for the object pair (o_3, o_4). Under $\mathcal{X}(o_3)$, the covering intervals for objects o_1, o_2, o_4, i.e., $\Theta_{31}, \Theta_{32}, \Theta_{34}$ are represented as segments in Fig. 4.

By traversing the o_i-bounded circles with angles in Φ_i in ascending order, we can obtain their sets of covered objects, where the difference between two adjacent sets is incurred by a specific event. Example 6 illustrates such procedure.

Example 6. Continuing with Example 5, three intervals Θ_{31}, Θ_{32} and Θ_{34} are drawn in Fig. 5. Interval Θ_{34} crosses the boundary between 0 and 2π, which is represented with a dashed line. We operate a ray originating from $o_i.\ell$ with the angle θ from 0 to 2π to rotate an o_3-bounded circle. Let $\mathcal{C}$ denote the set of spatial objects covered by $C_3(\theta)$. Then, when L meets θ_{31}^{start}, corresponding to C_{31}^{start} (the top green circle) in Fig. 3, o_1 is added into $\mathcal{C}$ ($\mathcal{C} = \{o_1, o_3\}$). Subsequently when L meets θ_{32}^{start} (the top pink circle), we have $\mathcal{C} = \{o_1, o_2, o_3\}$. When L meets θ_{31}^{end} (the bottom green circle), we have $\mathcal{C} = \{o_2, o_3\}$.

Given a collection of spatial objects $\mathcal{O}$, a reference point o_i and the corresponding event set Φ_i while rotating circle $C_i(\theta)$, we could track the current object set covered by $C_i(\theta)$ and the corresponding region score. Specifically, for a given monotonic function Score, we define two types of *optimal circles*:

- *Locally Optimal Circle* is the circle $C_i(\theta_i^*)$ whose region score is maximized. i.e., $\theta_i^* = \arg\max_{\theta \in \Phi_i} \mathsf{Score}(\mathcal{O}_{C_i(\theta)})$ where $\mathcal{O}_{C_i(\theta)} = \{o | o.\ell \in C_i(\theta) \land o \in \mathcal{O}\}$. Denote the local optimal circle for reference point o_i as C_i^{OPT}.
- *Globally Optima Circle* is the circle, denoted by C^{OPT}, with the highest region score among all local optimal circles, i.e., $C^{\mathsf{OPT}} = \arg\max_{C \in \{C_i^{\mathsf{OPT}}\}_{i=1}^n} \mathsf{Score}(\mathcal{O}_C)$.

With the concepts above, the connection between the circle rotation and RBR query result is shown in Theorem 3.

Theorem 3. *For the global optimal circle C^{OPT}, for arbitrary subset $\mathcal{O}' \subseteq \mathcal{O}$ satisfying the radius of $\mathsf{MCBR}(\mathcal{O}')$ is less than r_{max}, $\mathsf{Score}(\mathcal{O}_{C^{\mathsf{OPT}}}) \geq \mathsf{Score}(\mathcal{O}')$ where the equality holds i.f.f. $\mathcal{O}' = \mathcal{O}_{C^{\mathsf{OPT}}}$.*

Recall that each minimum bounding circle must have at least two points on its boundary. Theorem 3 implies that if the boundary of an RBR query result region contains point o_i, it can surely be found through the circle rotation process with o_i as the reference point. Thus, by repeatedly performing the circle rotation

---

Algorithm 2: BaseRotation

Input: A RBR query $\mathcal{Q} = \{\mathcal{O}, q, \mathsf{Score}, r_{max}\}$
Output: The result region of query $\mathcal{Q}$

1 $s^* \leftarrow 0, \mathcal{C}^* \leftarrow \{\}$;
2 **for** $i = 1, \cdots, |\mathcal{O}|$ **do**
3   $\quad \mathcal{N}(o_i) \leftarrow \{o_j| \ ||o_i.\ell - o_j.\ell||_2 < 2r_{max}, o_j \in \mathcal{O}, j \neq i\}$;
4   $\quad Q \leftarrow$ **new** priority queue, $\mathcal{C} \leftarrow \{\}$;
5   $\quad$ **for** $o' \in \mathcal{N}(o)$ **do**
6    $\quad\quad$ calculate θ_{ij}^{start} and θ_{ij}^{end} as Eq. (3);
7    $\quad\quad$ $Q.enqueue(\theta_{ij}^{start}), Q.enqueue(\theta_{ij}^{end})$;
8   $\quad$ **while** Q *is non-empty* **do**
9    $\quad\quad$ **if** $Q.pop()$ *is* θ_{ij}^{start} **then**
10    $\quad\quad\quad$ $\mathcal{C} \leftarrow \mathcal{C} \cup \{o_i\}$;
11    $\quad\quad\quad$ **if** $\mathsf{Score}(\mathcal{C}) > s^*$ **then**
12     $\quad\quad\quad\quad$ $s^* \leftarrow \mathsf{Score}(\mathcal{C}), \mathcal{C}^* \leftarrow \mathcal{C}$;
13    $\quad\quad$ **else if** $Q.pop()$ *is* θ_{ij}^{end} **then**
14     $\quad\quad\quad$ $\mathcal{C} \leftarrow \mathcal{C}/\{o_i\}$;
15 **return** $\mathsf{MCBR}(\mathcal{C}^*)$;

---

process with each object $o \in \mathcal{O}$ as reference point and tracking the region with the highest score, we can finally obtain the answer to an RBR query. Intuitively, the circle rotation based method is more efficient than PairEnum since it avoids computing $O(n^2)$ times range queries, and the update of the region with the highest score can be performed incrementally.

3.3   Algorithm: BaseRotation

Based on the circle rotation principle, we propose algorithm BaseRotation in Algorithm 2. Line 1 initializes s^* and $\mathcal{C}^*$, which represents the current highest score and the corresponding set of covered objects. Lines 2–14 repeatedly run the circle rotation procedure with the reference point o_i for $i = 1, \cdots, |\mathcal{O}|$. In each inner loop, Line 3 is a range query to find all objects within distance $2r_{max}$ from o_i, Line 4 creates a priority queue (Min-heap) Q, to store events for angle scanning and initializes an empty set $\mathcal{C}$ to track the current covered spatial objects, Lines 5–7 calculate $\theta_{ij}^{start}, \theta_{ij}^{end}$ for each $o_j \in \mathcal{N}(o_i)$ and insert them into Q, and Lines 8–14 scan the angles, which have been ordered in priority queue Q. Specifically, at each time we pop one element from Q, which simulates a ray scanning from 0 to 2π. For start events (i.e., θ_{ij}^{start}), the corresponding object o_j is added to a temporary set $\mathcal{C}$ which contains objects covered by current circle C_{ij}^{start}; whereas, for end events (i.e., θ_{ij}^{end}), we remove o_j from $\mathcal{C}$. s^* and $\mathcal{C}^*$ are updated if the current set $\mathcal{C}$ yields a higher score. Finally, we return the minimum bounding circle of $\mathcal{C}^*$ by invoking the subroutine MCBR.

Note that we ignore the corner cases in Algorithm 2 where intervals crossing $0/2\pi$, which can be handled by starting the scan from the first start event. There is a running example shown in Example 7.

Example 7. Suppose Score=COUNT, Fig. 6 illustrates the evolution of $\mathcal{C}$ regarding the scan in Fig. 5 where o_3 is the reference point. When the sweep-line meets θ_{32}^{start}, the corresponding circle C_{32}^{start} covers most objects. Thus, after this run of circle rotation, s^* and $\mathcal{C}^*$ are updated to 3 and $\{o_1, o_2, o_3\}$, respectively.

The correctness of Algorithm 2 can be easily derived from Theorem 3 since it gives every object a chance to be the reference point. We then analyze its time complexity. Algorithm 2 contains n range queries on spatial database $\mathcal{O}$ (Line 3), which takes time $O(n \log n)$ by adopting a common spatial index like R-tree. Since the region score function is required to be self-decomposable, Score$(\mathcal{C})$ could be simply updated using the previous value, which takes constant time. In each inner-loop, supposing there are c_i (i.e., $|\mathcal{N}(o_i)|$) objects in the result set of line 3, lines 4–14 involve $O(c_i)$ heap operations, which takes time $O(c_i \log c_i)$. Thus, the total time complexity of BaseRotation is $O(\sum_{i=1}^{n}(\log n + c_i \log c_i)) = O(n \log n + nC \log C)$ where $C = \max_{i=1,\cdots,n} c_i$. The worst case of Algorithm 2 occurs when the spatial objects are narrowly distributed and r_{max} is set to some large value. For such a case, $c_i = n$ and thus the time complexity becomes $O(n^2 \log n)$, which is near-optimal considering the 3SUM-hardness i.e., $\Omega(n^2)$.

3.4   Algorithm: OptRotation

The major overhead of BaseRotation lies in its $O(c_i)$ heap operations for rotating an o_i-bounded circle. To further improve the time efficiency, 1) we use another efficient data structure called *interval bucket* instead of the heap-based event queue; 2) we use several non-trivial pruning rules, observing that there are some unnecessary circle rotations.

Interval Bucket. Instead of using a single event queue at runtime, the *Interval Bucket* divides the angular space $[0, 2\pi]$ into k buckets and maintains a subqueue for each bucket to store its events (i.e., angles). The event scan is then conducted bucket by bucket. We suppose that the reference point is o_i and there are c_i objects within distance $2r_{max}$ from $o_i.\ell$. Then, by setting the bucket size $k = \Theta(c_i)$ (the ith bucket indicates the angular range $[2\pi i/k, 2\pi(i+1)/k)$ for $i = 0, \cdots, k-1$) and assuming the events are uniformly distributed into the buckets, the time complexity of event scan becomes $O(c_i)^2$, which improves from that of the pure heap-based approach, $O(c_i \log c_i)$. To make the events uniformly distributed into buckets, which would assure the efficiency of the method, we can use some histogram-base techniques [13] to split or merge some buckets.

Pruning and Lazy Evaluation. We further reduce the time cost of BaseRotation by avoiding unnecessary scanning events and adopting a *lazy evaluation* strategy. Two pruning rules are introduced below.

Lemma 1 (Pruning Rule 1). *Suppose the current reference point of circle rotation is o_i, if Score$(\mathcal{N}(o_i)) < s^*$ where s^* is the current highest region score, then it is sure that the optimal solution does not have o_i on its boundary.*

---

[2] The time complexity analysis is similar to that of *distribution sorting* [12].

Lemma 1 implies that we can use the current highest score to prune some reference points at runtime. This pruning rule can be effective if we achieve a large s^* at an early stage. To realize that, we start the circle rotation with the reference points in a relatively dense area, with the intuition that dense areas tend to have higher scores. To further reduce the time cost for a certain reference point o_i, we next show that the score variable s^* in BaseRotation can be updated lazily, which is stated in Lemma 2. We first introduce the concept of *Locally Maximal Set*.

Definition 8 (Locally Maximal Set). *Suppose the current reference point is o_i, the rotating circle $C_i(\theta)$ is called a locally maximal circle i.f.f. there exists no other $C_i(\theta')$ ($\theta \neq \theta'$) such that the objects covered by $C_i(\theta')$ is a superset of that of $C_i(\theta)$, i.e., $\mathcal{O}_{C_i(\theta)} \subset \mathcal{O}_{C_i(\theta')}$. The object set for a locally maximal circle is called a locally maximal set.*

Example 8. Continuing with the circle rotation result shown in Fig. 6, circle $C_i(\theta_{31}^{start})$ is not a locally maximal circle since $\{o_1, o_2, o_3\}$ is a superset of $\{o_1, o_2\}$, the set covered by $C_i(\theta_{31}^{start})$. On the other hand, $C_i(\theta_{32}^{start})$ is a locally maximal circle and $\{o_1, o_2, o_3\}$ is a local maximal set.

Lemma 2 (Pruning Rule 2). *Let C^{OPT} denote an optimal solution, then the object set covered by C^{OPT}, denoted by $\mathcal{O}_{C^{OPT}}$, must be a locally maximal set.*

Lemma 2 states that the object set covered by the optimal solution must be a locally maximal set. Thus, during the circle rotation process, we can safely skip some events if their corresponding circles do not cover a locally maximal set, which is expected to save much time from updating the current optimal score and the corresponding object set, i.e., s^* and $\mathcal{C}^*$ in BaseRotation. Since s^* and $\mathcal{C}^*$ are updated only if it is necessary, we call this pruning strategy as *lazy evaluation*. We have Lemma 3 to efficiently determine whether the current scanning event corresponds to a locally maximal set.

Lemma 3. *For a reference point o_i, we denote $\theta_{ij}^{t_j} \rightarrow \theta_{ik}^{t_k}$ as the transition from one scanning event to another where labels t_j, t_k can be either "start" or "end". Then, the object set covered by circle $C_i(\theta_{ij}^{t_j})$ is a locally maximal set i.f.f. $t_j = start, t_k = end$ (note that, j can be either equal to k or not).*

Example 9. By scanning the intervals shown in Fig. 6, the event θ_{32}^{start} corresponds to a locally maximal set $\{o_1, o_2, o_3\}$ since there is an event transition $\theta_{32}^{start} \rightarrow \theta_{31}^{end}$ satisfying the condition in Lemma 3. θ_{34}^{start} can be analyzed analogously. Thus, only θ_{32}^{start} and θ_{34}^{start} need to be checked, instead of all six events.

Based on the interval bucket and pruning rules introduced above, we propose OptRotation in Algorithm 3. Similar to BaseRotation, Line 1 initializes s^* and $\mathcal{C}^*$ to track the currently found optimum, and Lines 2–19 continuously perform circle rotation with $o_i \in \mathcal{O}$. Lines 4–5 implement Pruning Rule 1, which stops the current rotation if the score of $\mathcal{N}(o_i)$ is less than s^*. Lines 6–9 put the scanning events into the interval bucket B. Lines 10–19 scan the events one by one, which is similar to BaseRotation. The difference is that we record the last

---

Algorithm 3: OptRotation

Input: A RBR query $\mathcal{Q} = \{\mathcal{O}, q, \mathsf{Score}, \alpha, r_{max}\}$
Output: The result region of query $\mathcal{Q}$

1 $s^* \leftarrow 0, \mathcal{C}^* \leftarrow \{\}$;
2 **for** $i = 1, \cdots, |\mathcal{O}|$ **do**
3      $\mathcal{N}(o_i) \leftarrow \{o_j \mid \|o_i.\ell - o_j.\ell\|_2 < 2r_{max}, o_j \in \mathcal{O}, j \neq i\}$;
4   **if** $\mathsf{Score}(\mathcal{N}(o_i)) < s^*$ **then**
5   **continue**;   ▷ Pruning rule 1
6      $B \leftarrow$ **new** bucket interval with k buckets;
7   **for** $o' \in \mathcal{N}(o_i)$ **do**
8   calculate θ_{ij}^{start} and θ_{ij}^{end} as Eq. (3);
9   enqueue θ_{ij}^{start} and θ_{ij}^{end} into proper bucket of B;
10     $\mathcal{C} \leftarrow \{\}, prev \leftarrow NULL$;
11   **for** $j = 1, \cdots, k$ **do**
12        $Q \leftarrow$ queue stored in i^{th} bucket of B;
13   **while** Q *is non-empty* **do**
14   **if** $Q.pop()$ is θ_{ij}^{start} **then**
15              $\mathcal{C} \leftarrow \mathcal{C} \cup \{o_i\}, prev \leftarrow$ **true**;
16   **else if** $Q.pop()$ is θ_{ij}^{end} **then**
17   **if** $prev ==$ **true** *and* $\mathsf{Score}(\mathcal{C}) > s^*$ **then**
18                 $\mathcal{C}^* \leftarrow \mathcal{C}, s^* \leftarrow \mathsf{Score}(\mathcal{C})$;   ▷ Pruning rule 2
19              $\mathcal{C} \leftarrow \mathcal{C}/\{o_i\}, prev \leftarrow$ **false**;
20 **return** $\mathsf{MCBR}(\mathcal{C}^*)$;

---

scanning event type in $prev$, whose value is true for a start event, and s^* and $\mathcal{C}^*$ are updated only if it indicates a locally maximal set.

We then analyze the time complexity of Algorithm 3. In each rotation, Lines 3–5 retrieve a set of nodes $\mathcal{N}(o_i)$ through a range query and check whether Pruning Rule 1 holds. By adopting indexes designed for spatial aggregation query processing like [21], both the range query and the calculation of $\mathsf{Score}(\mathcal{N}(o_i))$ can be done in $O(\log n + c_i)$ where $c_i = |\mathcal{N}(o_i)|$. As we have analyzed before, for i^{th} rotation, the event scanning routine in Lines 11–19 takes $O(c_i)$ time in expectation. Let M be the number of conducted circle rotation operations. Then, the total time complexity is $O(n \log n + nC + MC) = O(n \log n + nC)$ where $C = \max c_i$. Compared with BaseRotation, Algorithm OptRotation reduces a factor of $\log C$ in time complexity by adopting the interval buckets. In practice, the efficiency improvement of OptRotation is significant since the two pruning rules can avoid most of the unnecessary circle rotation procedures.

4   Experimental Studies

In this section, we report experimental results of our proposed RBR query processing techniques. All the experiments were conducted on a Linux server with Intel(R) Xeon(R) CPU X5675 @ 3.07 GHz and 32 GB memory, and all the algorithms were implemented in Java using JDK10.

Table 1. Summary of real datasets.

Dataset	Category	#Objects	Avg. #Neighbors	Max. #Neighbors
Yelp	Real	192,610	2785.87	12133
Cal	Real	104,770	83.12	898

Table 2. Parameter setting (real).

Parameter	Values
p	[0.2, 0.4, 0.6, 0.8, <u>1.0</u>]
r_{max}	[0.5, 1, <u>5</u>, 10, 50]

Table 3. Parameter setting (synthetic).

Parameter	Values
n	[50K, 100K, <u>500K</u>, 1M, 5M]
σ	[50, 100, <u>500</u>, 1000, 5000]
r_{max}	[5, 10, <u>20</u>, 50, 100]

4.1 Experiment Setting

Compared Algorithms. We implement the baseline algorithm PairEnum (PE) and the two circle rotation based algorithms BaseRotation (BR) and OptRotation (OR). We also implement the state-of-the-art solution for the Maximum Range Sum (MaxRS) queries [11] for comparison.

Datasets. We use both real-world and synthetic datasets to evaluate the methods. Yelp is a dataset of 192,610 selected businesses [2], which are mostly restaurants, with multiple attributes including "stars" and "reviews". Cal is a dataset of 104,770 California POIs crawled by using Google Map Places API [1] where each POI is associated with multiple user-tagged textual labels (e.g., "nice barbershop"). We also adopt two synthetic datasets, Uniform and Mixture, generated by sampling from a uniform distribution and a Gaussian mixture distribution (with 5 identical Gaussian components), respectively. Table 1 presents some statistics of the adopted datasets. Specifically, for a dataset $\mathcal{O}$, Avg. #Neighbors$=\frac{1}{n}\sum_{i=1}^{n} c_i$ and Max. #Neighbors$=\max_{i=1,\cdots,n} c_i$ where c_i denotes the number of objects with distance less than 5 km from o_i. These two statistics depict the distribution and the locality of a spatial dataset.

Region Score and Queries. The score functions are set up as follows. For Yelp, the region score is Eq. (2) where Score_t($\cdot$) is the text relevance between query keywords and user reviews of restaurants, Score_w refers to the total sum of stars, and $\alpha = 0.5$. Note that, we normalize both text relevance and star value into range $[0, 1]$ before adding them together. For Cal, since the POIs do not have ratings, we only consider the text relevance, which means Score($\cdot$) = Score_t (i.e., $\alpha = 0$). For the two synthetic datasets, we set Score($\cdot$) = SUM and the objects weights are sampled from a uniform distribution $U(0, 1)$. Finally, for datasets Cal and Yelp which require keywords in the test queries, we randomly select 20 terms whose frequency in the whole text corpus is larger than 500 for each query.

Parameter Settings. The parameter settings are shown in Table 2 and Table 3 where the underlined values are the default settings. Specifically, p is the ratio for sampling the objects from the real datasets. n and σ are the number of objects and the variance of the distribution for generating the synthetic data. For both real and synthetic data, r_{max} is the maximum radius of the query region.

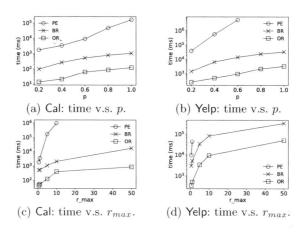

(a) Cal: time v.s. p.

(b) Yelp: time v.s. p.

(c) Cal: time v.s. r_{max}.

(d) Yelp: time v.s. r_{max}.

Fig. 7. Experimental results on real datasets by varying p and r_{max}.

4.2   Experiment Results

Influence of p. The scalability evaluation results w.r.t. p on real datasets are shown in Fig. 7a and Fig. 7b. On both Cal and Yelp, the running times of all three algorithms increase as p increases since both the number of spatial objects and the object density become larger as p increases. However, the increase of baseline PE is much faster than that of BR and OR, which coincides with the fact that the time complexity of PE is higher. This demonstrates the efficiency of our circle rotation-based techniques. More specifically, the most efficient algorithm, OR, is $10^2 \sim 10^3 \times$ faster than PE. Also, the growth of OR's running time keeps smooth over datasets of different locality levels.

Influence of n and σ. Similarly, we also conduct the scalability evaluation w.r.t. n, the number of samples, and σ, the variance of generated data, on the two synthetic datasets Uniform and Mix. The results are shown in Fig. 8a–Fig. 8e. For parameter n, the baseline PE yields "time-out" when n is higher than 1 million, whereas OR increases smoothly over the increase of n. We also find that the running times all grow faster on the synthetic data, compared with real data. That is because with the variance parameter σ, the data density increases very fast as n increases. Nonetheless, OR can finish in 10 s when n is set to 5 million, which is efficient enough to operate in real-time. The variance parameter σ is another parameter that affects the data distribution. As σ increases, the spatial objects tend to be more sparsely distributed. Consequently, the running times of the algorithms all decrease.

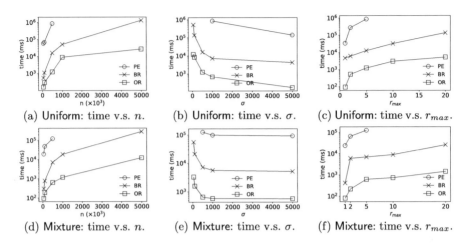

Fig. 8. Experimental results of parameters n and σ on synthetic datasets.

Influence of r_{max}. As an important query parameter, r_{max} not only constrains the maximum radius of the result region, but also influences the RBR query processing time. The results of running times w.r.t. r_{max} on both real and synthetic datasets are shown in Fig. 7c–Fig. 8f. The running times of all three algorithms increase since C increases as r_{max} grows. The running times are particularly large when the objects are densely distributed, e.g., dataset Yelp. However, algorithm OR still performs well even if r_{max} is set to some very large value (e.g., 50 km on Cal and Yelp).

5   Conclusion

In summary, this paper introduces a new type of spatial operator called *Radius Bounded ROI* (RBR) queries. Compared with existing studies on best region search over spatial databases, our RBR queries adopt a more flexible query geometry and a more generalized spatial data model enabling users to specify a large spectrum of region goodness measurements. We prove that answering RBR queries is 3SUM-hard and propose three algorithms PairEnum, BaseRotation and OptRotation based on non-trivial geometric observations. To demonstrate both efficiency and effectiveness of our proposed algorithms, we conduct extensive experimental studies on both real-world datasets and synthetic benchmarks, and the results show that OptRotation, our most efficient algorithm, beats the baseline algorithm by a factor of 3.

Acknowledgments. This work is partially supported by the Hong Kong RGC GRF Project 16207617, CRF Project C6030-18G, C1031-18G, C5026-18G, AOE Project AoE/E-603/18, China NSFC No. 61729201, Guangdong Basic and Applied Basic Research Foundation 2019B151530001, Hong Kong ITC ITF grants ITS/044/18FX and ITS/470/18FX, Microsoft Research Asia Collaborative Research Grant, Didi-HKUST joint research lab project, and Wechat and Webank Research Grants. Xiang Lian was supported by NSF OAC (No. 1739491) and Lian Startup (No. 220981) from Kent State University.

References

1. Places Google Map (2019). https://cloud.google.com/maps-platform/
2. Yelp Dataset (2019). https://www.yelp.com/dataset
3. Yelp Fusion (2019). https://www.yelp.com/developers/documentation/v3/
4. Amagata, D., Hara, T.: Monitoring MaxRS in spatial data streams. In: EDBT, pp. 317–328. OpenProceedings.org (2016)
5. Amagata, D., Hara, T.: A general framework for MaxRS and MaxCRS monitoring in spatial data streams. ACM TSAS **3**(1), 1:1–1:34 (2017)
6. Cao, X., Cong, G., Guo, T., Jensen, C.S., Ooi, B.C.: Efficient processing of spatial group keyword queries. ACM Trans. Database Syst. **40**(2), 13:1–13:48 (2015)
7. Cao, X., Cong, G., Jensen, C.S., Ooi, B.C.: Collective spatial keyword querying. In: SIGMOD Conference, pp. 373–384. ACM (2011)
8. Cao, X., Cong, G., Jensen, C.S., Yiu, M.L.: Retrieving regions of interest for user exploration. PVLDB **7**(9), 733–744 (2014)
9. Chen, Z., Liu, Y., Wong, R.C., Xiong, J., Cheng, X., Chen, P.: Rotating MaxRS queries. Inf. Sci. **305**, 110–129 (2015)
10. Choi, D., Chung, C., Tao, Y.: A scalable algorithm for maximizing range sum in spatial databases. PVLDB **5**(11), 1088–1099 (2012)
11. Choi, D., Chung, C., Tao, Y.: Maximizing range sum in external memory. ACM Trans. Database Syst. **39**(3), 21:1–21:44 (2014)
12. Cormen, T.H., Leiserson, C.E., Rivest, R.L., Stein, C.: Introduction to Algorithms. MIT Press, United States (2009)
13. Cormode, G., Garofalakis, M.N., Haas, P.J., Jermaine, C.: Synopses for massive data: samples, histograms, wavelets, sketches. Found. Trends Databases **4**(1–3), 1–294 (2012)
14. Feng, K., Cong, G., Bhowmick, S.S., Peng, W., Miao, C.: Towards best region search for data exploration. In: SIGMOD Conference, pp. 1055–1070. ACM (2016)
15. Feng, K., Guo, T., Cong, G., Bhowmick, S.S., Ma, S.: SURGE: continuous detection of bursty regions over a stream of spatial objects. In: ICDE, pp. 1292–1295. IEEE Computer Society (2018)
16. Gajentaan, A., Overmars, M.H.: On a class of $o(n^2)$ problems in computational geometry. Comput. Geom. **45**(4), 140–152 (2012)
17. Gao, Y., Zhao, J., Zheng, B., Chen, G.: Efficient collective spatial keyword query processing on road networks. IEEE Trans. Intell. Transp. Syst. **17**(2), 469–480 (2016)
18. Long, C., Wong, R.C., Wang, K., Fu, A.W.: Collective spatial keyword queries: a distance owner-driven approach. In: SIGMOD Conference, pp. 689–700. ACM (2013)
19. Megiddo, N.: Linear-time algorithms for linear programming in r̂3 and related problems. In: FOCS, pp. 329–338. IEEE Computer Society (1982)
20. Tao, Y., Hu, X., Choi, D., Chung, C.: Approximate MaxRS in spatial databases. In: PVLDB, vol. 6, no. 13, pp. 1546–1557 (2013)
21. Tao, Y., Papadias, D., Zhang, J.: Aggregate processing of planar points. In: Jensen, C.S., et al. (eds.) EDBT 2002. LNCS, vol. 2287, pp. 682–700. Springer, Heidelberg (2002). https://doi.org/10.1007/3-540-45876-X_42

22. Wang, K., Cao, X., Lin, X., Zhang, W., Qin, L.: Efficient computing of radius-bounded k-cores. In: ICDE, pp. 233–244. IEEE Computer Society (2018)
23. Wu, D., Jensen, C.S.: A density-based approach to the retrieval of top-k spatial textual clusters. In: CIKM, pp. 2095–2100. ACM (2016)
24. Zobel, J., Moffat, A.: Inverted files for text search engines. ACM Comput. Surv. **38**(2), 6 (2006)

An Attention-Based Bi-GRU for Route Planning and Order Dispatch of Bus-Booking Platform

Yucen Gao[1], Yuanning Gao[1], Yuhao Li[1], Xiaofeng Gao[1(✉)], Xiang Li[2], and Guihai Chen[1]

[1] Shanghai Key Laboratory of Scalable Computing and Systems,
Department of Computer Science and Engineering,
Shanghai Jiao Tong University, Shanghai, China
{guo_ke,gyuanning,Yggdrasils}@sjtu.edu.cn,
{gao-xf,gchen}@cs.sjtu.edu.cn
[2] Beijing University of Chemical Technology, Beijing, China
lixiang@mail.buct.edu.cn

Abstract. To cope with the high needs from passengers, especially for airports at night, we plan to develop a novel bus-booking platform, which can dispatch several passenger orders to one bus together. In this paper, we first give the formal definition of the Order Dispatch and Route Planning (ODRP) problem for the new bus-booking platform, and prove the ODRP problem is NP-hard. We then propose a new method based on attention mechanism and Bi-directional Gated Recurrent Unit (Bi-GRU) to realize the tasks of order dispatch and route planning simultaneously. To the best of our knowledge, this is the first method that uses main ideas of attention mechanism and Bi-GRU in order dispatch and route planning issues related to urban bus system. It can achieve the goal of increasing passenger number and reducing platform costs. Through experiments based on real-world data, we prove the effectiveness of the proposed method.

Keywords: Bus-booking platform · Order dispatch · Route planning

1 Introduction

With the rapid expansion of urban scale and population growth, the demand for urban transportation within city grows rapidly in recent years. However, all existing methods still cannot completely satisfy the needs from passengers, especially for some locations at special periods. For instance, according to the official

This work was supported by the National Key R&D Program of China [2020YFB1707903]; the National Natural Science Foundation of China [61872238, 71722007, 71931001], the Huawei Cloud [TC20201127009], the CCF-Tencent Open Fund [RAGR20200105], and the Tencent Marketing Solution Rhino-Bird Focused Research Program [FR202001].

© Springer Nature Switzerland AG 2021
C. S. Jensen et al. (Eds.): DASFAA 2021, LNCS 12681, pp. 609–624, 2021.
https://doi.org/10.1007/978-3-030-73194-6_40

statistics from Beijing Traffic Management Bureau (BTMB), the number of the passengers arrived at Beijing Capital International Airport is about 18,000 during midnight period (23:00–02:00). However, the local public transport capacity can only provide pick-up service for around 8,200 passengers, including 7,000 passengers by taxi and online car-hailing, and 1,200 passengers by airport shuttle buses (less than 7% of the total arrived passengers), leaving a large number of passengers waiting at the airport for a long duration [1].

To cope with this situation, we plan to develop a novel bus-booking platform. A bus-booking platform can dispatch several passenger orders to one bus together, compute a specially planned route cycle based on these requirements, and then ride the passengers to their destinations sequentially. However, its promotion faces many challenges. For example, to operate this new platform requires solving new variants of order dispatch, route planning and other issues. There is not much related work to study these issues.

Order dispatch refers to assigning multiple appropriate passenger orders to one bus. Didi Chuxing has done a lot of related work in modern taxi networks [2], but there has not been much reseach in bus networks. Route planning is another problem that needs to be solved urgently in the promotion of bus-booking platforms. Generally, route planning is a multi-objective optimization problem, where the optimization goal is to minimize transportation costs and maximize passenger number.

In this paper, we first investigate the previous work on the order dispatch and route planning problems of urban bus systems and cars. We then give a formal definition of the order dispatch and route planning (ODRP) problem for the bus-booking platform.

A method based on the attention mechanism [3] and Bi-directional Gated Recurrent Unit (Bi-GRU) [4] is then proposed to solve the ODRP problem. Specifically, we first preprocess the order information through the order information extraction module to obtain the vector representation of a series of order features. We then utilize the ideas used in the attention mechanism and give different degrees of attention to orders of different importance, where a higher attention weight indicates that the corresponding order has a higher importance level. This can meet the personalized needs of users and realize the goal of maximizing the profitability for the bus-booking platform. Finally, we use Bi-GRU to process station information, and dynamically provide the assigned order information and current station information as the context vector to the attention sub-network. Under the circular operation of attention sub-network and Bi-GRU, we can get the final order dispatch scheme and planned route.

We also evaluate the proposed method based on real data set. Simulation experiments under different order quantities, bus numbers and bus capacities are conducted. The results show that our method can achieve better results in terms of user experience and order satisfaction compared with the baseline methods.

2   Related Work

The new bus-booking platform is a novel system that can process passenger orders, thus there is little literature related to it. Because we study the route planning and order dispatch problem of the bus-booking platform, we investigate the literature related to route planning and order dispatch.

Route Planning: Around 2010, with the increase in urban traffic demand, researchers began to pay attention to the computational efficiency of the route planning algorithm, and combined a large amount of real data to conduct targeted analysis [5,6]. During the same period, some researchers also put forward some novel ideas. Khoa et al. [7] suggested that users can consider walking between two stations not far away to improve travel efficiency. Wang et al. [8] applied indexing technology to route planning. Since then, the relevant research on bus route planning has paid more and more attention to customer needs. Kong et al. [9] utilized machine learning models to accurately predict user travel needs with a dynamic programming method for route planning.

Order Dispatch: After the emergence of car-hailing platforms, researchers conducted an in-depth discussion on order dispatch based on the big data collected by car-hailing platforms and advanced research methods in recent years. Zhang et al. [10] used Bayesian framework to model the distribution of users and designed a combinatorial optimization method to maximize the global order success rate. Xu et al. [2] modeled the order dispatch problem as a large-scale sequential decision problem in a large-scale on-demand car-hailing platform, and aimed at optimal global resource utilization.

In addition to these studies that provide strategies for increasing the global order success rate, some scholars have also paid attention to the profits of platforms and drivers. On the basis of proving that the constrained optimization problem with platform profit as the optimization goal is NP-hard, Zheng et al. [11] proposed an approximation algorithm and carried out simulation experiments. Duan et al. [12] utilized a dynamic programming method and achieved a mutually beneficial win-win situation between car-hailing platform and passengers.

In this paper, we formally define a multi-objective optimization problem that considers both passenger experience and platform revenue, which is in line with the current research trend on order dispatch.

3   Problem Formulation

In this section, we first explain some concepts and describe the problem of order dispatch and route planning problem for the new bus-booking platform. Then, we prove the ODRP problem is NP-hard.

Definition 1. *(Time unit). A time unit refers to the time span of the bus-booking platform to dispatch existing orders to different buses and plan routes for these buses [11], after which the platform batches new orders until the next time unit.*

Definition 2. *(Order o_i). An order o_i is a quinary tuple $\langle d_i, t_i, v_i, np_i, p_i \rangle$, where d_i refers to the destination station of the order, t_i is the time when o_i is collected by the bus-booking platform, v_i represents the VIP level of o_i, np_i refers to the passenger number of the order, and p_i is the priority assigned to order o_i, which represents its importance. The relationship between p_i and other order attributes can be expressed as a function.*

$$p_i = f_1(t_i, v_i, np_i)$$

Definition 3. *(Departure station DS and destination station s_i). DS means the departure station, where all orders and buses start. s_i is a binary tuple $\langle longitude_i, latitude_i \rangle$, which represents the i_{th} destination station.*

Definition 4. *(Bus b_i). A bus b_i is a tuple $\langle route_i, pb_i, L, c_b, c_r, v_b \rangle$, where $route_i$ represents the route plan of the bus, pb_i refers to the numbe of passengers assigned to b_i, L means the capacity limit of the bus, c_b is the bus fuel cost per kilometer, c_r is the bus cost per route and v_b represents the bus speed per hour.*

Definition 5. *(Travel time $t(a,b)_i$). The travel time function $t(a,b)_i$ is the time from a to b in $route_i$.*

According to the definitions mentioned above, we can describe our optimization goals: the platform cost $C = \sum_{b_i \in B} C_i$ and the total priority of all accepted orders $P = \sum_{o_i \in O_{accepted}} p_i$. The cost of the i_{th} bus C_i can be obtained by $c_r + cost_i$, where $cost_i$ is the overall fuel cost of b_i.

Given an order set O, a bus set B, a departure station DS and a destination station set S in a time unit, the capacity constraint for the ODRP problem is:

$$\forall b_i \in B, pb_i \leq L$$

In order to give a formal programming form for this problem, we stipulate that $x_{ij} = 1$ represents the i_{th} order o_i is assigned to the j_{th} bus b_j. To ensure that an order is only assigned to one bus, we add constraints:

$$\sum_{b_j \in B} x_{ij} \leq 1, \forall o_i \in O \tag{1}$$

The capacity constraints mentioned above can be written as:

$$\sum_{o_i \in O} x_{ij} np_i \leq L, \forall b_j \in B \tag{2}$$

Our primary goal is to maximize benefits, and the secondary goal is to minimize costs, so the linear programming form of the problem is:

$$\max \sum_{o_i \in O} x_{ij} p_i$$

$$\min \sum_{b_j \in B} c_r + C_j$$

$$\text{s.t.Constraint } (1), (2)$$

Problem Hardness: We prove the ODRP problem is NP-hard by showing that the Hamiltonian cycle problem, an NPC problem, can be reduced to the ODRP problem.

Theorem 1. *The problem of maximizing priority and minimizing cost for the bus-booking platform is NP-hard.*

Given $G(V, E)$ with $|V| = n$, we define $cost(e) = 1, \forall e \in E$. Assume that there are $n - 1$ orders with different destinations, $c_b = 0$ and only one bus is at the departure station, which can accept all orders. To maximize the priority, the bus will accept all orders. Then we add edges E' to G to make G become a complete graph and define $cost(e') = 2, \forall e' \in E'$. We can do this operation in polynomial time. In the situation, we consider the decision version of the ODRP problem, which is "is there a route planning solution of cost at most n?". If the answer is yes, there must be a cycle that visits all nodes exactly once. At the same time, edges of the cycle are all from E. Therefore we can ensure that there is a Hamiltonian cycle in G. Since the Hamiltonian cycle problem is NP-Complete as we mentioned above, the decision version of the ODRP problem is NP-Complete, hence the ODRP problem is NP-hard.

4   Attention-Based Bi-GRU Method

4.1   Framework Overview

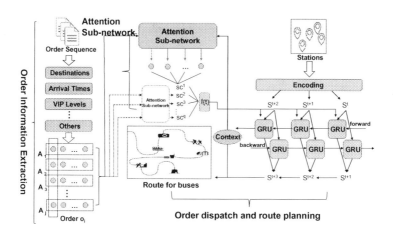

Fig. 1. Overview of the attention-based Bi-GRU method

The framework of our proposed method is shown in Fig. 1. The method can be divided into three parts, an order information extraction module, an attention sub-network module, and an order dispatch and route planning module. We will explain the details of each module in combination with mathematical symbols below.

4.2 Order Information Extraction

In this section, we will introduce the structure and function of the order information extraction module in detail. This module is mainly used to process the original information of orders to generate the corresponding attribute feature vector. In our discussion, given an order set $O = \{o_1, o_2, \cdots, o_n\}$, where $o_i = \langle d_i, t_i, v_i, np_i, p_i \rangle$, we can express the original attributes of the order set in another form, $V = \{V^1, V^2, V^3, V^4, V^5\}$, where $V^j = \{v_{o_1}^j, v_{o_2}^j, \cdots, v_{o_n}^j\}$ represents the original vector representation of the j_{th} attribute. Thus, we can get the order set O 's corresponding attribute feature vector sequence $A = \{A^1, A^2, A^3, A^4, A^5\}$, where $A^j = \{a_{o_1}^j, a_{o_2}^j, \cdots, a_{o_n}^j\}$ represents the vector representation of the j_{th} attribute obtained through information extraction. Because the order attributes are different, the corresponding information extraction methods are also different. We will then discuss in detail the information extraction methods of various order attributes considered in the paper.

- *Destination:* In general, the station information includes the station name and station location, where the station name is often Chinese or English characters, and the station location can be characterized by latitude and longitude coordinates. We will not consider the station location factor in this module in order to simplify the model.
- *Arrival Time:* The time when the order o_i is collected by the platform is called the arrival time t_i. In order to eliminate the influence of dimension, we use Min-max normalization method to process t_i.

$$a_i^2 = \frac{t_i}{max(V^2) - min(V^2)}$$

- *VIP Level:* VIP level is a parameter we introduce to measure the importance of users to the platform. Because online car-hailing platforms have not taken this into consideration at present, there is no VIP level entry in the data set. Hence, we use a randomization method to add a VIP level to each order. In fact, the proportion of VIP users to the total users of the platform is usually very small, so we think that the VIP level of most orders is zero. We set the VIP level to three levels: low, medium and high, and its priority is increased in turn. Therefore, $a_{o_i}^3$ can be $0, 1, 2$, or 3. Our randomization method can ensure that the total number of orders with a higher VIP level is less than the total number of orders with a lower VIP level, which is in line with reality.
- *Number of Passengers Included:* Unlike car-hailing services, orders of the bus-booking platform needs to indicate the number of passengers it contains in order to meet the capacity constraint when the order is dispatched.
- *Priority:* In fact, the priority should be related to the first four attributes instead of being obtained directly. The earlier the order arrives at the platform, the higher the VIP level is, and the larger the number of passengers included is, the higher the priority will be. However, because the attention sub-network will also consider the aggregation effect of the destination stations, we stipulate that the order priority p_i is only related to t_i, v_i and np_i

in order to facilitate the calculation. In our discussion, p_i can be calculated through the function $f_1(\cdot)$. We can adjust the weight coefficient to flexibly reflect the impact of the other factors on p_i.

$$a_{o_i}^4 = w_p^2(1 - a_{o_i}^2) + w_p^3 a_{o_i}^3 + w_p^4 a_{o_2}^4$$

It is worth noting that if the order has more attributes in the future, we can still convert the attributes into feature vectors by a suitable method, so that the order information extraction module can still be used to solve the problem. This structure ensures the scalability of the model.

4.3   Attention Sub-network

This module is used to process the attribute feature vectors of orders to get an aggregated representation of order set attributes. This aggregated representation can reflect the user's overall demand for different destination stations. Specifically, the attribute feature vector matrix A will be assigned to different attention sub-networks according to the destination station. Each attention sub-network then combines the assigned orders and the context vector c to calculate an aggregated representation of this part of the order attributes.

In fact, the attention mechanism used in our proposed method has some differences from the attention mechanism used for machine translation. We refer to the idea of providing different inputs with different attention and utilizing a variable context vector c to achieve the dynamic change of the weight matrix between the input and output used in the latter construction process. Such a mechanism will bring the following benefits:

- We provide different orders with different weights, which can make those important orders play significant roles in the aggregation representation.
- The impact of different attributes of an order on the final aggregate representation is also different.
- The impact of orders on the aggregated representation at different stages varies, which is related to the current order dispatch scheme and the current station at this stage.

We introduce attention weight $\alpha(i, j, k)$ to express the influence of the i_{th} order with s_j as the destination on aggregated representation sc^j at the planning stage k. For ease of presentation, we use o_i^j to represent the i_{th} order with s_j as the destination. According to the previous statement, if o_i^j is important to sc^j at the planning stage k, then $\alpha(i, j, k)$ should be high. The dynamic feature of the module is that when k takes different values, $\alpha(i, j, k)$ will also have different values according to the calculation based on the order dispatch and route planning situation at the time. We are able to detect changing order information and station information during the route by using the attention mechanism to learn $\alpha(i, j, k)$ dynamically, Next, we will introduce in detail how to obtain the context vector c and the internal structure of the attention sub-network.

In order to illustrate the construction of the context vector c, we use c_k^j to represent the context vector of the attention sub-network for processing orders whose destination is s_j at the planning stage k. c_k^j is related to the current order dispatch and route planning scheme. We think that $c_k^j = \{c_k^j[1], c_k^j[2]\}$, where $c_k^j[1]$ records the information of orders that has been dispatched, $c_k^j[2]$ records the relationship between the current station and s_j. In the paper, we assume that $c_k^j[2]$ mainly reflects distance information, and its value is related to the distance matrix between stations M_{dis}.

We introduce $A_{o_i^j}$ to represent the attribute characteristics of o_i^j to illustrate the internal structure of attention sub-network. We define $A_{o_i^j} = f_2(a_{o_i^j}^1, a_{o_i^j}^2, a_{o_i^j}^3, a_{o_i^j}^4, a_{o_i^j}^5)$. The attention self-network inputs the order attribute feature $A_{o_i^j}$ and the order information part of context vector $c_k^j[1]$ to a two-layer network, thereby obtaining the corresponding weight:

$$\alpha(i, j, k) = \omega^T \phi(W_c c_k^j[1] + W_A A_{o_i^j} + \epsilon_1) + \epsilon_2$$

where the first layer parameters are the matrices W_c, W_A and bias ϵ_1, the second layer parameters are the vector w and bias ϵ_2. Non-linear ReLU function $\phi(x) = max(0, x)$ is leveraged.

We use O^j to denote the collection of all orders with destination s_j. Then each attention sub-network can get an aggregated representation of the order information at planning stage k.

$$sc^j = \sum_{o_i^j \in O^j} \alpha(i, j, k) A_{o_1^j}$$

The overall group aggregated representation of all orders g_k at planning stage k can be obtained from a merge function $f(\cdot)$ and the station information part of context vector $c_k^j[2]$.

$$g_k = f(c_k^1[2]sc^1, c_k^2[2]sc^2, \cdots, c_k^1[2]sc^q)$$

The vector concatenation operation ($\oplus$) is used as the merge function $f(\cdot)$ for ease of operation,, which also means that the information of each station is treated fairly. Therefore, the overall group aggregated representation $g_k = c_k^1[2]sc^1 \oplus c_k^2[2]sc^2 \oplus \cdots \oplus c_k^q[2]sc^q$ in this paper.

4.4   Order Dispatch and Route Planning

In this section, we will explain the details of using Bi-GRU to complete the task of order dispatch and route planning. In car-hailing services, order dispatch and route planning are usually two separate issues. Even in urban bus system research, the two problems are usually handled with different methods. In this module, while using Bi-GRU to generate the bus route, we can complete the order dispatch task, which benefits from the detailed processing of the order information before.

The input of Bi-GRU in the planning stage k is a q-dimensional one-hot vector representing a certain station via encoding, where q is the total number of bus stops. At the same time, the group aggregated representation g_k obtained by attention sub-network and the hidden state of the previous stage will also be fed to Bi-GRU, and finally output a q-dimensional vector that can be decoded to the site.

It is worth noting that the group aggregated representation g_k at this moment contains the information of the context vector c_k, indicating that c_k will also affect the route planning of Bi-GRU. In addition, we reserve processing space for the feature vectors related to the current station in Bi-GRU so that we can take this information into account (for example, historical travel needs of the station).

Take the GRU with forward propagation at planning stage k as an example, the formulae are as follows:

$$r_k = \sigma(W_r x_k + W_{hr}(h_{k-1} \odot F_k) + b_r + W_a A)$$
$$z_k = \sigma(W_r x_k + W_{hz}(h_{k-1} \odot F_k) + b_z + W_a A)$$
$$h_k^* = tanh(W_{h_k^*} x_k + W_{hh_k^*}(r_k \odot (h_{k-1} \odot F_k)) + W_a A)$$
$$h_k = (1 - z_k) \odot (h_{k-1} A_k) + z_k \odot h_k^*$$
$$y_k = \sigma(W_o h_k)$$

where r_k is the reset gate, controlling some information in the memory state to be forgotten. On the contrary, z_k is the update gate, used to transfer information from the input to the memory state. h_{k-1} is the old memory state, and h_k^* is the temporary new memory state. In order to calculate the final memory state h_k at the planning stage k, the update gate needs to confirm which information is collected from h_k^* and the previous h_{k-1}. In addition, A is the attribute group aggregated representation vector, and W_a is the corresponding attribute weight matrix. F_k is the feature vector related to station information, and $\odot$ is an element-wise product operator. The remaining weight matrices W and biases b that appear in the formulae can be combined with subscripts to know their roles. These parameters need to be learned during the operation of Bi-GRU. Since the backward propagation GRU's formulae are similar to the above, it will not be repeated here.

From the above statement, we understand the advantages of Bi-GRU. It has a simple structure and only uses two gates to control the update and reset of the information in the memory state. By propagating the input and the previous hidden state into the gates, we can get the corresponding output station.

Finally, we can get the vector representation of the output station obtained at planning stage k.

$$r_k = \max(\sigma(y_k^f + y_k^b))$$

Among them, r_k can be decoded into the original station representation. Note that the moment the route planning is completed, the module will determine which orders are suitable for being dispatched to this bus. These dispatched orders and the current station information will be sent back to the attention sub-network's context vector c. The two modules run alternately to achieve the final order dispatch and route planning tasks.

5   Experiments and Analysis

5.1   Dataset and Evaluation Criteria

Dataset. The data set used in this section is the taxi GPS trajectory data from February 1, 2007 to March 1, 2007 in Shanghai. It contains various fields, such as record ID, longitude, latitude, as shown in Table 1. The data set contains about 357,000 GPS records of 1,367 taxis in Shanghai.

Data Preprocessing: Data preprocessing includes data filtering and data expansion. In term of the data filtering step, we first confirm the order information according to the status of taxis. We think that the status from idle to occupied indicates the starting point of an order, and the status from occupied to idle indicates the destination of an order, thus getting the basic information of an order: (Order ID, Source, Destination, Start time, End time). In order to adapt the data to our research scenario, we need to screen out orders that depart from the vicinity of Shanghai Pudong International Airport during midnight period. According to the map, we determine the latitude and longitude range of Shanghai Pudong International Airport and filter out the required order information. Because there are not many orders that meet our requirements in one day, data expansion is necessary. Based on the information of existing orders, we perform reasonable interpolation to generate more orders, thus expanding the effective order data set.

Table 1. Description of Taxi GPS track data

Field	Annotation
RecordID	ID of record
TaxiID	ID of taxi
Longitude	Longitude of taxi
Latitude	Latitude of taxi
Speed	Speed of taxi
Datetime	Time that GPS record was sent
Status	1: taxi is occupied while 0: taxi is vacant

Evaluation Criteria. In order to compare the performance of different methods, we specify three evaluation criteria based on the order attributes: average VIP level $\bar{V}$, total priority of all accepted orders P, and platform cost C.

$$\bar{V} = \frac{\sum_{o_i \in O_{accepted}} v_i}{\text{Size}(O_{accepted})}$$

$$P = \sum_{o_i \in O_{accepted}} p_i$$

$$C = \sum_{b_i \in B} (cost_i + c_b)$$

5.2   Destination Stations Selection

In this section, we first propose a DBSCAN-PAM hybrid clustering algorithm to cluster "hot" grid cells and determine destination stations.

DBSCAN-PAM Hybrid Clustering Algorithm. Our goal is to use the real taxi GPS data to study the travel hotspots of residents, so as to establish a proper bus destination station at the central point of the hot spot area. In order to solve this problem, a hybrid clustering algorithm based on DBSCAN algorithm and PAM algorithm is designed. The algorithm consists of two steps:

1. The DBSCAN algorithm is used to process the data of the get-off points, and the noise data is eliminated.
2. The PAM algorithm is used to cluster the data after noise elimination, and uses the class center point as a candidate station for the bus-booking platform.

We filter out suitable popular stations based on the destination parameter of valid order data. First, we need to rasterize the city map, which means dividing the city into $25\,m \times 25\,m$ grid cells. Grid cells with value greater than θ_{hot} are then marked as "hot" grid cells. Next, DBSCAN-PAM hybrid clustering algorithm is used to cluster and determine the center points of clusters. To conduct the hybrid algorithm, we need to set two key parameters, eps and $MinPts$. The setting of these two parameter values will directly affect the clustering effect.

Fig. 2. 50 selected hot destination stations

According to relevant regulations in China's "Urban Road Traffic Planning and Design Specification" GB50220-95, the distance between public transportation stations in the city's transportation system should be 500 m to 800 m, and the distance between stations in the suburban transportation system should be 800 m to 1000 m. Because the purpose of this paper is to design a new bus-booking platform that is applied to the urban transportation system, we set *eps* to 500 m. The value of *MinPts* affects the distance from the furthest points in the cluster to the center point. According to the distribution of "hot" grids cells, we set *MinPts* to 100. In this way, we can get the results of clustering and center point determination. At the same time, DBSCAN-PAM algorithm can identify some noise points and eliminate these noise data.

Shanghai Pudong International Airport is the departure station of the bus-booking platform in our discussion. Considering the total number of orders in each station cluster, the distance between other points and the center point, the distance between center points and the departure station, and some traditional rules for the location of bus stations, we finally selected 50 popular destination stations, as shown in Fig. 2.

5.3   Simulations

In this section, we first introduce the baseline methods for comparison. Simulation experiments are then conducted to investigate the performance of the baseline methods and our proposed method in terms of evaluation criteria.

Compared Methods. A genetic algorithm for solving classic CVRP problem and an advanced heuristic route planning method are chosen as baseline methods. Among them, the former represents the classic methods used for route planning, and the latter represents the novel methods proposed by scholars in recent years.

– **CVRP:** Capacitated Vehicle Routing Problem (CVRP) is to compute the travel lines for the vehicles with capacity so as to satisfy the requests of all the passengers and minimize the travel cost of all the vehicles [13–15]. In the simulation experiment, we use a genetic algorithm to solve the CVRP problem as one baseline method [14].
– **SubBus:** SubBus is a dynamic route planning method [9]. Candidate origins are first determined based on station information and passenger travel demands. Starting from the candidate origin set, a candidate route set is generated combined with road network information. A dynamic programming method is then designed with the optimization goal of minimizing the running distance to implement the dynamic route planning of multiple buses running at the same time.

Configurations. Our experiments are conducted on a Mi Notebook Pro with four processors (4× Intel Core i5-8250U, 1.60 GHz) and one graphics card

(NVIDIA GeForce MX250). The operating system of the device is Microsoft Windows 10 (64-bit). The code is mainly written with Python 3.5. Meanwhile, TensorFlow 1.2 is used as a machine learning framework for our experiments.

Order data is obtained from real-world data through data filtering and data expansion as mentioned above. The destination and arrival time attributes are inherent in the order, and the VIP and priority attributes are given by us manually, which will be discussed in detail below. The scale of experimental data can refer to Table 2.

Simulation for the ODRP Problem. In the simulation experiment, we use the control variable method to explore the advantages and disadvantages of each method and the effectiveness of each method under different conditions. For the order o_i, the attributes d_i and t_i can be obtained directly. However, v_i and np_i are not collected by the current car-hailing platform, so we need to give them values manually. For v_i, we use a random assignment method. At the same time, we ensure that most orders' v_i is 0. In the order group of $v_i > 0$, the order of large v_i is small, which can be achieved by setting the probability matrix. For np_i, we stipulate that $np_i = 1, 2$, or 3 in the simulation, and the ratio between the three meets 16:3:1. The advantage of this rule is that when we discuss the special case of $p_i = np_i$, the total number of orders and the total number of passengers can be used as controlled variables at the same time. So far, we have completed the discussion of attributes of order o_i in the simulation experiment.

We then discuss the control variables. In the simulation, we take $m \times L$, and number of orders n as the controlled variables. We set up five sets of experiments. In the first three sets of experiments, n remained unchanged, and $m \times L$ increased sequentially. In contrast, in the last three sets of experiments, $m \times L$ remained unchanged, and n increased in turn. Therefore, through the first three sets of experiments, we can observe changes in the effectiveness of different methods as the total carrying capacity of the platform increases. Through the last three sets of experiments, we can observe the changes with the total number of orders increases. For specific parameters of each variable in the five sets of experiments, please refer to Table 2. Next, we will specifically analyze the effect of each method on different evaluation criteria in conjunction with the figures showing the simulation results.

Table 2. Parameters in five experiments

Experiment	Bus number	Bus capacity	Order number	Passenger number
1	20	20	960	1200
2	30	20	960	1200
3	30	30	960	1200
4	30	30	1200	1500
5	30	30	1440	1800

We first analyze the evaluation criterion $\bar{V}$. As shown in Fig. 3, the $\bar{V}$s obtained by the first two baseline methods are basically in the range of 0.4–0.6, which reflects that these two methods do not deal with VIP level specially. However, the $\bar{V}$ obtained by the attention-based Bi-GRU method is in a dominant position and exhibits certain regularity. According to Experiment 1–3, when the total number of orders is the same, $\bar{V}$ decreases as the total bus carrying capacity increases. According to Experiment 3–5, when the total carrying capacity of the bus remains unchanged, $\bar{V}$ increases as the total number of orders increases. We speculate that the attention-based Bi-GRU method can absorb most of the VIP orders in the order set, so that we can derive a similar change trend as in Fig. 3.

We then analyze P, the total priority of $O_{accepted}$. In the simulation, we define the function $f_1(\cdots)$: $p_i = np_i \times \{0.5[1 - N(t_i)] + 0.5N(v_i)\}$, where $N(\cdot)$ indicates the normalization processing function. The simulation results are shown in Fig. 4. The reason why our method is better than SubBus is because we pay attention to t_i and v_i. Under this construction method of p_i, the P obtained by our method is close to the optimal maximum value that P can reach. To maximize P is the main optimization goal of the ODRP problem. Therefore, our method is very helpful to solve the problem.

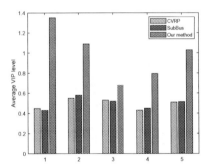

Fig. 3. Average VIP level

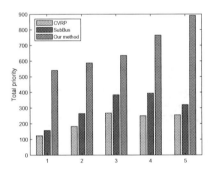

Fig. 4. Total priority of $O_{accepted}$

The special case of $p_i = np_i$ is also discusses, where P represents the total number of passengers accepted. According to Fig. 5, we can see that our method is still in a leading position, thanks to the global consideration of order information and the changing attention for orders in different planning stages. Compared with the baseline method, our method accepts significantly more passengers, which shows that our method has notable advantages.

The platform cost C is analyzed in the end. As shown in Fig. 6, CVRP always keeps the platform cost at a low level, thanks to the method only accepting few passengers. Our method is comparable to the SubBus method. From Experiment 1 to 2, the platform cost increases significantly, because 10 new buses are added.

From Experiment 2 to 5, although the total carrying capacity of buses and the number of orders increase, their impact on platform cost is not large. On the one hand, the simulation result shows that the three methods have achieved good results in route planning. On the other hand, it shows that reducing the number of buses used is one of the good ways to save platform cost. Although the effectiveness of our method in terms of minimizing the platform cost is not very satisfactory, accepting more passengers means that the platform can earn more profits as mentioned above. The platform can cover the cost through appropriate pricing strategies.

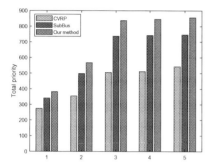

Fig. 5. P when $p_i = np_i$

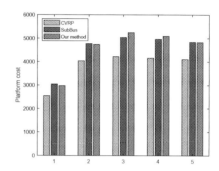

Fig. 6. Platform cost

In general, we prove the effectiveness of the attention-based Bi-GRU method to solve the ODRP problem through simulation experiments and comparison with the baseline methods. It maximizes the total priority of all accepted orders while keeping the platform cost as small as possible without violating constraints. Therefore, we believe that the attention-based Bi-GRU method has broad application prospects in the research field of the new bus-booking platform.

6   Conclusion

In this paper, we proposed an attention-based Bi-GRU method to solve the ODRP problem. The attention mechanism is mainly used to differentiate the order information. Bi-GRU gives the final order dispatch and route planning schemes.

Our contribution is mainly that the method not only uses the different order characteristics to describe the order information, but also dynamically learns the influence weight of each given order at different planning stages, combined with station information. In addition, this method is the first to combine the attention mechanism with Bi-GRU in the order dispatch and route planning problem related to urban buses. Experiments conducted on real-world data sets show that our proposed method has better performance than the baseline methods.

624   Y. Gao et al.

References

1. Zhou, H., Gao, Y., Gao, X., Chen, G.: Real-time route planning and online order dispatch for bus-booking platforms. In: Li, G., Yang, J., Gama, J., Natwichai, J., Tong, Y. (eds.) DASFAA 2019. LNCS, vol. 11447, pp. 748–763. Springer, Cham (2019). https://doi.org/10.1007/978-3-030-18579-4_44
2. Xu, Z., et al.: Large-scale order dispatch in on-demand ride-hailing platforms: a learning and planning approach. In: ACM SIGKDD International Conference on Knowledge Discovery & Data Mining (KDD), pp. 905–913 (2018)
3. Xia, B., Li, Y., Li, Q., Li, T.: Attention-based recurrent neural network for location recommendation. In: International Conference on Intelligent Systems and Knowledge Engineering (ISKE), pp. 1–6 (2017)
4. Li, L., Cai , G., Chen, N.: A rumor events detection method based on deep bidirectional GRU neural network. In: IEEE International Conference on Image, Vision and Computing (ICIVC), pp. 755–759 (2018)
5. Xian, O.Y., Chitre, M., Rus, D.: An approximate bus route planning algorithm. In: IEEE Symposium on Computational Intelligence in Vehicles & Transportation Systems (CIVTS), pp. 16–24 (2013)
6. Lei, S., Li, Z., Wu , B., Wang, H.: Research on multi-objective bus route planning model based on taxi GPS data. In: International Conference on Cyber-Enabled Distributed Computing and Knowledge Discovery (CyberC), pp. 249–255 (2016)
7. Khoa, V.D., Pham, T.V., Nguyen, H.T., Van Hoai, T.: Multi–criteria route planning in bus network. In: Saeed, K., Snášel, V. (eds.) CISIM 2014. LNCS, vol. 8838, pp. 535–546. Springer, Heidelberg (2014). https://doi.org/10.1007/978-3-662-45237-0_49
8. Wang, S., Lin, W., Yang, Y., Xiao, X., Zhou, S.: Efficient route planning on public transportation networks: a labelling approach. In: ACM SIGMOD International Conference on Management of Data (SIGMOD), pp. 967–982 (2015)
9. Kong, X., Li, M., Tang, T., Tian, K., Moreira-Matias, L., Xia, F.: Shared subway shuttle bus route planning based on transport data analytics. IEEE Trans. Autom. Sci. Eng. **15**(4), 1507–1520 (2018)
10. Zhang, L., et al.: A taxi order dispatch model based on combinatorial optimization. In: ACM SIGKDD International Conference on Knowledge Discovery and Data Mining (KDD), pp. 2151–2159 (2017)
11. Zheng, L., Chen, L., Ye, J.: Order dispatch in price-aware ridesharing. Int. Conf. Very Large Data Bases (VLDB) **11**(8), 853–865 (2018)
12. Duan, Y., Wang , N., Wu, J.: Optimizing order dispatch for ride-sharing systems. In: International Conference on Computer Communication and Networks (ICCCN), pp. 1–9 (2019)
13. Allahviranloo, M., Chow, J.Y., Recker, W.W.: Selective vehicle routing problems under uncertainty without recourse. Transp. Res. Part E Logistics Transp. Rev. **62**, 68–88 (2014)
14. Caceres-Cruz, J., Arias, P., Guimarans, D., Riera, D., Juan, A.A.: Rich vehicle routing problem: survey. ACM Comput. Surv. **47**(2), 1–28 (2014)
15. Ralphs, T.K., Kopman, L., Pulleyblank, W.R., Trotter, L.E.: On the capacitated vehicle routing problem. Math. Program. **94**, 343–359 (2003)

Top-k Closest Pair Queries over Spatial Knowledge Graph

Fangwei Wu[1], Xike Xie[1(✉)], and Jieming Shi[2]

[1] University of Science and Technology of China, Hefei, China
`wufw1995@mail.ustc.edu.cn`, `xkxie@ustc.edu.cn`
[2] The Hong Kong Polytechnic University, Hung Hom, Hong Kong
`jieming.shi@polyu.edu.hk`

Abstract. Recently, RDF data has been enriched with spatial semantics enabling spatial keyword search. Research spatial keyword search over spatial RDF data focus on finding the spatial entities rooted at subtrees which cover given query keywords. In this work, we study how relevant spatial entity pairs can be efficiently retrieved, where the relevance is determined according to both spatial distances and textual similarities. The retrieved top-k closest pairs are ranked and then returned to users for the interests of business intelligence and recommendation. We propose a branch-and-bound framework associated with effective lower and upper bound pruning techniques and early stopping conditions for efficiently retrieving relevant top-k closet pairs. The results demonstrate the high efficiency of our proposal compared to baseline solutions.

Keywords: Knowledge graph · Closest pair query · Spatial keyword search

1 Introduction

Due to the growth of knowledge-sharing communities and the development of automated information extraction technologies, large knowledge bases, have been used in various applications. For example, DBpedia [1] and YAGO [2] extract facts from Wikipedia automatically and store them in the Resources Description Framework (RDF) format to support structural queries. An RDF entity is linked to other entities and/or types and/or descriptions. Therefore, an RDF knowledge base can also be seen as a directed graph, where entities are entities or types or descriptions, and edges are predicates that describe the relationship between entities.

Recently, knowledge bases are enriched with additional spatial semantics. These knowledge bases make the location-based queries and information retrieval. The standard structured languages such as GeoSPARQL have been proposed by Open Geospatial Consortium (OGC) to support geospatial search operations. However, these structured languages often require users to be familiar with specific language and the apriori knowledge of the data domains, which

C. S. Jensen et al. (Eds.): DASFAA 2021, LNCS 12681, pp. 625–640, 2021.
https://doi.org/10.1007/978-3-030-73194-6_41

limits the access of common users. Given this, a keyword search on spatial knowledge bases model emerged [15], which is called kSP query and allows users to retrieve spatial entity without understanding the structure languages. As far as we know, the kSP query aims to find a subtree rooted at the spatial vertex in the knowledge graph, which covers all query keywords and minimizes the ranking scores. The ranking score is determined by the compactness of the subtree and the spatial distance to a retrieved query location. In other words, the size of a retrieved subtree can measure the relevance between the spatial entity and the query keywords.

In the paper, a spatial query that combines join and nearest neighbor queries based on knowledge graphs is exampled.

Example 1. Figure 1(a) shows the preprocessed graph representation of several triples extracted from DBpedia. Each node is an entity associated with a document. Squares corresponds to spatial vertices, for which the locations have been extracted and are shown in Fig. 1(b). Circles are non-spatial entities of the RDF graph. If one wants to find a park and a bus station, which are close to each other. According to [15], kSP can only find a bus station $\{s_1\}$ and parks $\{s_2, s_3\}$ since kSP aims to find a single spatial vertex that covers all the query keywords. In addition, kSP can not be leveraged to find two semantic spatial entities and measure the spatial distance between them. We call the pair of park and bus station as a spatial entity pair and investigate an efficient solution for finding top-k paired spatial entities.

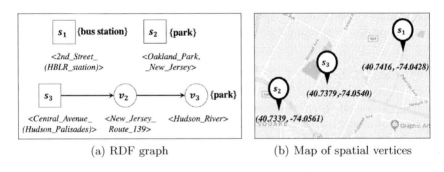

(a) RDF graph   (b) Map of spatial vertices

Fig. 1. RDF example

To address above mentioned problems, in the paper, we propose a novel way of searching on a spatial knowledge base, namely Top-k Closest Pair Queries, which returns k spatial entity pairs with minimum ranking scores. The ranking function aggregates the spatial distance between the paired entities and the textual relevance to query keywords.

Our contributions in the paper are summarized as follows.

- We define a general score function for ranking spatial entity pairs with respect to their locations and corresponding textual descriptions, and propose a basic solution (Sect. 3).
- We index the spatial RDF entities by an R-tree and derive the upper/lower bounds for the ranking function and an early stopping condition to speed up the ranking evaluation (Sect. 4).
- We conducted experiments on YAGO and DBpedia datasets to demonstrate the efficiency of our proposed algorithm (Sect. 5).

In addition, we introduce some preliminaries in Sect. 2 and related works in Sect. 6.

2   Preliminaries

A knowledge graph can be modeled as a directed graph $G = (V, E)$ where V indicates a collection of vertices and E indicates a collection of edges. Each vertex $v \in V$ refers to an entity and $v.\psi$ refers to a text description containing the entity's URI and literals. The knowledge graph adopted the RDF data model, which represents the data as collections of *<subject, predicate, object>* triples. For each triple, the description of the predicate is added to the document of the object entity for the purpose of the keyword search. In addition, some of the entities are associated with spatial attributes (e.g., coordinates). For the ease of presentation, we call these vertices *spatial vertices* in the graph and use s to represent a spatial vertex.

For a spatial vertex, a keyword-based retrieval model over knowledge graphs, such as [7,11,17], identifies a set of paths that reach vertices containing query keywords. In order to aggregate the proximity of the query keywords to the spatial vertex, the sum of the length of the paths denoted as *looseness*, is defined as follows:

Definition 1 (Looseness of Spatial Vertex [15]). *Assume a spatial vertex* s, *a knowledge graph* $G = (V, E)$ *and query keywords* $q.\psi = \{t_1, \ldots, t_m\}$. *Let* $d_g(s, t_i) = \min_{v \in V \wedge t_i \in v.\psi} d(s, v)$ *be the length of the shortest path from* s *to keyword* $t_i \in q.\psi$, *where* $d(s, v)$ *is the shortest path from* s *to* v. *The looseness of* s *to* $q.\psi$ *is defined as*

$$L(s, q.\psi) = 1 + \sum_{t_i \in q.\psi} \min(d_g(p, t_i), d_{\max}), \qquad (1)$$

where $d_{\max}$ *is the threshold of the maximum length of a shortest path.*

For each query keyword t_i of $q.\psi$, there must exist at least one path from s to a vertex v that contains t_i in G, so that s is relevant to t_i. In addition, the sum of the length of the shortest path from s to keyword t_i can be used as a relevance score. For normalization purposes, 1 is added to the sum of the paths to avoid the looseness being 0. The lower the looseness, the more relevant the spatial vertex is to the query keywords. $d_{\max}$ is the maximum length of the shortest path that a user can tolerate as proposed in [8]. In other words, when the length of a path from a spatial vertex is larger than $d_{\max}$, the relevance score of the keyword is rounded up to $d_{\max}$.

3 Problem Definition and Basic Algorithm

We define the query in Sect. 3.1, and give a basic algorithm in Sect. 3.2.

3.1 Problem Definition

A *top-k pair keyword queries on spatial knowledge graph* (*k*PKQ) *q* aims to find *k* spatial vertex pairs with the smallest *looseness* values, covering spatial distances and textual similarities, based on an aggregate function. The *k*PKQ query *q* consists of two arguments: the query keyword set ψ_R and the query keyword set ψ_B.

Thus, we can join the spatial vertices searched by ψ_R with the spatial vertex searched by ψ_R to get a set of spatial vertex pairs and returned the top-k results. The looseness of a spatial vertex pair $sp = (s_R, s_B)$ is as follows, according to Definition 1.

Definition 2 (Pair Looseness). *Assume a spatial vertex pair* $sp = (s_R, s_B)$, *a knowledge graph* $G = (V, E)$ *and query keywords* ψ_R *and* ψ_B. *The pair looseness of sp to* ψ_R *and* ψ_B *is defined as*

$$L_P(sp, q) = \max(L(s_R, \psi_R), L(s_B, \psi_B)). \tag{2}$$

The pair looseness is set to the maximum of two looseness values. It represents that each spatial vertex in the pair is compact and satisfies the corresponding query keywords so that the pair looseness is not smaller than any of the two both looseness values. Finally, we define the *k*PKQ problem as follows.

Definition 3 (*k*PKQ). *Given a kPKQ query q with two sets of query keywords* (ψ_R, ψ_B) *on a knowledge graph* $G = (V, E)$, *and a parameter k, the kPKQ query returns top-k spatial vertex pairs with smallest ranking scores* $f(L_P(sp, q), S(sp))$, *where* $S(sp)$ *is the spatial distance between the two entities of a pair.*

The *k*PKQ query aims to find *k* pairs of spatial vertices: (i) are spatially close to each other, (ii) have a looseness to the corresponding query keywords, respectively, and (iii) have a smaller pair looseness. Without loss of generality, Euclidean distance $S(sp) = |s_R, s_B|$ is used as spatial distance in our work. The ranking function $f(L_P(sp, q), S(sp))$ is described as follows.

$$f(L_P(sp, q), S(sp)) = L_P(sp, q) \times S(sp). \tag{3}$$

Example 2. Consider again the RDF data in Example 1 and assume an 1PKQ issued by a tourist who wants to find a bus station and a park which are close to each other, i.e., $\psi_R = \{busstation\}$ and $\psi_B = \{park\}$. Based on the RDF graph in Fig. 1(a), the spatial vertex pair $sp_1 = (s_1, s_2)$ has $L_P(sp_1, q) = \max(1, 1) = 1$ and $S(sp_1) = 0.0154$. $f(L_P(sp_1, q), S(sp_1)) = L_P(sp_1, q) \times S(sp_1) = 0.0154$. The spatial vertex pair $sp_2 = (s_1, s_3)$ has $L_P(sp_2, q) = \max(1, 3) = 3$ and $S(sp_2) = 0.0118$. $f(L_P(sp_2, q), S(sp_2)) = L_P(sp_2, q) \times S(sp_2) = 0.0384$. Therefore, sp_1 is returned as top-1 result.

3.2   Basic Solution

Assume that V_S indicates a collection of spatial vertices in the knowledge graph. Intuitively, the query can be answered by iterating over all spatial vertex pairs in $V_S \bowtie V_S$ and computing the ranking score for each pair. Meanwhile, a global min-heap H_k is used to maintain the current top-k results in terms of the ranking score. After all pairs in $V_S \bowtie V_S$ have been processed, the query returns the k pairs in top of H_k.

To compute the looseness, a straightforward way is to iterate over all spatial vertex pairs and use BFS to compute the pair looseness. As shown in Algorithm 1, there is a doubly nested loop in which both the outer loop and the inner loop iterate a spatial vertex in V_S with the corresponding query keywords. Apparently, the basic algorithm is expensive as it uses BFS for each spatial vertex pair in $V_S \bowtie V_S$. In the following section, we propose more efficient methods to compute L_P and to prune disqualified pairs.

Algorithm 1: Basic$(q, G, d_{\max})$

---

1   Minheap $H_k = \emptyset$, ordered by $f(L_P, S)$;
2   Threshold $\theta = +\infty$;
3   **while** $s_R =$ GetNext(V_S, ψ_R) **do**   ▷ Iterate over V_S with ψ_R
4   $\quad$ Compute $L(s_R, \psi_R)$ by BFS;
5   $\quad$ **while** $s_B =$ GetNext(V_S, ψ_B) **do**   ▷ Iterate over V_S with ψ_B
6   $\quad\quad$ Compute $L(s_B, \psi_B)$ by BFS;
7   $\quad\quad$ Compute the ranking score f of (s_R, s_B);
8   $\quad\quad$ **if** $f < \theta$ **then**
9   $\quad\quad\quad$ H_k.add$((s_R, s_B), f)$;
10 $\quad\quad\quad$ update θ;

11 **return** H_k;

---

4   Improved Solution

Obviously, the computation of looseness is much more expensive than the computation of spatial distance. Therefore, the key problem to be addressed is to reduce the number of BFS during the query. The main idea is to estimate the upper and lower bounds for the current ranking score to avoid looseness computation as much as possible. Suppose that all spatial vertices are spatially indexed by an R-tree $\mathcal{R}$. The doubly nested loop in Algorithm 1 can be improved to traverse R-tree node with the query keywords. When accessing an R-tree node, its minimum bounding rectangle can be used to estimate the upper and lower bounds, which can be used to stop BFS early.

In Sect. 4.1, we propose a vertex-join-node method in which the outer loop traverses spatial vertices in V_S and the inner loop traverses nodes in $\mathcal{R}$. After that, we propose a vertex-join-node with α-radius keywords index method in Sect. 4.2.

Algorithm 2: V2N

---

1 Minheap $H_k = \emptyset$, ordered by $f(L_P, S)$;
2 Threshold $\theta = +\infty$;
3 **while** $s_R =$ GETNEXT(V_S, ψ_R) **do**   ▷ Iterate V_S with ψ_R,
4   **if** $\theta = 0$ **then**
5   **break**;
6   **if** s_R *can not reach* ψ_R **then**   ▷ Pruning Rule 1
7   **continue**;
8   Compute $L(s_R, \psi_R)$ by BFS;
9   Minheap $H_e = \emptyset$, ordered by $|s_R, e_B|_{\min}$;
10   H_e.add($\mathcal{R}$, $|s_R, \mathcal{R}|_{\min}$);
11   **while** H_e *is not empty* **do**   ▷ Iterate $\mathcal{R}$ with ψ_B
12          $e_B = H_e$.pop();
13   **if** $L(s_R, \psi_R) \times |s_R, e_B|_{\min} \geq \theta$ **then**   ▷ Pruning Rule 2
14   **break**;
15   **if** e_B *is a non-leaf node* **then**
16   **foreach** *child node* e_B^c *in* e_B **do**
17   H_e.add(e_B^c, $|s_R, e_B^c|_{\min}$)
18   **else**   ▷ e_B is a leaf node, denoted as s_B
19   **if** s_B *can not reach* ψ_B **then**   ▷ Pruning Rule 1
20   **continue**;
21   Compute $L(s_B, \psi_B)$ by BFS$_P$;   ▷ Algorithm 3
22   Compute the ranking score f of (s_R, s_B);
23   **if** $f < \theta$ **then**
24   H_k.add((s_R, s_B), f);
25   update θ;

26 **return** H_k;

4.1 Vertex Join Node Method: V2N

In the basic method, it needs to compute the looseness to its corresponding query keywords for each spatial vertex. However, not every spatial vertex can reach vertices covering all the query keywords by BFS. For the spatial vertex that can not reach all the query keywords, it can not compute the looseness and can hence be pruned. We adopt the method in [15] to test whether a spatial vertex can reach a keyword. The pruning rule is formulated as follows.

Pruning Rule 1. Given a spatial vertex s, a knowledge graph and query keywords $q.\psi = \{t_1, \ldots, t_m\}$, if $\exists t_i \in q.\psi$, $d_g(s, t_i) = +\infty$, we say that the spatial vertex s can not reach the keyword t_i, so that s can be pruned.

Algorithm 2 shows the pseudo code of our Vertex Join Node (V2N) search method for evaluating kPKQ queries. Initially, we maintain a min-heap H_k for storing the current top-k ranking score and θ is the k-th scores in the heap (line 1–2). When the outer loop iterates to a spatial vertex s_R in V_S with the query keywords ψ_R, Pruning Rule 1 can be used before BFS to avoid unnecessary looseness computation (line 6–7). After $L(s_R, \psi_R)$ is computed, the inner loop begins to iterate node e_B in the R-tree. The accessing order with the query keywords ψ_B is in the ascending order of the minimum possible distances between s_R and the minimum bounding rectangle (MBR) of the node, denoted by $|s_R, e_B|_{\min}$ (line 9–10, 15–17). Based on the iteration order, the early stopping condition is given in Pruning Rule 2. If e_B is a leaf node, we denote it as s_B for symbol consistency (line 18). If s_B can reach all the query keywords ψ_B, BFS with Pruning function (BFS$_P$) is invoked to compute the looseness of s_B by adopting dynamic upper and lower bounds (line 21).

Pruning Rule 2. Let θ be the current top-k score in the heap. For an iterated spatial vertex s_R in the outer loop, and an iterated node e_B in the inner loop, if $L(s_R, \psi_R) \times |s_R, e_B|_{\min} \geq \theta$, then no spatial vertex pair containing s_R can be added to top-k result heap.

Proof. Suppose that s_B^r is one of the rest of non-iterated spatial vertices in the inner loop and $sp^r = (s_R, s_B^r)$ is the spatial vertex pair. As the order of iteration is in ascending order, the spatial distance from s_R to s_B^r is not less than the minimum possible distance from s_R to e_B, i.e., $|s_R, s_B^r| \geq |s_R, e_B|_{\min}$. In addition, $L_P(sp^r, q) \geq L(s_R, \psi_R)$. Hence, we can have $f(L_P(sp^r, q), S(sp)) = L_P(sp^r, q) \times |s_R, s_B^r| \geq L(s_R, \psi_R) \times |s_R, e_B|_{\min} \geq \theta$. Therefore, sp^r can not be added to top-k result heap, which means that the spatial vertex pair includes the rest of non-iterated spatial vertices in the inner loop can not be added to top-k result heap.   □

Function BFS$_P$ computes the looseness of s_B as BFS, except that dynamic lower and upper bounds are used during the process. We first derive the lower and upper bounds in Lemma 1.

Lemma 1. *Assume a set of query keywords* $\psi = \{t_1, \ldots, t_j, \ldots, t_m\}$ *and a distance threshold* $d_{\max}$, *without loss of generality. Also, without loss of generality, we assume that the function have already discovered the first* j *query keywords are reached with BFS depth* $d(s, v)$, *where* v *is the furthest vertex that BFS visits. A lower bound of the looseness* $L(s, \psi)$ *is* $L^-(s, \psi) = 1 + \sum_{i=1}^{j} d_g(s, t_i) + d(s, v) \times (m - j)$, *and an upper bound of the looseness* $L(s, \psi)$ *is* $L^+(s, \psi) = 1 + \sum_{i=1}^{j} d_g(s, t_i) + d_{\max} \times (m - j)$.

Proof. For the reached query keywords, the sum length of the shortest path is $\sum_{i=1}^{j} d_g(s, t_i)$. For the undiscovered query keywords, the depth of BFS is $d(s, v)$ where v is the encountering vertex, which means that keywords which are not reached cannot have a shorter distance from s to v, i.e., $d_g(s, t_k) \geq d(s, v)$, $j \leq k \leq m$. In addition, the distance cannot be larger than the distance threshold, i.e., $d_g(s, t_k) \geq d_{\max}$. Therefore, we can have $L^-(s, \psi) = 1 + \sum_{i=1}^{j} d_g(s, t_i) + d(s, v) \times (m - j) \leq 1 + \sum_{i=1}^{m} \min(d_g(p, t_i), d_{\max}) \leq 1 + \sum_{i=1}^{j} d_g(s, t_i) + d_{\max} \times (m - j) = L^+(s, \psi)$. □

For the iterated vertex s_R in the outer loop, the lower and upper bounds of the looseness $L^-(s_B, \psi_B)$ and $L^+(s_B, \psi_B)$ can be used for pruning and stopping. Based on Lemma 1, we introduce the dynamic bounds in Pruning Rule 3 and Pruning Rule 4.

Pruning Rule 3. Let θ be the current k-th score in the current heap. For an iterated spatial vertex s_R in the outer loop and an iterated spatial vertex s_B in the inner loop, as long as $L^-(s_B, \psi_B) \geq \theta/|s_B, s_R|$, the spatial pair (s_R, s_B) cannot be added to the result heap, and thus (s_R, s_B) can be pruned.

Proof. For the spatial vertex pair $sp = (s_R, s_B)$, its pair looseness $L_P(sp, q) = \max(L(s_R, \psi_R), L(s_B, \psi_B)) \geq L(s_B, \psi_B) \geq L^-(s_B, \psi_B)$, so the ranking score of $f(L_P(sp, q), S(sp)) = L_P(sp, q) \times S(sp) \geq L^-(s_B, \psi_B) \times |s_R, s_B| \geq \theta$, meaning that the ranking score of sp must be no smaller than the current k-th score in the result heap, then sp cannot be in the result. □

Pruning Rule 4. For an iterated spatial vertex s_R in the outer loop and an iterated spatial vertex s_B in the inner loop, as long as $L^+(s_B, \psi_B) \leq L(s_R, \psi_R)$, the spatial pair (s_R, s_B) can be added to the result heap, and thus the BFS$_P$ can be immediately stopped.

Proof. For the spatial vertex pair $sp = (s_R, s_B)$, because $L(s_B, \psi_B) \leq L^+(s_B, \psi_B) \leq L(s_R, \psi_R)$, its pair looseness $L_P = \max(L(s_R, \psi_R), L(s_B, \psi_B)) \leq \max(L(s_R, \psi_R), L^+(s_B, \psi_B)) \leq \max(L(s_R, \psi_R), L(s_R, \psi_R)) = L(s_R, \psi_R)$. In addition, $L_P(sp, q) \geq L(s_R, \psi_R)$. Hence, $L_P(sp, q) = L(s_R, \psi_R)$, which has already been computed in the outer loop. Thus, the computation can be immediately stopped and $L_R(s_R, \psi_R)$ can be returned for the computation of L_P. □

By applying the two dynamic bounds, we can design the algorithm BFS$_P$ in Algorithm 3. It computes the dynamic bounds for s_B when BFS encounters a new vertex (line 3–4). Pruning Rule 3 guarantees the ranking score of (s_R, s_B) is no less than k-th score in the result heap (line 5–6). Conversely, Pruning Rule 4 guarantees that it is no larger than k-th score in the result heap (line 7–8). Finally, the algorithm computes the looseness of s_B by Eq. 1 according to the depth of BFS and $d_{\max}$ (line 9–14).

4.2   Improved Vertex Join Node Method: V2N$_\alpha$

V2N can only apply pruning techniques at the vertex to the node level. Here, we develop techniques capable of pruning index node that cannot contribute to the result which is called Vode Join Node with α-radius keywords index (V2N$_\alpha$) method. We notice that in the processing of V2N, the bounds can only

---

Algorithm 3: BFS$_P$

1 $L_B = 1$;
2 Set $B = \psi_B$;
3 **while** $v = \text{BFS}(G, s_B)$ *and* $B \neq \emptyset$ **do**
4   Compute the dynamic bounds $L^+(s_B, \psi_B)$ and $L^-(s_B, \psi_B)$;
5   **if** $L^-(s_B, \psi_B) \geq \theta/|s_B, s_R|$ **then**   ▷ Pruning Rule 3
6   ⌊ **return** $+\infty$;
7   **if** $L^+(s_B, \psi_B) \leq L(s_R, \psi_R)$ **then**   ▷ Pruning Rule 4
8   ⌊ **return** $L(s_R, \psi_R)$;
9   **if** $B \cap v.\psi \neq \emptyset$ **then**
10         $B = B \setminus v.\psi$;
11         $L_B += |B \cap v.\psi| \times d(s_B, v)$;
12   **if** $d(s_B, v) \geq d_{\max}$ **then**   ▷ Distance threshold
13         $L_B += |B| \times d_{\max}$;
14   ⌊ **break**;
15 **return** L_B;

Algorithm 4: V2N$_\alpha$

1 Minheap $H_k = \emptyset$, ordered by $f(L_P, S)$;
2 Threshold $\theta = +\infty$;
3 **while** $s_R = \text{GetNext}(\mathcal{R}, \psi_R)$ **do**   ▷ Pruning Rule 1
4   **if** $\theta = 0$ **then**
5   ⌊ **break**;
6   **if** s_R *can not reach* ψ_R **then**   ▷ Pruning Rule 1
7   ⌊ **continue**;
8   Minheap $H_e = \emptyset$, ordered by $|s_R, e_B|_{\min}$;
9      $H_e.\text{add}(\mathcal{R}, |e_R, \mathcal{R}|_{\min})$;
10   **while** H_e *is not empty* **do**
11         $e_B = H_e.\text{pop}()$;
12   **if** $L_\alpha^-(s_R, \psi_R) \times |s_R, e_B|_{\min} \geq \theta$ **then**   ▷ Pruning Rule 5
13   ⌊ **break**;
14   **if** e_B *is a non-leaf node* **then**
15   **foreach** *child node* e_B^c *in* e_B **do**
16   **if** $L_\alpha^-(e_B^c, \psi_B) \times |s_R, e_B^c|_{\min} \geq \theta$ **then**   ▷ Pruning Rule 6
17   ⌊ **continue**;
18   **else**
19   ⌊ $H_e.\text{add}(s_R, e_B^c)$;
20   **else**   ▷ e_B is a leaf node, denoted as s_B
21   **if** s_B *can not reach* ψ_B **then**   ▷ Pruning Rule 1
22   ⌊ **continue**;
23   Compute $L(s_R, \psi_R)$ by BFS;
24   Compute $L(s_B, \psi_R)$ by BFS$_P$;
25   Compute the ranking score f of (s_R, s_B);
26   **if** $f < \theta$ **then**
27                 $H_k.\text{add}((s_R, s_B), f)$;
28   ⌊ update θ;
29 **return** H_k;

be computed in BFS_P. In order to compute the bound at the node level, we precompute the α-*radius keyword neighborhood* for each node in R-tree. We define the α-radius keyword neighborhood of spatial vertex in Definition 4.

Definition 4 (α-radius Keyword Neighborhood of Spatial Vertex). *For a spatial vertex s, its α-radius keyword neighborhood $KN_\alpha(s)$ contains a set of keyword-distance pairs $\{(t_i, d_g(s, t_i))\}$, where the length of the shortest path from the spatial vertex s to keyword t_i in $KN_\alpha(s)$ is no larger than α, i.e., $d_g(s.t_i) \leq \alpha$.*

Based on Definition 4, we define the α-radius keyword neighborhood of node in the R-tree in Definition 5.

Definition 5 (α-radius Keyword Neighborhood of Node). *For a node e of the R-tree $\mathcal{R}$, assuming that $\{s_j\}$ is the set of spatial vertices enclosed in e, its α-radius keyword neighborhood $KN_\alpha(e)$ contains a set of keyword-distance pairs $\{t_i, d_g(e, t_i)\}$, where the keywords in $KN_\alpha(e)$ is the union of the keywords in $KN_\alpha(s_j)$ for each s_j enclosed in e, and $d_g(e, t_i) = \min_{s_j \in e} d_g(s_j, t_i)$.*

Based on the α-radius keyword neighborhood of spatial vertex, we can derive the α-lower bound of the looseness for a spatial vertex based on Lemma 2, which can be added to Pruning Rule 3 to improve the efficiency of pruning BFS_P.

Lemma 2 (α-lower Bound of Looseness of Spatial Vertex). *Let $KN_\alpha(s)$ be the α-radius keyword neighborhood of spatial vertex s and $\psi = \{t_1, \ldots, t_j, \ldots, t_m\}$ be a set of keywords. Assume that the first j query keywords correspond to keywords in $KN_\alpha(s)$. The α-lower bound of the looseness $L(s, \psi)$ is $L_\alpha^-(s, \psi) = 1 + \sum_{i=i}^{j} d_g(s, t_i) + (\alpha + 1) \times (m - j)$.*

First, we introduce the α-lower bound of a node in R-tree.

Lemma 3 (α-lower Bound of Looseness of Node). *Let $KN_\alpha(e)$ be the α-radius keyword neighborhood of node e in the R-tree and $\psi = \{t_1, \ldots, t_j, \ldots, t_m\}$ be a set of keywords. Assume that the first j query keywords correspond to keywords in $KN_\alpha(s)$. The α-lower bound of the looseness $L(e, \psi)$ is $L_\alpha^-(e, \psi) = 1 + \sum_{i=i}^{j} d_g(s, t_i) + (\alpha + 1) \times (m - j)$ and $\forall s \in e$, $L_\alpha^-(e, \psi) \leq L(s, \psi)$.*

Based on Algorithm 2, $V2N_\alpha$ integrates the α-radius index and uses the Pruning Rule 5 for early stopping and Pruning Rule 6 for pruning the node whose pair looseness must be larger than the threshold. The proof of Pruning Rule 5 and Pruning Rule 6 is the same as Pruning Rule 2.

Pruning Rule 5. Let θ be the current top-k score in the result heap. For an iterated spatial vertex s_R in the outer loop and an iterated node e_B in the inner loop, as soon as $L_\alpha^-(s_R, \psi_R) \times |s_R, e_B|_{\min} \geq \theta$, then no spatial vertex pair includes s_R with query keywords ψ_R can be added to top-k result heap.

Pruning Rule 6. Let θ be the current top-k score in the result heap. For an iterated spatial vertex s_R in the outer loop and an iterated node e_B in the inner loop, and the child node of e_B is denoted as e_B^c, as soon as $L_\alpha^-(e_B^c, \psi_B) \times |s_R, e_B^c|_{\min} \geq \theta$, then the spatial vertex pair (s_R, e_B^c) can not be added into H_e.

5   Experiment

We conducted extensive experiments on a Yago dataset and a DBpedia dataset to test the performance and feasibility of the proposed algorithms: V2V (Sect. 3.2), V2N (Sect. 4.1) and V2N$_\alpha$ (Sect. 4.2).

5.1   Settings

Datasets. We extracted the data from widely-used knowledge bases: Yago (version 2.5) and DBpedia. For the purpose of keyword search, we simplify the graphs by collecting all the keywords from the URIs, types and literals to form a document for each vertex. The documents of all vertices are organized by an inverted index. In DBpedia, there are 8,099,955 vertices and 72,193,833 edges in the knowledge graph, and 2,927,026 unique keywords. Among all vertices, there are 883,665 spatial vertices with coordinates. In Yago, there are 8,091,179 vertices and 50,415,307 edges in the knowledge graph, and 3,778,457 unique keywords. Among all vertices, there are 4,774,479 spatial vertices with coordinates.

Platform. All methods were implemented in Java and evaluated on a 2.2 Ghz 20 cores machine running Ubuntu with 16 GBytes memory. All data was kept in memory.

Queries. In kPKQ problems, a join query is implemented to find top-k spatial vertex pairs, which means that the query process requires a lot of memory and time. To intuitively show the effect of the algorithms, we randomly set 1,000 spatial vertices from 883,665 spatial vertices in DBpedia and 2,000 spatial vertices from 4,774,479 spatial vertices in Yago for iteration, and other spatial vertices will not be iterated in the algorithm.

Preprocessing. We extracted the keywords from DBpedia and Yago and implemented the knowledge graph as adjacency lists for applying the BFS computation. Spatial vertices were indexed by R-tree. The keywords in vertices and α-radius keyword neighborhood were indexed using inverted indices. The storage cost is given in Table 1.

Table 1. Storage Cost (MB)

Dataset	Knowledge graph	R-tree	Inverted index	$\alpha(=1,3,5)$ index			TF-lable
DBpedia	607.95	0.63	2031.08	4.38	35.81	209.03	5048.48
Yago	454.81	1.20	508.93	0.58	6.32	22.98	1221.08

5.2   Efficiency Evaluation

V2V takes too long to finish for the kPKQ problem because (i) BFS wastes lots of computation for the spatial vertices that can not reach all the query keywords, and (ii) it does not have early stopping conditions or bounds for pruning. Hence, in our experiments, we set 1000 s as the maximum runtime for the queries using V2V and abort those that take a longer time.

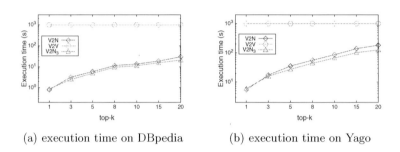

(a) execution time on DBpedia   (b) execution time on Yago

Fig. 2. Varying k

Figure 2 shows the execution time on DBpedia and Yago, respectively. As expected, the execution time increases as k increases, since a larger request number of spatial vertex pairs requires a larger search space. Note that, the execution time of V2V is always larger than the maximum runtime, which will affect the performance of the experiments. For the purpose of illustrating the details, the experiments of the following paper will not consider V2V.

Varying k. Figure 3 shows the test results of the proposed V2N and V2N$_\alpha$ ($\alpha = 1, 3, 5$), in terms of the number of execution time and pruning rate of dynamic bounds in BFS$_P$. The number of query keywords $|\psi_R|$ and $|\psi_B|$ is fixed to 5, respectively, and the distance threshold $d_{\max}$ is fixed to 5. In order to measure the pruning efficiency, we define the pruning rate of dynamic bounds as follows:

$$PR_d = \frac{N_3 + N_4}{N_{BFS_P}} \qquad (4)$$

where N_3 is the number of the spatial vertex pairs pruned by Pruning Rule 3 and N_4 is the number of the spatial vertex pairs pruned by Pruning Rule 4. N_{BFS_P} is the number of calls to BFS$_P$.

We count the number of BFS$_P$ that is called and the total number of Pruning Rule 3 and Pruning Rule 4 that are used in BFS$_P$. The larger PR_d indicates the pruning rate of dynamic bounds is better.

Figure 3(a) and (b) demonstrate that V2N$_\alpha$ is better than V2N with increasing k, and the execution time of V2N$_\alpha$ tends to be shorter with the value of α increasing.

Figure 3(c) and (d) show the PR_d in BFS_P by varying the value of k. As k increases, the k-th ranking score in iteration is more difficult to find, and Pruning Rule 2 can not limit the value of $L(s_R, \psi_R)$, which causes Pruning Rule 4 to prune more spatial vertex pairs.

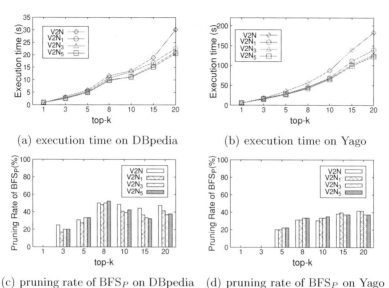

(a) execution time on DBpedia   (b) execution time on Yago

(c) pruning rate of BFS_P on DBpedia   (d) pruning rate of BFS_P on Yago

Fig. 3. Varying k

Varing $|\psi_R|$ and $|\psi_B|$. Figure 4(a) and (b) show the execution time of all method by varying the value of $|\psi_R|$ and $|\psi_B|$. The distance threshold d_{max} is fixed to 5 and k is fixed to 10. The execution time of kPKQ increases stably with $|\psi_R|$ and $|\psi_B|$, and decreases with α. The large $|\psi_R|$ and $|\psi_B|$ will lead to large cost of search space of BFS and BFS_P.

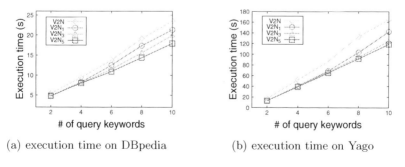

(a) execution time on DBpedia   (b) execution time on Yago

Fig. 4. Varying $|\psi_R| = |\psi_B|$

Varying $d_{\max}$. Figure 5(a) and (b) show the execution time of all method by varying the value of $d_{\max}$. The number of query keywords is fixed to $|\psi_R| = |\psi_B| = 5$, and k is fixed to 10. We observe that when $d_{\max}$ increases from 1 to 9, the execution time remains basically stable. On the other hand, we define the stopping rate as follows:

$$SR_d = \frac{N_{d_{\max}}}{N_{BFS_P}} \tag{5}$$

where $N_{d_{\max}}$ is the number of spatial vertex pairs whose traversal depth of BFS exceeds the distance threshold.

As shown in Fig. 5(c) and (d), when $d_{\max}$ is less than 5, it can limit the search space of BFS. However, $d_{\max}$ limits the size of pair looseness, so the current top-k score in the heap cannot be updated effectively, which makes Pruning Rule 2 not effective.

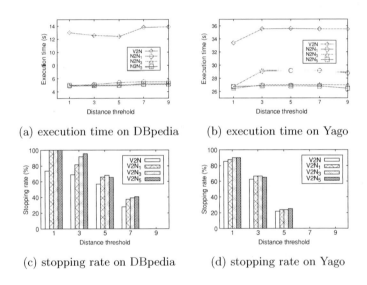

(a) execution time on DBpedia   (b) execution time on Yago

(c) stopping rate on DBpedia   (d) stopping rate on Yago

Fig. 5. Varying $d_{\max}$

6   Related Work

To the best of our knowledge, there is not any previous work on closest pair query using spatial keyword search over knowledge graph. Hereby, we discuss the related work about keyword search over knowledge graph and closest pair query.

Spatial Keyword Search over Knowledge Graph. Conventional spatial keyword search techniques focus on unstructured or semantic structured data, e.g., PoIs. Given a set of geo-textual objects, a spatial keyword search aims to find top-k objects w.r.t. keyword relevance and spatial distance[4,6,9,12,18–20,24]. Recently, knowledge graph have been enriched to include spatial information and it makes the spatial keyword search over knowledge graph feasible. For example, Jin et al. [10] propose a top-k collective keyword query to find a group of semantic places to collectively cover the query keywords. Shi et al. [15] propose the basic semantic place (BSP) and the semantic place retrieval with pruning (SPP) algorithms to compute looseness of a spatial vertex for the given query keywords. BSP is retrieved by invoking breadth-first search (BFS) from spatial vertex s until all the query keywords are searched.

Closest Pair Query. Closest pair query has been well-studied in computational geometry. Given two spatial databases, the closest pair query aims to find the k closest pairs in two set join. If both spatial databases are indexed by R-trees, the synchronous tree traversal employed by R-tree join [3] can be combined for query processing. [13,21,23] study index structures to improve the query. Except for two-dimensional, [16] gives an efficient method for the search in high dimensional space. [5] focus the problems on spatial networks. Besides, [14,22] considers the range-search of the closest query.

7 Conclusion

In this paper, we address the problem of answering top-k pair keyword queries on spatial knowledge graph. In order to solve the problem, we propose a basic solution. Based on it, we design a branch-and-bound framework associated with effective lower and upper bound pruning techniques and early stopping conditions for efficiently retrieving relevant top-k closet pairs. According to our experimental results, all techniques enables processing V2N$_\alpha$ method in less than a second for most settings and outperforms the basic method by orders of magnitude. About the future work, we plan to study how to further improve query performance with some approximates.

Acknowledgments. This work is supported by NSFC (No. 61772492, 62072428) and CAS Pioneer Hundred Talents Program. Jieming Shi is supported by the startup fund (1-BE3T) from Hong Kong Polytechnic University.

References

1. DBpedia. http://wiki.dbpedia.org
2. Yago. http://www.mpi-inf.mpg.de/departments/databases-and-information-syste ms/research/yago-naga/yago/
3. Beckmann, N., Kriegel, H.P., Schneider, R., Seeger, B.: The R*-tree: an efficient and robust access method for points and rectangles. In: SIGMOD, pp. 322–331 (1990)
4. Cao, X., Cong, G., Jensen, C.S.: Retrieving top-k prestige-based relevant spatial web objects. PVLDB **3**(1–2), 373–384 (2010)

5. Cheng, J., Huang, S., Wu, H., Fu, A.W.C.: TF-Label: a topological-folding labeling scheme for reachability querying in a large graph. In: SIGMOD, pp. 193–204 (2013)
6. Cong, G., Jensen, C.S., Wu, D.: Efficient retrieval of the top-k most relevant spatial web objects. PVLDB **2**(1), 337–348 (2009)
7. Elbassuoni, S., Blanco, R.: Keyword search over RDF graphs. In: CIKM, pp. 237–242 (2011)
8. Hristidis, V., Papakonstantinou, Y.: Discover: keyword search in relational databases. In: VLDB, pp. 670–681 (2002)
9. Hu, H., et al.: Top-k spatio-textual similarity join. TKDE **28**(2), 551–565 (2015)
10. Jin, X., Shin, S., Jo, E., Lee, K.H.: Collective keyword query on a spatial knowledge base. TKDE **31**(11), 2051–2062 (2018)
11. Le, W., Li, F., Kementsietsidis, A., Duan, S.: Scalable keyword search on large RDF data. TKDE **26**(11), 2774–2788 (2014)
12. Liu, Q., Feng, Z., Xie, X., Xu, J., Lin, X., Jensen, C.S.: iZone: efficient influence zone evaluation over geo-textual data. In: ICDE, pp. 1645–1648 (2018)
13. Lu, H., Yiu, M.L., Xie, X.: Querying spatial data by dominators in neighborhood. Inf. Syst. **77**, 71–85 (2018)
14. Shan, J., Zhang, D., Salzberg, B.: On spatial-range closest-pair query. In: SSTD, pp. 252–269 (2003)
15. Shi, J., Wu, D., Mamoulis, N.: Top-k relevant semantic place retrieval on spatial rdf data. In: SIGMOD, pp. 1977–1990 (2016)
16. Tao, Y., Yi, K., Sheng, C., Kalnis, P.: Efficient and accurate nearest neighbor and closest pair search in high-dimensional space. TODS **35**(3), 1–46 (2010)
17. Tran, T., Wang, H., Rudolph, S., Cimiano, P.: Top-k exploration of query candidates for efficient keyword search on graph-shaped (RDF) data. In: ICDE, pp. 405–416 (2009)
18. Wu, D., Yiu, M.L., Cong, G., Jensen, C.S.: Joint top-k spatial keyword query processing. TKDE **24**(10), 1889–1903 (2011)
19. Xie, X., Jin, P., Yiu, M.L., Du, J., Yuan, M., Jensen, C.S.: Enabling scalable geographic service sharing with weighted imprecise voronoi cells. TKDE **28**(2), 439–453 (2016)
20. Xie, X., Lin, X., Xu, J., Jensen, C.S.: Reverse keyword-based location search. In: ICDE, pp. 375–386 (2017)
21. Xie, X., Lu, H., Chen, J., Shang, S.: Top-k neighborhood dominating query. In: DASFAA, pp. 131–145 (2013)
22. Xue, J., Li, Y., Janardan, R.: Approximate range closest-pair queries. Comput. Geome. **90**, 101654 (2020)
23. Yang, C., Lin, K.I.: An index structure for improving closest pairs and related join queries in spatial databases. In: IDEAS, pp. 140–149 (2002)
24. Zheng, K., et al.: Interactive top-k spatial keyword queries. In: ICDE, pp. 423–434 (2015)

HIFI: Anomaly Detection for Multivariate Time Series with High-order Feature Interactions

Liwei Deng[1], Xuanhao Chen[1], Yan Zhao[2], and Kai Zheng[1(✉)]

[1] School of Computer Science and Engineering, University of Electronic Science and Technology of China, Chengdu, China
{deng_liwei,xhc}@std.uestc.edu.cn, zhengkai@uestc.edu.cn
[2] Aalborg University, Aalborg, Denmark
yanz@cs.aau.dk

Abstract. Monitoring complex systems results in massive multivariate time series data, and anomaly detection of these data is very important to maintain the normal operation of the systems. Despite the recent emergence of a large number of anomaly detection algorithms for multivariate time series, most of them ignore the correlation modeling among multivariate, which can often lead to poor anomaly detection results. In this work, we propose a novel anomaly detection model for multivariate time series with HIgh-order Feature Interactions (HIFI). More specifically, HIFI builds multivariate feature interaction graph automatically and uses the graph convolutional neural network to achieve high-order feature interactions, in which the long-term temporal dependencies are modeled by attention mechanisms and a variational encoding technique is utilized to improve the model performance and robustness. Extensive experiments on three publicly available datasets demonstrate the superiority of our framework compared with state-of-the-art approaches.

Keywords: Multivariate time series · Anomaly detection · Graph neural networks

1 Introduction

Complex systems such as servers [8] and aircrafts [1] are ubiquitous in the real world. Monitoring the behaviors of these systems generates huge amounts of multivariate time series data. A key task in managing complex systems is to detect system anomalies in a timely and accurate manner in order to reduce or avoid the losses caused by system anomalies. Due to the various of anomalies and lack enough labelled anomalies data, supervised anomaly detection methods are hard to adopt. In this work we will design an unsupervised model to detect system anomalies.

Recently, many unsupervised anomaly detection models [3,6,8] are proposed. EncDec-AD [6] uses autoencoder architecture with LSTM and treats the reconstruction errors as anomaly scores to detect anomaly. OmniAnomaly [8] adopts

© Springer Nature Switzerland AG 2021
C. S. Jensen et al. (Eds.): DASFAA 2021, LNCS 12681, pp. 641–649, 2021.
https://doi.org/10.1007/978-3-030-73194-6_42

advanced variational techniques to improve the ability of modeling complex time series. Although these methods can get better performance than their baselines for multivariate time series, they have some disadvantages.

Firstly, these studies ignore the correlation between multivariate, which is helpful for modeling complex temporal information [11]. Secondly, they generally adopt RNN or its variants to model temporal information, which hardly capture the long-term temporal dependencies [7]. Thirdly, some of them use deterministic models [1] which are unrobustness and weak representation ability [8].

To address the above-mentioned problems, we design a novel unsupervised model called HIFI (anomaly detection for multivariate time series with HIgh-order Feature Interactions). Specifically, a feature interaction graph is constructed automatically and then is delivered to GNN to model high-order feature interaction. To capture long-term dependencies and improve the robustness, attention-based time series modeling module and variational technique are used.

The main contributions of this paper are as follows:

- We design a multivariate feature interaction module, which uses the graph convolutional neural network to conduct high-order feature interactions on multi-dimensional temporal variables.
- We utilize the attention mechanism to model the long-term temporal dependence and variational encoding to improve the robustness of the model, which is critical for anomaly detection.
- We conduct extensive experiments on three publicly available datasets, which empirically demonstrate the advantages of our proposed model compared with the representative models.

2   Related Work

In this section, we will introduce some work related to our model, such as unsupervised time series anomaly detection, graph convolutional neural network, and time series modeling method based on attention mechanism.

EncDec-AD is proposed by [6], which adopts the structure of sequence to sequence and uses two independent LSTM models as encoder and decoder respectively. [1] develops the LSTM-NDT model, which is used to model the temporal information through LSTM. OmniAnomaly [8] aims to capture the normal patterns of multivariate time series by stochastic variable connections and planar normalizing flow. OED [3] is developed to use recurrent autoencoder ensembles to detect anomaly. However, none of them explicitly model the relationship between features. Therefore, in this work, we use the graph neural network to model the high-order interaction features.

Graph Convolutional network (GCN) is firstly proposed in [4], which shows a strong representation ability of unstructured graph in node classification task and attracts the attention of a large number of researchers. Then a series of variants of GCN are proposed. GAT [10] introduces the correlation between nodes through the attention mechanism. PPNP [5] uses personalized pagerank to extend the size of neighborhood and avoids oversmooth.

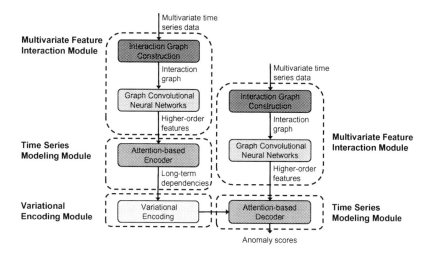

Fig. 1. Framework overview. Our model adopts encoder-decoder architecture in which three modules are involved such as multivariate feature interaction module, attention-based time series modeling module and variational encoding module.

Transformer [9] is the first time series modeling approach that completely abandons both recurrent neural network structures and convolutional neural network structures. Its performance reflects the superiority of its model structure. SASRec [2] also adopts the attention mechanism to model the sequential relation of items, and achieves the good performance in the field of sequential recommendation. However, few works adopt the attention mechanism in the field of anomaly detection.

3   Methodology

As shown in Fig. 1, our model consists of three type of parts, namely, the multivariate feature interaction module, the attention-based time series modeling module, and the variational encoding module. Each part will be elaborated in the following.

3.1   Multivariate Feature Interaction Module

In order to model the relationship between variables, the multivariate feature interaction module firstly constructs the interaction graph through feature embedding, and then gets the high-order interaction features through the graph convolutional neural network. In practice, it is hard to use graph convolutional neural network to get high-order features for multivariate time series because of the agnostic of relation graph among multivariate. So we construct the relation graph automatically in which each node represents a variable.

Inspired by MTGNN [11], we directly transform the original features $X = (x_{t-w+1}, x_{t-w+2}, \cdots, x_t) \in \mathbb{R}^{w \times d}$ to the hidden space $X^h = (x^h_{t-w+1},$

$x_{t-w+2}^h, \cdots, x_t^h) \in \mathbb{R}^{w \times d_1}$. In this way, we can control the size of the interaction graph. When the original features have a higher dimension, we can set a smaller d_1 to reduce the amount of calculation of the graph convolution.

$$x_i^h = W_h x_i + b_h \tag{1}$$

where $W_h \in \mathbb{R}^{d_1 \times d}$, $b_h \in \mathbb{R}^{d_1}$ are model parameters.

Then, in order to obtain the interaction graph with more expressive capability, we adopt an asymmetric construction method and define two independent feature embedding matrices E_1 and E_2, in which the i-th row of each matrix represents the embedding of i-th feature. Then we calculate the embedding similarity of each feature, and express the strength of feature correlation by the similarity as follow.

$$\begin{aligned} M_1 &= tanh(E_1 \Theta_1) \\ M_2 &= tanh(E_2 \Theta_2) \\ A &= ReLU(tanh(M_1 M_2^T - M_2^T M_1)) \end{aligned} \tag{2}$$

where $E_1 \in \mathbb{R}^{d_1 \times d_1}$, $E_2 \in \mathbb{R}^{d_1 \times d_1}$, $\Theta_1 \in \mathbb{R}^{d_1 \times d_2}$, $\Theta_2 \in \mathbb{R}^{d_1 \times d_2}$ are model weights, which are learnable during training. $A \in \mathbb{R}^{d_1 \times d_1}$ is the feature interaction graph.

Obviously, the interaction graph constructed by the above methods is a complete graph. In order to reduce the computation of graph convolution module, $topk$ is used to return the maximum k values in each row of the adjacency matrix, turning the complete graph into a sparse graph.

$$\mathbb{A} = topk(A) \tag{3}$$

Graph Convolution Module. We take the constructed feature interaction graph $\mathbb{A}$ and the hidden features X^h into the graph convolutional network [5] to obtain the higher-order interaction features. We first calculate $\hat{A} = \tilde{D}^{-\frac{1}{2}} \tilde{A} \tilde{D}^{-\frac{1}{2}}$, where $\tilde{A} = \mathbb{A} + I_N$ and I_N is identity matrix and $\tilde{D}_{ii} = \sum_j \tilde{A}_{ij}$.

$$H_{k+1} = (1 - \alpha)\hat{A}H_k + \alpha H_0 \tag{4}$$

where $H_k \in \mathbb{R}^{d_1 \times w}$ represents the high-order interaction features obtained by the k-th convolution. H_0 is the hidden features X^{h^T}. α is a hyperparameter defining the amount of information that retains the original feature at every convolution. It is important to note that the feature interaction does not interact across time steps. The higher-order interaction features of the current time step are completely calculated by its own hidden features.

Moreover, since the convolution depth required by each high-order feature may be different, we concatenate the output of the convolution of each step to obtain the final higher-order feature representation through linear transformation.

$$X^{ho} = Concat(H_0^T, H_1^T, \dots, H_K^T)W^{ho} \tag{5}$$

where $X^{ho} = (x_{t-w+1}^{ho}, x_{t-w+2}^{ho}, \cdots, x_t^{ho}) \in \mathbb{R}^{w \times d_1}$ is the high-order features corresponding to X^h. $Concat$ denote the concatenation operation. $W^{ho} \in \mathbb{R}^{Kd_1 \times d_1}$ is the model weights. K is the maximal graph convolutional depth.

3.2   Attention-Based Time Series Modeling Module

We use the attention mechanism to model temporal information and capture long-term temporal dependencies, which are hardly captured by RNN model.

Attention Layer. We use one of the most common attention mechanisms, scaled-dot attention [9], which can be described as follows.

$$Attention(Q, K, V) = softmax(\frac{QK^T}{\sqrt{d_k}})V \tag{6}$$

where $Q \in \mathbb{R}^{w \times d_k}$, $K \in \mathbb{R}^{w \times d_k}$, $V \in \mathbb{R}^{w \times d_v}$ are queries, keys and values respectively. $\sqrt{d_k}$ is the scaling factor. Intuitively, the nature of the attention mechanism is to take the weighted sum of all time steps according to the weights calculated. It allows information from any distance of time steps to flow directly to the current step, giving the attention mechanism the ability to capture long-term temporal dependencies. Multi-head attention [9], which can allow the model to jointly attend to information from different representation subspaces at different positions, is used in our work.

Nonlinear Layer. Although multi-head attention can aggregate the information of each time step through adaptive weights, it is still a linear model. This greatly limits the capability of the model, so we use two linear network with *ReLU* activation to introduce nonlinear information to the model.

$$NonLinear(X) = W_2^n ReLU(W_1^n X + b_1^n) + b_2^n \tag{7}$$

where $W_1^n \in \mathbb{R}^{d_3 \times d_1}$, $W_2^n \in \mathbb{R}^{d_1 \times d_3}$, $b_1^n \in \mathbb{R}^{d_3}$, $b_2^n \in \mathbb{R}^{d_1}$ are model parameters. Due to the nature of position invariance of attention layer, we follow [9] to introduce sinusoidal positional encoding $P \in \mathbb{R}^{w*d_1}$ to add the sequential information after multivariate feature interaction module. Specifically, $X^{in} = P + X^{ho} + X$ where X^{in} is the input of attention-based time series modeling module. Our encoder and decoder can be obtained by alternately stacking multi-head attention layer and nonlinear layer, in which layer normalization and residual connection are used to prevent overfitting and gradient disappearance respectively.

3.3   Variational Encoding Module

To improve the robustness and performance of the model, we model the deterministic encoding of the encoder $X^{eo} = (x_{t-w+1}^{eo}, x_{t-w+2}^{eo}, \cdots, x_t^{eo}) \in \mathbb{R}^{w \times d_1}$ as a normal distribution. We obtain the mean and the logarithm of variance of the normal distribution by two independent linear layers.

$$\begin{aligned} \mu_i &= W_\mu x_i^{eo} + b_\mu \\ log\, \sigma_i^2 &= W_\sigma x_i^{eo} + b_\sigma \end{aligned} \tag{8}$$

where $W_\mu \in \mathbb{R}^{d_1 \times d_1}$, $W_\sigma \in \mathbb{R}^{d_1 \times d_1}$, $b_\mu \in \mathbb{R}^{d_1}$, $b_\sigma \in \mathbb{R}^{d_1}$ are model weights. We then adopt the reparameterization trick to sample from the normal distribution and input the samples into the decoder. We fix the number of samples at 1.

$$z_i = \mu_i + \epsilon * \sigma_i \tag{9}$$

where ϵ is a sample from $N(0,1)$. μ_i and σ_i are the mean and the variance of normal distribution in i-th time step. The resampled Z is inputted to attention-based time series modeling module of decoder and treated as Q and K.

3.4 Model Training

We use the reconstruction error of the entire current window and the Kullback-Leibler divergence between variational encoding of sample and the standard normal distribution to train the model.

$$loss = \sum_{i=1}^{w} ||x_i - x_i^{rec}||_2 + \beta \sum_{i=1}^{w} KL(N(\mu_i, \sigma_i^2)||N(0,1))) \tag{10}$$

where β is a hyperparameter, which is used to balance the loss of the two parts.

In current window, we regard $||x_t - x_t^{rec}||_2$ as the anomaly score, based on which we can detect anomalies.

4   Experiments

In this section, we evaluate our model by comparing it with some state-of-the-art models. We begin by introducing the setup of the experiment, and then report and analyze the results of the experiment.

4.1   Experimental Setup

Datasets and Metrics. We conduct experiments on three publicly available datasets, i.e. SMD (Server Machine Dataset) [8], SMAP (Soil Moisture Active Passive satellite) and MSL (Mars Science Laboratory rover) [1], in two different domains. We use Precision (Pre), Recall (Rec) and F1-score ($F1$) to evaluate the performance of HIFI and baselines. We enumerate all possible anomaly thresholds to search for the best F1, denoted as $F1_{best}$. And a point-adjust [12] is adopted to get the final prediction which is same as [8].

Baselines. In our experiments, we select four representative models. LSTM-NDT [1] predicts the values of time step t and uses predictive error as anomaly scores of step t. EncDec-AD [6] adopts autoencoder architecture with LSTM and treats reconstruction error as anomaly score. OED-IF [3] employs multiple autoencoder with different connection structures to improve the performance of anomaly detection. OmniAnomaly [8] adopts advanced variational encoding techniques with autoencoder architecture to detect anomaly.

Implementation Details. All models take the sliding window data of the original data as input and we set the window size w to 100. We randomly select 30% from the training data as validation sets, set the batch size as 64 for training and run for 100 epochs. We save the model with the least loss of the validation set as the final test model. We use Adam optimizer for stochastic gradient descent with an initial learning rate of 0.005. For our model, we turn hyperparameters in validation set. Specifically, we set $d_1 = 64$, $d_2 = 64$, $d_k = 16$, $l = 2$, $\alpha = 0.2$, $\beta = 1$, $K = 3$ for all datasets. In MSL dataset, we set $d_3 = 256$ and in other datasets, we set $d_3 = 128$. For all of the baselines, if they are tested on the same dataset, we follow the settings of the original paper, or we tune the model to be optimal.

Table 1. The performance of HIFI with other baseline methods over three datasets

Methods	MSL			SMAP			SMD		
	$F1_{best}$	Pre	Rec	$F1_{best}$	Pre	Rec	$F1_{best}$	Pre	Rec
LSTM-NDT	0.8623	0.7830	0.9594	0.7852	0.6756	0.9373	0.7942	0.6865	0.9481
EncDec-AD	0.9039	0.8606	0.9520	0.8707	0.7737	0.9956	0.9491	0.9317	0.9673
OED-IF	0.9185	0.8754	0.9661	0.8458	0.7351	**0.9959**	0.9730	0.9685	**0.9777**
OmniAnomaly	0.9257	0.8802	0.9762	0.8966	0.8198	0.9893	0.9503	0.9337	0.9675
HIFI	**0.9546**	**0.9133**	**0.9998**	**0.9708**	**0.9475**	0.9952	**0.9773**	**0.9811**	0.9737

4.2   Overall Performance Comparison

Table 1 shows the performance of our model compared with the baseline models in three datasets. LSTM-NDT performs worst in all the baselines in terms of $F1_{best}$, which shows the disadvantage of the predictive model. In most cases, OED-IF can get better performance than EncDec-AD, which shows that model ensemble is a good strategy for anomaly detection. OmniAnomaly generally performs better than other baselines, which illustrates the variational encoding can improve the performance of anomaly detection. Last but not least, HIFI outperforms all baselines on three datasets in terms of $F1_{best}$. In particular, HIFI outperforms the best-performing state-of-the-art method by 2.89%, 7.42% and 0.43% on MSL, SMAP and SMD dataset respectively, which shows the effectiveness of the proposed model.

4.3   Ablation Study

We perform an ablation study at $F1_{best}$ to verify the effectiveness of each component of our model. We name different variants of HIFI as follows:

w/o FI: HIFI: without multivariate feature interaction module.
w/o VE: HIFI: without variational encoding module.
w/o FI+VE: HIFI without multivariate feature interaction module and stochastic variable embedding module.

w/o FI+VE+EN: HIFI only with encoder part. We stack 4 encoder layers to fair comparison.

The results of our model and its variants are shown in the Table 2. In most cases, HIFI can achieve the best results. In the SMD dataset, the performance of each model is similar. This is because the time series information is sufficient to detect anomalies, and introducing complex features does not improve the performance of the model. In MSL and SMAP datasets, we can clearly see that w/o FI and w/o VE outperform w/o FI+VE, which proves the effectiveness of the multivariate feature interaction module and variational encoding module. Comparing performance between w/o FI+VE and w/o FI+VE+EN, we can find that w/o FI+VE can get better performance, which shows the superiority of Encoder-Decoder model in anomaly detection. Finally, compared with the baselines, w/o FI+VE+EN can get better performance generally, which shows the importance of long-term temporal dependency modeling in anomaly detection.

Table 2. The $F1_{best}$ of HIFI and its variants

Methods	MSL	SMAP	SMD
HIFI	**0.9546**	**0.9708**	0.9773
w/o FI	0.9455	0.9674	0.9781
w/o VE	0.9423	0.8821	0.9777
w/o FI+VE	0.9415	0.8703	**0.9787**
w/o FI+VE+EN	0.9358	0.8295	0.9771

5   Conclusion

In this paper, we propose a model, namely HIFI, an unsupervised anomaly detection model for multivariate time series with high-order feature interaction. Extensive empirical study based on real datasets confirms the advantage of our proposed model compared with the state-of-the-art methods as well as the importance of different components of our model for multivariate anomaly detection.

Acknowledgements. This work is supported by NSFC (No. 61972069, 61836007, 61832017) and Sichuan Science and Technology Program under Grant 2020JDTD0007.

References

1. Hundman, K., Constantinou, V., Laporte, C., Colwell, I., Soderstrom, T.: Detecting spacecraft anomalies using LSTMs and nonparametric dynamic thresholding. In: KDD, pp. 387–395 (2018)
2. Kang, W.C., McAuley, J.: Self-attentive sequential recommendation. In: ICDM, pp. 197–206 (2018)

3. Kieu, T., Yang, B., Guo, C., Jensen, C.: Outlier detection for time series with recurrent autoencoder ensembles. In: IJCAI, pp. 2725–2732 (2019)
4. Kipf, T.N., Welling, M.: Semi-supervised classification with graph convolutional networks. ArXiv (2016)
5. Klicpera, J., Bojchevski, A., Günnemann, S.: Predict then propagate: graph neural networks meet personalized PageRank. In: ICLR (2019)
6. Malhotra, P., Ramakrishnan, A., Anand, G., Vig, L., Agarwal, P., Shroff, G.: LSTM-based encoder-decoder for multi-sensor anomaly detection. ArXiv (2016)
7. Qin, Y., Song, D., Chen, H., Cheng, W., Jiang, G., Cottrell, G.W.: A dual-stage attention-based recurrent neural network for time series prediction. In: IJCAI, pp. 2627–2633 (2017)
8. Su, Y., Zhao, Y., Niu, C., Liu, R., Sun, W., Pei, D.: Robust anomaly detection for multivariate time series through stochastic recurrent neural network. In: KDD, pp. 2828–2837 (2019)
9. Vaswani, A., et al.: Attention is all you need. ArXiv (2017)
10. Velickovic, P., Cucurull, G., Casanova, A., Romero, A., Liò, P., Bengio, Y.: Graph attention networks. ArXiv (2018)
11. Wu, Z., Pan, S., Long, G., Jiang, J., Chang, X., Zhang, C.: Connecting the dots: multivariate time series forecasting with graph neural networks. In: KDD (2020)
12. Xu, H., et al.: Unsupervised anomaly detection via variational auto-encoder for seasonal KPIs in web applications. In: WWW, pp. 187–196 (2018)

Incentive-aware Task Location in Spatial Crowdsourcing

Fei Zhu[1], Shushu Liu[2], Junhua Fang[1], and An Liu[1(✉)]

[1] School of Computer Science and Technology, Soochow University, Suzhou, China
20184227027@stu.suda.edu.cn, {jhfang,anliu}@suda.edu.cn
[2] Department of Communication and Networking, Aalto University, Espoo, Finland
liu.shushu@aalto.fi

Abstract. With the popularity of wireless network and mobile devices, spatial crowdsourcing has gained much attention from both academia and industry. One of the critical components in spatial crowdsourcing is task-worker matching, where workers are assigned to tasks to meet some pre-defined objectives. Previous works generally assume that the locations of tasks are known in advance. However, this does not always hold, since in many real world applications where to put tasks is not specific and needs to be determined on the fly. In this paper, we propose Incentive-aware Task Location (ITL), a novel problem in spatial crowdsourcing. Given a location-unspecific task with a fixed budget, the ITL problem seeks multiple locations to place the task and allocates the given budget to each location, such that the number of workers who are willing to participate the task is maximized. We prove that the ITL problem is NP-hard and propose three heuristic methods to solve it, including even clustering, uneven clustering and greedy location methods. Through extensive experiments on a real dataset, we demonstrate the efficiency and effectiveness of the proposed methods.

Keywords: Spatial crowdsourcing · Task location · Task assignment

1 Introduction

With the development of mobile Internet and GPS-equipped smart devices, spatial crowdsourcing has been booming in both academia [4,11,15–17] and industry [1]. In a typical workflow of spatial crowdsourcing, a requester specifies the functional requirement and a budget of a task, then the platform matches available workers to it, after which the selected workers physically move to the designated location to complete the task and gain the corresponding reward.

Existing studies mainly focus on the design of task assignment algorithms [5,6,8,9] where the platform needs to select the most appropriate workers according to some optimization objectives. A common assumption is that the location of a task is specified by its requester, and the platform takes location information as a given parameter to conduct task assignment. In practice, however, this

© Springer Nature Switzerland AG 2021
C. S. Jensen et al. (Eds.): DASFAA 2021, LNCS 12681, pp. 650–657, 2021.
https://doi.org/10.1007/978-3-030-73194-6_43

assumption is too strong, since in many crowdsourced spatiotemporal applications [2,3], tasks are location-unspecific, that is, requesters do not specify where these tasks should be completed. For example, foursquare [3] place artificial rewards at some points of interest to attract spatial users to check-in and write comments. In this application, how much rewards are placed at which locations are all determined by the platform whose objective is to achieve the highest user participation. Another example can be found in the famous game Pokemon GO [2], in which different pokemons are placed at multiple locations in the AR scenario and game players need to physically move in the real world to catch them.

As workers generally prefer nearby tasks, the locations of tasks can affect the enthusiasm of workers to participate. It is therefore of great importance to determine where to put these location-unspecific tasks. However, it is non-trivial due to the following observations:

- *Uneven distribution of workers.* In practice, workers are unevenly distributed in space, for example, placing tasks in dense areas is likely to attract more workers than placing them in sparse areas.
- *Rewarding duplication of tasks.* Some spatial task can be performed at multiple places and this deliberate duplication can bring more benefits. Taking foursquare as an example, the more points of interest can attract new users to check-in, the more comprehensive data the platform can obtain.
- *Correlation between budget and appeal.* The willingness of a worker to perform a task also depends on the reward of the task. Putting a task on multiple places will reduce its appeal since its budget fixed by the requester will also be distributed. From this aspect, task duplication (i.e., putting a task on multiple places) is not a good choice, which is contrary to the above observation.

Motivated by this, we propose a novel problem in spatial crowdsourcing, namely Incentive-aware Task Location (ITL) problem. Given a task with a fixed budget, ITL problem seeks multiple locations and allocates the given budget to them, with the optimization criterion being the maximization of the number of workers who are willing to perform the task on these locations. Unfortunately, previous achievements cannot be directly applied to solve this problem. Firstly, task assignment algorithms [5,7,8,12] focus on a different phase with us. That is, only after we determine the task locations and attracting as more workers to participate as possible, can assignment algorithms select appropriate workers for their own objectives. Secondly, incentive mechanisms [10,14] only deal with location-specific tasks and never specify locations where workers need to reach. Last but not least, the algorithms of location problem [13] can put each resource in an appropriate position to achieve an optimization goal, but they do not address the problem of allocating a given budget to multiple tasks at different locations. In a word, we make the following contributions:

- We formalize the ITL problem and prove its hardness by reducing it to the maximum coverage location problem (MCLP).

- We propose three efficient heuristic approaches, namely even clustering, uneven clustering and greedy location methods.
- We conduct extensive experiments on real data set and show the efficiency and effectiveness of our approaches.

2   Problem Statement

Definition 1. Location-unspecific Task. *A location-unspecific task t is defined as $\langle F, b \rangle$, where F is the functional requirement of the task and $t.b$ is the budget for the task.*

Definition 2. Location-specific Task. *A location-specific task (or task for short) t is defined as $\langle F, b, l \rangle$, where l is the place that the task is carried out, and F and $t.b$ are defined as above.*

Definition 3. Worker. *Let $W = \{w_1, \ldots, w_n\}$ denote a set of available workers. A worker w_i is defined as $\langle l, d(.), D, r, s \rangle$, where l is the current location of w with skill s, $d(t.b)$ is the distance w is willing to move given the reward $t.b$, D is the maximum distance w is willing to move no matter how much the reward is, and r is the minimum reward that makes w willing to perform a task.*

Definition 4. Valid Task-worker Pair. *A task-worker pair $\langle t, w \rangle$ is said to be valid if $\mathrm{dist}(t, w) \leq w.d(t.b)$, $\mathrm{dist}(t, w) \leq w.D$, $w.s = t.F$ and $t.b \geq w.r$.*

Definition 5. Participation Set. *The participation set of task t is defined as $p(t) = W_t$, where W_t is the set of the workers who can form a valid task-worker pair with task t, that is, $W_t = \{w | w \in W, \langle t, w \rangle \text{ is valid}\}$.*

Definition 6. Candidate Site. *Let $L = \{\mathrm{loc}_1, \ldots, \mathrm{loc}_m\}$ denote the set of m candidate sites where tasks can be located.*

Considering not all geographic locations are suitable for completing tasks, such as dangerous traffic intersections, the platform determines some suitable places as the candidate sites for placing tasks and stores these locations in advance. For each location-unspecific task, the candidate site set is the same.

Definition 7. Incentive-aware Task Location (ITL) Problem. *Given a set W of workers, a set L of candidate sites, and a location-unspecific task t, the ITL problem is to construct a set T of location-specific tasks, such that the participation set of T, i.e., $p(T) = \bigcup_{t_i \in T} p(t_i)$, is maximized, subject to the following constraints:*

$$t_i.F = t.F, \forall t_i \in T \tag{1}$$

$$t_i.l \in L, t_j.l \in L, t_i.l \neq t_j.l, \forall t_i, t_j \in T \tag{2}$$

$$\sum_{t_i \in T} t_i.b \leq t.b \tag{3}$$

3 Proposed Methods

3.1 Even Clustering Location Method

Workers in real world are often clustered in multiple business districts within a city, and rarely uniformly distributed. Therefore, we find worker-intensive areas by clustering, and then place tasks with the same budget near each cluster.

As a typical clustering method, K-means is easy to implement, simple in principle and fast in clustering. Since the number of clusters to be produced is unknown, we cannot apply K-means directly. We determine the size of K under the following consideration. If the number of clusters is too large, the budget evenly allocated to each task will be too small, even lower than the minimum accept reward of nearby workers, thus losing the attraction to workers. Therefore, we set the range as $\left[1, \frac{t.b}{max_{w_i \in W}(w_i.r)}\right]$, where $t.b$ is the budget of the given location-unspecific task and $max_{w_i \in W}(w_i.r)$ is the max value of the minimum acceptable rewards of all workers. As long as the number of clusters is less than $\frac{t.b}{max_{w_i \in W}(w_i.r)}$, the budget evenly allocated to each cluster is greater than $max_{w_i \in W}(w_i.r)$, which can meet all workers' minimum acceptable reward constraints.

We also design a filter process to delete those clusters without candidate sites nearby. If the distance between a cluster to its nearest candidate site exceeds $max_{w_i \in W}(w_i.D)$, the cluster will be filtered out. Otherwise, even if we place a task in the nearest candidate site, all the workers in this cluster will not be willing to move so far to complete it.

The complete process of the even clustering location method is as follows. For every integer K in range $\left[1, \frac{t.b}{max_{w_i \in W}(w_i.r)}\right]$, we use K-means to cluster workers spatially, then construct K location-specific tasks with the budget $\frac{t.b}{K}$ at the candidate sites closet to their clusters. After $\lfloor \frac{t.b}{max_{w_i \in W}(w_i.r)} \rfloor$ attempts, a location strategy that can make the most workers willing to participate is selected as the final result.

3.2 Uneven Clustering Location Method

Since the number of workers in each cluster may vary significantly, it seems wasteful to allocate equal rewards to clusters with fewer workers and those with far more workers, especially when the given budget is limited. Therefore, in the uneven location method, we not only place tasks at the candidate points closest to each effective cluster, but also allocate the budget according to the proportion of the population of each cluster in the total number of workers.

Note that, the K-means clustering methods cannot be applied in this uneven location method. Because the budget is allocated unevenly, we cannot ensure that the budget allocated to each cluster will exceed $max_{w_i \in W}(w_i.r)$ by limiting the number of clusters in advance. Hence, we introduce the hierarchical clustering that can divide workers into clusters without knowing the number of clusters as a priori, so that workers in the same cluster are close to each other, and those in different clusters are far away from each other. Specifically, it initializes every worker to their own individual cluster and then iteratively picks the two clusters that are the closest to each other to merge into a bigger cluster. The merging stops when the termination condition is satisfied.

We specify the termination condition of merging adjacent clusters to accommodate our ITL problem, rather than apply the clustering method directly. In our method, the distance between two clusters is calculated by the shortest distance from a worker of one cluster to another worker of another cluster. If the distance between two clusters is greater than twice the max value of the maximum moving distance of all workers, there is always a cluster of workers unwilling to move to complete the task placed near the other cluster. Therefore, we set the distance threshold DT to be $2 \cdot max_{w_i \in W}(w_i.d)$.

After hierarchical clustering, we obtain areas where the workers are close to each other. However, these clusters may include many with a high density but small number of workers. If we construct a location-specific location in each cluster and allocate the budget in proportion to the number of people in each cluster, the allocated budget of the clusters with few workers may not reach the lowest acceptable price of all workers. Therefore, we only place tasks near the clusters whose population exceeds the threshold $ST - \frac{max_{w_i \in W}(w_i.r) \cdot n}{t.b}$. We determine the ST value to ensure that each cluster exceeding it can attract participation. When the population of a cluster exceeds ST, it can receive a reward more than $\frac{ST}{n} \cdot t.b = max_{w_i \in W}(w_i.r)$, which is higher than all workers' minimum acceptable rewards. After screening out some clusters that do not meet the population threshold ST, we obtain several clusters where a large number of workers gather. The number of workers in each cluster is calculated, and budget is allocated to these valid clusters according to their proportions of the number of workers.

3.3   Uneven Greedy Location Method

Because the two methods above are based on clustering, their performances also highly depend on the clustering result, which means a poor clustering result will lead to very low participation of tasks. Therefore, we propose a greedy method to construct the location-specific task set, which iteratively assigns a unit of reward to a task that contributes to the highest participation increase. Before we present the greedy algorithm, we first define the increase Δp of $p(T)$, the number of workers willing to participate. It is noteworthy that the increase is calculated differently depending on whether the unit reward is added to a constructed task or to a newly constructed task.

If $t = \langle F, t.b, t.l \rangle$ is a previously constructed location-specific task in set T, the increase refers to the number of workers who newly involved in the task set T after adding a unit reward to the of task t.

If t is a task to be constructed, considering that if only one unit of reward is allocated, the minimum acceptable rewards of many workers may not be reached. Therefore, we assign t a reward $MT = max_{w_i \in W}(w_i.r)$, at one time when we just construct it, and then add t into T to calculate the average participation increase of each unit reward for T.

With the greedy principle, the whole process of the greedy location method is as follows. At first, we use L_new to denote the candidate sites set T includes all constructed tasks. Before the budget is used up, we may continuously allocate it in two ways: constructing a new task with a starting reward of MT and adding a unit reward to a constructed task in the set T. By comparing the unit reward participation increase Δp generated by the two strategies, we allocate the corresponding rewards, and update the relevant variables, including the remaining budget, the unused candidate sites set L_new and the constructed tasks set T.

4 Experimental Study

4.1 Experimental Methodology

Data Set. We use real data from Didi [1] to test our proposed ITL methods, which includes 5,590,861 order records of Chengdu, Sichuan province in 2016. From this dataset, we can get the real distribution and the relationship between travel distance and reward of workers. All pick-up and destination locations are regarded as candidate sites in our experiments, leaving the task locations to be determined by our proposed methods.

ITL Methods and Measures. To evaluate our three proposed methods, we need to compare with the global optimal solution. However, the ITL problem is NP-hard and thus infeasible to solve it optimally in polynomial time. Alternatively, we compare the effectiveness of our three methods with that of single task location method, which does not split the given budget and find the best candidate site to locate the single task. Table 1 depicts our experimental settings, where the default values of parameters are in bold font.

Table 1. Experiments settings.

Parameters	Values
Number of workers n	1K, **2K**, 5K, 7K, 10K
Budget range $[B^-, B^+]$	[5, 15], [15, 25], [25, 35], **[35, 45]**, [45, 55], [55, 65]

4.2 Experiments on Real Data

In this subsection, we show the effects of the ranges of budget, the number of workers, minimum acceptable reward and maximum moving distance.

Effect of the Number of Workers n. As shown in Fig. 1(a), when the range of n increases, the participation rates decrease. In real scenario, the more workers, the larger the geographical area they cover. Thus, when given the same default

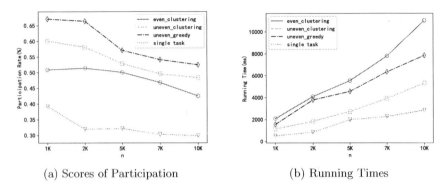

(a) Scores of Participation   (b) Running Times

Fig. 1. Effect of the number of workers n.

budget, the more workers are far away from the task location, the participation is lower. In Fig. 1(b), when the range of the number of workers n increases, the running times of our tested methods also increase, due to the cost of more valid task-worker pairs to be searched. Most the proposed methods can finish the ITL problem of 10K workers in 8 s, which shows the efficiency considering there is no real-time requirement in our problem.

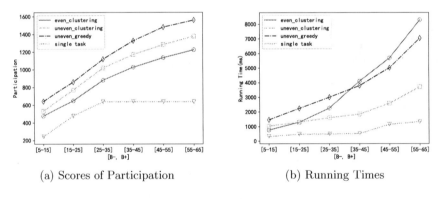

(a) Scores of Participation   (b) Running Times

Fig. 2. Effect of the range of budget $[B^-, B^+]$.

Effect of the Range of Budget $[B^-, B^+]$. In Fig. 2(a), the participation scores of all the four methods increase, when the value range of task budget gets larger. For single task location method, even if the reward of single task is very large, the workers who are far away are unable to participate due to the $w.D$ limit, which leads to the participation performance of this method remains unchanged. In any case, it has the lowest score, which proves that our proposed task location methods are more effective. From Fig. 2(b), we can observe that the running time of our methods increase which is owing to the fact of the increasing cost can cover more valid task-worker pairs.

5 Conclusion

In this paper, we formalize the incentive-aware task location (ITL) problem, which aims to construct a set of tasks with specified locations and allocate budget to maximize the participation under workers' constraints. We propose three heuristic methods (i.e., even clustering, uneven clustering, and greedy location methods), which can efficiently solve the NP-hard problem. Extensive experiments on real data set have shown the efficiency and effectiveness of them.

Acknowledgment. This paper is partially supported by Natural Science Foundation of China (Grant No. 61572336), Natural Science Research Project of Jiangsu Higher Education Institution (No. 18KJA520010), and a Project Funded by the Priority Academic Program Development of Jiangsu Higher Education Institutions.

References

1. https://gaia.didichuxing.com/
2. https://www.pokemongo.com/
3. https://foursquare.com/
4. Chen, Z., Cheng, P., Chen, L., Lin, X., Shahabi, C.: Fair task assignment in spatial crowdsourcing. Proc. VLDB Endow. **13**(11), 2479–2492 (2020)
5. Chen, Z., Cheng, P., Zeng, Y., Chen, L.: Minimizing maximum delay of task assignment in spatial crowdsourcing. In: ICDE, pp. 1454–1465. IEEE (2019)
6. Cheng, P., Chen, L., Ye, J.: Cooperation-aware task assignment in spatial crowdsourcing. In: ICDE, pp. 1442–1453. IEEE (2019)
7. Cheng, Y., Li, B., Zhou, X., Yuan, Y., Wang, G., Chen, L.: Real-time cross online matching in spatial crowdsourcing. In: ICDE, pp. 1–12 (2020)
8. Li, B., Cheng, Y., Yuan, Y., Wang, G., Chen, L.: Three-dimensional stable matching problem for spatial crowdsourcing platforms. In: KDD, pp. 1643–1653. ACM (2019)
9. Tao, Q., Tong, Y., Zhou, Z., Shi, Y., Chen, L., Xu, K.: Differentially private online task assignment in spatial crowdsourcing: a tree-based approach. In: ICDE, pp. 517–528. IEEE (2020)
10. Tong, Y., et al.: The simpler the better: a unified approach to predicting original taxi demands based on large-scale online platforms. In: SIGKDD, pp. 1653–1662 (2017)
11. Tong, Y., Zhou, Z., Zeng, Y., Chen, L., Shahabi, C.: Spatial crowdsourcing: a survey. VLDB J. **29**(1), 217–250 (2019)
12. Tran, L., To, H., Fan, L., Shahabi, C.: A real-time framework for task assignment in hyperlocal spatial crowdsourcing. TIST **9**(3), 37:1–37:26 (2018)
13. Wolf, G.W.: Facility location: concepts, models, algorithms and case studies. Int. J. Geogr. Inf. Sci. **25**(2), 331–333 (2011)
14. Xiao, M., et al.: SRA: secure reverse auction for task assignment in spatial crowdsourcing. TKDE **32**(4), 782–796 (2020)
15. Zhao, Y., Zheng, K., Cui, Y., Su, H., Zhu, F., Zhou, X.: Predictive task assignment in spatial crowdsourcing: a data-driven approach. In: ICDE, pp. 13–24 (2020)
16. Zheng, B., et al.: Online trichromatic pickup and delivery scheduling in spatial crowdsourcing. In: ICDE, pp. 973–984 (2020)
17. Zheng, B., et al.: Answering why-not group spatial keyword queries. TKDE **32**(1), 26–39 (2020)

Efficient Trajectory Contact Query Processing

Pingfu Chao[✉], Dan He, Lei Li, Mengxuan Zhang, and Xiaofang Zhou

School of Information Technology and Electrical Engineering,
University of Queensland, Brisbane, Australia
{p.chao,d.he,l.li3}@uq.edu.au, mengxuan.zhang@uqconnect.edu.au,
zxf@itee.uq.edu.au

Abstract. During an infectious disease outbreak, the contact tracing is regarded as the most crucial and effective way of disease control. As the users' trajectories are widely obtainable due to the ubiquity of positioning devices, the contact tracing can be achieved by examining trajectories of confirmed patients to identify other trajectories that are contacted either directly or indirectly. In this paper, we propose a generalised Trajectory Contact Search (TCS) query, which models the contact tracing problem as well as other similar trajectory-based problems. In addition, we answer the query by proposing an iterative algorithm that finds contacted trajectories progressively along the transmission chains, and we further optimise each iteration in terms of time and space efficiency by proposing a hop scanning algorithm and a grid-based time interval tree. Extensive experiments on large-scale real-world data demonstrate the effectiveness of our proposed solutions over baseline algorithms.

1 Introduction

Nowadays, our daily movements are collected into trajectory data by various devices like GPS, Bluetooth, cellular tower, etc. Among different trajectory query types, there is a lack of research on exploiting the trajectory's *contact* feature, which happens when two objects travel nearby each other over a period of time, such that information (contagious disease, chemical/radiative leaks, airborne materials, etc.) could spread from one to another. A direct application is the contact tracing during the COVID-19 pandemic. Massive literature emphasises the importance of contact tracing and several countries have established apps to track contact events through Bluetooth connection between devices. However, such a method is hardly practical as it requires users' strict participation and it is not capable of detecting contact events with different radius due to the hardware limitation. In contrast, contact tracing through trajectories is more applicable as the trajectory distance is measurable and the data is widely available.

Therefore, in this paper, we propose a new type of trajectory query, termed as *Trajectory Contact Search (TCS)*, which finds all trajectories that directly and indirectly contact the query trajectory. Specifically, when two trajectories

C. S. Jensen et al. (Eds.): DASFAA 2021, LNCS 12681, pp. 658–666, 2021.
https://doi.org/10.1007/978-3-030-73194-6_44

appear within a distance over a time period, we say they make a *contact* and one can *influence* the other. Then, given a query trajectory Tr_q, a distance ϵ, and the time step threshold k, we aim to find not only all trajectories R' it contacts, but also all the trajectories contacted by the influenced results R' subsequently.

In fact, trajectory contact tracing is non-trivial. As pre-computing all contact events is not viable due to the flexibility of contact definition (ϵ and k), the query can only be answered by searching direct contacts to the influenced trajectory recursively. Besides, as a contact event requires both spatial and temporal continuity, new index and scanning algorithm are required to store and retrieve timestamp-level trajectories efficiently. Overall, our contributions are as follows:

- We propose a *Trajectory Contact Search* query for contact tracing problem.
- We propose an iteration-based solution to answer the contact search query without redundancy. In addition, we propose a *hop scanning* algorithm and a *time interval grid* index to further improve the time and space efficiency.
- Extensive experiments on large-scale real-world dataset show that our methods can answer the *TCS* query more efficiently than existing methods.

2 Related Works

To the best of our knowledge, the *travelling group discovery* problem is closely related to *TCS*. It finds all groups of objects that move together over a period of time. Depending on how to define proximity (*distance-based* [1,2,4], or *density-based* [3,7,9]) and whether the time period is required to be consecutive [2,3,7] or not [5,9], various group patterns are identified. To discover the groups, the trajectories are first sliced into temporal snapshots, then a clustering algorithm or predefined criteria is applied to each snapshot to find groups. Finally, the clusters from adjacent snapshots are intersected and concatenated until forming a long time sequence satisfying travel requirements. Besides, to enable distance comparison in every timestamp, linear interpolation is introduced to ensure an object to appear in every snapshot it crosses, which greatly inflates the input size and the processing cost. To reduce the cost, [3] uses trajectory segments instead of points, and it further simplifies trajectories using *Douglas-Peucker* algorithm. Meanwhile, [7] proposes a *travelling buddy* structure to capture the minimal groups of objects and perform intersection on buddies instead.

Besides, *IMO* [8] is the only demo that works on a similar problem to ours. However, it focuses on simulating the disease spread and analysing the effectiveness of the policies in preventing transmission. The demo is integrated into an open-source spatial-temporal database and utilises the built-in spatial-temporal indices to reduce the search space of finding the infected trajectories. Although it is regarded as the first work that introduces the trajectory contact tracing problem, it doesn't work on the query optimisation, which is our main focus.

3   Problem Statement

Definition 1 (Trajectory). *A trajectory is a series of chronologically ordered points $Tr_o = \langle p_1 \rightarrow p_2 \rightarrow \cdots \rightarrow p_n \rangle$ representing the historical trace of an object o. Each point $p_i = \langle x, y, t \rangle$ indicates the location of o at time $p_i.t$.*

Note that, since the trajectories are sampled periodically, we define that the time interval between any two consecutive trajectory points is the multitude of Δt, i.e. $p_{n+1}.t - p_n.t = x * \Delta t, x \in \mathcal{N}$, which we call a *time step*.

As for a *contact event*, two objects are defined as contacted if their trajectories (1) are close to each other at a certain point in time, and (2) such proximity is kept for a continuous period of time, formally defined as follows:

Definition 2 (Contact Event). *Given a distance threshold ϵ and a duration k, objects a and b are directly contacted during $[t_u, t_v]$ if $\forall t_i \in [t_u, t_v], dist(a, b, t_i) \leq \epsilon$ and $t_v - t_u \geq k * \Delta t$, denoted as a contact event $C_{\epsilon,k}(a, b, [t_u, t_v])$.*

Subsequently, we define the *direct contact search* problem below:

Definition 3 (Direct Contact Search (DCS)). *Given a trajectory set R, a query trajectory Tr_q, a starting time t, a distance threshold ϵ and a duration k, a direct contact search $DCS(Tr_q, t, \epsilon, k)$ returns all trajectories Tr_o that satisfies: $\exists C_{\epsilon,k}(q, o, [t_u, t_v])$ where $t_u \geq t$ (direct contact).*

Note that, if not specified, the query starting time t is assumed to set to the starting time of Tr_q. Now we are ready to define the *trajectory contact search* which further capture the indirect contacts:

Definition 4 (Trajectory Contact Search (TCS)). *Given a trajectory set R, a query trajectory Tr_q, a distance threshold ϵ and a duration k, the trajectory contact search $TCS(Tr_q, \epsilon, k)$ returns all trajectories Tr_a which satisfy: there exists a sequence of trajectories $\langle Tr_0, Tr_1, ..., Tr_n \rangle$ where (1) $Tr_0 = Tr_q$, $Tr_n = Tr_a$, (2) $\forall i \in [1, n]$, Tr_i and Tr_{i-1} are contacted directly as $C_{\epsilon,k}(i-1, i, [ct_i, ct_i + k * \Delta t])$ and (3) $\forall i \in [2, n], ct_i \geq ct_{i-1} + k * \Delta t$.*

4   Iteration-Based Trajectory Contact Search

A direct solution to address *TCS* follows the same routine of the disease transmission process. Starting from the query trajectory, it performs a *DCS* on a contacted trajectory in each iteration. Then trajectories retrieved by *DCS* are regarded as the newly contacted trajectories. The algorithm terminates when all contacted trajectories are examined. Intuitively, the iteration process may follow either *Breath-First Search (BFS)* or *Depth-First Search (DFS)* order. However, both can retrieve the result correctly but they may incur redundant computation, as one trajectory may contact multiple reported trajectories. Motivated by this, we propose an *Iterative Direct Contact Search (Iterative-DCS)* algorithm that

Algorithm 1: Iterative-DCS Algorithm

Input: Trajectory Set R, Query Trajectory Tr_q, Contact Parameter $\langle \epsilon, k \rangle$
Output: A set of contacted trajectories R_q

1 Let Q_T be a priority queue to store contacted trajectory;
2 $Q_T.enqueue(\langle Tr_q, t_q \rangle)$;
3 **while** $Q_T \neq \emptyset$ **do**
4 $\quad$ $\langle Tr_i, t_i \rangle \leftarrow Q_T.dequeue()$;
5 $\quad$ $L \leftarrow \text{DCS}(Tr_i, t_i, \epsilon, k)$;
6 $\quad$ **for** $\langle Tr_j, t_j \rangle \in L$ and $Tr_j \notin R_q$ **do**
7 $\quad\quad$ **if** $\langle Tr_j, t'_j \rangle \in Q_T$ and $t_j < t'_j$ **then**
8 $\quad\quad\quad$ Update the entry of Tr_j with t_j;
9 $\quad\quad$ **else if** $\langle Tr_j, t'_j \rangle \notin Q_T$ **then**
10 $\quad\quad\quad$ $Q_T.enqueue(\langle Tr_j, t_j \rangle)$;
11 $\quad$ $R_q \leftarrow R_q \cup \{Tr_i\}$;
12 **return** R_q;

---

performs iterations through temporal order, which ensures that each contacted trajectory is observed and processed only once, demonstrated in Algorithm 1.

To enable the distance calculation between trajectories all the time, in the preprocessing step, we align all trajectories to the same frequency through linear interpolation, such that each trajectory has one point every time step Δt during its lifetime. In the algorithm, we maintain a priority queue Q_T that stores all contacted trajectories that have not been processed, where each entry consists of a contacted trajectory Tr_i and a time t_i indicating the time that this trajectory was first contacted, and the priority is defined by t_i of each entry. Initially, the given query trajectory Tr_q and its trajectory starting time are enqueued into Q_T (Line 2). In each iteration, we dequeue a trajectory from Q_T with smallest t_i and perform a DCS to retrieve a list L of contacted trajectories (Line 3–5). We then check if each item already exists in Q_T. If yes, we update the entry of Tr_j only if t_j is earlier than the existing contact time (Line 7–8). Otherwise, we enqueue the pair to Q_T (line 9–10) directly. Lastly, we add it to the result set R_q (Line 11). The algorithm terminates when Q_T is empty and we return R_q as the TCS result (Line 12). By doing so, the $Interactive\text{-}DCS$ guarantees that each contacted trajectories is processed only once.

Regarding the DCS algorithm, the main goal is to find trajectories which have no less than k consecutive points that are pair-wisely close to Tr_q. Therefore, we employ a slice-based grid for indexing. Specifically, for each timestamp t, we build a grid index G_t for all trajectory points at t, where each point is assigned with the trajectory id it belongs to. Meanwhile, we construct a B-tree for the temporal dimension to quickly retrieve the corresponding grid slice. The DCS algorithm works as follows: 1) For each point $p_i \in Tr_q$ $(1 \leq i \leq n)$, we perform a spatial range query on the grid index $G_{p_i.t}$ that contains p_i to retrieve the neighbour list $N_\epsilon(p_i)$ of points whose distance to p_i is no larger than ϵ. 2) We scan through all the neighbour lists chronologically and keep track of the continuous occurrence for each trajectory by maintaining a hash table. Specifically, for a trajectory who has consecutive points appearing in a sequence of neighbour lists, we record the first and last occurrence time step, and it resets whenever there is an interruption before reaching k time steps. 3) We report all the trajectories with no less than k consecutive points falling into the neighbour lists.

5   Advanced Contact Search Algorithm

In this section, we introduce two optimisation strategies on the DCS algorithm to improve the time and space efficiency, respectively. Recall that during one DCS in the *iterative-DCS* algorithm, we generate a neighbour list for every time step of the trajectory, and scan all of them chronologically. However, it is impossible for two trajectories to be contacted during a duration of $k * \Delta t$ if they fail to keep the distance at any moment amid the period. Thus, the full scan is unnecessary, motivated by which we propose a hop scanning algorithm.

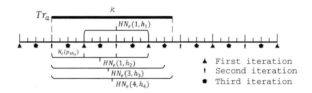

Fig. 1. Example of a hop scanning algorithm

5.1   Hop Scanning Algorithm

The idea of *hop scanning* is first generating a neighbour list N_ϵ for every certain number of points, such that two adjacent neighbour lists are far enough but still within a time duration $k * \Delta t$. Therefore, if the intersection result of them is empty, we can safely prune this time period. We repeat it for the remaining time periods iteratively by shrinking the gap size until all points are investigated. Specifically, given a query trajectory Tr_q, in the ith iteration, we first generate neighbour lists (if not exist) once every h_i points, i.e., p_{jh_i} where $j = 0, 1, 2, \dots$ and $h_i = \frac{k * \Delta t}{2^i}$, representing the current hop length. After generating the list $N_\epsilon(p_{jh_i})$, we create *hop neighbour lists*, denoted by $HN_\epsilon(j, h_i)$ to check if there is any possible contact event during this period, where $HN_\epsilon(j, h_i) = \bigcap_{l=0}^{2^i-1} N_\epsilon(p_{(j+l)h_i})$. After the intersection, objects in $HN_\epsilon(j, h_i)$ are the only objects that possibly contact with Tr_q during the period of $[jh_i, (j + 1)h_i]$. Besides, given that $h_i < k * \Delta t$ holds at all time, this result can further extend to $[jh_i, jh_i + k * \Delta t]$, meaning no contact event during k.

Figure 1 illustrates an example of hop scanning algorithm. In the figure, each line point represents a time step of the query trajectory, and $k = 16$ is given by the query. The top bold line lasting k time steps is a contact event between Tr_a and Tr_q. In the first iteration ($h_1 = 8 * \Delta t$), the time steps labelled by triangles are scanned to generate neighbour lists. When calculating the hop neighbour list, we can find that $Tr_a \in HN_\epsilon(1, h_1)$ as both $N_\epsilon(p_{h_1})$ and $N_\epsilon(p_{2h_1})$ contain it. However, from the second iteration ($h_2 = 4 * \Delta t$), the neighbour lists to be generated consist of both existing triangle ones and the new arrow-labelled ones, and the hop neighbour lists will not be built from scratch. For example, the $HN_\epsilon(1, h_2)$ is the intersection result of three components: (1) the calculated parent $HN_\epsilon(1, h_1)$, (2) $N_\epsilon(p_{3h_2})$ that falls within the time range of the parent $HN_\epsilon(1, h_1)$ and (3) an extra $N_\epsilon(p_{h_2})$ which is adjacent to $HN_\epsilon(1, h_1)$. In fact, for all subsequent hop neighbour list calculations, we will find that there is always one item in components (1) and (3), while the number of items in second component is decided by the iteration count, i.e. $2^{i-1} - 1$. Meanwhile, it is also worth noting that different hop neighbour lists may share the same parent. In general, the hop neighbour lists follow a binary tree structure, where each non-leaf node has two children (except the first and last HN in each iteration). Therefore, to avoid repetitive calculation, instead of generating hop neighbour lists sequentially, in i^{th} iteration, we first intersect each parent HN with all $2^{i-1} - 1$ neighbour lists that fall within its time range (component (2)), the result is then split into two HNs by intersecting left and right adjacent neighbour lists (component (3)), respectively. Hence, the algorithm ensures both children are generated by only processing their parent once. Meanwhile, even in the worst-case scenario when all neighbour lists are generated, each list is only generated once, which does not introduce extra cost over the full scan solution. Overall, it accelerates the process by efficiently pruning out the irrelevant time periods.

5.2   Time Interval Grid Index

In *Iterative-DCS* algorithm, we use slice-based grid to index all trajectory points. As the slice-based grid index maintains one index per time step, it aims for the optimal efficiency at the cost of massive space consumption. Note that the number of time steps is determined by the time span of the dataset and Δt. It is usually very large. Alternatively, by fitting each slice with multiple time steps, the index size can be significantly reduced at the cost of slower query speed. Therefore, we propose the *time interval grid (TIG)* index which reduces the number of the indexed points while keeping the query efficiency. The main idea of our index is to achieve spatial partitioning using grid index while constructing a time interval tree in each grid cell to enable efficient temporal searches.

Data Structure. The data structure consists of two layers. The first layer is a spatial grid index. When constructing the index, each trajectory from the dataset is split into a set of sub-trajectories, named as trajectory fragments, each of which is the longest sequence of consecutive trajectory points that falls in the same grid cell. The second layer is a time interval index. Following the idea from [6], for each grid cell i, we build a time interval index S_i to store all fragments located in it. Each fragment $f(j) = Tr_o(a, b)$ is represented by the timestamp

Table 1. Data specification

Name	# of objects	# of points	Time span (hours)	Avg length (sec)	Map size (km^2)	Point density (pts/(km^2 $*$ sec))
Beijing-M	196K	6.82M	120	398	57	3.18
Beijing-L	953K	50.82M	120	469	589	1.76

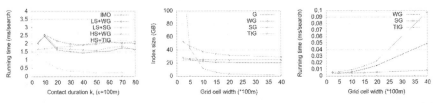

(a) Hop scanning efficiency (M)   (b) Index sizes comparison (L)   (c) Index query efficiency (L)

Fig. 2. Experimental results

of its left and right endpoints, i.e., $l_j = p_a.t$ and $r_j = p_b.t$ ($l_j \leq r_j$). Therefore, a fragment f_j is stored as a time interval $[l_j, r_j]$ in S_i. The construction process is omitted due to page limit and can be found in the original paper [6].

Neighbour Search Query. To find neighbours of a given trajectory point p_q at time t_q, we first search the grid index to find all cells intersecting the circular query region centred at p_q with the radius ϵ. For each grid cell I_i, we search the corresponding time interval index S_i by performing an interval stabbing query [6]. An important feature of the tree structure is that all intervals stabbed by the query q stay close, which ensures optimal query performance. Finally, we return the ids of trajectories that are stabbed by q. After the time interval tree search in each cell, we merge the results found from different cells and filter out the points whose actual distances to the query point p_q are larger than *epsilon* due to the low grid resolution. In general, since every time we compare an interval with q during traversal, we either add it to the output (if stabbed) or stop the subsequent search, the query complexity is $O(1 + m)$ (m is the number of output intervals), and the tree construction time and space complexity are both $O(n)$ (n is the number of intervals in S). Therefore, both search and space complexities are theoretically optimal. Compared with the basic slice-based grid index, the time interval tree only stores endpoints of each trajectory fragments instead of one point per time step. Although one trajectory may have multiple fragments, it is still more effective in terms of the storage cost (especially when grid cells are big), while its time efficiency is competitive.

6   Experiments

6.1   Experiment Settings

There is a lack of public trajectory dataset that has sufficiently large scale as well as high density, especially for pedestrian trajectories, so we conduct our experiments on real-world commercial data sets of taxi trajectories from Beijing, China. Besides, we set $\Delta t = 1$ s, and the data specification is listed in Table 1.

To ensure the correctness of evaluation results, we randomly select 25 trajectories from dataset as the query trajectories, and evaluate the total running time as well as the average running time taken to trace one contacted object (ms/rec). Our algorithms are implemented in C++, and all experiments are conducted on a single server. Our candidate solutions include our basic *iterative-DCS* solution with Linear Scanning (**LS**) and Slice-based Grid (**SG**) index, as well as the Hop Scanning (**HS**) algorithm and the Time Interval Grid (TIG) index. In addition, we introduce two more grid-based indices as references: (1) The typical spatial grid (**G**) which partitions the trajectory points only by their spatial location. (2) The Window-based Grid (**WG**) first partitions the temporal dimension into multiple windows and constructs a grid index for each window. In addition, we implement the only related work **IMO** [8] as another baseline. We test **IMO** on different grid indices, and we find that it achieves the best search performance when using **WG** with a 100 s window, which will be set as the default setting.

6.2   Effectiveness Study

Hop Scanning Algorithm. Figure 2a depicts the average running time of each DCS search using different scanning algorithms and indices. We can see that both hop scanning solutions (HS+WG & HS+TIG) outperform their linear scanning counterpart constantly, and the gap is more significant as k increases, which meets our expectation as a longer k enables more pruning opportunities. On the other hand, our advanced algorithm HS+TIG outperforms IMO significantly.

Time Interval Grid Index. We evaluate the performance of our proposed TIG index when $\epsilon = 100$ m and $k = 30$ s. Figures 2b and 2c illustrate its space and time efficiency on **Beijing-L**. As we expected, the slice-based grid consumes the largest space, which reduces steadily as the grid size increases due to the reduction of grid cell entries. In contrast, the grid cell size is crucial to TIG. A smaller grid cell means more trajectory fragments, resulting to more index entries. The figures show that the size of TIG is only $\frac{1}{3}$ of the others once the grid width reaches 1 km. Meanwhile, Fig. 2c shows that extremely small grid cells do not generally lead to better query performance. This is due to the excessive number of grid cells creates too many index entries, which slows down the entry search. In our experiments, by choosing a grid cell of 1 km × 1 km, we can achieve almost the best query performance and storage cost.

7   Conclusion

In this paper, we introduced a new trajectory contact search query to model the trajectory contact problem. We proposed an iteration-based algorithm which performs direct contact search iteratively to find all contacted trajectories. In addition, we further optimised it with a hop scanning algorithm and a time interval grid index, which improved the time and space efficiency, respectively. The experiments show that our proposal achieves both faster query speed and smaller space consumption.

References

1. Buchin, K., Buchin, M., van Kreveld, M., Speckmann, B., Staals, F.: Trajectory grouping structure. In: Dehne, F., Solis-Oba, R., Sack, J.-R. (eds.) WADS 2013. LNCS, vol. 8037, pp. 219–230. Springer, Heidelberg (2013). https://doi.org/10.1007/978-3-642-40104-6_19
2. Gudmundsson, J., van Kreveld, M., Speckmann, B.: Efficient detection of motion patterns in spatio-temporal data sets. In: Proceedings of the 12th Annual ACM International Workshop on Geographic Information Systems, pp. 250–257 (2004)
3. Jeung, H., Shen, H.T., Zhou, X.: Convoy queries in spatio-temporal databases. In: ICDE, pp. 1457–1459. IEEE (2008)
4. van Kreveld, M., Löffler, M., Staals, F., Wiratma, L.: A refined definition for groups of moving entities and its computation. Int. J. Comput. Geom. Appl 28(02), 181–196 (2018)
5. Li, Z., Ding, B., Han, J., Kays, R.: Swarm: mining relaxed temporal moving object clusters. PVLDB 3(1–2), 723–734 (2010)
6. Schmidt, J.M.: Interval stabbing problems in small integer ranges. In: Dong, Y., Du, D.-Z., Ibarra, O. (eds.) ISAAC 2009. LNCS, vol. 5878, pp. 163–172. Springer, Heidelberg (2009). https://doi.org/10.1007/978-3-642-10631-6_18
7. Tang, L.A., et al.: On discovery of traveling companions from streaming trajectories. In: ICDE, pp. 186–197. IEEE (2012)
8. Xu, J., Lu, H., Bao, Z.: IMO: a toolbox for simulating and querying "infected" moving objects. PVLDB 13(12), 2825–2828 (2020)
9. Zheng, K., Zheng, Y., Yuan, N.J., Shang, S., Zhou, X.: Online discovery of gathering patterns over trajectories. TKDE 26(8), 1974–1988 (2013)

STMG: Spatial-Temporal Mobility Graph for Location Prediction

Xuan Pan[1,3], Xiangrui Cai[2,3(✉)], Jiangwei Zhang[4], Yanlong Wen[1,3],
Ying Zhang[1,3], and Xiaojie Yuan[2,3]

[1] College of Computer Science, Nankai University, Tianjin, China
panxuan@dbis.nankai.edu.cn, {wenyl,yingzhang}@nankai.edu.cn
[2] College of Cyber Science, Nankai University, Tianjin, China
{caixr,yuanxj}@nankai.edu.cn
[3] Tianjin Key Laboratory of Network and Data Security Technology,
Nankai University, Tianjin, China
[4] Department of Computer Science, National University of Singapore,
Singapore, Singapore
A0054808@u.nus.edu

Abstract. Location-Based Social Networks (LBSNs) data reflects a
large amount of user mobility patterns. So it is possible to infer users'
unvisited Points of Interest (POIs) through the users' check-in records
in LBSNs. Existing location prediction approaches typically regard user
check-ins as sequences, while they ignore the spatial and temporal correlations between non-adjacent records. Moreover, the serialized form
is insufficient to analog user complex POI moving behaviors. In this
paper, we model user check-in records as a graph, named Spatial-
Temporal Mobility Graph (STMG), where the nodes and edges fuse
the spatial-temporal information in absolute and relative aspect respectively. Based on STMG, we propose a location prediction model named
Spatial-temporal Enhanced Graph Neural Network (SEGN). In SEGN,
the STMG nodes are encoded as the embeddings with specific time and
location semantics. Last but not the least, we introduce three kinds
of matrices, which completely depict the user moving behaviors among
POIs, as well as the relative relationships of time and location on STMG
edges. Extensive experiments on three real-world LBSNs datasets demonstrate that with specific time information, SEGN outperforms seven
state-of-the-art approaches on four metrics.

Keywords: Location-Based Social Network · User mobility · Graph
Neural Network · Location prediction

1 Introduction

While Location Based Social Networks (LBSNs) are getting popular, more and
more people share their locations with timestamps anytime and anywhere they
want. As a result, large number of user check-ins at Points of Interest (POIs) are
accumulated in LBSNs. As the user mobility is a significant property of people

© Springer Nature Switzerland AG 2021
C. S. Jensen et al. (Eds.): DASFAA 2021, LNCS 12681, pp. 667–675, 2021.
https://doi.org/10.1007/978-3-030-73194-6_45

behaviors, LBSNs data brings a great value in the study of human movement characteristics. Location prediction (also known as POI recommendation) is a typical task in user mobility prediction. Based on user check-in records, the task aims to predict the POIs that users have never been to. So the key issue of the task is understanding the complex user mobility via LBSNs data.

In fact, the user mobility is primarily manifested by two aspects of LBSNs data. On one hand, the mobility patterns such as sequentiality and periodicity can be reflected by the moving behaviors among large amount of POIs. On the other hand, the location habits are hidden behind the complex heterogeneous spatial-temporal contexts. In each user's check-in records, the visiting timestamp indicates the *absolute* time, and the time difference between every pair check-ins indicates the *relative* time. So as for locations, the POI coordinate means the *absolute* position, and the distance between each POI pair is the *relative* position. In this sense, absolute and relative information are equally important, but current location prediction methods are failed to cover them all.

To overcome the problem of insufficient attributes modeling on LBSNs data, we propose *Spatial-Temporal Mobility Graph* (STMG), a new kind of comprehensive representation for user mobility. Particularly, to enhance the POI moving behaviors, STMG connect both *adjacent* and *nonadjacent* check-in records. Besides, each node in the graph is expanded from a POI to a POI with the absolute time and location information. Meanwhile, edges fuse the relative information which consists of distances and time differences in the check-in pairs. Therefore, STMG carries a full spectrum of the information about time and locations. Then we propose the STMG embedding network with location prediction model, named *Spatial-temporal Enhanced Graph Neural Network* (SEGN). Gated Graph Neural Network [4] is used as the base model for graph embedding. The differences are the personalized node vectors with specific time and location semantics, as well as the extended node updating functions. With SEGN based on STMG, user mobility patterns with integrated spatial-temporal attributes can be explicitly understood. We conduct extensive experiments on three real-world LBSNs datasets. The experiment result show that with specific time information, SEGN outperforms seven state-of-the-art approaches on four metrics.

2   Related Work

Location Prediction Methods. Existing works can be roughly categorized into two groups. Matrix Factorization [3,5] based methods focus on looking for optimized factorized methods for user-location matrix. But they are not sufficient in establishing the user POI moving behaviors. Sequence-based approaches such as the Markov chain [15] and Recurrent Neural Networks [10,12] based models treat adjacent check-in records as serialized trajectories. But their inadequate exploitation of nonadjacent records is still a serious problem.

Graph Embedding. Recent location prediction models focus on developing customized Deep Neural Networks for learning POI, user or spatial-temporal contexts embedding vectors. Some embedding methods use graphs such as user-POI graphs [11], user-user graphs [14] and POI-POI graphs [16] as the model inputs to represent LBSNs data. However, none of above graphs integrate the spatial-temporal attributes both in absolute and relative aspects. Also, the graphs are not enough to indicate how user moves. So the above representations lack of the knowledge of time and space from a real-life perspective.

3 Proposed Model

3.1 Problem Definition

We define $P = \{p_1, p_2, \ldots, p_m\}$ as a set of POIs appearing in LBSNs data. Each of these POIs corresponds a site in real world, so every POI has a coordinate consisting of latitude and longitude. $U = \{u_1, u_2, \ldots, u_n\}$ is the user set. Let $R = \{R_1, R_2, \ldots, R_n\}$ denotes the set of user check-in records. R_i is a collection composed by all check-ins of user u_i. Each check-in record can be formally denoted as a tuple (t, p), where t is the timestamp, and p is the POI ID. Through a user's existing check-in records, the goal is to generate a prediction list of unvisited POIs which the user is likely to go to.

3.2 Spatial-temporal Mobility Graph Construction

A Spatial-Temporal Mobility Graph (STMG) is constructed on a user basis. Each user corresponds to a graph or a graph collection. A STMG is formally denoted as $G = \{V, E\}$, where $V = \{v_1, v_2, \ldots, v_k\}$ is the node set, and E is the edge set. In STMG, nodes represent POIs as well as the absolute spatial-temporal information. Therefore, each node means a POI in a specific time and locating at a certain place, rather than an isolated POI. For ease of representation in practice, the time is denoted as the time slot number where the timestamp is located, and the location is denoted as the region number gridded according to the geographic coordinate. STMG edges fuse the relative information of time and position. For an edge $(v_i, v_j) \in E$, it contains the time difference and the position distance between v_i and v_j. Particularly, the edges connect both of the adjacent and nonadjacent check-in records for each user. The directions of the edges depend on the time chronological order. So all nodes of a user's STMG can be associated with each other with the direction of time. Therefore, the edges hold the relative spatial-temporal attributes as well as analog the multi-connected POI moving behaviors. In the end, user mobility can be comprehensively represented by STMG.

3.3 Spatial-temporal Enhanced Graph Neural Network

In this section, we introduce Spatial-temporal Enhanced Graph Neural Network (SEGN for short) for STMG embedding and location prediction. The overall model is shown in Fig. 1. SEGN is a modified Gated Graph Neural Networks (GG-NNs) [4], which encodes STMG from two perspectives. Firstly, for STMG

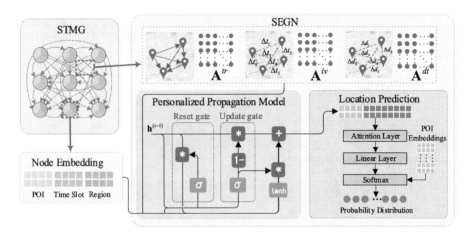

Fig. 1. The overall architecture of SEGN. The node embeddings consist of POI, time slot and region vectors. The connectivity as well as the information of the edges are converted to the matrices $\mathbf{A}^{tr}$, $\mathbf{A}^{iv}$ and $\mathbf{A}^{dt}$ respectively. After updating the node states, the prediction list is generated by the prediction network.

nodes, the absolute information consisting of the POI, time slot and region are embedded as three kinds of vectors, which are concatenated as the node vector $\mathbf{h}_v$. All three kinds of vectors can be learned through the network simultaneously. As a result, the node vectors have the semantics of POIs with the spatial-temporal context, which are rendered by users' real activities. Secondly, the nodes states are updated by spatial-temporal enhanced recurrence functions. To integrate the relative information and the connectivity of the edges, we extend the single $\mathbf{A}$ in vanilla recurrence function to three matrices. The first matrix is $\mathbf{A}^{tr} \in \mathbb{R}^{|V| \times 2|V|}$, where $|V|$ is the node number of STMG. It is used to represent the POI moving behaviors. Similar with SR-GNN [13], $\mathbf{A}^{tr}$ is a concatenation of two submatrices, which represent the outgoing and incoming edges respectively. The values in each submatrix denote the connection weights, which are decided by the directed relations in STMG. With the rich check-ins associations, $\mathbf{A}^{tr}$ hold more nodes transition relationships than serialized sessions in SR-GNN. So it is informative to simulate user transfer behaviors. The second matrix is $\mathbf{A}^{iv} \in \mathbb{R}^{|V| \times |V|}$. It is used to represent the time differences in node pairs which have edges in STMG. Values of $\mathbf{A}^{iv}$ are determined by the exponential decay function of the time interval between nodes. The third matrix is $\mathbf{A}^{dt} \in \mathbb{R}^{|V| \times |V|}$, which represents the distances of STMG node pairs. For measuring the geographic correlations, Here we adopt Gaussian Radial Basis Function (RBF) kernel according to coordinates of the POIs as [9,16] did. In this way, $\mathbf{A}^{tr}$ fuses the activations of POI transfers from edges in both directions. Meanwhile, $\mathbf{A}^{iv}$ and $\mathbf{A}^{dt}$ hold the relative information of time and locations. Then each of the three matrices passes information among nodes respectively. The spatial-temporal enhanced recurrence functions in SEGN are as following:

$$\mathbf{a}_v^{(t)tr} = \mathbf{A}_{v:}^{tr\top} \left[\mathbf{h}_1^{(t-1)\top} \dots \mathbf{h}_{|V|}^{(t-1)\top} \right]^{\top} + \mathbf{b}^{tr} \qquad (1)$$

$$\mathbf{a}_v^{(t)iv} = \mathbf{A}_{v:}^{iv\top} \left[\mathbf{h}_1^{(t-1)\top} \dots \mathbf{h}_{|V|}^{(t-1)\top} \right]^{\top} + \mathbf{b}^{iv} \qquad (2)$$

$$\mathbf{a}_v^{(t)dt} = \mathbf{A}_{v:}^{dt\top} \left[\mathbf{h}_1^{(t-1)\top} \dots \mathbf{h}_{|V|}^{(t-1)\top} \right]^{\top} + \mathbf{b}^{dt} \qquad (3)$$

$$\mathbf{a}_v^{(t)} = \mathbf{W}^{tr}\mathbf{a}_v^{(t)tr} + \mathbf{W}^{iv}\mathbf{a}_v^{(t)iv} + \mathbf{W}^{dt}\mathbf{a}_v^{(t)dt} + \mathbf{b} \qquad (4)$$

$$\mathbf{z}_v^t = \sigma \left(\mathbf{W}^z \mathbf{a}_v^{(t)} + \mathbf{U}^z \mathbf{h}_v^{(t-1)} \right) \qquad (5)$$

$$\mathbf{r}_v^t = \sigma \left(\mathbf{W}^r \mathbf{a}_v^{(t)} + \mathbf{U}^r \mathbf{h}_v^{(t-1)} \right) \qquad (6)$$

$$\widetilde{\mathbf{h}_v^{(t)}} = \tanh \left(\mathbf{W}\mathbf{a}_v^{(t)} + \mathbf{U} \left(\mathbf{r}_v^t \odot \mathbf{h}_v^{(t-1)} \right) \right) \qquad (7)$$

$$\mathbf{h}_v^{(t)} = \left(1 - \mathbf{z}_v^t \right) \odot \mathbf{h}_v^{(t-1)} + \mathbf{z}_v^t \odot \widetilde{\mathbf{h}_v^{(t)}}. \qquad (8)$$

Formula (1), (2) and (3) indicate that information passes between different nodes in three ways, then all the information is aggregated in formula (4). In the following GRU settings, $\mathbf{z}_v$ acts as the update gate, and $\mathbf{r}_v$ is the reset gate. $\sigma(x)$ is the logistic sigmoid function, and $\odot$ denotes element-wise multiplication. With the recurrence functions processing, each node receives message from the other nodes and from the previous time-step to update its hidden state. The final states of all nodes can be obtained until the network converges.

After the nodes updating process, the nodes hidden state $\mathbf{h}_v^{(t)}$ is used as the input of the location prediction model. $\mathbf{h}_v^{(t)}$ is firstly delivered to the attention layer. Each node embedding obtains the overall STMG weights as the global preference. Then after the linear layers processing, node embeddings multiply all POI embeddings to get the prediction probabilities. Finally, the probability distribution of all POIs is generated via a softmax function.

In the training set, each user's check-in records are split into two parts. One part is used for STMG constructing, the other part is used as the ground truths. For every STMG, we use cross-entropy as the object function as following:

$$\mathcal{L} = - \sum_{i=1}^{k} \mathbf{y}_i \log \left(\hat{\mathbf{y}}_i \right) + \left(1 - \mathbf{y}_i \right) \log \left(1 - \hat{\mathbf{y}}_i \right) \qquad (9)$$

Where k is the number of the unvisited POIs to current STMG, and $\mathbf{y}_i$ is the one hot vector of the ground truth i. In the prediction stage, with the user's known check-ins, the STMGs are generated as SEGN inputs. And in the model outputs, the unvisited POIs with higher probabilities are composed as the prediction list.

4   Experiments

4.1   Datasets and Settings

Datasets. We conduct the experiments on Gowalla[1], Foursquare[2] and Yelp[3], which are three real-world LBSNs datasets. The data version and the schemes of data filtering and splitting are as same as the previous work [7]. For each user,

---

[1] http://snap.stanford.edu/data/loc-gowalla.html.
[2] https://sites.google.com/site/yangdingqi/home/foursquare-dataset.
[3] https://www.yelp.com/dataset.

the earliest 70% check-in records are used for training. The latest 20% are for testing. And the remaining 10% are for validation.

Settings. For users with large check-in records, it is difficult to use only one STMG for the model to embed. Here we uniformly construct STMGs in weekly check-ins for each user. So every user have a graph collection for training. For Gowalla and Foursquare, we split one week as 24 * 7 time slots, and segment 7,500 regions according to the coordinates of POIs. Since Yelp only provides the check-in date instead of specific time, we can only split one week as 7 time slots, and segment 4500 regions for locations. We set the time decay coefficient to 0.03 for Gowalla and Foursquare, and 0.27 for Yelp. For all datasets, the geographical correlation level for RBF kernel is set to 60, the dimension of POI embedding and region embedding is set to 100 and 50 respectively for all datasets. And the dimension of time slot is 50 for Gowalla and Foursquare, and 10 for Yelp. In the prediction stage, we generate each user's STMG collection from the training set as the inputs, and the cumulative scores of the collection outputs are ranked as the prediction results. We use four metrics to evaluate our model. Precision (Pre@K) is the percentage of POIs among the top k prediction results has been visited by each user. Recall (Rec@K) is the percentage of each user's visiting POIs can emerge in the top k prediction results. Mean average precision (MAP@K) is the mean average precision in top k results. Normalized discounted cumulative gain (NDCG@K) considers the rank of the top k predictions by assigning higher score to the hits at higher POIs. All the metrics are with larger value, the better the performance. We choose seven location prediction models MGMPFM [1], LRT [2], IRenMF [8], GeoMF [5], RankGeoFM [3], GeoPFM [6] and SAE-NAD [9] for comparison.

4.2   Performance Comparison

The performances of SEGN and baselines are shown in Fig. 2, Fig. 3, Fig. 4 and Fig. 5. The observations are as following:

Spatial-temporal Context Mining. In Gowalla and Foursquare, SEGN achieves the best performance compared with all baselines. For example, with the metrics of average precision, recall, MAP and NDCG on Gowalla, SEGN has respectively reached 8.9%, 9.9%, 9.9% and 8.5% better than the second best model RankGeoFM. But in Yelp, SEGN doesn't outperform several models. The reason is that our method is strongly dependent on the spatial-temporal contexts. As mentioned above, Yelp does not provide the specific check-in time, so SEGN is inability to grasp user moving behaviors by hours. This is the direct cause of the performance degradation. Therefore, compared with other models, SEGN has obtained more absolute and relative spatial-temporal information respectively. It enables SEGN to learn the mobility patterns from user check-in records more completely than other baselines.

Semantic Richness. Compared to the embedding models SAE-NAD, SEGN embeds absolute spatial-temporal information as additional valuable knowledge

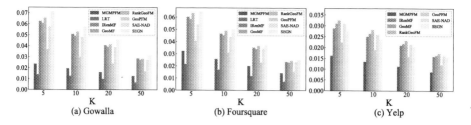

Fig. 2. The comparison of precision on the three datasets.

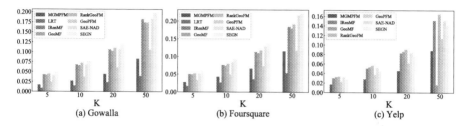

Fig. 3. The comparison of recall on the three datasets.

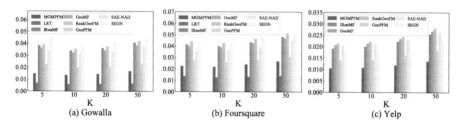

Fig. 4. The comparison of MAP on the three datasets.

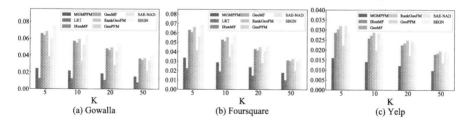

Fig. 5. The comparison of NDCG on the three datasets.

of POI, instead of the POI embeddings only. It makes the STMG nodes holding proper representation of the spatial-temporal semantics.

Moving Behaviors Modeling. The matrix factorization methods such as MGMPFM, LRT and IRenMF can not establish the users' POI moving behaviors. In contrast, SEGN captures the complex POI transferring relationships not only in the adjacent check-ins, but also in the nonadjacent ones. In a result, SEGN realizes a comprehensive understanding of user mobility patterns.

5   Conclusion

In this paper, we propose Spatial-Temporal Mobility Graph as a new kind of comprehensive representation of user mobility. It integrates the absolute and relative spatial-temporal contexts contained in user check-in records. Particularly, edges in the graph analog the complex moving behaviors. Then, we propose SEGN as the location prediction model. The specifics of STMG are fused in the embedding settings and node updating functions respectively. Relevant experiments indicate that SEGN brings the improvements in the location prediction task.

Acknowledgments. This work is supported by NSFC-General Technology Joint Fund for Basic Research (No. U1936206, No. U1836109), and National Natural Science Foundation of China (No. 62077031, No. U1903128).

References

1. Cheng, C., Yang, H., King, I., Lyu, M.R.: Fused matrix factorization with geographical and social influence in location-based social networks. In: AAAI, pp. 17–23 (2012)
2. Gao, H., Tang, J., Hu, X., Liu, H.: Exploring temporal effects for location recommendation on location-based social networks. In: RecSys, pp. 93–100. ACM (2013)
3. Li, X., Cong, G., Li, X., Pham, T.N., Krishnaswamy, S.: Rank-geoFM: a ranking based geographical factorization method for point of interest recommendation. In: SIGIR, pp. 433–442. ACM (2015)
4. Li, Y., Tarlow, D., Brockschmidt, M., Zemel, R.S.: Gated graph sequence neural networks. In: ICLR (2016)
5. Lian, D., Zhao, C., Xie, X., Sun, G., Chen, E., Rui, Y.: GeoMF: joint geographical modeling and matrix factorization for point-of-interest recommendation. In: SIGKDD, pp. 831–840. ACM (2014)
6. Liu, B., Xiong, H., Papadimitriou, S., Fu, Y., Yao, Z.: A general geographical probabilistic factor model for point of interest recommendation. IEEE Trans. Knowl. Data Eng. 27(5), 1167–1179 (2015)
7. Liu, Y., Pham, T., Cong, G., Yuan, Q.: An experimental evaluation of point-of-interest recommendation in location-based social networks. Proc. VLDB Endow. 10(10), 1010–1021 (2017)
8. Liu, Y., Wei, W., Sun, A., Miao, C.: Exploiting geographical neighborhood characteristics for location recommendation. In: CIKM, pp. 739–748. ACM (2014)
9. Ma, C., Zhang, Y., Wang, Q., Liu, X.: Point-of-interest recommendation: exploiting self-attentive autoencoders with neighbor-aware influence. In: CIKM, pp. 697–706. ACM (2018)
10. Manotumruksa, J., Macdonald, C., Ounis, I.: A contextual attention recurrent architecture for context-aware venue recommendation. In: SIGIR, pp. 555–564. ACM (2018)

11. Su, Y., et al.: HRec: heterogeneous graph embedding-based personalized point-of-interest recommendation. In: Gedeon, T., Wong, K.W., Lee, M. (eds.) ICONIP 2019. LNCS, vol. 11955, pp. 37–49. Springer, Cham (2019). https://doi.org/10.1007/978-3-030-36718-3_4

12. Sun, K., Qian, T., Chen, T., Liang, Y., Nguyen, Q.V.H., Yin, H.: Where to go next: Modeling long- and short-term user preferences for point-of-interest recommendation. In: AAAI, pp. 214–221 (2020)

13. Wu, S., Tang, Y., Zhu, Y., Wang, L., Xie, X., Tan, T.: Session-based recommendation with graph neural networks. In: AAAI, pp. 346–353 (2019)

14. Yang, C., Bai, L., Zhang, C., Yuan, Q., Han, J.: Bridging collaborative filtering and semi-supervised learning: A neural approach for POI recommendation. In: SIGKDD, pp. 1245–1254. ACM (2017)

15. Ying, H., et al.: Time-aware metric embedding with asymmetric projection for successive POI recommendation. World Wide Web **22**(5), 2209–2224 (2018). https://doi.org/10.1007/s11280-018-0596-8

16. Zhong, T., Zhang, S., Zhou, F., Zhang, K., Trajcevski, G., Wu, J.: Hybrid graph convolutional networks with multi-head attention for location recommendation. In: World Wide Web, pp. 1–27 (2020)

Author Index